Healthy Ageing and Longevity

Volume 3

Series editor

Suresh I.S. Rattan, Aarhus, Denmark

More information about this series at http://www.springer.com/series/13277

Alexander M. Vaiserman · Alexey A. Moskalev
Elena G. Pasyukova
Editors

Life Extension

Lessons from Drosophila

Springer

Editors
Alexander M. Vaiserman
Laboratory of Epigenetics
Institute of Gerontology
Kiev
Ukraine

Elena G. Pasyukova
Laboratory of Genome Variation
Institute of Molecular Genetics of RAS
Moscow
Russia

Alexey A. Moskalev
Laboratory of Molecular Radiobiology
and Gerontology
Institute of Biology of Komi Science
Center of Ural Branch of RAS
Syktyvkar, Komi Republic
Russia

ISSN 2199-9007 ISSN 2199-9015 (electronic)
Healthy Ageing and Longevity
ISBN 978-3-319-18325-1 ISBN 978-3-319-18326-8 (eBook)
DOI 10.1007/978-3-319-18326-8

Library of Congress Control Number: 2015938726

Springer Cham Heidelberg New York Dordrecht London

Printed on acid-free paper

Springer International Publishing AG Switzerland is part of Springer Science+Business Media
(www.springer.com)

Preface

Human life expectancy has nearly doubled over the last 100 years due, in part, to a wide range of novel medical technologies and treatments. The trend toward increased life expectancy in the developed countries is accompanied by the increased number of people surviving to an advanced age and having different chronic age-associated pathologies. This trend leads to the need to understand the genetic and physiological mechanisms underlying aging processes and particularly those that promote healthy aging. Moreover, in recent years, substantial evidence has emerged supporting the possibility of the radical human life extension, primarily due to the rapid development of genetic and stem cell-based technologies.

In the development of such technologies, several insect models may provide useful starting points prior to animal and human studies. The use of insect models seems particularly reasonable since, despite the large phylogenetic distance between insects and mammals, some metabolic processes and signaling pathways were shown to play an evolutionarily conserved role in aging across various insect and mammal species. Among them, the insulin/insulin growth factor signaling pathway, histone deacetylases, and genes involved in oxidative stress all exert evolutionarily conserved effects on aging and life span in a wide range of model organisms. These data suggest that aging itself is an evolutionarily conserved process and not simply an inevitable deterioration of biological systems. The high degree of conservation between diverse species in the genetic pathways that regulate longevity suggests that work in model organisms can expand the theoretical knowledge of aging, yield valuable insight into the molecular and cellular processes that underlie aging process, and perhaps provide new therapeutic targets for the treatment of age-related disorders.

Among the widespread model organisms, the fruit fly, *Drosophila melanogaster*, is likely one of the most appropriate model organisms to study biological mechanisms of aging due to its relatively short life span (60–80 days), convenient husbandry, and well-studied genetics. The *Drosophila* genome was one of the first to be sequenced. It has powerful systems for gene knockout and targeted mutagenesis. The large brood sizes also make it possible to measure survival in large numbers of individuals within each experimental cohort in controlled environments and to test the functional consequences of senescence either longitudinally in individuals or as sampled from the aging population. Furthermore, almost all cells in adult insects are postmitotic except a few cells in the malpighian tubules, gut, and gonads. Therefore, the age-related decline in cellular functions may be examined without interference from newly dividing cells. Certainly, not all senescent physiological changes revealed in flies can be simply translated to humans. However, flies and humans often show very similar age-related physiological phenotypes, suggesting that at least some of the basic biological properties and mechanisms that regulate longevity are conserved between flies and humans. In the last years, *Drosophila* models have been developed for a large variety of aging-related processes and diseases.

The goal of this book is to provide the reader with an overview of current research concerned with the use of the *Drosophila* experimental model as a tool for unraveling the genetic, molecular, and physiological mechanisms underlying the aging process and to search for life-extending remedies. This research field is currently a hot topic in biomedicine. Thereby, the present book, which is a collective work of the world's leading researchers in the field of biogerontology, may be of interest to a wide audience, ranging from academic researchers to the general public.

Finally, the editors would like to thank Prof. Suresh I.S. Rattan, the "Healthy Ageing and Longevity" book series editor, for his kind support and wise advices. We would also like to thank Oksana Zabuga for the valuable help in preparing the book manuscript.

Alexander M. Vaiserman
Alexey A. Moskalev
Elena G. Pasyukova

Contents

Part I
Genetic Regulation of Longevity

Chapter 1
Neuronal Genes and Developmental Neuronal Pathways in *Drosophila* Life Span Control

Elena Pasyukova, Alexander Symonenko, Natalia Roshina, Mikhail Trostnikov, Ekaterina Veselkina and Olga Rybina

Abstract The nervous system has long been suggested as a key tissue that defines life span. The identity of neuronal cell types is established during development and maintained throughout adulthood due to the expression of specific neuronal genes coding for ion channels, neurotransmitters and neuropeptides, G-protein-coupled receptors, motor proteins, recognition and adhesion molecules. In this paper, we review data on the role of neuronal genes in *Drosophila melanogaster* life span control. Several pathways responsible for life span regulation are also important for the development of the nervous system. Genes involved in insulin-like, Target of Rapamycin, Janus Kinase/Signal Transducer and Activator of Transcription and cell polarity pathways, a number of global regulators and transcription factors play key roles both in aging and longevity control and in shaping the nervous system as a network of specialized neuronal cells in early development. Is their impact on life span related, at least partially, to their developmental functions or is it explained by other pleiotropic influences later in life? In this paper, we address this question based on the published data and our own findings.

Keywords Nervous system · Neuronal genes · *Drosophila* · Life span · Transcription factors

1.1 Introduction

The nervous system has long been suggested as a key tissue that defines life span. The numerous and diverse interactions between the nervous system and life span are reciprocal and intimately linked. On the one hand, via different types of sensory neurons and a wide variety of environmental cues, the nervous system

E. Pasyukova (✉) · A. Symonenko · N. Roshina · M. Trostnikov · E. Veselkina · O. Rybina
Institute of Molecular Genetics of RAS, Moscow, Russia
e-mail: egpas@rambler.ru

A.M. Vaiserman et al. (eds.), *Life Extension*, Healthy Ageing and Longevity 3,
DOI 10.1007/978-3-319-18326-8_1

receives complex information from the environment and further processes, integrates and transforms it into various physiological outputs that have a major impact on life span; on the other hand, aging also affects the functional state of the nervous system and is associated with the development of age-associated neurodegenerative diseases.

The impact of the nervous system on *Drosophila* life span was initially indicated by accumulating data on the genetic control of life span. Primarily, overexpression of many genes only in the nervous system of transgenic flies resulted in an increase in life span (see, for example, Parkes et al. 1999; Seong et al. 2001; Ruan et al. 2002; Wang et al. 2003; Morrow et al. 2004; Bauer et al. 2005a, b; Orr et al. 2003; Fridell et al. 2005, 2009; Liao et al. 2008; Martínez-Azorín et al. 2008; Simonsen et al. 2008; Lee et al. 2009; Alic et al. 2011; Plyusnina et al. 2011; Rana et al. 2013). Multiple molecular and genetic mechanisms for the impact of the nervous system on aging and longevity were reported (for review, see Broughton and Partridge 2009; Alcedo et al. 2013). These include insulin-like signaling; stress-sensing pathways; antioxidative response mechanisms, reactive oxygen species (ROS) signaling and mitochondrial homeostasis; molecular chaperones, autophagy, lysosomal degradation; etc. Despite this progress, little is known whether genes that control specific functions of neuronal cells affect normal life span. Indeed, the abovementioned aging pathways are not specifically neuronal and function in several other tissues such as fat body, muscles, gonads, etc. (see, for example, Giannakou et al. 2004; Kapahi et al. 2004; Flatt et al. 2008; Biteau et al. 2010; Demontis and Perrimon 2010; Stenesen et al. 2013).

The identity of neuronal cell types is established during development and maintained throughout adulthood. In addition to housekeeping genes, a differentiated neuron is thought to express combinations of genes that define its functional properties (Hobert 2011). These genes code for: (1) ion channels (Potassium channels, Calcium channels, ligand-gated ion channels, etc.); (2) neurotransmitters and neuropeptides (their synthesis, transport, reuptake, and degradation); (3) G-protein-coupled receptors; (4) motor proteins and their associated complexes (kinesin, dynein and myosin motors); (5) recognition and adhesion molecules (immunoglobulin superfamily, cadherins, neurexins superfamily) and some others. Collectively such genes will be further referred to as neuronal genes, even though we fully realize that this term is conditional, given the pleiotropic nature of most genes. In this paper, we review data on the role of neuronal genes in *Drosophila melanogaster* life span control.

Several pathways responsible for life span regulation are also important for the development of the nervous system. Genes involved in Insulin/Insulin Receptor (InR), Target of Rapamycin (TOR), Janus Kinase/Signal Transducer and Activator of Transcription (JAK/STAT) and cell polarity pathways, a number of global regulators and transcription factors play key roles both in aging and longevity control and in shaping the nervous system as a network of specialized neuronal cells in early development (Table 1.1). Even though, according to the definition given above, these genes can not be regarded as neuronal genes, they exemplify another embodiment of the relationship between the nervous system and longevity. Is their

Table 1.1 Neuronal genes and developmental neuronal pathways in life span control

Gene and pathway	Protein	Function in the nervous system	Stage	References	Function in life span control	References
Shaker (*Sh*); **cAMP/PKA pathway?**	Alpha subunit of a voltage-dependent potassium channel	Synaptic activity	Larvae	Papazian et al. (1987), Budnik et al. (1990), Ueda and Wu (2006)	Mutation: decreased activity decreases LS	Trout and Kaplan (1970), Rogina and Helfand (1995)
Hyperkinetic (*Hk*); **cAMP/PKA pathway?**	Beta subunit of a voltage-dependent potassium channel	Synaptic activity	Larvae	Budnik et al. (1990), Wilson et al. (1998)	Mutation: decreased activity decreases LS	Trout and Kaplan (1970), Rogina and Helfand (1995)
Rutabaga (*rut*); **cAMP/PKA pathway**	Adenylyl cyclase	Learning and memory	Adults	McGuire et al. (2003)	Mutation decreases LS	Tong et al. (2007)
		Synaptic activity	Larvae	Renger et al. (2000), Baines (2004)		
Dunc (*dnc*); **cAMP/PKA pathway**	cAMP phosphordiesterase	Learning and memory	Adults	Dudai et al. (1976), Byers et al. (1981)	Mutation restores LS of NF1 mutants	Tong et al. (2007)
		Synaptic activity	Larvae	Renger et al. (2000), Zhong and Wu (2004)		
cAMP-dependent protein kinase 1 (*PKA-C1*); **cAMP/PKA pathway**	Catalytic subunit of cAMP-dependent protein kinase A (PKA)	Learning and memory	Adults	Drain et al. (1991), Skoulakis et al. (1993), Yamazaki et al. (2007), Gervas et al. (2010)	Mutation does not affect LS	Yamazaki et al. (2007)
		Synaptic activity	Larvae	Baines (2004)	Ubiquitously increased activity (mouse PKA) restores LS of *NF1* mutants	Tong et al. (2007)
Neurofibromin 1 (*NF1*); **cAMP/PKA pathway**	Ras GTPase activator	Learning and memory	Adults	Buchanan et al. (2000), Buchanan and Davis (2010)	Mutation decreases LS	Tong et al. (2007)
		Synaptic growth, synaptic activity	Larvae	Tsai et al. (2012)		

(continued)

Table 1.1 (continued)

Gene and pathway	Protein	Function in the nervous system	Stage	References	Function in life span control	References
Maleless (*mle*)	ATP-dependent double-stranded RNA helicase	Synaptic activity (via RNA editing of the Na^+ channel gene *paralytic*)	Larvae	Reenan et al. (2000), Zhong and Wu (2004)	Gain-of-function mutation: increased activity decreases LS	Reenan and Rogina (2008)
Paralytic (*para*)	Major voltage-gated sodium channel	Synaptic activity	Larvae	Reenan et al. (2000)	Increased dosage of *para* restores LF of mle mutants	Reenan and Rogina (2008)
Insulin-like peptide 2 (*Ilp2*); <u>InR/TOR pathway</u>	Insulin-like peptide	Insulin signaling	Larvae, adults	Grönke et al. (2010)	p53 dominant negative mutation: decreased activity increases LS	Bauer et al. (2005a, b, 2007, 2010)
					Loss-of-function mutation: decreased activity increases LS	Grönke et al. (2010)
					UCP3 overexpression: increased activity decreases LS	Humphrey et al. (2009)
					See other data and references in the text	
Insulin-like peptide 3 (*Ilp3*); <u>InR/TOR pathway</u>	Insulin-like peptide	Insulin signaling	Larvae, adults	Grönke et al. (2010)	mUCP1 or hUCP2 overexpression: decreased activity increases LS	Fridell et al. (2009)
					DR: decreased activity increases LS	Grönke et al. (2010)

(continued)

Table 1.1 (continued)

Gene and pathway	Protein	Function in the nervous system	Stage	References	Function in life span control	References
Insulin-like peptide 5 (Ilp5); InR/TOR pathway	Insulin-like peptide	Insulin signaling	Larvae, adults	Grönke et al. (2010)	DR: decreased activity increases LS	Min et al. (2008)
Catecholamines up (catsup)	Negative regulator of tyrosine hydroxylase activity	Neurotransmitter synthesis	Adults	Stathakis et al. (1999)	Mutational variation is associated with LS	Carbone et al. (2006), Roshina and Pasyukova (2007)
Dopa decarboxylase (Ddc)	Dopa decarboxylase	Neurotransmitter synthesis	Adults	Livingstone and Tempel (1983), Stathakis et al. (1995)	Mutational and molecular variation is associated with LS	De Luca et al. (2003)
Odorant receptor co-receptor (Orco)	Odorant receptor co-receptor protein	Sensory perception of smell	Adults	Mukunda et al. (2014)	Loss-of-function mutation: decreased activity increases LS	Libert et al. (2007)
Gustatory receptor 63a (Gr63a)	Gustatory receptor	Detection of carbon dioxide	Adults	Kwon et al. (2007)	Loss-of-function mutation: decreased activity increases LS	Poon et al. (2010)
Metabotropic GABA-B receptor subtype 2 (GABA-B-R2)	Metabotropic GABA-B receptor subtype 2 subunit	Synaptic activity, olfactory behavior	Adults	Root et al. (2008)	RNAi knockdown: decreased activity decreases LS	Enell et al. (2010)
Cyclin-dependent kinase 5 (Cdk5)	Cyclin-dependent kinase 5 (Cdk5)	Axon guidance	Embryo	Connell-Crowley et al. (2000), Connell-Crowley et al. (2007)	*p35* null mutation: decrease in activity decreases LS	Connell-Crowley et al. (2007)
p35	Cdk5 activating subunit	Synaptic growth	Larvae	Kissler et al. (2009)		

(continued)

Table 1.1 (continued)

Gene and pathway	Protein	Function in the nervous system	Stage	References	Function in life span control	References
Embryonic lethal abnormal vision (elav)	mRNA binding protein	Neuron development and functional integrity	Embryo, larvae, pupae, adults	Robinow and White (1988, 1991), Toba et al. (2010)	Hypomorphic mutation: decreased activity decreases LS	Toba et al. (2010)
Insulin-like receptor (InR); InR/TOR pathway	Insulin-like 1 receptor	Dendrite pruning	Larval-pupal transition	Wong et al. (2013)	Hypomorphic mutation: decreased activity increases LS	Tatar et al. (2001)
					Intestine: moderately decreased activity increases LS	Biteau et al. (2010)
Pi3K92E; InR/TOR pathway	Phosphatidylinositol-3-kinase	Dendrite growth	Larvae	Parrish et al. (2009)	Intestine: moderately decreased activity increases LS	Biteau et al. (2010)
		Synaptic growth	Larvae, pupae	Dimitroff et al. (2012)		
		Dendrite pruning	Larval-pupal transition	Wong et al. (2013)		
		Synapse number and maintenance	Larvae, adults	Martin-Pena et al. (2006)		
Pten; InR/TOR pathway	Phosphatidylinositol-3,4,5-trisphosphate 3-phosphatase	Dendrite growth	Larvae	Parrish et al. (2009)	Muscles: increased activity increases LS	Demontis and Perrimon (2010)
		Dendrite pruning	Larval-pupal transition	Wong et al. (2013)		
		Dendrite and axon regeneration	Larvae	Song et al. (2012)		

(continued)

Table 1.1 (continued)

Gene and pathway	Protein	Function in the nervous system	Stage	References	Function in life span control	References
Akt1; InR/TOR pathway	Protein kinase B (PKB)	Synaptic growth	Larvae	Natarajan et al. (2013)	Intestine: moderately decreased activity increases LS	Biteau et al. (2010)
		Dendrite growth	Larvae	Parrish et al. (2009)		
		Dendrite pruning	Larval-pupal transition	Wong et al. (2013)		
		Dendrite and axon regeneration	Larvae	Song et al. (2012)		
AMP-activated protein kinase alpha subunit (*AMPK-alpha*); InR/TOR pathway	AMP-activated protein kinase alpha subunit	Axon guidance	Larvae, pupae	Dimitroff et al. (2012)	Muscles, fat body: increased activity increases LS, decreased activity decreases LS	Stenesen et al. (2013)
Gigas (*gig*); InR/TOR pathway	dTSC2	Synaptic growth	Larvae	Natarajan et al. (2013)	Ubiquitously increased activity increases LS, muscles, fat body: increased activity increases LS, no effect in the nervous system	Kapahi et al. (2004)
		Dendrite pruning	Larval-pupal transition	Wong et al. (2013)		
Ras homolog enriched in brain (*Rheb*); InR/TOR pathway	Small GTPase	Synaptic growth, axon guidance	Larvae, pupae	Dimitroff et al. (2012)		Data not found

(continued)

Table 1.1 (continued)

Gene and pathway	Protein	Function in the nervous system	Stage	References	Function in life span control	References
Target of Rapamycin (*Tor*); InR/TOR pathway	TOR kinase	Synaptic growth, axon guidance	Larvae, pupae	Dimitroff et al. (2012)	Fat body: decreased activity increased LS	Kapahi et al. (2004)
		Dendritic growth and branching	Larvae	Koike-Kumagai et al. (2009)		
		Dendrite pruning	Larval-pupal transition	Wong et al. (2013)		
Raptor; InR/TOR pathway	Regulatory associated protein of TOR	Axon guidance	Pupae	Dimitroff et al. (2012)		Data not found
Rapamycin-insensitive companion of TOR (*Rictor*) *SAPK-interacting protein 1* (*Sin1*); InR/TOR pathway	Rapamycin-insensitive companion of TOR, stress-activated-proteinkinase-interacting protein 1	Dendritic tiling	Larvae	Koike-Kumagai et al. (2009)		Data not found
		Synaptic growth	Larvae	Dimitroff et al. (2012)		
RPS6-p70-protein kinase (*S6k*); InR/TOR pathway	S6 protein kinase	Axon guidance	Pupae	Dimitroff et al. (2012)	Ubiquitously decreased activity increases LS, ubiquitously increased activity decreases LS, fat body: decreased activity increased LS	Kapahi et al. (2004)
		Dendritic growth and branching	Larvae	Koike-Kumagai et al. (2009)		
		Dendrite pruning	Larval-pupal transition	Wong et al. (2013)		

(continued)

Table 1.1 (continued)

Gene and pathway	Protein	Function in the nervous system	Stage	References	Function in life span control	References
Thor; InR/TOR pathway	Eukaryotic translation initiation factor 4E binding protein (4E-BP)	Dendrite pruning	Larval-pupal transition	Wong et al. (2013)	Ubiquitously increased activity increases LS	Zid et al. (2009)
					Muscles: increased activity increases LS	Demontis and Perrimon (2010)
Shaggy (*sgg*); InR/TOR pathway **cell polarity pathway**	Glycogen syntase kinase 3 beta (GSK3-beta)	Axonal transport	Larvae	Mudher et al. (2004)	Nervous system, lifelong: both increased and decreased activity decreases LS	Trostnikov et al. (2014), Roshina et al., unpublished
		NMJ growth	Larvae	Franco et al. (2004)		
		Formation of neural precursor cells	Larvae	Kanuka et al. (2005)		
		Presynaptic neurotransmitter release, synaptic growth	Larvae	Franciscovich et al. (2008)		
		Asymmetric neuroblast division (?)	Embryos	Colosimo et al. (2010), Kaplan et al. (2011)		
		Dendrite pruning	Larvae	Wong et al. (2013)		
		Synaptic activity	Larvae	Trostnikov et al. (2014)		
aPKC; **cell polarity pathway**	Atypical protein kinase C	Asymmetric neuroblast division	Embryos	Betschinger et al. (2003), Cai et al. (2003)	Loss of function mutation: decreased expression increases LS	Symonenko et al., unpublished
		Asymmetric neuroblast division	Larvae	Atwood and Prehoda (2009)		
		Synaptic structure	Larvae	Ruiz-Canada et al. (2004), Chang et al. (2010), Colosimo et al. (2010)		

(continued)

Table 1.1 (continued)

Gene and pathway	Protein	Function in the nervous system	Stage	References	Function in life span control	References
Escargot (*esg*); **cell polarity pathway**	Transcription factor	Asymmetric neuroblast division	Embryos	Ashraf et al. (1999), Ashraf and Ip (2001), Cai et al. (2001)	Insertion mutation: decreased activity increases LS	Magwire et al. (2010), Zaitsev et al. (2010)
Inscuteable (*insc*); **cell polarity pathway**	Cytoskelet associated protein	Asymmetric neuroblast division	Embryos	Cai et al. (2003)	Insertion mutation increases LS	Symonenko et al., unpublished
		Neuron specification	Embryos	Udolph et al. (2009)		
Locomotion defects (*loco*); **cell polarity pathway cAMP-PKA pathway**	Guanine nucleotide dissociation inhibitor and GTPase-activating protein	Asymmetric neuroblast division	Embryos	Yu et al. (2005)	Heterozygous amorphic mutation: decreased activity increases LS, mild ubiquitous decrease in activity increases LS, ubiquitous increase in activity decreases LS	Lin et al. (2011a, b)
Rpd3; global regulators	Histone deacetylase activity	Axon and dendrite arborization, dendrite guidance	Larvae	Parrish et al. (2006), Tea et al. (2010)	Null mutation, hypomorphic mutant: decreased activity increases LS	Rogina et al. (2002)

(continued)

Table 1.1 (continued)

Gene and pathway	Protein	Function in the nervous system	Stage	References	Function in life span control	References
hopscotch (*hop*); JAK/STAT pathway global regulators	Receptor-associated janus kinase (JAK)	Nervous system development	Embryos	Li et al. (2003)	Loss-of-function mutation: decreased activity increased LS, gain-of-function mutation: increased activity decreased LS	Larson et al. (2012)
		Memory	Adults	Copf et al. (2011)		
Signal-transducer and activator of transcription protein at 92E (*Stat92E*); JAK/STAT pathway global regulators	STAT transcription factor	Nervous system development	Embryos	Li et al. (2003)	Mutation: decreased? activity decreased LS	Larson et al. (2012)
		Memory	Adults	Copf et al. (2011)		
Extra sexcombs (*esc*); global regulators	H3 binding protein	Dendrite arborization	Larvae	Parrish et al. (2006)	Null mutation, dominant-negative mutant: decreased activity increases LS	Siebold et al. (2010)
Mnt; global regulators	Mnt transcription factor	NS development	Embryo; larvae	Loo et al. (2005)	Null mutation, hypomorphic mutation: decreased activity decreases LS	Loo et al. (2005)
		Neuron morphogenesis	Cell culture	Sepp et al. (2008)		

(continued)

Table 1.1 (continued)

Gene and pathway	Protein	Function in the nervous system	Stage	References	Function in life span control	References
p53	p53 transcription factor	Neuroblast differentiation	Larvae	Ouyang et al. (2011)	Adult nervous system: decreased activity increases LS	Bauer et al. (2005a, b)
					Muscles, fat body: no effect	Bauer et al. (2007)
Lim3	Lim3 transcription factor	Neuron specification	Embryos	Thor et al. (1999), Certel and Thor (2004)	Mutational and molecular variation is associated with LS	Roshina and Pasyukova (2007), Rybina and Pasyukova (2010)
					Embryos: both increased and decreased activity decreases LS	Rybina and Pasyukova (2010), Rybina et al., unpublished
Tail up (*tup*)	Islet transcription factor	Neuron specification	Embryos	Thor et al. (1999), Certel and Thor (2004)	Mutational variation is associated with LS	Roshina and Pasyukova (2007), Symonneko et al., unpublished
Shuttle craft (*stc*)	Stc transcription factor	Axon guidance	Embryos	Stroumbakis et al. (1996)	Mutational variation is associated with LS	Pasyukova et al. (2004)
					Embryos: decreased transcription increases LS	Roshina et al. (2014)

LS life span; *UCP* uncoupling protein; *DR* diet restriction impact on life span related, at least partially, to developmental functions or is it explained by other pleiotropic influences later in life? Here we address this question based on the data found in publications and our own findings.

1.2 Neuronal Genes in Life Span Control

1.2.1 Genes Coding for Ion Channels

Many *Drosophila melanogaster* genes are implicated in establishing ion channels in the nervous system. *Sh* and *Hk* (Table 1.1) encode alpha and beta subunits of a voltage-dependent Potassium channel (Papazian et al. 1987; Wilson et al. 1998). Mutations in these genes affect synaptic activity and lead to hyperexcitability followed by enhanced ramification of larval nerve terminals (Budnik et al. 1990). Both *Sh* and *Hk* mutants show a decrease in life span as compared to matched controls. The decrease in life span is more pronounced for *Hk* than for *Sh* (Trout and Kaplan 1970; Rogina and Helfand 1995).

The effects of *Sh* and *Hk* have been suggested to be mediated by elevated cAMP levels in response to hyperneural activities, because *dnc* (Table 1.1) mutants with reduced phosphodiesterase activity, and hence higher cAMP levels, also cause enhanced nerve terminal arborization (see Zhong and Wu 2004 for references). Several genes involved in the cyclic AMP/Protein kinase A (cAMP/PKA) pathway affect both functions of the nervous system and life span. Mutations of the genes *rut* (Table 1.1) and *dnc* responsible for synthesis and degradation of cAMP, respectively, affect stability and fine tuning of synaptic structure and function (Baines 2004; Renger et al. 2000; Zhong and Wu 2004), learning, and memory (Dudai et al. 1976; Byers et al. 1981; McGuire et al. 2003). Mutations at *rut* and *dnc* also affect life span (Tong et al. 2007). Postsynaptic expression of a constitutively active form of PKA is sufficient to increase averaged synaptic current, while inhibition of this kinase results in a significant reduction in averaged synaptic current amplitude (Baines 2004). Strong modifications of the activity of the major PKA catalytic subunit encoded by *PKA-C1* (Table 1.1) also impair olfactory memory and learning (Drain et al. 1991; Skoulakis et al. 1993; Yamazaki et al. 2007; Gervas et al. 2010). Ubiquitously overexpressed mouse PKA was shown to affect *Drosophila* life span (Tong et al. 2007). cAMP/PKA pathway also mediates effects of *NF1* (Table 1.1) on longevity (Tong et al. 2007). Neurofibromin, the protein product of the *NF1* gene, affects synaptic growth, synaptic activity (Tsai et al. 2012), memory, and learning (Buchanan et al. 2000; Buchanan and Davis 2010). Multiple studies have indicated that *NF1*-dependent learning in *Drosophila* involves the cAMP pathway (see Buchanan and Davis 2010 for references): NF1 positively regulates the Ca^{2+}/calmodulin–sensitive adenylyl cyclase and, consequently, inactivation of NF1 results in downregulation of cAMP/PKA signaling. Thus, in this case, the same pathway is involved in both maintenance of specific neuronal functions and life span control.

Voltage-gated Sodium channels are necessary for initiation and propagated transmission of the action potentials which underlay many animal activities. In *Drosophila*, the gene *para* (Table 1.1) encodes the major voltage-gated Sodium channel and the gene *mle* (Table 1.1) encodes ATP-dependent double-stranded RNA helicase. Mle protein is required for adenosine-to-inosine RNA editing and proper expression of *para* (Reenan et al. 2000). Both genes are involved in regulation of synaptic activity (Reenan et al. 2000; Zhong and Wu 2004) and longevity control (Reenan and Rogina 2008). Decreased expression of Sodium channels in flies harboring the mutant *mle* allele confers rapid and reversible temperature dependent paralysis, because of failure of action potential propagation. The *mle* flies also exhibit decreased life span. Paralysis and decreased life span of *mle* mutants are partially rescued by increasing the dosage of *para* (Reenan and Rogina 2008).

1.2.2 Genes Coding for Neuropeptides

In *Drosophila melanogaster*, more than 40 genes encode precursors of neuropeptides, peptide hormones and protein hormones, and a large number of different neuropeptides has been identified in a variety of neuron types (for review, see Nassel and Winther 2010). The insulin-like peptides stay alone in the list of *Drosophila* neuropeptides since they are not necessarily always neuropeptides in other organisms. Seven genes encode insulin-like peptides (ILPs) in *Drosophila melanogaster*. Of these, ILP2, 3 and 5 display similarities to mammalian insulins and are found in the 14 insulin producing cells (IPCs) embedded in a cluster of median neurosecretory cells in the *pars intercerebralis* of the CNS (for review, see Nassel et al. 2013). Ablation of the IPCs significantly extends life span (Fig. 1.1, Broughton et al. 2005; Haselton et al. 2010). ILPs are secreted into hemolymph and spread between target tissues where they activate insulin-like signaling (Grönke et al. 2010).

ILP2, 3 and 5 are encoded by three genes, *ilp2*, *ilp3*, and *ilp5* (Table 1.1). Transcription level of *ilp2* is regulated by p53 transcription factor. Overexpression of the *p53* dominant-negative mutation in neurons reduces the amount of *ilp2* transcript, activity of insulin-like signaling pathway in the fat body of flies and increases life span (Fig. 1.1, Bauer et al. 2005a, b, 2007, 2010). In flies with a loss-of-functions mutation in *ilp2* life span is also increased (Fig. 1.1, Grönke et al. 2010). Inactivation of insulin-like signaling in the fat body of the head results in reduced transcription of *ilp2* in ICPs and increased life span (Fig. 1.1, Hwangbo et al. 2004). Similarly, the stress-responsive Jun kinase (JNK) in the IPCs promotes longevity by downregulating ILP2 through activation of FOXO (Fig. 1.1, Wang et al. 2003). Consistent with this, elevation of ILP2 amount in ICPs is associated with reduced life span (Fig. 1.1, Humphrey et al. 2009; Enell et al. 2010). All results presented above suggest that the level of expression of *ilp2* in ICPs plays a key role in life span control.

However, a number of other results contradict this hypothesis. For instance, RNAi knockdown of *ilp2* does not change life span (Fig. 1.1, Broughton et al. 2008). Also, overexpression of mouse uncoupling protein 1 and human uncoupling

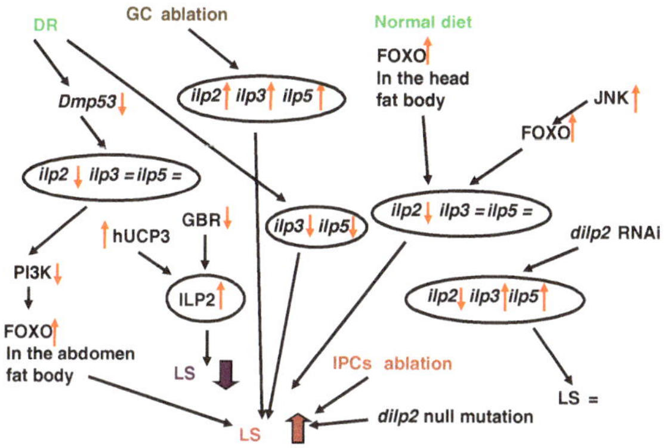

Fig. 1.1 *Drosophila melanogaster* insulin-like peptides (*ILPs*) in life span control. The effects of ILPs on life span are shown according to Hwangbo et al. (2004), Bauer et al. (2005a, b, 2007, 2010), Wang et al. (2003), Broughton et al. (2005, 2008), Flatt et al. (2008), Humphrey et al. (2009), Enell et al. (2010), Grönke et al. (2010), Haselton et al. (2010). *Yellow arrows* direction of changes in the amount of transcripts or proteins; *blue/red arrows* direction of changes in life span. *DR* dietrestriction; *GC* gonad cells; *FOXO* forkhead box protein; *PI3K* phosphatidylinositol-3 kinase; *hUCP3* human uncoupling protein 3; *GBR* gamma-aminobutyric acid B receptor; *ISPs* insulin producing cells; *LS* life span

protein 2 in *Drosophila* neurons increases life span but affects transcription of *ilp3*, not *ilp2* (Fridell et al. 2009). Diet restriction is associated with an increase in life span and with a reduction in the level of transcription of *ilp5* in adults (Min et al. 2008) and *ilp3* in larvae (Fig. 1.1, Grönke et al. 2010). Possibly, different ILPs synthesized in the central nervous system and slightly different in function can in some cases compensate for the function of each other. Indeed, a decrease in *ilp2* expression is accompanied by an increase in *ilp3* and *ilp5* expression, and both ILP6 produced in the fat body and ILP7 produced in gonads are also involved in complex interactions among ILPs (Broughton et al. 2008; Grönke et al. 2010). Interaction with ILP7 might explain why eliminating germ cells leads, in contrast to other data, to both an increase in life span and an increase in the level of expression of all three genes encoding ILPs in the CNS, (Fig. 1.1, Flatt et al. 2008).

The brain IPCs that produce DILP2, 3 and 5 are directly regulated by a few neurotransmitters and neuropeptides. Serotonin, octopamine, gamma-aminobutyric acid (GABA), short neuropeptide F, corazonin and tachykinin-related peptide have been identified in *Drosophila* as regulators of IPCs (for review, see Nässel et al. 2013). All these regulators might be involved in life span control, however, this has not yet been proved experimentally.

1.2.3 Genes Coding for Neurotransmitters Synthesis

Dopamine, a biogenic amine, plays a role as a neurotransmitter and a hormone in *Drosophila*. The conversion of tyrosine to l-DOPA catalyzed by tyrosine hydroxylase is the rate-limiting step in dopamine biosynthesis. *Catsup* (Table 1.1) encodes a negative regulator of tyrosine hydroxylase (Stathakis et al. 1995, 1999). DOPA is then converted to dopamine by dopa decarboxylase encoded by *Ddc* (Table 1.1). *Ddc* also decarboxylates 5-hydroxytryptophan and transforms it into serotonin, another neurotransmitter (Livingstone and Tempel 1983; Stathakis et al. 1995).

A genome-wide screen for genes affecting life span followed by quantitative complementation tests with deficiencies and mutations revealed *Catsup* and *Ddc* as candidate genes involved in life span regulation (Nuzhdin et al. 1997; Pasyukova et al. 2000; De Luca et al. 2003; Roshina and Pasyukova 2007). This result was confirmed by analyses of structural molecular variation at *Catsup* (Carbone et al. 2006) and *Ddc* (De Luca et al. 2003) in a natural population. *Catsup* and *Ddc* polymorphisms located within functional and 5′ regulatory gene regions are significantly associated with life span.

1.2.4 Genes Coding for Receptors

Sensory neurons perceive environmental cues and transmit them to non-neuronal tissues via neural circuits that consist of other types of neurons. Genes coding for receptors that are expressed in different types of sensory neurons (Table 1.1) have been found to affect life span in *Drosophila*.

Orco (*Or83b* in original paper) is broadly expressed throughout olfactory tissues and codes for co-receptor protein that interacts with conventional odorant receptors and is required for their localization to the neuronal dendrites; loss-of-function mutations in *Orco* reduce physiological and behavioral responses to a wide range of odorants (for references, see Libert et al. 2007). *Orco* participates in sensory perception of smell (Mukunda et al. 2014). The loss-of-function mutation of *Orco* significantly extends male and female life span (Libert et al. 2007). Mutation also produces severe olfactory defects, enhanced stress resistance, and altered adult metabolism.

Drosophila detect CO_2 using a small subpopulation of sensory neurons that are located in the antennae and innervate a single glomerulus in the antennal lobe; these neurons express at least two gustatory receptor genes, *Gr63a* and *Gr21a*, which together comprise a CO_2 odorant receptor (for references, see Poon et al. 2010) and participate in detection of carbon dioxide (Kwon et al. 2007). Loss-of-function mutation of *Gr63a* results in a lack of functional CO_2 receptor and extended longevity accompanied by improved resistance to environmental stresses (Poon et al. 2010). Unlike *Orco*, *Gr63a* is expressed in a highly specific population of neurons, which demonstrates that specific sensory cues and associated neural circuit have the ability to modulate fly life span.

Pheromone-sensing olfactory receptor neurons express a high level of the metabotropic GABA B receptor (GABA-B-R2) essential for synaptic activity and olfactory behavior (Root et al. 2008). Expression of GABA-B-R2 was also revealed in IPCs, suggesting that GABA is involved in regulation of these neurosecretory cells. RNAi knock down of *GABA-B-R2* specifically in the IPCs increased ILP amounts in these cells and shortened life span (Enell et al. 2010). Authors suggest that GABA-B-R2 is involved in inhibitory control of ILP production and release in adult flies and that this receptor mediates signals from brain interneurons that may convey environmental nutritional cues.

1.2.5 Neuron-Specific Genes

Several *Drosophila* genes that function almost exclusively in the nervous system are considered here.

In mammals, cyclin-dependent kinase 5 (Cdk5) is a member of the family of cyclin-dependent kinases, however, unlike the cell-cycle kinases, activity of Cdk5 is largely restricted to postmitotic neurons (for references, see Connell-Crowley et al. 2000, 2007). In *Drosophila*, Cdk5 encoded by the *Cdk5* gene (Table 1.1) and p35, a small protein necessary for the activation of Cdk5, expressed, in particular, in postmitotic neurons and encoded by the *p*35 gene (Table 1.1), form a complex that regulates the accuracy of neural wiring, the growth of axons and the formation of synapses (Kissler et al. 2009; Connell-Crowley et al. 2000). Cdk5 also participates in the phosphorylation of the tau protein connected with the development of certain pathologies of the nervous system (for review, see Noble et al. 2013). Flies that are mutant for the *dp*35 gene are short-lived and demonstrate age-dependent degradation of motor function (Connell-Crowley et al. 2007).

Elav protein encoded by the gene *elav* (Table 1.1) is a member of a family of mRNA binding proteins involved in the post-transcriptional regulation of gene expression, including alternative splicing and translation (for references, see Toba et al. 2010). *elav* is expressed exclusively in the nervous system, in every neuron at all developmental stages, and is involved in the development and maintenance of the functional integrity of neurons (Robinow and White 1988, 1991). A hypomorphic temperature-sensitive *elav* mutation decreases male life span (Toba et al. 2010).

1.2.6 Conclusions

Data on the role of neuronal genes in the control of longevity are rather sketchy, except for ILPs. Only in this later case the mechanism of life span-promoting effects is reasonably well understood and can be attributed to the well described evolutionary conserved insulin pathway also known to play a central role in regulating various aspects of growth, development, metabolism and reproduction. Not quite consistent with this, synthesis of ILPs in specialized neuronal cells is not conserved among taxa. However, in most species analyzed to date, neurons are involved in production and release of ILPs that regulate life span by influencing insulin-like signaling in remote tissues.

The empirical evidence at hand today indicate that another pathway that appears to mediate participation of neuronal genes in life span control is cAMP/PKA pathway, a signaling cascade used in cell communication. Several genes/proteins that are a part of this pathway affect life span, including the key enzyme, PKA. Multiple functions of PKA open up the possibility for much speculation regarding the alleged role of this enzyme in both interaction with neuronal genes and life span control. However, it remains to be further elucidated how exactly cAMP/PKA signaling might provide a cross-talk between neuronal genes and life span.

The sensory neurons constitute perceptual systems that continuously evaluate environment and provide information for a wide range of behavioral decisions essential for locomotor, feeding, mating and other life-asserting responses in the daily life of a fly. Evidently, this indicates a clear link between sensory cues, corresponding neural circuits and longevity programs. It is therefore not surprising that several observations strongly support the importance of proper receptor function for longevity.

Data presented above allow directly connecting longevity with specific functions of the nervous system. Not only neuronal genes themselves, but also genes providing general functions, yet exclusively in the nervous system, such as *Cdk5*, *p35* and *elav*, have an impact on aging. While much future work is needed for a detailed understanding of the underlying regulatory mechanisms, the available studies in *Drosophila* to date clearly show that different types of neuronal genes are important for life span determination.

1.3 Developmental Neuronal Pathways in Life Span Control

1.3.1 InR/TOR Pathway

Among the molecular pathways known to affect longevity, the InR and TOR pathways are perhaps the most important, mainly due to their major, evolutionarily conserved effects on life span in various model organisms (for review, see Tatar et al. 2003; Kapahi et al. 2010; Katewa and Kapahi 2011; Partridge et al. 2011). Signaling through the InR and TOR pathways can act both in parallel but also interact with each other (Fig. 1.2), though only few experiments have directly explored the interactions between these pathways in terms of life span (Katewa and Kapahi 2011). The role of the InR/TOR signaling in life span control was thoroughly reviewed elsewhere and is beyond the scope of this paper. Here we review data on the role of the InR/TOR signaling in the development and function of the nervous system.

Tissue-specific overexpression technique was used to understand how misregulation of the InR/TOR signaling affects neural and behavioral development (Dimitroff et al. 2012). Neural overexpression of *Pi3K*, a principal mediator of growth factor inputs to TOR, causes synapse overgrowth, while neural overexpression of RHEB, the direct downstream target inhibited by Gigas (Table 1.1, Fig. 1.2), produces significant synapse overgrowth, axon misrouting, and phototaxis deficits. Overexpression of *AMPK* (Table 1.1 and Fig. 1.2), coding for a component of the cellular energy sensing pathway, rescues behavioral and axon guidance deficits, but is not able to rescue synapse overgrowth. Altering the function of Raptor, a TOR complex 1 (TORC1) component, or a TORC1 downstream element S6K (Table 1.1 and Fig. 1.2) affects axon guidance and behavior. Reducing the function of TOR complex 2 (TORC2) components Rictor or Sin1 (Table 1.1 and Fig. 1.2) suppresses synapse overgrowth. According to Dimitroff et al. (2012), these findings demonstrate that different inputs to InR/TOR signaling have specific activities in the nervous system development, and that *Tor* provides a connection between nutrient-energy sensing systems and patterning of the nervous system. The role of TPRC2 in regulating proper synaptic growth was further confirmed by (Natarajan et al. 2013). *Gigas* and *Rictor* mutants show increased synaptic growth, whereas *Raptor* knockdown has no effect on this trait. Furthermore, in *gigas* mutants the levels of phosphorylated Akt are dramatically decreased and *Akt* mutants phenocopy *gigas* mutants, leading to the conclusion that *gigas* and *Akt* work via the same genetic pathway to regulate synaptic growth.

The InR/TOR pathway also regulates dendrite growth. Early in development dendrites must grow in concert with animal growth to maintain proper connectivity of tissues. This phenomenon, referred to as scaling growth of dendrites, requires the function of the microRNA *bantam* (*ban*) in the epithelial cells, which affects Akt activity in adjacent neurons (Parrish et al. 2009). *Akt* expression is increased in neurons but reduced in epithelial cells of *ban* mutants, and the amount of active, phosphorylated Akt and phosphorylated S6K, a downstream target of Akt (Fig. 1.2) is reduced. Activation of *Pi3k* leads to activation of Akt (Fig. 1.2) and a significant increase in dendrite coverage, while overexpression of Akt antagonist, *Pten* (Fig. 1.2) in neurons causes a significant reduction in dendrite coverage (Parrish et al. 2009). Components of the TORC2: TOR, Rictor and Sin1 are also required for dendritic tiling. TORC2 components physically and genetically interact with the kinase Tricornered: TORC2 is essential for its phosphorylation on a residue that is critical for Tricornered activity (Koike-Kumagai et al. 2009).

The selective removal of unnecessary or exuberant neuronal processes without loss of neurons, referred to as pruning, is a central event in the maturation of the nervous system during animal development. Several genes that are a part of InR/TOR pathway: *InR*, *Pi3K*, *Pten*, *Akt1*, *gig*, *TOR*, *S6k*, *Thor* (Table 1.1) were shown to participate in the control of dendrite pruning during larval-pupal transition (Wong et al. 2013). The F-box protein Slimb forms a complex with Akt, an activator of the InR/TOR pathway, and promotes Akt ubiquitination. Authors claim that activation of the InR/TOR pathway is sufficient to inhibit dendrite pruning.

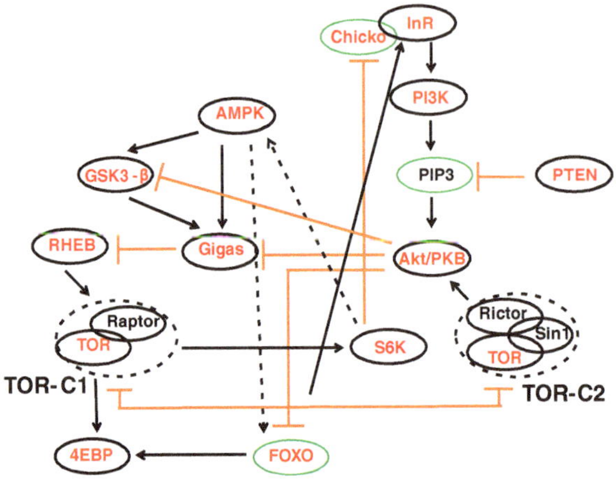

Fig. 1.2 Insulin/TOR signaling in *Drosophila melanogaster*: interacting genes and proteins. Gene (protein) interactions are shown according to Blagosklonny (2009), Feng (2010), Kapahi et al. (2010), Madeo et al. (2010), Partridge et al. (2011), Van Dam et al. (2011), Troulinaki and Bano (2012). Genes (proteins) involved in life span control are in *red* (for references, see Table 1.1). *Dashed ellipses* protein complexes. *Green ellipses* genes were not shown to participate in the nervous system development and function (see text for explanations). *Black arrows* activation; *yellow lines* inactivation; *dashed lines* interactions were shown for other organisms. *InR* insulin-like receptor; *PI3K* phosphatidylinositol-3 kinase; *PTEN* phosphatidylinositol-3,4,5-trisphosphate phosphatase; *PIP3* phosphatidylinositol-3,4,5-trisphosphate; *Akt*/PKB protein kinase B; *AMPK* AMP-activated protein kinase; *Gigas* (*dTSC2*) tuberosis sclerosis complex 2; *Rheb* RAS homolog enriched in brain; *TOR* target of rapamycin kinase; *Raptor* regulatory associated protein of TOR; *Rictor* rapamycin-insensitive companion of TOR; *Sin1* stress-activated protein kinase interacting protein 1; *S6K* S6 kinase; *4EBP* 4E binding protein; *FOXO* forkhead box protein

Importantly, the InR/TOR signaling also has specific neuronal functions in adult flies. For example, the levels of Pi3K regulate synapse number in adult brain projection neurons and is necessary for synapse maintenance (Martin-Pena et al. 2006). Moreover, Pi3K activation induces synaptogenesis in aged adult neurons as well. Remarkably, *Pten* loss-of-function as well as *Akt* overexpression result in increased dendrite branching and regeneration (Song et al. 2012). Altogether, InR/TOR signaling participates in axon and dendrite maintenance and regeneration in the CNS (Martin-Pena et al. 2006; Song et al. 2012).

1.3.2 Glycogen Syntase Kinase 3 Beta

Serine-threonine protein kinase Glycogen syntase kinase 3 beta (GSK3-beta) encoded by the gene *sgg* (Table 1.1) is another participant in the InR/TOR cascade. When active, GSK-beta has been shown to directly phosphorylate and activate Gigas when primed by AMPK-dependent phosphorylation (Fig. 1.2, for references, see Inoki and Guan 2006). Therefore, inactivation of GSK3-beta alleviates Gigas-driven inhibition of Rheb and results in TOR activation.

GSK3-beta has multiple roles in the nervous system development and function. In some cases, GSK3-beta effects are mediated by the InR/TOR signaling. Like this, GSK3-beta participates in the control of dendrite pruning during larval-pupal transition (Wong et al. 2013). However, other pathways also mediate *sgg* effects. For example, GSK3-beta is cleaved by the Dark-dependent caspase, and this cleavage converts it to an active kinase, which contributes to the development of neural precursor cells (Kanuka et al. 2005). *sgg* is able to negatively influences synaptic growth by modulating the Jun-N-terminal kinase pathway, and also regulates presynaptic neurotransmitter release at the larval neuromuscular junction (Franciscovich et al. 2008). GSK3-beta also negatively controls the neuromuscular junctions growth and the microtubule cytoskeleton dynamics in motoneurons (Franco et al. 2004). GSK3-beta directly phosphorylates tau protein (for review, see Noble et al. 2013). Overexpression of tau disrupts axonal transport causing vesicle aggregation, and co-overexpression of constitutively active GSK3-beta enhances the effects (Mudher et al. 2004). GSK3-beta also directly phosphorylates atypical Protein kinase C (aPKC) (Fig. 1.3, Colosimo et al. 2010), a key component ensuring asymmetric neuroblast division during *Drosophila* early development.

We addressed the question whether *sgg* affects life span in *Drosophila*. Panneuronal overexpression of *sgg* decreases the life span of both males and, at a lesser extent, females (Trostnikov et al. 2014). On the contrary, life span is increased in sgg^{EP1576} and $sgg^{EY02862}$ mutant females compared to controls (Fig. 1.4a). Of note, both mutations are caused by insertions of the vector constructs P{EP} or P{EPgy2} into the first intron of the gene and were not activated by a driver. Accordingly, in these experiments, the effects on life span are associated with gene disruptions but not with *sgg* overexpression. Overexpression of sgg^{EP1576} demonstrates an intrinsic role for GSK3-beta in rhythmic locomotor activity (Stoleru et al. 2007), and overexpression of $sgg^{EY02862}$ in differentiated post mitotic neurons suppresses synaptic growth (Franciscovich et al. 2008).

1.3.3 Cell Polarity Pathway

The above studies demonstrating the role of *sgg* in life span control attracted our attention to other genes involved in asymmetric neuroblast division as to potential aging genes.

Delaminated neuroblasts in *Drosophila* function as stem cells during embryonic central nervous system development. They go through repeated asymmetric divisions to generate multiple ganglion mother cells, which divide only once more to produce postmitotic neurons. Polarity of neuroblasts is controlled by a protein complex consisting of Bazooka (Baz), PAR-6 and aPKC (Fig. 1.3, Wodarz et al. 1999, 2000; Cai et al. 2003). aPKC is the main signaling component of this complex that functions by phosphorylating downstream targets, while the PDZ domain proteins Baz and PAR-6 control the subcellular localization and kinase activity of aPKC. GSK3-beta directly phosphorylates aPKC, which likely promotes its ubiquitin-mediated proteosomal degradation (Fig. 1.3, Colosimo et al. 2010). Apically localized aPKC is required for segregation of neuronal differentiation factors such as Prospero and Numb and the adaptor proteins Miranda and Partner of Numb to the basal cortical domain, to ensure their segregation into the basal daughter cell (Fig. 1.3). Numb and Miranda are polarized by direct aPKC phosphorylation (Smith et al. 2007; Atwood and Prehoda 2009), with the help of the cytoskeletal protein Lethal (2) giant larvae that is also phosphorylated by aPKC (Fig. 1.3) and thus released from its association with membranes and the actin cytoskeleton (Betschinger et al. 2003). Of note, as it was demonstrated recently, aPKC also affects synaptic structure in larvae (Ruiz-Canada et al. 2004; Chang et al. 2010; Colosimo et al. 2010).

When neuroblasts delaminate from the epithelial layer, Baz and Inscuteable (Insc) (Kraut and Campos-Ortega 1996; Kraut et al. 1996; Wodarz et al. 1999) concentrate in the apical stalk, maintaining the polarity cues for neuroblasts. Partnerof-Inscuteable (Pins), Locomotion defects (Loco), and Gαi protein join Insc (Fig. 1.3), forming an

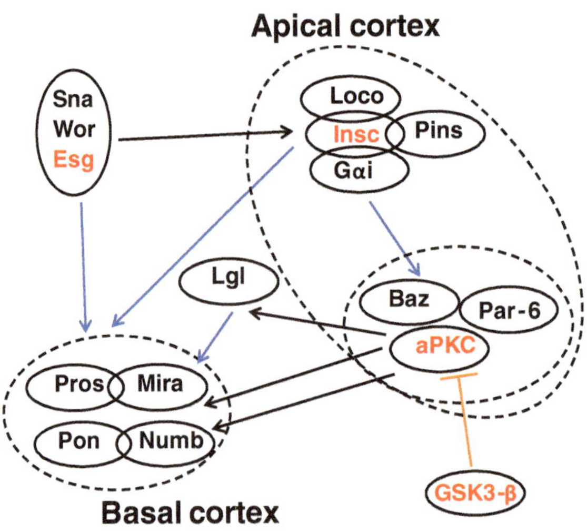

Fig. 1.3 Asymmetric neuroblast division in *Drosophila melanogaster*: interacting genes and proteins. Gene (protein) interactions are shown according to Ashraf and Ip (2001), Cai et al. (2001), Betschinger et al. (2003), Atwood and Prehoda (2009), Colosimo et al. (2010). Genes (proteins) involved in life span control are in *red* (for references, see Table 1.1). *Dashed ellipses* protein complexes. *Blue arrows* instability or mislocalization; *black arrows* activation; *yellow lines* inactivation. *Sna* snail; *Wor* worrier; *Esc* escargot; *Insc* inscrutable; *Pins* partner of inscutable; *Loco* locomotion defects; *Gαi* G protein; *Baz* Bazooka; *Par-6* Parkin-6; *aPKC* atypical protein kinase C; *GSK3- β* Glycogen syntase kinase 3 β; *Pros* Prospero; *Mira* Miranda; *Pon* Partner of numb; *Lgl* Lethal (2) giant larvae

apically localized functional complex before the neuroblast enters mitosis (Schaefer et al. 2000; Yu et al. 2000, 2003, 2005). Pins, Loco, and Gαi function redundantly with the Baz/aPKC/PAR-6 complex in regulating spindle geometry. Loco and Pins, through their GoLoco motifs, acts as a guanine nucleotide dissociation inhibitors (GDI) for Gαi protein. Furthermore, the regulator of G protein signaling (RGS) domain of Loco can also activate the GTPase activity of Gαi protein to regulate the equilibrium between the GDP- and the GTP-bound forms of Gαi. Thus, Loco can potentially regulate heterotrimeric G-protein signaling via two distinct modes of action during *Drosophila* neuroblast asymmetric divisions (Yu et al. 2005).

Three snail family genes *snail*, *worniu*, and *esg* (Table 1.1) encode related zinc finger transcription factors that have redundant and essential functions in modulating asymmetric neuroblast division: they activate *insc* transcription and translation (Fig. 1.3, Ashraf and Ip 2001; Cai et al. 2001).

We assessed the life span of females heterozygous for $aPKC^{k06403}$ loss-of-function mutation. Decreased function of aPKC results in a significantly increased life span (Fig. 1.4). Effects on life span were also revealed for mutations in *insc* and *esg*. $insc^{EY09668}$ and $esg^{BG01042}$ mutant females live longer than their matched controls (Fig. 1.4c, d). Experiments with $esg^{BG01042}$ mutant flies also allowed to directly prove the causal association between changes in life span and *esg* mutation (Magwire et al. 2010) and an association of increased life span with decreased

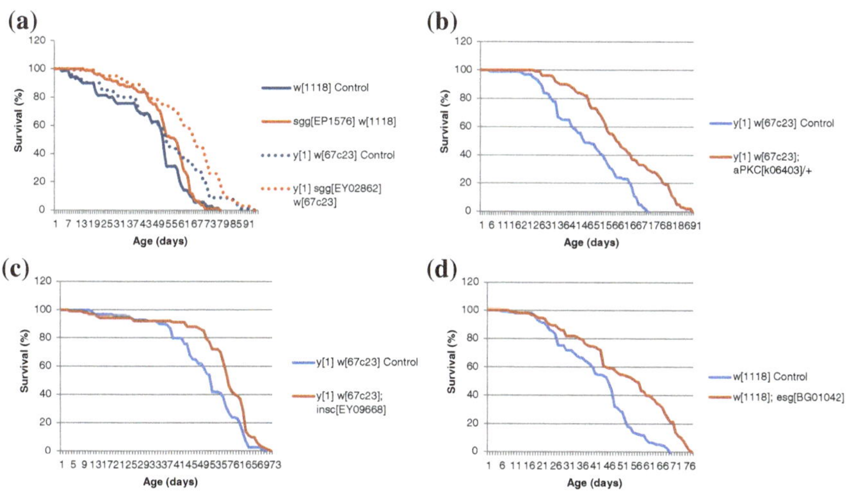

Fig. 1.4 Mutations in neuronal genes affect female life span. Lines used: **a** w^{1118} (control) and $P\{EP\}sgg^{EP1576}$ w^{1118}; y^1 w^{67c23} (control) and y^1 $P\{EPgy2\}sgg^{EY02862}$ w^{67c23}; **b** y^1 w^{67c23} (control) and y^1 w^{67c23}; $P\{lacW\}aPKC^{k06403}/CyO$ (life span was measured in hybrids between control and mutant females); **c** y^1 w^{67c23} (control) and y^1 w^{67c23}; $P\{EPgy2\}insc^{EY09668}$; **d** w^{1118} (control) and w^{1118}; $esg^{BG01042}$. Five virgin females of the same genotype, all collected on the same day from cultures with moderate density, were placed in replicate vials and maintained at 25 °C. Flies were transferred to vials with fresh food containing approximately 5 mL of standard medium without live yeast on the surface weekly. Dead flies were recorded daily. Sample sizes were 100–150 flies/genotype

esg transcription (Zaitsev et al. 2010), consistent with the results obtained for *aPKC* and *insc*.

Reduced expression of Loco results in a longer life span and a stronger resistance to different stressors. In addition, the reduction in Loco expression decreases cAMP levels. In contrast, overexpression of both genomic and cDNA *loco* gene significantly shortens the life span with weaker stress resistance. Deletion analysis of the Loco demonstrates that its RGS domain is required for the regulation of longevity (Lin et al. 2011a, b). Accordingly, Lin et al. (2011a, b) proposed that Loco activates inhibitory Gαi•GTP protein to reduce activity of adenylate cyclase and may regulate stress resistance and longevity as an activator of cAMP-PKA pathway.

1.3.4 Global Regulators

Global regulators of gene expression programs play multiple roles in maintaining genome integrity and providing epigenetic modulation of expression patterns during development and throughout life span. In 1997, Villeponteau proposed a heterochromatin loss model of aging. According to this model, heterochromatin domains, which are set up early in embryogenesis, are gradually lost with aging, resulting in derepression of silenced genes and aberrant gene expression patterns (Villeponteau 1997). Experimental tests of the role of heterochromatin formation in aging, however, have produced controversial results. It was shown that mutations at *Su(var)2-1* and *Su(var)205* loci have virtually no effect on life span (Rogina et al. 2002; Frankel and Rogina 2005). At the same time, increasing and decreasing HP1 levels shortens and prolongs life span, respectively (Larson et al. 2012). Here we review data on longevity effects of global regulators that were also shown to affect the development and function of the nervous system.

The histone deacetilase encoded by the gene *Rpd3* (Table 1.1) was shown to affect axon and dendrite arborization and dendrite guidance (Parrish et al. 2006; Tea et al. 2010). Interestingly, overexpression of Pros can suppress *Rpd3* mutant phenotypes (Tea et al. 2010). Authors suggest a specific function for the general chromatin remodeling factor Rpd3 in regulating neuron development, largely through the postmitotic action of the Pros transcription factor. It was also shown that heterozygous hypomorphic or loss-of-function *Rpd3* mutants have extended life span (Rogina et al. 2002). Authors present evidences that the effect on life span is not associated with the overall increase in histone acetylation.

JAK/STAT signaling have two roles: in the canonical pathway, JAK/STAT directly regulates target gene expression; in the non-canonical pathway, unphosphorylated STAT is involved in heterochromatin formation. In the non-canonical function, therefore, loss of STAT has the same effects as JAK overactivation, causing heterochromatin destabilization (for references, see Larson et al. 2012). In *Drosophila*, JAK is encoded by the gene *hop* (Table 1.1) and STAT is encoded by the gene *Stat* (Table 1.1). Lack of both JAK and STAT activation causes defects in the growth and organization of axonal projections in embryos (Li et al. 2003), indicating the role of canonical signaling in the development of the nervous system. Cytokine signaling through the JAK/STAT pathway is also required for long-term memory in *Drosophila* (Copf et al. 2011). On the contrary, non-canonical JAK/STAT pathway was shown to affect life span (Larson et al. 2012). Flies heterozygous for a loss-of-function *hop* allele have longer life span compared to controls, while flies heterozygous for a gain-of-function *hop* allele and flies heterozygous for a *Stat* mutation are short-lived. It is not known yet whether non-canonical JAK/STAT pathway plays a role in the nervous system.

Polycomb Group (PcG) and Trithorax Group proteins are key epigenetic regulators of global transcription. The H3 binding protein, a core subunit of the PcG complex, encoded by the gene *esc* (Table 1.1), was revealed in a genome-wide analysis as a factors required for proper morphogenesis of *Drosophila* sensory neuron dendrites (Parrish et al. 2006). It is not yet known whether its partner, another core subunit of the PcG complex, the histone H3 lysine 27-specific methyltransferase encoded by the gene *E(Z)*, also plays a role in neuron morphogenesis. However, both genes were shown to affect life span and stress-resistance (Siebold et al. 2010). Mutations both increase longevity and stress-resistance and reduce adult levels of trimethylated H3K27, which allowed authors to suggest that both the longevity and stress resistance phenotypes of PcG mutants are specifically due to violation of Polycomb silencing.

The Myc-Max-Mad-Mnt conserved transcription factors of the basic helix-loop-helix zipper class are supposed to function together as a molecular module to transcriptionally regulate cell growth, proliferation, and differentiation (for references, see Loo et al. 2005). In *Drosophila*, Mnt was identified and shown to affect the nervous system development in embryos (Loo et al. 2005) and the neuron morphogenesis in cell culture (Sepp et al. 2008). Null *Mnt* mutants have a decreased life span and an increased body size (Loo et al. 2005).

1.3.5 Transcription Factors

Transcriptional cascades have long been considered as important regulators of many biological processes including the nervous system development (Skeath and Thor 2003). Individual transcription factors may also play a crucial role in regulation of development, homeostasis, metabolism, growth, and life span, as exemplified, for instance, by the transcription factor Forkhead Box Protein (for review, see Puig and Mattila 2011). Several transcription factors have already been considered above in the context of their neuronal functions and participation in the life span control via ILPs signaling (p53) and alleged association with asymmetric neuroblas division (Esg). Here we review data on several other transcription factors that are involved in the development of the nervous system and longevity control.

In *Drosophila, Lim3* and *tup* (Table 1.1) encode two RNA polymerase II transcription factors required for development and function of neurons. With Drifter and Apterous, Lim3 and Tup constitute a "combinatorial code" that generates distinct motor neuron identities during embryonic development (Thor et al. 1999; Certel and Thor 2004). Mutations at *Lim3* locus and natural polymorphisms at the regulatory region of the *Lim3* main transcript are associated with life span variation (Roshina and Pasyukova 2007; Rybina and Pasyukova 2010). A naturally occurring structural variation causes a sixfold change in gene transcription and a 25 % change in life span. We hypothesized that polymorphic markers associated with *Lim3A* expression are located within the binding sites for proteins that regulate gene function, and provide general rather than tissue-specific regulation of transcription, and that intermediate levels of *Lim3A* expression confer a selective advantage and longer life span. Analysis of a reporter gene expression governed by different polymorphic variants of the *Lim3* regulatory region in transgenic flies confirmed the functionality of naturally occurring structural variation (Rybina et al. manuscript in preparation). Mutations at *tup* locus are associated with life span variation (Roshina and Pasyukova 2007).Our preliminary data also indicate that an insertion at the 5′ regulatory region of *tup* increases male and female life span, and precise reversion of the mutation leads to the restoration of the life span phenotype specific to control flies, indicating the causal association between changes in life span and *tup* mutation (Symonenko et al. unpublished).

stc (Table 1.1) encodes an RNA polymerase II transcription factor homologous to human transcription factor NF-X1 (Stroumbakis et al. 1996). In *Drosophila*, *stc* is expressed throughout all developmental stages and in adults. In embryos, *stc* is expressed in the central nervous system, where it is required to maintain the proper morphology of motoneuronal axon nerve routes (Stroumbakis et al. 1996). In adults, *stc* expression is highest in ovaries and provides essential maternal contributions to early development (Tolias and Stroumbakis 1998). Mutations at *stc* locus are associated with life span variation (Pasyukova et al. 2004). In particular, one of the *stc* mutations was shown to increase the life span of unmated females and decrease the life span of mated females, without affecting males (Roshina et al. 2014). Precise reversions of the mutation lead to the restoration of the life span specific to control females, indicating the causal association between changes in life span and *stc* mutation. No differences in *stc* transcription are observed between whole bodies, ovaries, and brains of mutant and control females of different ages, either unmated or mated. The amount of *stc* transcript is substantially decreased in mutant embryos (Roshina et al. 2014).

1.3.6 Conclusions

The empirical data at hand today indicate that many genes are involved in both the nervous system development and the life span control. Do the same pathways mediate an impact of a gene on the neuron development and longevity? Is an impact of a gene on life span related, at least partially, to its developmental functions or is it explained by other pleiotropic influences later in life?

It is well established that InR/TOR signaling pathway integrates signals originating from changes in growth factors, nutrient availability, energy status and various physiological stresses and conveys them to downstream outputs. Recently, InR/TOR signaling have been found to influence the development and functions of the nervous system via controlling synapse growth, axon guidance, dendrite pruning, etc. It was shown that axon guidance and behavioral phenotypes are affected by altering functions of TORC1 involved in the control of growth and life span. These findings demonstrate that TOR signaling provides a critical link between metabolic factors, axon development, and longevity. Synapse overgrowth was only suppressed by reducing the function of TORC2. These findings demonstrate that regulation of synapse overgrowth probably occurs via growth-and life span- independent signaling mechanisms.

InR and TOR cascades were at first regarded as separate pathways, however, later, many new genes/proteins belonging to these cascades were discovered and new interactions among them were revealed. Now, InR and TOR cascades make up a single extensive network of interacting elements. Eventually, many other pathways may be appended to the network, creating an even more complicated system. Which exactly parts of this interacting network will be assigned to regulate life span is so far unclear. Evidently, at present, the puzzle is not yet complete. However, some promising links are indicated.

sgg belongs to many pathways, including InR/TOR signaling and asymmetric cell division pathway; GSK3-beta participates in phosphorylation of tau protein and thus is associated with the development of age-related neurodegenerative diseases. Thereby, *sgg* effects on longevity, most probably, are mediated by sophisticated interactions among several pathways. Cdk5 is also involved in phosphorylation of tau, and GSK3-beta and Cdk5 effects on longevity may be mediated by partially overlapping mechanisms.

Loco is involved in asymmetric neuroblast division and may regulate both this process and longevity as an activator of cAMP-PKA pathway. cAMP-PKA pathway also mediates effects of ion channel genes on life span and, in this case, the same pathway is also involved in both maintenance of the integrity of specific neuronal functions and the life span control. These examples indicate that effects of seemingly unrelated genes may converge on the same pathway.

In certain cases, experimental data clearly point out that the effect on the life span is not mediated by the main functional activity of a gene/protein. For example, Rpd3 is supposed to affect longevity independently of histone deacetylation. The role of canonical JAK/STAT signaling in the development of the nervous system is indicated, while the non-canonical pathway plays a role the life span regulation. Conversely, the longevity phenotype of PcG mutants appears specifically due to violation of Polycomb silencing.

Accumulating data suggest that several key life span regulators such as mitochondrial electron transport chain enzymes, microRNAs, and the transcription factors HSF-1 and FOXO affect life span predominantly during early larval development and early adulthood (for review, see Alcedo et al. 2013). We demonstrated that only *stc* transcription in embryos is altered in long-lived *stc* mutants. It is tempting to suggest that life span might depend on gene function during early development. We hypothesized that the long-term, carry-over effects of the *stc* mutation might be epigenetically inherited in cell lineages. Alternatively, the STC transcription factor might participate in transcriptional cascades that predetermine structural and therefore functional properties of the adult nervous system. Similarly, several genes belonging to the asymmetric neuroblast division pathway affect life span, and it remain to be understood if neuroblast division that occurs during early embryonic development, may have a carry-over effect on aging and longevity.

1.4 Perspectives

It may be that virtually all genes are involved in life span control, the strength of their effects being different in different environments (variations in food supplies, oxygen, temperature, humidity, etc.), in different sexes, at different ages. The most intriguing and difficult puzzles remained to be solved in future work concern interactions among different genes and pathways that provide an optimal level of integrity and functionality of all organs and tissues, at all ages. At the level of the whole organism, pleiotropic effects of genes and tissue-, temporal-, sex-, age-specific activities of pathways comprise an extremely complicated functional network. Considering neuronal genes and developmental neuronal pathways in this context, the ambitious and exciting goals would be, among others: to establish pathways that mediate neuronal inputs in aging and longevity; to reveal feed-back and feed-forward mechanism underlying interactions between the nervous system and other tissues in life span control; to understand how the nervous system mediates an impact of environmental cues on longevity; to discover the basis of undoubted sex-specificity of neuronal inputs in life span.

Acknowledgments Authors are supported by the Presidium of the Russian Academy of Sciences Program "Biodiversity of natural systems" and the Russian Foundation for Basic Research grants #14-04-01464-a and #15-04-05797-a.

References

Alcedo J, Flatt T, Pasyukova EG (2013) Neuronal inputs and outputs of aging and longevity. Front Genet 4:71

Alic N, Hoddinott MP, Vinti G, Partridge L (2011) Lifespan extension by increased expression of the *Drosophila homologue* of the IGFBP7 tumour suppressor. Aging Cell 10:137–147

Ashraf SI, Ip YT (2001) The Snail protein family regulates neuroblast expression of inscuteable and string, genes involved in asymmetry and cell division in *Drosophila*. Development 128:4757–4767

Ashraf SI, Hu X, Roote J, Ip YT (1999) The mesoderm determinant snail collaborates with related zinc-finger proteins to control *Drosophila* neurogenesis. EMBO J 18:6426–6638

Atwood SX, Prehoda KE (2009) aPKC phosphorylates Miranda to polarize fate determinants during neuroblast asymmetric cell division. Curr Biol 19:723–729

Baines RA (2004) Synaptic strengthening mediated by bone morphogenetic protein-dependent retrograde signaling in the *Drosophila* CNS. J Neurosci 24:6904–6911

Bauer JH, Chang C, Morris SN, Hozier S, Andersen S, Waitzman JS, Helfand SL (2005a) Expression of dominant-negative Dmp53 in the adult fly brain inhibits insulin signaling. Proc Natl Acad Sci USA 104:13355–13360

Bauer JH, Poon PC, Glatt-Deeley H, Abrams JM, Helfand SL (2005b) Neuronal expression of p53 dominant-negative proteins in adult *Drosophila melanogaster* extends life span. Curr Biol 15:2063–2068

Bauer JH, Chang C, Morris SN, Hozier S, Andersen S, Waitzman JS, Helfand SL (2007) Expression of dominant-negative Dmp53 in the adult fly brain inhibits insulin signaling. Proc Natl Acad Sci USA 104:13355–13360

Bauer JH, Chang C, Bae G, Morris SN, Helfand SL (2010) Dominant-negative Dmp53 extends life span through the dTOR pathway in *D. melanogaster*. Mech Ageing Dev 131:193–201

Betschinger J, Mechtler K, Knoblich JA (2003) The Par complex directs asymmetric cell division by phosphorylating the cytoskeletal protein Lgl. Nature 422:326–330

Biteau B, Karpac J, Supoyo S, Degennaro M, Lehmann R, Jasper H (2010) Lifespan extension by preserving proliferative homeostasis in *Drosophila*. PLoS Genet 6:e1001159

Blagosklonny MV (2009) Validation of anti-aging drugs by treating age-related diseases. Aging (Albany NY) 1:281–288

Broughton S, Partridge L (2009) Insulin/IGF-like signalling, the central nervous system and aging. Biochem J 418:1–12

Broughton SJ, Piper MD, Ikeya T, Bass TM, Jacobson J, Driege Y, Martinez P, Hafen E, Withers DJ, Leevers SJ, Partridge L (2005) Longer lifespan, altered metabolism, and stress resistance in *Drosophila* from ablation of cells making insulin-like ligands. Proc Natl Acad Sci USA 102:3105–3110

Broughton S, Alic N, Slack C, Bass T, Ikeya T, Vinti G, Tommasi AM, Driege Y, Hafen E, Partridge L (2008) Reduction of DILP2 in *Drosophila* triages a metabolic phenotype from lifespan revealing redundancy and compensation among DILPs. PLoS ONE 3:e3721

Buchanan ME, Davis RL (2010) A distinct set of *Drosophila* brain neurons required for neurofibromatosis type 1-dependent learning and memory. J Neurosci 30:10135–10143

Buchanan ME, Guo HF, Tong J, Hannan F, Luo L, Zhong Y (2000) A *neurofibromatosis-1*-regulated pathway is required for learning in *Drosophila*. Nature 403:895–898

Budnik V, Zhong Y, Wu C-F (1990) Morphological plasticity of motor axons in *Drosophila* mutants with altered excitability. J Neurosci 10:3754–3768

Byers D, Davis RL, Kiger JA Jr (1981) Defect in cyclic AMP phosphodiesterase due to the dunce mutation of learning in Drosophila melanogaster. Nature 289:79–81

Cai Y, Chia W, Yang XA (2001) A family of snail-related zinc finger proteins regulates two distinct and parallel mechanisms that mediate *Drosophila* neuroblast asymmetric divisions. EMBO J 20:1704–1714

Cai Y, Yu F, Lin S, Chia W, Yang X (2003) Apical complex genes control mitotic spindle geometry and relative size of daughter cells in *Drosophila* neuroblast and pI asymmetric divisions. Cell 112(1):51–62

Carbone MA, Jordan KW, Lyman RF, Harbison ST, Leips J, Morgan TJ, DeLuca M, Awadalla P, Mackay TF (2006) Phenotypic variation and natural selection at catsup, a pleiotropic quantitative trait gene in *Drosophila*. Curr Biol 16:912–919

Certel SJ, Thor S (2004) Specification of *Drosophila* motoneuron identity by the combinatorial action of POU and LIM-HD factors. Development 131:5429–5439

Chang KC, Garcia-Alvarez G, Somers G, Sousa-Nunes R, Rossi F, Lee YY, Soon SB, Gonzalez C, Chia W, Wang H (2010) Interplay between the transcription factor Zif and aPKC regulates neuroblast polarity and self-renewal. Dev Cell 19:778–785

Colosimo PF, Liu X, Kaplan NA, Tolwinski NS (2010) GSK3beta affects apical-basal polarity and cell-cell adhesion by regulating aPKC levels. Dev Dyn 239:115–125

Connell-Crowley L, Le Gall M, Vo DJ, Giniger E (2000) The cyclin-dependent kinase Cdk5 controls multiple aspects of axon patterning in vivo. Curr Biol 10:599–602

Connell-Crowley L, Vo D, Luke L, Giniger E (2007) *Drosophila* lacking the Cdk5 activator, p35, display defective axon guidance, age-dependent behavioral deficits and reduced lifespan. Mech Dev 124:341–349

Copf T, Goguel V, Lampin-Saint-Amaux A, Scaplehorn N, Preat T (2011) Cytokine signaling through the JAK/STAT pathway is required for long-term memory in *Drosophila*. Proc Natl Acad Sci USA 108:8059–8064

De Luca M, Roshina NV, Geiger-Thornsberry GL, Lyman RF, Pasyukova EG, Mackay TFC (2003) *Dopa decarboxylase* (*Ddc*) affects variation in *Drosophila* longevity. Nat Genet 34:429–433

Demontis F, Perrimon N (2010) FOXO/4E-BP signaling in *Drosophila* muscles regulates organism-wide proteostasis during aging. Cell 143:813–825

Dimitroff B, Howe K, Watson A, Campion B, Lee HG, Zhao N, O'Connor MB, Neufeld TP, Selleck S (2012) Diet and energy-sensing inputs affect TorC1-mediated axon misrouting but not TorC2-directed synapse growth in a *Drosophila* model of tuberous sclerosis. PLoS ONE 7:e30722

Drain P, Folkers E, Quinn WG (1991) cAMP-dependent protein kinase and the disruption of learning in transgenic flies. Neuron 6:71–82

Dudai Y, Jan YN, Byers D, Quinn WG, Benzer S (1976) *dunce*, a mutant of *Drosophila* deficient in learning. Proc Natl Acad Sci USA 73:1684–1688

Enell LE, Kapan N, Söderberg JAE, Kahsai L, Nässel DR (2010) Insulin signaling, lifespan and stress resistance are modulated by metabotropic GABA receptors on insulin producing cells in the brain of *Drosophila*. PLoS ONE 5:e15780

Feng Z (2010) p53 regulation of the IGF-1/AKT/mTOR pathways and the endosomal compartment. Cold Spring Harb Perspect Biol 2:a001057

Flatt T, Min KJ, D'Alterio C, Villa-Cuesta E, Cumbers J, Lehmann R, Jones DL, Tatar M (2008) *Drosophila* germ-line modulation of insulin signaling and lifespan. Proc Natl Acad Sci USA 105:6368–6373

Franciscovich AL, Mortimer AD, Freeman AA, Gu J, Sanyal S (2008) Overexpression screen in *Drosophila* identifies neuronal roles of GSK-3 beta/shaggy as a regulator of AP-1-dependent developmental plasticity. Genetics 180:2057–2071

Franco B, Bogdanik L, Bobinnec Y, Debec A, Bockaert J, Parmentier ML, Grau Y (2004) Shaggy, the homolog of glycogen synthase kinase 3, controls neuromuscular junction growth in *Drosophila*. J Neurosci 24:6573–6577

Frankel S, Rogina B (2005) *Drosophila* longevity is not affected by heterochromatin-mediated gene silencing. Aging Cell 4:53–56

Fridell Y-WC, Sanchez-Blanco A, Silvia BA, Helfand SL (2005) Targeted expression of the human uncoupling protein 2 (hUCP2) to adult neurons extends life span in the fly. Cell Metab 1:145–152

Fridell YW, Hoh M, Kréneisz O, Hosier S, Chang C, Scantling D, Mulkey DK, Helfand SL (2009) Increased uncoupling protein (UCP) activity in *Drosophila* insulin-producing neurons attenuates insulin signaling and extends lifespan. Aging (Albany NY) 1:699–713

Gervas N, Tchénio P, Preat T (2010) PKA dynamics in a *Drosophila* learning center: coincidence detection by *rutabaga* adenylyl cyclase and spatial regulation by *dunce* phosphodiesterase. Neuron 65:516–529

Giannakou ME, Goss M, Jünger MA, Hafen E, Leevers SJ, Partridge L (2004) Long-lived *Drosophila* with overexpressed dFOXO in adult fat body. Science 305:361

Grönke S, Clarke DF, Broughton S, Andrews TD, Partridge L (2010) Molecular evolution and functional characterization of *Drosophila* insulin-like peptides. PLoS Genet 6:e1000857

Haselton A, Sharmin E, Schrader J, Sah M, Poon P, Fridell YW (2010) Partial ablation of adult *Drosophila* insulin-producing neurons modulates glucose homeostasis and extends life span without insulin resistance. Cell Cycle 9:3063–3071

Hobert O (2011) Regulation of terminal differentiation programs in the nervous system. Annu Rev Cell Dev Biol 27:681–696

Humphrey DM, Toivonen JM, Giannakou M, Partridge L, Brand MD (2009) Expression of human uncoupling protein-3 in *Drosophila* insulin-producing cells increases insulin-like peptide (DILP) levels and shortens lifespan. Exp Gerontol 44:316–327

Hwangbo DS, Gershman B, Tu MP, Tatar M, Palmer M (2004) *Drosophila* dFOXO controls lifespan and regulates insulin signalling in brain and fat body. Nature 429:562–566

Inoki K, Guan KL (2006) Complexity of the TOR signaling network. Trends Cell Biol 16:206–212

Kanuka H, Kuranaga E, Takemoto K, Hiratou T, Okano H, Miura M (2005) *Drosophila* caspase transduces Shaggy/GSK-3beta kinase activity in neural precursor development. EMBO J 24:3793–3806

Kapahi P, Zid BM, Harper T, Koslover D, Sapin V, Benzer S (2004) Regulation of lifespan in *Drosophila* by modulation of genes in the TOR signaling pathway. Curr Biol 14:885–890

Kapahi P, Chen D, Rogers AN, Katewa SD, Li PW, Thomas EL, Kockel L (2010) With TOR, less is more: a key role for the conserved nutrient-sensing TOR pathway in aging. Cell Metab 11:453–465

Kaplan NA, Colosimo PF, Liu X, Tolwinski NS (2011) Complex interactions between GSK3 and aPKC in *Drosophila* embryonic epithelial morphogenesis. PLoS ONE 6:e18616

Katewa SD, Kapahi P (2011) Role of TOR signaling in aging and related biological processes in *Drosophila melanogaster*. Exp Gerontol 46:382–390

Kissler AE, Pettersson N, Frölich A, Sigrist SJ, Suter B (2009) *Drosophila cdk5* is needed for locomotive behavior and NMJ elaboration, but seems dispensable for synaptic transmission. Dev Neurobiol 69:365–377

Koike-Kumagai M, Yasunaga KI, Morikawa R, Kanamori T, Emoto K (2009) The target of rapamycin complex 2 controls dendritic tiling of *Drosophila* sensory neurons through the Tricornered kinase signalling pathway. EMBO J 28:3879–3892

Kraut R, Campos-Ortega JA (1996) *Inscuteable*, a neural precursor gene of *Drosophila*, encodes a candidate for a cytoskeleton adaptor protein. Dev Biol 174:65–81

Kraut R, Chia W, Jan LY, Jan YN, Knoblich JA (1996) Role of *inscuteable* in orienting asymmetric cell divisions in *Drosophila*. Nature 383:50–55

Kwon JY, Dahanukar A, Weiss LA, Carlson JR (2007) The molecular basis of CO_2 reception in *Drosophila*. Proc Natl Acad Sci USA 104:3574–3578

Larson K, Yan SJ, Tsurumi A, Liu J, Zhou J, Gaur K, Guo D, Eickbush TH, Li WX (2012) Heterochromatin formation promotes longevity and represses ribosomal RNA synthesis. PLoS Genet 8:e1002473

Lee KS, Iijima-Ando K, Iijima K, Lee WJ, Lee JH, Yu K, Lee DS (2009) JNK/FOXO-mediated neuronal expression of fly homologue of peroxiredoxin II reduces oxidative stress and extends life span. J Biol Chem 284:29454–29461

Li J, Li W, Calhoun HC, Xia F, Gao FB, Li W (2003) Patterns and functions of STAT activation during *Drosophila* embryogenesis. Mech Dev 120:1455–1468

Liao PC, Lin HY, Yuh CH, Yu LK, Wang HD (2008) The effect of neuronal expression of heat shock proteins 26 and 27 on lifespan, neurodegeneration, and apoptosis in *Drosophila*. Biochem Biophys Res Commun 376:637–641

Libert S, Zwiener J, Chu X, Vanvoorhies W, Roman G, Pletcher SD (2007) Regulation of *Drosophila* life span by olfaction and food-derived odors. Science 315:1133–1137

Lin YR, Kim K, Yang Y, Ivessa A, Sadoshima J, Park Y (2011a) Regulation of longevity by regulator of G-protein signaling protein, Loco. Aging Cell 10:438–447

Lin YR, Parikh H, Park Y (2011b) Loco signaling pathway in longevity. Small GTPases 2:158–161

Livingstone MS, Tempel BL (1983) Genetic dissection of monoamine neurotransmitter synthesis in *Drosophila*. Nature 303:67–70

Loo LW, Secombe J, Little JT, Carlos LS, Yost C, Cheng PF, Flynn EM, Edgar BA, Eisenman RN (2005) The transcriptional repressor dMnt is a regulator of growth in *Drosophila melanogaster*. Mol Cell Biol 25:7078–7091

Madeo F, Tavernarakis N, Kroemer G (2010) Can autophagy promote longevity? Nat Cell Biol 12:842–846

Magwire MM, Yamamoto A, Carbone MA, Roshina NV, Symonenko AV, Pasyukova EG, Morozova TV, Mackay TFC (2010) Quantitative and molecular genetic analyses of mutations increasing *Drosophila* life span. PLoS Genet 6:e1001037

Martínez-Azorín F, Calleja M, Hernández-Sierra R, Farr CL, Kaguni LS, Garesse R (2008) Over-expression of the catalytic core of mitochondrial DNA (mtDNA) polymerase in the nervous system of *Drosophila melanogaster* reduces median life span by inducing mtDNA depletion. J Neurochem 105:165–176

Martin-Pena A, Acebes A, Rodriguez JR, Sorribes A, de Polavieja GG, Fernandez-Funez P, Ferrus A (2006) Age-independent synaptogenesis by phosphoinositide 3 kinase. J Neurosci 26:10199–10208

McGuire SE, Le PT, Osborn AJ, Matsumoto K, Davis RL (2003) Spatiotemporal rescue of memory dysfunction in *Drosophila*. Science 302:1765–1768

Min KJ, Yamamoto R, Buch S, Pankratz M, Tatar M (2008) *Drosophila* lifespan control by dietary restriction independent of insulin-like signaling. Aging Cell 7:199–206

Morrow G, Samson M, Michaud S, Tanguay RM (2004) Overexpression of the small mitochondrial Hsp22 extends *Drosophila* lifespan and increases resistance to oxidative stress. FASEB J 18:598–609

Mudher A, Shepherd D, Newman TA, Mildren P, Jukes JP, Squire A, Mears A, Drummond JA, Berg S, MacKay D, Asuni AA, Bhat R, Lovestone S (2004) GSK-3beta inhibition reverses axonal transport defects and behavioural phenotypes in *Drosophila*. Mol Psychiatry 9:522–530

Mukunda L, Miazzi F, Kaltofen S, Hansson BS, Wicher D (2014) Calmodulin modulates insect odorant receptor function. Cell Calcium 55:191–199

Nässel DR, Winther AM (2010) *Drosophila* neuropeptides in regulation of physiology and behavior. Prog Neurobiol 92:42–104

Nässel DR, Kubrak OI, Liu Y, Luo J, Lushchak OV (2013) Factors that regulate insulin producing cells and their output in *Drosophila*. Front Physiol 4:252

Natarajan R, Trivedi-Vyas D, Wairkar YP (2013) Tuberous sclerosis complex regulates *Drosophila* neuromuscular junction growth via the TORC2/Akt pathway. Hum Mol Genet 22:2010–2023

Noble W, Hanger DP, Miller CC, Lovestone S (2013) The importance of tau phosphorylation for neurodegenerative diseases. Front Neurol 4:83

Nuzhdin SV, Pasyukova EG, Dilda CL, Zeng Z-B, Mackay TFC (1997) Sex-specific quantitative trait loci affecting longevity in *Drosophila melanogaster*. Proc Natl Acad Sci USA 94:9734–9739

Orr WC, Mockett RJ, Benes JJ, Sohal RS (2003) Effects of overexpression of copper-zinc and manganese superoxide dismutases, catalase, and thioredoxin reductase genes on longevity in *Drosophila melanogaster*. J Biol Chem 278:26418–26422

Ouyang Y, Song Y, Lu B (2011) dp53 restrains ectopic neural stem cell formation in the *Drosophila* brain in a non-apoptotic mechanism involving Archipelago and cyclin E. PLoS ONE 6:e28098

Papazian DM, Schwarz TL, Tempel BL, Jan YN, Jan LY (1987) Cloning of genomic and complementary DNA from *Shaker*, a putative potassium channel gene from *Drosophila*. Science 237:749–753

Parkes TL, Hilliker AJ, Phillips JP (1999) Motorneurons, reactive oxygen, and life span in *Drosophila*. Neurobiol Aging 20:531–535

Parrish JZ, Kim MD, Jan LY, Jan YN (2006) Genome-wide analyses identify transcription factors required for proper morphogenesis of *Drosophila* sensory neuron dendrites. Genes Dev 20:820–835

Parrish JZ, Xu P, Kim CC, Jan LY, Jan YN (2009) The microRNA bantam functions in epithelial cells to regulate scaling growth of dendrite arbors in *Drosophila* sensory neurons. Neuron 63:788–802

Partridge L, Alic N, Bjedov I, Piper MD (2011) Ageing in *Drosophila*: the role of the insulin/Igf and TOR signalling network. Exp Gerontol 46:376–381

Pasyukova EG, Vieira C, Mackay TFC (2000) Deficiency mapping of quantitative trait loci affecting longevity in *Drosophila melanogaster*. Genetics 156:1129–1146

Pasyukova EG, Roshina NV, Mackay TFC (2004) Shuttle craft: a candidate quantitative trait gene for *Drosophila* lifespan. Aging Cell 3:297–307

Plyusnina EN, Shaposhnikov MV, Moskalev AA (2011) Increase of *Drosophila melanogaster* lifespan due to D-GADD45 overexpression in the nervous system. Biogerontology 12:211–226

Poon PC, Kuo TH, Linford NJ, Roman G, Pletcher SD (2010) Carbon dioxide sensing modulates lifespan and physiology in *Drosophila*. PLoS Biol 8:e1000356

Puig O, Mattila J (2011) Understanding Forkhead box class O function: lessons from *Drosophila melanogaster*. Antioxid Redox Signal 14:635–647

Rana A, Rera M, Walker DW (2013) Parkin overexpression during aging reduces proteotoxicity, alters mitochondrial dynamics, and extends lifespan. Proc Natl Acad Sci USA 110:8638–8643

Reenan RA, Rogina B (2008) Acquired temperature-sensitive paralysis as a biomarker of declining neuronal function in aging *Drosophila*. Aging Cell 7:179–186

Reenan RA, Hanrahan CJ, Ganetzky B (2000) The *mle* (*napts*) RNA helicase mutation in *Drosophila* results in a splicing catastrophe of the *para* Na+ channel transcript in a region of RNA editing. Neuron 25:139–149

Renger JJ, Ueda A, Atwood HL, Govind CK, Wu CF (2000) Role of cAMP cascade in synaptic stability and plasticity: ultrastructural and physiological analyses of individual synaptic boutons in *Drosophila* memory mutants. J Neurosci 20:3980–3992

Robinow S, White K (1988) The locus *elav* of *Drosophila melanogaster* is expressed in neurons at all developmental stages. Dev Biol 126:294–303

Robinow S, White K (1991) Characterization and spatial distribution of the ELAV protein during *Drosophila melanogaster* development. J Neurobiol 22:443–461

Rogina B, Helfand SL (1995) Regulation of gene expression is linked to life span in adult *Drosophila*. Genetics 141:1043–1048

Rogina B, Helfand SL, Frankel S (2002) Longevity regulation by *Drosophila* Rpd3 deacetylase and caloric restriction. Science 298:1745

Root CM, Masuyama K, Green DS, Enell LE, Nässel DR, Lee CH, Wang JW (2008) A presynaptic gain control mechanism fine-tunes olfactory behavior. Neuron 59:311–321

Roshina NV, Pasyukova EG (2007) Genes regulating the development and functioning of the nervous system determine life span in *Drosophila melanogaster*. Russ J Genet 43:356–362

Roshina NV, Symonenko AV, Krementsova AV, Trostnikov MV, Pasyukova EG (2014) Embryonic expression of *shuttle craft*, a *Drosophila* gene involved in neuron development, is associated with adult lifespan. Aging (Albany NY) 6:1076–1093

Ruan H, Tang XD, Chen ML, Joiner ML, Sun G, Brot N, Weissbach H, Heinemann SH, Iverson L, Wu CF, Hoshi T (2002) High-quality life extension by the enzyme peptide methionine sulfoxide reductase. Proc Natl Acad Sci USA 99:2748–2753

Ruiz-Canada C, Ashley J, Moeckel-Cole S, Drier E, Yin J, Budnik V (2004) New synaptic bouton formation is disrupted by misregulation of microtubule stability in aPKC mutants. Neuron 42:567–580

Rybina OY, Pasyukova EG (2010) A naturally occurring polymorphism at *Drosophila melanogaster Lim3* locus, a homolog of human *LHX3/4*, affects *Lim3* transcription and fly lifespan. PLoS ONE 5:e12621

Schaefer M, Shevchenko A, Shevchenko A, Knoblich JA (2000) A protein complex containing Inscuteable and the Galpha-binding protein pins orients asymmetric cell divisions in *Drosophila*. Curr Biol 10:353–362

Seong KH, Matsuo T, Fuyama Y, Aigaki T (2001) Neural-specific overexpression of *Drosophila* plenty of SH3s (DPOSH) extends the longevity of adult flies. Biogerontology 2:271–281

Sepp KJ, Hong P, Lizarraga SB, Liu JS, Mejia LA, Walsh CA, Perrimon N (2008) Identification of neural outgrowth genes using genome-wide RNAi. PLoS Genet 4:e1000111

Siebold AP, Banerjee R, Tie F, Kiss DL, Moskowitz J, Harte PJ (2010) Polycomb repressive complex 2 and Trithorax modulate *Drosophila* longevity and stress resistance. Proc Natl Acad Sci USA 107:169–174

Simonsen A, Cumming RC, Brech A, Isakson P, Schubert DR, Finley KD (2008) Promoting basal levels of autophagy in the nervous system enhances longevity and oxidant resistance in adult *Drosophila*. Autophagy 4:176–184

Skeath JB, Thor S (2003) Genetic control of *Drosophila* nerve cord development. Curr Opin Neurobiol 13:8–15

Skoulakis EM, Kalderon D, Davis RL (1993) Preferential expression in mushroom bodies of the catalytic subunit of protein kinase A and its role in learning and memory. Neuron 11:197–208

Smith CA, Lau KM, Rahmani Z, Dho SE, Brothers G, She YM, Berry DM, Bonneil E, Thibault P, Schweisguth F, Le Borgne R, McGlade CJ (2007) aPKC-mediated phosphorylation regulates asymmetric membrane localization of the cell fate determinant Numb. EMBO J 26:468–480

Song Y, Ori-McKenney KM, Zheng Y, Han C, Jan LY, Jan YN (2012) Regeneration of *Drosophila* sensory neuron axons and dendrites is regulated by the Akt pathway involving Pten and microRNA bantam. Genes Dev 26:612–1625

Stathakis DG, Pentz ES, Freeman ME, Kullman J, Hankins GR, Pearlson NJ, Wright TR (1995) The genetic and molecular organization of the *Dopa decarboxlyase* gene cluster of *Drosophila melanogaster*. Genetics 141:629–655

Stathakis DG, Burton DY, McIvor WE, Krishnakumar S, Wright TR, O'Donnell JM (1999) The catecholamines up (Catsup) protein of *Drosophila melanogaster* functions as a negative regulator of tyrosine hydroxylase activity. Genetics 153:361–382

Stenesen D, Suh JM, Seo J, Yu K, Lee KS, Kim JS, Min KJ, Graff JM (2013) Adenosine nucleotide biosynthesis and AMPK regulate adult life span and mediate the longevity benefit of caloric restriction in flies. Cell Metab 17:101–112

Stoleru D, Nawathean P, Fernandez MP, Menet JS, Ceriani MF, Rosbash M (2007) The *Drosophila* circadian network is a seasonal timer. Cell 129:207–219

Stroumbakis ND, Li Z, Tolias PP (1996) A homolog of human transcription factor NF-X1 encoded by the *Drosophila shuttle craft* gene is required in the embryonic central nervous system. Mol Cell Biol 16:192–201

Tatar M, Kopelman A, Epstein D, Tu MP, Yin CM, Garofalo RS (2001) A mutant *Drosophila* insulin receptor homolog that extends life-span and impairs neuroendocrine function. Science 292:107–110

Tatar M, Bartke A, Antebi A (2003) The endocrine regulation of aging by insulin-like signals. Science 299:1346–1351

Tea JS, Chihara T, Luo L (2010) Histone deacetylase Rpd3 regulates olfactory projection neuron Dendrite targeting via the transcription factor Prospero. J Neurosci 30:9939–9946

Thor S, Andersson SGE, Tomlinson A, Thomas JB (1999) A LIM-homodomain combinatorial code for motorneuron pathway selection. Nature 397:76–80

Toba G, Yamamoto D, White K (2010) Life-span phenotypes of *elav* and *Rbp9* in *Drosophila* suggest functional cooperation of the two ELAV-family protein genes. Arch Insect Biochem Physiol 74:261–265

Tolias PP, Stroumbakis ND (1998) The *Drosophila* zygotic lethal gene *shuttle craft* is required maternally for proper embryonic development. Dev Genes Evol 208:274–282

Tong JJ, Schriner SE, McCleary D, Day BJ, Wallace DC (2007) Life extension through neurofibromin mitochondrial regulation and antioxidant therapy for *neurofibromatosis-1* in *Drosophila melanogaster*. Nat Genet 39:476–485

Trostnikov MV, Roshina NV, Symonenko AV, Pasyukova EG (2014) GSK-3 beta affects survival and synaptic function in *Drosophila melanogaster*. In: Abstract book of the 3rd international conference "genetics of aging and longevity", Russia, Sochi, 6–10 Apr, p 54

Troulinaki K, Bano D (2012) Mitochondrial deficiency: a double-edged sword for aging and neurodegeneration. Front Genet 3:244

Trout WE, Kaplan WD (1970) A relation between longevity, metabolic rate, and activity in *Shaker* mutants of *Drosophila melanogaster*. Exp Gerontol 5:83–92

Tsai PI, Wang M, Kao HH, Cheng YJ, Walker JA, Chen RH, Chien CT (2012) Neurofibromin mediates FAK signaling in confining synapse growth at *Drosophila* neuromuscular junctions. J Neurosci 32:16971–16981

Udolph G, Rath P, Tio M, Toh J, Fang W, Pandey R, Technau GM, Chia W (2009) On the roles of *Notch, Delta, kuzbanian*, and *inscuteable* during the development of *Drosophila* embryonic neuroblast lineages. Dev Biol 336:156–168

Ueda A, Wu CF (2006) Distinct frequency-dependent regulation of nerve terminal excitability and synaptic transmission by IA and IK potassium channels revealed by *Drosophila Shaker* and *Shab* mutations. J Neurosci 26:6238–6248

van Dam TJ, Zwartkruis FJ, Bos JL, Snel B (2011) Evolution of the TOR pathway. J Mol Evol 73:209–220

Villeponteau B (1997) The heterochromatin loss model of aging. Exp Gerontol 32:383–394

Wang MC, Bohmann D, Jasper H (2003) JNK signaling confers tolerance to oxidative stress and extends lifespan in *Drosophila*. Dev Cell 5:811–816

Wilson GF, Wang Z, Chouinard SW, Griffith LC, Ganetzky B (1998) Interaction of the K channel beta subunit, hyperkinetic, with eag family members. J Biol Chem 273:6389–6394

Wodarz A, Ramrath A, Kuchinke U, Knust E (1999) Bazooka provides an apical cue for Inscuteable localization in *Drosophila* neuroblasts. Nature 402:544–547

Wodarz A, Ramrath A, Grimm A, Knust E (2000) *Drosophila* atypical protein kinase C associates with Bazooka and controls polarity of epithelia and neuroblasts. J Cell Biol 150:1361–1374

Wong JJ, Li S, Lim EK, Wang Y, Wang C, Zhang H, Kirilly D, Wu C, Liou YC, Wang H, Yu F (2013) A Cullin1-based SCF E3 ubiquitin ligase targets the InR/PI3K/TOR pathway to regulate neuronal pruning. PLoS Biol 11:e1001657

Yamazaki D, Horiuchi J, Nakagami Y, Nagano S, Tamura T, Saitoe M (2007) The *Drosophila* DCO mutation suppresses age-related memory impairment without affecting lifespan. Nat Neurosci 10:478–484

Yu F, Morin X, Cai Y, Yang X, Chia W (2000) Analysis of *partner of inscuteable*, a novel player of *Drosophila* asymmetric divisions, reveals two distinct steps in inscuteable apical localization. Cell 100:399–409

Yu F, Cai Y, Kaushik R, Yang X, Chia W (2003) Distinct roles of Galphai and Gbeta13F subunits of the heterotrimeric G protein complex in the mediation of *Drosophila* neuroblast asymmetric divisions. J Cell Biol 162:623–633

Yu F, Wang H, Qian H, Kaushik R, Bownes M, Yang X, Chia W (2005) Locomotion defects, together with Pins, regulates heterotrimeric G-protein signaling during *Drosophila* neuroblast asymmetric divisions. Genes Dev 19:1341–1353

Zaitsev AA, Symonenko AV, Roshina NV, Pasyukova EG (2010) Involvement of the *escargot* gene of *Drosophila melanogaster* in lifespan control. Proc Tomsk State Univ 275:402–404

Zhong Y, Wu CF (2004) Neuronal activity and adenylyl cyclase in environment-dependent plasticity of axonal outgrowth in *Drosophila*. J Neurosci 24:1439–1445

Zid BM, Rogers AN, Katewa SD, Vargas MA, Kolipinski MC, Lu TA, Benzer S, Kapahi P (2009) 4E-BP extends lifespan upon dietary restriction by enhancing mitochondrial activity in *Drosophila*. Cell 139:149–160

Chapter 2
Gadd45 Proteins in Aging and Longevity of Mammals and *Drosophila*

Alexey A. Moskalev, Ekaterina N. Proshkina and Mikhail V. Shaposhnikov

Abstract Proteins of the GADD45 family play an essential role in the integration of cellular response to a wide variety of stressors and maintenance of homeostasis at the level of a cell, a tissue and an organism. The basic homeostatic processes are implicated in the determination of the progression of aging and development of major age-related disorders. Moreover, GADD45s mediate several well-known aging-associated signaling pathways through the interaction with such proteins as FOXO, p53, ATM, ATR, SIRT1, mTOR and some other. These reasons point out the role of the GADD45 proteins in the aging and life span regulation. Indeed, we have shown that constitutive and conditional (mifepristone-inducible) *D-GADD45* overexpression in *Drosophila melanogaster* nervous system extends median and maximum life span, and increases the resistance to genotoxic, oxidative, thermal stress, and starvation. The life span-extending effect was apparently due to more efficient recognition and repair of DNA damage, because the spontaneous DNA damage in the larva neuroblasts with *D-GADD45* overexpression was reduced. However, data obtained for flies with conditional ubiquitous *D-GADD45* overexpression demonstrates a negative effect of this intervention on the life span and stress resistance.

Keywords Aging · Age-related disease · Gadd45 proteins · Longevity · Stress resistance

---

A.A. Moskalev (✉)
The Engelhardt Institute of Molecular Biology, Moscow, Russia
e-mail: amoskalev@list.ru

A.A. Moskalev · E.N. Proshkina · M.V. Shaposhnikov
Institute of Biology of Komi Science Center of Ural Branch of RAS, Syktyvkar, Russia

A.A. Moskalev
Moscow Institute of Physics and Technology (State University), Dolgoprudny, Russia

A.A. Moskalev · E.N. Proshkina · M.V. Shaposhnikov
Syktyvkar State University, Syktyvkar, Russia

A.M. Vaiserman et al. (eds.), *Life Extension*, Healthy Ageing and Longevity 3,
DOI 10.1007/978-3-319-18326-8_2

2.1 Introduction

Proteins of the GADD45 family (Growth Arrest and DNA Damage-inducible) play an essential role in the integration of cellular response to a wide variety of stressors and maintenance of homeostasis at the level of a cell, a tissue and an organism (Liebermann and Hoffman 1994; Zhang et al. 1999; Fornace et al. 2002; Moskalev et al. 2012b; Salvador et al. 2013) as well as in the determination of aging-related processes and longevity (for review, see Moskalev et al. 2012b).

Majority of gerontogenes (genes whose activity determines organism life span) are members of conserved biological pathways across different groups of species (for review, see Moskalev et al. 2014). Likewise, GADD45 orthologs first appeared in molluscs, and were also found in anemones, polychaete worms, insects, fish, amphibians and mammals. The number of GADD45 proteins in each species varies from one in lower organisms to 5–6 in fish, and decreases to 2–3 in amphibians and mammals (*GADD45*α, *GADD45*β and *GADD45*γ). Additionally, several isoforms were described, which are generated as a result of alternative mRNA splicing (Flicek et al. 2011). This indicates that the GADD45 family is relatively "young" and has undergone duplications and deletions in the course of evolution (Moskalev et al. 2012b). In *Drosophila* there is a single ortholog of the *GADD45* family *D-GADD45* (*CG11086*) (Peretz et al. 2007).

The gerontogenes are classified as life span regulators, mediators, effectors, housekeeping genes, genes involved in mitochondrial function, and genes regulating cellular senescence and apoptosis (for review, see Moskalev et al. 2014). *GADD45s* can be relevant to the last category of regulatory genes. Protein products of *GADD45* genes are small (about 18–20 kDa), acidic (pH 4.0–4.2) proteins with high level of homology (55–57 % of identic aminoacids). Mainly, they have nucleus location and are associated with ribonucleoprotein speckles (Abdollahi et al. 1991; Zhang et al. 1999; Vairapandi et al. 2002; Sytnikova et al. 2011). GADD45 proteins form homo- and hetero-dimers and olygomers (Kovalsky et al. 2001; Schrag et al. 2008). The median half-life of the *GADD45* mRNA is unusually short (less than 1 h), suggesting a regulatory rather than metabolic function for GADD45 proteins (Sharova et al. 2009). They don't exercise enzymatic properties, and function through the interactions with other proteins and RNA (Sytnikova et al. 2011), or by way of modification of DNA/RNA accessibility for enzymes (Carrier et al. 1999; Moskalev et al. 2012b). The expression of the GADD45 proteins is detected in different tissues, including heart, brain, spleen, lungs, liver, skeletal muscles, kidneys, testicles, placenta (Zhang et al. 1999).

Most of the longevity genes described are related to stress response (for review, see Moskalev et al. 2014). GADD45 proteins are highly expressed after exposure to different physical, chemical and biological agents, and physiological factors. During stress response, they control such processes as DNA repair, cell cycle regulation, cellular senescence, apoptosis, inflammatory response, maintaining of the stem cell pool and cell differentiation. However, their inducibility is reduced with age (Edwards et al. 2004) which can be a reason of age-dependent descent of resistance to spontaneous and induced stress influences, and organism survivability. Processes that provide basic homeostasis reactions are implicated in the determination of the progression of aging and age-related diseases (ARDs) (for review, see Moskalev et al. 2012b). Indeed, there are evidences that GADD45 proteins are involved in the development of major ARDs, including cancer, metabolic, cardiovascular and autoimmunity disorders (Budovsky et al. 2009; Wolfson et al. 2009; Moskalev et al. 2012b). Moreover, GADD45 activity is responsible for embryogenesis and ontogenesis, and its imbalance is implicated in the development of such pathologic reactions as preeclampsia (Xiong et al. 2009; Geifman-Holtzman et al. 2013). GADD45s mediate several well-known aging-associated signaling pathways through the interaction with such proteins as FOXO, p53, ATM, ATR, SIRT1, mTOR and some other (Furukawa-Hibi et al. 2002; Kobayashi et al. 2005; Bortoff et al. 2010; Moskalev et al. 2012b; Salvador et al. 2013). All these reasons point out the role of the GADD45 proteins in the aging and life span regulation. Recently, we have shown the life span-extending effects and stress resistance stimulation due to neuro-specific *D-GADD45* overexpression using the fruit fly *Drosophila melanogaster* (Plyusnina et al. 2011; Moskalev et al. 2012b; Plyusnina et al. 2012). This data confirm the involvement of GADD45 proteins in longevity determination, and suggest their ectopic expression can appear as beneficial method for the organism life span extension.

GADD45 family is evolutionary conserved in multicellular animals. In this chapter we considered participation of *GADD45* genes in different aging-related processes, based on the data obtained in different model objects. At the same time we focused on the results of our investigations, which demonstrate the role of *D-GADD45* gene in *Drosophila melanogaster* longevity.

2.2 Involvement of GADD45 Proteins in Stress Resistance Regulation

2.2.1 GADD45 in DNA Repair and Epigenetic Regulation

The expression of GADD45s is one of the critical conditions at the early stages of the DNA damage response with following activation of DNA repair machinery. The GADD45 promoters contain binding motifs for FOXO (AFX/FOXO4, FKHRL1/FOXO3A, and AKT/FOXO1 transcription factors), p53, p33 (ING1), p73, OCT-1, BRCA1. Their activation under stress conditions leads to GADD45 upregulation with subsequent stimulation of DNA repair, blockage of cell cycle in the G1/S and G2/M checkpoints, and apoptosis (Kastan et al. 1992; Guillouf et al. 1995; Vairapandi et al. 1996; el-Deiry 1998; De Laurenzi and Melino 2000; Jin et al. 2000; MacLachlan et al. 2000; Cheung et al. 2001; Jin et al. 2001; Furukawa-Hibi et al. 2002; Tran et al. 2002; Ju et al. 2014).

The role of GADD45 in DNA repair is supported by the studies on in vitro and in vivo models. *GADD45α*-null mouse embryo fibroblasts and *GADD45α*-deficient human colon cancer cells exhibited slow base excision repair and delays removal of AP sites after treatment with methyl methanesulfonate (MMS) (Jung et al. 2007). The *GADD45a*$^{-/-}$ mice demonstrate genomic instability, reduced nucleotide excision repair, increased level of mutations and high susceptibility to chemical oncogenes (Hollander et al. 1999, 2001). The disrupted expression of *GADD45β* caused by the hepatitis C viral infection suppresses the DNA excision repair as well (Higgs et al. 2010). GADD45 proteins participate in base and nucleotide excision repair through the links with DNA repair enzymes and regulation of their activity. GADD45α and GADD45β interact with the apurinic/apyrimidinic endonuclease 1 (APE1) (Jung et al. 2007), xeroderma pigmentosum proteins XPC and XPG (Hartman and Ford 2002; Chang et al. 2003; Ma et al. 2009; Le May et al. 2010; Schäfer et al. 2010), and proliferating cell nuclear antigen (PCNA) (Smith et al. 1994; Hall et al. 1995; Vairapandi et al. 2000).

Another way of GADD45 involvement in DNA repair is its involvement in the repair-mediated active DNA demethylation (Barreto et al. 2007; Rai et al. 2008; Cortellino et al. 2011; Schomacher 2013). A possible GADD45-mediated demethylation mechanism involves nucleotide excision repair associated with the endonuclease activity of XPG protein. Specifically, 5-methylcytosine containing nucleotides could be recognized and removed through GADD45–XPG complex, ultimately resulting in the demethylation of CpG dinucleotides (Ma et al. 2009; Schmitz et al. 2009; Le May et al. 2010; Schäfer et al. 2010; Schomacher 2013). Additionally, GADD45 and XPG are involved in base excision repair, which could be another DNA repair mechanism associated with removal of methylated DNA (Jung et al. 2007). Additionally, GADD45 is able to bind histones and modify accessibility of damaged DNA for repair enzymes, and participates in chromatin decondensation (Carrier et al. 1999; Ma et al. 2009; Schomacher 2013). Thus, GADD45 recruits nucleotide and/or base excision repair factors to gene-specific loci and acts as an adapter between repair factors and chromatin, thereby creating a nexus between epigenetics and DNA repair (for review, see Niehrs and Schäfer 2012).

Recently, GADD45α was shown to bind RNA, forming ribonucleoprotein particles. GADD45 was detected inside nuclear speckles which are sites of active transcription, RNA splicing and processing (Sytnikova et al. 2011). Thus, GADD45 could exert its epigenetic effects both through active DNA demethylation, chromatin remodeling and post-transcriptional RNA regulation.

2.2.2 GADD45 in Cell Cycle Regulation and Cellular Senescence

ATM- and p53-mediated activation of GADD45 is essential for a DNA damage-induced G1 and G2/M cell cycle arrest during stress response (Wang et al. 1999). Thus, human and mouse GADD45-dificient fibroblasts and lymphocytes failed to arrest at G2/M after exposure to stress stimulus (Wang et al. 1999). Microinjection of the *GADD45α* expression vector into human primary fibroblasts arrests the cells in G2/M phase (Wang et al. 1999). On the same hand, ectopic expression of GADD45α, GADD45β or GADD45γ in cancer cells (M1 human myeloblastic leukemia and H1299 lung carcinoma) leads to accumulation of cells arrested in the G1 phase that later underwent apoptosis (Zhang et al. 2001). Additionally, it was found that GADD45α expression and G1 cell cycle arrest are activated by c-Jun N-terminal kinase (SAPK/JNK) under fucoxanthin treatment of LNCap prostate cancer cells, while the inhibition of SAPK/JNK attenuated the induction of G1 arrest and GADD45α expression (Satomi 2012). To achieve cell cycle arrest, GADD45 proteins interact with the protein kinase cell division cycle 2 (Cdc2), Cyclin B1, PCNA and cyclin-dependent kinase inhibitor p21 (Liebermann and Hoffman 2003). The interaction of GADD45α and GADD45β with the Cdc2/Cyclin B1 kinase complex leads to its dissociation and following G2/M cell cycle arrest as well as the inhibition of Cdc2 kinase activity (Zhan et al. 1999; Zhang et al. 1999; Vairapandi et al. 2002; Hsu et al. 2014). The interaction of all three GADD45 proteins with p21 induces both the G1 and G2/M arrest (Smith et al. 1994; el-Deiry 1998; Xiong et al. 2009; Zhang et al. 2014a).

GADD45 mediates cellular senescence in the case of unrepaired DNA damages, through the cell cycle arrest in the G1 phase with following unresponsiveness to growth factors (for review, see Moskalev et al. 2012b). A significant increase in *GADD45α* expression was observed upon stress-induced cellular senescence triggered by hydrogen dioxide (Duan et al. 2005). On the other hand, ectopic expression of GADD45γ robustly elicited senescence in hepatocellular carcinoma cells and suppressed tumor growth in vivo (Zhang et al. 2014b). Induction of GADD45 expression with subsequent cellular senescence can be activated by p53-dependent and JAK/STAT3 signaling pathways (Jackson and Pereira-Smith 2006; Zhang et al. 2014b). GADD45-mediated cellular senescence involves an increased expression of p21, mitochondrial dysfunction and generation of reactive oxygen species through the GADD45/p38 MAPK/GRB2/TGFBR2/TGFβ signaling pathway (Passos et al. 2010; Zhang et al. 2014a).

2.2.3 GADD45 Role in Cell Death and Survival

GADD45 family members play a dual role during mediation of apoptosis associated with two major components—p38/JNK mitogen-activated kinase (MAPK) and NF-κB signaling pathways (Takekawa and Saito 1998; Harkin et al. 1999; Lu et al. 2001; Hildesheim et al. 2002; Yoo et al. 2003; Tront et al. 2006). In fact, the MAPK/GADD45/NF-κB axis responds to a variety of extracellular stimuli, converting them to intracellular responses (Yang et al. 2009; Moskalev et al. 2012b). It is noteworthy that GADD45 proteins and stress kinases form a feedback regulatory loop: the expression of GADD45 is also under the control from p38 and JUNK MAPKs. GADD45γ and GADD45β bind to MEK kinase 4 (MEKK4) and promote phosphorylation and activation of the p38 and JNK MAP kinases by

MEKK4 (Takekawa and Saito 1998). However, specific inhibitor of p38 MAPK SB202190 suppresses the expression of all three *GADD45* genes (Oh-Hashi et al. 2001). NF-κB and *GADD45s* form a positive feedback regulatory loop as well (Gupta et al. 2006).

In the case of irreparable damages, GADD45 proteins exert a pro-apoptotic function. For example, the GADD45 proteins mediate the endoplasmic reticulum stress-induced apoptosis in mouse liver cells (Ji et al. 2005). Ectopic expression of GADD45 triggers apoptosis via the TGFβ/MEKK4/p38/JNK pathway in human leukemic cells or in mouse hepatocytes (Selvakumaran et al. 1994; Yoo et al. 2003). At the same time, blocking of early expression of GADD45β suppresses the apoptosis induced by TGFβ in myeloid leukemia cells (Selvakumaran et al. 1994).

However, GADD45α and GADD45β also can fulfil an anti-apoptotic function. For example, their activity increases hematopoietic cell survival under UV-irradiation or treatment with certain chemotherapeutic drugs (Gupta et al. 2005). Bone marrow cells obtained from $GADD45\alpha^{-/-}$ and $GADD45\beta^{-/-}$ mice show an impaired ability for differentiation and increased sensitivity to the induction of apoptosis after being stimulated by cytokines (Gupta et al. 2006). *GADD45α*-deficient E1A + Ras cells treated with HDAC inhibitors demonstrated a higher level of pro-apoptotic signals, whereas the anti-apoptotic program is suppressed (Igotti Abramova et al. 2014). Anti-apoptotic function of GADD45 is realized through two mechanisms: activation of the p38/NF-κB anti-apoptotic pathway by GADD45α (Zhang et al. 2005; Gupta et al. 2006) and inhibition of the MKK7/JNK pro-apoptotic pathway by GADD45β (Papa et al. 2004b; Tornatore et al. 2008). Additionally, interactions of GADD45 with PCNA may promote cell survival, apoptosis inhibition together with DNA repair (Vairapandi et al. 2000; Azam et al. 2001).

It should be noted that the MAPK-mediated effect of GADD45 activation on the apoptosis onset is cell type specific. For example, activation of p38 and JNK kinases by GADD45 is associated with apoptosis in endothelial and epithelial cells (Harkin et al. 1999; Hildesheim et al. 2002), whereas it increases survival of hematopoietic cells (Platanias 2003). Additionally, induction of GADD45β by NF-κB downregulates pro-apoptotic JNK signaling in mouse embryonic fibroblasts (De Smaele et al. 2001) and in hepatocytes during liver regenerations after partial hepatectomy (Papa et al. 2008).

2.2.4 GADD45 in Antioxidant System Regulation and Heat Shock Response

GADD45 proteins can be involved in the prevention of cellular damages and heat shock response induction. Indeed, it was found that *D-GADD45* gene was upregulated in fly heads after treatment with free radical inductor paraquat and high temperatures. Furthermore, flies with *D-GADD45* overexpression in the nervous system were more resistant to this stressor compared with ones without overexpression (Moskalev et al. 2012a). In mammalian cells GADD45β-mediated activation of the protein kinases MEKK4 and JNK (Takekawa and Saito 1998; Papa et al. 2004a) increases the level of the ASK1 (Ko et al. 2001), which acts in opposition to the SOD1 protein. The GADD45 proteins affect the expression level of the transcription factors PPARγ (Hamza et al. 2009) and RXRA (Wu et al. 2004) as a part of JAK and MEK kinase signaling cascades, and participate in the induction of downstream antioxidant systems (SOD1, thioredoxin and glutaredoxin enzymes). Thus, in the process of oxidative stress response, GADD45 family proteins are involved in the control of the activity and maintenance of the balance between antioxidant enzymes and determine the fate of cells (Moskalev et al. 2012a).

Additionally, the transcription factors PPAR-γ and RXRA increase the expression of the heat shock protein HSP22 (Hamza et al. 2009). HSP22 is responsible for activating another heat shock protein, HSP27 (Sun et al. 2004) and following HSP70 induction (Whitlock et al. 2005). For example, GADD45 proteins can participate in the heat shock response through PPAR-γ and RXRA. Another way of involvement of the GADD45 family members in the heat shock protein activation is mediated by its interaction with the CDK1 protein kinase. All three GADD45 proteins bind and inhibit CDK1, which phosphorylates the transcription factor SP1 (Chuang et al. 2012). The tanscription factor SP1 activates the expression of heat shock proteins HSP27 and HSP60 (Reed et al. 2008; Friedman et al. 2009). Finally, GADD45s activates the transcription factor HSF1 by inhibiting the p38 protein kinase (Moskalev et al. 2012a). HSF1 is a key element for pathways activating heat shock proteins, such as HSP60, HSP90, HSPA4, HSP70, HSP27 and HSP22.

2.2.5 GADD45 in Inflammatory Response and Immunity

GADD45 proteins also contribute to cellular inflammation response and organism survival by modulation of the immune response (for review, see Schmitz 2013). It was shown using the *Drosophila* model, that the inflammation induced by bacterial infection increases both the level of *D-GADD45* mRNA and protein (Peretz et al. 2007; Moskalev et al. 2012a). The *GADD45* genes are induced by the pro-inflammatory transcription factor NF-κB (Balliet et al. 2001), cytokines including interleukins TNFα, TNFβ, IL-2, IL-6, IL-8, IL-12, IL-18 (Fan et al. 1999; Zhang et al. 1999; Yang et al. 2001; Salerno et al. 2012) and oncostatin M (Nakayama et al. 1999). The main function of GADD45 in the inflammation response is determined by its interactions with mitogen-activated protein kinase p38, cyclin-dependent kinase p34 (Yang et al. 2000), and PCNA (Smith et al. 1994). For example, after IL-12 and IL-18 treatment GADD45β activated p38 and selectively increased cytokine-induced interferon γ (IFNγ) production (Yang et al. 2001). Additionally, the GADD45 proteins affect the transcription of IFNγ by interacting with PCNA-p300 family (Nakayama et al. 2001). Regulation of this pathway is mediated by interaction of GADD45s with the transcription factors PPARα, C/EBPβ and c-Jun (Moskalev et al. 2012a).

The GADD45β and GADD45γ proteins provide proliferation of T helper 1 (Th1) cells and induce the production of IFNγ in these cells (Yang et al. 2001). The essential role of GADD45 in elevating the level of IFNγ is evidenced by the absence of this process in *GADD45γ*- (Lu et al. 2001) and *GADD45β*-deficient mice (Ju et al. 2009). Furthermore, GADD45 proteins play an important role in the process of Th1-mediated anti-tumor immune responses (Ju et al. 2009) and in autoimmunity reaction development (Liu et al. 2005).

It was found that GADD45α and GADD45β are also essential for differentiation of bone marrow cells into macrophage and granulocyte. The *GADD45α*- and *GADD45β*-deficient mice were characterized by increased apoptosis during differentiation and reduced clonogenicity (Gupta et al. 2006). Additionally, GADD45s activation of the p38 kinase is implicated in the response of granulocytes to lipopolysaccharide (a component of gramm-negative bacterial cells) mediated chemotaxis, whereas Gadd45α and Gadd45β curtailment of JNK activation was linked to chemotaxis of macrophages in response to this inflammatory stimulus (Salerno et al. 2012). Moreover, Gadd45β regulates the autophagy process, a catabolic pathway that also degrades intracellular pathogens (Schmitz 2013). The Gadd45β-MEKK4 pathway specifically directs p38 to autophagosomes and mediate phosphorylation of the autophagy regulator autophagy-related 5 (ATG5) protein. This process results in an accumulation of autophagosomes through the p38-mediated inhibition of lysosome fusion (Keil et al. 2013).

2.3 Role of GADD45 Proteins in Aging and Life Span Regulation

The GADD45 family members are deeply implicated in the maintaining of cellular, tissue and organism homeostasis which determines the aging rate and longevity. Indeed, some well-known regulators of aging-associated processes and longevity are partner proteins for GADD45s (Budovsky et al. 2009; Wolfson et al. 2009; Moskalev et al. 2012b). The GADD45 proteins contain the FOXO- and p53-binding motifs (Kastan et al. 1992; Guillouf et al. 1995; el-Deiry 1998; Furukawa-Hibi et al. 2002; Tran et al. 2002; Ju et al. 2014), and are activated by the ATM- and ATR-dependent way (Kastan et al. 1992; O'Prey et al. 2003; Jang et al. 2010). In particular, RNA interference of FOXO3a leads to inhibition of *GADD45* stress-induced expression (Tran et al. 2002). Another example demonstrated that in human epithelial cells an inhibitor of ATM/ATR prevented induction of GADD45 and growth arrest by flavonoid treatment (O'Prey et al. 2003). The SIRT1 histone deacetylase is another key longevity regulator (Guarente 2011; Satoh et al. 2013; Hubbard and Sinclair 2014), that is involved in the GADD45 functioning regulation (Kobayashi et al. 2005; Scuto et al. 2013). The FOXO-mediated GADD45 induction was markedly impaired in cells, which depleted

SIRT1 expression by RNA-interference (Kobayashi et al. 2005). Additionally, there are evidences that *GADD45* expression is linked with the target of rapamycin (TOR) signaling. Insulin induces *GADD45β* transcription by activating the mTOR pathway (Bortoff et al. 2010), well known for its association with aging, longevity, and ARDs (for review, see Blagosklonny 2008; Zoncu et al. 2011). These examples of the relationship between *GADD45* family members and longevity-associated genes confirm their immixture in the life span control.

Another way of GADD45 participation in the aging and longevity determination at the tissue level is subjected by the role in stem cell pool maintenance (for review, see Moskalev et al. 2012b). The Gadd45 proteins participate also in the maintenance of the pool of myeloid quiescent stem cells. *Gadd45α* or *Gadd45β* deletion was shown to suppress the quiescent stem cell population or lower the survival rate of progenitor cells, leading to the depletion of the stem cell compartment, and to affect the clonogenic potential of these cells (Hoffman and Liebermann 2007). Gadd45γ is involved in stem cell pool maintenance as well. A possible regulatory mechanism of stem cell pool maintenance is mediated by the Nucleus accumbens-1 (NAC-1) protein which is important for self-regeneration and pluripotency of embryonic stem cells, negatively regulates the expression of Gadd45γ-interacting protein 1 (Gadd45γ-ip1), preventing its suppressive activity towards Gadd45γ (Jinawath et al. 2009).

Aging negatively affects the ability of cells to express GADD45 proteins in response to stress conditions. For example, treatment of cardiomyocytes of young mice with free radical inducer paraquat led to the significantly increased expression of GADD45 isoforms, but did not stimulate its expression in the myocardium of old animals (Edwards et al. 2004). Decreased inducibility of GADD45 members may have far-reaching consequences including genome instability, accumulation of DNA damage, and disorders in cellular homeostasis—all of which may eventually contribute to the aging process (for review, see Moskalev et al. 2012b).

The *GADD45* family members as well as longevity-associated genes are concerned with ARDs. One of the main implications of the GADD45 proteins in the ARDs is associated with the cancerogenesis determination (for review, see Liebermann et al. 2011; Hoffman and Liebermann 2013). It was shown that the *GADD45α*-deficient mice were characterized by genomic instability, increased sensitivity to cancerogenes, and high aptitude to ovarian, hepatocellular and vascular tumors (Hollander et al. 1999, 2001; Tront et al. 2006). Mice with the *GADD45β* gene knockout are more susceptible to ionizing radiation and chemical carcinogens, and display a lower immune response against implanted melanoma cells (Ju et al. 2009). In the in vivo model of Ras-overexpressing mice with different GADD45α expression levels (*Ras/GADD45α*$^{+/+}$, Ras/GADD45α$^{+/-}$, and Ras/GADD45α$^{-/-}$), it was shown that Ras-driven genesis and growth of breast tumors is a GADD45α-dependent process (Tront et al. 2006). Clinical patients with solid and hematopoietic cancers including breast, lung, nasopharyngeal, brain, liver, prostate cancer, and lymphoma showed disruption in *GADD45* expression pattern (Hoggard et al. 1995; Sun et al. 2003; Jiang and Wang 2004; Qiu et al. 2004; Ying et al. 2005; Cretu et al. 2009; Na et al. 2010; Liebermann et al. 2011; Hoffman and Liebermann 2013).

The main cause of *GADD45* expression loss in cancers is epigenetic modifications, particularly, DNA methylation (for review, see Moskalev et al. 2012b). For example, the methylation of the *GADD45γ* promoter was significantly higher in different types of cancers than in normal tissues (Zhang et al. 2010). On the other hand, treatment of cancer cells with DNA methyltransferase inhibitors restored *GADD45β* expression to its level in the non-tumorous cells (Qiu et al. 2004). One of the pathways that determine *GADD45* methylation is NF-κB signaling. This is supported by the effect of NF-κB inhibition in cancer cells which leads to the GADD45α- and γ-dependent induction of apoptosis and reduction in tumor growth (Zerbini et al. 2004). The NF-κB transcription factor induces the expression of proto-oncogene c-Myc, which binds to the GC-rich sites of the *GADD45* promoters and significantly reduces the *GADD45* inducibility in response to genotoxic stress (Amundson et al. 1998; Zerbini et al. 2004; Zerbini and Libermann 2005). It is known that hipermethylation of gene promoters provides the aging process as well (Muñoz-Najar and Sedivy 2011), thus methylation of the *GADD45* promoters can be involved in the age-dependent reduction of its expression and inducibility.

Conversely, ectopic expression of the GADD45 members blocks cell growth by arresting the cells in the G2/M phase (Zhu et al. 2009) and G1/G0 phase (Higgins et al. 2009), and/or induces apoptosis in human tumor cell lines (Zhan et al. 1994; Vairapandi et al. 1996; Zhang et al. 2001; Sun et al. 2003; Jiang and Wang 2004; Ying et al. 2005). For example, *GADD45β* overexpression in LβT2 mouse gonadotrope cells blocked tumor cell proliferation and increased rates of apoptosis in response to growth factor withdrawal (Michaelis et al. 2011). Anti-cancer activity of the GADD45 proteins is conditioned by its role in apoptosis and cell cycle control as well as in negative regulation of oncogenes, such as p63 and β-catenin (Hildesheim et al. 2004).

However, in some cases, GADD45α may exert a pro-cancer action, depending on the type of the oncogenic stimuli. For example, the Myc-driven breast cancer is promoted by GADD45α activity, which dramatically decreased level of the enzyme MMP10 and led to angiogenesis. In Myc expressing tumors loss of GADD45α was accompanied by apoptosis or cellular senescence (Tront et al. 2010).

Additionally, other ARDs are associated with changes in *GADD45* expression. The GADD45 proteins participate in the development of the nervous system during ontogenesis and provide long-term memory formation, as well as their deregulation results in neuronal pathologies including brain cancers, ischemia, insults, seizures, memory decline, autism, Alzheimer's disease, psychosis (for review, see Sultan and Sweatt 2013). For example, Alzheimer's disease patients are characterized by highly increased level of *GADD45* expression in neurons, that prevents neuronal cells from apoptosis induced by accumulation of β-amyloid (Torp et al. 1998; Santiard-Baron et al. 1999, 2001). The upregulation of GADD45 was also observed in the in vitro model (human neuroblastoma cells) of dopamine-induced neurotoxicity, which is a part of some neurodegenerative disorders (for example, Parkinson's disease) and normal brain aging (Stokes et al. 2002). The same changes were found in cultures of human endothelial cells derived from atherosclerotic aorta or coronary arteries, as well as in the mouse model of atherosclerosis (Thum and Borlak 2008). Thus, the GADD45 proteins apparently are induced during neurodegenerative processes and atherosclerosis providing a vicarious protective mechanism.

Chronic inflammation is largely attributed to an age-related increase in pro-inflammatory cytokines TNFα, IL-1β, IL-6 and NF-κB (Finch 1990; Chung et al. 2009; Coppé et al. 2010), that induce GADD45 proteins. For example, the induction of GADD45 was observed in the course of liver inflammation (Gant et al. 2003). Additionally, the GADD45 proteins can be involved in the process of the epithelial to mesenchymal transition (EMT). The EMT is a crucial process in the development of different tissues in the embryo and its reactivation in the adult is a part of inflammatory responses useful for the healing damaged tissue. However, abnormality of its control leads to tumorogenesis and organ fibrosis development (López-Novoa and Nieto 2009). GADD45s closely cooperate with key EMT regulators NF-κB, β-catenin, and matrix metalloproteases (Moskalev et al. 2012b).

Aging-dependent induction of oxidative stress and inflammation processes contributes to a process known as immunosenescence. Immunosenescence manifests in a decreased immune responsiveness to foreign and self-antigens, leading to an increased susceptibility to infection, cancer and autoimmune diseases. A decreased ability to maintain tolerance against self-antigens may result in autoimmune disorders (for review, see Moskalev et al. 2012b). Mice with deficiency in *GADD45β* and *GADD45γ* spontaneously develop signs of autoimmune lymphoproliferative syndrome and systemic lupus erythematosus. The reduced inducibility of GADD45s in immune cells is one of the possible factors that increase frequency of autoimmune conditions in aging (Liu et al. 2005).

High GADD45 expression may sustain the age-related immune dysfunctions, particularly, rheumatoid arthritis (for review, see Lindstrom and Robinson 2010). It is known that infiltrated Th1 cells in the synovial fluid of patients with rheumatoid arthritis are resistant to apoptosis. This resistance is accompanied by the high levels of GADD45β resulting from stimulation by pro-inflammatory cytokines TNFα and IL-12 (Du et al. 2008). The activated Th1 cells avoid utilization, which leads to chronic inflammation and tissue destruction. The important role of GADD45β in this process also follows from the finding that silencing of GADD45β by RNA interference abolished the anti-apoptotic effect of rheumatoid arthritis synovial fluid (for review, see Moskalev et al. 2012b). Another disorder associated with elevated expression of GADD45α protein is preeclampsia. Inflammatory and immune activation in preeclampsia may function in a feedback loop to maintain elevated expression of GADD45α protein (Geifman-Holtzman et al. 2013). GADD45α activates Mkk3-p38 and/or JNK signaling that leads to immunological and inflammatory changes as well as to triggering the production of circulating factors such as sFlt-1 (Xiong et al. 2009; Geifman-Holtzman et al. 2013; Xiong et al. 2013).

In hepatocytes both injury and growth stimulation remarkably increase the expression of the GADD45β protein. In liver cancer, promoter methylation frequently silences GADD45β, demonstrating a suppressive proapoptotic function. This contrasts with normal hepatocytes, where GADD45β facilitates cell survival, growth, and proliferation. GADD45β protects the liver through two ways: binding MKK7 to block damaging signal transduction or binding CAR to coactivate anabolic transcription (Tian et al. 2011; Tian and Locker 2013). Furthermore, the GADD45γ protein deregulation may be a reason of liver hypertrophy and liver tumor as well through the interaction with cyclins and cyclin-dependent kinase inhibitors (Ozawa et al. 2011).

Thus, GADD45 proteins are involved in major aging-associated conditions including oxidative stress, chronic inflammation, immunosenescence and fibroproliferative repair that contribute to the development of ARDs and aging progression (for review, see Moskalev et al. 2012b).

Aging-related changes in organism fertility can be caused by GADD45 expression alterations as well. In a model mice with deficiency of GADD45 isoforms, it was shown that GADD45s determine male fertility, testis development, and primary sex determination (Johnen et al. 2013).

In a view of its functions, it seems reasonable that *GADD45* overexpression might promote longevity, in particular, by increasing the efficiency of DNA repair (for review, see Moskalev et al. 2012b). Recently, we have confirmed this hypothesis in the *Drosophila melanogaster* model and have shown a life span-extending effects of *D-GADD45* overactivation in the nervous system (Plyusnina et al. 2011; Moskalev et al. 2012a; Plyusnina et al. 2012).

2.4 Life Span and Stress Resistance in Fruit Flies with *D-Gadd45* Overexpression

Research of the life span and stress resistance in model animals such as fruit fly *Drosophila melanogaster* with overexpressed investigated genes is a promising method to reveal their life span-extending properties. Therethrough, we investigated the effect of conditional and constitutive overexpression of the GADD45 gene both in the nervous system and the whole body on the life span and some age-dependent physiological parameters. To realize this aim the UAS/GAL4 system was used.

It was obtained that despite the fact that the overall spontaneous activity of the *D-GADD45* gene increases with age, the level of expression of this gene in the nervous system is practically eliminated (Moskalev et al. 2012a). The dramatic decrease of its activity within the nervous system with age might be one of the reasons for the decrease in the organism's stress resistance. Moreover, flies with constitutive *D-GADD45* overexpression in the nervous system showed a reduction in mRNA levels of *D-GADD45* with age as well (Moskalev et al. 2012a) that may indicate an epigenetic cause of the low level of activity of this gene in old flies.

Peretz et al. (2007) have shown that *D-GADD45* overexpression in *Drosophila melanogaster* has diverse phenotypic manifestations depending on the target tissue. Ubiquitous overexpression of this gene in *Drosophila* flies from the first stages of life cycle is lethal. Our investigations revealed that flies with conditional (mifepristone-inducible) ubiquitous *D-GADD45* overexpression in the imaginal developmental stage are viablc, but characterized by 22–46 % decreased life span (Fig. 2.1) (our unpublished data) and low resistance to the acute γ-irradiation and oxidative stress induced by paraquat treatment. The reason of this effect may be an insufficient epigenetic regulation of *D-GADD45* activity. For example, in human fibroblasts increased activity of DNA repair genes slows the replicative senescence only under simultaneous overexpression of histone deacetylase SIRT6 (Mao et al. 2012). The *GADD45* expression is depends on the sirtuins activity as well (Kobayashi et al. 2005; Scuto et al. 2013). Another reason could be associated with a high energy rate of repair processes (Halmosi et al. 2001). Ubiquitous *D-GADD45* overexpression can lead to excessive energy expenditure, which violates other processes.

Opposite to ubiquitous overexpression, tissue-specific *D-GADD45* overexpression in the nervous system leads to life span prolongation. Our data indicate that constitutive *D-GADD45* overexpression in the *Drosophila* nervous system leads to median life span extension (by 22–77 %) in comparison with flies without overexpression (Fig. 2.2a). Furthermore, the maximum life span was also increased, which is an evidence of slowing down aging (Plyusnina et al. 2011). To avoid effect of heterosis we also studied the life span effects of the conditional (mifepristone-inducible) *D-GADD45* overexpression in the nervous system. We found that the median life span of individuals with conditional overexpression of the *D-GADD45* gene in the nervous system was higher in comparison with animals with the same genotype kept on a medium without mifepristone (by 3–102 %) (Fig. 2.2b) (Plyusnina et al. 2011). It must be noted that the life span-extending effect of *D-GADD45* overexpression in *Drosophila* nervous

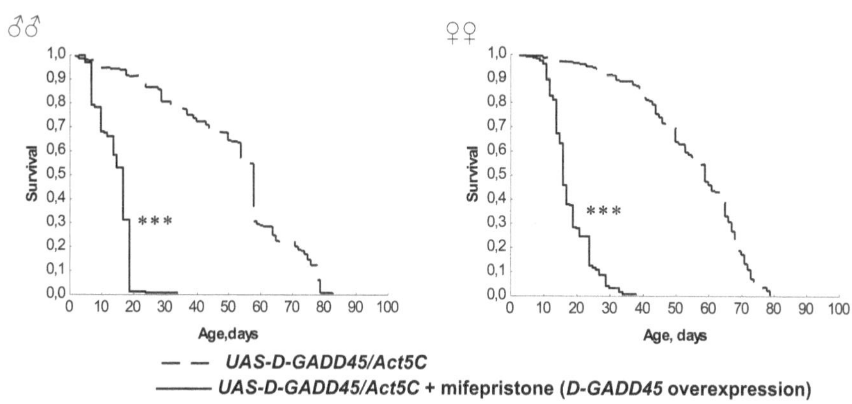

Fig. 2.1 Survival curves for *Drosophila melanogaster* males (♂♂) and females (♀♀) with the *UAS-D-GADD45/Act5C* genotype not treated with mifepristone and treated with mifepristone (conditional ubiquitous *D-GADD45* overexpression) (combined results of two replications), *** $p < 0.001$ (Kolmogorov–Smirnov test)

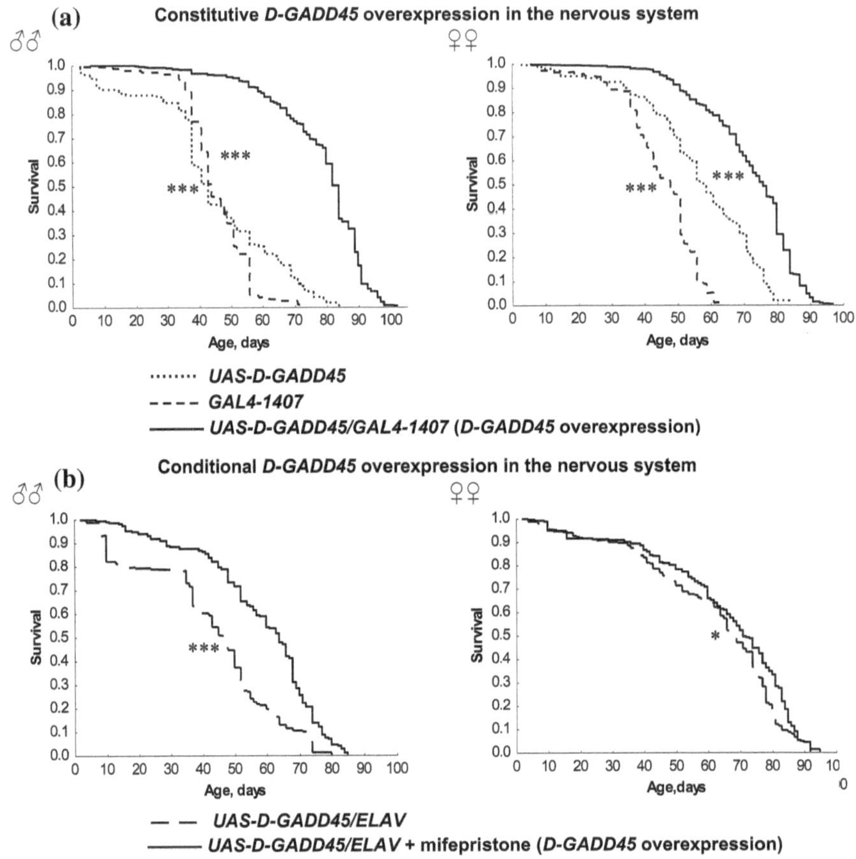

Fig. 2.2 Survival curves for *Drosophila melanogaster* flies with and without *D-GADD45* overexpression in the nervous system: **a** Survival curves for *Drosophila melanogaster* males (♂♂) and females (♀♀) with the parental *UAS-D-GADD45* and *GAL4-1407* genotypes and constitutive *D-GADD45* overexpression in the nervous system (combined results of three replications), **b** Survival curves for *Drosophila melanogaster* males (♂♂) and females (♀♀) with the *UAS-D-GADD45/ELAV* genotype not treated with mifepristone and treated with mifepristone (conditional *D-GADD45* overexpression in the nervous system) (combined results of two replications), * $p < 0.05$, *** $p < 0.001$ (Kolmogorov–Smirnov test) (Plyusnina et al. 2011)

system was not accompanied by decreases in fertility and locomotor activity parameters (Plyusnina et al. 2011). We proposed that *D-GADD45* overexpression causes more effective functioning of stress response mechanisms. Indeed, the DNA comet assay showed that neuroblasts of third-instar larvae with *D-GADD45* overexpression had the decreased DNA damage level (by 21–27 %). Therefore, overexpression of the *D-GADD45* gene in *Drosophila* nervous system results in more efficient recognition and elimination of spontaneous DNA damage caused by physiological processes and environmental factors (Plyusnina et al. 2011).

In further experiment we found additional evidences of increased resistance of *Drosophila melanogaster* individuals with constitutive and conditional *D-GADD45* overexpression in the nervous system. In most cases, these flies are characterized by increased survival under conditions of genotoxic stress (chronic and acute γ-irradiation), oxidative stress (paraquat), hyperthermia and starvation (Figs. 2.3, 2.4 and 2.5) (Moskalev et al. 2012a). Additionally, the involvement of the *D-GADD45* gene in the formation of biological responses to γ-irradiation was shown in the experiment on the fruit flies with homozygous and heterozygous *D-GADD45* mutation. Our results revealed the effects of hormesis and radio-adaptive response for the wild-type flies irradiated by the low 40 cGy dose. Over against, the *D-GADD45* mutations led to elimination of these reactions (Moskalev et al. 2012a).

Thus, ubiquitous *D-GADD45* overexpression leads to decrease of life span and stress resistance. At the same time, neuron-specific overexpression of the

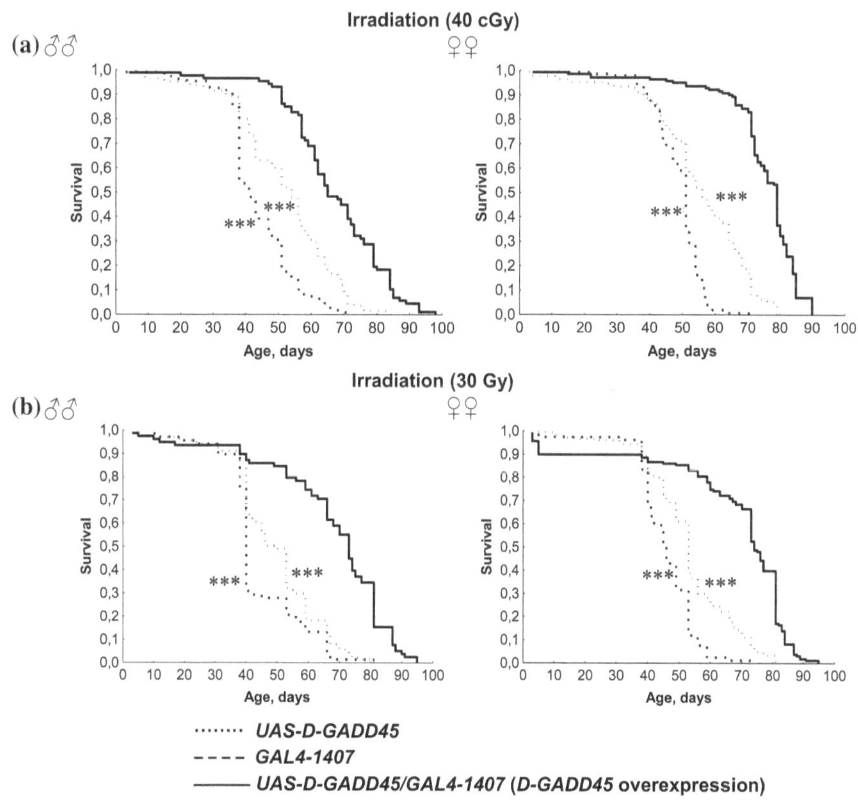

Fig. 2.3 Survival curves of *Drosophila* males (♂♂) and females (♀♀) with the parental *UAS-D-GADD45* and *GAL4-1407* genotypes and constitutive *D-GADD45* overexpression in the nervous system under different irradiation conditions: **a** chronic 40 cGy γ-irradiation, **b** acute 30 Gy γ-irradiation, *** $p < 0.001$ (Kolmogorov–Smirnov test) (Moskalev et al. 2012a)

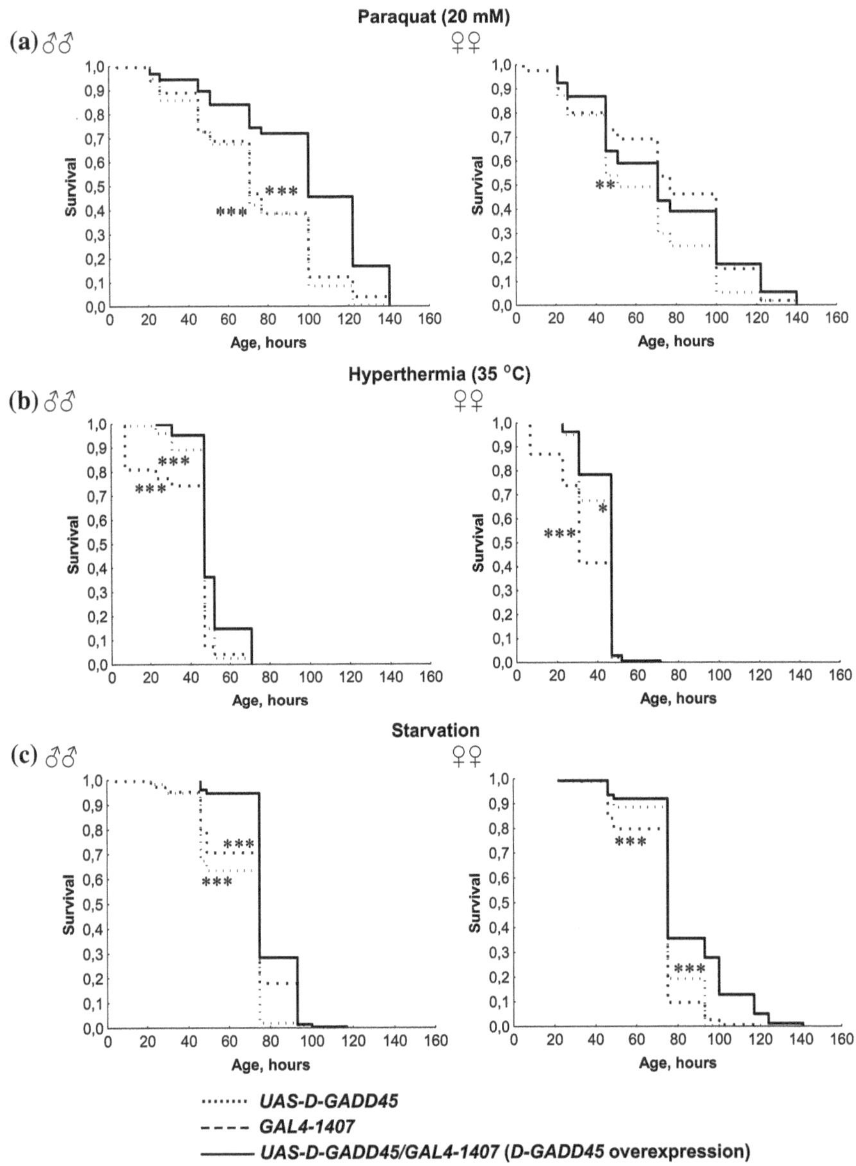

Fig. 2.4 Survival curves of *Drosophila* males (♂♂) and females (♀♀) with the parental *UAS-D-GADD45* and *GAL4-1407* genotypes and constitutive *D-GADD45* overexpression in the nervous system under different stress conditions: **a** paraquat (20 mM), **b** hyperthermia (35 °C), **c** starvation, * $p < 0.05$, ** $p < 0.01$, *** $p < 0.001$ (Kolmogorov–Smirnov test) (Moskalev et al. 2012a)

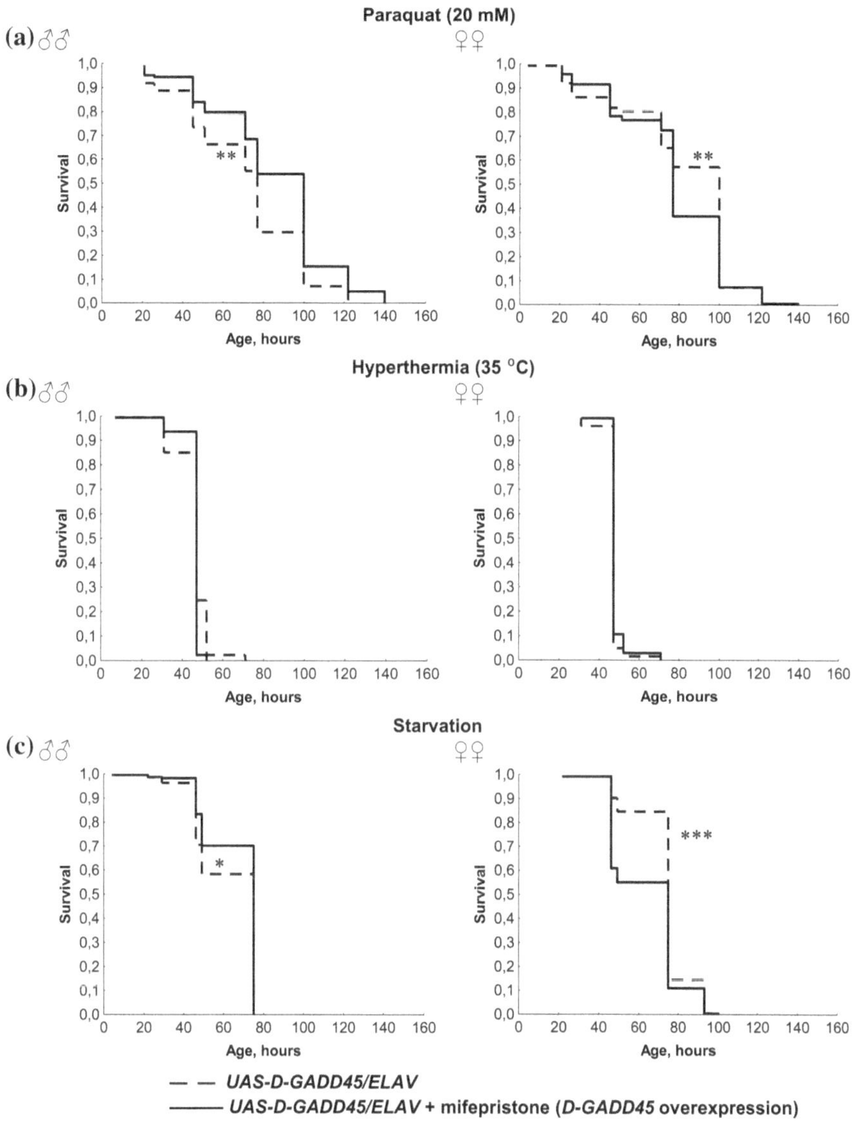

Fig. 2.5 Survival curves for *Drosophila melanogaster* males (♂♂) and females (♀♀) with the *UAS-D-GADD45/ELAV* genotype not treated with mifepristone and treated with mifepristone (conditional *D-GADD45* overexpression in the nervous system) under different stress conditions: **a** paraquat (20 mM), **b** hyperthermia (35 °C), **c** starvation, * $p < 0.05$, ** $p < 0.01$, *** $p < 0.001$ (Kolmogorov–Smirnov test) (Moskalev et al. 2012a)

D-GADD45 gene demonstrates a high life span-extending potential of controlled manipulation with this gene.

2.5 Concluding Remarks

Proteins of the GADD45 family are essential for stress resistance, and display antiaging and prolongevity activities (Fig. 2.6). Particularly, GADD45s provide a maintenance of basic homeostatic reactions and regulate a balance between cell (DNA) repair, eliminating (apoptosis) or preventing the expansion of potentially dangerous cells (cell cycle arrest, cellular senescence), maintaining of the stem cell pool and cellular differentiation. These processes provide survival of cells from different tissues and contribute to tissue regeneration. In turn, a decreased inducibility of the GADD45 family members may have far reaching consequences including genome instability, accumulation of DNA damage, and disorders in cellular homeostasis. All these negative processes may eventually lead to the age-dependent decline of organism system and organ functioning, aging progression, and promotion of carcinogenesis and development of other ARDs. It must be noted, that the GADD45 protein members are deeply involved in major longevity-associated signaling pathways, which confirm their role in aging and longevity determination.

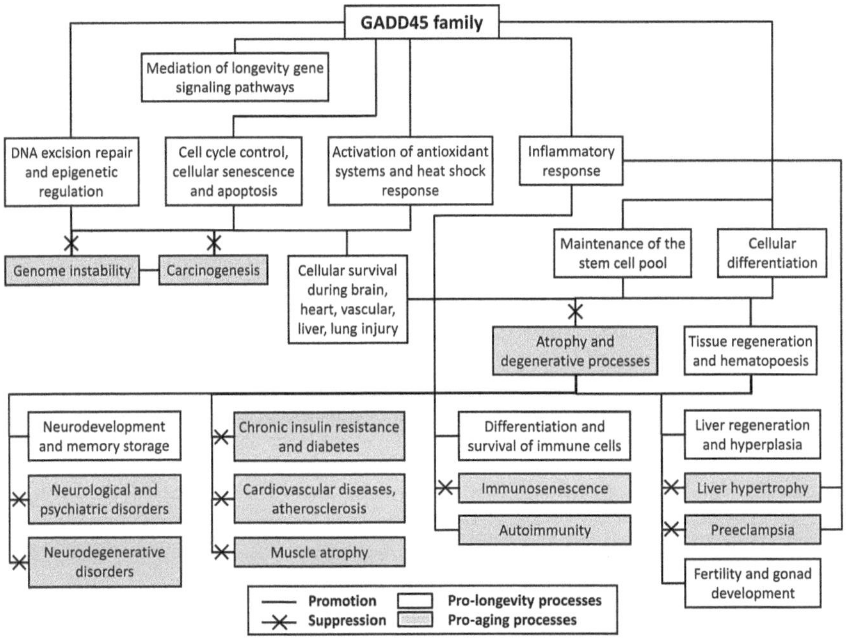

Fig. 2.6 The main anti-aging and pro-longevity activities of Gadd45 family

Investigations carried out in *Drosophila melanogaster* model disclosed the life span-extending activity of the *D-GADD45* gene due to its neuron-specific overactivation (Plyusnina et al. 2011, 2012) but not ubiquitous overexpression. It was shown that both constitutive and conditional *D-GADD45* overactivation in *Drosophila* nervous system extends median and maximum life span without negative changes in fertility and locomotor activity. This effect is apparently conditioned by elevated efficiency of recognition and elimination of spontaneous DNA damages. Additionally, neuron-specific *D-GADD45* overexpression can stimulate the resistance to different stress agents including genotoxic (γ-radiation), oxidative (paraquat) and thermal stressor as well as starvation.

Obtained results suggest that controlled manipulations of GADD45s and its interacting partners may also bring benefits to humans. Indeed, the increased GADD45 expression can be induced by several anti-tumor, anti-oxidant, anti-inflammatory pharmacological agents with potential life span-extending action, such as troglitazone (Yin et al. 2004), arsenic trioxide (Li et al. 2003), cucurbitacin E (Hsu et al. 2014), xanthatin (Takeda et al. 2011, 2013), quercetin (Yoshida et al. 2005), fucoxanthin (Kumar et al. 2013), epicatechin (Saha et al. 2010), ibuprofen (Bonelli et al. 2011). Thus, future studying the GADD45 family may provide prosperous therapeutic targets for promoting longevity and combating ARDs, as well as for stimulation of organism stress-resistance and enhancement of survivability.

Acknowledgments This work was supported by Russian Science Foundation grant N 14-50-00060.

References

Abdollahi A, Lord KA, Hoffman-Liebermann B, Liebermann DA (1991) Sequence and expression of a cDNA encoding *MyD118*: a novel myeloid differentiation primary response gene induced by multiple cytokines. Oncogene 6(1):165–167

Amundson SA, Zhan Q, Penn LZ, Fornace AJ Jr (1998) Myc suppresses induction of the growth arrest genes *gadd34*, *gadd45*, and *gadd153* by DNA-damaging agents. Oncogene 17(17):2149–2154

Azam N, Vairapandi M, Zhang W, Hoffman B, Liebermann DA (2001) Interaction of CR6 (GADD45γ) with proliferating cell nuclear antigen impedes negative growth control. J Biol Chem 276(4):2766–2774

Balliet AG, Hatton KS, Hoffman B, Liebermann DA (2001) Comparative analysis of the genetic structure and chromosomal location of the murine *MyD118* (*Gadd45beta*) gene. DNA Cell Biol 20(4):239–247

Barreto G, Schäfer A, Marhold J, Stach D, Swaminathan SK, Handa V, Döderlein G, Maltry N, Wu W, Lyko F, Niehrs C (2007) Gadd45a promotes epigenetic gene activation by repair-mediated DNA demethylation. Nature 445(7128):671–675

Blagosklonny MV (2008) Prevention of cancer by inhibiting ageing. Cancer Biol Ther 7(10):1520–1524

Bonelli P, Tuccillo FM, Calemma R, Pezzetti F, Borrelli A, Martinelli R, De Rosa A, Esposito D, Palaia R, Castello G (2011) Changes in the gene expression profile of gastric cancer cells in response to ibuprofen: a gene pathway analysis. Pharmacogenomics J 11(6):412–428

Bortoff KD, Keeton AB, Franklin JL, Messina JL (2010) Anti-inflammatory action of insulin via induction of *Gadd45-β* transcription by the mTOR signaling pathway. Hepat Med 2001(2):79–85

Budovsky A, Tacutu R, Yanai H, Abramovich A, Wolfson M, Fraifeld V (2009) Common gene signature of cancer and longevity. Mech Ageing Dev 130(1–2):33–39

Carrier F, Georgel PT, Pourquier P, Blake M, Kontny HU, Antinore MJ, Gariboldi M, Myers TG, Weinstein JN, Pommier Y, Fornace AJJ (1999) Gadd45, a p53-responsive stress protein, modifies DNA accessibility on damaged chromatin. Mol Cell Biol 19(3):1673–1685

Chang HC, Tsai J, Guo YL, Huang YH, Tsai HN, Tsai PC, Huang W (2003) Differential UVC-induced *gadd45* gene expression in xeroderma pigmentosum cells. Biochem Biophys Res Commun 305(4):1109–1115

Cheung KJJr, Mitchell D, Lin P, Li G (2001) The tumor suppressor candidate p33^{ING1} mediates repair of UV-damaged DNA. Cancer Res 61(13):4974–4977

Chuang JY, Wang SA, Yang WB, Yang HC, Hung CY, Su TP, Chang WC, Hung JJ (2012) Sp1 phosphorylation by cyclin-dependent kinase 1/cyclin B1 represses its DNA-binding activity during mitosis in cancer cells. Oncogene 31(47):4946–4959

Chung HY, Cesari M, Anton S, Marzetti E, Giovannini S, Seo AY, Carter C, Yu BP, Leeuwenburgh C (2009) Molecular inflammation: underpinnings of ageing and age-related diseases. Ageing Res Rev 8(1):18–30

Coppé JP, Desprez PY, Krtolica A, Campisi J (2010) The senescence-associated secretory phenotype: the dark side of tumor suppression. Annu Rev Pathol 5:99–118

Cortellino S, Xu J, Sannai M et al (2011) Thymine DNA glycosylase is essential for active DNA demethylation by linked deamination-base excision repair. Cell 146(1):67–79

Cretu A, Sha X, Tront J, Hoffman B, Liebermann DA (2009) Stress sensor *Gadd45* genes as therapeutic targets in cancer. Cancer Ther 7(A):268–276

De Laurenzi V, Melino G (2000) Evolution of functions within the p53/p63/p73 family. Ann N Y Acad Sci 926:90–100

De Smaele E, Zazzeroni F, Papa S, Nguyen DU, Jin R, Jones J, Cong R, Franzoso G (2001) Induction of *gadd45beta* by NF-kappaB downregulates pro-apoptotic JNK signalling. Nature 414(6861):308–313

Du F, Wang L, Zhang Y, Jiang W, Sheng H, Cao Q, Wu J, Shen B, Shen T, Zhang JZ, Bao C, Li D, Li N (2008) Role of GADD45 beta in the regulation of synovial fluid T cell apoptosis in rheumatoid arthritis. Clin Immunol 128(2):238–247

Duan J, Zhang Z, Tong T (2005) Irreversible cellular senescence induced by prolonged exposure to H_2O_2 involves DNA-damage-and-repair genes and telomere shortening. Int J Biochem Cell Biol 37(7):1407–1420

Edwards MG, Sarkar D, Klopp R, Morrow JD, Weindruch R, Prolla TA (2004) Impairment of the transcriptional responses to oxidative stress in the heart of aged C57BL/6 mice. Ann N Y Acad Sci 1019:85–95

el-Deiry WS (1998) Regulation of p53 downstream genes. Semin Cancer Biol 8(5):345–357

Fan W, Richter G, Cereseto A, Beadling C, Smith KA (1999) Cytokine response gene 6 induces p21 and regulates both cell growth and arrest. Oncogene 18(47):6573–6582

Finch CE (1990) Longevity, senescence, and the genome. University of Chicago Press, Chicago

Flicek P, Amode MR, Barrell D et al (2011) Ensembl 2011. Nucleic Acids Res 39(Database issue):D800–D806

Fornace AJ Jr, Amundson SA, Do KT, Meltzer P, Trent J, Bittner M (2002) Stress-gene induction by low-dose gamma irradiation. Mil Med 167(2 Suppl):13–15

Friedman MJ, Li S, Li XJ (2009) Activation of gene transcription by heat shock protein 27 may contribute to its neuronal protection. J Biol Chem 284(41):27944–27951

Furukawa-Hibi Y, Yoshida-Araki K, Ohta T, Ikeda K, Motoyama N (2002) FOXO forkhead transcription factors induce G(2)-M checkpoint in response to oxidative stress. J Biol Chem 277(30):26729–26732

Gant TW, Baus PR, Clothier B, Riley J, Davies R, Judah DJ, Edwards RE, George E, Greaves P, Smith AG (2003) Gene expression profiles associated with inflammation, fibrosis, and cholestasis in mouse liver after griseofulvin. EHP Toxicogenomics 111(1T):37–43

Geifman-Holtzman O, Xiong Y, Holtzman EJ (2013) Gadd45 stress sensors in preeclampsia. Adv Exp Med Biol 793:121–129

Guarente L (2011) Sirtuins, ageing, and metabolism. Cold Spring Harb Symp Quant Biol 76:81–90

Guillouf C, Grana X, Selvakumaran M, De Luca A, Giordano A, Hoffman B, Liebermann DA (1995) Dissection of the genetic programs of p53-mediated G1 growth arrest and apoptosis: blocking p53-induced apoptosis unmasks G1 arrest. Blood 85(10):2691–2698

Gupta M, Gupta SK, Balliet AG, Hollander MC, Fornace AJ, Hoffman B, Liebermann DA (2005) Hematopoietic cells from *Gadd45a*- and *Gadd45b*-deficient mice are sensitized to genotoxic-stress-induced apoptosis. Oncogene 24(48):7170–7179

Gupta M, Gupta SK, Hoffman B, Liebermann DA (2006) Gadd45a and Gadd45b protect hematopoietic cells from UV-induced apoptosis via distinct signaling pathways, including p38 activation and JNK inhibition. J Biol Chem 281(26):17552–17558

Hall PA, Kearsey JM, Coates PJ, Norman DG, Warbrick E, Cox LS (1995) Characterisation of the interaction between PCNA and Gadd45. Oncogene 10(12):2427–2433

Halmosi R, Berente Z, Osz E, Toth K, Literati-Nagy P, Sumegi B (2001) Effect of poly(ADP-ribose) polymerase inhibitors on the ischemia-reperfusion-induced oxidative cell damage and mitochondrial metabolism in Langendorff heart perfusion system. Mol Pharmacol 59(6):1497–1505

Hamza MS, Pott S, Vega VB, Thomsen JS, Kandhadayar GS, Ng PW, Chiu KP, Pettersson S, Wei CL, Ruan Y, Liu ET (2009) De-novo identification of PPARgamma/RXR binding sites and direct targets during adipogenesis. PLoS ONE 4(3):e4907

Harkin DP, Bean JM, Miklos D, Song YH, Truong VB, Englert C, Christians FC, Ellisen LW, Maheswaran S, Oliner JD, Haber DA (1999) Induction of GADD45 and JNK/SAPK-dependent apoptosis following inducible expression of BRCA1. Cell 97(5):575–586

Hartman AR, Ford JM (2002) BRCA1 induces DNA damage recognition factors and enhances nucleotide excision repair. Nat Genet 32(1):180–184

Higgins S, Wong SH, Richner M, Rowe CL, Newgreen DF, Werther GA, Russo VC (2009) Fibroblast growth factor 2 reactivates G1 checkpoint in SK-N-MC cells via regulation of p21, inhibitor of differentiation genes (*Id1-3*), and epithelium-mesenchyme transition-like events. Endocrinology 150(9):4044–4055

Higgs MR, Lerat H, Pawlotsky JM (2010) Downregulation of Gadd45beta expression by hepatitis C virus leads to defective cell cycle arrest. Cancer Res 70(12):4901–4911

Hildesheim J, Belova GI, Tyner SD, Zhou X, Vardanian L, Fornace AJJ (2004) Gadd45a regulates matrix metalloproteinases by suppressing DeltaNp63alpha and beta-catenin via p38 MAP kinase and APC complex activation. Oncogene 23(10):1829–1837

Hildesheim J, Bulavin DV, Anver MR, Alvord WG, Hollander MC, Vardanian L, Fornace AJJ (2002) Gadd45a protects against UV irradiation-induced skin tumors, and promotes apoptosis and stress signaling via MAPK and p53. Cancer Res 62(24):7305–7315

Hoffman B, Liebermann DA (2007) Role of gadd45 in myeloid cells in response to hematopoietic stress. Blood Cells Mol Dis 39(3):344–347. doi:S1079-9796(07)00127-1

Hoffman B, Liebermann DA (2013) Gadd45 in modulation of solid tumors and leukemia. Adv Exp Med Biol 793:21–33

Hoggard N, Hey Y, Brintnell B, James L, Jones D, Mitchell E, Weissenbach J, Varley JM (1995) Identification and cloning in yeast artificial chromosomes of a region of elevated loss of heterozygosity on chromosome 1p31.1 in human breast cancer. Genomics 30(2):233–243

Hollander MC, Sheikh MS, Bulavin DV, Lundgren K, Augeri-Henmueller L, Shehee R, Molinaro TA, Kim KE, Tolosa E, Ashwell JD, Rosenberg MP, Zhan Q, Fernandez-Salguero PM, Morgan WF, Deng CX, Fornace AJ Jr (1999) Genomic instability in *Gadd45a*-deficient mice. Nat Genet 23(2):176–184

Hollander MC, Kovalsky O, Salvador JM, Kim KE, Patterson AD, Haines DC, Fornace AJ Jr (2001) Dimethylbenzanthracene carcinogenesis in *Gadd45α*-null mice is associated with decreased DNA repair and increased mutation frequency. Cancer Res 61(6):2487–2491

Hsu YC, Huang TY, Chen MJ (2014) Therapeutic ROS targeting of GADD45γ in the induction of G2/M arrest in primary human colorectal cancer cell lines by cucurbitacin E. Cell Death Dis 5:e1198

Hubbard BP, Sinclair DA (2014) Small molecule SIRT1 activators for the treatment of ageing and age-related diseases. Trends Pharmacol Sci 35(3):146–154

Igotti Abramova MV, Pojidaeva AK, Filippova EA, Gnedina OO, Svetlikova SB, Pospelov VA (2014) HDAC inhibitors induce apoptosis but not cellular senescence in *Gadd45α*-deficient E1A+ Ras cells. Int J Biochem Cell Biol 51:102–110

Jackson JG, Pereira-Smith OM (2006) p53 is preferentially recruited to the promoters of growth arrest genes p21 and GADD45 during replicative senescence of normal human fibroblasts. Cancer Res 66(17):8356–8360

Jang ER, Choi JD, Park MA, Jeong G, Cho H, Lee JS (2010) ATM modulates transcription in response to histone deacetylase inhibition as part of its DNA damage response. Exp Mol Med 42(3):195–204

Ji C, Mehrian-Shai R, Chan C, Hsu YH, Kaplowitz N (2005) Role of CHOP in hepatic apoptosis in the murine model of intragastric ethanol feeding. Alcohol Clin Exp Res 29(8):1496–1503

Jiang F, Wang Z (2004) Gadd45gamma is androgen-responsive and growth-inhibitory in prostate cancer cells. Mol Cell Endocrinol 213(2):121–129. doi:10.1016/j.mce.2003.10.050

Jin S, Zhao H, Fan F, Blanck P, Fan W, Colchagie AB, Fornace AJ Jr, Zhan Q (2000) BRCA1 activation of the GADD45 promoter. Oncogene 19(35):4050–4057

Jin S, Fan F, Fan W, Zhao H, Tong T, Blanck P, Alomo I, Rajasekaran B, Zhan Q (2001) Transcription factors Oct-1 and NF-YA regulate the p53-independent induction of the GADD45 following DNA damage. Oncogene 20(21):2683–2690

Jinawath N, Vasoontara C, Yap KL, Thiaville MM, Nakayama K, Wang TL, Shih IM (2009) NAC-1, a potential stem cell pluripotency factor, contributes to paclitaxel resistance in ovarian cancer through inactivating Gadd45 pathway. Oncogene 28(18):1941–1948

Johnen H, González-Silva L, Carramolino L, Flores JM, Torres M, Salvador JM (2013) Gadd45g is essential for primary sex determination, male fertility and testis development. PLoS ONE 8(3):e58751

Ju S, Zhu Y, Liu L, Dai S, Li C, Chen E, He Y, Zhang X, Lu B (2009) Gadd45b and Gadd45g are important for anti-tumor immune responses. Eur J Immunol 39(11):3010–3018

Ju Y, Xu T, Zhang H, Yu A (2014) FOXO1-dependent DNA damage repair is regulated by JNK in lung cancer cells. Int J Oncol 44(4):1284–1292

Jung HJ, Kim EH, Mun JY, Park S, Smith ML, Han SS, Seo YR (2007) Base excision DNA repair defect in *Gadd45a*-deficient cells. Oncogene 26(54):7517–7525

Kastan MB, Zhan Q, el-Deiry WS, Carrier F, Jacks T, Walsh WV, Plunkett BS, Vogelstein B, Fornace AJ Jr (1992) A mammalian cell cycle checkpoint pathway utilizing p53 and GADD45 is defective in ataxia-telangiectasia. Cell 71(4):587–597

Keil E, Höcker R, Schuster M, Essmann F, Ueffing N, Hoffman B, Liebermann DA, Pfeffer K, Schulze-Osthoff K, Schmitz I (2013) Phosphorylation of Atg5 by the Gadd45β-MEKK4-p38 pathway inhibits autophagy. Cell Death Differ 20(2):321–332

Ko YG, Kang YS, Park H, Seol W, Kim J, Kim T, Park HS, Choi EJ, Kim S (2001) Apoptosis signal-regulating kinase 1 controls the proapoptotic function of death-associated protein (Daxx) in the cytoplasm. J Biol Chem 276(42):39103–39106

Kobayashi Y, Furukawa-Hibi Y, Chen C, Horio Y, Isobe K, Ikeda K, Motoyama N (2005) SIRT1 is critical regulator of FOXO-mediated transcription in response to oxidative stress. Int J Mol Med 16(2):237–243

Kovalsky O, Lung FD, Roller PP, Fornace AJ Jr (2001) Oligomerization of human Gadd45a protein. J Biol Chem 276(42):39330–39339

Kumar SR, Hosokawa M, Miyashita K (2013) Fucoxanthin: a marine carotenoid exerting anticancer effects by affecting multiple mechanisms. Mar Drugs 11(12):5130–5147
Le May N, Egly JM, Coin F (2010) True lies: the double life of the nucleotide excision repair factors in transcription and DNA repair. J Nucleic Acids 2010:616342
Li X, Ding X, Adrian TE (2003) Arsenic trioxide induces apoptosis in pancreatic cancer cells via changes in cell cycle, caspase activation, and GADD expression. Pancreas 27(2):174–179
Liebermann DA, Hoffman B (1994) Differentiation primary response genes and proto-oncogenes as positive and negative regulators of terminal hematopoietic cell differentiation. Stem Cells 12(4):352–369
Liebermann DA, Hoffman B (2003) Myeloid differentiation (*MyD*) primary response genes in hematopoiesis. Blood Cells Mol Dis 31(2):213–228
Liebermann DA, Tront JS, Sha X, Mukherjee K, Mohamed-Hadley A, Hoffman B (2011) Gadd45 stress sensors in malignancy and leukemia. Crit Rev Oncog 16(1–2):129–140
Lindstrom TM, Robinson WH (2010) Rheumatoid arthritis: a role for immunosenescence? J Am Geriatr Soc 58(8):1565–1575
Liu L, Tran E, Zhao Y, Huang Y, Flavell R, Lu B (2005) Gadd45 beta and Gadd45 gamma are critical for regulating autoimmunity. J Exp Med 202(10):1341–1347
López-Novoa JM, Nieto MA (2009) Inflammation and EMT: an alliance towards organ fibrosis and cancer progression. EMBO Mol Med 1(6–7):303–314
Lu B, Yu H, Chow C, Li B, Zheng W, Davis RJ, Flavell RA (2001) GADD45gamma mediates the activation of the p38 and JNK MAP kinase pathways and cytokine production in effector TH1 cells. Immunity 14(5):583–590
Ma DK, Guo JU, Ming GL, Song H (2009) DNA excision repair proteins and Gadd45 as molecular players for active DNA demethylation. Cell Cycle 8(10):1526–1531
MacLachlan TK, Somasundaram K, Sgagias M, Shifman Y, Muschel RJ, Cowan KH, El-Deiry WS (2000) BRCA1 effects on the cell cycle and the DNA damage response are linked to altered gene expression. J Biol Chem 275(4):2777–2785
Mao Z, Tian X, Van Meter M, Ke Z, Gorbunova V, Seluanov A (2012) Sirtuin 6 (SIRT6) rescues the decline of homologous recombination repair during replicative senescence. Proc Natl Acad Sci U S A 109(29):11800–11805
Michaelis KA, Knox AJ, Xu M, Kiseljak-Vassiliades K, Edwards MG, Geraci M, Kleinschmidt-DeMasters BK, Lillehei KO, Wierman ME (2011) Identification of growth arrest and DNA-damage-inducible gene beta (*GADD45beta*) as a novel tumor suppressor in pituitary gonadotrope tumors. Endocrinology 152(10):3603–3613
Moskalev A, Plyusnina E, Shaposhnikov M, Shilova L, Kazachenok A, Zhavoronkov A (2012a) The role of *D-GADD45* in oxidative, thermal and genotoxic stress resistance. Cell Cycle 11(22):4222–4241
Moskalev A, Smit-McBride Z, Shaposhnikov M, Plyusnina E, Zhavoronkov A, Budovsky A, Tacutu R, Fraifeld VE (2012b) Gadd45 proteins: relevance to ageing, longevity and age-related pathologies. Ageing Res Rev 11(1):51–66
Moskalev A, Aliper A, Smit-McBride Z, Buzdin A, Zhavoronkov A (2014) Genetics and epigenetics of ageing and longevity. Cell Cycle 13(7):1063–1077. doi:10.4161/cc.28433
Muñoz-Najar U, Sedivy JM (2011) Epigenetic control of ageing. Antioxid Redox Signal 14(2):241–259
Na YK, Lee SM, Hong HS, Kim JB, Park JY, Kim DS (2010) Hypermethylation of growth arrest DNA-damage-inducible gene 45 in non-small cell lung cancer and its relationship with clinicopathologic features. Mol Cells 30(1):89–92
Nakayama K, Hara T, Hibi M, Hirano T, Miyajima A (1999) A novel oncostatin M-inducible gene *OIG37* forms a gene family with *MyD118* and *GADD45* and negatively regulates cell growth. J Biol Chem 274(35):24766–24772
Nakayama A, Kawasaki H, Jin C, Munekata E, Taira K, Yokoyama KK (2001) Transcriptional regulation of interferon gamma gene by p300 co-activator. Nucleic Acids Res 1:89–90

Niehrs C, Schäfer A (2012) Active DNA demethylation by Gadd45 and DNA repair. Trends Cell Biol 22(4):220–227

O'Prey J, Brown J, Fleming J, Harrison PR (2003) Effects of dietary flavonoids on major signal transduction pathways in human epithelial cells. Biochem Pharmacol 66(11):2075–2088

Oh-Hashi K, Maruyama W, Isobe K (2001) Peroxynitrite induces GADD34, 45, and 153 VIA p38 MAPK in human neuroblastoma SH-SY5Y cells. Free Radic Biol Med 30(2):213–221

Ozawa S, Gamou T, Habano W, Inoue K, Yoshida M, Nishikawa A, Nemoto K, Degawa M (2011) Altered expression of *GADD45* genes during the development of chemical-mediated liver hypertrophy and liver tumor promotion in rats. J Toxicol Sci 36(5):613–623

Papa S, Zazzeroni F, Bubici C, Jayawardena S, Alvarez K, Matsuda S, Nguyen DU, Pham CG, Nelsbach AH, Melis T, De Smaele E, Tang WJ, D'Adamio L, Franzoso G (2004a) Gadd45 beta mediates the NF-kappa B suppression of JNK signalling by targeting MKK7/JNKK2. Nat Cell Biol 6(2):146–153

Papa S, Zazzeroni F, Pham CG, Bubici C, Franzoso G (2004b) Linking JNK signaling to NF-kappaB: a key to survival. J Cell Sci 117(Pt 22):5197–5208

Papa S, Zazzeroni F, Fu YX, Bubici C, Alvarez K, Dean K, Christiansen PA, Anders RA, Franzoso G (2008) Gadd45beta promotes hepatocyte survival during liver regeneration in mice by modulating JNK signaling. J Clin Invest 118(5):1911–1923

Passos JF, Nelson G, Wang C, Richter T, Simillion C, Proctor CJ, Miwa S, Olijslagers S, Hallinan J, Wipat A, Saretzki G, Rudolph KL, Kirkwood TB, von Zglinicki T (2010) Feedback between p21 and reactive oxygen production is necessary for cell senescence. Mol Syst Biol 6:347

Peretz G, Bakhrat A, Abdu U (2007) Expression of the *Drosophila melanogaster GADD45* homolog (*CG11086*) affects egg asymmetric development that is mediated by the c-Jun N-terminal kinase pathway. Genetics 177(3):1691–1702

Platanias LC (2003) Map kinase signaling pathways and hematologic malignancies. Blood 101(12):4667–4679

Plyusnina EN, Shaposhnikov MV, Moskalev AA (2011) Increase of *Drosophila melanogaster* lifespan due to *D-GADD45* overexpression in the nervous system. Biogerontology 12(3):211–226

Plyusnina EN, Shaposhnikov MV, Moskalev AA (2012) Geroprotective effects of activation of *D-GADD45* DNA reparation gene in *Drosophila melanogaster* nervous system. Bull Exp Biol Med 152(3):340–343

Qiu W, Zhou B, Zou H, Liu X, Chu PG, Lopez R, Shih J, Chung C, Yen Y (2004) Hypermethylation of growth arrest DNA damage-inducible gene 45 beta promoter in human hepatocellular carcinoma. Am J Pathol 165(5):1689–1699

Rai K, Huggins IJ, James SR, Karpf AR, Jones DA, Cairns BR (2008) DNA demethylation in zebrafish involves the coupling of a deaminase, a glycosylase, and gadd45. Cell 135(7):1201–1212

Reed BD, Charos AE, Szekely AM, Weissman SM, Snyder M (2008) Genome-wide occupancy of SREBP1 and its partners NFY and SP1 reveals novel functional roles and combinatorial regulation of distinct classes of genes. PLoS Genet 4(7):e1000133

Saha A, Kuzuhara T, Echigo N, Suganuma M, Fujiki H (2010) New role of (-)-epicatechin in enhancing the induction of growth inhibition and apoptosis in human lung cancer cells by curcumin. Cancer Prev Res (Phila) 3(8):953–962

Salerno DM, Tront JS, Hoffman B, Liebermann DA (2012) Gadd45a and Gadd45b modulate innate immune functions of granulocytes and macrophages by differential regulation of p38 and JNK signaling. J Cell Physiol 227(11):3613–3620

Salvador JM, Brown-Clay JD, Fornace AJJ (2013) Gadd45 in stress signaling, cell cycle control, and apoptosis. Adv Exp Med Biol 793:1–19

Santiard-Baron D, Gosset P, Nicole A, Sinet PM, Christen Y, Ceballos-Picot I (1999) Identification of beta-amyloid-responsive genes by RNA differential display: early induction of a DNA damage-inducible gene, *gadd45*. Exp Neurol 158(1):206–213

Santiard-Baron D, Lacoste A, Ellouk-Achard S, Soulie C, Nicole A, Sarasin A, Ceballos-Picot I (2001) The amyloid peptide induces early genotoxic damage in human preneuron NT2. Mutat Res 479(1–2):113–120

Satoh A, Brace CS, Rensing N, Cliften P, Wozniak DF, Herzog ED, Yamada KA, Imai S (2013) Sirt1 extends life span and delays ageing in mice through the regulation of Nk2 homeobox 1 in the DMH and LH. Cell Metab 18(3):416–430

Satomi Y (2012) Fucoxanthin induces GADD45A expression and G1 arrest with SAPK/JNK activation in LNCap human prostate cancer cells. Anticancer Res 32(3):807–813

Schäfer A, Schomacher L, Barreto G, Döderlein G, Niehrs C (2010) Gemcitabine functions epigenetically by inhibiting repair mediated DNA demethylation. PLoS ONE 5(11):e14060

Schmitz KM, Schmitt N, Hoffmann-Rohrer U, Schäfer A, Grummt I, Mayer C (2009) *TAF12* recruits Gadd45a and the nucleotide excision repair complex to the promoter of rRNA genes leading to active DNA demethylation. Mol Cell 33(3):344–353

Schmitz I (2013) Gadd45 proteins in immunity. Adv Exp Med Biol 793:51–68. doi:10.1007/978-1-4614-8289-5_4

Schomacher L (2013) Mammalian DNA demethylation: multiple faces and upstream regulation. Epigenetics 8(7):679–684

Schrag JD, Jiralerspong S, Banville M, Jaramillo ML, O'Connor-McCourt MD (2008) The crystal structure and dimerization interface of GADD45gamma. Proc Natl Acad Sci U S A 105(18):6566–6571

Scuto A, Kirschbaum M, Buettner R, Kujawski M, Cermak JM, Atadja P, Jove R (2013) SIRT1 activation enhances HDAC inhibition-mediated upregulation of *GADD45G* by repressing the binding of NF-κB/STAT3 complex to its promoter in malignant lymphoid cells. Cell Death Dis 4:e635

Selvakumaran M, Lin HK, Sjin RT, Reed JC, Liebermann DA, Hoffman B (1994) The novel primary response gene *MyD118* and the proto-oncogenes *myb, myc*, and *bcl-2* modulate transforming growth factor beta 1-induced apoptosis of myeloid leukemia cells. Mol Cell Biol 14(4):2352–2360

Sharova LV, Sharov AA, Nedorezov T, Piao Y, Shaik N, Ko MS (2009) Database for mRNA half-life of 19 977 genes obtained by DNA microarray analysis of pluripotent and differentiating mouse embryonic stem cells. DNA Res 16(1):45–58

Smith ML, Chen IT, Zhan Q, Bae I, Chen CY, Gilmer TM, Kastan MB, O'Connor PM, Fornace AJ Jr (1994) Interaction of the p53-regulated protein Gadd45 with proliferating cell nuclear antigen. Science 266(5189):1376–1380

Stokes AH, Freeman WM, Mitchell SG, Burnette TA, Hellmann GM, Vrana KE (2002) Induction of GADD45 and GADD153 in neuroblastoma cells by dopamine-induced toxicity. Neurotoxicology 23(6):675–684

Sultan FA, Sweatt JD (2013) The role of the Gadd45 family in the nervous system: a focus on neurodevelopment, neuronal injury, and cognitive neuroepigenetics. Adv Exp Med Biol 793:81–119

Sun L, Gong R, Wan B, Huang X, Wu C, Zhang X, Zhao S, Yu L (2003) *GADD45gamma*, down-regulated in 65 % hepatocellular carcinoma (HCC) from 23 chinese patients, inhibits cell growth and induces cell cycle G2/M arrest for hepatoma Hep-G2 cell lines. Mol Biol Rep 30(4):249–253

Sun X, Fontaine JM, Rest JS, Shelden EA, Welsh MJ, Benndorf R (2004) Interaction of human HSP22 (HSPB8) with other small heat shock proteins. J Biol Chem 279(4):2394–2402

Sytnikova YA, Kubarenko AV, Schäfer A, Weber AN, Niehrs C (2011) Gadd45a is an RNA binding protein and is localized in nuclear speckles. PLoS ONE 6(1):e14500

Takeda S, Matsuo K, Yaji K, Okajima-Miyazaki S, Harada M, Miyoshi H, Okamoto Y, Amamoto T, Shindo M, Omiecinski CJ, Aramaki H (2011) (–)-Xanthatin selectively induces *GADD45γ* and stimulates caspase-independent cell death in human breast cancer MDA-MB-231 cells. Chem Res Toxicol 24(6):855–865

Takeda S, Nishimura H, Koyachi K, Matsumoto K, Yoshida K, Okamoto Y, Amamoto T, Shindo M, Aramaki H (2013) (–)-Xanthatin induces the prolonged expression of c-Fos through an N-acetyl-L-cysteine (NAC)-sensitive mechanism in human breast cancer MDA-MB-231 cells. J Toxicol Sci 38(4):547–557

Takekawa M, Saito H (1998) A family of stress-inducible GADD45-like proteins mediate activation of the stress-responsive MTK1/MEKK4 MAPKKK. Cell 95(4):521–530

Thum T, Borlak J (2008) LOX-1 receptor blockade abrogates oxLDL-induced oxidative DNA damage and prevents activation of the transcriptional repressor Oct-1 in human coronary arterial endothelium. J Biol Chem 283(28):19456–19464. doi:M708309200

Tian J, Locker J (2013) Gadd45 in the liver: signal transduction and transcriptional mechanisms. Adv Exp Med Biol 793:69–80

Tian J, Huang H, Hoffman B, Liebermann DA, Ledda-Columbano GM, Columbano A, Locker J (2011) Gadd45β is an inducible coactivator of transcription that facilitates rapid liver growth in mice. J Clin Invest 121(11):4491–4502

Tornatore L, Marasco D, Dathan N, Vitale RM, Benedetti E, Papa S, Franzoso G, Ruvo M, Monti SM (2008) Gadd45beta forms a homodimeric complex that binds tightly to MKK7. J Mol Biol 378(1):97–111

Torp R, Su JH, Deng G, Cotman CW (1998) GADD45 is induced in Alzheimer's disease, and protects against apoptosis in vitro. Neurobiol Dis 5(4):245–252

Tran H, Brunet A, Grenier JM, Datta SR, Fornace AJ Jr, DiStefano PS, Chiang LW, Greenberg ME (2002) DNA repair pathway stimulated by the forkhead transcription factor FOXO3a through the Gadd45 protein. Science 296(5567):530–534

Tront JS, Hoffman B, Liebermann DA (2006) Gadd45a suppresses Ras-driven mammary tumorigenesis by activation of c-Jun NH2-terminal kinase and p38 stress signaling resulting in apoptosis and senescence. Cancer Res 66(17):8448–8454

Tront JS, Huang Y, Fornace AJJ, Hoffman B, Liebermann DA (2010) Gadd45a functions as a promoter or suppressor of breast cancer dependent on the oncogenic stress. Cancer Res 70(23):9671–9681

Vairapandi M, Balliet AG, Fornace AJ Jr, Hoffman B, Liebermann DA (1996) The differentiation primary response gene *MyD118*, related to *GADD45*, encodes for a nuclear protein which interacts with PCNA and p21WAF1/CIP1. Oncogene 12(12):2579–2594

Vairapandi M, Azam N, Balliet AG, Hoffman B, Liebermann DA (2000) Characterization of *MyD118*, Gadd45, and proliferating cell nuclear antigen (PCNA) interacting domains. PCNA impedes *MyD118* AND Gadd45-mediated negative growth control. J Biol Chem 275(22):16810–16819

Vairapandi M, Balliet AG, Hoffman B, Liebermann DA (2002) GADD45b and GADD45g are cdc2/cyclinB1 kinase inhibitors with a role in S and G2/M cell cycle checkpoints induced by genotoxic stress. J Cell Physiol 192(3):327–338

Wang XW, Zhan Q, Coursen JD, Khan MA, Kontny HU, Yu L, Hollander MC, O'Connor PM, Fornace AJ Jr, Harris CC (1999) GADD45 induction of a G2/M cell cycle checkpoint. Proc Natl Acad Sci USA 96(7):3706–3711

Whitlock NA, Lindsey K, Agarwal N, Crosson CE, Ma JX (2005) Heat shock protein 27 delays Ca^{2+}-induced cell death in a caspase-dependent and -independent manner in rat retinal ganglion cells. Invest Ophthalmol Vis Sci 46(3):1085–1091

Wolfson M, Budovsky A, Tacutu R, Fraifeld V (2009) The signaling hubs at the crossroad of longevity and age-related disease networks. Int J Biochem Cell Biol 41(3):516–520

Wu Y, Zhang X, Bardag-Gorce F, Robel RC, Aguilo J, Chen L, Zeng Y, Hwang K, French SW, Lu SC, Wan YJ (2004) Retinoid X receptor alpha regulates glutathione homeostasis and xenobiotic detoxification processes in mouse liver. Mol Pharmacol 65(3):550–557

Xiong Y, Liebermann DA, Tront JS, Holtzman EJ, Huang Y, Hoffman B, Geifman-Holtzman O (2009) Gadd45a stress signaling regulates sFlt-1 expression in preeclampsia. J Cell Physiol 220(3):632–639

Xiong Y, Liebermann DA, Holtzman EJ, Jeronis S, Hoffman B, Geifman-Holtzman O (2013) Preeclampsia-associated stresses activate Gadd45a signaling and sFlt-1 in placental explants. J Cell Physiol 228(2):362–370

Yang Q, Manicone A, Coursen JD, Linke SP, Nagashima M, Forgues M, Wang XW (2000) Identification of a functional domain in a GADD45-mediated G2/M checkpoint. J Biol Chem 275(47):36892–36898

Yang J, Zhu H, Murphy TL, Ouyang W, Murphy KM (2001) IL-18-stimulated GADD45 beta required in cytokine-induced, but not TCR-induced *IFN-gamma* production. Nat Immunol 2(2):157–164

Yang Z, Song L, Huang C (2009) Gadd45 proteins as critical signal transducers linking NF-kappaB to MAPK cascades. Curr Cancer Drug Targets 9(8):915–930

Yin F, Bruemmer D, Blaschke F, Hsueh WA, Law RE, Herle AJ (2004) Signaling pathways involved in induction of *GADD45* gene expression and apoptosis by troglitazone in human MCF-7 breast carcinoma cells. Oncogene 23(26):4614–4623

Ying J, Srivastava G, Hsieh WS, Gao Z, Murray P, Liao SK, Ambinder R, Tao Q (2005) The stress-responsive gene *GADD45G* is a functional tumor suppressor, with its response to environmental stresses frequently disrupted epigenetically in multiple tumors. Clin Cancer Res 11(18):6442–6449

Yoo J, Ghiassi M, Jirmanova L, Balliet AG, Hoffman B, Fornace AJJ, Liebermann DA, Bottinger EP, Roberts AB (2003) Transforming growth factor-beta-induced apoptosis is mediated by Smad-dependent expression of GADD45b through p38 activation. J Biol Chem 278(44):43001–43007

Yoshida T, Maeda A, Horinaka M, Shiraishi T, Nakata S, Wakada M, Yogosawa S, Sakai T (2005) Quercetin induces *gadd45* expression through a p53-independent pathway. Oncol Rep 14(5):1299–1303

Zerbini LF, Wang Y, Czibere A, Correa RG, Cho JY, Ijiri K, Wei W, Joseph M, Gu X, Grall F, Goldring MB, Zhou JR, Libermann TA (2004) NF-kappa B-mediated repression of growth arrest- and DNA-damage-inducible proteins 45alpha and gamma is essential for cancer cell survival. Proc Natl Acad Sci U S A 101(37):13618–13623

Zerbini LF, Libermann TA (2005) Life and death in cancer. GADD45 alpha and gamma are critical regulators of NF-kappaB mediated escape from programmed cell death. Cell Cycle 4(1):18–20

Zhan Q, Lord KA, Alamo I Jr, Hollander MC, Carrier F, Ron D, Kohn KW, Hoffman B, Liebermann DA, Fornace AJ Jr (1994) The *gadd* and *MyD* genes define a novel set of mammalian genes encoding acidic proteins that synergistically suppress cell growth. Mol Cell Biol 14(4):2361–2371

Zhan Q, Kontny U, Iglesias M, Alamo I Jr, Yu K, Hollander MC, Woodworth CD, Fornace AJ Jr (1999) Inhibitory effect of Bcl-2 on p53-mediated transactivation following genotoxic stress. Oncogene 18(2):297–304

Zhang W, Bae I, Krishnaraju K, Azam N, Fan W, Smith K, Hoffman B, Liebermann DA (1999) CR6: A third member in the *MyD118* and *Gadd45* gene family which functions in negative growth control. Oncogene 18(35):4899–4907

Zhang W, Hoffman B, Liebermann DA (2001) Ectopic expression of *MyD118/Gadd45/CR6* (*Gadd45beta/alpha/gamma*) sensitizes neoplastic cells to genotoxic stress-induced apoptosis. Int J Oncol 18(4):749–757

Zhang N, Ahsan MH, Zhu L, Sambucetti LC, Purchio AF, West DB (2005) NF-kappaB and not the MAPK signaling pathway regulates GADD45beta expression during acute inflammation. J Biol Chem 280(22):21400–21408

Zhang W, Li T, Shao Y, Zhang C, Wu Q, Yang H, Zhang J, Guan M, Yu B, Wan J (2010) Semi-quantitative detection of GADD45-gamma methylation levels in gastric, colorectal and pancreatic cancers using methylation-sensitive high-resolution melting analysis. J Cancer Res Clin Oncol 136(8):1267–1273

Zhang L, Yang Z, Liu Y (2014a) GADD45 proteins: roles in cellular senescence and tumor development. Exp Biol Med (Maywood) 239(7):773–778

Zhang L, Yang Z, Ma A, Qu Y, Xia S, Xu D, Ge C, Qiu B, Xia Q, Li J, Liu Y (2014b) Growth arrest and DNA damage 45G down-regulation contributes to Janus kinase/signal transducer and activator of transcription 3 activation and cellular senescence evasion in hepatocellular carcinoma. Hepatology 59(1):178–189

Zhu N, Shao Y, Xu L, Yu L, Sun L (2009) Gadd45-alpha and Gadd45-gamma utilize p38 and JNK signaling pathways to induce cell cycle G2/M arrest in Hep-G2 hepatoma cells. Mol Biol Rep 36(8):2075–2085

Zoncu R, Efeyan A, Sabatini DM (2011) mTOR: from growth signal integration to cancer, diabetes and ageing. Nat Rev Mol Cell Biol 12(1):21–35

Chapter 3
Superoxide Dismutase (SOD) Genes and Aging in *Drosophila*

John Tower

Abstract Since their discovery by McCord and Fridovich the superoxide dismutase (SOD) enzymes have been of particular interest to the field of aging. The *Drosophila* SOD genes are required for normal oxidative stress resistance and life span, and have been targets for investigation of mechanisms of aging. The ability of SOD genes to affect *Drosophila* life span is dependent upon the genetic background, including the sex of the animal, as well as the dietary environment. There is increasing understanding of the role of the SODs in signaling pathways that modulate aging. The Cu/ZnSOD is important in linking diet to life span, and MnSOD can activate the mitochondrial unfolded protein response and increase life span. The SOD genes also modulate survival in *Drosophila* models of human disease. The conservation of SOD genes and functions in *Drosophila* combined with the availability of powerful genetic and transgenic technologies promises to keep *Drosophila* at the forefront of research on aging and the role of SOD.

Keywords Superoxide dismutase genes · Cu/ZnSOD · MnSOD · Oxidative stress · Life span · Aging

3.1 Aging and Oxidative Stress

Aging is thought to be the result of antagonistic pleiotropy of gene function between developmental stages and the sexes (Hughes and Reynolds 2005; Tower 2006). Gene alleles may be selected for beneficial effects on growth and reproduction despite having deleterious effects at late ages. As a consequence, genetic interventions in aging may be dependent on the stage of the life cycle and/or the sex of the animal, and this has been observed upon manipulation of the SOD genes. The mitochondria appear

J. Tower (✉)
Molecular and Computational Biology Program, Department of Biological Sciences, University of Southern California, Los Angeles, CA 90089-2910, USA
e-mail: jtower@usc.edu

A.M. Vaiserman et al. (eds.), *Life Extension*, Healthy Ageing and Longevity 3,
DOI 10.1007/978-3-319-18326-8_3

to be particularly susceptible to damage and malfunction during aging (Cho et al. 2011). The mitochondria are the primary source of reactive oxygen species (ROS) in the cell, and increased production of ROS is associated with aging in both mammals and *Drosophila* (Salmon et al. 2010). In most organisms where it has been examined, aging is associated with increased abundance of oxidatively damaged macromolecules, including lipids, nucleic acids, carbohydrates and proteins (Yu 1993; Meli et al. 2003; Stadtman and Levine 2003; Negre-Salvayre et al. 2010; Salmon et al. 2010; Simm 2013). This holds true for aging *Drosophila*, where increased levels of protein carbonyls, lipid peroxidation, and glycation have been reported (Schwarze et al. 1998; Jacobson et al. 2010). ROS also function as cellular signaling molecules in both normal and transformed cells (Campos et al. 2013; Bauer 2014). Because the SOD enzymes play a key role in regulating cellular ROS levels they are implicated in both normal ROS signaling and in modulating ROS levels during aging.

3.2 SOD Genes, Enzymes, and Functions

The major families of eukaryotic SOD genes are each represented by a single member in *Drosophila* (Table 3.1) (McCord and Fridovich 1969; Parker et al. 2004; Landis and Tower 2005). The gene *Sod* encodes the canonical cytoplasmic

Table 3.1 *Drosophila* SOD and related genes

Gene	Symbol	Identifiers	Protein	Localization	Function	References
Superoxide dismutase	*Sod*	CG11793	Cu/ZnSOD	Cytoplasm, outer mito space	SOD enzyme	Seto et al. (1989) and Phillips et al. (1989)
Superoxide dismutase 2 (Mn)	*Sod2*	CG8905	MnSOD	Inner mito space	SOD enzyme	Duttaroy et al. (1997)
Superoxide dismutase 3	*Sod3*	CG9027	ecSOD (Cu/Zn)	Extracellular	SOD enzyme	Landis and Tower (2005) and Parker et al. (2004)
Copper chaperone for superoxide dismutase	*Ccs*	CG17753	Copper chaperone	Cytoplasm, outer mito space	Copper binding	Kirby et al. (2008)
Sodesque	*Sodq*	CG5948	Related to Cu/ZnSOD	Extracellular (predicted)	Unknown (Lacks active site)	Landis and Tower (2005)
Related to SOD	*Rsod*	CG31028	Related to Cu/ZnSOD	Cytoplasm (predicted)	Unknown (Lacks active site)	Landis and Tower (2005)

Mito mitochondria

Cu/ZnSOD enzyme (Phillips 1989; Seto et al. 1989). Cu/ZnSOD functions as a homodimer and is found both in the cytoplasm and the outer mitochondrial space (Okado-Matsumoto and Fridovich 2001; Sturtz et al. 2001). The gene *Sod2* encodes the MnSOD enzyme (Duttaroy et al. 1997). MnSOD functions as a tetramer and is localized to the inner mitochondrial space (Weisiger and Fridovich 1973). The gene *Sod3* encodes the extracellular form of Cu/ZnSOD, sometimes called ecSOD (Fukai et al. 2002; Oberley-Deegan et al. 2009). Phylogenetic analysis of SOD gene sequences indicates that within each phylum the ecSOD is more related to the cytoplasmic Cu/ZnSOD than to the ecSODs from other phyla, suggesting that ecSOD has evolved multiple times by the addition of a signal peptide to cytoplasmic Cu/ZnSOD (Landis and Tower 2005). The gene *Ccs* encodes the conserved copper chaperone that donates copper to Cu/ZnSOD (Kirby et al. 2008). In addition, there are two genes of unknown function that are distantly related to Cu/ZnSOD and that lack the active site (Table 3.1) (Landis and Tower 2005). The Cu/ZnSOD and MnSOD enzymes carry out the same reaction and convert superoxide to hydrogen peroxide and oxygen. The enzymes catalase and peroxiredoxin convert the hydrogen peroxide to water and oxygen (Radyuk et al. 2001). The SOD enzymes acting in concert with catalase and the peroxiredoxins comprise one of the major antioxidant systems of the cell.

The mitochondrial genome is associated with several proteins in a structure called the nucleoid. In mammals MnSOD was reported to be an integral component of the nucleoid complex and to act there to protect mitochondrial DNA and associated proteins from oxidative damage (Bakthavatchalu et al. 2012). The data suggest that MnSOD might have a protective structural function in addition to its protective enzymatic function.

3.3 SOD Gene Expression and Aging

During aging the expression of RNA for *Drosophila* MnSOD is reduced, whereas the RNA for Cu/ZnSOD is more constant (Landis et al. 2004, 2012; Radyuk et al. 2004). Analyses of enzyme activities in extracts of aging flies have reported both increases and decreases (Sohal et al. 1990), whereas strains genetically selected for increased life span are reported to have increased SOD enzyme activity (Hari et al. 1998). The gene *Darkener of apricot* (*Doa*) encodes a LAMMER-type kinase that is implicated as a negative regulator of Cu/ZnSOD gene expression (James et al. 2009). The gene *rolled* encodes a p38 MAPK and has been identified as a positive regulator of MnSOD gene expression (Duttaroy et al. 1997). It was reported that the p38 MAPK activates the transcription factor Mef2, which in turn activates expression of MnSOD (Vrailas-Mortimer et al. 2011). In contrast, the regulator of G-protein signaling encoded by the *loco* gene is reported to negatively regulate both MnSOD gene expression and longevity (Lin et al. 2011).

3.4 Reduced SOD Function and Aging

The role of the SOD genes in *Drosophila* aging has been studied by reducing their function, using both classical mutations and RNAi approaches. Null mutation of Cu/ZnSOD (n108 allele) caused reduced adult life span and reduced fertility, and also partly reduced viability during late development (Phillips et al. 1989). The null adult flies were hypersensitive to ionizing radiation and to the oxidative stressors paraquat and CuSO4. These phenotypes demonstrate the importance of Cu/ZnSOD for normal oxidative stress resistance.

When MnSOD gene expression was reduced during development and adulthood using an RNAi approach (da-GAL4 driver), development proceeded relatively normally and abundant adults were produced (Kirby et al. 2002). The adults had greatly reduced MnSOD enzyme activity and increased endogenous oxidative stress, as indicated by reduced activity of aconitase and other mitochondrial enzymes. The flies also had increased sensitivity to the oxidative stressor paraquat (Kirby et al. 2002). Using RNAi to knockdown MnSOD specifically in muscle tissue was sufficient to reduce oxidative stress resistance and life span, underscoring the importance of MnSOD in muscle (Godenschwege et al. 2009; Martin et al. 2009b). A strong mutation of the MnSOD gene caused greatly shortened life span and neuropathology, indicating the importance of MnSOD in the nervous system (Paul et al. 2007; Celotto et al. 2012). A MnSOD null mutation (homozygous n283 allele) produced the most severe phenotype, and flies died at or shortly after eclosion (Duttaroy et al. 2003; Mukherjee et al. 2011). These results indicate the importance of MnSOD for normal oxidative stress resistance and mitochondrial function, and suggest a relatively greater requirement in the adult stage.

Intriguingly, the effects of MnSOD and Cu/ZnSOD on oxidative stress resistance differed in the adult (Missirlis et al. 2003). The deleterious effects of hyperoxia on MnSOD RNAi flies were transient and reversible upon transfer to a hypoxic environment, whereas the effects of hyperoxia on Cu/ZnSOD RNAi flies were cumulative and could not be reversed by hypoxia (Wicks et al. 2009). A partial spontaneous reversibility in the negative effect of the MnSOD null mutation on heart rate and stimulus response was also reported (Piazza et al. 2009). The data suggest that Cu/ZnSOD may be more involved in preventing some type of cumulative damage, whereas MnSOD may be more involved in preventing acute mortality. Knock-down of gene expression specifically in the renal tubules indicated that MnSOD was required for survival upon dessication stress, whereas Cu/ZnSOD was not (Soderberg et al. 2011).

Finally, using an RNAi approach the ecSOD gene *Sod3* was recently reported to be required for normal oxidative stress resistance and life span (Jung et al. 2011). *Sod3* was also found to be induced in flies over-expressing the human Alzheimer's disease protein fragment Abeta, and to negatively affect survival of the mutant flies (Favrin et al. 2013).

3.5 Enhanced SOD Function and Aging

One method that has been used extensively to study the role of the SOD genes in *Drosophila* has been to engineer over-expression in transgenic flies. Early experiments utilized the gene's native promoters, or constitutive heterologous promoters, to over-express Cu/ZnSOD and/or catalase, and some conflicting results were obtained regarding effects on life span (Seto et al. 1990; Stavely et al. 1990; Reveillaud et al. 1991; Griswold et al. 1993; Orr and Sohal 1992, 1993, 1994). One of these studies concluded that coincident over-expression of Cu/ZnSOD and catalase might be required to observe life span increase (Orr and Sohal 1994). However, several authors reviewing these results concluded the studies lacked sufficient controls for genetic background effects on life span, and this masked any possible effects of the transgenes (Tower 1996; Kaiser et al. 1997; Tatar 1999). Subsequent studies using a larger number of lines confirmed that over-expression of the Cu/ZnSOD, MnSOD and catalase genes using their native promoters does not increase life span, regardless of whether the transgenes are co-over-expressed, or whether they are expressed in short-lived starting strains (Mockett et al. 1999; Orr et al. 2003). In contrast, several groups have reported life span increase when Cu/ZnSOD and MnSOD are over-expressed using tissue-specific and/or conditional transgenic systems, where the transgenes are expressed using heterologous (non-native) promoters.

In the binary "GAL4/UAS" system one transgenic construct contains *Drosophila* gene regulatory sequences that drive tissue-specific expression of the yeast transcription factor GAL4. The GAL4 in turn activates a second transgenic construct where a promoter containing the GAL4 binding site (called UAS) drives expression of a gene of interest (Brand and Perrimon 1993), such as a SOD gene. The GAL4/UAS system was used to target over-expression of human Cu/ZnSOD to the nervous system of male *Drosophila*, during developmental and adult stages, and a life span increase was observed (Parkes et al. 1998). A subsequent analysis of one of these transgenes in several long-lived genetic backgrounds reported life span increase that depended on genotype and sex, with greater effect observed in females (Spencer et al. 2003). Finally, a different analysis of one of these transgenes using male flies reported life span increase using tissue-general over-expression, but no life span increase upon nervous system over-expression (Martin et al. 2009a). The exact reasons for the variability in results across these studies is not clear, but they are generally consistent with the conclusion that over-expression of human Cu/ZnSOD in *Drosophila* can increase life span, and suggest that the life span increase depends upon the specific genetic background, including the sex of the animal (Fig. 3.1).

The binary transgenic system called "FLP-out" employs conditional somatic recombination to activate gene expression (Struhl and Basler 1993). In one transgenic construct a heat-activated promoter is used to drive expression of the

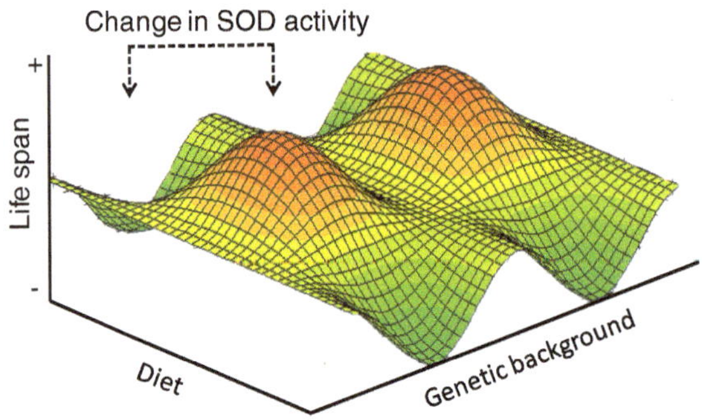

Fig. 3.1 Model for effect of altered SOD gene expression on life span. A hypothetical surface plot is diagrammed to indicate a complex relationship between diet, genetic background, and life span. A change in SOD gene activity is hypothesized to shift the relationship from one position on the plot to another. In this way altered SOD gene activity can have either a positive or negative effect on life span depending on the specific combination of diet and genetic background

yeast FLP-recombinase protein. In the second transgenic construct the FLP recombinase causes excision of a transcriptional stop signal, so that the gene of interest becomes actively transcribed from an upstream promoter. In this way a brief heat pulse to the developing or adult animal activates expression of the gene of interest, such as SOD, from that point in time onwards. FLP-out was used to create tissue-general over-expression of *Drosophila* Cu/ZnSOD specifically in adult animals, and this approach allowed for controlled analysis of interactions with the genetic background (Sun and Tower 1999). This study found that Cu/ZnSOD over-expression in adult males and females could increase life span, and that life span increase was dependent upon the genetic background. Similarly, over-expression of MnSOD in adult flies using FLP-out caused increased life span in male and female flies (Sun et al. 2002). Co-incident over-expression of Cu/ZnSOD and MnSOD was found to have partially additive effects (Sun et al. 2004). In contrast, catalase over-expression was found to increase resistance to hydrogen peroxide toxicity, but to have small negative effects on life span, both alone, and in combination with either Cu/ZnSOD or MnSOD (Sun and Tower 1999; Sun et al. 2002). The results with the FLP-out system again indicated that over-expression of Cu/ZnSOD and MnSOD could increase life span depending on genetic background, and the negative effects of catalase suggested the possibility that this might involve a retrograde signal of hydrogen peroxide.

The "Tet-on" conditional transgenic system is also binary (Bieschke et al. 1998). In one transgenic construct the tissue-general *Drosophila Actin5C* gene promoter drives expression of the artificial transcription factor called "rtTA". The rtTA becomes active upon binding to tetracycline or the tetracycline derivative doxycycline. The activated rtTA binds to a binding site called Tet-O in the promoter of the second transgenic construct, and drives expression of the gene of interest, such as SOD. In this way simple feeding of doxycycline to either larvae or adult flies yields conditional over-expression of SOD. The Tet-on system was used to over-express MnSOD in adult males, and this was also found to increase life span (Curtis et al. 2007). The life span increase resulted from a change in initial mortality rate, arguing against a mechanism involving prevention of accumulated oxidative damage, and instead suggested a mechanism involving retrograde ROS signaling. Consistent with this idea, transcriptional profiling revealed that MnSOD over-expression causes up-regulation of a conserved set of genes associated with increased life span (Curtis et al. 2007). These genes include ones involved in energy metabolism, purine biosynthesis, apoptotic pathways, endocrine signals and the detoxification and excretion of metabolites. In addition, MnSOD over-expression caused up-regulation of the Hsp22 gene and several other genes associated with the mitochondria unfolded protein response (UPRmt), as discussed further below.

3.6 Human ALS-Associated Cu/ZnSOD Over-Expression in *Drosophila*

The human disease amyotrophic lateral sclerosis (ALS) is sometimes caused by inherited mutations in the human Cu/ZnSOD gene (Rosen et al. 1993). Various possible mechanisms have been proposed for the toxicity of mutant Cu/ZnSOD, including misfolding and aggregation (Liu et al. 2012), disruptions of intracellular transport, disruptions of metal homeostasis (Lovejoy and Guillemin 2014), and disruptions of mitochondrial function (Hitchler and Domann 2014; Muyderman and Chen 2014). Transgenic *Drosophila* have been used in several studies to examine the effect of expression of human Cu/ZnSOD, including transgenes bearing human disease mutations. Interestingly, expression of one such mutant human transgene (G41S mutation) in the fly motorneurons was found to have no toxic effects, and to increase stress resistance and life span similar to the wild-type transgenes (Elia et al. 1999). In contrast, expression in the motorneurons of a human Cu/ZnSOD engineered to be Zinc-deficient caused toxicity including motor impairment and mitochondrial disruption (Bahadorani et al. 2013). Similarly, another study found that both the human wild-type and the ALS-associated mutant (A4V, G85R) caused locomotor defects and disruptions of neural circuit physiology when expressed in the motorneurons, whereas the wild-type *Drosophila* Cu/ZnSOD did not (Watson et al. 2008; Islam et al. 2012; Kumimoto et al. 2013). Transcriptional profiling of *Drosophila* expressing an ALS-mutant form of Cu/ZnSOD identified stress response genes, including oxidative stress response genes and proteasome subunit genes, that were also altered in ALS model mammals (Kumimoto et al. 2013). The data indicate that *Drosophila* can be used to model several aspects of human ALS, but may not reveal the toxicity of all disease-associated alleles.

3.7 MnSOD Over-Expression and the UPRmt

The UPRmt is a response to protein folding stress in the mitochondrial compartment. A UPRmt has been described in mammals, *C. elegans* and *Drosophila* (Baker and Haynes 2011; Haynes et al. 2013). The UPRmt is characterized by the induction of nuclear genes encoding Hsps, and the targeting of these Hsps to the mitochondria. These Hsps targeted to the mitochondria include Hsp60, the mitochondrial form of Hsp70, and the small heat shock protein Hsp22. In *Drosophila*, the UPRmt has been experimentally induced by targeting a mutant form of ornithine-transcarbamylase to the mitochondria to cause unfolded protein stress (Pimenta de Castro et al. 2012). The *Drosophila* UPRmt has also been experimentally induced in *Drosophila* flight muscle using RNAi to an ETC component, and in this case a benefit for life span was observed (Owusu-Ansah et al. 2013). Interestingly, the benefit of the UPRmt for life span was blocked by coincident over-expression of catalase, suggesting a possible retrograde hydrogen peroxide signal generated by SOD, however effects on Hsp induction were not examined. The *Drosophila tko* gene encodes a mitochondrial ribosomal protein, and a partial loss of function mutation induced expression of the mitochondrial chaperone Hsp22, consistent with a UPRmt, however increased Hsp60 and mitochondrial Hsp70 was not observed. The different Hsps observed in these different mitochondrial perturbations may indicate different UPRmt responses, and/or may reflect differences in the degree of the stress.

Over-expression of MnSOD causes induction of Hsp22, consistent with induction of the UPRmt (Curtis et al. 2007). MnSOD over-expression caused Hsp22 induction preferentially in the oenocytes (liver-like cells), and also reduced fly oxygen consumption and reduced the accumulation of age pigment in the oenocytes, indicating reduced oenocyte metabolism (Fig. 3.2) (Tower et al. 2013). Moreover, tissue-general over-expression of Hsp22 also caused induction of the endogenous Hsp22 gene preferentially in the oenocytes, consistent with a UPRmt (Shen and Tower 2013; Tower et al. 2013). The life span increase caused by MnSOD over-expression was reduced by coincident over-expression of catalase, again suggesting a possible retrograde hydrogen peroxide signal (Sun et al. 2002; Curtis et al. 2007). The ability of *Drosophila* MnSOD over-expression to induce the UPRmt and increase life span might therefore require the enzymatic activity of MnSOD and a retrograde hydrogen peroxide signal, or conceivably a signal involving reduced superoxide. The alternative is the UPRmt results from a non-enzymatic activity of MnSOD over-expression in inducing the UPRmt. This might be related to a normal structural function of MnSOD in the mitochondrial nucleoid, or alternatively the UPRmt may result from disruption of normal mitochondrial protein import and folding homeostasis due to the abnormally high level of expression of MnSOD. Because normal aging in *Drosophila* is associated with a progressive induction of UPRmt genes, including Hsp22 in the oenocytes and other cells, it suggests that the MnSOD life span intervention is an example of hormesis (Shen and Tower 2013; Tower et al. 2013). Determining the precise mechanisms for UPRmt induction and life span extension by MnSOD will be an important area for future research.

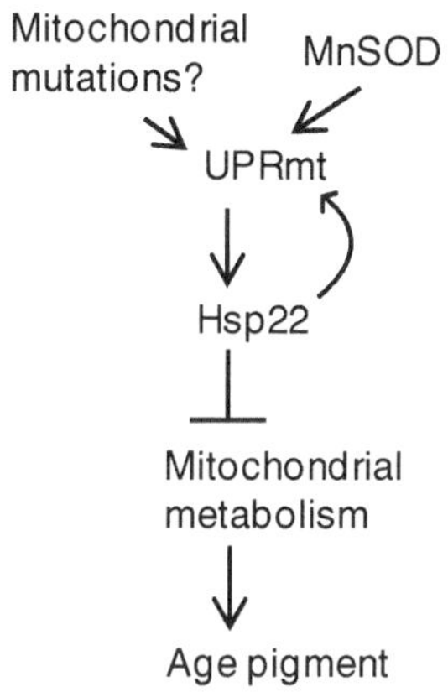

Fig. 3.2 Induction of the UPRmt by MnSOD. Over-expression of MnSOD caused induction of Hsp22 preferentially in the oenocytes (liver-like cells), indicating induction of the UPRmt. Hsp22 over-expression also induced Hsp22 expression in the oenocytes. MnSOD over-expression reduces oxygen consumption, and both MnSOD over-expression and Hsp22 over-expression reduced age pigment accumulation in the oenocytes, consistent with reduced metabolism in these cells. The UPRmt, as indicated by Hsp22 expression, is induced during normal aging of the oenocytes in a cell-specific and cell lineage-specific pattern, suggesting the possible involvement of mitochondrial mutations

3.8 SOD as Modifier of Aging Phenotypes

The SOD genes have been shown to modify several aging-related phenotypes. For example, over-expression of Cu/ZnSOD in the female gonad was reported to delay the loss of germ-line stem cells during aging (Pan et al. 2007). Over-expression of Cu/ZnSOD in the muscle and other tissues (twist-GAL4 driver) was reported to rescue the life-span shortening effect of a mutation in the *parkin* gene (Saini et al. 2010), and over-expression of MnSOD was reported to reduce the developmental defects caused by a PINK1 gene mutation (Koh et al. 2012). Over-expression of either Cu/ZnSOD or MnSOD was found to reduce the toxic effects of poly-Q-containing proteins in the *Drosophila* heart (Melkani et al. 2013), and to reduce the toxic effects of disrupted mitochondrial fusion in the heart (Dorn et al. 2011; Bhandari et al. 2014). Over-expression of MnSOD increased survival of flies expressing the mutant protein associated with human spinal cerebellar ataxia type 12 (Wang et al. 2011). These results suggest the importance of the SOD genes in resistance to stress associated with human disease genes.

An RNAi approach was used to examine the possible role of Cu/ZnSOD in mediating life span extension caused by protein restriction (Sun et al. 2012). Interestingly, Cu/ZnSOD was required for increased life span upon protein restriction, but only under conditions of high dietary sugar, demonstrating a complex diet-dependent interaction between Cu/ZnSOD activity and life span (Fig. 3.1). Consistent with a link to carbohydrate metabolism, Cu/ZnSOD null flies showed evidence of down-regulated carbohydrate metabolism enzyme activity (Bernard et al. 2011). Studies in yeast demonstrate the role of Cu/ZnSOD in repressing metabolism in response to glucose (Reddi and Culotta 2013). Similarly, *Drosophila* MnSOD gene activity was found to be required for the longevity-promoting effects of cranberry extract (Sun et al. 2014), supporting a diet-dependent interaction between MnSOD and life span (Fig. 3.1).

3.9 Conservation of SOD Functions in Aging

Both *Drosophila* Cu/ZnSOD and MnSOD have been found to modulate life span in a way that is dependent upon the genetic background, including the sex of the animal. Moreover, these effects were highly dependent upon the dietary environment (Fig. 3.1). Interestingly, over-expression of both Cu/ZnSOD and MnSOD has been shown to increase life span in the nematode *C. elegans* (Cabreiro et al. 2011), suggesting that the role of these genes in aging may be conserved across species. Consistent with this idea, reduced activity of the insulin-like pathway can increase life span in both *C. elegans* and *Drosophila*, and in *C. elegans* this involves up-regulated activity of MnSOD (Honda and Honda 1999; Zarse et al. 2012). In mammals, the Sirtuin SIRT3 regulates MnSOD acetylation and activity, including in response to dietary restriction (Qiu et al. 2010; Chen et al. 2011). These results support a conserved role for the SOD genes in modulating life span in response to the genetic background and the dietary environment (Fig. 3.1). The powerful genetics of *Drosophila* combined with the high degree of conservation of SOD genes and pathways promises to continue to provide insight into basic mechanisms of aging.

Acknowledgments This work was supported by a grant from the Department of Health and Human Services (AG011833) and by pilot project funds from the Southern California Environmental Health Sciences Center (5P30ES007048).

References

Bahadorani S, Mukai ST, Rabie J, Beckman JS, Phillips JP, Hilliker AJ (2013) Expression of zinc-deficient human superoxide dismutase in *Drosophila* neurons produces a locomotor defect linked to mitochondrial dysfunction. Neurobiol Aging 34:2322–2330

Baker BM, Haynes CM (2011) Mitochondrial protein quality control during biogenesis and aging. Trends Biochem Sci 36:254–261

Bakthavatchalu V, Dey S, Xu Y, Noel T, Jungsuwadee P, Holley AK, Dhar SK, Batinic-Haberle I, St Clair DK (2012) Manganese superoxide dismutase is a mitochondrial fidelity protein that protects Polγ against UV-induced inactivation. Oncogene 31:2129–2139

Bauer G (2014) Targeting extracellular ROS signaling of tumor cells. Anticancer Res 34:1467–1482

Bernard KE, Parkes TL, Merritt TJ (2011) A model of oxidative stress management: moderation of carbohydrate metabolizing enzymes in SOD1-null *Drosophila melanogaster*. PLoS One 6:e24518

Bhandari P, Song M, Chen Y, Burelle Y, Dorn GW 2nd (2014) Mitochondrial contagion induced by Parkin deficiency in *Drosophila* hearts and its containment by suppressing mitofusin. Circ Res 114:257–265

Bieschke ET, Wheeler JC, Tower J (1998) Doxycycline-induced transgene expression during *Drosophila* development and aging. Mol Gen Genet MGG 258:571–579

Brand AH, Perrimon N (1993) Targeted gene expression as a means of altering cell fates and generating dominant phenotypes. Development 118:401–415

Cabreiro F, Ackerman D, Doonan R, Araiz C, Back P, Papp D, Braeckman BP, Gems D (2011) Increased life span from overexpression of superoxide dismutase in *Caenorhabditis elegans* is not caused by decreased oxidative damage. Free Radic Biol Med 51:1575–1582

Campos JC, Gomes KM, Ferreira JC (2013) Impact of exercise training on redox signaling in cardiovascular diseases. Food Chem Toxicol 62:107–119 (an international journal published for the British Industrial Biological Research Association)

Celotto AM, Liu Z, Vandemark AP, Palladino MJ (2012) A novel *Drosophila* SOD2 mutant demonstrates a role for mitochondrial ROS in neurodevelopment and disease. Brain Behav 2:424–434

Chen Y, Zhang J, Lin Y, Lei Q, Guan KL, Zhao S, Xiong Y (2011) Tumour suppressor SIRT35 deacetylates and activates manganese superoxide dismutase to scavenge ROS. EMBO Rep 12:534–541

Cho J, Hur JH, Walker DW (2011) The role of mitochondria in *Drosophila* aging. Exp Gerontol 46:331–334

Curtis C, Landis GN, Folk D, Wehr NB, Hoe N, Waskar M, Abdueva D, Skvortsov D, Ford D, Luu A, Badrinath A, Levine RL, Bradley TJ, Tavare S, Tower J (2007) Transcriptional profiling of MnSOD-mediated lifespan extension in *Drosophila* reveals a species-general network of aging and metabolic genes. Genome Biol 8:R262

Dorn GW 2nd, Clark CF, Eschenbacher WH, Kang MY, Engelhard JT, Warner SJ, Matkovich SJ, Jowdy CC (2011) MARF and Opa1 control mitochondrial and cardiac function in *Drosophila*. Circ Res 108:12–17

Duttaroy A, Parkes T, Emtage P, Kirby K, Boulianne GL, Wang X, Hilliker AJ, Phillips JP (1997) The manganese superoxide dismutase gene of *Drosophila*: structure, expression, and evidence for regulation by MAP kinase. DNA Cell Biol 16:391–399

Duttaroy A, Paul A, Kundu M, Belton A (2003) A Sod2 null mutation confers severely reduced adult life span in *Drosophila*. Genetics 165:2295–2299

Elia AJ, Parkes TL, Kirby K, St George-Hyslop P, Boulianne GL, Phillips JP, Hilliker AJ (1999) Expression of human FALS SOD in motorneurons of *Drosophila*. Free Radic Biol Med 26:1332–1338

Favrin G, Bean DM, Bilsland E, Boyer H, Fischer BE, Russell S, Crowther DC, Baylis HA, Oliver SG, Giannakou ME (2013) Identification of novel modifiers of Aβ toxicity by transcriptomic analysis in the fruitfly. Sci Rep 3:3512

Fukai T, Folz RJ, Landmesser U, Harrison DG (2002) Extracellular superoxide dismutase and cardiovascular disease. Cardiovasc Res 55:239–249

Godenschwege T, Forde R, Davis CP, Paul A, Beckwith K, Duttaroy A (2009) Mitochondrial superoxide radicals differentially affect muscle activity and neural function. Genetics 183:175–184

Griswold CM, Matthews AL, Bewley KE, Mahaffey JW (1993) Molecular characterization of rescue of acatalasemic mutants of *Drosophila melanogaster*. Genetics 134:731–788

Hari R, Burde V, Arking R (1998) Immunological confirmation of elevated levels of CuZn superoxide dismutase protein in an artificially selected long-lived strain of *Drosophila melanogaster*. Exp Gerontol 33:227–237

Haynes CM, Fiorese CJ, Lin YF (2013) Evaluating and responding to mitochondrial dysfunction: the mitochondrial unfolded-protein response and beyond. Trends Cell Biol 23:311–318

Hitchler MJ, Domann FE (2014) Regulation of CuZnSOD and its redox signaling potential: implications for amyotrophic lateral sclerosis. Antioxid Redox Signal 20:1590–1598

Honda Y, Honda S (1999) The daf-2 gene network for longevity regulates oxidative stress resistance and Mn-superoxide dismutase gene expression in *Caenorhabditis elegans*. FASEB J 13:1385–1393

Hughes KA, Reynolds RM (2005) Evolutionary and mechanistic theories of aging. Annu Rev Entomol 50:421–445

Islam R, Kumimoto EL, Bao H, Zhang B (2012) ALS-linked SOD1 in glial cells enhances β-N-Methylamino L-Alanine (BMAA)-induced toxicity in *Drosophila*. F1000Res 1:47

Jacobson J, Lambert AJ, Portero-Otin M, Pamplona R, Magwere T, Miwa S, Driege Y, Brand MD, Partridge L (2010) Biomarkers of aging in *Drosophila*. Aging Cell 9(4):466–477

James BP, Staatz WD, Wilkinson ST, Meuillet E, Powis G (2009) Superoxide dismutase is regulated by LAMMER kinase in *Drosophila* and human cells. Free Radic Biol Med 46:821–827

Jung I, Kim TY, Kim-Ha J (2011) Identification of *Drosophila* SOD3 and its protective role against phototoxic damage to cells. FEBS Lett 585:1973–1978

Kaiser M, Gasser M, Ackermann R, Stearns SC (1997) P element inserts in transgenic flies: a cautionary tale. Heredity 78:1–11

Kirby K, Hu J, Hilliker AJ, Phillips JP (2002) RNA interference-mediated silencing of Sod2 in *Drosophila* leads to early adult-onset mortality and elevated endogenous oxidative stress. Proc Natl Acad Sci USA 99:16162–16167

Kirby K, Jensen LT, Binnington J, Hilliker AJ, Ulloa J, Culotta VC, Phillips JP (2008) Instability of superoxide dismutase 1 of *Drosophila* in mutants deficient for its cognate copper chaperone. J Biol Chem 283:35393–35401

Koh H, Kim H, Kim MJ, Park J, Lee HJ, Chung J (2012) Silent information regulator 2 (Sir2) and Forkhead box O (FOXO) complement mitochondrial dysfunction and dopaminergic neuron loss in *Drosophila* PTEN-induced kinase 1 (PINK1) null mutant. J Biol Chem 287:12750–12758

Kumimoto EL, Fore TR, Zhang B (2013) Transcriptome profiling following neuronal and glial expression of ALS-linked SOD1 in *Drosophila*. G3 (Bethesda) 3:695–708

Landis GN, Tower J (2005) Superoxide dismutase evolution and life span regulation. Mech Ageing Dev 126:365–379

Landis GN, Abdueva D, Skvortsov D, Yang J, Rabin BE, Carrick J, Tavare S, Tower J (2004) Similar gene expression patterns characterize aging and oxidative stress in *Drosophila melanogaster*. Proc Natl Acad Sci USA 101:7663–7668

Landis G, Shen J, Tower J (2012) Gene expression changes in response to aging compared to heat stress, oxidative stress and ionizing radiation in *Drosophila melanogaster*. Aging (Albany NY) 4:768–789

Lin YR, Kim K, Yang Y, Ivessa A, Sadoshima J, Park Y (2011) Regulation of longevity by regulator of G-protein signaling protein, Loco. Aging Cell 10:438–447

Liu HN, Tjostheim S, Dasilva K, Taylor D, Zhao B, Rakhit R, Brown M, Chakrabartty A, McLaurin J, Robertson J (2012) Targeting of monomer/misfolded SOD1 as a therapeutic strategy for amyotrophic lateral sclerosis. J Neurosci 32:8791–8799

Lovejoy DB, Guillemin GJ (2014) The potential for transition metal-mediated neurodegeneration in amyotrophic lateral sclerosis. Front Aging Neurosci 6:173

Martin I, Jones MA, Grotewiel M (2009a) Manipulation of Sod1 expression ubiquitously, but not in the nervous system or muscle, impacts age-related parameters in *Drosophila*. FEBS Lett 583:2308–2314

Martin I, Jones MA, Rhodenizer D, Zheng J, Warrick JM, Seroude L, Grotewiel M (2009b) Sod2 knock-down in the musculature has whole organism consequences in *Drosophila*. Free Radic Biol Med 47(6):803–813

McCord JM, Fridovich I (1969) Superoxide dismutase. An enzymic function for erythrocuprein (hemocuprein). J Biol Chem 244:6049–6055

Meli M, Frey J, Perier C (2003) Native protein glycoxidation and aging. J Nutr Health Aging 7:263–266

Melkani GC, Trujillo AS, Ramos R, Bodmer R, Bernstein SI, Ocorr K (2013) Huntington's disease induced cardiac amyloidosis is reversed by modulating protein folding and oxidative stress pathways in the *Drosophila* heart. PLoS Genet 9:e1004024

Missirlis F, Hu J, Kirby K, Hilliker AJ, Rouault TA, Phillips JP (2003) Compartment-specific protection of iron-sulfur proteins by superoxide dismutase. J Biol Chem 278:47365–47369

Mockett RJ, Orr WC, Rahmandar JJ, Benes JJ, Radyuk SN, Klichko VI, Sohal RS (1999) Overexpression of Mn-containing superoxide dismutase in transgenic *Drosophila melanogaster*. Arch Biochem Biophys 371:260–269

Mukherjee S, Forde R, Belton A, Duttaroy A (2011) SOD2, the principal scavenger of mitochondrial superoxide, is dispensable for embryogenesis and imaginal tissue development but essential for adult survival. Fly (Austin) 5:39–46

Muyderman H, Chen T (2014) Mitochondrial dysfunction in amyotrophic lateral sclerosis—a valid pharmacological target? Br J Pharmacol 171:2191–2205

Negre-Salvayre A, Auge N, Ayala V, Basaga H, Boada J, Brenke R, Chapple S, Cohen G, Feher J, Grune T, Lengyel G, Mann GE, Pamplona R, Poli G, Portero-Otin M, Riahi Y, Salvayre R, Sasson S, Serrano J, Shamni O, Siems W, Siow RC, Wiswedel I, Zarkovic K, Zarkovic N (2010) Pathological aspects of lipid peroxidation. Free Radic Res 44:1125–1171

Oberley-Deegan RE, Regan EA, Kinnula VL, Crapo JD (2009) Extracellular superoxide dismutase and risk of COPD. COPD 6:307–312

Okado-Matsumoto A, Fridovich I (2001) Subcellular distribution of superoxide dismutases (SOD) in rat liver: Cu, Zn-SOD in mitochondria. J Biol Chem 276:38388–38393

Orr WC, Sohal RS (1992) The effects of catalase gene overexpression on life span and resistance to oxidative stress in transgenic *Drosophila melanogaster*. Arch Biochem Biophys 297:35–41

Orr WC, Sohal RJ (1993) Effects of Cu-Zn superoxide dismutase overexpression on life span and resistance to oxidative stress in transgenic *Drosophila melanogaster*. Arch Biochem Biophys 301:34–40

Orr WC, Sohal RS (1994) Extension of life-span by overexpression of superoxide dismutase and catalase in *Drosophila melanogaster*. Science 263:1128–1130

Orr WC, Mockett RJ, Benes JJ, Sohal RS (2003) Effects of overexpression of copper-zinc and manganese superoxide dismutases, catalase, and thioredoxin reductase genes on longevity in *Drosophila melanogaster*. J Biol Chem 278:26418–26422

Owusu-Ansah E, Song W, Perrimon N (2013) Muscle mitohormesis promotes longevity via systemic repression of insulin signaling. Cell 155:699–712

Pan L, Chen S, Weng C, Call G, Zhu D, Tang H, Zhang N, Xie T (2007) Stem cell aging is controlled both intrinsically and extrinsically in the *Drosophila* ovary. Cell Stem Cell 1:458–469

Parker JD, Parker KM, Keller L (2004) Molecular phylogenetic evidence for an extracellular Cu Zn superoxide dismutase gene in insects. Insect Mol Biol 13:587–594

Parkes TL, Elia AJ, Dickinson D, Hilliker AJ, Phillips JP, Boulianne GL (1998) Extension of *Drosophila* lifespan by overexpression of human SOD1 in motorneurons. Nat Genet 19:171–174

Paul A, Belton A, Nag S, Martin I, Grotewiel MS, Duttaroy A (2007) Reduced mitochondrial SOD displays mortality characteristics reminiscent of natural aging. Mech Ageing Dev 128:706–716

Phillips JP, Campbell SD, Michaud D, Charbonneau M, Hilliker AJ (1989) Null mutation of copper/zinc superoxide dismutase in *Drosophila* confers hypersensitivity to paraquat and reduced longevity. Proc Natl Acad Sci USA 86:2761–2765

Piazza N, Hayes M, Martin I, Duttaroy A, Grotewiel M, Wessells R (2009) Multiple measures of functionality exhibit progressive decline in a parallel, stochastic fashion in *Drosophila* Sod2 null mutants. Biogerontology 10:637–648

Pimenta de Castro I, Costa AC, Lam D, Tufi R, Fedele V, Moisoi N, Dinsdale D, Deas E, Loh SH, Martins LM (2012) Genetic analysis of mitochondrial protein misfolding in *Drosophila* melanogaster. Cell Death Differ 19:1308–1316

Qiu X, Brown K, Hirschey MD, Verdin E, Chen D (2010) Calorie restriction reduces oxidative stress by SIRT3-mediated SOD2 activation. Cell Metab 12:662–667

Radyuk SN, Klichko VI, Spinola B, Sohal RS, Orr WC (2001) The peroxiredoxin gene family in *Drosophila melanogaster*. Free Radic Biol Med 31:1090–1100

Radyuk SN, Klichko VI, Orr WC (2004) Profiling Cu, Zn-superoxide dismutase expression in Drosophila melanogaster—a critical regulatory role for intron/exon sequence within the coding domain. Gene 328:37–48

Reddi AR, Culotta VC (2013) SOD1 integrates signals from oxygen and glucose to repress respiration. Cell 152:224–235

Reveillaud I, Niedzwiecki A, Bensch KG, Fleming JE (1991) Expression of bovine superoxide dismutase in *Drosophila melanogaster* augments resistance of oxidative stress. Mol Cell Biol 11:632–640

Rosen DR, Siddique T, Patterson D, Figlewicz DA, Sapp P, Hentati A, Donaldson D, Goto J, O'Regan JP, Deng HX et al (1993) Mutations in Cu/Zn superoxide dismutase gene are associated with familial amyotrophic lateral sclerosis. Nature 362:59–62

Saini N, Oelhafen S, Hua H, Georgiev O, Schaffner W, Bueler H (2010) Extended lifespan of *Drosophila* parkin mutants through sequestration of redox-active metals and enhancement of anti-oxidative pathways. Neurobiol Dis 40:82–92

Salmon AB, Richardson A, Perez VI (2010) Update on the oxidative stress theory of aging: does oxidative stress play a role in aging or healthy aging? Free Radic Biol Med 48:642–655

Schwarze SR, Weindruch R, Aiken JM (1998) Oxidative stress and aging reduce COX I RNA and cytochrome oxidase activity in *Drosophila*. Free Radic Biol Med 25:740–747

Seto NO, Hayashi S, Tener GM (1989) Cloning, sequence analysis and chromosomal localization of the Cu-Zn superoxide dismutase gene of *Drosophila melanogaster*. Gene 75:85–92

Seto NOL, Hayashi S, Tener GM (1990) Overexpression of Cu-Zn superoxide dismutase in *Drosophila* does not affect life span. Proc Nat Acad Sci (USA) 87:4270–4274

Shen J, Tower J (2013) Aging, MnSOD, and hormesis mechanisms converge on liver mUPR. Cell Cycle 12:3237–3238

Simm A (2013) Protein glycation during aging and in cardiovascular disease. J Proteomics 92:248–259

Soderberg JA, Birse RT, Nassel DR (2011) Insulin production and signaling in renal tubules of *Drosophila* is under control of tachykinin-related peptide and regulates stress resistance. PLoS One 6:e19866

Sohal RS, Arnold L, Orr WC (1990) Effect of age on superoxide dismutase, catalase, glutathione reductase, inorganic peroxides, TBA-reactive material, GSH/GSSG, NADPH/NADP+ and NADH/NAD+ in *Drosophila melanogaster*. Mech Ageing Dev 56:223–235

Spencer CC, Howell CE, Wright AR, Promislow DE (2003) Testing an 'aging gene' in long-lived drosophila strains: increased longevity depends on sex and genetic background. Aging Cell 2:123–130

Stadtman ER, Levine RL (2003) Free radical-mediated oxidation of free amino acids and amino acid residues in proteins. Amino Acids 25:207–218

Stavely BE, Phillips JP, Hilliker AJ (1990) Phenotypic consequences of copper/zinc superoxide dismutase overexpression in *Drosophila melanogaster*. Genome 33:867–872 National Research Council Canada (Genome/Conseil national de recherches Canada)

Struhl G, Basler K (1993) Organizing activity of wingless protein in *Drosophila*. Cell 72:527–540

Sturtz LA, Diekert K, Jensen LT, Lill R, Culotta VC (2001) A fraction of yeast Cu, Zn-superoxide dismutase and its metallochaperone, CCS, localize to the intermembrane space of mitochondria. A physiological role for SOD1 in guarding against mitochondrial oxidative damage. J Biol Chem 276:38084–38089

Sun J, Tower J (1999) FLP recombinase-mediated induction of Cu/Zn-superoxide dismutase transgene expression can extend the life span of adult *Drosophila melanogaster* flies. Mol Cell Biol 19:216–228

Sun J, Folk D, Bradley TJ, Tower J (2002) Induced overexpression of mitochondrial Mn-superoxide dismutase extends the life span of adult *Drosophila melanogaster*. Genetics 161:661–672

Sun J, Molitor J, Tower J (2004) Effects of simultaneous over-expression of Cu/ZnSOD and MnSOD on *Drosophila melanogaster* life span. Mech Ageing Dev 125:341–349

Sun X, Komatsu T, Lim J, Laslo M, Yolitz J, Wang C, Poirier L, Alberico T, Zou S (2012) Nutrient-dependent requirement for SOD1 in lifespan extension by protein restriction in *Drosophila melanogaster*. Aging Cell 11:783–793

Sun Y, Yolitz J, Alberico T, Sun X, Zou S (2014) Lifespan extension by cranberry supplementation partially requires SOD2 and is life stage independent. Exp Gerontol 50:57–63

Tatar M (1999) Transgenes in the analysis of life span and fitness. Am Nat 154(supplement):S67–S81

Tower J (1996) Aging mechanisms in fruit files. BioEssays 18:799–807

Tower J (2006) Sex-specific regulation of aging and apoptosis. Mech Ageing Dev 127:705–718

Tower J, Landis G, Gao R, Luan A, Lee J, Sun Y (2013) Variegated expression of Hsp22 transgenic reporters indicates cell-specific patterns of aging in *Drosophila* oenocytes. J Gerontol A Biol Sci Med Sci 55:109–118

Vrailas-Mortimer A, del Rivero T, Mukherjee S, Nag S, Gaitanidis A, Kadas D, Consoulas C, Duttaroy A, Sanyal S (2011) A muscle-specific p38 MAPK/Mef2/MnSOD pathway regulates stress, motor function, and life span in *Drosophila*. Dev Cell 21:783–795

Wang YC, Lee CM, Lee LC, Tung LC, Hsieh-Li HM, Lee-Chen GJ, Su MT (2011) Mitochondrial dysfunction and oxidative stress contribute to the pathogenesis of spinocerebellar ataxia type 12 (SCA12). J Biol Chem 286:21742–21754

Watson MR, Lagow RD, Xu K, Zhang B, Bonini NM (2008) A drosophila model for amyotrophic lateral sclerosis reveals motor neuron damage by human SOD1. J Biol Chem 283:24972–24981

Weisiger RA, Fridovich I (1973) Mitochondrial superoxide simutase. Site of synthesis and intramitochondrial localization. J Biol Chem 248:4793–4796

Wicks S, Bain N, Duttaroy A, Hilliker AJ, Phillips JP (2009) Hypoxia rescues early mortality conferred by superoxide dismutase deficiency. Free Radic Biol Med 46:176–181

Yu BP (1993) Oxidative damage by free radicals and lipid peroxidation in aging. In: Yu BP (ed) Free radicals in aging. CRC Press Inc., Boca Raton, pp 57–88

Zarse K, Schmeisser S, Groth M, Priebe S, Beuster G, Kuhlow D, Guthke R, Platzer M, Kahn CR, Ristow M (2012) Impaired insulin/IGF1 signaling extends life span by promoting mitochondrial L-proline catabolism to induce a transient ROS signal. Cell Metab 15:451–465

Chapter 4
Effect of *Wolbachia* Infection on Aging and Longevity-Associated Genes in *Drosophila*

Oleksandr M. Maistrenko, Svitlana V. Serga, Alexander M. Vaiserman and Iryna A. Kozeretska

Abstract Microbiota is known to interact with metabolic and regulatory networks of the host affecting its fitness. The composition of microbiota was shown to change throughout the host aging. Such changes can be likely caused by aging process or, vice versa, changes in microbiota composition can impact the aging process. It is suggested that microbiota plays an important role in life span determination. Several species from the genus *Drosophila*, especially *D. melanogaster*, are powerful models to study many biological processes including microbiota functioning and its effects on the host aging. The host fitness can be substantially affected by endosymbiotic bacteria such as *Wolbachia* that infects up to two-thirds of insects taxa, including *Drosophila*. *Wolbachia* was shown to significantly affect *Drosophila* aging and life span. However, the molecular mechanisms underlying interactions between *Wolbachia* and *Drosophila* remain mostly unknown. In this chapter, we summarize data suggesting that *Wolbachia-Drosophila* molecular cross-talk associated with life span determination and aging can occur through the immune deficiency pathway, stress-induced JNK pathway, insulin/IGF signaling pathway, ecdysteroid biosynthesis and signaling pathway, as well as through the heat shock and autophagy-specific genes/proteins.

Keywords Microbiota · *Wolbachia* · *D. melanogaster* · Infection · Aging · Longevity-associated genes

---

O.M. Maistrenko (✉) · S.V. Serga · I.A. Kozeretska
Taras Shevchenko National University of Kyiv, Kyiv, Ukraine
e-mail: o.m.maistrenko@gmail.com

A.M. Vaiserman
Laboratory of Epigenetics, D. F. Chebotarev State Institute of Gerontology NAMS of Ukraine, Kyiv, Ukraine

A.M. Vaiserman et al. (eds.), *Life Extension*, Healthy Ageing and Longevity 3,
DOI 10.1007/978-3-319-18326-8_4

4.1 Introduction

Symbiosis is a biological phenomenon, referring to coexistence of different organisms that form persistent associations and are called symbionts (Douglas 2010). Symbiosis typically results in broad range of effects in physiology, metabolism, structure and other biological characteristics. Symbiotic interactions are commonly characterized as mutualism, commensalism and parasitism (Martin and Schwab 2013). This chapter mainly focuses on how microbiota and especially endosymbiotic bacteria *Wolbachia* affect life span and aging of host organism—*Drosophila*.

Insects' organisms may harbour the amount of cells of microorganisms which exceed the number of their own cells (Dilon and Dilon 2004). The role of symbiotic organisms in many aspects of host organism functioning was, however, underestimated until recently (Russell et al. 2014). Microbial symbionts are shown to substantially affect growth, development, nutrition and immunological defence of hosts (reviewed in McFall-Ngai et al. 2013). Thus, multicellular animal is rather a consortium of organisms (a "holobiont") that becomes a functionally integrated "whole" incorporating the zoological organism per se, along with its persistent microbial symbionts (Gilbert 2014). A variety of holobiont's microbial symbionts forms a new "organ system", thereby becoming integrated into its metabolism and development. Moreover, there are numerous reciprocal interactions between the cells of the host organism and its microorganisms that alter gene expression in both sets of cells (Gilbert 2014). A totality of host and microbial genomes is defined as "hologenome" (Rosenberg et al. 2009). Animal fitness is believed to be highly dependent on its microbiota, and a hologenome is supposed to be a unit of natural selection (Zilber-Rosenberg and Rosenberg 2008). In physiological terms, microbiota plays an important role in the host's nutrient digestion, response to environmental stresses, and also in normal development of the host individuals (Costello et al. 2012; McFall-Ngai et al. 2013). Furthermore, it is substantially implicated in the maintenance of homeostasis and, thus, in the determination of longevity of the organism.

It is suggested that microbiota composition may influence the aging of holobiont (Heintz and Mair 2014). Age-related changes in microbiota are observed in human (Biagi et al. 2011; Ottaviani et al. 2011; Cheng et al. 2013), *Drosophila melanogaster* and *Caenorhabditis elegans* (reviewed in Ottaviani et al. 2011; Heintz and Mair 2014). However, it is still unclear whether the age related changes in host organism or bacterial community fluctuations or both are the driving force for the bacterial community development (Erkosar et al. 2013). Symbiosis is also believed to play a crucial role in the eukaryotic evolution (Douglas 2014). Much evidence supports the conclusion that mitochondria originated from eubacterial (specifically alpha-proteobacterial) ancestor (Margulis et al. 2000; Gray et al. 2001). This is an important point because mitochondria are essential for aging/senescence of the organism (Jacobs 2003). It can be suggested, therefore, that highlighting the symbiotic interactions may expand our understanding of aging and longevity.

4.2 Microbiota and Aging of *Drosophila*

Several species from the genus *Drosophila*, especially *D. melanogaster*, are powerful models to study many biological processes including microbiota functioning and its effects on host/holobiont aging (Broderick and Lemaitre 2012). The most studied reservoir of microbes in Drosophila is digestive tract, especially gut and intestine. According to Erkosar et al. (2013), the *Drosophila* bacterial communities include *Lactobacilus plantarum*, *Lactobacilus brevis*, *Enterococcus faecalis*, *Acetobacter pomorum*. Most of these bacteria are presented in larva and adult flies in all analyzed laboratory stocks. *E. faecalis* in known to be presented only in adult flies. Other minor representatives of bacterial microbiota include the genera *Gluconoacetobacter*, *Gluconobacter*, *Enterobacter*, *Commensalibacter* and *Acetobacter*. It was shown that microbiota in digestive tract can vary across different species of *Drosophila* (Corby-Harris et al. 2007; Chandler et al. 2011; Wong et al. 2013). These variations depend on the diet of flies and other environmental conditions (reviewed in Broderick and Lemaitre 2012; Erkosar et al. 2013). It was also shown that microbiota varies between laboratory and wild flies (Chandler et al. 2011; Staubach et al. 2013).

Bacterial microbiota composition is known to change throughout both developmental and adult life stages of the host. In *Drosophila*, it has been demonstrated that amount of *Lactobacillus fructivorans* tends to decrease and amount of *Lactobacillus plantarum* tends to increase throughout the larval development (Wong et al. 2011a, 2013). At the pupal stage, *Acetobacter tropicalis* becomes the dominant species. *L. fructivorans* and *Acetobacter pomorum* are dominant species in young adult flies and in aged flies, respectively. Species of the genus *Drosophila* are frequently infected by endosymbiotic bacteria, namely, *Wolbachia* and *Spiroplasma* (Chandler et al. 2011).

The link between the bacterial community and flies' longevity was studied repeatedly (Brummel et al. 2004; Ren et al. 2007; Ridley et al. 2012). Brummel et al. (2004) showed that germ-free (GF) flies have shorter life span indicating that microbiota is required for manifestation of "wild type" longevity phenotype. However, it has been shown that antibiotic treatment of aged flies may increase life span (Brummel et al. 2004). This discrepancy can likely be explained by the deleterious effects of bacterial load in late life. Moreover, life extension in *Drosophila* caused by microbiota may be associated in some cases with maintenance of intestinal stem cells homeostasis (Biteau et al. 2010; O'Brien et al. 2011). In addition, bacteria in some researches may not affect flies' longevity (Ren et al. 2007).

Microbiota may affect longevity and aging through interaction with pathways that are known to be involved in the control of life span, such as the immunity deficiency (IMD) and the insulin/IGF (IIS) pathways. For example, Shin et al. (2011) have shown that *Acetobacter pomorum* that inhabits *Drosophila* digestive tract and also is detected in the rest part of the body is able to induce activation of IIS-pathway through *Drosophila* insulin like peptides (dIlp). Bacteria with mutation in gene of pyrroloquinoline quinone–dependent alcohol dehydrogenase (PQQ-ADH) were not able to enhance host insulin/IGF signaling when comparing with wild type bacteria. Flies that harbored mutant *A. pomorum* had smaller body size, altered development time, elevated circulating sugar and triacylglycerides content, all effects similar to those of the mutations in the IIS pathway. Similarly, Storelli et al. (2011) showed that infection by *Lactobacillus plantarum* correlates with upregulated IIS pathway when comparing to GF flies. *L. plantarum* exhibits growth-promoting effect in *L. plantarum* monoassosiated *D. melanogaster* (Storelli et al. 2011). This indicates that *L. plantarum* may be required for normal body size in *D. melanogaster*. Furthermore, *L. plantarum* was found to interact with TOR pathway that controls the ecdysone (Ecd) production, thereby mediating IIS growth phenotype.

Described above evidences of *Drosophila*-microbiota interactions on molecular level result in upregulation of IIS pathway that allows concluding that bacteria is expected to decrease life span of flies. On other hand GF flies had shorter life span than flies with intact microbiota (e.g. Brummel et al. 2004). Mechanisms of interactions between mentioned and other pathways with microbiota are complex and multidirectional, explaining the high diversity and contradictions of effects of microbiota on host life span in different experiments.

Immunity protects organism against pathogenic bacteria that can directly cause death and also regulates normal bacterial community that is necessary for survival. Generally, bacterial load increases throughout the lifetime, being accompanied by upregulation of antimicrobial effector genes (*Cecropins*, *Attacins*, *Defensin* and *Relish*) (Pletcher et al. 2002), *Diptericin* (Seroude et al. 2002), peptidoglican recognition proteins (PGRPs) (Seroude et al. 2002), and IMD pathway target genes (Eleftherianos and Castillo 2012; Combe et al. 2014).

Probably, the capability of the *Drosophila* immune system to eliminate bacteria does not change with age and flies can tolerate high amounts of bacteria in late age (reviewed in Eleftherianos and Castillo 2012). These findings, however, contradict the aforementioned study by Brummel et al. (2004) where life extension in *Drosophila* was observed when bacteria were removed after the age of 4 weeks. Such contradictory data on impact of microbiota on the flies' life span may likely be explained, at least partially, by different methods of obtaining of the GF flies (Ridley et al. 2013); as well as by high variability of microbiota among different labs and wild populations and by high molecular complexity of host-bacteria interactions.

Eukaryotic microbiota is likely essential for the development and longevity in *Drosophila* as well. It is reported that yeast associated with *D. melanogaster* can affect development and nutrition by providing flies with sterols and B vitamins. Symbiosis between *Drosophila* and yeast is rather mutualistic (Broderick and Lemaitre 2012). *D. melanogaster* can be infected by microsporidia that cause decrease in various fitness traits, e.g., fecundity (Futerman et al. 2006). *D. melanogaster* may also be the host for several intra- and extra-cellular trypanosomatids parasite species (reviewed in Keebaugh and Schlenke 2014). The eukaryotic members of the *Drosophila* microbial consortium are, however, poorly characterized. Putative complex interactions between the prokaryotic and eukaryotic members of *Drosophila* microbiota could provide another explanation why data on the use of antibiotics in fruit fly may differ among studies.

4.3 *Wolbachia* Infection in *Drosophila*

Wolbachia is the most widespread heritable endosymbiont of arthropods (Hingelbroeker et al. 2008). *Wolbachia* is transmitted maternally and causes host reproductive outputs such as male-killing, parthenogenesis induction, cytoplasmic incompatibility (CI) and feminization that promote spreading of bacteria (Werren 1997; Werren et al. 2008). Phenotype manifestation depends on the bacteria strain and host genotype (Werren et al. 2008). A summary of the impacts of *Wolbachia* infection on biology of some species of *Drosophila* genus is present in Table 4.1.

Table 4.1 *Wolbachia* effects on *Drosophila*

Species[a]	*Wolbachia* strain	*Wolbachia* effect on host	References
Drosophila annanasae	*w*Ana[b]	Partial CI	Bourtzis et al. (1996)
D. auraria	*w*Ri, *w*Dau[b]	Partial CI	Bourtzis et al. (1996)
D. bifasciata	*w*Dbif[b]	Weak male-killing (MK)	Hurst et al. (2000)
		Weak CI	Hurst et al. (2000)
D. borealis	*w*Bor[b]	MK	Sheeley and McAllister (2009)
D. innubila	*w*Din[b]	MK	Dyer and Jaenike (2004)
		Increased fecundity in infected females	Unckless and Jaenike (2012)
		Increased survival after Flock house virus (FHV) injection	Unckless and Jaenike (2012)
D. melanogaster	*w*Mel	Weak CI	Hoffmann (1988)
		Altered survival in cold/hot conditions	Versace et al. (2014)
		Altered fitness and fecundity	Olsen et al. (2001), Fry et al. (2004), Serga et al. (2014)
		Increased tolerance to RNA viruses	Hedges et al. (2008), Teixeira et al. (2008)
		Altered mating behaviour	Markov et al. (2009), Sharon et al. (2010)
		Increased life span in females infected by *Wolbachia* compared with tetracycline-treated	E.g. Alexandrov et al. (2007)
		Restored fertility in *Wolbachia*-infected *Sxl* mutants	Starr and Cline (2002)
	*w*MelPop	Degradation of nervous and muscles tissues: decrease of life span	Min and Benzer (1997)

(continued)

Table 4.1 (continued)

Species[a]	*Wolbachia* strain	*Wolbachia* effect on host	References
D. paulistorum	*w*Au-like, *w*Mel-like	Increased fecundity in infected females compared to tetracycline treated	Miller et al. (2010)
		Premating isolation between *D. paulistorum* semispecies	Miller et al. (2010)
		CI between *D. paulistorum* semispecies	Miller et al. (2010)
D. recens	Strain unknown	CI in matings between infected *D. recens* males and uninfected *D. subquinaria* females	Shoemaker et al. (1999)
	Strain unknown	CI	Werren and Jaenike (1995)
D. sechellia	*w*Ha[b], *w*Sn[b], *w*Sh[b]	Partial CI	Giordano et al. (1995), Bourtzis et al. (1996)
D. simulans	*w*Ri	CI	Hoffmann et al. (1986), Ballard (2004)
		Resistance to RNA viruses	Osborne et al. (2009)
		Increased fecundity in infected females compared to uninfected	Kriesner et al. (2013)
	*w*Ha	CI	O'Neill and Karr (1990)
	*w*No	CI	Mercot et al. (1995)
	*w*Au	Resistance to RNA viruses	Osborne et al. (2009)
D. subquinaria	Strain unknown (same strain as in *Drosophila recens*)	MK	Jaenike (2007)

[a]*Drosophila* species were not included in the table if phenotypic effects of *Wolbachia* are unknown or absent

[b]This strain is presented in populations of given species, however reference article in column 4 does not provide name of *Wolbachia* strain that is responsible for inducing phenotype listed in the column 3 of table

In several *Drosophila* species including *D. neotestacea* (Jaenike et al. 2010), *D. mauritania* (Giordano et al. 1995), *D. santomea*, *D. teissieri*, *D. yakuba* (Zabalou et al. 2004; Hughes and Rasgon 2014), *Wolbachia* causes no reproductive distortions such as CI. In *D. simulans*, naturally infecting strains *w*Cof, *w*Au (Hoffmann et al. 1996), and *w*Ma (Charlat et al. 2003) do not exhibit CI as well. In these species, the mechanisms of *Wolbachia* maintenance in host populations remain unknown.

D. melanogaster natural populations are infected by *Wolbachia* at relatively high frequency (Hoffmann et al. 1998; Serga et al. 2014). However, *Wolbachia* that naturally infects populations of *D. melanogaster* is known to induce low levels of CI. In case of high prevalence of CI inducing *Wolbachia* in host population it gives reproductive advantage to infected females because in crosses between uninfected females and infected males the progeny is not viable. However, low CI may be unsufficient to keep high infection rate of *Wolbachia* in *D. melanogaster* population. Many papers report other kinds of effects (see Table 4.1) by which *Wolbachia* increases hosts fitness or increases chances to be transmitted in next generation that finally results in propagation of *Wolbachia* infection in natural populations of *D. melanogaster* (reviewed in Serga and Kozeretskaya 2013).

Wolbachia was shown to affect *Drosophila* fitness by a variety of mechanisms. In *D. melanogaster*, *Wolbachia* was shown to mitigate the phenotypic manifestation of mutations in IIS pathway (Ikeya et al. 2009) and to restore fertility in *Sxl* mutants (Starr and Cline 2002). In *D. paulistorum*, these bacteria are considered as an obligate mutualist because of their ability to suppress female lethality through interaction with IIS pathway (Miller et al. 2010). These observations suggest ability of *Wolbachia* to rescue lethal mutations or mitigate the manifestation of the mutant phenotype.

To study *Wolbachia*—host interactions, the transinfection technique are widely applied (reviewed in Hughes and Rasgon 2014).

CI is the most frequent reproductive manipulation phenotype in transinfection experiments. For instance, *Wolbachia* (strain *w*Cer2 that induces CI in natural host) from *Rhagolesi cerasi* transifected to *D. simulans*, was shown to induce CI in *Drosophila* (Riegler et al. 2004). *Wolbachia* can acquire reproductive manipulation ability in the novel host. For example, *Wolbachia* strains *w*Yak, *w*Tei, and *w*San do not cause any observable effects in natural hosts, but cause CI in transinfected *D. simulans* (Zabalou et al. 2008). However, in some cases, the reproductive manipulation phenotype specific for natural hosts may not manifest in transinfected species. For example, transinfection of *Wolbachia* *w*Din strain from *D. innubila* (Table 4.1) that induces male-killing in natural host to *D. simulans* and *D. melanogaster* flies does not cause male-killing (Veneti et al. 2012). Surprisingly, one of the two transinfected lines tested also showed the increased longevity. In general, the transinfection method reveals that reproductive manipulation phenotypes may be inconstant in novel hosts species.

In transinfection experiments, *Wolbachia* also affected fecundity and viral resistance of novel host species. In particular, *w*Cer2 *Wolbachia* strain in *D. simulans* reduced the female fecundity by 10 % (Riegler et al. 2004), and *w*Mel strain in *D. simulans* increased the viral resistance (Osborne et al. 2009).

4.4 Effects of *Wolbachia* on Life Span in *D. melanogaster*

D. melanogaster harbours at least two endosymbiotic bacteria: *Wolbachia* and *Spiroplasma* (Mateos et al. 2006). It was observed that number of *Spiroplasma* cells increase through aging of fly, causing the enhancement of male-killing phenotype manifestation (reviewed in Haselkorn 2010). To date, however, there is a little information on effects of *Spiroplasma* on *D. melanogaster* longevity. In this chapter, we will focus on the effects of another endosymbiont, *Wolbachia pipientis*, on aging and life span in *D. melanogaster*.

Aging and life span are fitness related traits that are probably regulated by *Wolbachia*. An association between *Wolbachia* infection and *Drosophila* longevity has been observed repeatedly. In the study by Alexandrov et al. (2007) infected females had extended life span when compared with the tetracycline-treated (uninfected) female flies.

The effects of the infection on life span of infected and tetracycline-treated flies have been found to be strongly dependent on host genetic background (Fry and Rand 2002; Fry et al. 2004). To investigate the inheritance of the survival phenotype and its dependence on the host sex and genotype, Fry and Rand (2002) used reciprocal hybrid crosses between the strains, one that lived longer with *Wolbachia* (Z53) and one that did not (Z2). The positive effects of infection were found to be more pronounced in hybrids than in parental flies. These effects were more apparent in the single-sex cages, where courtship and mating were not permitted. In the subsequent study by the same group, infected females from three different strains also showed altered survival or fecundity associated with *Wolbachia* infection, compared to uninfected flies (Fry et al. 2004). Z53 and Ftf1 strains lived longer when they were infected with *Wolbachia*, and the life span of Wj9 strain was decreased. Other experimental fly strains, Z2 and Ftf100, demonstrated no survival effects associated with *Wolbachia*. In the study by Min and Benzer (1997), the *Wolbachia popcorn* (*wMelPop*) strain was shown to be strongly virulent, causing degeneration of brain, retina, and muscle tissues, and culminating in life-shortening of flies. Tetracycline treatment eliminated both the bacteria and the degeneration, restoring normal longevity.

4.5 Molecular Mechanisms Underlying the Interaction Between *Wolbachia* and *Drosophila*

Wolbachia like other intracellular bacteria can secret different factors via the Type IV Secretion System to mediate the interactions with the host (Masui et al. 2000). Among the candidate factors, the ankyrin domain containing proteins seem to play a key role. *Wolbachia* genome contains 23 genes encoding proteins with ankyrin domains (Wu et al. 2004). The products of genes WD0285, WD0636, and WD0637 are the most probable candidates to contribute to the *Wolbachia-Drosophila* interactions (Wu et al. 2004, Foster et al. 2005). Ankyrin repeats, tandem motifs of about 33 amino acids, are known to mediate protein-protein interactions in eukaryotes (Caturegli et al. 2000). These proteins are found in obligate and facultative endosymbiotic bacteria (Siozios et al. 2013). Proteins with ankyrin repeat from *Wolbachia* are involved in CI induction in *D. melanogaster* and *D. simulans* (Papafotiou et al. 2011). Capability of ankyrin containing proteins to mediate *Wolbachia*-induced CI makes them candidates to play role in other aspects of *Wolbachia-Drosophila* interactions including aging.

Wolbachia may likely influence the levels of both mRNA expression and protein synthesis of some factors involved in the aging of fruit fly. The associations between *Wolbachia*-induced transcriptional profiles and flies' aging process, however, have not been systematically analyzed until now.

We summarized in this chapter genes potentially implicated in the aging process in *Drosophila*, and, among them, those influenced by *Wolbachia*. We searched available research data (published before August 2014) for evidences of any interactions with *Wolbachia* for each of 159 genes involved in *D. melanogaster* aging [according to FlyBase Gene Ontology (ID: GO: 0007568, the release date: July 21, 2014)]. The list of *Drosophila* genes potentially implicated in aging and longevity which are affected by *Wolbachia* are presented in Table 4.2.

Table 4.2 Genes affected by *Wolbachia* that are involved in aging and life span determination of *D. melanogaster*

Gene name	Model on which *Wolbachia-Drosophila* interactions were studied	Reference
Atg8a (*Autophagy-related 8a*)[a]	Adult females	Voronin et al. (2012)
chico	Larval testes	Zheng et al. (2011)
EcR (*Ecdysone receptor*)	N/A[b]	Negri and Pellecchia (2012)
Hsp22 (*Heat shock protein 22*)	S2 cell line	Xi et al. (2008)
	Larval testes	Zheng et al. (2011)
Hsp26	S2 cell line	Xi et al. (2008)
Hsp27	S2 cell line	Xi et al. (2008)
Hsp68	S2 cell line	Xi et al. (2008)
Hsp70	S2 cell line	Xi et al. (2008)
Ilp (*Insulin-like peptide*)	Adult males and females	Gronke et al. 2010
Indy ("*I'm not dead yet*")	Larval testes	Zheng et al. (2011)
	Adult males	Toivonen et al. (2007)
InR (*Insulin receptor*)	Adult males and females	Ikeya et al. (2009)
mld (*molting defective* (DTS-3)	S2 cell line	Xi et al. (2008)
mthl5 (*methuselah-like 5*)	S2 cell line	Xi et al. (2008)
mys (*myospheroid*)	S2 cell line	Xi et al. (2008)
PGRP-LE (*Peptidoglycan recognition protein-LE*)	N/A[b]	Kaneko et al. (2006), Lemaitre and Hoffmann (2007)
PGRP-LF (*Peptidoglycan recognition protein LF*)	S2 cell line	Xi et al. (2008)
puc (*puckered*)	S2 cell line	Xi et al. (2008)
Sod (*Superoxide dismutase*)	Adult males and females	Wang et al. (2012)

[a]Gene names are in line with http://flybase.org/ (the release date: November 21, 2014)

[b]Direct experimental research was not performed. Statement about interaction of this gene with *Wolbachia* is rather hypothetical

4.5.1 Immune Deficiency Pathway

Peptidoglycan recognition proteins (PGRPs) are innate immunity proteins that play a key role in defense against pathogenic organisms (Gupta 2008). PGRPs' functions include bacterial cell wall peptidoglycan recognition and activation of two central immune pathways in *Drosophila*, namely, Toll and Immune Deficiency (IMD) signaling pathways (Royet and Dziarski 2007). A member of the PGRP family, PGRP-LE protein, plays a key role in peptidoglycan recognition and activation of IMD pathway (Fig. 4.1). Unlike another activator of IMD, transmembrane protein PGRP-LC, PGRP-LE is located in cytoplasm (Kaneko et al. 2006); thus, this protein seems the most likely candidate to interact with *Wolbachia* (Lemaitre and Hoffmann 2007). Overexpression of PGRP-LE in fat body resulted in constitutive up-regulation of the immune response and enhanced pathogen resistance; the chronic activation of PGRP-LE, however, led to an inflammatory state and reduced fly life span (Libert et al. 2006). Paik et al. (2012) also suggest that high bacterial infection load and chronically over-activated immune response can lead to life span shortening. Other negative regulator of IMD pathway, PGRP-LF, likely acts through inhibition of PGRP-LC by concurrent interaction with bacterial peptidoglycans (Maillet et al. 2008). Paik et al. (2012) suggested that over activation of *PGRP-LF* can likely cause life extension. Remarkably, *PGRP-LF* has been shown to be up-regulated (1.26 fold change of gene expression) in S2 cell lines infected with *Wolbachia* (Xi et al. 2008) compared to uninfected. Since the activity of IMD and other immune pathways is obviously important for survival of fruit fly, the effects of *Wolbachia* or other associated bacteria on both PGRP-LF and/or PGRP-LE activity at either gene expression and/or protein levels may likely affect the *Drosophila* longevity. Several previous studies, however, have failed to show any effects of *Wolbachia* on *D. melanogaster* survival after infection by other bacteria (Wong et al. 2011b; Rottschaefer and Lazzaro 2012), despite the fact that up-regulation of IMD and Toll pathways genes along with up-regulation of antimicrobial peptides genes have been shown in flies infected with *Wolbachia* (Eleftherianos et al. 2013).

4.5.2 Stress-Induced JNK Pathway

In *Drosophila*, Jun N-terminal Kinase (JNK) signaling pathway is responsible for induction of autophagy by oxidative stress (Wu et al. 2009), heat stress (Gonda et al. 2012), and antibacterial responses (Maillet et al. 2008) and known to play central roles in aging and life span determination. Overactivation of JNK pathway through overexpression of its components such as JNKK/Hep (Hemipterous) (Libert et al. 2008) and JNK/Bsk (Basket) (Biteau et al. 2010) has been shown to be able to extend fly's life span both under normal and stress conditions. JNK/Bsk targets AP1 (Fig. 4.1) (protein dimer of the products of the *Drosophila*

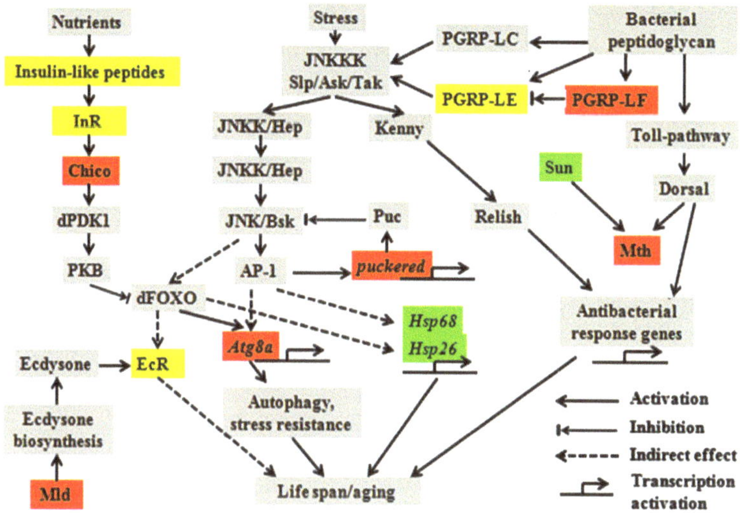

Fig. 4.1 Interactions between *Wolbachia* and host regulatory network contributing to longevity in *Drosophila*. Genes up-regulated in the presence of *Wolbachia* are marked in red, down-regulated—in *green*, proteins and/or genes that interact with *Wolbachia* in unknown or complicated way are marked in *yellow*

proto-oncongenes *dJun* (Jun-related antigen) and *dFos* (also known as *kayak*)) (Kockel et al. 2001), and *Drosophila* Forkhead transcription factor O (dFOXO) that activates genes involved in stress response (Alic et al. 2014).

The *puckered* gene encoding a VH1-like phosphatase has been shown to play a key role in negative regulation JNK activity in *Drosophila* (Fig. 4.1) (Martin-Blanco et al. 1998). In flies heterozygous for loss-of-function alleles, *pucA251.1* or *pucE69*, a significant life span extension was observed (Wang et al. 2003, 2005).

In the S2 cell system, the *Wolbachia* infection was shown to be able to up-regulate the *puckered* expression (1.30 fold change) (Xi et al. 2008) thereby potentially causing life extension in vivo. Up-regulation of *dJun* was observed (1.3 fold change) in infected S2 cells, potentially inducing transcriptional up-regulation of *puckered* which is target of AP1 complex that contains dJun (Xi et al. 2008).

4.5.3 Heat Shock Protein Genes

Many studies show that up-regulation of heat shock protein (*Hsp*) genes is associated with increased life span while down-regulation results in life shortening in *Drosophila* (reviewed in Tower 2011). The primary function of chaperone proteins coded by these genes is assistance in the folding and refolding of other proteins, particularly under stress conditions, including aging.

Down-regulation of 11 of 27 studied *Hsp* genes (from −1.2 to −1.7 fold change) was observed in *Wolbachia*-infected S2 cells (Xi et al. 2008). Zheng et al. (2011) also showed downregulation of *Hsp22* (−1.55 fold change) in larval testes. Among these down-regulated genes *Hsp22*, *Hsp68* and *Hsp70* were repeatedly found to affect life span in *D. melanogaster* (Tatar et al. 1997; Wang et al. 2003; Morrow et al. 2004). Interestingly that in the fruit fly, JNK signaling pathway up-regulates the transcription of *Hsp68* and *Hsp26* genes through dFOXO (Fig. 4.1) (Wang et al. 2003, 2005). Because of up-regulation of *puckered* that negatively regulates JNK pathway *Wolbachia* infection can result in decreased stress resistance through down-regulation of *Hsp* genes, contributing to the life-shortening effect.

4.5.4 Autophagy-Specific Genes

Autophagy refers to the normal self-cleaning process responsible for elimination of damaged cellular components. Several recent findings demonstrate that autophagy can likely contribute to many life-extending manipulations (Madeo et al. 2010). In *D. melanogaster*, autophagy is regulated by crosstalk between IMD, JNK, TOR and IIS pathways (Gelino and Hansen 2012). One of the autophagy-related genes, *autophagy-specific gene 8a* (*Atg8a*) encodes protein that is necessary to control the intracellular *Wolbachia* density in many invertebrates (Voronin et al. 2012). It was shown that autophagy gene *Atg8a* was three times overexpressed in *D. melanogaster* infected by pathogenic *Wolbachia* strain *wMel-Pop* when compared to uninfected flies (Voronin et al. 2012). Increased expression of *Atg8a* in the nervous system of adult flies can increase life span due oxidative stress resistance and elimination of damaged cell components (Simonsen et al. 2008). It may be hypothesized that *Wolbachia*-induced autophagy can also cause the removal of damaged cellular components and 'rejuvenation' of cell population.

4.5.5 Insulin/IGF Signaling Pathway

The insulin/IGF signaling (IIS) pathway is thought to play a central role in growth, stress resistance, reproduction and metabolism as well as in determination of life span of all multicellular organisms including *D. melanogaster* (Wang et al. 2014). The moderate tissue-specific and/or whole-organism reduction of IIS pathway activity was found to be associated with life extension in fruit fly (Broughton and Partridge 2009). The *Drosophila* insulin-like peptides (DILPs) are triggers of insulin signaling cascade that act through binding to insulin receptor (*InR*). There are 7 DILPs that are expressed in *D. melanogaster* in a tissue- and developmental-specific manner (Brogiolo et al. 2001). In the Grönke et al. (2010) study, homozygous *dilp2* null mutants and homozygous *dilp2,3-* null double mutants had significantly extended median life span and homozygous *dilp2-3,5* null triple mutants had slightly extended maximum life span compared to control flies. The extended mean life span was also observed in flies with ablated median neurosecretory cells that produce DILPs 2, 3 and 5 (Broughton et al. 2005).

Grönke et al. (2010) by examining how *Wolbachia* interacts with *Drosophila* IIS pathway, found that loss of DILPs produced in the brain significantly extended life span but only in the presence of the *Wolbachia*. Specifically, *wDah dilp2-3,5* mutants that carried *Wolbachia* had increased median and maximum life span when comparing with *wDah* wild type lines with *Wolbachia* and tetracycline-treated *wDah* wild type lines and *wDah dilp2-3,5* mutants. However, *Wolbachia* infection was not contributing to the observed life span extension in *dilp2* single mutant and *dilp2-3* double mutant flies. *Wolbachia* infection also contributed to DDT resistance of *dilp2-3,5* triple mutant but had no effect on survival of flies under starvation and peroxide treatment. Authors suggest that moderate down-regulation of IIS can cause life extension. Simultaneous loss of DILPs 2, 3, 5 may, however, lead to deleterious phenotypic effects, whereas the *Wolbachia* infection can likely attenuate the manifestation of *dilp2-3,5* mutant phenotype through the increased IIS signaling.

In the Ikeya et al. (2009) study, the dominant negative reduction of insulin receptor (InRDN) activity in precense of *Wolbachia* led to reduced growth and fecundity phenotypes and extended life span when compared to infected and uninfected control flies. In uninfected InRDN flies, the extreme dwarfism, sterility, increased fat content and decreased life span were observed compared to infected and uninfected control flies and flies with mutant receptor (InRDN) and with *Wolbachia*. Removal of *Wolbachia* from control flies caused moderate reduction in weight and fecundity but did not affect the life span. Expression of InRDN in the fat body had no effect on life span in *Wolbachia*-infected flies, whereas removal of *Wolbachia* resulted in life span extension. These data suggest that *Wolbachia* may interact with IIS downstream of InR.

No differences in expression of IIS downstream target, 4E-BP, were, however, observed in infected and uninfected *dilp2-3,5* flies in the Grönke et al. (2010) study. Moreover, there were no differences in egg-to-adult survival, development time, energy storage, fecundity and stress resistance between infected and uninfected *dilp2-3*,5 mutants. Taken together, these findings suggest that life extension in infected flies could proceed through other pathways than the IIS. Generally, interaction of *Wolbachia* with components of IIS pathway is unclear but presence of *Wolbachia* in many cases tends to "smooth" the mutant phenotypes of IIS pathway.

Reducing the gene dose of *chico* that encodes ligand of InR may also cause life extension. Flies carrying loss-of-function allele *chico1* exhibited extended median life span (Clancy et al. 2001). The transcription of *chico* was found to be up-regulated (1.53 fold change) in larval testes of *D. melanogaster* infected by *Wolbachia* (Zheng et al. 2011). Thus, some *Wolbachia*-induced longevity phenotypes may be mediated by *chico*.

Transcription factor dFOXO is maintained in cytoplasm through its phosphorylation by IIS. While dFOXO is localized in cytoplasm JNK-activated stress response genes are repressed.

Thus, IIS-mediated phosphorylation works antagonistically to JNK-mediated phosphorylation (Partridge and Bruning 2008). Overactivation of dFOXO causes life extension in flies (Giannakou et al. 2004).

By examining the effects of reduced IIS in the absence of dFOXO in *Drosophila*, using a newly generated null allele of *dfoxo*, Slack et al. (2011) found that the removal of dFOXO almost completely blocks IIS-dependent life span extension. However, unlike in *C. elegans*, removal of dFOXO does not suppress the body size, fecundity, or oxidative stress resistance phenotypes of IIS-compromised flies.

Down-regulation of JNK-signaling and up-regulation of insulin/IGF signaling, expectedly, will reduce life span. However, non-pathogenic strains of *Wolbachia* cause life span extension of *D. melanogaster*. Consequently, *Wolbachia* possibly extend life span through interaction with other pathways.

4.5.6 Ecdysteroid Biosynthesis and Signaling

Several findings have demonstrated that ecdysteroid signaling pathway is involved in life span determination in *Drosophila*. For example, flies heterozygous for the mutation in the gene encoding ecdysone receptor, *EcRV559fs*, exhibit increased life span and resistance to various stresses compared to controls without the mutation, with no apparent deficit in fertility or activity (Simon et al. 2003). It may be hypothesized that *Wolbachia* produces regulators able to interact directly or indirectly with ecdysone receptor resulting in modulation of ecdysteroid signaling (Negri and Pellecchia 2012).

Female DTS-3/+ flies that are mutant for *moulting defective (mld)* gene involved in ecdysone biosynthesis also had increased life span when were cultivated at 29 °C. The up-regulation (1.67 fold change) of this gene was observed in *Wolbachia* infected S2 cell lines as well (Xi et al. 2008).

Ecdysteroid signaling pathway is involved in manifestation of *Wolbachia* reproductive manipulation phenotypes (Negri et al. 2010, 2012). Thus, the ecdysteroid pathway may likely be implicated in *Wolbachia*-mediated life span modulations in *D. melanogaster*.

4.5.7 Indy Gene

The *Indy* (“*I’m not dead yet*”) gene is one of the most known life-extending genes. This gene encodes a transporter of Krebs cycle intermediates including citrate which plays a role in energy regulation by affecting fatty acid synthesis and glycolysis. The decreased expression of *Indy* results in metabolic changes similar to those induced by calorie restriction, such as decreased levels of lipids, changes in carbohydrate metabolism and increased mitochondrial biogenesis (Frankel and Rogina 2012), as well as with an extended longevity in *Drosophila* (Rogina and Helfand 2013). The variations of this gene in natural populations were shown to confer fitness advantage and extension of life span through transposon insertion (Zhu et al. 2014). Specifically, heterozygous flies for *Hoppel* transposone insertion significantly outlive homozygous flies lacking the insertion in *Indy* gene. A mutation of this gene can lead to doubling the life span of fruit fly through metabolic changes, while maintaining high levels of functioning and fertility (Zhu et al. 2014).

Remarkably, in CS-Indy^{206} fly line, the extended longevity phenotype was removed by tetracycline treatment (Toivonen et al. 2007). Authors suggest that life span extension phenotype was induced by the presence of *Wolbachia* or other bacteria that can be removed by tetracycline.

4.5.8 Myospheroid Gene

The *myospheroid (mys)* gene encodes the *Drosophila* beta-integrin (beta-PS), one of the cell surface receptors that mediate linkage between the extracellular matrix and cytoskeleton (MacKrell et al. 1988). The loss-of-function mutations in *mys* were reported to extend median life span in *D. melanogaster* (Goddeeris et al. 2003). In the Xi et al. (2008) study, it has been demonstrated that infection of S2 cells with *Wolbachia* leads to up-regulation (1.29 fold change) of the *mys* gene.

4.5.9 Methuselah Gene

The *methuselah* (*mth*) gene, encoding G-protein-coupled receptor, is known to be associated with longevity and stress resistance in *Drosophila* (Petrosyan et al. 2014). Mutations decreasing the activity of this gene are known to extend the life span in flies (Lin et al. 1998). *The methuselah-like (mthl)* genes are also assumed to play a role in life span determination based on their homology with *mth* (Araújo 2012). *mth* was not expressed at different levels in *Wolbachia*-infected and uninfected S2 cells cultures; one of the *mth* genes, *mthl5*, was, however, up-regulated (1.26 fold change). *stunted* (*sun*) gene, that encodes a ligand/agonist of *mth* product and whose mutation displays an extension of life span (Cvejic et al. 2004), was down-regulated (−1.18 fold change) in infected S2 cells (Xi et al. 2008).

4.5.10 Superoxide Dismutase

Superoxide dismutase (SOD) is an important enzyme that catalytically converts superoxide radical to oxygen and hydrogen peroxide (Fukai and Ushio-Fukai

2011). This enzyme is a major player in the organismal antioxidant defense system since it is involved in protecting the cells from superoxide radicals generated during aerobic metabolism. Overexpression of SOD has been repeatedly found to be associated with reduced oxidative damage and extended life span in *Drosophila* under both normal metabolism and environmental stress (Fleming et al. 1992; Landis and Tower 2005). *Wolbachia* likely may be implicated in these associations. Wang et al. (2012) demonstrated that *Wolbachia* infection decreased activity of SOD compared to tetracycline treated flies.

4.6 Conclusion

Analysis of data from the studies included in this systematic review suggests that, to ensure their persistence within the host, *Wolbachia* may affect expression of key genes controlling different pathways implicated in immune response, stress resistance, metabolic processes, antioxidant defense system, autophagy and other important survival functions, thus influencing longevity in *Drosophila*. It remains unclear, however, why the *Wolbachia* infection in some cases leads to life extension, while in other cases can shorten the life span of the flies. These effects of infection by *Wolbachia* must be thereby taken into account by using the *Drosophila* model system to investigate the aging process. Further researches are also required for the identification of molecular pathways by which *Wolbachia* infection affects the aging process and longevity in *Drosophila*.

Acknowledgments Authors thank Dr. Zhiyong Xi and Dr. Stephen Dobson from Department of Entomology of University of Kentucky for providing additional information about genes that changed expression in case of *Wolbachia* infection in *Drosophila* S2 cell line. Authors acknowledge Dr. Elena Pasyukova for critical review of manuscript draft.

References

Alexandrov ID, Alexandrova MV, Goryacheva II et al (2007) Removing endosymbiotic *Wolbachia* specifically decreases life span of females and competitiveness in a laboratory strain of *Drosophila melanogaster*. Russian Journal of Genetics 43(10):1147–1152

Alic N, Tullet JM, Niccoli T et al (2014) Cell-nonautonomous effects of dFOXO/DAF-16 in aging. Cell Reports 6(4):608–616

Araújo A (2012) Are all mth-like genes involved in life span determination? Master thesis, ICBAS and FCUP

Ballard JWO (2004) Sequential evolution of a symbiont inferred from the host: *Wolbachia* and *Drosophila simulans*. Mol Biol Evol 21(3):428–442

Biagi E, Candela M, Franceschi C, Brigidi P (2011) The aging gut microbiota: new perspectives. Ageing Res Rev 10(4):428–429

Biteau B, Karpac J, Supoyo S et al (2010) Life span extension by preserving proliferative homeostasis in *Drosophila*. PLoS Genet 6(10):e1001159

Bourtzis K, Nirgianaki A, Markakis G, Savakis C (1996) Wolbachia infection and cytoplasmic incompatibility in *Drosophila* species. Genetics 144(3):1063–1073

Broderick NA, Lemaitre B (2012) Gut-associated microbes of *Drosophila melanogaster*. Gut Microbes 3(4):307–321

Brogiolo W, Stocker H, Ikeya T et al (2001) An evolutionarily conserved function of the *Drosophila* insulin receptor and insulin-like peptides in growth control. Curr Biol 11(4):213–221

Broughton SJ, Piper MDW, Ikeya T et al (2005) Longer life span, altered metabolism and stress resistance in *Drosophila* from ablation of cells making insulin-like ligands. Proc Natl Acad Sci USA 102:3105–3110

Broughton S, Partridge L (2009) Insulin/IGF-like signalling, the central nervous system and aging. Biochem J 418:1–12

Brummel T, Ching A, Seroude L et al (2004) *Drosophila* life span enhancement by exogenous bacteria. Proc Natl Acad Sci USA 101:12974–12979

Caturegli P, Asanovich KM, Walls JJ et al (2000) ankA: an *Ehrlichia phagocytophila* group gene encoding a cytoplasmic protein antigen with ankyrin repeats. Infect Immun 68(9):5277–5283

Chandler JA, Lang JM, Bhatnagar S et al (2011) Bacterial communities of diverse *Drosophila* species: ecological context of a host–microbe model system. PLoS Genet 7(9):e1002272

Charlat S, Le Chat L, Mercot H (2003) Characterization of non-cytoplasmic incompatibility inducing *Wolbachia* in two continental African populations of *Drosophila simulans*. Heredity 90(1):49–55

Cheng J, Palva AM, de Vos WM, Satokari R (2013) Contribution of the intestinal microbiota to human health: from birth to 100 years of age. In: Between Pathogenicity and Commensalism, vol 358, pp 323–346

Clancy DJ, Gems D, Harshman LG et al (2001) Extension of life-span by loss of CHICO, a *Drosophila* insulin receptor substrate protein. Science 292(5514):104–106

Combe BE, Defaye A, Bozonnet N et al (2014) *Drosophila* microbiota modulates host metabolic gene expression via IMD/NF-κB signaling. PLoS ONE 9(4):e94729

Corby-Harris V, Pontaroli AC, Shimkets LJ et al (2007) Geographical distribution and diversity of bacteria associated with natural populations of *Drosophila melanogaster*. Appl Environ Microbiol 73(11):3470–3479

Costello EK, Stagaman K, Dethlefsen L et al (2012) The application of ecological theory toward an understanding of the human microbiome. Science 336(6086):1255–1262

Cvejic S, Zhu Z, Felice SJ et al (2004) The endogenous ligand Stunted of the GPCR Methuselah extends life span in *Drosophila*. Nat Cell Biol 6(6):540–546

Dillon RJ, Dillon VM (2004) The gut bacteria of insects: nonpathogenic interactions. Annu Rev Entomol 49:71–92

Douglas AE (2010) The symbiotic habit. Princeton University Press, Princeton

Douglas AE (2014) Symbiosis as a general principle in eukaryotic evolution. Cold Spring Harb Perspect Biol 6(2):a016113

Dyer KA, Jaenike J (2004) Evolutionarily stable infection by a male-killing endosymbiont in *Drosophila innubila* molecular evidence from the host and parasite genomes. Genetics 168(3):1443–1455

Eleftherianos I, Castillo JC (2012) Molecular mechanisms of aging and immune system regulation in *Drosophila*. Int J Mol Sci 13(8):9826–9844

Eleftherianos I, Atri J, Accetta J, Castillo JC (2013) Endosymbiotic bacteria in insects: guardians of the immune system? Front Physiol 4:46

Erkosar B, Storelli G, Defaye A, Leulier F (2013) Host-intestinal microbiota mutualism: "learning on the fly". Cell Host Microbe 13(1):8–14

Fleming JE, Reveillaud I, Niedzwiecki A (1992) Role of oxidative stress in *Drosophila* aging. Mutat Res/DNAging 275(3):267–279

Foster J, Ganatra M, Kamal I et al (2005) The *Wolbachia* genome of *Brugia malayi*: endosymbiont evolution within a human pathogenic nematode. PLoS Biol 3:0599–0614

Frankel S, Rogina B (2012) Indy mutants: live long and prosper. Genet Aging 3:13

Fry AJ, Palmer MR, Rand DM (2004) Variable fitness effects of *Wolbachia* infection in *Drosophila melanogaster*. Heredity 93:379–389

Fry AJ, Rand DM (2002) *Wolbachia* interactions that determine *Drosophila melanogaster* survival. Evolution Int J Org Evol 56:1976–1981

Fukai T, Ushio-Fukai M (2011) Superoxide dismutases: role in redox signaling, vascular function, and diseases. Antioxid Redox Signal 15(6):1583–1606

Futerman PH, Layen SJ, Kotzen ML et al (2006) Fitness effects and transmission routes of a microsporidian parasite infecting *Drosophila* and its parasitoids. Parasitology 132(04):479–492

Gelino S, Hansen M (2012) Autophagy—an emerging anti-aging mechanism? J Clin Exp Pathol S 4:006. doi:10.4172/2161-0681.S4-006

Gilbert SF (2014) Symbiosis as the way of eukaryotic life: the dependent co-origination of the body. J Biosci 39(2):201–209

Giannakou ME, Goss M, Jünger MA et al (2004) Long-lived *Drosophila* with overexpressed dFOXO in adult fat body. Science 305(5682):361

Giordano R, O'Neill SL, Robertson HM (1995) *Wolbachia* infections and the expression of cytoplasmic incompatibility in *Drosophila sechellia* and *D. mauritiana*. Genetics 140(4):1307–1317

Gray MW, Burger G, Lang BF (2001) The origin and early evolution of mitochondria. Genome Biol 2(6):1018–1021

Grönke S, Clarke DF, Broughton S et al (2010) Molecular evolution and functional characterization of *Drosophila* insulin-like peptides. PLoS Genet 6(2):e1000857

Goddeeris MM, Cook-Wiens E, Horton WJ et al (2003) Delayed behavioural aging and altered mortality in *Drosophila* β integrin mutants. Aging Cell 2(5):257–264

Gonda RL, Garlena RA, Stronach B (2012) *Drosophila* heat shock response requires the JNK pathway and phosphorylation of mixed lineage kinase at a conserved serine-proline motif. PLoS ONE 7(7):e42369

Gupta D (2008) Peptidoglycan recognition proteins-maintaining immune homeostasis and normal development. Cell Host Microbe 3:273–274

Haselkorn TS (2010) The *Spiroplasma* heritable bacterial endosymbiont of *Drosophila*. Fly 4(1):80–87

Hedges LM, Brownlie JC, O'Neill SL, Johnson KN (2008) *Wolbachia* and virus protection in insects. Science 322(5902):702

Heintz C, Mair W (2014) You are what you host: microbiome modulation of the aging process. Cell 156(3):408–411

Hilgenboecker K, Hammerstein P, Schlattmann P et al (2008) How many species are infected with *Wolbachia*?—A statistical analysis of current data. FEMS Microbiol Lett 281(2):215–220

Hoffmann AA, Turelli M, Simmons GM (1986) Unidirectional incompatibility between populations of *Drosophila simulans*. Evolution 40(4):692–701

Hoffmann AA (1988) Partial cytoplasmic incompatibility between two Australian populations of *Drosophila melanogaster*. Entomol Exp Appl 48(1):61–67

Hoffmann AA, Clancy D, Duncan J (1996) Naturally-occurring *Wolbachia* infection in *Drosophila simulans* that does not cause cytoplasmic incompatibility. Heredity 76(1):1–8

Hoffmann AA, Hercus M, Dagher H (1998) Population dynamics of the *Wolbachia* infection causing cytoplasmic incompatibility in *Drosophila melanogaster*. Genetics 148(1):221–231

Hughes GL, Rasgon JL (2014) Transinfection: a method to investigate *Wolbachia*-host interactions and control arthropod-borne disease. Insect Mol Biol 23(2):141–151

Hurst GD, Johnson AP, vd Schulenburg JHG, Fuyama Y (2000) Male-killing *Wolbachia* in *Drosophila*: a temperature-sensitive trait with a threshold bacterial density. Genetics 156(2):699–709

Ikeya T, Broughton S, Alic N et al (2009) The endosymbiont *Wolbachia* increases insulin/IGF-like signalling in *Drosophila*. Proc R Soc Lond B Biol Sci 276:3799–3807

Jacobs HT (2003) The mitochondrial theory of aging: dead or alive? Aging Cell 2(1):11–17

Jaenike J (2007) Spontaneous emergence of a new *Wolbachia* phenotype. Evolution 61(9):2244–2252

Jaenike J, Stahlhut JK, Boelio LM, Unckless RL (2010) Association between *Wolbachia* and *Spiroplasma* within *Drosophila neotestacea*: an emerging symbiotic mutualism?. Mol. Ecol. 19(2):414–425

Kaneko T, Yano T, Aggarwal K et al (2006) PGRP-LC and PGRP-LE have essential yet distinct functions in the *Drosophila* immune response to monomeric DAP-type peptidoglycan. Nat Immunol 7:715–723

Keebaugh ES, Schlenke TA (2014) Insights from natural host–parasite interactions: the *Drosophila* model. Dev Comp Immunol 42(1):111–123

Kockel L, Homsy JG, Bohmann D (2001) *Drosophila* AP-1: lessons from an invertebrate. Oncogene 20:2347–2364

Kriesner P, Hoffmann AA, Lee SF et al (2013) Rapid sequential spread of two *Wolbachia* variants in *Drosophila simulans*. PLoS Pathog 9(9):e1003607

Landis GN, Tower J (2005) Superoxide dismutase evolution and life span regulation. Mech Ageing Dev 126(3):365–379

Lemaitre B, Hoffmann J (2007) The host defense of *Drosophila melanogaster*. Annu Rev Immunol 25:697–743

Libert S, Chao Y, Chu X, Pletcher SD (2006) Trade-offs between longevity and pathogen resistance in *Drosophila melanogaster* are mediated by NFκB signaling. Aging Cell 5(6):533–543

Libert S, Chao Y, Zwiener J, Pletcher SD (2008) Realized immune response is enhanced in long-lived puc and chico mutants but is unaffected by dietary restriction. Mol Immunol 45(3):810–817

Lin YJ, Seroude L, Benzer S (1998) Extended life-span and stress resistance in the *Drosophila* mutant methuselah. Science 282(5390):943–946

Maillet F, Bischoff V, Vignal C et al (2008) The *Drosophila* peptidoglycan recognition protein PGRP-LF blocks PGRP-LC and IMD/JNK pathway activation. Cell Host Microbe 3(5):293–303

MacKrell AJ, Blumberg B, Haynes SR, Fessler JH (1988) The lethal myospheroid gene of *Drosophila* encodes a membrane protein homologous to vertebrate integrin beta subunits. Proc Natl Acad Sci USA 85:2633–2637

Madeo F, Tavernarakis N, Kroemer G (2010) Can autophagy promote longevity? Nat Cell Biol 12(9):842–846

Margulis L, Dolan MF, Guerrero R (2000) The chimeric eukaryote: origin of the nucleus from the karyomastigont in amitochondriate protists. Proc Natl Acad Sci 97(13):6954–6959

Markov AV, Lazebny OE, Goryacheva II et al (2009) Symbiotic bacteria affect mating choice in *Drosophila melanogaster*. Anim Behav 77(5):1011–1017

Martin BD, Schwab E (2013) Current usage of symbiosis and associated terminology. Int J Biol 5:32–45

Martín-Blanco E, Gampel A, Ring J et al (1998) puckered encodes a phosphatase that mediates a feedback loop regulating JNK activity during dorsal closure in *Drosophila*. Genes Dev 12(4):557–570

Masui S, Sasaki T, Ishikawa H (2000) Genes for the type IV secretion system in an intracellular symbiont, *Wolbachia*, a causative agent of various sexual alterations in arthropods. J Bacteriol 182(22):6529–6531

Mateos M, Castrezana SJ, Nankivell BJ et al (2006) Heritable endosymbionts of *Drosophila*. Genetics 174(1):363–376

McFall-Ngai M, Hadfield MG, Bosch TC et al (2013) Animals in a bacterial world, a new imperative for the life sciences. Proc Natl Acad Sci 110(9):3229–3236

Mercot H, Llorente B, Jacques M et al (1995) Variability within the Seychelles cytoplasmic incompatibility system in *Drosophila simulans*. Genetics 141(3):1015–1023

Miller WJ, Ehrman L, Schneider D (2010) Infectious speciation revisited: impact of symbiont-depletion on female fitness and mating behavior of *Drosophila paulistorum*. PLoS Pathog 6(12):e1001214

Min KT, Benzer S (1997) *Wolbachia*, normally a symbiont of *Drosophila*, can be virulent, causing degeneration and early death. Proc Natl Acad Sci 94(20):10792–10796

Morrow G, Samson M, Michaud S, Tanguay RM (2004) Overexpression of the small mitochondrial Hsp22 extends *Drosophila* life span and increases resistance to oxidative stress. FASEB J 18(3):598–599

Negri I (2012) *Wolbachia* as an "infectious" extrinsic factor manipulating host signaling pathways. Front Endocrinol (Lausanne) 2:115

Negri I, Pellecchia M, Grève P et al (2010) Sex and stripping: the key to the intimate relationship between *Wolbachia* and host. Commun Integr Biol 3(2):110–115

Negri I, Pellecchia M (2012) Sex steroids in insects and the role of the endosymbiont *Wolbachia*: a new perspective. Sex Hormones 353–374

Olsen K, Reynolds KT, Hoffmann AA (2001) A field cage test of the effects of the endosymbiont *Wolbachia* on *Drosophila melanogaster*. Heredity 86(6):731–737

O'Brien LE, Soliman SS, Li X, Bilder D (2011) Altered modes of stem cell division drive adaptive intestinal growth. Cell 147:603–614

O'Neill SL, Karr TL (1990) Bidirectional incompatibility between conspecific populations of *Drosophila simulans*. Nature 348(6297):178–180

Osborne SE, San Leong Y, O'Neill SL, Johnson KN (2009) Variation in antiviral protection mediated by different *Wolbachia* strains in *Drosophila simulans*. PLoS Pathog 5(11):e1000656

Ottaviani E, Ventura N, Mandrioli M et al (2011) Gut microbiota as a candidate for life span extension: an ecological/evolutionary perspective targeted on living organisms as metaorganisms. Biogerontology 12(6):599–609

Paik D, Jang YG, Lee YE et al (2012) Misexpression screen delineates novel genes controlling *Drosophila* life span. Mech Ageing Dev 133(5):234–245

Partridge L, Brüning JC (2008) Forkhead transcription factors and ageing. Oncogene 27(16):2351–2363

Papafotiou G, Oehler S, Savakis C, Bourtzis K (2011) Regulation of *Wolbachia* ankyrin domain encoding genes in *Drosophila* gonads. Res Microbiol 162(8):764–772

Petrosyan A, Gonçalves ÓF, Hsieh IH, Saberi K (2014) Improved functional abilities of the life-extended *Drosophila* mutant Methuselah are reversed at old age to below control levels. Age 36(1):213–221

Pletcher SD, Macdonald SJ, Marguerie R et al (2002) Genome-wide transcript profiles in aging and calorically restricted *Drosophila melanogaster*. Curr Biol 12(9):712–723

Ren C, Webster P, Finkel SE, Tower J (2007) Increased internal and external bacterial load during *Drosophila* aging without life-span trade-off. Cell Metab 6:144–152

Ridley EV, Wong AC, Westmiller S, Douglas AE (2012) Impact of the resident microbiota on the nutritional phenotype of *Drosophila melanogaster*. PLoS ONE 7(5):e36765

Ridley EV, Wong AC, Douglas AE (2013) Microbe-dependent and nonspecific effects of procedures to eliminate the resident microbiota from *Drosophila melanogaster*. Appl Environ Microbiol 79(10):3209–3214

Riegler M, Charlat S, Stauffer C, Merçot H (2004) *Wolbachia* transfer from *Rhagoletis cerasi* to *Drosophila simulans*: investigating the outcomes of host-symbiont coevolution. Appl Environ Microbiol 70(1):273–279

Rogina B, Helfand SL (2013) Indy mutations and *Drosophila* longevity. Front Genet 4:47

Rosenberg E, Sharon G, Zilber-Rosenberg I (2009) The hologenome theory of evolution contains Lamarckian aspects within a Darwinian framework. Environ Microbiol 11(12):2959–2962

Rottschaefer SM, Lazzaro BP (2012) No effect of *Wolbachia* on resistance to intracellular infection by pathogenic bacteria in *Drosophila melanogaster*. PLoS ONE 7(7):e40500. doi:10.1371/journal.pone.0040500

Royet J, Dziarski R (2007) Peptidoglycan recognition proteins: pleiotropic sensors and effectors of antimicrobial defences. Nat Rev Microbiol 5(4):264–277

Russell JA, Dubilier N, Rudgers JA (2014) Nature's microbiome: introduction. Mol Ecol 23(6):1225–1237

Serga SV, Kozeretskaya IA (2013) The puzzle of *Wolbachia* spreading out through natural populations of *Drosophila melanogaster*. Zh Obshch Biol 74(2):99–111

Serga S, Maistrenko O, Rozhok A et al (2014) Fecundity as one of possible factors contributing to the dominance of the *wMel* genotype of *Wolbachia* in natural populations of *Drosophila melanogaster*. Symbiosis 63(1):11–17. doi:10.1007/s13199-014-0283-1

Seroude L, Brummel T, Kapahi P, Benzer S (2002) Spatio-temporal analysis of gene expression during aging in *Drosophila melanogaster*. Aging Cell 1(1):47–56

Sharon G, Segal D, Ringo JM et al (2010) Commensal bacteria play a role in mating preference of *Drosophila melanogaster*. Proc Natl Acad Sci 107(46):20051–20056

Sheeley SL, McAllister BF (2009) Mobile male-killer: similar *Wolbachia* strains kill males of divergent *Drosophila* hosts. Heredity 102(3):286–292

Shin SC, Kim SH, You H et al (2011) *Drosophila* microbiome modulates host developmental and metabolic homeostasis via insulin signaling. Science 334(6056):670–674

Shoemaker DD, Katju V, Jaenike J (1999) *Wolbachia* and the evolution of reproductive isolation between *Drosophila recens* and *Drosophila subquinaria*. Evolution 1157–1164

Simon AF, Shih C, Mack A, Benzer S (2003) Steroid control of longevity in *Drosophila melanogaster*. Science 299(5611):1407–1410

Simonsen A, Cumming RC, Brech A et al (2008) Promoting basal levels of autophagy in the nervous system enhances longevity and oxidant resistance in adult *Drosophila*. Autophagy 4(2):176

Siozios S, Ioannidis P, Klasson L et al (2013) The diversity and evolution of *Wolbachia* Ankyrin repeat domain genes. PLoS ONE 8(2):e55390. doi:10.1371/journal.pone.0055390

Slack C, Giannakou ME, Foley A et al (2011) dFOXO-independent effects of reduced insulin-like signaling in *Drosophila*. Aging Cell 10(5):735–748

Starr DJ, Cline TW (2002) A host-parasite interaction rescues *Drosophila* oogenesis defects. Nature 418(6893):76–79

Staubach F, Baines JF, Künzel S et al (2013) Host species and environmental effects on bacterial communities associated with *Drosophila* in the laboratory and in the natural environment. PLoS ONE 8(8):e70749. doi:10.1371/journal.pone.0070749

Storelli G, Defaye A, Erkosar B et al (2011) *Lactobacillus plantarum* promotes *Drosophila* systemic growth by modulating hormonal signals through TOR-dependent nutrient sensing. Cell Metab 14(3):403–414

Tatar M, Khazaeli AA, Curtsinger JW (1997) Chaperoning extended life. Nature 390(6655):30

Teixeira L, Ferreira Á, Ashburner M (2008) The bacterial symbiont *Wolbachia* induces resistance to RNA viral infections in *Drosophila melanogaster*. PLoS Biol 6(12):e1000002

Toivonen JM, Walker GA, Martinez-Diaz P et al (2007) No influence of Indy on life span in *Drosophila* after correction for genetic and cytoplasmic background effects. PLoS Genet 3(6):e95. doi:10.1371/journal.pgen.0030095

Tower J (2011) Heat shock proteins and *Drosophila* aging. Exp Gerontol 46(5):355–362

Unckless RL, Jaenike J (2012) Maintenance of a male-killing *Wolbachia* in *Drosophila innubila* by male-killing dependent and male-killing independent mechanisms. Evolution 66(3):678–689

Veneti Z, Zabalou S, Papafotiou G et al (2012) Loss of reproductive parasitism following transfer of male-killing *Wolbachia* to *Drosophila melanogaster* and *Drosophila simulans*. Heredity 109(5):306–312

Versace E, Nolte V, Pandey RV et al (2014) Experimental evolution reveals habitat-specific fitness dynamics among *Wolbachia* clades in *Drosophila melanogaster*. Mol Ecol 23(4):802–814

Voronin D, Cook DA, Steven A, Taylor MJ (2012) Autophagy regulates *Wolbachia* populations across diverse symbiotic associations. Proc Natl Acad Sci 109(25):E1638–E1646

Wang MC, Bohmann D, Jasper H (2003) JNK signaling confers tolerance to oxidative stress and extends life span in *Drosophila*. Dev Cell 5(5):811–816

Wang MC, Bohmann D, Jasper H (2005) JNK extends life span and limits growth by antagonizing cellular and organism-wide responses to insulin signaling. Cell 121(1):115–125

Wang L, Karpac J, Jasper H (2014) Promoting longevity by maintaining metabolic and proliferative homeostasis. J Exp Biol 217(1):109–118

Wang L, Zhou C, He Z (2012) *Wolbachia* infection decreased the resistance of *Drosophila* to lead. PLoS ONE 7(3):e32643. doi:10.1371/journal.pone.0032643

Werren JH (1997) Biology of *Wolbachia*. Annu Rev Entomol 42(1):587–609

Werren JH, Jaenike J (1995) *Wolbachia* and cytoplasmic incompatibility in mycophagous *Drosophila* and their relatives. Heredity 75(3):320–326

Werren JH, Baldo L, Clark ME (2008) *Wolbachia*: master manipulators of invertebrate biology. Nat Rev Microbiol 6:741–751

Wong AC, Chaston JM, Douglas AE (2013) The inconstant gut microbiota of *Drosophila* species revealed by 16S rRNA gene analysis. ISME J 7(10):1922–1932

Wong CN, Ng P, Douglas AE (2011a) Low-diversity bacterial community in the gut of the fruitfly *Drosophila melanogaster*. Environ Microbiol 13:1889–1900

Wong ZS, Hedges LM, Brownlie JC, Johnson KN (2011b) *Wolbachia*-mediated antibacterial protection and immune gene regulation in *Drosophila*. PLoS ONE 6(9):e25430. doi:10.1371/journal.pone.0025430

Wu M, Sun LV, Vamathevan J et al (2004) Phylogenomics of the reproductive parasite *Wolbachia* pipientis *wMel*: a streamlined genome overrun by mobile genetic elements. PLoS Biol 2:0327–0341

Wu H, Wang MC, Bohmann D (2009) JNK protects *Drosophila* from oxidative stress by trancriptionally activating autophagy. Mech Dev 126(8):624–637

Xi Z, Gavotte L, Xie Y, Dobson SL (2008) Genome-wide analysis of the interaction between the endosymbiotic bacterium *Wolbachia* and its *Drosophila* host. BMC Genom 9(1):1

Zabalou S, Apostolaki A, Pattas S et al (2008) Multiple rescue factors within a *Wolbachia* strain. Genetics 178(4):2145–2160

Zabalou S, Charlat S, Nirgianaki A et al (2004) Natural *Wolbachia* infections in the *Drosophila yakuba* species complex do not induce cytoplasmic incompatibility but fully rescue the *w*Ri modification. Genetics 167:827–834

Zilber-Rosenberg I, Rosenberg E (2008) Role of microorganisms in the evolution of animals and plants: the hologenome theory of evolution. FEMS Microbiol Rev 32(5):723–735

Zheng Y, Wang JL, Liu C et al (2011) Differentially expressed profiles in the larval testes of *Wolbachia* infected and uninfected *Drosophila*. BMC Genom 12(1):595

Zhu CT, Chang C, Reenan RA, Helfand SL (2014) Indy gene variation in natural populations confers fitness advantage and life span extension through transposon insertion. Aging (Albany NY) 6(1):58

Part II
Drosophila Models for Aging-Associated Diseases

Chapter 5
Skeletal Muscle Homeostasis and Aging in *Drosophila*

Melissa J. Puppa and Fabio Demontis

Abstract A debilitating feature of aging in humans is the progressive loss of skeletal muscle mass and function termed sarcopenia. A variety of intrinsic and extrinsic factors that are induced by aging contribute to sarcopenia, which in turn is a risk factor for many other age-related diseases. While widely studied in human and rodent models, sarcopenia has been identified also in the common fruit fly *Drosophila melanogaster*. *Drosophila* is emerging as a powerful system to study the mechanisms underlying sarcopenia, as it shares many of the same skeletal muscle characteristics as mammalian models. Decreased protein homeostasis, mitochondrial dysfunction, increased apoptosis, and alterations in transcription are just a few of the features of sarcopenia that are shared between mammals and *Drosophila*. Given its short life span compared to mammals and the ease in conducting genetic manipulations, including genome-wide muscle-specific transgenic screens, *Drosophila* offers unique advantages for studying the fundamental mechanisms of skeletal muscle aging and may provide potential therapeutic targets to combat sarcopenia in humans.

Keywords *Drosophila* · Skeletal muscle · Sarcopenia · Protein homeostasis · Mitochondrial dysfunction · Apoptosis · Aging

5.1 Introduction

During their life span many animal species, including nematodes, flies, rodents, and primates, develop sarcopenia, the progressive loss of skeletal muscle mass and strength with age (Herndon et al. 2002; Wolkow 2006; Augustin and Partridge 2009; Demontis et al. 2013a, b). Sarcopenia can be attributed to muscle-intrinsic and extrinsic defects that lead to a gradual decrease in the capacity to maintain

M.J. Puppa · F. Demontis (✉)
Division of Developmental Biology, Department of Developmental Neurobiology, St. Jude Children's Research Hospital, Memphis, TN 38105, USA
e-mail: Fabio.Demontis@stjude.org

A.M. Vaiserman et al. (eds.), *Life Extension*, Healthy Ageing and Longevity 3,
DOI 10.1007/978-3-319-18326-8_5

skeletal muscle function and mass during aging. While sarcopenia is progressive and it is not fully reversible, muscle atrophy that occurs in response to catabolic stimuli such as fasting is rapid (a matter of days), generally reversible, and typically it does not entail intrinsic defects. There are many models available to study skeletal muscle agingand *Drosophila* has become a useful one to examine the cellular processes and genetic pathways responsible for functional decrements of skeletal muscle with aging (Augustin and Partridge 2009; Demontis et al. 2013a, b).

The organization and metabolic properties of skeletal muscle fibers of *Drosophila melanogaster* are similar to those of mammals (Sink 2006). Both *Drosophila* and mammalian skeletal muscles are composed of tandem arrays of sarcomeres containing thin filaments, composed of actin, and thick filaments, composed of myosin (Sink 2006). Release of calcium from the sarcoplasmic reticulum results in binding of the myosin head to the actin filament, which leads to generating the force of contraction. After the calcium is reabsorbed by the sarcoplasmic reticulum, the muscle is prepared for another contraction. Alternatively, some skeletal muscle in *Drosophila,* such as the indirect flight muscles, do not rely on extensive calcium recycling to maintain contractions. Instead, mechanical stimuli, through stretching and shortening of the muscle, are used to maintain the high frequency of contractions needed for flight (Dickinson 2006; Vigoreaux 2006; Tregear 2011).

Similar to mammals, *Drosophila* skeletal myofibers appear to be either glycolytic or oxidative (Sink 2006). The flight muscles, both direct and indirect, have been proposed to have an oxidative phenotype, as they are fatigue resistant, similar to the *soleus* muscle in mammals. Conversely, the leg muscles of adult flies and the body wall muscles of larvae, which are used intermittently, are thought to be glycolytic, similar to the *tibialis anterior* muscle in mammals (Sink 2006).

The developmental origin of *Drosophila* skeletal muscle is similar to vertebrates, as the somatic, visceral and cardiac musclesarise from the mesoderm. The commonality in structure and developmental origin makes *Drosophila* an excellent model for studying skeletal muscle differentiation, growth, aging, and disease. During the embryonic stage muscle progenitor cells differentiate and fuse to form individual myofibers. Embryonic muscle development is completed within one day (Fig. 5.1). *Drosophila* embryonic development provides a useful system for the examination of the cellular mechanisms and genetic pathways regulating myoblast fusion and myofiber differentiation, which may be relevant for understanding muscle regeneration and satellite cell function during aging in mammals. Subsequently, during the ~5 days of larval development (Fig. 5.1), larval muscles (each composed by a single myofiber) grow 50-fold in size (Demontis and Perrimon 2009). The profound changes in muscle mass observed during larval development provide a sensitized setting for the identification of the cellular mechanisms and genetic pathways regulating muscle growth and atrophy.

After the larval stage, developing *Drosophila* undergo pupal metamorphosis (Fig. 5.1) during which most of the larval muscles are degraded in a process called "histolysis" and replaced with adult muscles composed of multiple fibers similar to vertebrates (Patel et al. 2002). Analysis of the pupal stage of muscle development can further the understanding of myofiber degeneration and loss in

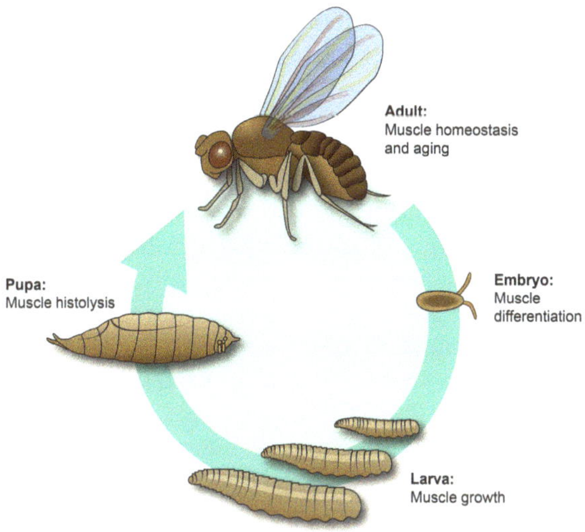

Fig. 5.1 The *Drosophila* life cycle: muscle differentiation, growth, remodeling, and aging. During embryonic development myoblasts fuse to form myofibers, which undergo differentiation. Subsequently, myofibers grow up to 50-fold in size during 4–5 days of larval development. During the pupal stage, most myofibers undergo rapid atrophy and degradation ("histolysis") while a few myofibers escape histolysis and form the template for the formation of the adult musculature by adult muscle precursor cells (AMPs). Finally in the adult fly, muscles must maintain their functional and structural integrity despite decreased homeostatic capacity and progressive incidence of muscle-extrinsic and muscle-intrinsic age-related defects

mammalian models of age-related diseases. Importantly, because some myofibers are spared and do not undergo histolysis, studies on muscle pupal development may shed light on the genetic and metabolic properties that can provide protection from different atrophic stimuli. For example, glycolytic muscles preferentially undergo atrophy during cachexia, sarcopenia and starvation in mammals (Li et al. 2007; Yu et al. 2008; Yamada ct al. 2012; Wang and Pessin 2013), while oxidative muscles undergo wasting during disuse and immobilization in rodents and humans (Edstrom 1970; Appell 1990; Thomason and Booth 1990). However, the mechanistic basis for this differential sensitivity to atrophic stimuli is largely unknown and studies in *Drosophila* may provide some clues on the fundamental mechanisms involved.

After around 10 days of development, the adult flies eclose. Examination of skeletal muscle in the adult allows for the analysis of the role of organelle turnover and proteinhomeostasis in regulating skeletal muscle homeostasis and function during aging (Fig. 5.1). In fact, although there is currently no evidence for muscle mass loss in old flies, there is an abundance of literature demonstrating age-related muscle intrinsic defects leading to functional declines including decreased flight, climbing, and walking ability (Grotewiel et al. 2005; Dickinson 2006; Martinez et al. 2007; Miller et al. 2008). Here below we describe how *Drosophila* is emerging as a convenient model organism to study sarcopenia and related processes.

5.2 Developmental Skeletal Muscle Growth and Histolysis in Drosophila: A Model for Mammalian Muscle Atrophy and Hypertrophy

There are similarities in the mechanisms regulating skeletal muscle growth and atrophy in *Drosophila* and mammals. In mammals signaling from contraction, nutrients, and hormones through the IGF/Akt/TOR pathway leads to robust hypertrophy (Schiaffino and Mammucari 2011). Importantly, the hypertrophic response to nutrient and contraction stimuli can be attenuated in mice by inhibition of the TOR pathway through muscle specific knockouts or through activation/suppression of TOR regulators such as TSC1/2 and Raptor (Bentzinger et al. 2008; Schiaffino and Mammucari 2011; Bentzinger et al. 2013; Sandri et al. 2013). Similar to mammals, the rapid muscle growth during *Drosophila* larval development is heavily dependent on nutrient sensing through the insulin/Akt/TOR pathway. Signaling through the transcription factors FoxO, Myc, and Mnt are also required for normal muscle growth in *Drosophila*. Overexpression of *FoxO* can inhibit muscle growth at least in part through inhibition of Myc activity, and overexpression of *Mnt*, a Myc antagonist, also inhibits muscle growth (Demontis and Perrimon 2009). The extensive genetic resources available in *Drosophila* allow for rapidly testing the role of any given gene and signaling pathway in the process of muscle growth in the larva, providing comprehensive insight into signaling pathways that may regulate muscle hypertrophy in mammals.

Similar to hypertrophy, many different stimuli including disuse, starvation, denervation, aging, stress signals, and reactive oxygen species can induce atrophy in mammalian skeletal muscle. In *Drosophila*, muscle atrophy followed by cell death is observed during histolysis. Larval muscles undergo two fates during metamorphosis. While a few muscles change morphology and become the adult muscles, most are degraded and replaced with new muscles formed by adult muscle precursor cells (AMPs). One of the signals regulating the rapid degradation of muscle during histolysis is a class of steroid hormones, called ecdysteroids, which induce similar gene expression changes as glucocorticoids when applied to mammalian cells (Christopherson et al. 1992; Jindra 1993). These steroid molecules cause the rapid breakdown of myofibrils leading to severe atrophy. Similar to the mammalian response to glucocorticoids, not all muscles undergo histolysis when exposed to ecdysteroids (Goldberg and Goodman 1969; Hegstrom and Truman 1996). Interestingly, the muscles that do not undergo histolysis are used as scaffolding for the growth and development of the adult skeletal muscles (Chinta et al. 2012). Taken together, although muscle growth and atrophy in *Drosophila* is limited to specific stages of development, the vast similarities in the signaling pathways regulating muscle size in *Drosophila* and mammals makes flies a good model for examining the regulation of muscle hypertrophy and atrophy.

While there are no stem cells in the muscles of adult flies, there are some similarities between mammalian satellite cells and stem cell-like adult muscle precursors (AMPs) in *Drosophila*. The AMPs are responsible for the formation of adult musculature during pupal metamorphosis (Figeac et al. 2007). AMPs, like quiescent satellite cells, can be regulated by epidermal growth factor (EGF) signaling and undergo fusion to promote muscle growth and new fiber formation. In *Drosophila*, over-expression of *EGFR* leads to increased number of AMPs in embryos (Bidet et al. 2003). Although different in many regards, AMPs can be a good model system for better understanding the cellular processes and signaling pathways regulating muscle stem cell behavior. Unlike mammals, there is no apparent regeneration of *Drosophila* muscles in adulthood. Therefore any muscle damage and intrinsic defects may rapidly lead to decreased muscle function given the lack of compensatory mechanisms.

In summary, different steps of *Drosophila* development allow for the investigation of the cellular and molecular mechanisms responsible for muscle growth and atrophy, regeneration, and muscle stem cell function, processes that are relevant for modulating muscle aging in mammals.

5.3 Age-Related Skeletal Muscle Functional Decay in Drosophila: A Model for Mammalian Sarcopenia

Sarcopenia has recently been defined as "an aging-related condition that normally manifests during or after the 4th decade of life where the overall quality of skeletal muscle decreases, ultimately leading to muscle weakness" (Brotto and Abreu 2012). Much attention has been given to the analysis of sarcopenia in mammals through the use of monkeys and rodents. The relatively long life span and the high costs associated with aging studies with these animal models as well as the inability to conduct large scale genetic screens suggests that simpler organisms may complement research in mammals and provide further insight into the mechanisms regulating sarcopenia.

Drosophila melanogaster undergoes dramatic age-related muscle deterioration, recalling the age-related decline in muscle function observed in humans (Doherty 2003). Defects in climbing, flight, and spontaneous movement are clearly discernable with aging in *Drosophila* over the course of its short life span of around 2–3 months (Grotewiel et al. 2005; Martinez et al. 2007; Miller et al. 2008). In *Drosophila* the decline in muscle function is largely due to a decrease in muscle strength. Decrements in muscle ultrastructure identified in old flies include mitochondrial degeneration, accumulation of misfolded protein and lysosomal dysfunction, and disorganization of the sarcoplasmic reticulum and sarcomeres (Takahashi et al. 1970; Hunt and Demontis 2013) (Fig. 5.2), defects that are also seen in mammalian models of sarcopenia (Tomonaga 1977).

The age-related decline in muscle function in humans correlates with an overall decrease in skeletal muscle mass, and muscle cross sectional area has been shown to decrease by 40 % between the ages of 20 and 60 years (Vandervoort 1994; Porter et al. 1995). Some of the loss in muscle mass could be attributed to disuse atrophy as the elderly display the lowest levels of physical activity of any age group (Nelson

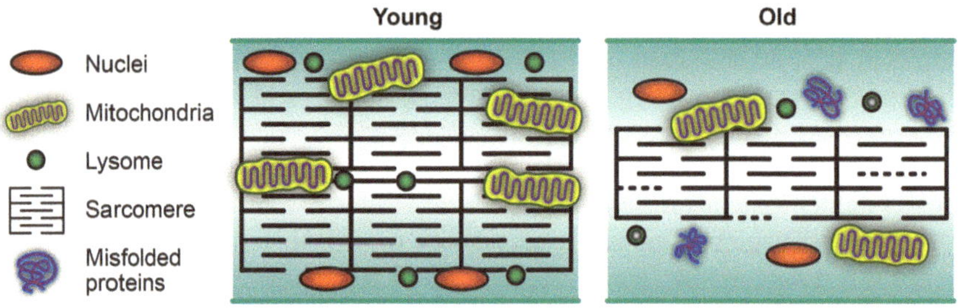

Fig. 5.2 Intracellular changes in *Drosophila* skeletal muscle during aging. During aging, skeletal muscle function declines. This is associated with accumulation of misfolded protein aggregates, dysfunctional organelles (including mitochondria and lysosomes), decreased number of nuclei (which are lost via syncytial apoptosis), decreased organization and contractile properties of sarcomeres, and many other degenerative changes in several cellular functions. Altogether, these changes contribute to the decreased muscle function observed during aging

et al. 2007). However, the decrease in muscle strength is 3-fold greater than the loss in muscle mass indicating that the loss in mass alone is not sufficient to explain the decrements in muscle function (Goodpaster et al. 2006). Contrary to humans, disuse atrophy and age related decreases in muscle mass have not been described in the adult *Drosophila* (Piccirillo et al. 2014), whereby age-related changes in cellular homeostatic systems and organelles appear to be largely responsible for the decline in muscle function observed during aging. Alternatively, the lack of evidence for muscle loss with aging in *Drosophila* could be attributable to the small size of the muscle making it difficult to detect small changes in muscle mass that may occur with aging. Although this may be a potential difference in the occurrence of sarcopenia in *Drosophila* compared to mammals, there are many benefits to using *Drosophila* for investigating the intrinsic changes that are responsible for decreased muscle function during aging. Along with the wide array of genetic tools that allow for muscle specific and genome wide analysis of gene function in *Drosophila*, large cohorts of flies of a given age and/or genotype can be analyzed for environmental, pharmacological, and dietary interventions. Additionally, laboratory mice are typically housed in small cages that limit physical activity, a condition that clearly associates with decreased muscle mass and function, whereas flies have proportionally larger housing, which provides the possibility for being more active, and thus age in a manner closer to fly populations in the wild.

5.4 Protective Role of Balanced Protein Homeostasis

Skeletal muscle mass and function is maintained by constant turnover of proteins and organelles. A shift in the balance between protein synthesis and degradation can lead to severe myopathies and skeletal muscle atrophy. Protein degradation is vital to proper muscle growth and homeostasis, as the inhibition of protein degradation pathways such as autophagy leads to the accumulation of dysfunctional proteins and organelles, ultimately leading to decreased function, atrophy, and apoptosis of myofibers (Masiero et al. 2009). The importance of muscle protein homeostasis is well established in humans and in rodent models of aging. During aging, multiple stress resistance pathways such as the unfolded protein response are activated to cope with the progressive accumulation of damaged proteins and organelle dysfunction (Haigis and Yankner 2010; Aoi and Sakuma 2011).

In *Drosophila*, increased gene expressions of the ubiquitin proteasome system components and antioxidant response pathways have been observed in aging skeletal muscle (Girardot et al. 2006). These responses include increased levels of heat shock protein 70 (Hsp70) (Wheeler et al. 1995), a cytosolic chaperone that aids in refolding proteins, as well as increased expression of JNK pathway components including *Jra* (the homologue of *Jun*) and factors involved in cytoskeletal rearrangements such as Myo31DF and Rac2 (Girardot et al. 2006). This response is also seen in the muscle of mice during aging (Clavel et al. 2006). While overexpression of *Hsp70* does not prevent age-induced accumulation of poly-ubiquitin aggregates in flies, Hsp70 does associate with these aggregates (Demontis and Perrimon 2010) presumably suppressing the proteotoxicity of the ubiquitinated misfolded proteins. In addition to the increase in cytosolic chaperones, mitochondrial chaperones Hsp22 and Hsp23 are also induced in aged *Drosophila* muscle (Wheeler et al. 1995). In mice, overexpression of *Hsp10*, another mitochondrial heat shock protein, attenuates both the decline in muscle mass and the decrease in muscle strength during aging (Kayani et al. 2010). In addition to heat shock proteins, the unfolded protein response in the endoplasmic reticulum (UPR^{ER}) is increased during aging in mouse skeletal muscle. The UPR^{ER} induces phosphorylation of eIF-2α leading to the suppression of muscle protein synthesis in mammals (Hasten et al. 2000) and *Drosophila* (Webster et al. 1980).

Although protein synthesis is a necessary component of the homeostatic balance that maintains skeletal muscle function during aging, its role is paradoxical. In humans, aged muscle is characterized by anabolic resistance, i.e. decreased protein synthesis in response to anabolic stimuli such as amino acid ingestion (Guillet et al. 2004; Cuthbertson et al. 2005). While dietary supplementation of amino acids and subsequent increases in protein synthesis may offset in the short-term some of the decreases in skeletal muscle mass observed with aging (Fujita and Volpi 2006), caloric restriction (which decreases protein synthesis) has been shown to delay age-associated muscle dysfunction in mammals and flies (Colman et al. 2008; Altun et al. 2010; Katewa et al. 2012). The suppression of protein synthesis during dietary restriction is mediated in part by the inhibition of IGF-1/TOR signaling (Mercken et al. 2013) and is associated with decreased muscle accrual of damaged proteins in mice (Lass et al. 1998) and preservation of muscle mass in rats (Altun et al. 2010). Moreover, activation of the protein synthesis machinery by insulin signaling and TOR activation has been associated with decreased longevity in both mammals and *Drosophila* (Garofalo 2002; Bartke 2008; Harrison et al. 2009), further indicating that a partial decrease in protein synthesis with age may be a protective mechanism elicited in the muscle. Recent studies have indeed suggested that moderate suppression of protein synthesis may in part explain the life span extension induced by caloric restriction or inhibition of TOR signaling through rapamycin administration in both vertebrates and invertebrates (Masoro 2005; Bjedov et al. 2010). Moreover, a recent paper has shown that overexpression of the transcription factor *Mnt* in skeletal muscle can reduce age-related muscle climbing defects and extend life span by reducing the function of the nucleolus, the site of rRNA transcription and ribosome biogenesis (Demontis et al. 2014) which determines the capacity for protein synthesis in the muscle. In addition to regulating protein synthesis, TOR activity has been linked to the inhibition of the autophagy pathway (Castets et al. 2013). Excessive activation of autophagy in mice has been associated with muscle wasting conditions in response to fasting and other catabolic stimuli (Bonaldo and Sandri 2013). However, activation of the autophagy pathway is necessary to maintain muscle mass and muscle quality (Masiero et al. 2009). Inhibition of autophagy through ablation of muscle Atg7 in mice results in abnormal mitochondria, accumulation of protein aggregates, and misaligned sarcomeres, which lead to decreased muscle function (Masiero et al. 2009). Muscle knockout of Atg7 in mice activates several stress responses including up-regulation of chaperones and ubiquitin proteasome system components such as the muscle E3 ligases atrogin-1 and MuRF1 (Masiero et al. 2009). Similar to the muscle ablation of Atg7, muscle Atg5 knockout mice accumulate protein aggregates leading to decreased muscle mass (Raben et al. 2008). A transcription factor regulating the expression of autophagy genes is FoxO. As demonstrated by Sandri et al. in mice and by Demontis et al. in *Drosophila*, overexpression of *FoxO* can lead to the induction of the autophagy pathway (Sandri et al. 2004; Demontis and Perrimon 2009). In *Drosophila*, muscle overexpression of *FoxO* delays the accumulation of poly-ubiquinated protein aggregates during muscle aging while *FoxO* null flies demonstrate accelerated accumulation of these aggregates. FoxO increases the expression of *Hsp70* as well as several genes regulating the autophagy/lysosomal pathway, such as *Atg* genes and *Lamp1*. Furthermore the knockdown of Atg7 in muscle with *FoxO* overexpression results in an increase in protein aggregate accumulation compared with age-matched *FoxO* overexpressing controls. These data demonstrate that overexpression of wild type *FoxO* in muscle of *Drosophila* prevents the age-related decrease in muscle protein homeostasis and maintains muscle function during aging by preserving the functionality of the autophagy pathway (Demontis and Perrimon 2010).

In addition to its role in autophagy, FoxO is also a mediator of protein degradation through the ubiquitin proteasome system (UPS). Activation of the UPS helps to maintain muscle integrity through selected degradation of poly-ubiquitinated proteins. The UPS requires the attachment of ubiquitin to targeted substrates. Ubiquitinated proteins are then preferentially degraded by the 26S proteolytic complex eliminating damaged and misfolded proteins and allowing for the recycling of amino acids under stress conditions. Overexpression of *FoxO* can increase expression of the E3 ubiquitin ligase *atrogin-1* in mice and in cultures of C2C12 myotubes (Sandri et al. 2004; Zhao et al. 2007). Sandri et al. demonstrated FoxO-mediated regulation of *atrogin-1* expression by examining mouse skeletal muscle transfected with constitutively active FoxO, which lead to a 20-fold increase in atrogin-1 promoter luciferase activity. Additionally, inhibition of *FoxO* through RNAi in mouse skeletal muscle prevented fasting-induced atrogin-1 promoter luciferase activity. These data led to the identification of FoxO binding sites at the 5′ end of the *atrogin-1* gene that are necessary for its expression by FoxO (Sandri et al. 2004). In addition, constitutively active FoxO induces several autophagy-related genes including *LC3B*, *Gabarapl1* and *Atg12* in C2C12 myotubes through direct binging to their promoters (Zhao et al. 2007). While removal of damaged proteins is vital to the maintenance of muscle function with aging, over activation of protein degradation pathways including the UPS and autophagy is known to be deleterious and can lead to atrophy in mice and C2C12 myotubes.

In addition to its regulation through the ubiquitination of proteins, the proteasome is regulated in part by the composition of the 20S catalytic core as well as the 19S regulatory cap (Ciechanover 1994). The composition of the proteasome is altered during aging in mammals (Husom et al. 2004; Ferrington et al. 2005). Recently, a decrease in 26S proteasome activity and a decrease in proteasome assembly were reported in the course of *Drosophila* aging (Vernace et al. 2007). The decrease in proteasome assembly and activity was associated with an age-related reduction in ATP levels in the fly skeletal muscle, and given the ATP-dependent nature of the proteasome this may well contribute to proteasome dysfunction in aged muscle (Vernace et al. 2007). In rats, increased expression of the alternative version of the proteasome (the immunoproteasome) and increases in proteasome subunit oxidation may explain the overall decrease in proteasomal activity observed during aging in skeletal muscle (Husom et al. 2004; Ferrington et al. 2005). However, while some studies have demonstrated decreased activation of the proteasome in aged muscle (Low 2011), others reported increased function (Carrard et al. 2002; Altun et al. 2010). For example, proteasome peptidase activity was increased in the skeletal muscle of 30-month-old rats, indicating an increase in proteasome capacity, which was reduced by caloric restriction (Altun et al. 2010). These conflicting data on the regulation and role of the proteasome in skeletal muscle aging demonstrate the need for further research in this area. Although there is much debate over the role of the UPS in aging skeletal muscle, there is clear evidence that dysregulation of the systems controlling protein homeostasis can lead to accumulation of dysfunctional proteins and organelles as well as to unselective loss of muscle mass, ultimately leading to decreased muscle function.

In summary, the specificity and activity of protein degradation pathways needs to be tightly controlled for ensuring the maintenance of skeletal muscle mass and function during aging.

5.5 Systemic Aging and Life Span Determination in Response to Exercise and Signals from Skeletal Muscle

The human body is comprised by nearly 50 % skeletal muscle that mediate the body's movements. There is extensive epidemiological evidence indicating that muscle contraction leads to organism-wide responses following different exercise regimens, at least in part due to the high nutrient/metabolic demand of contracting muscle. A common example of muscle's systemic effects is the observation of cellular and metabolic responses following muscle contraction. For example, exercise can reduce whole body glucose levels as well as increase lipolysis in the adipose tissue. Exercise also protects from neurodegeneration (Ahlskog 2011; Revilla et al. 2014), reduces the risk of developing many types of cancers (Brown et al. 2012), improves cardiac and endothelial function (Shephard and Balady 1999), and potentially increases life span (Piazza et al. 2009).

Similar to mammals, studies in *Drosophila* also demonstrate muscle adaptations to exercise. For example, climbing (negative geotaxis) exercise increases mitochondrial function and preserves motor capacity in flies (Piazza et al. 2009) at least in part via the PGC-1α/β homolog *spargel* (Tinkerhess et al. 2012). While exercise has potent health benefits, muscle disuse and physical inactivity have been linked with increased health risk and mortality. Interestingly, studies in both flies and mammals have shown an increase in physical activity that is associated with caloric restriction (Holloszy and Schechtman 1991; Giustina et al. 1997; Weed et al. 1997; Katewa et al. 2012). Moreover, disuse of skeletal muscle induced through wing ablation can prevent the life span extension caused by caloric restriction in *Drosophila* (Katewa et al. 2012). These data demonstrate that the protective effects of some life span-extending interventions such as caloric restriction appear to be directly tied to skeletal muscle function and physical activity.

The complex interplay between exercise, metabolic homeostasis, and life span clearly requires further studies. A recent avenue of investigation suggests that skeletal muscle is an important endocrine tissue with the capacity to influence whole-organism metabolism via the secretion of muscle-derived cytokines and growth factors known as "myokines" (Pedersen and Febbraio 2008; Pratesi et al. 2013). The release of myokines from muscle may explain how exercise and perturbations in skeletal muscle signaling can lead to alterations in organism-wide physiological homeostasis and aging. Specifically, there is evidence for myokine-based crosstalk of skeletal muscle with several organs and tissues such as the liver, endothelium, pancreas, adipose tissue, and perhaps the brain during both healthy and disease states (Pedersen and Hojman 2012).

A notable recent case is irisin, a myokine that is secreted by skeletal muscle during endurance exercise following the cleavage of its transmembrane precursor (FNDC5, fibronectin type III domain containing 5). Once released, irisin can then induce the browning of adipose tissue making it more metabolically active than the white adipose tissue. The browning of the adipose tissue may indeed result in increased metabolic substrate utilization and may thus mimic (phenocopy) caloric restriction. The health benefits of irisin are still under investigation; however, irisin is closely linked to metabolic homeostasis and is emerging as a possible therapeutic target for the treatment of obesity and diabetes (Elbelt et al. 2013). Treatment of C2C12 myotubes with irisin can lead to an increase in mitochondrial biogenesis and increased expression of Glut4 glucose transporter (Vaughan et al. 2014). Additionally, irisin has been shown to increase IGF-1 and suppress myostatin in primary human myocytes (Huh et al. 2014). A recent study suggested that irisin may also play a role in the aging process as elevated plasma irisin in humans was associated with increased telomere length, which declines during cellular aging (Rana et al. 2014).

Other myokines such as Myostatin (MSTN) and interleukin-6 (IL-6) can also regulate whole body metabolism. Myostatin is a negative regulator of muscle mass; however, it has recently been shown to act on non-muscle tissues such as the adipose and other tissues, as reviewed by Argiles et al. (2012). For example, exposure of mesenchymal stem cells to MSTN leads to the differentiation of immature adipocytes that protect from obesity and metabolic diseases (Feldman et al. 2006). Interestingly Myoglianin, the *Drosophila* homolog of human Myostatin and GDF11, has been recently shown to extend life span and delay systemic aging by acting on muscle, adipocytes, and possibly other tissues (Demontis et al. 2014). These effects were not due to feeding or changes in muscle mass (Demontis et al. 2014), suggesting that *Drosophila* may be a convenient system for testing the direct signaling roles of GDF11/Myostatin signaling without the indirect confounding effects deriving from the increased muscle mass observed in Myostatin (MSTN) knock-out mice. In fact, MSTN knock-out mice have increased insulin sensitivity and glucose oxidation and decreased whole body adiposity at least in part due to the higher metabolic demand deriving from the doubling in muscle mass, which leads to higher nutrient utilization in muscle and reduced nutrient availability for other tissues (Guo et al. 2009).

Another prominent myokine is Interleukin-6, IL-6, which can induce lipolysis and increase insulin sensitivity when elevated for a short term, such as during exercise. While acute elevations in IL-6 appear beneficial, long-term increases in IL-6 levels are associated with muscle wasting (Haddad et al. 2005). Although paradoxical, IL-6 is a myokine with the potential to affect the aging process through inter-tissue crosstalk. For example, IL-6 increases glucose uptake in the muscle, it also signals for the secretion of insulin from the pancreas (Ellingsgaard et al. 2011).

Although little is known on myokine signaling in *Drosophila*, many putative evolutionarily conserved myokines are encoded by the *Drosophila* genome. Several studies indicate systemic regulation of agingfollowing muscle-specific genetic interventions in *Drosophila* (Demontis et al. 2013b) and signals released by muscle (such as myokines) may play a role. For example, muscle-specific *FoxO* overexpression increases autophagy/lysosomal activities locally in the muscle but also systemically in the brain, retina, and adipose tissue via the organism-wide induction of FoxO/4E-BP signaling (Demontis and Perrimon 2010). Interestingly, muscle specific overexpression of FoxO increases life span and also preserves muscle function, decreases feeding behavior, and lowers glycemia (Demontis and Perrimon 2010). FoxO activity in muscle can improve muscle function and systemic proteostasis at least in part by decreasing the expression of *dawdle*, an activin-related secreted factor which is a direct FoxO target gene (Bai et al. 2013). Thus, FoxO-regulated myokines released by the muscle may be responsible at least in part for the regulation of systemic aging and life span.

A recent study has described a transgenic RNAi screen for myokines regulating life span in *Drosophila* (Demontis et al. 2014). Among the myokines identified, there was Myoglianin, the *Drosophila* homolog of human Myostatin and the related factor GDF11. Overexpression of *myoglianin* in muscle extended life span and reduced the number of flies that displayed climbing defects in old age (Demontis et al. 2014). Conversely, *myoglianin* RNAi in the muscle lead to accelerated muscle aging and shorter life span. Myoglianin regulates aging by reducing the activity of the nucleolus (which is the key site for ribosome biogenesis and thus protein synthesis) and by activating p38 MAPK (Demontis et al. 2014), a regulator of aging in multiple species and a signal transduction component of non-canonical TGF-beta signaling in vertebrates. Thus, myokine signaling appears to be an important determinant of systemic aging and life span. *Drosophila* may be a valuable system for studying how muscle-specific genetic interventions can regulate life span, the role of myokines in mediating the crosstalk between muscle and other tissues, and the cellular and molecular responses induced in distant tissues.

Other studies have emphasized a protective role of skeletal muscle against whole body oxidative stress. Muscle specific suppression of super-oxide dismutase-2 (Sod-2), p38 MAPK, and AMPK expression can reduce the resistance of the organism to oxidative stress while overexpression of *p38 MAPK* in muscle can increase life span and stress resistance in *Drosophila* (Vrailas-Mortimer et al. 2011). Additionally, p38 has been recently implicated in the regulation of skeletal muscle protein translation through its interaction with the scaffold protein Receptor of activated protein kinase C-1 (Rack1) (Belozerov et al. 2014). Taken together, these data demonstrate that muscle-specific activation of signaling pathways can alter organism-wide aging and stress resistance through modulation of systemic metabolism, myokine signaling, and perhaps also neuronal interactions (Fig. 5.3).

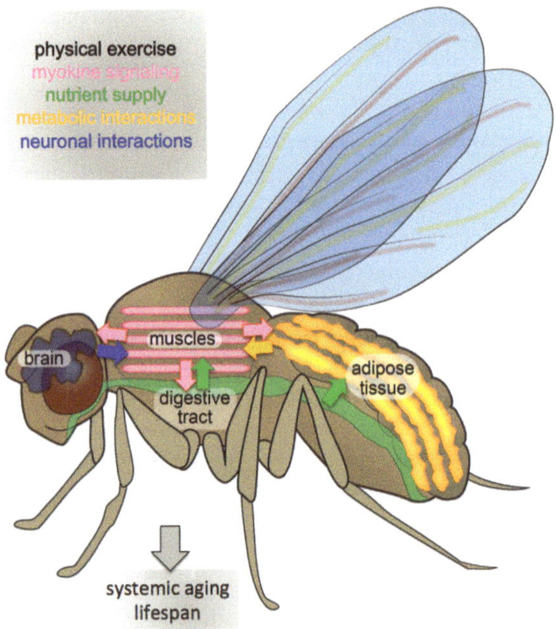

Fig. 5.3 Systemic regulation of aging and lifespan in response to signals from skeletal muscle. Skeletal muscle can interact with a host of tissues and organs to regulate systemic aging and lifespan. Myokine signaling, nutrient demand of contracting muscle, metabolic homeostasis, and crosstalk with neuronal circuits all intersect with skeletal muscle and may impact lifespan and systemic aging in *Drosophila*

5.6 Methods for Studying Skeletal Muscle Homeostasis and Aging in *Drosophila*

Drosophila melanogaster has been a model organism for developmental research for over a century. Recently it has emerged as a promising model for the study of skeletal muscle homeostasis during aging because of its short life span, the evolutionary conservation of signaling pathways, and the wide array of genetic and molecular interventions available. In particular, the ability to quickly generate muscle-specific genetic mutations that encompass all of the skeletal muscle has allowed for the analysis of muscle signaling on organismal wide aging and homeostasis. While techniques to induce genetic mutations in the muscle of rodents are available through electroporation of overexpression and knockdown plasmids, typically only few myofibers are successfully transfected, thus limiting the experimental tissue available for analysis. It is well known that electroporation allows for the monitoring of transgene expression only for a few weeks rather than over many months, as it would be required for the analysis of sarcopenia progression. Alternative methods, such as the Cre-LoxP system, allow for the modulation of transgene expression in the entire tissue and for a longer period of time. However, the breeding and development of tissue specific transgenic mice can be expensive and slow. Additionally, variation in the genetic background in mice can be a confounder in life span and aging studies. For example, Liao et al. demonstrated that caloric restriction only extended life span in 9 mouse strains out of 42, and that caloric restriction even shortened life span in four strains (Liao et al. 2010). These findings demonstrate the inconsistency across different mouse strains and the need for careful control of the genetic background in aging studies. Given the short lifecycle of *Drosophila*, fly strains carrying different transgenes or classical mutations can be easily isogenized by backcrossing them over many generations against the same genetic background to avoid any confounding effects due to background mutations.

In addition to minimization of variation deriving from differences in the genetic background, gene knockdown through RNA interference (RNAi) has allowed for genome-wide screens to be conducted in the adult bypassing developmental effects (Duffy 2002), such as gene mutations causing embryonic and larval lethality. In *Drosophila* RNAi is cell autonomous (Van Roessel et al. 2002). When used in combination with the GAL4/UAS expression system, tissue/cell specific gene function can be studied (Brand and Perrimon 1993). For example several GAL4 drivers are available that are specific to skeletal muscle including Mef2-GAL4 and Mhc-GAL4. Additionally, the availability of drug- and temperature-controlled GAL4 systems allow for gene inactivation and transgene expression during specific stages of the lifecycle and tissue-specifically (Duffy 2002). Importantly, the availability of multiple genome wide transgenic RNAi libraries has made it possible to conduct systematic RNAi screens in specific tissue types during different stages of development (Dietzl et al. 2007). Additionally, clustered regularly interspaced short palindromic repeats (CRISPR) and TALEN technologies are emerging as complementary approaches for the generation of *Drosophila* lines containing novel mutations in targeted genes (Bassett and Liu 2014).

In addition to a vast array of technologies for modulating gene expression and function during aging, multiple tests can be used in *Drosophila* to examine muscle activity. Spontaneous locomotion, climbing, jumping and flight assays have all been used to analyze the functional decrements of different muscle subsets during aging and thus provide a physiological readout of any given genetic intervention. Climbing assays utilize the flies' negative geotaxis behavior, e.g. the innate instinct of the flies to move away from the earth, while flight and jumping assays probe the function of other muscle groups upon stimulation. Analysis of single muscle fibers from *Drosophila* can also be conducted to determine alterations in muscle tension, power output, and calcium levels in the muscle (Miller et al. 2008). Along with the host of genetic and functional assays available, *Drosophila* is also amenable to many routine molecular and cellular assays used in mammals for the investigation of gene function in muscle and for probing the effect of drugs and dietary regimens and for assessing their interaction with genetic interventions.

5.7 Conclusion

Drosophila melanogaster is a promising model for the study of skeletal muscle homeostasis and aging. During aging, *Drosophila* muscle display profound deterioration and dysfunction, a key characteristic of mammalian sarcopenia. In addition to age-associated muscle dysfunction, systemic aging is evident in *Drosophila* and the examination of muscle-specific genetic alterations has lead to striking findings on the role of muscle in regulating aging in other tissues and life span. Further work is still necessary to determine the signaling factors released by muscle and regulating organismal aging and to dissect the fundamental muscle-intrinsic mechanisms responsible for sarcopenia.

References

Ahlskog JE (2011) Does vigorous exercise have a neuroprotective effect in Parkinson disease? Neurology 77(3):288–294. doi:10.1212/WNL.0b013e318225ab66

Altun M, Besche HC, Overkleeft HS, Piccirillo R, Edelmann MJ, Kessler BM, Goldberg AL, Ulfhake B (2010) Muscle wasting in aged, sarcopenic rats is associated with enhanced activity of the ubiquitin proteasome pathway. J Biol Chem 285(51):39597–39608. doi:10.1074/jbc.M110.129718

Aoi W, Sakuma K (2011) Oxidative stress and skeletal muscle dysfunction with aging. Curr Aging Sci 4(2):101–109

Appell H-J (1990) Muscular atrophy following immobilisation. Sports Med 10(1):42–58. doi:10.2165/00007256-199010010-00005

Argiles JM, Orpi M, Busquets S, Lopez-Soriano FJ (2012) Myostatin: more than just a regulator of muscle mass. Drug Discov Today 17(13–14):702–709. doi:10.1016/j.drudis.2012.02.001

Augustin H, Partridge L (2009) Invertebrate models of age-related muscle degeneration. Biochim Biophys Acta 1790(10):1084–1094. doi:10.1016/j.bbagen.2009.06.011

Bai H, Kang P, Hernandez AM, Tatar M (2013) Activin signaling targeted by insulin/dFOXO regulates aging and muscle proteostasis in *Drosophila*. PLoS Genet 9(11):e1003941. doi:10.1371/journal.pgen.1003941

Bartke A (2008) Impact of reduced insulin-like growth factor-1/insulin signaling on aging in mammals: novel findings. Aging Cell 7(3):285–290. doi:10.1111/j.1474-9726.2008.00387.x

Bassett A, Liu JL (2014) CRISPR/Cas9 mediated genome engineering in *Drosophila*. Methods 69(2):128–136. doi:10.1016/j.ymeth.2014.02.019

Belozerov VE, Ratkovic S, McNeill H, Hilliker AJ, McDermott JC (2014) In vivo interaction proteomics reveal a novel p38 mitogen-activated protein kinase/Rack1 pathway regulating proteostasis in *Drosophila* muscle. Mol Cell Biol 34(3):474–484. doi:10.1128/MCB.00824-13

Bentzinger CF, Lin S, Romanino K, Castets P, Guridi M, Summermatter S, Handschin C, Tintignac LA, Hall MN, Ruegg MA (2013) Differential response of skeletal muscles to mTORC1 signaling during atrophy and hypertrophy. Skelet Muscle 3(1):6. doi:10.1186/2044-5040-3-6

Bentzinger CF, Romanino K, Cloetta D, Lin S, Mascarenhas JB, Oliveri F, Xia J, Casanova E, Costa CF, Brink M, Zorzato F, Hall MN, Ruegg MA (2008) Skeletal muscle-specific ablation of raptor, but not of rictor, causes metabolic changes and results in muscle dystrophy. Cell Metab 8(5):411–424. doi:10.1016/j.cmet.2008.10.002

Bidet Y, Jagla T, Da Ponte JP, Dastugue B, Jagla K (2003) Modifiers of muscle and heart cell fate specification identified by gain-of-function screen in Drosophila. Mech Dev 120(9):991–1007

Bjedov I, Toivonen JM, Kerr F, Slack C, Jacobson J, Foley A, Partridge L (2010) Mechanisms of life span extension by rapamycin in the fruit fly *Drosophila melanogaster*. Cell Metab 11(1):35–46. doi:10.1016/j.cmet.2009.11.010

Bonaldo P, Sandri M (2013) Cellular and molecular mechanisms of muscle atrophy. Dis Models Mech 6(1):25–39. doi:10.1242/dmm.010389

Brand AH, Perrimon N (1993) Targeted gene expression as a means of altering cell fates and generating dominant phenotypes. Development 118(2):401–415

Brotto M, Abreu EL (2012) Sarcopenia: pharmacology of today and tomorrow. J Pharmacol Exp Ther 343(3):540–546. doi:10.1124/jpet.112.191759

Brown JC, Winters-Stone K, Lee A, Schmitz KH (2012) Cancer, physical activity, and exercise. Compr Physiol 2(4):2775–2809. doi:10.1002/cphy.c120005

Carrard G, Bulteau AL, Petropoulos I, Friguet B (2002) Impairment of proteasome structure and function in aging. Int J Biochem Cell Biol 34(11):1461–1474

Castets P, Lin S, Rion N, Di Fulvio S, Romanino K, Guridi M, Frank S, Tintignac LA, Sinnreich M, Ruegg MA (2013) Sustained activation of mTORC1 in skeletal muscle inhibits constitutive and starvation-induced autophagy and causes a severe, late-onset myopathy. Cell Metab 17(5):731–744. doi:10.1016/j.cmet.2013.03.015

Chinta R, Tan JH, Wasser M (2012) The study of muscle remodeling in *Drosophila* metamorphosis using in vivo microscopy and bioimage informatics. BMC Bioinformatics 13(Suppl 17):S14. doi:10.1186/1471-2105-13-S17-S14

Christopherson KS, Mark MR, Bajaj V, Godowski PJ (1992) Ecdysteroid-dependent regulation of genes in mammalian cells by a *Drosophila* ecdysone receptor and chimeric transactivators. Proc Natl Acad Sci USA 89(14):6314–6318

Ciechanover A (1994) The ubiquitin-proteasome proteolytic pathway. Cell 79(1):13–21

Clavel S, Coldefy AS, Kurkdjian E, Salles J, Margaritis I, Derijard B (2006) Atrophy-related ubiquitin ligases, atrogin-1 and MuRF1 are up-regulated in aged rat Tibialis Anterior muscle. Mech Ageing Dev 127(10):794–801. doi:10.1016/j.mad.2006.07.005

Colman RJ, Beasley TM, Allison DB, Weindruch R (2008) Attenuation of sarcopenia by dietary restriction in rhesus monkeys. J Gerontol A Biol Sci Med Sci 63(6):556–559

Cuthbertson D, Smith K, Babraj J, Leese G, Waddell T, Atherton P, Wackerhage H, Taylor PM, Rennie MJ (2005) Anabolic signaling deficits underlie amino acid resistance of wasting, aging muscle. FASEB J (official publication of the Federation of American Societies for Experimental Biology) 19(3):422–424. doi:10.1096/fj.04-2640fje

Demontis F, Patel VK, Swindell WR, Perrimon N (2014) Intertissue control of the nucleolus via a myokine-dependent longevity pathway. Cell Rep 7(5):1481–1494. doi:10.1016/j.celrep.2014.05.001

Demontis F, Perrimon N (2009) Integration of Insulin receptor/Foxo signaling and dMyc activity during muscle growth regulates body size in *Drosophila*. Development 136(6):983–993. doi: 10.1242/dev.027466

Demontis F, Perrimon N (2010) FOXO/4E-BP signaling in *Drosophila* muscles regulates organism-wide proteostasis during aging. Cell 143(5):813–825. doi:10.1016/j.cell.2010.10.007

Demontis F, Piccirillo R, Goldberg AL, Perrimon N (2013a) Mechanisms of skeletal muscle aging: insights from *Drosophila* and mammalian models. Dis Models Mech 6(6):1339–1352. doi:10.1242/dmm.012559

Demontis F, Piccirillo R, Goldberg AL, Perrimon N (2013b) The influence of skeletal muscle on systemic aging and lifespan. Aging Cell 12(6):943–949. doi:10.1111/acel.12126

Dickinson MH (2006) Insect flight. Curr Biol 16(9):R309–R314

Dietzl G, Chen D, Schnorrer F, Su KC, Barinova Y, Fellner M, Gasser B, Kinsey K, Oppel S, Scheiblauer S, Couto A, Marra V, Keleman K, Dickson BJ (2007) A genome-wide transgenic RNAi library for conditional gene inactivation in *Drosophila*. Nature 448(7150):151–156. doi:10.1038/nature05954

Doherty TJ (2003) Invited review: aging and sarcopenia. J Appl Physiol 95(4):1717–1727. doi:10.1152/japplphysiol.00347.2003

Duffy JB (2002) GAL4 system in *Drosophila*: a fly geneticist's Swiss army knife. Genesis 34(1–2):1–15. doi:10.1002/gene.10150

Edstrom L (1970) Selective atrophy of red muscle fibres in the quadriceps in long-standing knee-joint dysfunction. Injuries to the anterior cruciate ligament. J Neurol Sci 11(6):551–558

Elbelt U, Hofmann T, Stengel A (2013) Irisin: what promise does it hold? Curr Opin Clin Nutr Metab Care 16(5):541–547. doi:10.1097/MCO.0b013e328363bc65

Feldman BJ, Streeper RS, Farese RV Jr, Yamamoto KR (2006) Myostatin modulates adipogenesis to generate adipocytes with favorable metabolic effects. Proc Natl Acad Sci USA 103(42):15675–15680. doi:10.1073/pnas.0607501103

Ellingsgaard H, Hauselmann I, Schuler B, Habib AM, Baggio LL, Meier DT, Eppler E, Bouzakri K, Wueest S, Muller YD, Hansen AM, Reinecke M, Konrad D, Gassmann M, Reimann F, Halban PA, Gromada J, Drucker DJ, Gribble FM, Ehses JA, Donath MY (2011) Interleukin-6 enhances insulin secretion by increasing glucagon-like peptide-1 secretion from L cells and alpha cells. Nat Med 17(11):1481–1489. doi:10.1038/nm.2513

Ferrington DA, Husom AD, Thompson LV (2005) Altered proteasome structure, function, and oxidation in aged muscle. FASEB J (official publication of the Federation of American Societies for Experimental Biology) 19(6):644–646. doi:10.1096/fj.04-2578fje

Figeac N, Daczewska M, Marcelle C, Jagla K (2007) Muscle stem cells and model systems for their investigation. Dev Dyn (an official publication of the American Association of Anatomists) 236(12):3332–3342. doi:10.1002/dvdy.21345

Fujita S, Volpi E (2006) Amino acids and muscle loss with aging. J Nutr 136(1 Suppl):277S–280S

Garofalo RS (2002) Genetic analysis of insulin signaling in *Drosophila*. Trends Endocrinol Metab TEM 13(4):156–162

Girardot F, Lasbleiz C, Monnier V, Tricoire H (2006) Specific age-related signatures in *Drosophila* body parts transcriptome. BMC Genom 7:69. doi:10.1186/1471-2164-7-69

Giustina A, Desenzani P, Bossoni S, Perini P (1997) Growth hormone treatment in aging: state of the art and perspectives. Aging 9(4 Suppl):73–74

Goldberg AL, Goodman HM (1969) Relationship between cortisone and muscle work in determining muscle size. J Physiol 200(3):667–675

Goodpaster BH, Park SW, Harris TB, Kritchevsky SB, Nevitt M, Schwartz AV, Simonsick EM, Tylavsky FA, Visser M, Newman AB (2006) The loss of skeletal muscle strength, mass, and quality in older adults: the health, aging and body composition study. J Gerontol A Biol Sci Med Sci 61(10):1059–1064

Grotewiel MS, Martin I, Bhandari P, Cook-Wiens E (2005) Functional senescence in *Drosophila melanogaster*. Ageing Res Rev 4(3):372–397. doi:10.1016/j.arr.2005.04.001

Guillet C, Prod'homme M, Balage M, Gachon P, Giraudet C, Morin L, Grizard J, Boirie Y (2004) Impaired anabolic response of muscle protein synthesis is associated with S6K1 dysregulation in elderly humans. FASEB J (official publication of the Federation of American Societies for Experimental Biology) 18(13):1586–1587. doi:10.1096/fj.03-1341fje

Guo T, Jou W, Chanturiya T, Portas J, Gavrilova O, McPherron AC (2009) Myostatin inhibition in muscle, but not adipose tissue, decreases fat mass and improves insulin sensitivity. PLoS ONE 4(3):e4937. doi:10.1371/journal.pone.0004937

Haddad F, Zaldivar F, Cooper DM, Adams GR (2005) IL-6-induced skeletal muscle atrophy. J Appl Physiol 98(3):911–917. doi:10.1152/japplphysiol.01026.2004

Haigis MC, Yankner BA (2010) The aging stress response. Mol Cell 40(2):333–344. doi:10.1016/j.molcel.2010.10.002

Harrison DE, Strong R, Sharp ZD, Nelson JF, Astle CM, Flurkey K, Nadon NL, Wilkinson JE, Frenkel K, Carter CS, Pahor M, Javors MA, Fernandez E, Miller RA (2009) Rapamycin fed late in life extends lifespan in genetically heterogeneous mice. Nature 460(7253):392–395. doi:10.1038/nature08221

Hasten DL, Pak-Loduca J, Obert KA, Yarasheski KE (2000) Resistance exercise acutely increases MHC and mixed muscle protein synthesis rates in 78-84 and 23-32 yr olds. Am J Physiol Endocrinol Metab 278(4):E620–E626

Hegstrom CD, Truman JW (1996) Steroid control of muscle remodeling during metamorphosis in Manduca sexta. J Neurobiol 29(4):535–550. doi:10.1002/(SICI)1097-4695(199604)29:4<535:AID-NEU9>3.0.CO;2-9

Herndon LA, Schmeissner PJ, Dudaronek JM, Brown PA, Listner KM, Sakano Y, Paupard MC, Hall DH, Driscoll M (2002) Stochastic and genetic factors influence tissue-specific decline in ageing *C. elegans*. Nature 419(6909):808–814. doi:10.1038/nature01135

Holloszy JO, Schechtman KB (1991) Interaction between exercise and food restriction: effects on longevity of male rats. J Appl Physiol 70(4):1529–1535

Huh JY, Dincer F, Mesfum E, Mantzoros CS (2014) Irisin stimulates muscle growth-related genes and regulates adipocyte differentiation and metabolism in humans. Int J Obes (Lond) 38(12):1538–1544. doi:10.1038/ijo.2014.42

Hunt LC, Demontis F (2013) Whole-mount immunostaining of *Drosophila* skeletal muscle. Nat Protoc 8(12):2496–2501. doi:10.1038/nprot.2013.156

Husom AD, Peters EA, Kolling EA, Fugere NA, Thompson LV, Ferrington DA (2004) Altered proteasome function and subunit composition in aged muscle. Arch Biochem Biophys 421(1):67–76

Jindra M (1993) Gene regulation by sex hormones: Vertebrates and insects. Eur J Entomol 91:163–187

Katewa SD, Demontis F, Kolipinski M, Hubbard A, Gill MS, Perrimon N, Melov S, Kapahi P (2012) Intramyocellular fatty-acid metabolism plays a critical role in mediating responses to dietary restriction in *Drosophila melanogaster*. Cell Metab 16(1):97–103. doi:10.1016/j.cmet.2012.06.005

Kayani AC, Close GL, Dillmann WH, Mestril R, Jackson MJ, McArdle A (2010) Overexpression of HSP10 in skeletal muscle of transgenic mice prevents the age-related fall in maximum tetanic force generation and muscle Cross-Sectional Area. Am J Physiol Regul Integr Comp Physiol 299(1):R268–R276. doi:10.1152/ajpregu.00334.2009

Lass A, Sohal BH, Weindruch R, Forster MJ, Sohal RS (1998) Caloric restriction prevents age-associated accrual of oxidative damage to mouse skeletal muscle mitochondria. Free Radic Biol Med 25(9):1089–1097

Li P, Waters RE, Redfern SI, Zhang M, Mao L, Annex BH, Yan Z (2007) Oxidative phenotype protects myofibers from pathological insults induced by chronic heart failure in mice. Am J Pathol 170(2):599–608. doi:10.2353/ajpath.2007.060505

Liao CY, Rikke BA, Johnson TE, Diaz V, Nelson JF (2010) Genetic variation in the murine lifespan response to dietary restriction: from life extension to life shortening. Aging Cell 9(1):92–95. doi:10.1111/j.1474-9726.2009.00533.x

Low P (2011) The role of ubiquitin-proteasome system in ageing. Gen Comp Endocrinol 172(1):39–43. doi:10.1016/j.ygcen.2011.02.005

Martinez VG, Javadi CS, Ngo E, Ngo L, Lagow RD, Zhang B (2007) Age-related changes in climbing behavior and neural circuit physiology in *Drosophila*. Dev Neurobio 67(6):778–791. doi:10.1002/dneu.20388

Masiero E, Agatea L, Mammucari C, Blaauw B, Loro E, Komatsu M, Metzger D, Reggiani C, Schiaffino S, Sandri M (2009) Autophagy is required to maintain muscle mass. Cell Metab 10(6):507–515. doi:10.1016/j.cmet.2009.10.008

Masoro EJ (2005) Overview of caloric restriction and ageing. Mech Ageing Dev 126(9):913–922. doi:10.1016/j.mad.2005.03.012

Mercken EM, Crosby SD, Lamming DW, JeBailey L, Krzysik-Walker S, Villareal DT, Capri M, Franceschi C, Zhang Y, Becker K, Sabatini DM, de Cabo R, Fontana L (2013) Calorie restriction in humans inhibits the PI3 K/AKT pathway and induces a younger transcription profile. Aging Cell 12(4):645–651. doi:10.1111/acel.12088

Miller MS, Lekkas P, Braddock JM, Farman GP, Ballif BA, Irving TC, Maughan DW, Vigoreaux JO (2008) Aging enhances indirect flight muscle fiber performance yet decreases flight ability in *Drosophila*. Biophys J 95(5):2391–2401. doi:10.1529/biophysj.108.130005

Nelson ME, Rejeski WJ, Blair SN, Duncan PW, Judge JO, King AC, Macera CA, Castaneda-Sceppa C (2007) Physical activity and public health in older adults: recommendation from the American College of Sports Medicine and the American Heart Association. Med Sci Sports Exerc 39(8):1435–1445. doi:10.1249/mss.0b013e3180616aa2

Patel K, Christ B, Stockdale FE (2002) Control of muscle size during embryonic, fetal, and adult life. Results Probl Cell Differ 38:163–186

Pedersen BK, Febbraio MA (2008) Muscle as an endocrine organ: focus on muscle-derived interleukin-6. Physiol Rev 88(4):1379–1406. doi:10.1152/physrev.90100.2007

Pedersen L, Hojman P (2012) Muscle-to-organ cross talk mediated by myokines. Adipocyte 1(3):164–167. doi:10.4161/adip.20344

Piazza N, Gosangi B, Devilla S, Arking R, Wessells R (2009) Exercise-training in young *Drosophila* melanogaster reduces age-related decline in mobility and cardiac performance. PLoS ONE 4(6):e5886. doi:10.1371/journal.pone.0005886

Piccirillo R, Demontis F, Perrimon N, Goldberg AL (2014) Mechanisms of muscle growth and atrophy in mammals and *Drosophila*. Dev Dyn 243(2):201–215. doi:10.1002/dvdy.24036

Porter MM, Vandervoort AA, Lexell J (1995) Aging of human muscle: structure, function and adaptability. Scand J Med Sci Sports 5(3):129–142

Pratesi A, Tarantini F, Di Bari M (2013) Skeletal muscle: an endocrine organ. Clin Cases Mineral Bone Metab (the official journal of the Italian Society of Osteoporosis, Mineral Metabolism, and Skeletal Diseases) 10(1):11–14. doi:10.11138/ccmbm/2013.10.1.011

Raben N, Hill V, Shea L, Takikita S, Baum R, Mizushima N, Ralston E, Plotz P (2008) Suppression of autophagy in skeletal muscle uncovers the accumulation of ubiquitinated proteins and their potential role in muscle damage in Pompe disease. Hum Mol Genet 17(24):3897–3908. doi:10.1093/hmg/ddn292

Rana KS, Arif M, Hill EJ, Aldred S, Nagel DA, Nevill A, Randeva HS, Bailey CJ, Bellary S, Brown JE (2014) Plasma irisin levels predict telomere length in healthy adults. Age (Dordr) 36(2):995–1001. doi:10.1007/s11357-014-9620-9

Revilla S, Sunol C, Garcia-Mesa Y, Gimenez-Llort L, Sanfeliu C, Cristofol R (2014) Physical exercise improves synaptic dysfunction and recovers the loss of survival factors in 3xTg-AD mouse brain. Neuropharmacology 81:55–63. doi:10.1016/j.neuropharm.2014.01.037

Sandri M, Barberi L, Bijlsma AY, Blaauw B, Dyar KA, Milan G, Mammucari C, Meskers CG, Pallafacchina G, Paoli A, Pion D, Roceri M, Romanello V, Serrano AL, Toniolo L, Larsson L, Maier AB, Munoz-Canoves P, Musaro A, Pende M, Reggiani C, Rizzuto R, Schiaffino S (2013) Signalling pathways regulating muscle mass in ageing skeletal muscle: the role of the IGF1-Akt-mTOR-FoxO pathway. Biogerontology 14(3):303–323. doi:10.1007/s10522-013-9432-9

Sandri M, Sandri C, Gilbert A, Skurk C, Calabria E, Picard A, Walsh K, Schiaffino S, Lecker SH, Goldberg AL (2004) Foxo transcription factors induce the atrophy-related ubiquitin ligase atrogin-1 and cause skeletal muscle atrophy. Cell 117(3):399–412

Schiaffino S, Mammucari C (2011) Regulation of skeletal muscle growth by the IGF1-Akt/PKB pathway: insights from genetic models. Skelet Muscle 1(1):4. doi:10.1186/2044-5040-1-4

Shephard RJ, Balady GJ (1999) Exercise as cardiovascular therapy. Circulation 99(7):963–972

Sink H (2006) Muscle development in *Drosophila*. In: Landes R (ed) Molecular biology intelligence unit. Springer, New York (Landes Bioscience/Eurekah.com)

Takahashi A, Philpott DE, Miquel J (1970) Electron microscope studies on aging *Drosophila melanogaster*. 3. Flight muscle. J Gerontol 25(3):222–228

Thomason DB, Booth FW (1990) Atrophy of the soleus muscle by hindlimb unweighting. J Appl Physiol 68(1):1–12

Tinkerhess MJ, Healy L, Morgan M, Sujkowski A, Matthys E, Zheng L, Wessells RJ (2012) The *Drosophila* PGC-1alpha homolog spargel modulates the physiological effects of endurance exercise. PLoS ONE 7(2):e31633. doi:10.1371/journal.pone.0031633

Tomonaga M (1977) Histochemical and ultrastructural changes in senile human skeletal muscle. J Am Geriatr Soc 25(3):125–131

Tregear RT (2011) Physiology of insect flight muscle. In: Comprehensive physiology. Source: Supplement 27: Handbook of Physiology, Skeletal Muscle, pp 487–506

Yamada E, Bastie CC, Koga H, Wang Y, Cuervo AM, Pessin JE (2012) Mouse skeletal muscle fiber-type-specific macroautophagy and muscle wasting are regulated by a Fyn/STAT3/Vps34 signaling pathway. Cell Rep 1(5):557–569. doi:10.1016/j.celrep.2012.03.014

Yu Z, Li P, Zhang M, Hannink M, Stamler JS, Yan Z (2008) Fiber type-specific nitric oxide protects oxidative myofibers against cachectic stimuli. PLoS ONE 3(5):e2086. doi:10.1371/journal.pone.0002086

Van Roessel P, Hayward NM, Barros CS, Brand AH (2002) Two-color GFP imaging demonstrates cell-autonomy of GAL4-driven RNA interference in *Drosophila*. Genesis 34(1–2):170–173. doi:10.1002/gene.10146

Vandervoort AA (1994) Aging and muscle performance. Phys Ther 74(5):509

Vaughan RA, Gannon NP, Barberena MA, Garcia-Smith R, Bisoffi M, Mermier CM, Conn CA, Trujillo KA (2014) Characterization of the metabolic effects of irisin on skeletal muscle in vitro. Diab Obes Metab 16(8):711–718. doi:10.1111/dom.12268

Vernace VA, Arnaud L, Schmidt-Glenewinkel T, Figueiredo-Pereira ME (2007) Aging perturbs 26S proteasome assembly in Drosophila melanogaster. FASEB J (official publication of the Federation of American Societies for Experimental Biology) 21(11):2672–2682. doi:10.1096/fj.06-6751com

Vigoreaux JO (2006) Nature's versatile engine: insect flight muscle inside and out. In: Landes R (ed) Molecular biology intelligence unit. Springer, New York (Landes Bioscience/Eurekah.com)

Vrailas-Mortimer A, del Rivero T, Mukherjee S, Nag S, Gaitanidis A, Kadas D, Consoulas C, Duttaroy A, Sanyal S (2011) A muscle-specific p38 MAPK/Mef2/MnSOD pathway regulates stress, motor function, and life span in *Drosophila*. Dev Cell 21(4):783–795. doi:10.1016/j.devcel.2011.09.002

Wang Y, Pessin JE (2013) Mechanisms for fiber-type specificity of skeletal muscle atrophy. Curr Opin Clin Nutri Metab Care 16(3):243–250. doi:10.1097/MCO.0b013e328360272d

Webster GC, Beachell VT, Webster SL (1980) Differential decrease in protein synthesis by microsomes from aging *Drosophila melanogaster*. Exp Gerontol 15(5):495–497

Weed JL, Lane MA, Roth GS, Speer DL, Ingram DK (1997) Activity measures in rhesus monkeys on long-term calorie restriction. Physiol Behav 62(1):97–103

Wheeler JC, Bieschke ET, Tower J (1995) Muscle-specific expression of *Drosophila* hsp70 in response to aging and oxidative stress. Proc Natl Acad Sci USA 92(22):10408–10412

Wolkow CA (2006) Identifying factors that promote functional aging in *Caenorhabditis elegans*. Exp Gerontol 41(10):1001–1006. doi:10.1016/j.exger.2006.06.033

Zhao J, Brault JJ, Schild A, Cao P, Sandri M, Schiaffino S, Lecker SH, Goldberg AL (2007) FoxO3 coordinately activates protein degradation by thc autophagic/lysosomal and proteasomal pathways in atrophying muscle cells. Cell Metab 6(6):472–483. doi:10.1016/j.cmet.2007.11.004

Chapter 6
Drosophila Models of Cardiac Aging and Disease

Alyson Sujkowski and Robert Wessells

Abstract Cardiac dysfunction is a critical problem in the aging population, with heart disease the number one cause of death in humans. While much has been learned about the characteristics of the aging heart from human longitudinal and cohort studies, as well as from vertebrate animal models, substantial logistical obstacles exist to the study of these phenomena in long-lived animals or humans. The emergence of *Drosophila* as a short-lived model system for studying cardiac function across ages has thus been an important factor in boosting understanding of conserved changes during cardiac aging. Here we discuss established and emerging methodology for assessment of cardiac function in *Drosophila* and review conserved changes to function during normal aging that have been observed in flies. We also review genetic factors contributing to cardiac aging that have been identified and studied using these techniques, including genes involved in stress response, contractile function, ion exchange, and nutrient sensing. Further, we discuss the use of *Drosophila* to study longitudinal effects of environmental interventions, such as exercise, on cardiac function. Lastly, we compare transcriptional changes induced by various methods of longevity extension in *Drosophila* and point out common pathways induced by selective breeding, exercise and dietary restriction.

Keywords Cardiac function · Aging · *Drosophila* model · Stress response · Diet · Exercise · Contractile function

---

A. Sujkowski · R. Wessells (✉)
Department of Physiology, School of Medicine, Wayne State University, Detroit, MI 48201, USA
e-mail: rwessell@med.wayne.edu

A.M. Vaiserman et al. (eds.), *Life Extension*, Healthy Ageing and Longevity 3,
DOI 10.1007/978-3-319-18326-8_6

6.1 Introduction

The use of invertebrate models to understand the genetics of pattern formation during development has been one of the greatest success stories of the latter twentieth century. More recently, the same model systems have begun to be used to understand the genetic regulation of various physiological processes and how the functions of various organ systems are integrated. The problems of how organ function declines during normal aging, and how this process is regulated genetically, are questions where invertebrate models are particularly well suited for examination.

Some of the same advantages that invertebrates offer to developmental genetics are equally valuable for the study of progressive changes to organ function, including (1) large numbers, (2) short life span, and (3) extensive conservation of genetic regulatory factors. Until recently, the major obstacle to use of invertebrate models for functional research has been the lack of suitable assays to track organ performance in these smaller organisms. However, within the last decade, several useful methods for analyzing performance of cardiac function in *Drosophila* have been developed, making the fly model system (Fig. 6.1a) available for study of progressive effects of aging or disease on cardiac function (summarized in Table 6.1).

6.2 Methods for Analyzing Heart Performance in *Drosophila*

6.2.1 M-Mode

High-speed cameras applied to dissected, beating hearts in physiologically accurate solution capture video images of heart movements in two dimensions. These images can be processed to focus on a single slice of the heart, then make a sequential print of the edges of the heart in that single region across time up to a minute (Fink et al. 2009). Several important functional assessments can be calculated from these M-mode transcriptions, including heart period, rhythmicity, systolic and diastolic diameter, and fractional shortening (Fig. 6.1c).

6.2.2 Optical Coherence Tomography (OCT)

This ultrasound-like technology has been adapted and scaled for use on small organisms, such as flies (Wolf et al. 2006). This method provides sequential recordings of the size and shape of the heart cavity and has been effectively used to generate similar data as M-mode transcriptions from fully intact, unanesthetized flies (Fig. 6.1b). Recent adjustments to the technology now also allow 3D diagrams of the thickness of the heart wall itself along the full length of the fly as well as in cross-section (Li et al. 2013).

6.2.3 GFP Tracing

Stable integration of a cardiac-specific GFP-expressing element allows tracing of cardiac movements using brightness to identify the edges of the heart in two dimensions (Monnier et al. 2012). This allows measurement of rate, rhythm and diameter of the heart and does not require dissection (Fig. 6.1e).

6.2.4 Atomic Force Microscopy

This method allows detection of resistance to physical perturbation by monitoring the force displacement when the heart is indented with a cantilevered piece of plastic (Kaushik et al. 2011). This method allows the measurement of mechanical resistance and stiffness, an important factor in aging hearts.

6.2.5 Electrical Pacing

This method uses a small current to externally pace the heart to an increased heart rate. It serves as a measure of stress tolerance, an important early indicator of progressive heart dysfunction before overt pathology is manifested. This method has been used in two ways, (1) to measure the maximal heart rate that the animal can tolerate (Paternostro et al. 2001) and (2) to measure the changes in the tolerance to a single protocol during aging or genetic alteration (Fig. 6.1f) (Wessells and Bodmer 2004).

6.2.6 Bioinformatics

In addition to physical experimentation, databases have been created that catalog the total expressed proteome of the *Drosophila* heart (Cammarato et al. 2011) and the total transcriptome of the heart at multiple ages (Monnier et al. 2012).

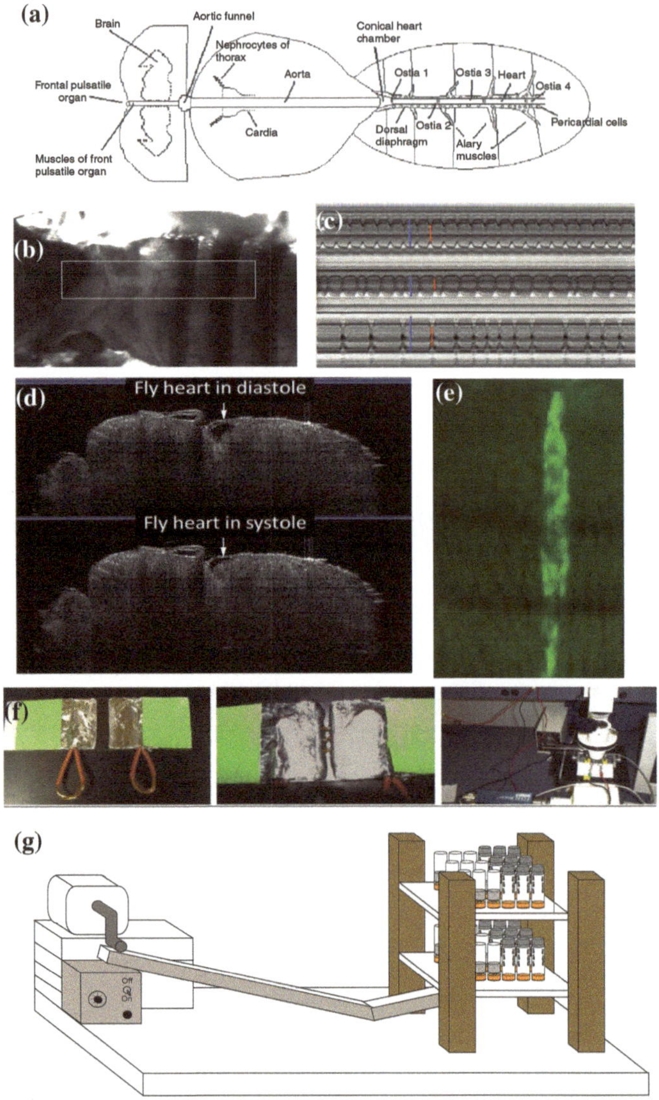

Fig. 6.1 Methods to assess cardiac function in *Drosophila*. **a** Structural anatomy of the adult *Drosophila* heart (Piazza and Wessells 2011). **b** Photograph of a live *Drosophila* on a bright field dissecting microscope. Located within the white box is a visible portion of the heart tube from which live images can be taken. **c** M-mode image constructed from sequential photographs of a beating heart to display contractile function over time in a linear fashion (Wessells et al. 2009). **d** Two OCT images of a live *Drosophila* heartbeat. Images are a kind gift from Gerald Dorn. **e** Fluorescent image of a live *Drosophila* heart made possible by stable integration of a heart-specific GFP driving construct (Wessells et al. 2004). **f** Electrical pacing apparatus used to pass an electrical current through the fly to temporarily increase heart rate. This assay is used to test cardiac stress tolerance. **g** The Power Tower apparatus takes advantage of the flies' natural instinct for negative geotaxis. Fly vials are repetitively dropped, knocking the flies to the bottom of their vial, causing them to rapidly climb. Repeating this process allows for (1) measurement of acute endurance or for (2) an automated daily exercise training regime (Tinkerhess et al. 2012a, b)

Table 6.1 *Drosophila* genes associated with adult cardiac function, with their respective human orthologs, and known associated human disease models

Drosophila gene	Human ortholog	Associated human disease/ disorder
Wingless *wg*	Wingless-type MMTV integration site family, member 1 *WNT1*	Ischemic and hypertrophic cardiomyopathies Reentrant ventricular tachycardia
Decapentaplegic *dpp*	Bone morphogenetic protein 2 *BMP2*	Congenital heart disease Mechanical failure of calcified heart valves
Notch *N*	notch 1 *NOTCH1*	Congenital heart disease Aortic valve disease
Hedgehog *hh*	Sonic hedgehog *shh*	Abnormal heart development
Abdominal A *abd-A*	Homeobox *HOX*	Congenital heart disease
Tinman *tin*	NK2 homeobox 5 *NKX2-5*	Congenital heart disease Atrial septal defect Hypoplastic left heart syndrome
Myocyte enhancer factor *Mef2*	Myocyte enhancer factor 2C *MEF2C*	Congenital heart disease Impaired cardiac remodeling
Ras oncogene *Ras85D*	Kirsten rat sarcoma viral oncogene homolog *KRAS*	Congenital heart disease Cardiofaciocutaneous syndrome
Rhomboid *rho*	Rhomboid, veinlet-like 2 *RHBDL2*	Enlarged cardiac chamber
Scalloped *sd*	TEA domain family member 1 *TEAD1*	Cardiomyopathy Heart failure
Actin 57B *Act57B*	Actin, beta *ACTB*	Cardiomyopathy, idiopathic dilated
Dystrophin *dys*	dystrophin *DMD*	Becker muscular dystrophy, cardiomyopathy, dilated, X-linked, duchenne muscular dystrophy
Heartless *htl*	Fibroblast growth factor receptor 1 *FGFR1*	Venous malformations
Band4.1 inhibitor LRP interactor *Bili*	FERM domain containing 8 *FRMD8*	Cavernous angiomatous malformations
wings up A *wupA*		Cardiomyopathy, familial hypertrophic, 3
Target of rapamycin *Tor*	Mechanistic target of rapamycin *MTOR*	Cardiac hypertrophy Distinctive ventricular electrophysiologic abnormality

(continued)

Table 6.1 (continued)

Drosophila gene	Human ortholog	Associated human disease/ disorder
Upheld *up*		Cardiomyopathy, familial hypertrophic, 2
Muscle LIM protein at 84B *Mlp84B*	cysteine and glycine-rich protein 2 *CSRP2*	
KCNQ *KCNQ*	KCNQ1 *KCNQ1*	Jervell and Lange–Nielsen syndrome
Mitochondrial assembly regulatory factor *Marf*	mitofusin 2 *MFN2*	
Myosin heavy chain *mhc*	Myosin, heavy chain 6 and 7 (alpha and beta) MYH6,7	Cardiomyopathy, familial hypertrophic, 1
Zipper *zip*	Myosin, heavy chain 11 *MYH11*	Cardiomyopathy, familial hypertrophic, 1
Myosin light chain cytoplasmic *Mlc-c*	Myosin, light chain 3 *MYL3*	Cardiomyopathy, hypertrophic, midventicular chamber type (3)
Spaghetti squash *sqh*	Myosin, light chain 2, regulatory, cardiac, slow *MYL2*	Cardiomyopathy, hypertrophic, midleft ventricular chamber type (3)
Bent *bt*	Titin *TTN*	Cardiomyopathy, familial hypertrophic, 9
Sallimus *sls*	Myosin binding protein C *MYBPC3*	Cardiomyopathy, familial hypertrophic, 4
Tropomyosin 1 *Tm1*	Tropomyosin 1 (alpha) *TPM1*	Cardiomyopathy, familial hypertrophic, 3
Actin 5C *Act5C*	Actin, alpha, cardiac muscle 1 ACTC1	Cardiomyopathy, idiopathic dilated
Actin 88F *Act88F*	Actin, alpha, cardiac muscle 1 ACTC1	Cardiomyopathy, idiopathic dilated
Lamin *Lam*	Lamin A/C *LMNA*	Cardiomyopathy (1), myopathy, desminopathic

(continued)

Table 6.1 (continued)

Drosophila gene	Human ortholog	Associated human disease/disorder
Seizure *sei*	Potassium voltage-gated channel, subfamily H (eag-related), member 2, 6, 6 *KCNH2,6,7*	Long QT syndrome-2
Paralytic *para*	Sodium channel, voltage-gated, type I, alpha subunit *SCN1A*	Long QT syndrome-3
Ventral nervous system defective *vnd*	NK2 homeobox 1-6,8 *NK2-1-6,8*	Artial septal defect with atrioventricular conduction defects
Ryanodine receptor *RyR*	ryanodine receptor 1-3 *RYR1-3*	Stress-induced polymorphic ventricular tachycardia Arrhythmogenic right ventricular dysplasia
Angiotensin converting enzyme *Ance*	Angiotensin I converting enzyme ACE	Myocardial infarction
G protein alpha o subunit *Galphao*	Guanine nucleotide binding protein (G protein), alpha activating activity polypeptide O *GNAO1*	Idiopathic ventricular arrhythmias
Acetyl Coenzyme A synthase *AcCoAS*	Acyl-CoA synthetase short-chain family member 1/Acyl-CoA synthetase short-chain family member 2 ACSS1/ACSS2	Hypertension, essential

More genes associated with cardiac function continue to be identified in this active field of research. More information can be found at www.Flybase.org and www.OMIM.org, in addition to further reviews (Piazza and Wessells 2011; Wolf and Rockman 2011; Qian and Bodmer 2012; Seyres et al. 2012; Whitaker et al. 2014)

6.2.7 Exercise

The development of an automated exercise machine for flies (Piazza et al. 2009a, b) has allowed heart performance to be measured under conditions of chronic exercise and to measure fatigue tolerance (Fig. 6.1g) (Tinkerhess et al. 2012a, b).

6.3 Phenotypic Changes During Normal Cardiac Heart Aging

Young flies exhibit a regular, highly rhythmic heartbeat, reported variously from 3–6 Hz, depending on measurement conditions (Wessells and Bodmer 2004; Wolf et al. 2006; Monnier et al. 2012). Flies undergo a progressive decline in most measures of function between one and five weeks of age, at which point the rate of decline levels off as flies enter the log phase of their survival curve (Wessells et al. 2004).

In resting hearts, the heart rate at steady temperature declines progressively with age (Wessells et al. 2004), as does the maximal heart rate to which flies can be stimulated by external electrical current (Paternostro et al. 2001). Fly hearts also exhibit a progressive deterioration of rhythmic rigor, with increasing episodes of arrhythmic beating or pauses (Ocorr et al. 2007a, b). These changes are reflected in measurements of calcium transients in semi-intact preparations (Santalla et al. 2014). Contractile strength also deteriorates, as reflected in reduced fractional shortening with age (Wolf et al. 2006).

Resistance to acute stress is also dramatically reduced with age. Flies aged to five weeks, for example, enter arrest or fibrillation at three times the rate of one-week old flies, when subjected to transient electrical pacing (Wessells et al. 2004). Aging hearts are also more sensitive to acute hypoxic stress and take longer to recover to a normal resting heart rate following a bout of hypoxia (Coquin et al. 2008).

These changes to resting function are reflected in anatomical degradation in the myocardium of aged flies. Electron microscopy of related *Drosophila* species reveals alterations to myofibril structure and mitochondrial degradation in older wild type flies (Burch et al. 1970). Staining actin filaments with fluorescent markers reveals the gradual disorganization of myofiber arrays with age, likely contributing to reduced contractility (Taghli-Lamallem et al. 2008). In addition, the physical stiffness of fibers gradually increases with age (Kaushik et al. 2011), probably also contributing to lack of elasticity during contraction and relaxation.

6.4 Role of Stress Response Genes in Cardiac Aging

A decreased ability to tolerate stress is a major component of cardiac aging and aging in general. In this section, we will survey the genetic models that have been utilized to model various stress responses in the fly heart, and report findings of relevance to vertebrate research.

6.4.1 Oxidative Stress

Oxidative stress has been proposed to be a conserved mechanism contributing to heart failure in aging populations. In the fly model, transcripts related to oxidative stress resistance, including the JNK signaling pathway, are upregulated in the heart with age, suggesting that heart tissue may indeed be increasingly subjected to oxidative stress during aging (Monnier et al. 2012).

In support of this proposed mechanism, flies mutant for the reactive oxygen species scavenging protein Superoxide Dismutase 2 (SOD2) have drastically reduced fractional shortening, rhythmicity and pacing tolerance (Piazza et al. 2009a, b). Conversely, cardiac overexpression of *MafS*, a rate-limiting cofactor for the fly homolog of the oxidative stress responsive transcription factor Nrf2, leads to protection against age-related increases in arrhythmias and heart period (Rahman et al. 2013).

Taken together, these observations suggest that increasing oxidative stress and/or decreasing tolerance for oxidative stress are likely to be significant contributing factors to late-stage cardiac aging.

6.4.2 Stress-Induced Hypertrophy

Chronic stress can result in stress-induced cardiac hypertrophy in vertebrates. A fly model for this process has recently been described, generating increased wall thickness, abnormal fiber morphology and decreased end diastolic lumen dimensions (Yu et al. 2013). These phenotypes were shown to be entirely dependent on the presence of activated Raf in the cardiomyocytes, and activation of Raf in cardiac tissue was sufficient to produce the hypertrophy-like state.

6.4.3 Proteostatic Control

Maintenance of protein quality in the cell has been proposed to be a driving factor in the aging process. Recent evidence supports the role of protein quality control in *Drosophila* cardiac aging.

Age-associated atrial fibrillation is an important consequence of normal aging in humans. The remodeling events associated with atrial fibrillation are thought to involve loss of proteostasis. Consequently, gene products that promote effective proteostasis could ameliorate some of the consequences of atrial fibrillation-induced remodeling. Indeed, overexpression of Hsp proteins has been shown to have protective effects in this context in humans (Brundel et al. 2006).

Using a fly model for atrial fibrillation caused by tachypacing (Zhang et al. 2011), it was shown that induction of fly Hsp proteins by either hormetic treatments or by transgene expression could reduce arrhythmias caused by aging or tachypacing (Zhang et al. 2011).

6.5 Role of Developmental Genes in Cardiac Aging

A common theme in adult cardiac dysfunction is the reiteration of developmental programs to generate pathological states of postnatal growth, or hypertrophy. The fly system has been instrumental in elucidating the role of several important developmental pathways in maintenance of adult cardiac function over time. For a review, see (Qian and Bodmer 2012).

Several genes necessary for cardiac specification have also been demonstrated to be necessary for function in adult flies. The Nkx2.5 homolog *tinman*, for example, plays a key role, as cardiac-specific disruption of the *tin* product causes disruption of the heart period and deficits in heart size (Zaffran et al. 2006). Likewise, adult-specific disruption of the GATA4 homolog *pannier* in flies causes increased arrhythmias and decreased tolerance of pacing stress (Qian and Bodmer 2009). Similar phenotypes are seen in hearts with adult-specific disruption of the T-box factor *neuromancer*, which also exhibit increased arrhythmias, decreased tolerance of external pacing, reduced heart rate, and progressive structural disruption of myofibrils (Qian and Bodmer 2009).

These genes seem to work cooperatively, as strong genetic interactions were seen between trans-heterozygotes of *tinman* and *neuromancer* (Qian and Bodmer 2009), as well as *tinman* and the *Drosophila Cdc42* (Qian et al. 2011). These phenotypes are likely to be relevant to normal aging, since chronic overexpression of either *tinman* or *neuromancer* offers significant protection against age-related decline in heart rate and stress resistance.

Embryonic induction of cardiac specification genes such as *tinman* is dependent on proper regulation of Wnt signaling (Venkatesh et al. 2000). In adult flies, cardiac knockdowns of several components of the Wnt signaling pathway have mild cardiac defects. By contrast, the transcriptional cofactor and Wnt signaling pathway component *Pygopus* is essential for maintenance of adult cardiac performance (Tang et al. 2013). *Pygopus* is specifically expressed in adult heart cells, and adult-specific knockdowns cause progressive functional and structural defects in the myocardium (Tang et al. 2013). The *Pygopus* adult phenotype, perhaps surprisingly, is not dependent on interactions with Wnt signaling, however, but on Wnt-independent interactions with Calcium/Calmodulin-Dependent Protein Kinase 2 (CaMKII) (Tang et al. 2013).

The Wnt-antagonizing transcription factor *Sox102F* is a functional homolog of the vertebrate SOX5. Cardiac knockdowns of *Sox102F* result in increased arrhythmias, increased thickness of cardiac walls, leading to a restricted cardiomyopathy type phenotype (Li et al. 2013). Anatomical disruptions are also seen in these knockdowns with disruptions in both myofibril array and mitochondrial integrity (Li et al. 2013). This phenotypic array may reflect the results of deregulated Wnt signaling on adult function.

In addition to modeling the broad impact of transcription factors and stress response on aging, flies have also been used to dissect more specific gene interactions affecting subsets of the characteristic age-related declines in cardiac function. In the next sections, we will discuss the roles of ion channels in regulating progressive dysrhythmia and the role of contractile proteins in maintaining functional integrity.

6.6 Role of Ion Channels in Regulating Dysrhythmias

The *Drosophila KCNQ* gene encodes the alpha subunit of a potassium channel responsible for the slow repolarizing current in cardiac muscle (Ocorr et al. 2007a, b). Cardiac expression of *KCNQ* declines progressively with age, suggesting a role in normal age-related decline. Both *KCNQ* mutant flies and flies with cardiac-specific knockdown of *KCNQ* exhibit arrhythmias that worsen progressively, while cardiac overexpression of *KCNQ* is sufficient to rescue the mutant phenotype. Overexpression in otherwise wild-type flies can also protect hearts from age-related increases in arrhythmias and tolerance for external pacing (Ocorr et al. 2007a, b).

Another critical channel-encoding gene is *dSur*. *dSur* is highly expressed in the fly myocardium under the direction of Nkx2.5 and GATA factor homologs, and expression has been observed to decline progressively with age, concurrent with increased arrhythmias (Akasaka et al. 2006). *dSur* encodes an essential component of a K^{ATP} channel, and knockdowns have defects in rhythmic control, tolerance of pacing stress and resistance to pacing stress (Akasaka et al. 2006). These defects are similar to a subset of defects seen in normal aging.

Taken together, these observations indicate that age-related changes in expression and availability of channel proteins play an important role in loss of rhythmic homeostasis during aging.

6.7 Role of Contractile Proteins in Maintaining Functional Heart Integrity

Mutations in structural proteins that are directly involved in forming the contractile machinery have been associated with dystrophic phenotypes in both skeletal and cardiac muscle (Spletter and Schnorrer 2014). Similar phenotypes have been observed in several fly models for dystrophies, and these progressive phenotypes may also serve as models for gradual dysfunction with age.

For example, the fly homolog of the dystrophin gene, *Dys*, has been causally associated with dilated cardiomyopathy in flies (Taghli-Lamallem et al. 2008). These phenotypes progressively worsen with age, as they do in humans (Corrado et al. 2002). Dys proteins have two conserved roles, serving as mechanical links to a complex that acts to stabilize the sarcolemma, and also acting as a signaling molecule that modulates the activity of multiple proteins, including nNos (Brenman et al. 1996).

A recent study takes advantage of prior work on the structure/function of vertebrate dystrophin to express variant constructs of the human Dys in the fly heart and break down the relative importance of these two roles in aging heart function. Transgenes were created that express forms of *Dys* that lack either the mechanical role or the signaling role. Both forms of *Dys* were able to provide substantial, though incomplete, rescue of systolic diameter and contractility, suggesting that both roles are important and required for full function (Taghli-Lamallem et al. 2014). In addition, Nos was identified as an important target of Dys signaling in flies, as *Nos* gain of function had protective effects in Dys signaling-impaired hearts (Taghli-Lamallem et al. 2014).

As a side note, these experiments, as well as others (e.g. Grossman et al. 2011), validate the approach of expressing vertebrate constructs directly in the fly heart to assess the phenotypic consequences of different protein variants.

Flies have also been used to detail the effects of mutations to the Myosin Heavy Chain (MHC), where the ATPase activity of MHC was found to be a key factor in predicting the consequences of specific mutations. Lesions that increase ATPase activity of MHC cause restrictive cardiomyopathy, whereas mutations that decrease ATPase activity cause dilated cardiomyopathy (Cammarato et al. 2008).

A series of deletion mutants in delta-sarcoglycan have also been examined in flies. Heart function was quite tolerant of smaller deletions in the lengthy coding region, but larger deletions caused dilated cardiomyopathy and poor fractional shortening, along with progressive motor impairment, suggestive of accompanying somatic muscle phenotypes (Allikian et al. 2007). In another example, mutation to the tropomyosin-binding region of the fly homolog of Troponin T was observed to cause prolonged systole, impaired relaxation and increased myocardial stiffness (Viswanathan et al. 2014). These phenotypes were dependent on aberrant localization of tropomyosin, as verified by genetics and electron microscopy.

The Muscle Lim Protein (MLP) homolog Mlp84B was shown to localize to the Z-disc of cardiac sarcomeres in flies (Mery et al. 2008), as it does in vertebrates (Arber et al. 1997). Mutant hearts or hearts with cardiac-specific knockdown of *Mlp84B* showed arrhythmias and reduced heart rate, leading to a reduced life span. However, the fly mutant did not show the sarcomeric disarray shown in the mouse model, suggesting an alternative mechanism for LIM proteins in the development of cardiomyopathy (Mery et al. 2008).

Taken together, these results emphasize the highly conserved nature of the contractile machinery and the value of invertebrate research in investigating complex interactions between structural proteins during the aging process.

6.8 Calcium Signaling and Cardiac Disorders

Dysregulation of calcium signaling or calcium storage is a component of the etiology of several cardiac disorders. For example, constitutive activation of the calcium-dependent phosphatase calcineurin induces pathological hypertrophy in vertebrates (Molkentin et al. 1998; Wilkins and Molkentin 2002). Conversely, inhibition of calcineurin suppresses induction of hypertrophy (Sussman et al. 1998; Taigen et al. 2000).

Recently, similar constitutively active calcineurin constructs have been employed to model this effect in the fly heart, where activated calcineurin led to enlargement of heart walls and reduced fractional shortening (Lee et al. 2014a, b). Excitingly, the fly model was employed to identify novel modifiers of this effect. Knockdown of the galactokinase gene in the myocardium was sufficient to rescue enlargement of the heart wall and fractional shortening in the presence of activated calcineurin (Lee et al. 2014a, b).

Presenilin is an important factor in the etiology of Alzheimer's and age-related cardiac dysfunction in humans. Knockdown of the fly homolog of presenilin in the heart caused a prematurely reduced heart rate, reduced expression of *sarco/endoplasmic reticulum calcium-ATPase* (*SERCA*), and reduced symptoms of dilated cardiomyopathy, indicating that Presenilin may have a conserved role in the onset of age-related dilated cardiomyopathy (Li et al. 2011). Another study took advantage of a temperature sensitive mutation in *SERCA* to show that inactivation of *SERCA* causes reduction of heart rate with no effect on fractional shortening (Abraham and Wolf 2013). Cardiac-specific expression of a mutant form of *SERCA* instead caused dilation of the heart tube and reduced fractional shortening, with no effect on heart rate. These phenotypes were dependent on calcium flux, as they could be rescued by a mutation in the ryanodine receptor homolog (Abraham and Wolf 2013). Interestingly, overexpression of fly *SERCA* was capable of rescuing a fly model of cardiomyopathy, lending further credence to the fly as an inexpensive model for testing potential interventions.

6.9 Degenerative Disease Models

Degenerative diseases, such as Huntington's disease or Parkinson's disease, often manifest cardiac pathologies, which in many cases, are the proximate cause of death in patients (Ziemssen and Reichmann 2010; Jain and Goldstein 2012; Abildtrup and Shattock 2013). Therefore, modeling the development of cardiac dysfunction in patients with progressive diseases is an important research priority. For example, Huntington's disease causes inclusion bodies to form in cardiac tissue as well as neuronal tissue (Gunawardena and Goldstein 2005; Abildtrup and Shattock 2013). Inclusion bodies are thought to result from aggregation of proteins with Poly-glutamine sequences of various lengths (Finkbeiner 2011; Arrasate and Finkbeiner 2012).

Recently, ectopic Poly-Q sequences were expressed in fly hearts and shown to cause increased arrhythmias, cardiac dilation, myofibril disorganization, mitochondrial fragmentation and consequent decreased fractional shortening (Melkani et al. 2013). The severity of these phenotypes was dependent on the length of the Poly-Q sequences in the expressed construct. The mechanism by which aggregations cause dysfunction was explored genetically and pharmacologically.

In support of an oxidative stress-based mechanism, co-expression of antioxidant SOD proteins or feeding with the antioxidant resveratrol was able to partially rescue cardiac function (Melkani et al. 2013). In support of a protein folding/endoplasmic reticulum ER stress-based mechanism, co-expression of the chaperone Unc45 was also able to partially rescue (Melkani et al. 2013). Taken together, these results indicate that both oxidative stress and protein misfolding are likely to play a role in development of cardiac dysfunction in Huntington's hearts, and therapies aimed at either mechanism have palliative potential.

The role of the ubiquitin ligase Parkin in wild type or diseased hearts has been ambiguous, due to compensatory upregulation of other ubiquitin ligases in the hearts of vertebrate knockdown models. Due to reduced complexity in the fly genome, the cardiac role of Parkin was recently studied in the fly heart model. Heart-specific knockdown of *dparkin* caused the formation of dysmorphic mitochondria with abnormal depolarization. This, in turn, led to increased reactive oxygen species ROS in the myocardium and reduced fractional shortening (Bhandari et al. 2014). Cardiac dysfunction in this model was ROS-dependent, as co-expression of SOD proteins or RNAi against Reactive Oxygen Species Modulator One *ROMO1* was able to rescue function (Bhandari et al. 2014). Interestingly, the same study found that inhibition of mitochondrial fusion by blocking an ortholog of mitofusin was able to rescue the cardiac phenotype, suggesting that the effects of reduced mitochondrial clearance can be ameliorated by preventing damaged mitochondria from fusing with undamaged ones (Bhandari et al. 2014).

These results highlight the utility of the *Drosophila* model for the study of degenerative cardiac phenotypes across ages. The next section will examine recent findings related to lipid metabolism in the fly heart, and its relationship to environmental factors such as diet and exercise.

6.10 Lipid Metabolism

Lipodystrophies are an important factor in cardiac performance across species. A screen for mutants that alter the profile of aging heart performance identified two loci encoding genes involved in lipid metabolism that strongly regulate adult cardiac performance, *easily shocked* (*eas*) and the fatty acid transporter *dFatp*.

Eas encodes an enzyme necessary for biosynthesis of phosphatdiylethanolamine (PE), which in turn induces biosynthesis of Sterol regulatory element binding protein (SREBP), a conserved lipogenic transcription factor. Flies mutant for *eas* accumulate SREBP in the nucleus of cardiomyocytes, leading to the activation of lipogenic genes (Lim et al. 2011). Either *eas* mutants or flies overexpressing *SREBP* in the heart show elevated triglycerides in the heart, resulting in arrhythmias, relaxation defects, increased sensitivity to pacing stress and increased heart rate (Lim et al. 2011). The cardiac phenotype of mutants can be completely rescued by interfering downstream with lipid biosynthesis, either by knocking down *SREBP* itself or its targets *acetyl-coA carboxylase* (*acc*) or *fatty acid synthase* (*fas*) (Lim et al. 2011), demonstrating that the defects are completely dependent on the cardiac steatosis.

dFatp encodes a conserved fatty acid transporter protein with extensive homology to the vertebrate Fatp family. Although homozygous null mutants for *dFatp* are lethal, heterozygotes display accumulations of triglycerides in haemolymph and in cardiac muscle (Sujkowski et al. 2012). As a result, heterozygous hearts have defects very similar to those seen in *eas* mutants, with increased heart rate, relaxation defects, and decreased tolerance for pacing stress (Sujkowski et al. 2012). This phenotype is completely dependent on triglyceride accumulation, as interventions eliminating steatosis such as lipase overexpression or endurance exercise are capable of rescuing the cardiac phenotype (Sujkowski et al. 2012).

Interestingly, both of these mutations act at the level of lipid processing and have effects that are tissue-specific and diet-independent, raising the possibility of interventions downstream of diet that can ameliorate the effects of high-fat diets. Fly models for the effects of diet and exercise modulation will be discussed below in a separate section.

In addition to the direct role of lipid accumulation in the heart on aging, there also exist important signaling roles of lipid processing and/or biosynthesis with critical effects on aging heart function. For example, the effects of the drug family known as statins on maintenance of cardiac function during aging are an important subject of current research and clinical guidelines. In particular, the possible cholesterol-independent benefits of statins for cardiac function are a topic of great interest. Since flies are auxotrophic for cholesterol, they make an excellent model to test cholesterol-independent functions of statins.

In addition to impairing cholesterol biosynthesis, statins impair production of isoprenoids, which are required for protein farnesylation. Farnesylation, in turn, is necessary for various small molecules to be anchored to cell membranes, including important signaling molecules such as Ras GTPases. The altered localization of signaling GTPases has been proposed to account for cholesterol-independent protective effects of statins in vertebrates (Liao and Laufs 2005).

Simvistatin has recently been demonstrated to extend the life span of flies in a mechanism dependent on prenylation of Ras and Rab4 (Spindler et al. 2012). Importantly, this intervention also decreases arrhythmias in aging flies, suggesting a possible conserved role for prenylation in maintenance of cardiac function during normal aging.

On the other side of the equation from metabolism of dietary constituents are the downstream signaling effects that occur when nutrient intake is sensed by the animal. These nutrient-sensing pathways and their exciting therapeutic potential for aging hearts are discussed in the following section.

6.11 Nutrient-Sensing Pathways

Reduction in signaling through nutrient-sensing pathways has been well established as a method of life span extension in model organisms (Greer et al. 2008; Greer and Brunet 2009; Tatar et al. 2014). However, the mechanisms by which these pathways act at the level of organ function is less well understood. In flies, mutations that reduce signaling through the insulin receptor or reduce the secretion of insulin-like peptide protect cardiac function to very advanced ages, as measured by heart rate and resistance to pacing stress (Wessells et al. 2004). Importantly, the heart itself is an important target tissue for these signaling events, since heart-specific knockdown of insulin signaling also protects against cardiac aging, to the same extent as genomic mutation (Wessells et al. 2004). This protection is dependent on the conserved transcription factor *dFoxo*, as manipulation of *dFoxo* expression levels is sufficient to drive the protective phenotype.

The related and interdependent Target of Rapamycin (TOR) kinase signaling pathway (Chong et al. 2011, O' Neill 2013) also plays an important role in cardiac aging. Viable TOR transheterozygotes are protected against age-related decrease in heart rate and resistance to pacing stress (Luong et al. 2006). This phenotype, too, is organ-autonomous, as heart-specific expression of *tuberous sclerosis complex 1* and *2* (*TSC1* and *TSC2*), which act together to reduce TOR activity, provides protection to a degree similar to TOR transheterozygotes. Conversely, cardiac overexpression of *TOR* itself hastens age-related decline (Wessells et al. 2009).

Interestingly, the cardioprotection provided by these knockdowns is not dependent on changes to circulating glucose levels, since *TOR* knockdowns have reduced circulating glucose (Luong et al. 2006), while insulin signaling knockdowns have increased glucose levels (Clancy et al. 2001; Tatar et al. 2001).

Instead, the key downstream factor seems to be the regulation of cap-dependent translation by the eukaryotic initiation factor binding protein d4eBP. 4eBP is a conserved protein, regulated at the transcriptional level by Foxo-dependent insulin signaling, and at the post-transcriptional level by TOR signaling (Grewal 2009). Cardiac overexpression of *d4eBP* is sufficient to protect against age-related decline in heart rate, rhythmicity, and resistance to pacing stress (Wessells et al. 2009). Furthermore, *d4eBP* appears to be fully epistatic to TOR in this process, as co-overexpression of *Tor* and *d4eBP* protects to the same extent as expressing *d4eBP* alone. Co-overexpression of *dfoxo* with *d4eBP* also protects to the same extent as *d4eBP* alone, indicating that Foxo and 4eBP have no additive effect in this process.

An important modulator of these pathways is the TOR antagonist Sestrin (Sesn). *dSesn* is both a target and a feedback regulator of TOR signaling. Mutants display several phenotypes associated with overactive TOR, including increased lipid accumulation, and decreased cardiac performance (Lee et al. 2010). *dSesn* mutant hearts have increased arrhythmias, decreased heart rate and expansion of the diastolic period, similar to the dilated cardiomyopathy phenotype seen in *TOR*-overexpressing hearts. These phenotypes can be rescued by administration of the TOR inhibitor rapamycin or by the AMP-activated protein kinase (AMPK) activator AICAR, suggesting that dSestrin is dependent on both TOR and AMPK for its effects (Lee et al. 2010).

The cardiac phenotype of *dSesn* mutants is similar to that of cardiac knockdowns of essential autophagy genes, suggesting that TOR-mediated autophagy may be a critical intermediate in the Sesn-mediated control of cardiac aging. Upregulation of AMPK and autophagy are also characteristics of endurance exercise and caloric restriction, raising the idea that Sestrin may be an important link between these processes. In the next section, we discuss data using the fly model to examine external interventions in the diet or exercise environment and their effect on long-term cardiac function.

6.12 Diet and Exercise

Laboratory flies are typically fed a very simple diet, composed of a sugar source (sucrose or molasses/corn meal) and yeast (either killed or live baker's yeast). This simplicity offers some advantages for calculating the effects of varying relative amounts of these components in diet on long-term performance. Multiple groups have observed that the relative composition of the diet has a bigger impact on life span and other markers of aging than does the total amount of calories ingested (Lee et al. 2008; Skorupa et al. 2008; Simpson and Raubenheimer 2009).

Cardiac performance is particularly sensitive to dietary composition in flies. When a high-fat dietary preparation using oil as the primary food source was fed to wild-type flies, they exhibited dramatic increases in whole-body triglycerides and accumulated deposits of lipid in the myocardium (Birse et al. 2010). The steatosis seen in these hearts led, in turn, to increased arrhythmias, decreased heart rate and decreased fractional shortening. Such hearts had increased expression of *fas* and decreased expression of the triglyceride lipase *brummer*. This process was dependent on increased signaling through TOR kinase, as heart specific reduction of TOR signaling was able to rescue the effect (Birse et al. 2010).

It is noteworthy that the reduced fractional shortening and increased arrhythmias caused by this diet are similar to the phenotypes seen in mutants with increased myocardial lipid deposits, such as *eas* and *dFatp* (see above).

In an attempt to model acquired diabetic cardiomyopathy, high sugar diets have also been formulated and fed to adult flies. Flies on such diets have high circulating sugars and exhibit progressive insulin resistance, as measured by phosporylation levels of insulin targets (Na et al. 2013). Hearts in these flies exhibit progressively worsening arrhythmias and decreased fractional shortening, without changes to heart rate. An increase in pericardin expression in heart tissue was also observed, suggesting the development of a fibrosis like phenotype. Like flies on a high-fat diet, flies on the high-sugar formulation also accumulated triglyceride deposits in the myocardium (Na et al. 2013).

When sugar and yeast (source of lipid and protein) were varied inversely in a matrix of different dietary components, it was found that cardiac performance was strongly responsive to more subtle variations in diet as well. As was found for life span, cardiac performance was more dependent on ratio of dietary components than on total caloric intake (Bazzell et al. 2013). Fractional shortening and resistance to pacing stress were best on low-calorie diets with a balanced amount of sugar and yeast, while the diets that promoted the worst cardiac function were those with high sugar/yeast ratios. In this context, the requirements for running speed seem to be different than those for cardiac performance, as high sugar diets promote increased performance in speed assays, although not in endurance assays (Bazzell et al. 2013).

Although mutants that reduce insulin signaling are protective against cardiac aging on normal diets (see above), mutants in the IRS homolog *chico* enhance the cardiac dysfunction of flies on a high-sugar diet (Na et al. 2013). This highlights the importance of testing model organisms under multiple conditions to get a clear picture of a mutation's effects. In addition to diet, varying exercise conditions can have an enormous impact on the expression of mutant phenotypes.

Wild-type flies that complete a three-week endurance training program show delayed age-related decline in fractional shortening (Tinkerhess et al. 2012a, b), resistance to pacing stress (Piazza et al. 2009a, b), maintenance of myofibril integrity (Bazzell et al. 2013), and mitochondrial integrity (Laker et al. 2014). As in vertebrates (Ehsani et al. 2003; Evans et al. 2005), these improvements to healthspan do not cause a significant increase in maximal life span (Sujkowski et al. 2012). However, there is likely to be substantial overlap in mechanism between exercise and life span-extending interventions. In support of this idea, recently performed microarrays compared changes in gene expression following exercise to changes in gene expression caused by selective breeding for longevity. About 70 % of gene expression changes found in long-lived flies were also found in wild-type flies after three weeks of endurance exercise (Fig. 6.2a, b) (Sujkowski and Wessells, unpublished).

Invertebrate exercise seems to work through conserved pathways. For example, the fly *spargel*, a homolog of the well-studied vertebrate exercise response gene peroxisome proliferator-activated receptor-gamma coactivator 1 alpha (PGC-1 alpha), is also necessary for flies to receive the full benefits of exercise (Tinkerhess et al. 2012a, b). Conversely, *spargel* overexpression in somatic and cardiac muscle is sufficient to improve both endurance and cardiac resistance to pacing stress.

Exercise induces some of the same downstream genes as nutrient-signaling pathways, including *sestrin* and *d4eBP* (Lee and Wessells, unpublished), suggesting that further study of the exercise response in flies may shed important light on mechanisms of cardiac healthspan. Furthermore, exercise can restore normal function to mutant flies with steatosis phenotypes (Sujkowski et al. 2012), emphasizing again the synergistic importance of diet and exercise to stimulate pro-healthspan genetic pathways.

Exercise is a powerful stimulator of mitochondrial biogenesis in cardiac tissue in flies (Tinkerhess et al. 2012a, b; Laker et al. 2014). Using a marker of mitochondrial stress, it has been observed that older flies that have been exposed to chronic exercise have a higher percentage of healthy mitochondria and fewer that show signs of exposure to oxidative stress (Laker et al. 2014). It is probable that this is a result of directed mitophagy in conjunction with biogenesis to rejuvenate the mitochondrial population in cardiac muscle.

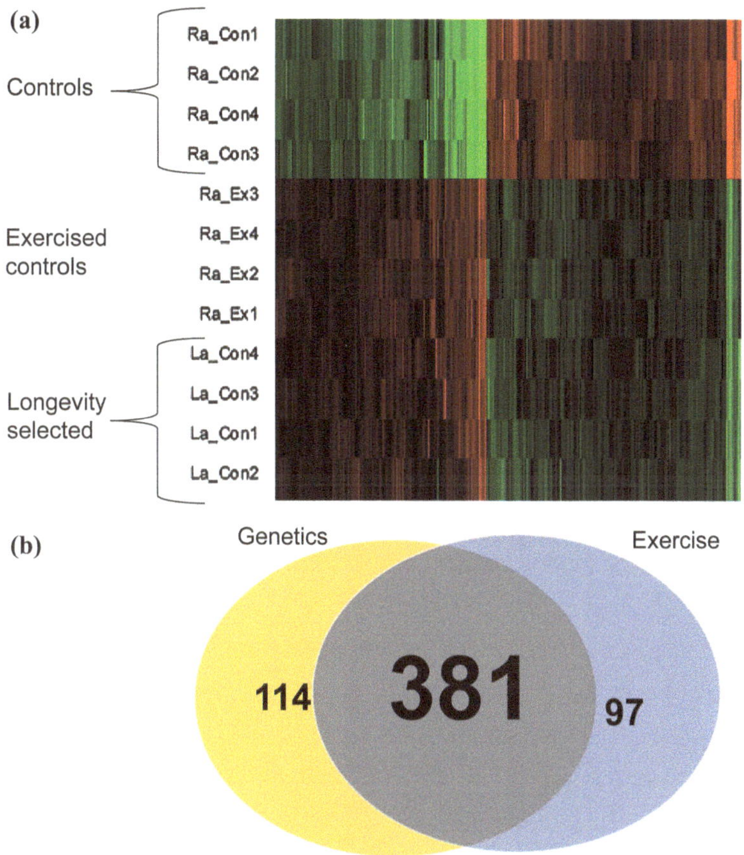

Fig. 6.2 Endurance exercise causes transcriptional changes similar to those caused by selective breeding for longevity. **a** Heat map of control, exercised control, and longevity-selected flies. After three weeks of daily exercise, exercised flies have transcriptional profiles more like long-lived flies. **b** Venn diagram showing overlap in transcriptional changes induced by selective breeding for longevity and exercise

The induction of secreted factors from cardiac and skeletal muscle by diet or exercise is an important emerging topic in metabolic research. Flies have been an important model for discovery of such factors. For example, a fly screen for factors mediating the protective role of reduced insulin signaling uncovered the TGF-beta ligand *dawdle* as an important downstream factor whose expression in muscle is sufficient to increase muscle autophagy and life span (Bai et al. 2013). Whether *dawdle* secretion is activated by exercise in addition to reduced insulin signaling is a topic of ongoing study.

In support of this potential link, it has been demonstrated that perturbation of the mitochondrial electron transport chain (ETC) in *Drosophila* muscle induces a mitohormesis effect that extends life span and muscle performance (Owusu-Ansah et al. 2013). The beneficial effects of this treatment are dependent on induction of the mitochondrial unfolded protein response, as has also been observed for life span-extending perturbations of the ETC in *C. elegans* (Durieux et al. 2011). As endurance exercise is known to generate transient oxidative stress in *Drosophila* muscle mitochondria (Laker et al. 2014), this supports a model where endurance exercise can protect muscle healthspan through a mitohormesis mechanism.

Muscle-specific factors have also recently been shown to be critical for susceptibility to obesity in flies. Cardiac overexpression of the transcriptional regulatory proteins Mediator 12 and Mediator 13 plays a protective role against the development of obesity in mice (Grueter et al. 2012). Using the fly model, it has recently been discovered that this effect is mediated by induction of Wnt signaling through the fly gene *wingless*, and that activation of Wnt signaling in muscle leads to both tissue-autonomous Wg pathway activity and secretion of the Wg protein, which can potentially be sensed in adipose tissue (Lee et al. 2014a, b).

The advantages of the fly model for combining genetic manipulation with longitudinal physiology in large numbers position the fly heart model as an important system for analyzing genetics, diet and exercise on aging and disease in ever greater detail in the years to come.

Acknowledgments We thank Gerald Dorn for use of unpublished images, and Rolf Bodmer and Leah Cannon for helpful and generous discussion prior to publication.

References

Abildtrup M, Shattock M (2013) Cardiac dysautonomia in Huntington's disease. J Huntingtons Dis 2(3):251–261

Abraham DM, Wolf MJ (2013) Disruption of sarcoendoplasmic reticulum calcium ATPase function in *Drosophila* leads to cardiac dysfunction. PLoS ONE 8(10):e77785

Akasaka TS, Klinedinst K, Ocorr et al (2006) The ATP-sensitive potassium (KATP) channel-encoded dSUR gene is required for *Drosophila* heart function and is regulated by tinman. Proc Natl Acad Sci U S A 103(32):11999–12004

Allikian MJ, Bhabha G, Dospoy P et al (2007) Reduced life span with heart and muscle dysfunction in *Drosophila* sarcoglycan mutants. Hum Mol Genet 16(23):2933–2943

Arber S, Hunter JJ, Ross J Jr et al (1997) MLP-deficient mice exhibit a disruption of cardiac cytoarchitectural organization, dilated cardiomyopathy, and heart failure. Cell 88(3):393–403

Arrasate M, Finkbeiner S (2012) Protein aggregates in Huntington's disease. Exp Neurol 238(1):1–11

Bai H, Kang P, Hernandez AM et al (2013) Activin signaling targeted by insulin/dFOXO regulates aging and muscle proteostasis in *Drosophila*. PLoS Genet 9(11):e1003941

Bazzell B, Ginzberg S, Healy L et al (2013) Dietary composition regulates *Drosophila* mobility and cardiac physiology. J Exp Biol 216(Pt 5):859–868

Bhandari P, Song M, Chen Y et al (2014) Mitochondrial contagion induced by Parkin deficiency in *Drosophila* hearts and its containment by suppressing mitofusin. Circ Res 114(2):257–265

Birse RT, Choi J, Reardon K et al (2010) High-fat-diet-induced obesity and heart dysfunction are regulated by the TOR pathway in *Drosophila*. Cell Metab 12(5):533–544

Brenman JE, Chao DS, Gee SH et al (1996) Interaction of nitric oxide synthase with the postsynaptic density protein PSD-95 and alpha1-syntrophin mediated by PDZ domains. Cell 84(5):757–767

Brundel BJ, Henning RH, Ke L et al (2006) Heat shock protein upregulation protects against pacing-induced myolysis in HL-1 atrial myocytes and in human atrial fibrillation. J Mol Cell Cardiol 41(3):555–562
Burch GE, Sohal RS, Fairbanks LD (1970) Senescent changes in the heart of *Drosophila* repleta Wollaston. Nature 225(5229):286–288
Cammarato A, Ahrens CH, Alayari NN et al (2011) A mighty small heart: the cardiac proteome of adult *Drosophila melanogaster*. PLoS ONE 6(4):e18497
Cammarato A, Dambacher CM, Knowles AF et al (2008) Myosin transducer mutations differentially affect motor function, myofibril structure, and the performance of skeletal and cardiac muscles. Mol Biol Cell 19(2):553–562
Chong ZZ, Shang YC, Maiese K (2011) Cardiovascular disease and mTOR signaling. Trends Cardiovasc Med 21(5):151–155
Clancy DJ, Gems D, Harshman LG et al (2001) Extension of life-span by loss of CHICO, a *Drosophila* insulin receptor substrate protein. Science 292(5514):104–106
Coquin L, Feala JD, McCulloch AD et al (2008) Metabolomic and flux-balance analysis of age-related decline of hypoxia tolerance in *Drosophila* muscle tissue. Mol Syst Biol 4:233
Corrado G, Lissoni A, Beretta S et al (2002) Prognostic value of electrocardiograms, ventricular late potentials, ventricular arrhythmias, and left ventricular systolic dysfunction in patients with Duchenne muscular dystrophy. Am J Cardiol 89(7):838–841
Durieux J, Wolff S, Dillin A (2011) The cell-non-autonomous nature of electron transport chain-mediated longevity. Cell 144(1):79–91
Ehsani AA, Spina RJ, Peterson LR et al (2003) Attenuation of cardiovascular adaptations to exercise in frail octogenarians. J Appl Physiol (1985) 95(5):1781–1788
Evans EM, Racette SB, Peterson LR et al (2005) Aerobic power and insulin action improve in response to endurance exercise training in healthy 77–87 yr olds. J Appl Physiol (1985) 98(1):40–45
Fink M, Callol-Massot C, Chu A et al (2009) A new method for detection and quantification of heartbeat parameters in *Drosophila*, zebrafish, and embryonic mouse hearts. Biotechniques 46(2):101–113
Finkbeiner S (2011) Huntington's Disease. Cold Spring Harb Perspect Biol 3(6):a007476
Greer EL, Brunet A (2009) Different dietary restriction regimens extend lifespan by both independent and overlapping genetic pathways in *C. elegans*. Aging Cell 8(2):113–127
Greer ER, Perez CL, Van Gilst MR et al (2008) Neural and molecular dissection of a *C. elegans* sensory circuit that regulates fat and feeding. Cell Metab 8(2):118–131
Grewal SS (2009) Insulin/TOR signaling in growth and homeostasis: a view from the fly world. Int J Biochem Cell Biol 41(5):1006–1010
Grossman TR, Gamliel A, Wessells RJ et al (2011) Over-expression of DSCAM and COL6A2 cooperatively generates congenital heart defects. PLoS Genet 7(11):e1002344
Grueter CE, van Rooij E, Johnson BA et al (2012) A cardiac microRNA governs systemic energy homeostasis by regulation of MED13. Cell 149(3):671–683
Gunawardena S, Goldstein LS (2005) Polyglutamine diseases and transport problems: deadly traffic jams on neuronal highways. Arch Neurol 62(1):46–51
Jain S, Goldstein DS (2012) Cardiovascular dysautonomia in Parkinson disease: from pathophysiology to pathogenesis. Neurobiol Dis 46(3):572–580
Kaushik G, Fuhrmann A, Cammarato A et al (2011) In situ mechanical analysis of myofibrillar perturbation and aging on soft, bilayered *Drosophila* myocardium. Biophys J 101(11):2629–2637
Laker RC, Xu P, Ryall KA et al (2014) A novel MitoTimer reporter gene for mitochondrial content, structure, stress, and damage in vivo. J Biol Chem 289(17):12005–12015
Lee JH, Bassel-Duby R, Olson EN (2014a) Heart- and muscle-derived signaling system dependent on MED13 and Wingless controls obesity in *Drosophila*. Proc Natl Acad Sci U S A 111(26):9491–9496
Lee JH, Budanov AV, Park EJ et al (2010) Sestrin as a feedback inhibitor of TOR that prevents age-related pathologies. Science 327(5970):1223–1228

Lee KP, Simpson SJ, Clissold FJ et al (2008) Lifespan and reproduction in *Drosophila*: new insights from nutritional geometry. Proc Natl Acad Sci U S A 105(7):2498–2503

Lee TE, Yu L, Wolf MJ et al (2014b) Galactokinase is a novel modifier of calcineurin-induced cardiomyopathy in *Drosophila*. Genetics [Epub ahead of print]. PubMed ID: 25081566

Li A, Ahsen OO, Liu JJ et al (2013) Silencing of the *Drosophila* ortholog of SOX5 in heart leads to cardiac dysfunction as detected by optical coherence tomography. Hum Mol Genet 22(18):3798–3806

Li A, Zhou C, Moore J et al (2011) Changes in the expression of the Alzheimer's disease-associated presenilin gene in *Drosophila* heart leads to cardiac dysfunction. Curr Alzheimer Res 8(3):313–322

Liao JK, Laufs U (2005) Pleiotropic effects of statins. Annu Rev Pharmacol Toxicol 45:89–118

Lim HY, Wang W, Wessells RJ et al (2011) Phospholipid homeostasis regulates lipid metabolism and cardiac function through SREBP signaling in *Drosophila*. Genes Dev 25(2):189–200

Luong N, Davies CR, Wessells RJ et al (2006) Activated FOXO-mediated insulin resistance is blocked by reduction of TOR activity. Cell Metab 4(2):133–142

Melkani GC, Trujillo AS, Ramos R et al (2013) Huntington's disease induced cardiac amyloidosis is reversed by modulating protein folding and oxidative stress pathways in the *Drosophila* heart. PLoS Genet 9(12):e1004024

Mery A, Taghli-Lamallem O, Clark KA et al (2008) The *Drosophila* muscle LIM protein, Mlp84B, is essential for cardiac function. J Exp Biol 211(Pt 1):15–23

Molkentin JD, Lu JR, Antos CL et al (1998) A calcineurin-dependent transcriptional pathway for cardiac hypertrophy. Cell 93(2):215–228

Monnier V, Iche-Torres M, Rera M et al (2012) dJun and Vri/dNFIL3 are major regulators of cardiac aging in *Drosophila*. PLoS Genet 8(11):e1003081

Na J, Musselman LP, Pendse J et al (2013) A *Drosophila* model of high sugar diet-induced cardiomyopathy. PLoS Genet 9(1):e1003175

Ocorr K, Akasaka T, Bodmer R (2007a) Age-related cardiac disease model of *Drosophila*. Mech Ageing Dev 128(1):112–116

Ocorr K, Reeves NL, Wessells RJ et al (2007b) KCNQ potassium channel mutations cause cardiac arrhythmias in *Drosophila* that mimic the effects of aging. Proc Natl Acad Sci U S A 104(10):3943–3948

O' Neill (2013) PI3-kinase/Akt/mTOR signaling: impaired on/off switches in aging, cognitive decline and Alzheimer's disease. Exp Gerontol 48(7):647–653

Owusu-Ansah E, Song W, Perrimon N (2013) Muscle mitohormesis promotes longevity via systemic repression of insulin signaling. Cell 155(3):699–712

Paternostro G, Vignola C, Bartsch DU et al (2001) Age-associated cardiac dysfunction in *Drosophila melanogaster*. Circ Res 88(10):1053–1058

Piazza N, Gosangi B, Devilla S et al (2009a) Exercise-training in young *Drosophila melanogaster* reduces age-related decline in mobility and cardiac performance. PLoS ONE 4(6):e5886

Piazza N, Hayes M, Martin I et al (2009b) Multiple measures of functionality exhibit progressive decline in a parallel, stochastic fashion in *Drosophila* Sod2 null mutants. Biogerontology 10(5):637–648

Piazza N, Wessells RJ (2011) *Drosophila* models of cardiac disease. Prog Mol Biol Transl Sci 100:155–210

Qian L, Bodmer R (2009) Partial loss of GATA factor Pannier impairs adult heart function in *Drosophila*. Hum Mol Genet 18(17):3153–3163

Qian L, Bodmer R (2012) Probing the polygenic basis of cardiomyopathies in *Drosophila*. J Cell Mol Med 16(5):972–977

Qian L, Wythe JD, Liu J et al (2011) Tinman/Nk2–5 acts via miR-1 and upstream of Cdc42 to regulate heart function across species. J Cell Biol 193(7):1181–1196

Rahman MM, Sykiotis GP, Nishimura M et al (2013) Declining signal dependence of Nrf2-MafS-regulated gene expression correlates with aging phenotypes. Aging Cell 12(4):554–562

Santalla M, Valverde CA, Harnichar E et al (2014) Aging and CaMKII alter intracellular Ca2 + transients and heart rhythm in *Drosophila melanogaster*. PLoS ONE 9(7):e101871

Seyres D, Roder L, Perrin L (2012) Genes and networks regulating cardiac development and function in flies: genetic and functional genomic approaches. Brief Funct Genomics 11(5):366–374

Simpson SJ, Raubenheimer D (2009) Macronutrient balance and lifespan. Aging (Albany NY) 1(10):875–880

Skorupa DA, Dervisefendic A, Zwiener J et al (2008) Dietary composition specifies consumption, obesity, and lifespan in *Drosophila melanogaster*. Aging Cell 7(4):478–490

Spindler SR, Li R, Dhahbi JM et al (2012) Statin treatment increases lifespan and improves cardiac health in *Drosophila* by decreasing specific protein prenylation. PLoS ONE 7(6):e39581

Spletter ML, Schnorrer F (2014) Transcriptional regulation and alternative splicing cooperate in muscle fiber-type specification in flies and mammals. Exp Cell Res 321(1):90–98

Sujkowski A, Saunders S, Tinkerhess M et al (2012) dFatp regulates nutrient distribution and long-term physiology in *Drosophila*. Aging Cell 11(6):921–932

Sussman MA, Lim HW, Gude N et al (1998) Prevention of cardiac hypertrophy in mice by calcineurin inhibition. Science 281(5383):1690–1693

Taghli-Lamallem O, Akasaka T, Hogg G et al (2008) Dystrophin deficiency in *Drosophila* reduces lifespan and causes a dilated cardiomyopathy phenotype. Aging Cell 7(2):237–249

Taghli-Lamallem O, Jagla K, Chamberlain JS et al (2014) Mechanical and non-mechanical functions of Dystrophin can prevent cardiac abnormalities in *Drosophila*. Exp Gerontol 49:26–34

Taigen T, De LJ, Windt H, Lim W et al (2000) Targeted inhibition of calcineurin prevents agonist-induced cardiomyocyte hypertrophy. Proc Natl Acad Sci U S A 97(3):1196–1201

Tang M, Yuan W, Fan X et al (2013) Pygopus maintains heart function in aging *Drosophila* independently of canonical Wnt signaling. Circ Cardiovasc Genet 6(5):472–480

Tatar M, Kopelman A, Epstein D et al (2001) A mutant *Drosophila* insulin receptor homolog that extends life-span and impairs neuroendocrine function. Science 292(5514):107–110

Tatar M, Post S, Yu K (2014) Nutrient control of *Drosophila* longevity. Trends Endocrinol Metab 25(10):509–517

Tinkerhess MJ, Ginzberg S, Piazza N et al (2012a) Endurance training protocol and longitudinal performance assays for *Drosophila melanogaster*. J Vis Exp 61:3786

Tinkerhess MJ, Healy L, Morgan M et al (2012b) The *Drosophila* PGC-1alpha homolog spargel modulates the physiological effects of endurance exercise. PLoS ONE 7(2):e31633

Venkatesh TV, Park M, Ocorr K et al (2000) Cardiac enhancer activity of the homeobox gene tinman depends on CREB consensus binding sites in *Drosophila*. Genesis 26(1):55–66

Viswanathan MC, Kaushik G, Engler AJ et al (2014) A *Drosophila melanogaster* model of diastolic dysfunction and cardiomyopathy based on impaired troponin-T function. Circ Res 114(2):e6–17

Wessells RJ, Bodmer R (2004) Screening assays for heart function mutants in *Drosophila*. Biotechniques 37(1):58–60

Wessells RJ, Fitzgerald E, Cypser JR et al (2004) Insulin regulation of heart function in aging fruit flies. Nat Genet 36(12):1275–1281

Wessells R, Fitzgerald E, Piazza N et al (2009) d4eBP acts downstream of both dTOR and dFoxo to modulate cardiac functional aging in *Drosophila*. Aging Cell 8(5):542–552

Whitaker R, Gil MP, Ding F et al (2014) Dietary switch reveals fast coordinated gene expression changes in *Drosophila melanogaster*. Aging (Albany NY) 6(5):355–368

Wilkins BJ, Molkentin JD (2002) Calcineurin and cardiac hypertrophy: where have we been? Where are we going? J Physiol 541(Pt 1):1–8

Wolf MJ, Amrein H, Izatt JA et al (2006) *Drosophila* as a model for the identification of genes causing adult human heart disease. Proc Natl Acad Sci U S A 103(5):1394–1399

Wolf MJ, Rockman HA (2011) *Drosophila*, genetic screens, and cardiac function. Circ Res 109(7):794–806

Yu L, Daniels J, Glaser AE et al (2013) Raf-mediated cardiac hypertrophy in adult *Drosophila*. Dis Model Mech 6(4):964–976

Zaffran S, Reim I, Qian L et al (2006) Cardioblast-intrinsic Tinman activity controls proper diversification and differentiation of myocardial cells in *Drosophila*. Development 133(20):4073–4083

Zhang D, Ke L, Mackovicova K et al (2011) Effects of different small HSPB members on contractile dysfunction and structural changes in a *Drosophila melanogaster* model for atrial fibrillation. J Mol Cell Cardiol 51(3):381–389

Ziemssen T, Reichmann H (2010) Cardiovascular autonomic dysfunction in Parkinson's disease. J Neurol Sci 289(1 2):74–80

Chapter 7
Parallels Between Mammals and Flies in Inflammatory Bowel Disease

Christofi Theodoulakis and Yiorgos Apidianakis

Abstract The intestinal hologenome lies at the center of intestinal homeostasis and disease. It is the sum of the host genome and the intestinal microbiome. During the last decade we came to appreciate the pivotal role of the microbiome in intestinal inflammation and concomitant diseases. Moreover environmental factors, predominantly diet, affect the microbiome and the host immune responses. Thus manipulation of intestinal microbiota and environmental factors are rightfully the focus of current research. The study of the hologenome necessitates not only clinical assessments but also the use of model organisms, for example, mice and flies. Despite the limitations imposed by the evolutionary distance between flies and mammals, *Drosophila* research provided us during the last few years with a wealth of information regarding intestinal inflammation and the role of microbiota. Conserved aspects of intestinal homeostasis and disease between flies and mice, for example in signalling pathways, the intestinal defence responses and the role of microbiota, consolidate and may advance the principles that govern intestinal inflammation in humans.

Keywords Intestinal microbiota · Microbiome · Inflammation · Immune response · Inflammatory bowel disease · *Drosophila*

7.1 Introduction

Inflammatory bowel disease (IBD) is a condition caused by inflammationof the small or large intestine. It is increasingly prevalent in Western countries, and impacts on the patients' quality of life and the risk for colorectal cancer (CRC)

An erratum of this chapter can be found under DOI 10.1007/978-3-319-18326-8_16

C. Theodoulakis · Y. Apidianakis (✉)
Department of Biological Sciences, University of Cyprus, Nicosia, Cyprus
e-mail: apidiana@ucy.ac.cy

C. Theodoulakis
e-mail: christofi.theodoulakis@ucy.ac.cy

A.M. Vaiserman et al. (eds.), *Life Extension*, Healthy Ageing and Longevity 3,
DOI 10.1007/978-3-319-18326-8_7

(Molodecky et al. 2012). A number of disorders are classified as IBD, including the most common ones, Crohn's disease (CD) and ulcerative colitis (UC). Disease classification has been first proposed in 1988 (Greenstein et al. 1988), but over the years new information has modified the criteria based on the disease location, clinical behaviour and age of onset (Silverberg et al. 2005; Levine et al. 2011). CD can occur discontinuously anywhere along the digestive tract and throughout the intestinal wall. It is usually associated with granulomas and can develop fistulas and intestinal stenosis (Pal 2014). UC generally affects the large intestine causing inflammation continuously along the colon, and is usually limited to the mucosa (Abraham and Cho 2009). Epidemiological data showed in 2013 that in the United States 1,171,000 people have IBD, 565,000 of which are CD and 593,000 UC patients (Kappelman et al. 2013). In Europe 0.3 % of the population, that is, 2.5–3 million people, suffer from IBD. Of those approximately 1.1 million suffer from CD and 1.5 million from UC (Burisch et al. 2013). This imposes a substantial public health burden because IBD is usually chronic and cannot be easily treated.

While the etiopathology of IBD is not yet clear, accumulating evidence indicates that genetic predisposition leads to an aberrant immune response against enteric microbes, which may be triggered by environmental factors, such as western world diets (Sartor 2006). The gut houses many trillions of microorganisms of >500 different aerobic and anaerobic species (Human Microbiome Project Consortium 2012). Microbiota metabolise nutrients that the host cannot digest, provide vitaminsand protect against harmful microbes (Backhed et al. 2005). Microbiota also helps to maintain the intestinal tissue homeostasis through a crosstalk between the intestinal epithelial and immune cells (Mortha et al. 2014). However, genetic and environmental factors may cause a persistent microbial infection or a shift in the relative abundance of microbial species and may lead to an imbalance in the host defence. Consequently bacteria may invade the intestinal mucosa or activate an abnormal immune response, leading to innate and adaptive immune cell infiltration, tissue damage and regeneration (Xavier and Podolsky 2007). On a chronic basis intestinal inflammation increases the risk for colorectal cancer (CRC) (Pohl et al. 2000). The precise risk levels of cancer in IBD can be difficult to quantify due to the variation of methodological approaches. Evidence nevertheless shows that there is an at least two fold increase in the prevalence of colorectal cancer in patients with UC whereas patients with CD are at increased risk for small bowel adenocarcinoma (Pohl et al. 2000). One in five patients with UC develops CRC within 30 years of the disease onset and more than half of them will die from colitis-associated cancer (CAC) (Lakatos and Lakatos 2008). This is because CRC is caused by the accumulation of gene mutations in oncogenes and tumor suppressor genes, such as APC, K-Ras and p53, which lead to abnormal cell proliferation, tissue invasion and metastasis (Fearon and Vogelstein 1990). In IBD, the persistent production of pro-inflammatorycytokines and reactive oxygen species may induce genomic instability and cancer promoting mutations (Neurath 2014).

Below we summarize prominent risk and associated factors, clinical assessments and treatments for IBD and introduce a comparative view of the literature

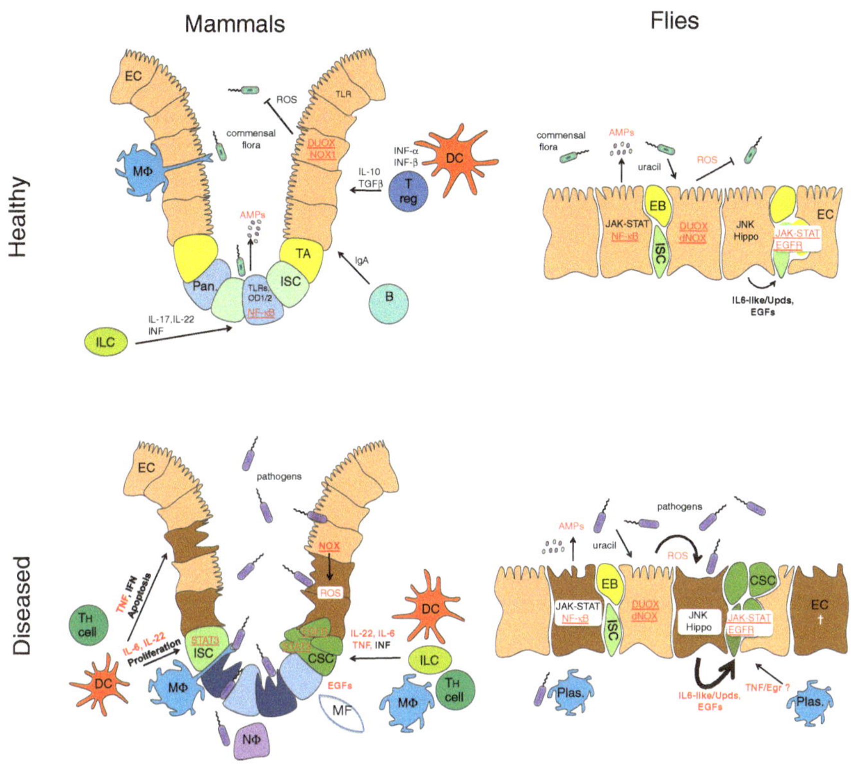

Fig. 7.1 Parallels in intestinal homeostasis and inflammation-driven tumorigenesis between mammals and flies. In IBD patients and animal models cytokines and growth factors play a role in intestinal homeostasis, inflammation and cancer. **In mammals**, Toll-like (*TLRs*) and NOD receptors in epithelial cells, recognise bacterial components to induce anti-microbial peptides (AMPs) via NF-κB activation. Macrophages (*MΦ*), dendritic cells (*DC*), innate lymphoid cells (*ILC*), B cells (*B*) and T regulatory (T_{reg}) cells maintain the immune balance by secreting IgA and anti-nflammatory cytokines and the induction of AMPs. Intestinal stem cells (*ISCs*) repair the intestinal barrier through STAT3 signalling. Similarly, **in the *Drosophila*** intestinal enterocytes (*EC*), the NF-κB pathway Imd senses potentially pathogenic bacteria and induces AMP secretion. EC also recognise bacterial-derived uracil and secrete reactive oxygen species (*ROS*) to regulate bacterial numbers but also stimulate intestinal regeneration upon EC damage. Microbiota and/or ROS mediated EC stress induce stem cell proliferation and intestinal regeneration through the JAK-STAT and EGFR pathways. **During inflammation in mammals**, immune cells secrete many pro-inflammatorycytokines, such as IL-6, IL-22, TNF and INF, which cause EC apoptosisand ISC proliferation. Myofibroblasts (*MF*) may secrete EGFs, which also induce ISC proliferation. In addition NOX induces ROS, which may lead to mutations in oncogenes and tumor suppressor genes capable of inducing cancer stem cell (*CSC*)-mediated tumorigenesis. **In bacterially infected fly epithelia**, innate immune phagocytes (plasmatocytes) regulate bacteria that pass through the epithelium via phagocytosis. Plasmatocytes (Plas.) may be recruited and may produce pro-inflammatory cytokines, such as TNF/Egr to induce apoptosis or proliferation of epithelial cells depending on the genetic background. In addition, damaged EC and microenvironment cells secrete IL6-like/Upds and EGFs to induce ISC proliferation. In the presence of oncogenic mutations signals from damaged enterocytes drive ISC proliferation towards CSC-like tumorigenesis. *EC* enterocytes, *Pan.* Paneth cells, *DC* dendritic cell, *ILC* innate lymphoid cell, *MΦ* macrophage, *NΦ* neutrophils, T_{reg} T regulatory cells, *TA* transit amplifying cell, *ISC* intestinal stem cell, *EB* enteroblast, *CSC* cancer stem cell, *B* B cell, †: damaged cell. *Red (light color) lettering* indicates conserved or analogous pathways and factors between mammals and flies. Intraepithelial lymphocytes, Peyer's patches and other components on the mammalian intestinal immune system are not described for simplicity

regarding mouse and fly models of human intestinal inflammationand cancer. Modelling of intestinal inflammation and inflammation-driven tumorigenesis in flies and mice has been fruitful because despite the differences with humans, some parallels among them are striking (Fig. 7.1). Mouse models of IBD are traditionally used due to the high degree of conservation between the human and the mouse intestinal tract (Wirtz and Neurath 2007; Gkouskou et al. 2014). Nevertheless, key principles of mammalian inflammatory bowel disease pathogenesis can be recapitulated and new aspects of disease have been proposed using *Drosophila* models (Apidianakis and Rahme 2011; Panayidou and Apidianakis 2013).

7.2 Risk and Associated Factors in IBD

7.2.1 Genetic Predisposition

The inheritance of IBD has been the focus of many epidemiologists because of the increased risk for IBD among relatives (Halme et al. 2006). According to Halme and colleagues, 2–14 % of CD patients reported a family history of CD, 7–11 % of UC patients a UC history in their family and 10 % of either UC or CD patients a family history of some type of IBD (Halme et al. 2006). Moreover, identical (monozygotic) versus non-identical (dizygotic) twins studies provide a strong confirmation for the genetic basis of IBD. For example, the concordance of CD and UC is much higher in monozygotic versus dizygotic twins, which can be explained by genetic rather than environmental factors (Brant 2011).

Various genome wide association studies (GWAS)on UC and CD have examined the genetic makeup of affected versus healthy individuals (Table 7.1). Many of them belong to the epithelial barrier function group, such as mucins and matrix metalloproteinases (Buisine et al. 1999; von Lampe et al. 2000; Makitalo et al. 2009, 2010; Giuffrida et al. 2014), or oxidative stress related genes, such as ascorbate transporters SLC23A1 polymorphisms (Amir Shaghaghi et al. 2014). Innate and adaptive immunity, regeneration and autophagy genes are described below in more detail. Nevertheless, the gene lists provided here are not exhaustive. Moreover, many of the listed genes, such as neuropeptide S receptor 1 (NPSR1) are associated with IBD susceptibility but their function is poorly understood (D'Amato et al. 2007).

Genes controlling inflammation, such as anti-inflammatory cytokines of the IL10 family (IL-10/19/26), IL-2, IL-27 and pro-inflammatory cytokines(IL-3/12b/21) and receptors (IL1R2, IL7R, IL8RA/B, IL17REL, 23R) have been strongly associated with risk for IBD indicating the crucial role of immune system responses at the site of inflammation (Franke et al. 2010; Anderson et al. 2011; Jostins et al. 2012). The magnitude of inflammatory signals expressed by the immune system (ILC, T cells, macrophages, dendritic cells) or the intestinal cells associated with IBD can be expanded to include tumor necrosis factors (TNFSF) 8/9/11/14/15 and receptors

Table 7.1 Processes and genes affecting susceptibility to intestinal infection, stress or inflammation (in humans, mice and flies)

	Human SNPs/genes	Mouse SNPs/genes	Fly genes
Autophagy	ATG16L1 (Prescott et al. 2007); IRGM1 (McCarroll et al. 2008); ULK1 (Henckaerts et al. 2011); MTMR3 (Henderson et al. 2011); VAMP3, DAP (Franke et al. 2010); LRRK2, CUL2, PARK7, NOD2 (Khor et al. 2011); RIPK3 (Gunther et al. 2011)	ATG4B (Cabrera et al. 2013); IRGM1 (Henry et al. 2007); ATG16L1 (Saitoh et al. 2008); RIPK3 (Gunther et al. 2011)	ATG1, ATG6 (Wu et al. 2009a)
Innate immunity	NF-kB, TNF-α (Plevy et al. 1997); COX-2 (Hendel and Nielsen 1997); TLR (Bank et al. 2014); TLR-4 (Franchimont et al. 2004); IL18RAP (Zhernakova et al. 2008); NOD2 (Ogura et al. 2001); β-defensins; HBD2; HBD3; HBD4 (Wehkamp et al. 2003); HBD5; HBD6 (Wehkamp et al. 2005); JAK; TYK2 (Franke et al. 2010); IL-1 (Casini-Raggi et al. 1995); SLC11A1, FCGR2A/B, REL, CARD9 (Khor et al. 2011); MIF (Griga et al. 2007); FOXO3 (Lee et al. 2013a)	TNFR, NF-kB, TLR (Araki et al. 2005) TLR5 (Uematsu et al. 2006); IFN-γ, TNF-α (Powrie et al. 1994); IL-12 (Kullberg et al. 1998); NOD1 (Chatterjee and Chaudhuri 2013); NOD2 (Kobayashi et al. 2005); PI3 K/AKT (Khan et al. 2013); FOXO3 (Snoeks et al. 2009)	IMD/Relish (Buchon et al. 2009b, Lemaitre et al. 1995); Jak-Stat (Buchon et al. 2009b); Ras (Ragab et al. 2011); Toll (Lemaitre et al. 1996); bbg (Bonnay et al. 2013); PGRP-LB (Zaidman-Remy et al. 2006); PGRP-LC (Schmidt et al. 2008); PGRP-LE (Bosco-Drayon et al. 2012); PLCβ-DUOX, p38 MAPK (Lee et al. 2013b); dome-less, Upds (Buchon et al. 2009b); Eiger (Schneider et al. 2007); FOXO (Karpac et al. 2013)

(continued)

Table 7.1 (continued)

	Human SNPs/genes	Mouse SNPs/genes	Fly genes
Adaptive immunity	PRDM1 (Franke et al. 2010); LSP1, Smad3, Smad7, TGF-β (Monteleone et al. 2001); TNFRSF 6/9/14 (Anderson et al. 2011); IL4/6/10/12B/13/17A/18/21/22, IL1R/7R/8R/12R, IL17R//23R (Franke et al. 2010; Anderson et al. 2011); TNF-α, IFN-γ, TNFRSF6B (Kugathasan et al. 2008); VNN1 (Gensollen et al. 2013); TNFSF8/11/15 (Franke et al. 2010); CCR3 (Manousou et al. 2010); CCR9 (Linton et al. 2012); HLA-DR1/2, HLA-DQw5 (Toyoda et al. 1993); CXCR3/4 (Hosomi et al. 2011); CXCL1 (Alzoghaibi et al. 2008); IL-5. GATA3 (Neurath et al. 2002); ERAP2, DENND1B, LNPEP (Khor et al. 2011)	CXCL1 (Shea-Donohue et al. 2008); CXCR3, CCR2/5 (Tokuyama et al. 2005); TNF-α (Kojouharoff et al. 1997); TNFRFSF9 (Maerten et al. 2006); IL-1/6/10/12/22/23, IFN-γ (Grivennikov et al. 2009; Zenewicz et al. 2008; Sturlan et al. 2001); Smad3 (Maggio-Price et al. 2006); VNN1 (Pouyet et al. 2010); INF-α (Katakura et al. 2005)	
Regeneration	Wnt, Notch (van Es and Clevers 2005); APC (Groden et al. 1993); Stat1 (Schreiber et al. 2002); Stat3 (Jiang et al. 2013); Stat4 (Ohtani et al. 2010); Stat5 (Connelly et al. 2013); JAK2 (Anderson et al. 2011); KRas (Vogelstein et al. 1988); PI3 K/AKT (Khan et al. 2013)	Wnt, Notch (Dahan et al. 2011); BMP (Haramis et al. 2004); Stat3 (Pickert et al. 2009); JNK (Jones et al. 2008); Stat4 (Simpson et al. 1998); Stat6 (Elrod et al. 2005); p 53, Ras/MAPK (Valentin-Vega et al. 2008); PI3 K, APC (Deming et al. 2014)	JNK (Buchon et al. 2009a); Jak/Stat (Jiang et al. 2009); Warts, Yorkie (Staley and Irvine 2010); EGF (Biteau and Jasper 2011); Wingless (Xu et al. 2011); vein, spitz, Keren, MKP3, Ras (Jiang et al. 2011a); Wnt, APC (Wang et al. 2013)
Oxidative stress	NOS2 (Singer et al. 1996; Dhillon et al. 2014); Duox2 (Wu et al. 2013); ADO, SLC22A4, GPX1/4, UTS2, PEX13, PARK7, DLD, BACH2, NOD2, LRRK2, PRDX5, HSPA6, CARD9, (Khor et al. 2011); SLC23A1 (Amir et al. 2014); Nox1 (Szanto et al. 2005)	iNOS (Krieglstein et al. 2001); GPX1/2 (Khor et al. 2011); Nox1 (Jones et al. 2013)	dDuox (Ha et al. 2009); dNox (Jones et al. 2013)

(continued)

Table 7.1 (continued)

	Human SNPs/genes	Mouse SNPs/genes	Fly genes
Epithelial barrier	MUC1/3/4/5B (Buisine et al. 1999); HNF4A, CDH1/3, LAMB1, GNA12 (UK IBD Genetics Consortium et al. 2009); CDH11 (Costello et al. 2005); CDH1, ERRFI1, MUC19, ITLN1 (Khor et al. 2011); MMP-1/3/7/9/10/12/13/14/26, TIMP-1/2/3, (Makitalo et al. 2009, 2010; von Lampe et al. 2000; Giuffrida et al. 2014); DLG5 (Stoll et al. 2004); MLCK (Blair et al. 2006)	MUC1 (McAuley et al. 2007); MUC2 (Velcich et al. 2002); XBP1 (Kaser et al. 2008); MMP3 (Li et al. 2004); MLCK (Su et al. 2009)	Myo1B (Hegan et al. 2007); Drosocrystallin (Kuraishi et al. 2011); bbg (Bonnay et al. 2013); βν integrin (Okumura et al. 2014)

(TNFRSF) 6/9/14. Upstream and downstream regulators of these cytokines, such as TYK2, IRF5, DENND1B, PRDM1, LSP1, JAK2, PTPN2 and STAT3 have also been identified within the IBD risk loci, signifying the importance of these pathways in IBD pathogenesis. Implication of these pathways in IBD has been further validated in animal models where investigation has proven their link to IBD. For example, IL-10 deficient mice are widely used for IBD modeling. These mice are predisposed for pro-inflammatory signal production from macrophages, such as TNF-α, through the JAK-STAT pathway (Riley et al. 1999). Moreover, a number of single nucleotide polymorphisms (SNPs) in NOD2 and CARD9, suggest that recognition of bacteria is essential in maintaining intestinal homeostasis (Ogura et al. 2001; Khor et al. 2011; Beaudoin et al. 2013). NOD2 is primarily associated with IBD because it can promote both innate and adaptive immune responses. Upon bacterial recognition, NOD2 stimulates production of AMPs, antigen presentation and autophagy demonstrating a central role in intestinal protection (Khor et al. 2011). Macrophage migration inhibitor factor (MIF), have also been associated with CD predisposition and the need for higher steroid treatment in patients, because for example, MIF induces the production of pro-inflammatorycytokines and counteracts the anti-inflammatory effects of glucocorticoids (Griga et al. 2007).

Genes of the signal transducers and activators of transcription (STATs) pathways are implicated in both innate immunity and intestinal regeneration in mammals and in flies (Panayidou and Apidianakis 2013) and SNPs in many of them have been linked to IBD (Table 7.1). In fact, STAT3 pathway lies at the intersection of inflammationand regeneration because innate immune cells secrete its ligands, IL6 and IL22, to induce epithelial regeneration in the nearby cells (Grivennikov et al. 2009; Kirchberger et al. 2013; Kryczek et al. 2014). Intestinal regeneration is driven by stem cells and is also linked to intestinal tumorigenesis via many of the same genes that drive CRC, such as APC, WNT/Wg, K-Ras/Ras1 and PI3 K (Table 7.1).

Lastly, autophagy has been implicated in the pathogenesis of IBD. Autophagy is the process where the cell catabolises unnecessary or foreign components through lysosomal degradation. Dysfunction of this mechanism either through polymorphisms or mutations is believed to increase the chance of IBD (Scharl and Rogler 2012). A threonine to alanine substitution (T300A) in the autophagy related 16 like (ATG16L1) gene has been associated with an increased risk for CD and UC (Prescott et al. 2007). Similarly, a non-synonymous SNP in ATG16L1 has been shown to increase the risk of CD due to loss of Paneth cell function and morphology (Prescott et al. 2007). Another autophagy-related protein is IRGM, which is associated with mitochondrial function, apoptosisand CD (McCarroll et al. 2008; Singh et al. 2010). The role of autophagy is crucial for the correct immune response. ATG16L1 and IRGM have been shown to be important for the elimination of pathogens present in the mucus layer, such as adherent invasive *Escherichia coli* and *Salmonella typhimurium* (Henry et al. 2007; Brest et al. 2011), but also those found in the lumen. VAMP3 is a SNARE protein that mediates exocytosis of vesicles into the extracellular medium and is involved in cell migration and integrin trafficking (Luftman et al. 2009) and SCAMP3 is a key ingredient in the intestinal epithelium mucus barrier, both of which are related to autophagy and have been reported to be associated with intestinal inflammation(Campbell et al. 2001; Franke et al. 2010). DAP encodes a death associated protein and is a negative regulator of apoptosis. While previously associated with CD, expression of DAP kinase is linked to autophagy and is elevated in UC (Kuester et al. 2010).

7.2.2 Geographic and Dietary Associations

The prevalence of UC and CD varies widely—from 0 to more than 20 per 100,000 people worldwide (Molodecky et al. 2012)—probably because IBD is highly related to western societies' lifestyle. For example, there is a higher incidence of IBD in North America compared to Asian countries (Molodecky et al. 2012). Furthermore, within Europe there is a gradient of IBD incidence from East to West and from North to South (Shivananda et al. 1996; Burisch et al. 2014). Even within individual countries, such as France and Finland, a North to South geographical distribution is clear (Nerich et al. 2006; Jussila et al. 2013). This phenomenon may be due to many factors, including vitamin D and lactase levels, sunshine and latitude (Szilagyi et al. 2014). However a combination of environmental factors, dietary habits and genetic background would give a more complete explanation of disease prevalence.

One of the pivotal factors in IBD is diet (Table 7.2). Dietary changes have been investigated in a number of studies on immigrant populations. Traditional diets of South Asian populations consist of mostly complex carbohydrates and minimal fats or sugars (Misra et al. 2009). However when foreign populations are exposed to western diets high in fat and sugar, in their lifetime or in subsequent generations, a significantly higher incidence of IBD is observed (Pinsk et al. 2007). A systematic review on food intake suggests a correlation between specific nutrients (fat, proteins, carbohydrates) and food groups (meats, vegetables, fruits) and the prevalence of IBD (Hou et al. 2011). For example, consumption of meat, saturated and total fat, and polyunsaturated fatty acids (PUFAS), such as low omega-3 to omega-6 ratio, correlate with increased risk for IBD. In contrast, the high intake of fiber, vegetables and fruits has the opposite effect (Hou et al. 2011).

7.2.3 Intestinal Microbiota and Dysbiosis

The human intestinal tract contains 10 times more bacterial cells than the total body cells (Backhed et al. 2005). The human microbiome project revealed that bacterial species vary among individuals, but Bacteroides and Firmicutes and to a lesser extent Proteobacteria and Actinobacteria phyla prevail in the human intestine (Human Microbiome Project Consortium 2012). Bacterial species interact

Table 7.2 Dietary factors that influence IBD and intestinal homeostasis [increase (↑), decrease (↓) or change (↕)]

Diet	Organism (Human, Mice, Flies)	Effect
Probiotic	H (Bibiloni et al. 2005; Parvez et al. 2006; Kanai et al. 2014), M (Matsumoto et al. 2001; Mennigen et al. 2009), F (Ryu et al. 2008; Lee et al. 2013b)	↓IBD, ↓CRC, ↕homeostasis
Fermented food	H (Parvez et al. 2006; Kanai et al. 2014), M (Matsumoto et al. 2001)	↓IBD, ↓dysbiosis
N-6 PUFAs	H (Hou et al. 2011, Andersen et al. 2012), M (Ghosh et al. 2013)	↑CD, ↑UC, ↑colitis
N-3 PUFAs/UFAs	H (Andersen et al. 2012), M (Vilaseca et al. 1990; Hudert et al. 2006; Devkota et al. 2012)	↓UC, ↓CRC, ↓inflammation
Meat (red)	H (Maconi et al. 2010; Hou et al. 2011)	↑CD, ↑UC
Fat	H (Hou et al. 2011), M (Devkota et al. 2012; Paik et al. 2013)	↑IBD, ↑colitis, ↑dysbiosis
Vegetables	H (Hou et al. 2011)	↓UC, ↓CD, ↓inflammation
Fiber	H (Hou et al. 2011), M (Bassaganya-Riera et al. 2011), F (Shin et al. 2011)	↓CD, ↓inflammation, ↕homeostasis
Fruits	H (Hou et al. 2011), M (Fujisawa et al. 2005; Kohno et al. 2006)	↓CD, ↓colitis, ↓inflammation
Proteins (meat/veg.)	H (Jantchou et al. 2010; Andersen et al. 2012), M (MacDonald and Przybyszewski 2008; Jiang et al. 2011b), F (Chandler et al. 2011)	↑colitis, ↑IBD/↓colitis, ↕microbiome
Milk	H (Gudmand-Hoyer and Jarnum 1970; Pittschieler 1990), M (Madsen et al. 2002; Lara-Villoslada et al. 2006; Fuhrer et al. 2010)	↑IBD/↓colitis
Dairy products	H (Asakura et al. 2008; Maconi et al. 2010)	↑CD,↑UC
Sugars	H (Reif et al. 1997; Sakamoto et al. 2005), M (Turnbaugh et al. 2009; Fuhrer et al. 2010), F (Chandler et al. 2011)	↑dysbiosis, ↑IBD, ↕microbiome
Vitamins	H (Todhunter et al. 2005), M (Song et al. 2000; Carrier et al. 2003; Lagishetty et al. 2010; Benight et al. 2011)	↕colitis

dynamically and are usually in balance with each other, the host and the environment. Bacterial communities in the gut provide a physical barrier to incoming pathogens through direct and indirect competition. Antagonism for food sources or attachment sites and production of various antimicrobial substances protect the host from invading pathogens. *Ruminococcus brommi* and other colonic bacteria degrade indigestible nutrients, such as dietary fibber and resistant starch, to produce acetic acid and other physiologically important short-chain fatty acids (Wong et al. 2006; Ze et al. 2012). Similarly, *Acetobacter pomorum* affect host metabolism via acetic acid production in the *Drosophila* gut (Shin et al. 2011). Bifidobacteria and lactic acid bacteria produce necessary vitamins that cannot be synthesized by human cells (LeBlanc et al. 2013). Another important aspect of the gut microbiome is its ability to modulate intestinal homeostasis. This is crucial for a healthy gut since it is a cell-damaging environmentwhere the need for constant cell renewal is vital for proper gut function. *Bacteroides thetaiotaomicron* has been found to modulate genes involved in intestinal maturation and mucosal barrier fortification (Hooper et al. 2001). Microbiota also facilitates gut immune system development and modulate immune responses to control inflammation (Round and Mazmanian 2009; Hooper et al. 2012). For example, Polysacharide A (PSA) produced by *Bacteroides fragilis* directs maturation of the immune system by expanding and differentiating splenic CD4 + T cells (Mazmanian et al. 2005). In addition, PSA protects mice against induction of experimental IBD by decreasing the levels of pro-inflammatorycytokinesvia IL-10 expression by CD4 + T cells (Mazmanian et al. 2008). Regulatory T cells accumulate in the mucosa after administration of specific indigenous *Clostridium* species protecting against systemic immune responses and colitis development (Atarashi et al. 2011). Similarly, *Bifidobacterium infantis* protects against *Salmonela typhimirium* via regulatory T cells (O'Mahony et al. 2008).

On the contrary, when microbiota is depleted as in germ free mice, a number of bacterial pathogens can colonize and induce inflammationin mice (Cahill et al. 1997; Balish and Warner 2002; Kim et al. 2007a, b; Tlaskalova-Hogenova et al. 2011). Similarly, IBD in CD patients may relapse if beneficial bacteria, such as *Faecalibacterium prausnitzii*, are reduced (Sokol et al. 2008). Such imbalances in bacterial diversity are common among IBD patients and may have a major impact in humans. Studies investigating the bacterial composition of IBD and control patients have identified differences in the microbiota of UC and UC patients compared to controls with a characteristic depletion of commensal bacteria belonging to the Bacteroidetes and Firmicutes phyla compared to the family of Enterobacteriaceae (Ott et al. 2004; Frank et al. 2007; Kaakoush et al. 2012; Morgan et al. 2012). Whether the rise in Enterobacteriaceae species, such as *Escherichia coli,* is a cause or a consequence of IBD is a subject of intensive research.

7.2.4 Dysbiosis upon Bacterial Infection or Antibiotics Treatment

The gastrointestinal track is exposed to a plethora of ingested microbes and therefore is prone to microbial infections. Even though a few potentially pathogenic strains reside in the gut, these are dormant and cause no health issues. Some bacterial species however (Table 7.3), such as the adherent and invasive *E. coli* (AIEC) are directly associated with IBD pathogenesis (Barnich et al. 2007). AIEC has been found to colonize the ileal mucosa of CD patients by attaching to carcinoembryonic antigen-related cell adhesion molecule 6 (CEACAM6) present on CD patients' ileal enterocytes. IFN-γ and TNF-α cytokines produced upon AIEC infection induce CEACAM6 expression in enterocytes, indicating that these bacteria can promote their own colonization (Barnich et al. 2007). In immunocompromised or genetically predisposed animal models a number of potential pathogens have been implicated in the pathogenesis of IBD. *Helicobacter hepaticus, Enterococcus faecalis* and *Clostridium difficile* species have been linked to IBD (Cahill et al. 1997; Balish and Warner 2002; Khanna and Pardi 2012). A more complex scheme was observed in gnotobiotic transgenic mice where infection with two commensal bacteria, a nonpathogenic *Enterococcus faecalis* or a nonpathogenic *Escherichia coli* strain, act additively to induce duodenal inflammation and aggressive pancolitis, showing the cooperation of different bacteria in the pathogenesis of IBD (Kim et al. 2007a, b). Similarly, in genetically predisposed mouse models of IBD, intestinal inflammation can be cured with the administration of broad-spectrum antibiotics, indicating that some intestinal microbes may induce IBD (Garrett et al. 2007). The same study showed that wild type mice developed colitis when co-housed with the diseased transgenic mice, which was explained by pathogenic bacteria transferred between mice.

Nevertheless, administration of antibiotics in healthy individuals may induce microbial changes favouring the growth of potentially harmful bacteria or opportunistic pathogens. For example, *Clostridium difficile* is a bacterium that causes diarrhea and pseudomembranous colitis in patients using antibiotics (George et al. 1978). *Clostridium difficile* can take advantage of the antibiotic-mediated reduction in microbial diversity to infect the adult gut (Chang et al. 2008). Nevertheless, the effect of antibiotics can be even more pronounced during early childhood. Antibiotic use in early postnatal period can modulate the relative abundance of intestinal microbiota, decreasing Bifidobacteria while increasing Enterobaceriaceae (Tanaka et al. 2009). A study of 36 IBD patients and 360 controls showed that administration of antibiotics to children less than 1 year old triples their chance to develop IBD (Shaw et al. 2010). A population based cohort study in the UK assessed over 1 million individuals 748 of whom developed IBD and found that antianaerobic antibiotics promote the development of IBD in a number of dose- and young age-dependent manner (Kronman et al. 2012).

7.2.5 Others Risk Factors

Exogenous factors also play a role in the prevalence and development of IBD. Smoking is a known risk factor for a number of diseases. In IBD, smoking has been found to be a risk factor for CD but interestingly protective against UC (Jang et al. 2006; Mahid et al. 2006). In addition, appendectomy may lower the risk

Table 7.3 Bacterial species linked to intestinal damage or inflammation

Phyla	Species—host (Human, Mice, Flies)	Effect
Firmicutes	*Lactobacillus plantarum* (H, M, F) (Schultz et al. 2002)	↓inflammation, ↓IFN-γ, ↓IL-12, ↓CRC
	Lactobacillus salivarius (H, M) (Macho Fernandez et al. 2011)	↓inflammation, ↑IL-10
	Lactobacillus casei Shirota (H, M) (Matsumoto et al. 2005)	↓inflammation
	Lactobacillus acidophilus (H, M) (Mohamadzadeh et al. 2011)	↓inflammation, ↓IL-12, ↓TNF-α, ↑IL-10
	Lactobacillus bulgaricus (H, M) (Hunter et al. 2009)	↕systemic immune responses
	Lactobacillus reuteri (H, M) (Jones and Versalovic 2009)	↓inflammation, ↓TNF, ↑antimicrobials
	Lactobacillus fermentum (H, M) (Peran et al. 2006)	↑glutathion, ↓iNOS, ↓TNF-α
	Lactobacillus rhamnosus (H, M) (Ma et al. 2004; Tao et al. 2006; Lin et al. 2009)	↓inflam, ↓TNF-α, ↑cytoprotection, ↑NGF
	Lactobacillus crispatus (H, M) (Zhou et al. 2012)	↑IL-1β, ↑IL-6, ↑TNF-a, ↓IL-10
	Lactobacillus johnsonii (H, M) (Sgouras et al. 2005)	↓inflammation, ↓infection
	Lactobacillus delbrueckii lactis (H, M) (Santos Rocha et al. 2014)	↓inflammation, ↓colitis
	Lactobacillus brevis (H, M, F) (Ueno et al. 2011)	↕intestinal homeostasis, ↓inflammation
	Bacillus subtilis (H, M) (Selvam et al. 2009)	↓inflammation, ↓colitis
	Enterococcus faecalis (H, M, F) (Balish and Warner 2002)	↑inflammation, ↑IBD
	Streptococcus thermophilus (H, M) (Menard et al. 2004)	↓inflammation, ↓TNF-α
	Staphylococcus sp. (H, M, F) (Sibley et al. 2008; Edwards et al. 2012)	↑inflammation
	Clostridium difficile (H, M) (George et al. 1978)	↑diarrhea, ↑colitis, ↑IBD
	Faecalibacterium prausnitzii (H, M) (Sokol et al. 2008)	↓inflammation
	Ruminococcus gnavus (H, M) (Menard et al. 2004)	↓inflammation, ↓TNF-α
Bacteroides	*Bacteroides thetaiotaomicron* (H, M) (Bloom et al. 2011)	↕intestinal homeostasis, ↑IBD
	Bacteroides fragilis (H, M) (Ruiz-Perez et al. 2005; Mazmanian et al. 2008; Round and Mazmanian 2010)	↓inflammation
	enterotoxigenic Bacteroides fragilis (H, M) (Toprak et al. 2006; Rhee et al. 2009)	↑inflammation, ↑CRC
	Bacteroides vulgatus (H) (Shiba et al. 2003; Bloom et al. 2011)	↑IBD
	Bacteroides ovatus (H) (Saitoh et al. 2002)	↑inflammation

(continued)

Table 7.3 (continued)

Phyla	Species—host (Human, Mice, Flies)	Effect
Proteobacteria	*Acetobacter pomorum* (F) (Shin et al. 2011)	↕intestinal homeostasis
	Helicobacter hepaticus (H, M) (Kullberg et al. 1998; Mazmanian et al. 2008)	↑inflam., ↑IFN-γ, ↑IL-12, ↑IBD
	Helicobacter muridarum (H, M) (Jiang et al. 2002; Monceaux et al. 2013)	↑disease activity
	Helicobacter bilis (H, M) (Javed et al. 2013)	↑inflammation, ↑ROS, ↑IL-8
	adherent invasive E. coli (H, M) (Small et al. 2013)	↑inflammation, ↑fibrosis
	Pseudomonas entomophila (F) (Liehl et al. 2006)	↑inflammation
	Pseudomonas aeruginosa (H, M, F) (Apidianakis et al. 2009; Wagner et al. 2013)	↑inflammation
	Cronobacter sakazakii (H, M) (Hunter et al. 2009)	↑Nitric oxide, ↑inflam./IL-6
	Escherichia coli Nissle 1917 (H, M) (Kruis et al. 2004)	↕UC remission
	Salmonella typhimurium (H, M) (Stecher et al. 2006)	↑inflammation, ↑colitis
	Serratia marcescens (H, M, F) (Nehme et al. 2007)	↑inflammation
	Vibrio cholerae (H, M, F) (Blow et al. 2005; Ou et al. 2009)	↑inflammation, ↑IL-8
	Neisseria sp. (H, M, F) (Sibley et al. 2008)	↑inflammation
Actinobacteria	*Bifidobacterium infantis* (H, M) (Shiba et al. 2003)	↕immune response
	Bifidobacterium longum (H, M) (Ocon et al. 2013)	↓inflammation, ↓iNOS
	Bifidobacterium breve (H, M) (Menard et al. 2004)	↓inflammation, ↑IL-10, ↓TNF-α
	Bifidobacterium bifidum (H, M) (Kim et al. 2007a, b; Philippe et al. 2011b)	↓inflammation, ↓IFN-γ, ↓IL-6
	Bifidobacterium adolescentis (Kawabata et al. 2013)	↓inflammation, ↓NO
	Bifidobacterium lactis (H, M) (Philippe et al. 2011a)	↓inflammation, ↓TNF-α, ↓IL-6
	Mycobacterium paratuberculosis (H, M) (Momotani et al. 2012)	↑inflammation, ↑colitis
	Propionibacterium freudenreichii (H, M) (Uchida et al. 2007)	↑bacterial homeostasis, ↓IBD
	Propionibacterium acnes (H, M, F) (Sibley et al. 2008)	↑inflammation

for UC even though it may increase the risk for CD (Radford-Smith et al. 2002; Kaplan et al. 2008). Also, psychological factors affect the onset and progression of the disease, such as anxiety and depression that exacerbate IBD (Goodhand et al. 2012). While IBD patients may find it difficult to exercise (Bilski et al. 2014), recent evidence shows that exercise is associated with anti-inflammatory effects and prevention of colon cancer (Aoi et al. 2013).

7.3 Treatments

7.3.1 Diagnosis and Prognosis

The world gastroenterology organisation has released practice guidelines for the diagnosis and management of IBD (Bernstein et al. 2010). Currently, the use of video capsule endoscopy provides additional clinical information, treatment adjustments and better outcomes in children with IBD (Min et al. 2013). IBD or colon cancer history in the family is an indicator of possible genetic predisposition or environmental factors within the family environment. Intestinal inflammation symptoms vary from mild to severe and are attributed to UC if there is continuous inflammation in the colonic mucosa accompanied by frequent bloody diarrhea. In CD there is discontinuous transmural inflammation in any part of the alimentary canal, diarrhea with abdominal pain and tissue defects like ulcers and deep fissures. Two serological antibodies, perinuclear anti-neutrophil cytoplasmic antibody (p-ANCA) and anti-Saccharomyces cerevisiae antibody (ASCA) indicative of heightened immune response (Konrad et al. 2004; Walker et al. 2011) are certified as IBD diagnostic tools, however the number of biomarkers is rising (Meuwis et al. 2007; Han et al. 2013). Proteomic studies investigate biomarkers that aid in early diagnosis and classification of IBD through a selection of methodologies in sera, tissue and fecal samples (Bennike et al. 2014). Such simple non-invasive tests could identify IBD and possibly help prognosis and the development of personalised treatments.

Due to the multivariable nature of the disease, it is difficult to predict how IBD will progress. As with most diseases, the earlier the stage of IBD diagnosis the better the prognosis. However, the disease type and manifestation can provide some clues for the patient's prognosis. The presence of perianal disease is associated with 3–4 fold increased risk for severe disease (Tarrant et al. 2008). Furthermore, if the disease affects the small bowel instead of the colon, there is faster progression to complicated disease (Tarrant et al. 2008). Mucosal healing is a sign of good prognosis in both UC and CD and is linked with disease remission and decreased need for active therapy (Froslie et al. 2007; Baert et al. 2010). Serum biomarkers are another method of investigating the disease prognosis. Levels of the bacterial antigen I2, ASCA and anti-*E. coli* outer membrane porin protein C antibodies have been found to be correlated with a more perforating disease, strictures, fibrostenosis and the need for small bowel surgery (Mow et al. 2004). C-reactive protein

(CRP) is an indicator of prognosis in a number of diseases and in inflammatory bowel disease (Keshet et al. 2009). Elevated CRP indicates active disease and is associated with a more severe clinical course, with better response in infliximab treatment (Solem et al. 2005; Koelewijn et al. 2008). Other studies have also validated a panel of serological markers with antibody response to be associated with disease progression. For example, anti-neutrophil cytoplasmic (ANCA), anti-CBir1 Flagellin and anti-glycan antibodies have been used to distinguish disease type and been correlated with disease prognosis (Papp et al. 2008; Dotan 2010; Kovacs et al. 2014). Anti-CBir1 and pANCA expression have been correlated with ileal pouch anal anastomosis and the pouchitis in UC patients (Fleshner et al. 2008). ASCA antibody reactivity is associated with stenosing and penetrating CD behaviour with increased need for surgery in children and adults (Dubinsky et al. 2008). Pancreatic autoantibodies for glycoprotein 2 (GP2) and CUB/zona pellucida-like domain-containing protein 1 (CUZD1) have been recently implicated in the diagnosis of UC and CD but further research is needed for their correlation with any prognostic value (Roggenbuck et al. 2009; Komorowski et al. 2013).

7.3.2 Drugs and Surgery

Management of UC and CD have been proposed by the European evidence-based consensus panel of the European Crohn's and Colitis Organisation in 2010 and 2012 (Dignass et al. 2010, 2012). Drug treatment of IBD is focused on the reduction of inflammation in the gastrointestinal tract. Aminosalicylates, such as sulfasalazine and mesalazine, are usually the first anti-inflammatory drugs used for the treatment of active and quiescent UC, but not CD (Nielsen and Munck 2007). Corticosteroids are commonly used as anti-inflammatory agents in both CD and UC. They act by inhibiting various inflammatory pathways and induce remission and minimize the chances of a relapse. These can be delivered orally, intravenously or through suppositories for direct absorption. However, steroids have a number of side effects therefore an initial high dose is administered in conjunction with other regiments to counteract the side effects, and a steady dose reduction until complete withdrawal from steroids (Mowat et al. 2011). If steroids are ineffective, then intensification of treatment or surgical intervention is considered. Up to 1 in 4 patients with UC and up to 70 % of patients with CD will require surgery (Hoie et al. 2007). This is usually required if the disease progresses without responding to therapies, upon extensive damage with chronic severe symptoms or increased risk of malignancy. Half of operated CD patients will need to do another surgery due to disease relapse (de Buck van Overstraeten et al. 2012). However it is mostly preferred to maintain a remission status through alternative treatments before proceeding to any surgical intervention.

The use of non-steroidal anti-inflammatory drugs (NSAIDS) is not recommended because they may cause gastrointestinal toxicity and mucosal injury and have been associated with the onset or relapse of IBD (Klein and Eliakim 2010).

Azathioprine (AZA) and 6-mercaptopurine are routinely used as non-steroids for the treatment of UC and CD, despite the concerns about toxicity (Ardizzone et al. 2006; Prefontaine et al. 2009; Chande et al. 2013). Antibiotics are important in the treatment of secondary complications of IBD as they control bacterial overgrowth and abscess formation (Castiglione et al. 2003). Ciprofloxacin, metronidazole, rifaximin and ornidazole are antibiotics with antimicrobial action that have been found to be beneficial for the treatment of perianal CD and pouchitis (Kale-Pradhan et al. 2013). Calcineurin and methotrexate are some other drugs that have been found to be beneficial in treating UC and CD respectively (Mowat et al. 2011). Patients that have failed to respond to standard immunosuppression therapies are given antibodies against Tumor Necrosis Factor α (anti-TNFα), namely Infliximab or Adalimumab. TNF can promote chronic colitis by suppressing glucocorticoid (anti-inflammatory steroid hormones related with decreased IBD severity) synthesis in the intestine (Huang et al. 2014). Infliximab and Adalimumab block effectively the pro-inflammatory effects of TNFα in both UC and CD, but patients might experience other inflammatory disorders as a side effect (Rutgeerts et al. 2005; Hanauer et al. 2006; Niess and Danese 2014).

7.3.3 Diet and Probiotics

Malnutrition has been observed in up to 50 % of IBD hospitalised patients and is an indicator of disease severity (Nguyen et al. 2008). Therefore, frequent analysis of nutrient uptake should be performed to assess if nutrient supplements (vitamins, iron, calcium, proteins) should be provided especially when significant weight loss is observed. Diet may directly affect bacterial diversity in the gut, shape the abnormal microbiota composition and restore intestinal balance. For example, semi-vegetarian diet has been investigated in CD patients that achieved remission and found to significantly prevent relapse in 15 out of 16 patients (Chiba et al. 2010). Probiotics, like *Lactobacillus* and *Bifidobacterium* species, have been found to prevent pouchitis and alleviate symptoms by inducing and maintain remission of ulcerative colitis (Bibiloni et al. 2005; Gionchetti et al. 2003). A probiotic cocktail of four strains of *Lactobacillus*, three strains of *Bifidobacterium* and *Streptococcus salivaris* has been tested in humans and mice and found to alleviate IBD by inducing IL-10 and TGFβ expressing cells (Bibiloni et al. 2005; Di Giacinto et al. 2005). Additional studies corroborating these findings show that *Lactobacillus* species may decrease pro-inflammatory cytokines and intestinal inflammation (Schultz et al. 2002; Ma et al. 2004; Foligne et al. 2007). Pathogen induced inflammation can also be controlled by probiotics. For example, *Bifidobacterium infantis* decreases the levels of pro-inflammatory cytokines and inflammation upon Salmonella infection by regulating T cells induction and NF-κB activation in mice (O'Mahony et al. 2008). Fecal microbiota transplantation is a newer approach with promising results as it is associated with beneficial changes in fecal bacterial diversity and it was successful in treating IBD in a number of cases (van Nood et al. 2013; Kao et al. 2014; Kunde et al. 2013).

7.3.4 Clinical Trials

A number of human clinical trials are on-going around the world to investigate promising biological agents and various drugs towards their efficacy in IBD. For UC and CD children that do not respond to conventional therapies a clinical trial (Identifier: NCT02150551) has been initiated that will assess intravenously allogeneic bone marrow-derived mesenchymal stromal cells for their ability to locate inflammation, decrease pro-inflammatory cytokines and promote tissue repair. Moreover, new drugs and biomarkers are being tested in humans, such as AST-120 (Identifier: NCT00321412) and serum calprotectin (Meuwis et al. 2013). Humanised monoclonal antibodies against cytokines, such as anti-IL-2 receptor are investigated against moderate to severe UC (Identifier: NCT00073047). Several studies are using fecal microbiota transplantation or bacteriotherapy in treating IBD (Identifiers: NCT01560819, NCT02108821, NCT01793831, NCT01757964).

7.4 Advances and Overlapping Findings in Ibd and Cancer Research with the Use of Mouse and Fly Models

7.4.1 Mouse Models

In recent years animal models of inflammatory bowel disease have been developed to assist researchers in investigating IBD. Mouse is a very good model organism because it retains much of the complexity of human biology while being experimentally fairly manageable. There are two groups of inflammatory bowel disease models, the chemically-induced and the genetically-predisposed mouse models. Intestinal inflammation can be triggered by chemicals like dextran sodium sulphate (DSS), TNBS/DNBS and Oxazolone. DSS induced colitis is the most widely used chemical model due to its ability to induce both acute or chronic colitis depending on the protocol (Okayasu et al. 1990). Feeding mice with DSS in the drinking water for several days induces acute colitis accompanied by ulcerations, bloody diarrhea and infiltration with granulocytes (Wirtz and Neurath 2007). Various mouse but also human studies have shown that underneath the gut epithelium, lamina propria immune cells, such as macrophages, dendritic cells and T-cells, respond to microbial products and tissue damage to produce cytokines, such as TNFα, and attract neutrophils and innate lymphoid cells (ILCs) and activate further macrophages, T cells and other tissue cells (Rimoldi et al. 2005; Zaph et al. 2007; Coccia et al. 2012; Kang et al. 2012; Peterson and Artis 2014). For example, *Helicobacter hepaticus* has been found to enhance recruitment of granulocytes and ILCs in the gut causing intestinal inflammation through IL-23 and IL-1β signalling (Kullberg et al. 2006; Coccia et al. 2012). TNFα stimulates angiogenesis, epithelial cell death, DNA damage through reactive oxygen species and promotes cellular transformation (Yan et al. 2006). IL-6 and IL-22 cytokines, are secreted by macrophages and dendritic cells or T cells and ILCs respectively to activate the transcription factor STAT3. Under physiological conditions, activated STAT3 maintains intestinal regeneration of epithelial cells and activates the production of mucus and antimicrobial peptides, such as defensins and regenerating islet derived (REG) proteins (Zindl et al. 2013). However, activation of the transcription factor STAT3 in genetically predisposed epithelial cells by IL-6 and IL-22 may also promote stemness and colorectal cancer development (Grivennikov et al. 2009; Kirchberger et al. 2013; Kryczek et al. 2014).

In conjunction with a carcinogen, such as azoxymethane (AOM), chronic inflammation in mice leads to the development of colorectal cancer (Tanaka et al. 2003). As an alternative, transgenic mice provide a sensitised genetic background that can be used to directly investigate particular molecular mechanisms of inflammation and its contribution to carcinogenesis. Interleukin-10 (IL-10) deficient mice, for example, exhibit spontaneous colitis and adenocarcinoma associated with aberrant cytokine production and T cell responses (Berg et al. 1996). To investigate the pathogenesis of intestinal cancer, adenomatous polyposis coli gene (APC) mutant mice are one of the first models to be used for the development of intestinal adenomas (Oshima et al. 1995). These models have also been used in combination with bacterial infections to study the role of intestinal microbiota in IBD because inflammation and cancer do not develop in germ free mice (Gkouskou et al. 2014). Increased Enterobacteriaceae and Bacteroides levels are correlated with intestinal inflammation in DSS and IL-10 deficient mice (Bloom et al. 2011; Arthur et al. 2012; Hakansson et al. 2014). Moreover, *Fusobacterium nucleatum* and enterotoxigenic *Bacteroides fragilis* have the potential to induce colon cancer (Wu et al. 2009b; Kostic et al. 2013). Genotoxic and adherent invasive *Escherichia coli* have also been linked with colon cancer where they were found to cause genomic instability through their polyketide synthase (pks) genotoxic island and promote invasive carcinoma in a mouse model (Arthur et al. 2012; Sears and Garrett 2014). The extent to which inflammation plays a role in the models of intestinal carcinogenesis it is still an open question.

7.4.2 Fly Models

Drosophila research is facilitated by a wealth of genetic tools for the investigation of innate immunity and intestinal pathophysiology. It is a relatively fast, cheap and easy to use organism with significant similarities to the digestive tract of humans. The digestive tract of mammals goes through the esophagus and into the stomach for digestion; the nutrients are absorbed in the small and large intestine and are excreted from the rectum. In flies a similar process follows the consumption of food, as it passes through the foregut and into the crop; absorption takes place in the midgut and hindgut and waste products eventually exit through the anus (Apidianakis and Rahme 2011). Also tissue architecture and regeneration in both flies and mice depend on intestinal stem cells that give rise to enteroblasts or transit amplifying cells, which develop into enterocytes or secretory cells through conserved signaling pathways (Apidianakis and Rahme 2011). The simpler immune system of the fly reduces the complexity and enables the easier study of human intestinal microbes and their relation to the host defense responses that lead to disease (Panayidou et al. 2014). Furthermore, infectious disease and cancer studies in *Drosophila* are facilitated by established techniques for large scale in vivo RNAi and drug testing (Tzelepis et al. 2013).

Drosophila intestinal homeostasis and pathology upon infection or aging is guided by conserved signaling pathways (Apidianakis and Rahme 2011) (Fig. 7.1). *Drosophila* lacks adaptive immunity but its innate immune system is well conserved (Lemaitre and Hoffmann 2007). Microbicidal reactive oxygen species (ROS) are produced by the NADPH oxidase Duox forming a line of defense in the intestinal mucosa (Ha et al. 2005). The NF-κB/Imd and JAK-STAT serve as an additional level of defense controlling intestinal antimicrobial peptide (AMP) production (Buchon et al. 2009b; Lee and Lee 2014). *Drosophila* gut regeneration serves as a third line of defense against intestinal infection and stress (Panayidou et al. 2014). While stem cell-mediated proliferation and differentiation maintain intestinal homeostasis (Micchelli and Perrimon 2006; Ohlstein and Spradling 2006), intestinal infection or enterocyte stress or damage or aging induce regenerative inflammatory signaling via the JNK and JAK-STAT signaling pathway (Apidianakis et al. 2009; Biteau et al. 2008; Buchon et al. 2009b; Cronin et al. 2009; Jiang et al. 2009; Panayidou and Apidianakis 2013).

Flies constitutively activating the NF-κB/Imd pathway in the gut exhibit an intestinal inflammation-like condition, accompanied by dysbiosis that leads to gut cell apoptosis and premature death (Ryu et al. 2008). These flies produce AMPs excessively, which lead to dysbiosis, characterized by pathobiont proliferation and intestinal damage (Lee and Lee 2014). Interestingly, inhibition of the NF-κB/Imd pathway results in high colonization with gut commensals, which leads to faster intestinal regeneration and hyperplasia because of concomitant excessive ROS production by Duox (Buchon et al. 2009a). Similarly, Duox loss of function in transgenic flies increases death rate after a rather mild bacterial infection, while Duox overexpression leads to enterocyte damage via excessive ROS production (Ha et al. 2005). A recent study showed that chronic activation of Foxo, a transcription factor found to repress peptidoglycan recognition protein SC2 (homolog of human anti-inflammatory molecule PGLYRP1-4 and negative regulator of the NF-κB/Imd pathway), induces immunosenescence in aging enterocytes causing commensal dysbiosis and intestinal epithelial dysplasia (Guo et al. 2014). Furthermore overexpression of PGRP-SC2 in enterocytes prevents dysbiosis and extends life span, emphasizing the importance of bacterial balance and gut homeostasis in host longevity (Guo et al. 2014). Similarly to mice, germ free flies reverse the effect of genetic- or aging-driven and host defense-mediated intestinal damage (Guo et al. 2014; Lee and Lee 2014).

Genetically predisposed flies have also been investigated in the context of intestinal defense response-driven tumorigenesis. Infection with *Pseudomonas aeruginosa*, a Gram-negative opportunistic pathogen, induces enterocyte apoptosis and mediates intestinal stem cell regeneration via c-Jun N-terminal kinase pathway activation (JNK) (Apidianakis et al. 2009). When flies are predisposed with a latent Ras1 oncogene, *P. aeruginosa* infection stimulates extreme stem cell proliferation and intestinal dysplasia, illustrating a synergy between bacterial infection and genetic predisposition (Apidianakis et al. 2009). Furthermore, persistent bacterial infection activates the Imd-dTab2-dTak1 innate immune pathway, which synergizes with the Ras1 ^{V12} oncogene to induce extracellular matrix degradation, basal invasion and hindgut cell dissemination to distant sites (Bangi et al. 2012; Christofi and Apidianakis 2012).

7.4.3 Perspectives on the Parallels Observed Between Mammals and Flies

Inflammatory bowel diseases have been on the rise during the last decades. There are a number of etiologies for IBD, such as, the westernization of life style, genetic predisposition, dysbiosis and prolonged use of antibiotics. Treatment strategies have also been updated with new drugs and probiotic cocktails that seem to alleviate the disease. However, the diverse responses to these treatments suggest the need for more personalized therapies tailored for each patient. The use of models organisms is invaluable for the study of this multifactorial disease, opening a new era for UC and CD understanding and treatment. Nevertheless many questions remain unanswered: Which synergisms among diet, microbiota and host genetics trigger IBD and inflammation-driven tumorigenesis? Consequently, which combinations of diets and microbiota may eliminate dysbiosis, IBD and inflammation-driven tumorigenesis? Given the differences between humans and mice in inflammatory response (Seok et al. 2013), to what extent are mouse and fly data translatable into therapies for humans?

While answers to these questions may come from further studies on humans, mice and flies, model organisms already provide some consensus on the signals that drive intestinal inflammation-driven tumorigenesis (Fig. 7.1) that should be relevant to human disease. Lessons from mice during the last few years indicate among others that innate immune cells (ILCs, dendritic cells, macrophages and neutrophils) secrete IL6/IL22 to induce intestinal epithelium cell proliferation and innate immunity (mucus, AMPs) via STAT3 signaling (Takeda et al. 1999; Grivennikov et al. 2009; Pickert et al. 2009; Wittkopf et al. 2011; Murano et al. 2014). Also EGFR pathway is induced in epithelial cells by ligands (Epiregulin, Amphiregulin) expressed in epithelial cells and myofibroblasts (Nishimura et al. 2008) and intestinal defense response NADPH oxidases, Nox1 induces epithelial regeneration (Jones et al. 2013). Interestingly, innate and adaptive immune cells produce TNF and IFN to induce enterocyte apoptosis, but these same signals promote tumorigenesis upon genetic predisposition (Neurath 2014).

Some striking parallels come from fly studies of the last few years (Fig. 7.1). *Drosophila* intestinal cells (enterocytes, progenitor cells and intestinal muscle) induce regenerative inflammatory signaling and AMPs via JAK-STAT signaling (Apidianakis et al. 2009; Buchon et al. 2009b; Jiang et al. 2009). Also EGFR pathway is induced in intestinal stem cells (ISCs) by ligands (Vein, Spitz, Keren) expressed in neighboring muscle and epithelial cells (Biteau and Jasper 2011; Xu et al. 2011) and intestinal defense response NADPH oxidase, dNox induces epithelial regeneration (Jones et al. 2013). Lastly, *Drosophila* TNF/Egr is induced by phagocytes (plasmatocytes) inducing apoptosis in epithlelial cells, but promotes proliferation of tumorigenic cells (Cordero et al. 2010). Thus modeling in *Drosophila* is not only poised to identify new genes pertinent to intestinal inflammation and inflammation-driven tumorigenesis, but may also provide insights and reinforce observations on some basic principles of disease development.

References

Abraham C, Cho JH (2009) Inflammatory bowel disease. N Engl J Med 361(21):2066–2078

Alzoghaibi MA, Al-Mofleh IA, Al-Jebreen AM (2008) Neutrophil chemokines GCP-2 and GRO-alpha in patients with inflammatory bowel disease. J Dig Dis 9(3):144–148

Amir Shaghaghi M, Bernstein CN, Serrano Leon A, El-Gabalawy H, Eck P (2014) Polymorphisms in the sodium-dependent ascorbate transporter gene SLC23A1 are associated with susceptibility to Crohn disease. Am J Clin Nutr 99(2):378–383

Andersen V, Olsen A, Carbonnel F, Tjonneland A, Vogel U (2012) Diet and risk of inflammatory bowel disease. Dig Liver Dis 44(3):185–194

Anderson CA, Boucher G, Lees CW, Franke A, D'Amato M, Taylor KD et al (2011) Meta-analysis identifies 29 additional ulcerative colitis risk loci, increasing the number of confirmed associations to 47. Nat Genet 43(3):246–252

Aoi W, Naito Y, Takagi T, Tanimura Y, Takanami Y, Kawai Y et al (2013) A novel myokine, secreted protein acidic and rich in cysteine (SPARC), suppresses colon tumorigenesis via regular exercise. Gut 62(6):882–889

Apidianakis Y, Pitsouli C, Perrimon N, Rahme L (2009) Synergy between bacterial infection and genetic predisposition in intestinal dysplasia. Proc Natl Acad Sci U S A 106(49):20883–20888

Apidianakis Y, Rahme LG (2011) Drosophila melanogaster as a model for human intestinal infection and pathology. Dis Model Mech 4(1):21–30

Araki A, Kanai T, Ishikura T, Makita S, Uraushihara K, Iiyama R et al (2005) MyD88-deficient mice develop severe intestinal inflammation in dextran sodium sulfate colitis. J Gastroenterol 40(1):16–23

Ardizzone S, Maconi G, Russo A, Imbesi V, Colombo E, Bianchi Porro G (2006) Randomised controlled trial of azathioprine and 5-aminosalicylic acid for treatment of steroid dependent ulcerative colitis. Gut 55(1):47–53

Arthur JC, Perez-Chanona E, Muhlbauer M, Tomkovich S, Uronis JM, Fan TJ et al (2012) Intestinal inflammation targets cancer-inducing activity of the microbiota. Science 338(6103):120–123

Asakura H, Suzuki K, Kitahora T, Morizane T (2008) Is there a link between food and intestinal microbes and the occurrence of Crohn's disease and ulcerative colitis? J Gastroenterol Hepatol 23(12):1794–1801

Atarashi K, Tanoue T, Shima T, Imaoka A, Kuwahara T, Momose Y et al (2011) Induction of colonic regulatory T cells by indigenous *Clostridium* species. Science 331(6015):337–341

Backhed F, Ley RE, Sonnenburg JL, Peterson DA, Gordon JI (2005) Host-bacterial mutualism in the human intestine. Science 307(5717):1915–1920

Baert F, Moortgat L, Van Assche G, Caenepeel P, Vergauwe P, De Vos M et al (2010) Mucosal healing predicts sustained clinical remission in patients with early-stage Crohn's disease. Gastroenterology 138(2):463–468

Balish E, Warner T (2002) Enterococcus faecalis induces inflammatory bowel disease in interleukin-10 knockout mice. Am J Pathol 160(6):2253–2257

Bangi E, Pitsouli C, Rahme LG, Cagan R, Apidianakis Y (2012) Immune response to bacteria induces dissemination of Ras-activated Drosophila hindgut cells. EMBO Rep 13(6):569–576

Bank S, Skytt Andersen P, Burisch J, Pedersen N, Roug S, Galsgaard J et al (2014) Polymorphisms in the inflammatory pathway genes TLR2, TLR4, TLR9, LY96, NFKBIA, NFKB1, TNFA, TNFRSF1A, IL6R, IL10, IL23R, PTPN22, and PPARG are associated with susceptibility of inflammatory bowel disease in a Danish Cohort. PLoS ONE 9(6):98815

Barnich N, Carvalho FA, Glasser AL, Darcha C, Jantscheff P, Allez M et al (2007) CEACAM6 acts as a receptor for adherent-invasive *E. coli*, supporting ileal mucosa colonization in Crohn disease. J Clin Invest 117(6):1566–1574

Bassaganya-Riera J, DiGuardo M, Viladomiu M, de Horna A, Sanchez S, Einerhand AW et al (2011) Soluble fibers and resistant starch ameliorate disease activity in interleukin-10-deficient mice with inflammatory bowel disease. J Nutr 141(7):1318–1325

Beaudoin M, Goyette P, Boucher G, Lo KS, Rivas MA, Stevens C et al (2013) Deep resequencing of GWAS loci identifies rare variants in CARD9, IL23R and RNF186 that are associated with ulcerative colitis. PLoS Genet 9(9):e1003723

Benight NM, Stoll B, Chacko S, da Silva VR, Marini JC, Gregory JF et al (2011) B-vitamin deficiency is protective against DSS-induced colitis in mice. Am J Physiol Gastrointest Liver Physiol 301(2):G249–G259

Bennike T, Birkelund S, Stensballe A, Andersen V (2014) Biomarkers in inflammatory bowel diseases: current status and proteomics identification strategies. World J Gastroenterol 20(12):3231–3244

Berg DJ, Davidson N, Kuhn R, Muller W, Menon S, Holland G et al (1996) Enterocolitis and colon cancer in interleukin-10-deficient mice are associated with aberrant cytokine production and CD4(+) TH1-like responses. J Clin Invest 98(4):1010–1020

Bernstein CN, Fried M, Krabshuis JH, Cohen H, Eliakim R, Fedail S et al (2010) World gastroenterology organization practice guidelines for the diagnosis and management of IBD in 2010. Inflamm Bowel Dis 16(1):112–124

Bibiloni R, Fedorak RN, Tannock GW, Madsen KL, Gionchetti P, Campieri M et al (2005) VSL#3 probiotic-mixture induces remission in patients with active ulcerative colitis. Am J Gastroenterol 100(7):1539–1546

Bilski J, Brzozowski B, Mazur-Bialy A, Sliwowski Z, Brzozowski T (2014) The role of physical exercise in inflammatory bowel disease. Biomed Res Int 2014:429031

Biteau B, Hochmuth CE, Jasper H (2008) JNK activity in somatic stem cells causes loss of tissue homeostasis in the aging *Drosophila* gut. Cell Stem Cell 3(4):442–455

Biteau B, Jasper H (2011) EGF signaling regulates the proliferation of intestinal stem cells in *Drosophila*. Development 138(6):1045–1055

Blair SA, Kane SV, Clayburgh DR, Turner JR (2006) Epithelial myosin light chain kinase expression and activity are upregulated in inflammatory bowel disease. Lab Invest 86(2):191–201

Bloom SM, Bijanki VN, Nava GM, Sun L, Malvin NP, Donermeyer DL et al (2011) Commensal *Bacteroides* species induce colitis in host-genotype-specific fashion in a mouse model of inflammatory bowel disease. Cell Host Microbe 9(5):390–403

Blow NS, Salomon RN, Garrity K, Reveillaud I, Kopin A, Jackson FR et al (2005) *Vibrio cholerae* infection of Drosophila melanogaster mimics the human disease cholera. PLoS Pathog 1(1):e8

Bonnay F, Cohen-Berros E, Hoffmann M, Kim SY, Boulianne GL, Hoffmann JA et al (2013) Big bang gene modulates gut immune tolerance in *Drosophila*. Proc Natl Acad Sci U S A 110(8):2957–2962

Bosco-Drayon V, Poidevin M, Boneca IG, Narbonne-Reveau K, Royet J, Charroux B (2012) Peptidoglycan sensing by the receptor PGRP-LE in the *Drosophila* gut induces immune responses to infectious bacteria and tolerance to microbiota. Cell Host Microbe 12(2):153–165

Brant SR (2011) Update on the heritability of inflammatory bowel disease: the importance of twin studies. Inflamm Bowel Dis 17(1):1–5

Brest P, Lapaquette P, Souidi M, Lebrigand K, Cesaro A, Vouret-Craviari V et al (2011) A synonymous variant in IRGM alters a binding site for miR-196 and causes deregulation of IRGM-dependent xenophagy in Crohn's disease. Nat Genet 43(3):242–245

Buchon N, Broderick NA, Chakrabarti S, Lemaitre B (2009a) Invasive and indigenous microbiota impact intestinal stem cell activity through multiple pathways in *Drosophila*. Genes Dev 23(19):2333–2344

Buchon N, Broderick NA, Poidevin M, Pradervand S, Lemaitre B (2009b) *Drosophila* intestinal response to bacterial infection: activation of host defense and stem cell proliferation. Cell Host Microbe 5(2):200–211

Buisine MP, Desreumaux P, Debailleul V, Gambiez L, Geboes K, Ectors N et al (1999) Abnormalities in mucin gene expression in Crohn's disease. Inflamm Bowel Dis 5(1):24–32

Burisch J, Jess T, Martinato M, Lakatos PL, ECCO-EpiCom (2013) The burden of inflammatory bowel disease in Europe. J Crohns Colitis 7(4):322–337

Burisch J, Pedersen N, Cukovic-Cavka S, Brinar M, Kaimakliotis I, Duricova D et al (2014) East-West gradient in the incidence of inflammatory bowel disease in Europe: the ECCO-EpiCom inception cohort. Gut 63(4):588–597

Cabrera S, Fernandez AF, Marino G, Aguirre A, Suarez MF, Espanol Y et al (2013) ATG4B/autophagin-1 regulates intestinal homeostasis and protects mice from experimental colitis. Autophagy 9(8):1188–1200

Cahill RJ, Foltz CJ, Fox JG, Dangler CA, Powrie F, Schauer DB (1997) Inflammatory bowel disease: an immunity-mediated condition triggered by bacterial infection with *Helicobacter hepaticus*. Infect Immun 65(8):3126–3131

Campbell BJ, Yu LG, Rhodes JM (2001) Altered glycosylation in inflammatory bowel disease: a possible role in cancer development. Glycoconj J 18(11–12):851–858

Carrier J, Medline A, Sohn KJ, Choi M, Martin R, Hwang SW et al (2003) Effects of dietary folate on ulcerative colitis-associated colorectal carcinogenesis in the interleukin 2-and beta(2)-microglobulin-deficient mice. Cancer Epidemiol Biomarkers Prev 12(11 Pt 1):1262–1267

Casini-Raggi V, Kam L, Chong YJ, Fiocchi C, Pizarro TT, Cominelli F (1995) Mucosal imbalance of IL-1 and IL-1 receptor antagonist in inflammatory bowel disease. A novel mechanism of chronic intestinal inflammation. J Immunol 154(5):2434–2440

Castiglione F, Rispo A, Di Girolamo E, Cozzolino A, Manguso F, Grassia R et al (2003) Antibiotic treatment of small bowel bacterial overgrowth in patients with Crohn's disease. Aliment Pharmacol Ther 18(11–12):1107–1112

Chande N, Tsoulis DJ, MacDonald JK (2013) Azathioprine or 6-mercaptopurine for induction of remission in Crohn's disease. Cochrane Database Syst Rev 4:CD000545

Chandler JA, Lang JM, Bhatnagar S, Eisen JA, Kopp A (2011) Bacterial communities of diverse *Drosophila* species: ecological context of a host-microbe model system. PLoS Genet 7(9):e1002272

Chang JY, Antonopoulos DA, Kalra A, Tonelli A, Khalife WT, Schmidt TM et al (2008) Decreased diversity of the fecal Microbiome in recurrent *Clostridium difficile*-associated diarrhea. J Infect Dis 197(3):435–438

Chatterjee D, Chaudhuri K (2013) *Vibrio cholerae* O395 outer membrane vesicles modulate intestinal epithelial cells in a NOD1 protein-dependent manner and induce dendritic cell-mediated Th2/Th17 cell responses. J Biol Chem 288(6):4299–4309

Chiba M, Abe T, Tsuda H, Sugawara T, Tsuda S, Tozawa H et al (2010) Lifestyle-related disease in Crohn's disease: relapse prevention by a semi-vegetarian diet. World J Gastroenterol 16(20):2484–2495

Christofi T, Apidianakis Y (2012) Ras-oncogenic *Drosophila* hindgut but not midgut cells use an inflammation-like program to disseminate to distant sites. Gut Microbes 4(1):1–6

Coccia M, Harrison OJ, Schiering C, Asquith MJ, Becher B, Powrie F et al (2012) IL-1beta mediates chronic intestinal inflammation by promoting the accumulation of IL-17A secreting innate lymphoid cells and CD4(+) Th17 cells. J Exp Med 209(9):1595–1609

Connelly TM, Koltun WA, Berg AS, Hegarty JP, Brinton D, Deiling S et al (2013) A single nucleotide polymorphism in the STAT5 gene favors colonic as opposed to small-bowel inflammation in Crohn's disease. Dis Colon Rectum 56(9):1068–1074

Cordero JB, Macagno JP, Stefanatos RK, Strathdee KE, Cagan RL, Vidal M (2010) Oncogenic Ras diverts a host TNF tumor suppressor activity into tumor promoter. Dev Cell 18(6):999–1011

Costello CM, Mah N, Hasler R, Rosenstiel P, Waetzig GH, Hahn A et al (2005) Dissection of the inflammatory bowel disease transcriptome using genome-wide cDNA microarrays. PLoS Med 2(8):e199

Cronin SJ, Nehme NT, Limmer S, Liegeois S, Pospisilik JA, Schramek D et al (2009) Genome-wide RNAi screen identifies genes involved in intestinal pathogenic bacterial infection. Science 325(5938):340–343

D'Amato M, Bruce S, Bresso F, Zucchelli M, Ezer S, Pulkkinen V et al (2007) Neuropeptide s receptor 1 gene polymorphism is associated with susceptibility to inflammatory bowel disease. Gastroenterology 133(3):808–817

Dahan S, Rabinowitz KM, Martin AP, Berin MC, Unkeless JC, Mayer L (2011) Notch-1 signaling regulates intestinal epithelial barrier function, through interaction with CD4 + T cells, in mice and humans. Gastroenterology 140(2):550–559

de Buck van Overstraeten A, Wolthuis A, D'Hoore A (2012) Surgery for Crohn's disease in the era of biologicals: a reduced need or delayed verdict? World J Gastroenterol 18(29):3828–3832

Deming DA, Leystra AA, Nettekoven L, Sievers C, Miller D, Middlebrooks M et al (2014) PIK3CA and APC mutations are synergistic in the development of intestinal cancers. Oncogene 33(17):2245–2254

Devkota S, Wang Y, Musch MW, Leone V, Fehlner-Peach H, Nadimpalli A et al (2012) Dietary-fat-induced taurocholic acid promotes pathobiont expansion and colitis in Il10-/- mice. Nature 487(7405):104–108

Dhillon SS, Mastropaolo LA, Murchie R, Griffiths C, Thoni C, Elkadri A et al (2014) Higher activity of the inducible nitric oxide synthase contributes to very early onset inflammatory bowel disease. Clin Transl Gastroenterol 5:e46

Di Giacinto C, Marinaro M, Sanchez M, Strober W, Boirivant M (2005) Probiotics ameliorate recurrent Th1-mediated murine colitis by inducing IL-10 and IL-10-dependent TGF-beta-bearing regulatory cells. J Immunol 174(6):3237–3246

Dignass A, Eliakim R, Magro F, Maaser C, Chowers Y, Geboes K et al (2012) Second European evidence-based consensus on the diagnosis and management of ulcerative colitis part 1: definitions and diagnosis. J Crohns Colitis 6(10):965–990

Dignass A, Van Assche G, Lindsay JO, Lemann M, Soderholm J, Colombel JF et al (2010) The second European evidence-based Consensus on the diagnosis and management of Crohn's disease: Current management. J Crohns Colitis 4(1):28–62

Dotan I (2010) New serologic markers for inflammatory bowel disease diagnosis. Dig Dis 28(3):418–423

Dubinsky MC, Kugathasan S, Mei L, Picornell Y, Nebel J, Wrobel I et al (2008) Increased immune reactivity predicts aggressive complicating Crohn's disease in children. Clin Gastroenterol Hepatol 6(10):1105–1111

Edwards LA, O'Neill C, Furman MA, Hicks S, Torrente F, Perez-Machado M et al (2012) Enterotoxin-producing staphylococci cause intestinal inflammation by a combination of direct epithelial cytopathy and superantigen-mediated T-cell activation. Inflamm Bowel Dis 18(4):624–640

Elrod JW, Laroux FS, Houghton J, Carpenter A, Ando T, Jennings MH et al (2005) DSS-induced colitis is exacerbated in STAT-6 knockout mice. Inflamm Bowel Dis 11(10):883–889

Fearon ER, Vogelstein B (1990) A genetic model for colorectal tumorigenesis. Cell 61(5):759–767

Fleshner P, Ippoliti A, Dubinsky M, Vasiliauskas E, Mei L, Papadakis KA et al (2008) Both pre-operative perinuclear antineutrophil cytoplasmic antibody and anti-CBir1 expression in ulcerative colitis patients influence pouchitis development after ileal pouch-anal anastomosis. Clin Gastroenterol Hepatol 6(5):561–568

Foligne B, Nutten S, Grangette C, Dennin V, Goudercourt D, Poiret S et al (2007) Correlation between in vitro and in vivo immunomodulatory properties of lactic acid bacteria. World J Gastroenterol 13(2):236–243
Franchimont D, Vermeire S, El Housni H, Pierik M, Van Steen K, Gustot T et al (2004) Deficient host-bacteria interactions in inflammatory bowel disease? The toll-like receptor (TLR)-4 Asp299gly polymorphism is associated with Crohn's disease and ulcerative colitis. Gut 53(7):987–992
Frank DN, St Amand AL, Feldman RA, Boedeker EC, Harpaz N, Pace NR (2007) Molecular-phylogenetic characterization of microbial community imbalances in human inflammatory bowel diseases. Proc Natl Acad Sci U S A 104(34):13780–13785
Franke A, McGovern DP, Barrett JC, Wang K, Radford-Smith GL, Ahmad T et al (2010) Genome-wide meta-analysis increases to 71 the number of confirmed Crohn's disease susceptibility loci. Nat Genet 42(12):1118–1125
Froslie KF, Jahnsen J, Moum BA, Vatn MH, IBSEN Group (2007) Mucosal healing in inflammatory bowel disease: results from a Norwegian population-based cohort. Gastroenterology 133(2):412–422
Fuhrer A, Sprenger N, Kurakevich E, Borsig L, Chassard C, Hennet T (2010) Milk sialyllactose influences colitis in mice through selective intestinal bacterial colonization. J Exp Med 207(13):2843–2854
Fujisawa M, Oguchi K, Yamaura T, Suzuki M, Cyong JC (2005) Protective effect of hawthorn fruit on murine experimental colitis. Am J Chin Med 33(2):167–180
Garrett WS, Lord GM, Punit S, Lugo-Villarino G, Mazmanian SK, Ito S et al (2007) Communicable ulcerative colitis induced by T-bet deficiency in the innate immune system. Cell 131(1):33–45
Gensollen T, Bourges C, Rihet P, Rostan A, Millet V, Noguchi T et al (2013) Functional polymorphisms in the regulatory regions of the VNN1 gene are associated with susceptibility to inflammatory bowel diseases. Inflamm Bowel Dis 19(11):2315–2325
George RH, Symonds JM, Dimock F, Brown JD, Arabi Y, Shinagawa N et al (1978) Identification of clostridium difficile as a cause of pseudomembranous colitis. Br Med J 1(6114):695
Ghosh S, DeCoffe D, Brown K, Rajendiran E, Estaki M, Dai C et al (2013) Fish oil attenuates omega-6 polyunsaturated fatty acid-induced dysbiosis and infectious colitis but impairs LPS dephosphorylation activity causing sepsis. PLoS ONE 8(2):e55468
Gionchetti P, Rizzello F, Helwig U, Venturi A, Lammers KM, Brigidi P et al (2003) Prophylaxis of pouchitis onset with probiotic therapy: a double-blind, placebo-controlled trial. Gastroenterology 124(5):1202–1209
Giuffrida P, Biancheri P, MacDonald TT (2014) Proteases and small intestinal barrier function in health and disease. Curr Opin Gastroenterol 30(2):147–153
Gkouskou KK, Deligianni C, Tsatsanis C, Eliopoulos AG (2014) The gut microbiota in mouse models of inflammatory bowel disease. Front Cell Infect Microbiol 4:28
Goodhand JR, Wahed M, Mawdsley JE, Farmer AD, Aziz Q, Rampton DS (2012) Mood disorders in inflammatory bowel disease: relation to diagnosis, disease activity, perceived stress, and other factors. Inflamm Bowel Dis 18(12):2301–2309
Greenstein AJ, Lachman P, Sachar DB, Springhorn J, Heimann T, Janowitz HD et al (1988) Perforating and non-perforating indications for repeated operations in Crohn's disease: evidence for two clinical forms. Gut 29(5):588–592
Griga T, Wilkens C, Wirkus N, Epplen J, Schmiegel W, Klein W (2007) A polymorphism in the macrophage migration inhibitory factor gene is involved in the genetic predisposition of Crohn's disease and associated with cumulative steroid doses. Hepatogastroenterology 54(75):784–786
Grivennikov S, Karin E, Terzic J, Mucida D, Yu GY, Vallabhapurapu S et al (2009) IL-6 and Stat3 are required for survival of intestinal epithelial cells and development of colitis-associated cancer. Cancer Cell 15(2):103–113

Groden J, Gelbert L, Thliveris A, Nelson L, Robertson M, Joslyn G et al (1993) Mutational analysis of patients with adenomatous polyposis: identical inactivating mutations in unrelated individuals. Am J Hum Genet 52(2):263–272
Gudmand-Hoyer E, Jarnum S (1970) Incidence and clinical significance of lactose malabsorption in ulcerative colitis and Crohn's disease. Gut 11(4):338–343
Gunther C, Martini E, Wittkopf N, Amann K, Weigmann B, Neumann H et al (2011) Caspase-8 regulates TNF-alpha-induced epithelial necroptosis and terminal ileitis. Nature 477(7364):335–339
Guo L, Karpac J, Tran SL, Jasper H (2014) PGRP-SC2 promotes gut immune homeostasis to limit commensal dysbiosis and extend lifespan. Cell 156(1–2):109–122
Ha EM, Lee KA, Seo YY, Kim SH, Lim JH, Oh BH et al (2009) Coordination of multiple dual oxidase-regulatory pathways in responses to commensal and infectious microbes in drosophila gut. Nat Immunol 10(9):949–957
Ha EM, Oh CT, Bae YS, Lee WJ (2005) A direct role for dual oxidase in *Drosophila* gut immunity. Science 310(5749):847–850
Hakansson A, Tormo-Badia N, Baridi A, Xu J, Molin G, Hagslatt ML et al (2014) Immunological alteration and changes of gut microbiota after dextran sulfate sodium (DSS) administration in mice. Clin Exp Med 1:107–120
Halme L, Paavola-Sakki P, Turunen U, Lappalainen M, Farkkila M, Kontula K (2006) Family and twin studies in inflammatory bowel disease. World J Gastroenterol 12(23):3668–3672
Han NY, Choi W, Park JM, Kim EH, Lee H, Hahm KB (2013) Label-free quantification for discovering novel biomarkers in the diagnosis and assessment of disease activity in inflammatory bowel disease. J Dig Dis 14(4):166–174
Hanauer SB, Sandborn WJ, Rutgeerts P, Fedorak RN, Lukas M, MacIntosh D et al (2006) Human anti-tumor necrosis factor monoclonal antibody (adalimumab) in Crohn's disease: the CLASSIC-I trial. Gastroenterology 130(2):323–333
Haramis AP, Begthel H, van den Born M, van Es J, Jonkheer S, Offerhaus GJ et al (2004) De novo crypt formation and juvenile polyposis on BMP inhibition in mouse intestine. Science 303(5664):1684–1686
Hegan PS, Mermall V, Tilney LG, Mooseker MS (2007) Roles for *Drosophila melanogaster* myosin IB in maintenance of enterocyte brush-border structure and resistance to the bacterial pathogen Pseudomonas entomophila. Mol Biol Cell 18(11):4625–4636
Henckaerts L, Cleynen I, Brinar M, John JM, Van Steen K, Rutgeerts P et al (2011) Genetic variation in the autophagy gene ULK1 and risk of Crohn's disease. Inflamm Bowel Dis 17(6):1392–1397
Hendel J, Nielsen OH (1997) Expression of cyclooxygenase-2 mRNA in active inflammatory bowel disease. Am J Gastroenterol 92(7):1170–1173
Henderson P, van Limbergen JE, Wilson DC, Satsangi J, Russell RK (2011) Genetics of childhood-onset inflammatory bowel disease. Inflamm Bowel Dis 17(1):346–361
Henry SC, Daniell X, Indaram M, Whitesides JF, Sempowski GD, Howell D et al (2007) Impaired macrophage function underscores susceptibility to Salmonella in mice lacking Irgm1 (LRG-47). J Immunol 179(10):6963–6972
Hoie O, Wolters FL, Riis L, Bernklev T, Aamodt G, Clofent J et al (2007) Low colectomy rates in ulcerative colitis in an unselected European cohort followed for 10 years. Gastroenterology 132(2):507–515
Hooper LV, Littman DR, Macpherson AJ (2012) Interactions between the microbiota and the immune system. Science 336(6086):1268–1273
Hooper LV, Wong MH, Thelin A, Hansson L, Falk PG, Gordon JI (2001) Molecular analysis of commensal host-microbial relationships in the intestine. Science 291(5505):881–884
Hosomi S, Oshitani N, Kamata N, Sogawa M, Okazaki H, Tanigawa T et al (2011) Increased numbers of immature plasma cells in peripheral blood specifically overexpress chemokine receptor CXCR3 and CXCR4 in patients with ulcerative colitis. Clin Exp Immunol 163(2):215–224

Hou JK, Abraham B, El-Serag H (2011) Dietary intake and risk of developing inflammatory bowel disease: a systematic review of the literature. Am J Gastroenterol 106(4):563–573

Huang SC, Lee CT, Chung BC (2014) Tumor necrosis factor suppresses NR5A2 activity and intestinal glucocorticoid synthesis to sustain chronic colitis. Sci Signal 7(314):ra20

Hudert CA, Weylandt KH, Lu Y, Wang J, Hong S, Dignass A et al (2006) Transgenic mice rich in endogenous omega-3 fatty acids are protected from colitis. Proc Natl Acad Sci U S A 103(30):11276–11281

Human Microbiome Project Consortium (2012) Structure, function and diversity of the healthy human microbiome. Nature 486(7402):207–214

Hunter CJ, Williams M, Petrosyan M, Guner Y, Mittal R, Mock D et al (2009) *Lactobacillus bulgaricus* prevents intestinal epithelial cell injury caused by Enterobacter sakazakii-induced nitric oxide both in vitro and in the newborn rat model of necrotizing enterocolitis. Infect Immun 77(3):1031–1043

Jang JY, Kim HJ, Jung JH, Chae MJ, Kim NH, Lee SK et al (2006) The role of smoking as a risk factor in inflammatory bowel diseases: single center study in Korea. Korean J Gastroenterol 47(3):198–204

Jantchou P, Morois S, Clavel-Chapelon F, Boutron-Ruault MC, Carbonnel F (2010) Animal protein intake and risk of inflammatory bowel disease: the E3 N prospective study. Am J Gastroenterol 105(10):2195–2201

Javed S, Mejias-Luque R, Kalali B, Bolz C, Gerhard M (2013) *Helicobacter bilis* gamma-glutamyltranspeptidase enhances inflammatory stress response via oxidative stress in colon epithelial cells. PLoS ONE 8(8):e73160

Jiang H, Grenley MO, Bravo MJ, Blumhagen RZ, Edgar BA (2011a) EGFR/Ras/MAPK signaling mediates adult midgut epithelial homeostasis and regeneration in *Drosophila*. Cell Stem Cell 8(1):84–95

Jiang HQ, Kushnir N, Thurnheer MC, Bos NA, Cebra JJ (2002) Monoassociation of SCID mice with Helicobacter muridarum, but not four other enterics, provokes IBD upon receipt of T cells. Gastroenterology 122(5):1346–1354

Jiang H, Patel PH, Kohlmaier A, Grenley MO, McEwen DG, Edgar BA (2009) Cytokine/Jak/Stat signaling mediates regeneration and homeostasis in the Drosophila midgut. Cell 137(7):1343–1355

Jiang H, Przybyszewski J, Mitra D, Becker C, Brehm-Stecher B, Tentinger A et al (2011b) Soy protein diet, but not *Lactobacillus rhamnosus* GG, decreases mucin-1, trefoil factor-3, and tumor necrosis factor-alpha in colon of dextran sodium sulfate-treated C57BL/6 mice. J Nutr 141(7):1239–1246

Jiang R, Wang H, Deng L, Hou J, Shi R, Yao M et al (2013) IL-22 is related to development of human colon cancer by activation of STAT3. BMC Cancer 13:59

Jones RM, Luo L, Ardita CS, Richardson AN, Kwon YM, Mercante JW et al (2013) Symbiotic lactobacilli stimulate gut epithelial proliferation via Nox-mediated generation of reactive oxygen species. EMBO J 32(23):3017–3028

Jones RM, Wu H, Wentworth C, Luo L, Collier-Hyams L, Neish AS (2008) Salmonella AvrA coordinates suppression of host immune and apoptotic defenses via JNK pathway blockade. Cell Host Microbe 3(4):233–244

Jones SE, Versalovic J (2009) Probiotic Lactobacillus reuteri biofilms produce antimicrobial and anti-inflammatory factors. BMC Microbiol 9:35

Jostins L, Ripke S, Weersma RK, Duerr RH, McGovern DP, Hui KY et al (2012) Host-microbe interactions have shaped the genetic architecture of inflammatory bowel disease. Nature 491(7422):119–124

Jussila A, Virta LJ, Salomaa V, Maki J, Jula A, Farkkila MA (2013) High and increasing prevalence of inflammatory bowel disease in Finland with a clear North-South difference. J Crohns Colitis 7(7):e256–e262

Kaakoush NO, Day AS, Huinao KD, Leach ST, Lemberg DA, Dowd SE et al (2012) Microbial dysbiosis in pediatric patients with Crohn's disease. J Clin Microbiol 50(10):3258–3266

Kale-Pradhan PB, Zhao JJ, Palmer JR, Wilhelm SM (2013) The role of antimicrobials in Crohn's disease. Expert Rev Gastroenterol Hepatol 7(3):281–288

Kanai T, Matsuoka K, Naganuma M, Hayashi A, Hisamatsu T (2014) Diet, microbiota, and inflammatory bowel disease: lessons from Japanese foods. Korean J Intern Med 29(4):409–415

Kang S, Okuno T, Takegahara N, Takamatsu H, Nojima S, Kimura T et al (2012) Intestinal epithelial cell-derived semaphorin 7A negatively regulates development of colitis via alphavbeta1 integrin. J Immunol 188(3):1108–1116

Kao D, Hotte N, Gillevet P, Madsen K (2014) Fecal microbiota transplantation inducing remission in Crohn's colitis and the associated changes in fecal microbial profile. J Clin Gastroenterol 48(7):625–628

Kaplan GG, Jackson T, Sands BE, Frisch M, Andersson RE, Korzenik J (2008) The risk of developing Crohn's disease after an appendectomy: a meta-analysis. Am J Gastroenterol 103(11):2925–2931

Kappelman MD, Moore KR, Allen JK, Cook SF (2013) Recent trends in the prevalence of Crohn's disease and ulcerative colitis in a commercially insured US population. Dig Dis Sci 58(2):519–525

Karpac J, Biteau B, Jasper H (2013) Misregulation of an adaptive metabolic response contributes to the age-related disruption of lipid homeostasis in *Drosophila*. Cell Rep 4(6):1250–1261

Kaser A, Lee AH, Franke A, Glickman JN, Zeissig S, Tilg H et al (2008) XBP1 links ER stress to intestinal inflammation and confers genetic risk for human inflammatory bowel disease. Cell 134(5):743–756

Katakura K, Lee J, Rachmilewitz D, Li G, Eckmann L, Raz E (2005) Toll-like receptor 9-induced type I IFN protects mice from experimental colitis. J Clin Invest 115(3):695–702

Kawabata K, Sugiyama Y, Sakano T, Ohigashi H (2013) Flavonols enhanced production of anti-inflammatory substance(s) by *Bifidobacterium adolescentis*: prebiotic actions of galangin, quercetin, and fisetin. Biofactors 39(4):422–429

Keshet R, Boursi B, Maoz R, Shnell M, Guzner-Gur H (2009) Diagnostic and prognostic significance of serum C-reactive protein levels in patients admitted to the department of medicine. Am J Med Sci 337(4):248–255

Khan MW, Keshavarzian A, Gounaris E, Melson JE, Cheon EC, Blatner NR et al (2013) PI3 K/AKT signaling is essential for communication between tissue-infiltrating mast cells, macrophages, and epithelial cells in colitis-induced cancer. Clin Cancer Res 19(9):2342–2354

Khanna S, Pardi DS (2012) IBD: Poor outcomes after Clostridium difficile infection in IBD. Nat Rev Gastroenterol Hepatol 9(6):307–308

Khor B, Gardet A, Xavier RJ (2011) Genetics and pathogenesis of inflammatory bowel disease. Nature 474(7351):307–317

Kim N, Kunisawa J, Kweon MN, Eog Ji G, Kiyono H (2007a) Oral feeding of Bifidobacterium bifidum (BGN4) prevents CD4(+) CD45RB(high) T cell-mediated inflammatory bowel disease by inhibition of disordered T cell activation. Clin Immunol 123(1):30–39

Kim SC, Tonkonogy SL, Karrasch T, Jobin C, Sartor RB (2007b) Dual-association of gnotobiotic IL-10-/- mice with 2 nonpathogenic commensal bacteria induces aggressive pancolitis. Inflamm Bowel Dis 13(12):1457–1466

Kirchberger S, Royston DJ, Boulard O, Thornton E, Franchini F, Szabady RL et al (2013) Innate lymphoid cells sustain colon cancer through production of interleukin-22 in a mouse model. J Exp Med 210(5):917–931

Klein A, Eliakim R (2010) Non steroidal anti-inflammatory drugs and inflammatory bowel disease. Pharmaceuticals 3(4):1084–1092

Kobayashi KS, Chamaillard M, Ogura Y, Henegariu O, Inohara N, Nunez G et al (2005) Nod2-dependent regulation of innate and adaptive immunity in the intestinal tract. Science 307(5710):731–734

Koelewijn CL, Schwartz MP, Samsom M, Oldenburg B (2008) C-reactive protein levels during a relapse of Crohn's disease are associated with the clinical course of the disease. World J Gastroenterol 14(1):85–89

Kohno H, Suzuki R, Curini M, Epifano F, Maltese F, Gonzales SP et al (2006) Dietary administration with prenyloxycoumarins, auraptene and collinin, inhibits colitis-related colon carcinogenesis in mice. Int J Cancer 118(12):2936–2942

Kojouharoff G, Hans W, Obermeier F, Mannel DN, Andus T, Scholmerich J et al (1997) Neutralization of tumour necrosis factor (TNF) but not of IL-1 reduces inflammation in chronic dextran sulphate sodium-induced colitis in mice. Clin Exp Immunol 107(2):353–358

Komorowski L, Teegen B, Probst C, Aulinger-Stocker K, Sina C, Fellermann K et al (2013) Autoantibodies against exocrine pancreas in Crohn's disease are directed against two antigens: the glycoproteins CUZD1 and GP2. J Crohns Colitis 7(10):780–790

Konrad A, Rutten C, Flogerzi B, Styner M, Goke B, Seibold F (2004) Immune sensitization to yeast antigens in ASCA-positive patients with Crohn's disease. Inflamm Bowel Dis 10(2):97–105

Kostic AD, Chun E, Robertson L, Glickman JN, Gallini CA, Michaud M et al (2013) *Fusobacterium nucleatum* potentiates intestinal tumorigenesis and modulates the tumor-immune microenvironment. Cell Host Microbe 14(2):207–215

Kovacs M, Muller KE, Papp M, Lakatos PL, Csondes M, Veres G (2014) New serological markers in pediatric patients with inflammatory bowel disease. World J Gastroenterol 20(17):4873–4882

Krieglstein CF, Cerwinka WH, Laroux FS, Salter JW, Russell JM, Schuermann G et al (2001) Regulation of murine intestinal inflammation by reactive metabolites of oxygen and nitrogen: divergent roles of superoxide and nitric oxide. J Exp Med 194(9):1207–1218

Kronman MP, Zaoutis TE, Haynes K, Feng R, Coffin SE (2012) Antibiotic exposure and IBD development among children: a population-based cohort study. Pediatrics 130(4):e794–e803

Kruis W, Fric P, Pokrotnieks J, Lukas M, Fixa B, Kascak M et al (2004) Maintaining remission of ulcerative colitis with the probiotic *Escherichia coli* Nissle 1917 is as effective as with standard mesalazine. Gut 53(11):1617–1623

Kryczek I, Lin Y, Nagarsheth N, Peng D, Zhao L, Zhao E et al (2014) IL-22(+)CD4(+) T cells promote colorectal cancer stemness via STAT3 transcription factor activation and induction of the methyltransferase DOT1L. Immunity 40(5):772–784

Kuester D, Guenther T, Biesold S, Hartmann A, Bataille F, Ruemmele P et al (2010) Aberrant methylation of DAPK in long-standing ulcerative colitis and ulcerative colitis-associated carcinoma. Pathol Res Pract 206(9):616–624

Kugathasan S, Baldassano RN, Bradfield JP, Sleiman PM, Imielinski M, Guthery SL et al (2008) Loci on 20q13 and 21q22 are associated with pediatric-onset inflammatory bowel disease. Nat Genet 40(10):1211–1215

Kullberg MC, Jankovic D, Feng CG, Hue S, Gorelick PL, McKenzie BS et al (2006) IL-23 plays a key role in Helicobacter hepaticus-induced T cell-dependent colitis. J Exp Med 203(11):2485–2494

Kullberg MC, Ward JM, Gorelick PL, Caspar P, Hieny S, Cheever A et al (1998) *Helicobacter hepaticus* triggers colitis in specific-pathogen-free interleukin-10 (IL-10)-deficient mice through an IL-12- and gamma interferon-dependent mechanism. Infect Immun 66(11):5157–5166

Kunde S, Pham A, Bonczyk S, Crumb T, Duba M, Conrad HJr et al (2013) Safety, tolerability, and clinical response after fecal transplantation in children and young adults with ulcerative colitis. J Pediatr Gastroenterol Nutr 56(6):597–601

Kuraishi T, Binggeli O, Opota O, Buchon N, Lemaitre B (2011) Genetic evidence for a protective role of the peritrophic matrix against intestinal bacterial infection in *Drosophila melanogaster*. Proc Natl Acad Sci U S A 108(38):15966–15971

Lagishetty V, Misharin AV, Liu NQ, Lisse TS, Chun RF, Ouyang Y et al (2010) Vitamin D deficiency in mice impairs colonic antibacterial activity and predisposes to colitis. Endocrinology 151(6):2423–2432

Lakatos PL, Lakatos L (2008) Risk for colorectal cancer in ulcerative colitis: changes, causes and management strategies. World J Gastroenterol 14(25):3937–3947

Lara-Villoslada F, Debras E, Nieto A, Concha A, Galvez J, Lopez-Huertas E et al (2006) Oligosaccharides isolated from goat milk reduce intestinal inflammation in a rat model of dextran sodium sulfate-induced colitis. Clin Nutr 25(3):477–488

LeBlanc JG, Milani C, de Giori GS, Sesma F, van Sinderen D, Ventura M (2013) Bacteria as vitamin suppliers to their host: a gut microbiota perspective. Curr Opin Biotechnol 24(2):160–168

Lee JC, Espeli M, Anderson CA, Linterman MA, Pocock JM, Williams NJ et al (2013a) Human SNP links differential outcomes in inflammatory and infectious disease to a FOXO3-regulated pathway. Cell 155(1):57–69

Lee KA, Kim SH, Kim EK, Ha EM, You H, Kim B et al (2013b) Bacterial-derived uracil as a modulator of mucosal immunity and gut-microbe homeostasis in *Drosophila*. Cell 153(4):797–811

Lee KA, Lee WJ (2014) *Drosophila* as a model for intestinal dysbiosis and chronic inflammatory diseases. Dev Comp Immunol 42(1):102–110

Lemaitre B, Hoffmann J (2007) The host defense of *Drosophila melanogaster*. Annu Rev Immunol 25:697–743

Lemaitre B, Kromer-Metzger E, Michaut L, Nicolas E, Meister M, Georgel P et al (1995) A recessive mutation, immune deficiency (imd), defines two distinct control pathways in the *Drosophila* host defense. Proc Natl Acad Sci U S A 92(21):9465–9469

Lemaitre B, Nicolas E, Michaut L, Reichhart JM, Hoffmann JA (1996) The dorsoventral regulatory gene cassette spatzle/toll/cactus controls the potent antifungal response in *Drosophila* adults. Cell 86(6):973–983

Levine A, Griffiths A, Markowitz J, Wilson DC, Turner D, Russell RK et al (2011) Pediatric modification of the Montreal classification for inflammatory bowel disease: the Paris classification. Inflamm Bowel Dis 17(6):1314–1321

Li CK, Pender SL, Pickard KM, Chance V, Holloway JA, Huett A et al (2004) Impaired immunity to intestinal bacterial infection in stromelysin-1 (matrix metalloproteinase-3)-deficient mice. J Immunol 173(8):5171–5179

Liehl P, Blight M, Vodovar N, Boccard F, Lemaitre B (2006) Prevalence of local immune response against oral infection in a *Drosophila/Pseudomonas* infection model. PLoS Pathog 2(6):e56

Lin PW, Myers LE, Ray L, Song SC, Nasr TR, Berardinelli AJ et al (2009) *Lactobacillus rhamnosus* blocks inflammatory signaling in vivo via reactive oxygen species generation. Free Radical Biol Med 47(8):1205–1211

Linton L, Karlsson M, Grundstrom J, Hjalmarsson E, Lindberg A, Lindh E et al (2012) HLA-DR(hi) and CCR9 define a pro-inflammatory monocyte subset in IBD. Clin Transl Gastroenterol 3:e29

Luftman K, Hasan N, Day P, Hardee D, Hu C (2009) Silencing of VAMP3 inhibits cell migration and integrin-mediated adhesion. Biochem Biophys Res Commun 380(1):65–70

Ma D, Forsythe P, Bienenstock J (2004) Live *Lactobacillus rhamnosus* is essential for the inhibitory effect on tumor necrosis factor alpha-induced interleukin-8 expression. Infect Immun 72(9):5308–5314

MacDonald RS, Przybyszewski J (2008) Soy protein diet suppresses COX-2 induction by dextran sodium sulfate in CD-1 mouse colon. FASEB J 22:885

Macho Fernandez E, Valenti V, Rockel C, Hermann C, Pot B, Boneca IG et al (2011) Anti-inflammatory capacity of selected lactobacilli in experimental colitis is driven by NOD2-mediated recognition of a specific peptidoglycan-derived muropeptide. Gut 60(8):1050–1059

Maconi G, Ardizzone S, Cucino C, Bezzio C, Russo AG, Bianchi Porro G (2010) Pre-illness changes in dietary habits and diet as a risk factor for inflammatory bowel disease: a case-control study. World J Gastroenterol 16(34):4297–4304

Madsen KL, Fedorak RN, Tavernini MM, Doyle JS (2002) Normal breast milk limits the development of colitis in IL-10-deficient mice. Inflamm Bowel Dis 8(6):390–398

Maerten P, Kwon BS, Shen C, De Hertogh G, Cadot P, Bullens DM et al (2006) Involvement of 4-1BB (CD137)-4-1BBligand interaction in the modulation of CD4 T cell-mediated inflammatory colitis. Clin Exp Immunol 143(2):228–236

Maggio-Price L, Treuting P, Zeng W, Tsang M, Bielefeldt-Ohmann H, Iritani BM (2006) *Helicobacter* infection is required for inflammation and colon cancer in SMAD3-deficient mice. Cancer Res 66(2):828–838

Mahid SS, Minor KS, Soto RE, Hornung CA, Galandiuk S (2006) Smoking and inflammatory bowel disease: a meta-analysis. Mayo Clin Proc 81(11):1462–1471

Makitalo L, Kolho KL, Karikoski R, Anthoni H, Saarialho-Kere U (2010) Expression profiles of matrix metalloproteinases and their inhibitors in colonic inflammation related to pediatric inflammatory bowel disease. Scand J Gastroenterol 45(7–8):862–871

Makitalo L, Sipponen T, Karkkainen P, Kolho KL, Saarialho-Kere U (2009) Changes in matrix metalloproteinase (MMP) and tissue inhibitors of metalloproteinases (TIMP) expression profile in Crohn's disease after immunosuppressive treatment correlate with histological score and calprotectin values. Int J Colorectal Dis 24(10):1157–1167

Manousou P, Kolios G, Valatas V, Drygiannakis I, Bourikas L, Pyrovolaki K et al (2010) Increased expression of chemokine receptor CCR3 and its ligands in ulcerative colitis: the role of colonic epithelial cells in in vitro studies. Clin Exp Immunol 162(2):337–347

Matsumoto S, Hara T, Hori T, Mitsuyama K, Nagaoka M, Tomiyasu N et al (2005) Probiotic Lactobacillus-induced improvement in murine chronic inflammatory bowel disease is associated with the down-regulation of pro-inflammatory cytokines in lamina propria mononuclear cells. Clin Exp Immunol 140(3):417–426

Matsumoto S, Watanabe N, Imaoka A, Okabe Y (2001) Preventive effects of *Bifidobacterium*-and *Lactobacillus*-fermented milk on the development of inflammatory bowel disease in senescence-accelerated mouse P1/Yit strain mice. Digestion 64(2):92–99

Mazmanian SK, Liu CH, Tzianabos AO, Kasper DL (2005) An immunomodulatory molecule of symbiotic bacteria directs maturation of the host immune system. Cell 122(1):107–118

Mazmanian SK, Round JL, Kasper DL (2008) A microbial symbiosis factor prevents intestinal inflammatory disease. Nature 453(7195):620–625

McAuley JL, Linden SK, Png CW, King RM, Pennington HL, Gendler SJ et al (2007) MUC1 cell surface mucin is a critical element of the mucosal barrier to infection. J Clin Invest 117(8):2313–2324

McCarroll SA, Huett A, Kuballa P, Chilewski SD, Landry A, Goyette P et al (2008) Deletion polymorphism upstream of IRGM associated with altered IRGM expression and Crohn's disease. Nat Genet 40(9):1107–1112

Menard S, Candalh C, Bambou JC, Terpend K, Cerf-Bensussan N, Heyman M (2004) Lactic acid bacteria secrete metabolites retaining anti-inflammatory properties after intestinal transport. Gut 53(6):821–828

Mennigen R, Nolte K, Rijcken E, Utech M, Loeffler B, Senninger N et al (2009) Probiotic mixture VSL#3 protects the epithelial barrier by maintaining tight junction protein expression and preventing apoptosis in a murine model of colitis. Am J Physiol Gastrointest Liver Physiol 296(5):G1140–G1149

Meuwis MA, Fillet M, Geurts P, de Seny D, Lutteri L, Chapelle JP et al (2007) Biomarker discovery for inflammatory bowel disease, using proteomic serum profiling. Biochem Pharmacol 73(9):1422–1433

Meuwis MA, Vernier-Massouille G, Grimaud JC, Bouhnik Y, Laharie D, Piver E et al (2013) Serum calprotectin as a biomarker for Crohn's disease. J Crohns Colitis 7(12):e678–e683

Micchelli CA, Perrimon N (2006) Evidence that stem cells reside in the adult *Drosophila* midgut epithelium. Nature 439(7075):475–479

Min SB, Le-Carlson M, Singh N, Nylund CM, Gebbia J, Haas K et al (2013) Video capsule endoscopy impacts decision making in pediatric IBD: a single tertiary care center experience. Inflamm Bowel Dis 19(10):2139–2145

Misra A, Rastogi K, Joshi SR (2009) Whole grains and health: perspective for Asian Indians. J Assoc Physicians India 57:155–162

Mohamadzadeh M, Pfeiler EA, Brown JB, Zadeh M, Gramarossa M, Managlia E et al (2011) Regulation of induced colonic inflammation by *Lactobacillus acidophilus* deficient in lipoteichoic acid. Proc Natl Acad Sci U S A 108(Suppl 1):4623–4630

Molodecky NA, Soon IS, Rabi DM, Ghali WA, Ferris M, Chernoff G et al (2012) Increasing incidence and prevalence of the inflammatory bowel diseases with time, based on systematic review. Gastroenterology 142(1):46–54

Momotani E, Ozaki H, Hori M, Yamamoto S, Kuribayashi T, Eda S et al (2012) *Mycobacterium avium* subsp. *paratuberculosis* lipophilic antigen causes Crohn's disease-type necrotizing colitis in mice. Springerplus 1(1):47

Monceaux CP, Testerman TL, Boktor M, Jordan P, Adegboyega P, McGee DJ et al (2013) Helicobacter infection decreases basal colon inflammation, but increases disease activity in experimental IBD. Open J Gastroenterology 3:177

Monteleone G, Kumberova A, Croft NM, McKenzie C, Steer HW, MacDonald TT (2001) Blocking Smad7 restores TGF-beta1 signaling in chronic inflammatory bowel disease. J Clin Invest 108(4):601–609

Morgan XC, Tickle TL, Sokol H, Gevers D, Devaney KL, Ward DV et al (2012) Dysfunction of the intestinal microbiome in inflammatory bowel disease and treatment. Genome Biol 13(9):R79

Mortha A, Chudnovskiy A, Hashimoto D, Bogunovic M, Spencer SP, Belkaid Y et al (2014) Microbiota-dependent crosstalk between macrophages and ILC3 promotes intestinal homeostasis. Science 343(6178):1249288

Mow WS, Vasiliauskas EA, Lin YC, Fleshner PR, Papadakis KA, Taylor KD et al (2004) Association of antibody responses to microbial antigens and complications of small bowel Crohn's disease. Gastroenterology 126(2):414–424

Mowat C, Cole A, Windsor A, Ahmad T, Arnott I, Driscoll R et al (2011) Guidelines for the management of inflammatory bowel disease in adults. Gut 60(5):571–607

Murano T, Okamoto R, Ito G, Nakata T, Hibiya S, Shimizu H et al (2014) Hes1 promotes the IL-22-mediated antimicrobial response by enhancing STAT3-dependent transcription in human intestinal epithelial cells. Biochem Biophys Res Commun 443(3):840–846

Nehme NT, Liegeois S, Kele B, Giammarinaro P, Pradel E, Hoffmann JA et al (2007) A model of bacterial intestinal infections in *Drosophila melanogaster*. PLoS Pathog 3(11):e173

Nerich V, Monnet E, Etienne A, Louafi S, Ramee C, Rican S et al (2006) Geographical variations of inflammatory bowel disease in France: a study based on national health insurance data. Inflamm Bowel Dis 12(3):218–226

Neurath MF (2014) Cytokines in inflammatory bowel disease. Nat Rev Immunol 14(5):329–342

Neurath MF, Weigmann B, Finotto S, Glickman J, Nieuwenhuis E, Iijima H et al (2002) The transcription factor T-bet regulates mucosal T cell activation in experimental colitis and Crohn's disease. J Exp Med 195(9):1129–1143

Nguyen GC, Munsell M, Harris ML (2008) Nationwide prevalence and prognostic significance of clinically diagnosable protein-calorie malnutrition in hospitalized inflammatory bowel disease patients. Inflamm Bowel Dis 14(8):1105–1111

Nielsen OH, Munck LK (2007) Drug insight: aminosalicylates for the treatment of IBD. Nat Clin Pract Gastroenterol Hepatol 4(3):160–170

Niess JH, Danese S (2014) Anti-TNF and skin inflammation in IBD: a new paradox in gastroenterology? Gut 63(4):533–535

Nishimura T, Andoh A, Inatomi O, Shioya M, Yagi Y, Tsujikawa T et al (2008) Amphiregulin and epiregulin expression in neoplastic and inflammatory lesions in the colon. Oncol Rep 19(1):105–110

Ocon B, Anzola A, Ortega-Gonzalez M, Zarzuelo A, Suarez MD, Sanchez de Medina F et al (2013) Active hexose-correlated compound and Bifidobacterium longum BB536 exert symbiotic effects in experimental colitis. Eur J Nutr 52(2):457–466

Ogura Y, Bonen DK, Inohara N, Nicolae DL, Chen FF, Ramos R et al (2001) A frameshift mutation in NOD2 associated with susceptibility to Crohn's disease. Nature 411(6837):603–606

Ohlstein B, Spradling A (2006) The adult *Drosophila* posterior midgut is maintained by pluripotent stem cells. Nature 439(7075):470–474

Ohtani K, Ohtsuka Y, Ikuse T, Baba Y, Yamakawa Y, Aoyagi Y et al (2010) Increased mucosal expression of GATA-3 and STAT-4 in pediatric ulcerative colitis. Pediatr Int 52(4):584–589

Okayasu I, Hatakeyama S, Yamada M, Ohkusa T, Inagaki Y, Nakaya R (1990) A novel method in the induction of reliable experimental acute and chronic ulcerative colitis in mice. Gastroenterology 98(3):694–702

Okumura T, Takeda K, Taniguchi K, Adachi-Yamada T (2014) betanu integrin inhibits chronic and high level activation of JNK to repress senescence phenotypes in *Drosophila* adult midgut. PLoS ONE 9(2):e89387

O'Mahony C, Scully P, O'Mahony D, Murphy S, O'Brien F, Lyons A, Sherlock G, MacSharry J, Kiely B, Shanahan F, O'Mahony L (2008) Commensal-induced regulatory T cells mediate protection against pathogen-stimulated NF-kappaB activation. PLoS Pathog 4(8):e1000112

Oshima M, Oshima H, Kitagawa K, Kobayashi M, Itakura C, Taketo M (1995) Loss of Apc heterozygosity and abnormal tissue building in nascent intestinal polyps in mice carrying a truncated Apc gene. Proc Natl Acad Sci U S A 92(10):4482–4486

Ott SJ, Musfeldt M, Wenderoth DF, Hampe J, Brant O, Folsch UR et al (2004) Reduction in diversity of the colonic mucosa associated bacterial microflora in patients with active inflammatory bowel disease. Gut 53(5):685–693

Ou G, Rompikuntal PK, Bitar A, Lindmark B, Vaitkevicius K, Wai SN et al (2009) *Vibrio cholerae* cytolysin causes an inflammatory response in human intestinal epithelial cells that is modulated by the PrtV protease. PLoS ONE 4(11):e7806

Paik J, Fierce Y, Treuting PM, Brabb T, Maggio-Price L (2013) High-fat diet-induced obesity exacerbates inflammatory bowel disease in genetically susceptible Mdr1a-/- male mice. J Nutr 143(8):1240–1247

Pal A (2014) Crohn's disease. InnovAiT Educ Inspiration Gen Pract 7(1):43–54

Panayidou S, Apidianakis Y (2013) Regenerative inflammation: lessons from *Drosophila* intestinal Epithelium in health and disease. Pathogens 2(2):209–231

Panayidou S, Ioannidou E, Apidianakis Y (2014) Human pathogenic bacteria, fungi, and viruses in *Drosophila*: disease modeling, lessons, and shortcomings. Virulence 5(2):253–269

Papp M, Altorjay I, Dotan N, Palatka K, Foldi I, Tumpek J et al (2008) New serological markers for inflammatory bowel disease are associated with earlier age at onset, complicated disease behavior, risk for surgery, and NOD2/CARD15 genotype in a Hungarian IBD cohort. Am J Gastroenterol 103(3):665–681

Parvez S, Malik KA, Ah Kang S, Kim HY (2006) Probiotics and their fermented food products are beneficial for health. J Appl Microbiol 100(6):1171–1185

Peran L, Camuesco D, Comalada M, Nieto A, Concha A, Adrio JL et al (2006) *Lactobacillus* fermentum, a probiotic capable to release glutathione, prevents colonic inflammation in the TNBS model of rat colitis. Int J Colorectal Dis 21(8):737–746

Peterson LW, Artis D (2014) Intestinal epithelial cells: regulators of barrier function and immune homeostasis. Nat Rev Immunol 14(3):141–153

Philippe D, Favre L, Foata F, Adolfsson O, Perruisseau-Carrier G, Vidal K et al (2011a) Bifidobacterium lactis attenuates onset of inflammation in a murine model of colitis. World J Gastroenterol 17(4):459–469

Philippe D, Heupel E, Blum-Sperisen S, Riedel CU (2011b) Treatment with Bifidobacterium bifidum 17 partially protects mice from Th1-driven inflammation in a chemically induced model of colitis. Int J Food Microbiol 149(1):45–49

Pickert G, Neufert C, Leppkes M, Zheng Y, Wittkopf N, Warntjen M et al (2009) STAT3 links IL-22 signaling in intestinal epithelial cells to mucosal wound healing. J Exp Med 206(7):1465–1472

Pinsk V, Lemberg DA, Grewal K, Barker CC, Schreiber RA, Jacobson K (2007) Inflammatory bowel disease in the South Asian pediatric population of British Columbia. Am J Gastroenterol 102(5):1077–1083

Pittschieler K (1990) Cow's milk protein-induced colitis in the breast-fed infant. J Pediatr Gastroenterol Nutr 10(4):548–549

Plevy SE, Landers CJ, Prehn J, Carramanzana NM, Deem RL, Shealy D et al (1997) A role for TNF-alpha and mucosal T helper-1 cytokines in the pathogenesis of Crohn's disease. J Immunol 159(12):6276–6282

Pohl C, Hombach A, Kruis W (2000) Chronic inflammatory bowel disease and cancer. Hepatogastroenterology 47(31):57–70

Pouyet L, Roisin-Bouffay C, Clement A, Millet V, Garcia S, Chasson L et al (2010) Epithelial vanin-1 controls inflammation-driven carcinogenesis in the colitis-associated colon cancer model. Inflamm Bowel Dis 16(1):96–104

Powrie F, Leach MW, Mauze S, Menon S, Caddle LB, Coffman RL (1994) Inhibition of Th1 responses prevents inflammatory bowel disease in scid mice reconstituted with CD45RBhi CD4 + T cells. Immunity 1(7):553–562

Prefontaine E, Macdonald JK, Sutherland LR (2009) Azathioprine or 6-mercaptopurine for induction of remission in Crohn's disease. Cochrane Database Syst Rev (4):CD000545

Prescott NJ, Fisher SA, Franke A, Hampe J, Onnie CM, Soars D et al (2007) A nonsynonymous SNP in ATG16L1 predisposes to ileal Crohn's disease and is independent of CARD15 and IBD5. Gastroenterology 132(5):1665–1671

Radford-Smith GL, Edwards JE, Purdie DM, Pandeya N, Watson M, Martin NG et al (2002) Protective role of appendicectomy on onset and severity of ulcerative colitis and Crohn's disease. Gut 51(6):808–813

Ragab A, Buechling T, Gesellchen V, Spirohn K, Boettcher AL, Boutros M (2011) *Drosophila* Ras/MAPK signalling regulates innate immune responses in immune and intestinal stem cells. EMBO J 30(6):1123–1136

Reif S, Klein I, Lubin F, Farbstein M, Hallak A, Gilat T (1997) Pre-illness dietary factors in inflammatory bowel disease. Gut 40(6):754–760

Rhee KJ, Wu S, Wu X, Huso DL, Karim B, Franco AA et al (2009) Induction of persistent colitis by a human commensal, enterotoxigenic Bacteroides fragilis, in wild-type C57BL/6 mice. Infect Immun 77(4):1708–1718

Riley JK, Takeda K, Akira S, Schreiber RD (1999) Interleukin-10 receptor signaling through the JAK-STAT pathway. Requirement for two distinct receptor-derived signals for anti-inflammatory action. J Biol Chem 274(23):16513–16521

Rimoldi M, Chieppa M, Salucci V, Avogadri F, Sonzogni A, Sampietro GM et al (2005) Intestinal immune homeostasis is regulated by the crosstalk between epithelial cells and dendritic cells. Nat Immunol 6(5):507–514

Roggenbuck D, Hausdorf G, Martinez-Gamboa L, Reinhold D, Buttner T, Jungblut PR et al (2009) Identification of GP2, the major zymogen granule membrane glycoprotein, as the autoantigen of pancreatic antibodies in Crohn's disease. Gut 58(12):1620–1628

Round JL, Mazmanian SK (2009) The gut microbiota shapes intestinal immune responses during health and disease. Nat Rev Immunol 9(5):313–323

Round JL, Mazmanian SK (2010) Inducible Foxp3 + regulatory T-cell development by a commensal bacterium of the intestinal microbiota. Proc Natl Acad Sci U S A 107(27):12204–12209

Ruiz-Perez B, Chung DR, Sharpe AH, Yagita H, Kalka Moll WM, Sayegh MH et al (2005) Modulation of surgical fibrosis by microbial zwitterionic polysaccharides. Proc Natl Acad Sci U S A 102(46):16753–16758

Rutgeerts P, Sandborn WJ, Feagan BG, Reinisch W, Olson A, Johanns J et al (2005) Infliximab for induction and maintenance therapy for ulcerative colitis. N Engl J Med 353(23):2462–2476

Ryu JH, Kim SH, Lee HY, Bai JY, Nam YD, Bae JW et al (2008) Innate immune homeostasis by the homeobox gene caudal and commensal-gut mutualism in *Drosophila*. Science 319(5864):777–782

Saitoh T, Fujita N, Jang MH, Uematsu S, Yang BG, Satoh T et al (2008) Loss of the autophagy protein Atg16L1 enhances endotoxin-induced IL-1beta production. Nature 456(7219):264–268

Saitoh S, Noda S, Aiba Y, Takagi A, Sakamoto M, Benno Y et al (2002) Bacteroides ovatus as the predominant commensal intestinal microbe causing a systemic antibody response in inflammatory bowel disease. Clin Diagn Lab Immunol 9(1):54–59

Sakamoto N, Kono S, Wakai K, Fukuda Y, Satomi M, Shimoyama T et al (2005) Dietary risk factors for inflammatory bowel disease: a multicenter case-control study in Japan. Inflamm Bowel Dis 11(2):154–163

Santos Rocha C, Gomes-Santos AC, Garcias Moreira T, de Azevedo M, Diniz Luerce T, Mariadassou M et al (2014) Local and systemic immune mechanisms underlying the anti-colitis effects of the dairy bacterium *Lactobacillus delbrueckii*. PLoS ONE 9(1):e85923

Sartor RB (2006) Mechanisms of disease: pathogenesis of Crohn's disease and ulcerative colitis. Nat Clin Pract Gastroenterol Hepatol 3(7):390–407

Scharl M, Rogler G (2012) Inflammatory bowel disease: dysfunction of autophagy? Dig Dis 30(Suppl 3):12–19

Schmidt RL, Trejo TR, Plummer TB, Platt JL, Tang AH (2008) Infection-induced proteolysis of PGRP-LC controls the IMD activation and melanization cascades in *Drosophila*. FASEB J 22(3):918–929

Schneider DS, Ayres JS, Brandt SM, Costa A, Dionne MS, Gordon MD et al (2007) *Drosophila* eiger mutants are sensitive to extracellular pathogens. PLoS Pathog 3(3):e41

Schreiber S, Rosenstiel P, Hampe J, Nikolaus S, Groessner B, Schottelius A et al (2002) Activation of signal transducer and activator of transcription (STAT) 1 in human chronic inflammatory bowel disease. Gut 51(3):379–385

Schultz M, Veltkamp C, Dieleman LA, Grenther WB, Wyrick PB, Tonkonogy SL et al (2002) Lactobacillus plantarum 299V in the treatment and prevention of spontaneous colitis in interleukin-10-deficient mice. Inflamm Bowel Dis 8(2):71–80

Sears CL, Garrett WS (2014) Microbes, microbiota, and colon cancer. Cell Host Microbe 15(3):317–328

Selvam R, Maheswari P, Kavitha P, Ravichandran M, Sas B, Ramchand CN (2009) Effect of *Bacillus subtilis* PB6, a natural probiotic on colon mucosal inflammation and plasma cytokines levels in inflammatory bowel disease. Indian J Biochem Biophys 46(1):79–85

Seok J, Warren HS, Cuenca AG, Mindrinos MN, Baker HV, Xu W et al (2013) Genomic responses in mouse models poorly mimic human inflammatory diseases. Proc Natl Acad Sci U S A 110(9):3507–3512

Sgouras DN, Panayotopoulou EG, Martinez-Gonzalez B, Petraki K, Michopoulos S, Mentis A (2005) Lactobacillus johnsonii La1 attenuates Helicobacter pylori-associated gastritis and reduces levels of proinflammatory chemokines in C57BL/6 mice. Clin Diagn Lab Immunol 12(12):1378–1386

Shaw SY, Blanchard JF, Bernstein CN (2010) Association between the use of antibiotics in the first year of life and pediatric inflammatory bowel disease. Am J Gastroenterol 105(12):2687–2692

Shea-Donohue T, Thomas K, Cody MJ, Aiping Z, Detolla LJ, Kopydlowski KM et al (2008) Mice deficient in the CXCR2 ligand, CXCL1 (KC/GRO-alpha), exhibit increased susceptibility to dextran sodium sulfate (DSS)-induced colitis. Innate Immun 14(2):117–124

Shiba T, Aiba Y, Ishikawa H, Ushiyama A, Takagi A, Mine T et al (2003) The suppressive effect of bifidobacteria on Bacteroides vulgatus, a putative pathogenic microbe in inflammatory bowel disease. Microbiol Immunol 47(6):371–378

Shin SC, Kim SH, You H, Kim B, Kim AC, Lee KA et al (2011) *Drosophila* microbiome modulates host developmental and metabolic homeostasis via insulin signaling. Science 334(6056):670–674

Shivananda S, Lennard-Jones J, Logan R, Fear N, Price A, Carpenter L et al (1996) Incidence of inflammatory bowel disease across Europe: is there a difference between north and south? Results of the European collaborative study on inflammatory bowel disease (EC-IBD). Gut 39(5):690–697

Sibley CD, Duan K, Fischer C, Parkins MD, Storey DG, Rabin HR et al (2008) Discerning the complexity of community interactions using a *Drosophila* model of polymicrobial infections. PLoS Pathog 4(10):e1000184

Silverberg MS, Satsangi J, Ahmad T, Arnott ID, Bernstein CN, Brant SR et al (2005) Toward an integrated clinical, molecular and serological classification of inflammatory bowel disease: report of a working party of the 2005 montreal world congress of gastroenterology. Can J Gastroenterol 19(Suppl A):5A–36A

Simpson SJ, Shah S, Comiskey M, de Jong YP, Wang B, Mizoguchi E et al (1998) T cell-mediated pathology in two models of experimental colitis depends predominantly on the interleukin 12/Signal transducer and activator of transcription (Stat)-4 pathway, but is not conditional on interferon gamma expression by T cells. J Exp Med 187(8):1225–1234

Singer II, Kawka DW, Scott S, Weidner JR, Mumford RA, Riehl TE et al (1996) Expression of inducible nitric oxide synthase and nitrotyrosine in colonic epithelium in inflammatory bowel disease. Gastroenterology 111(4):871–885

Singh SB, Ornatowski W, Vergne I, Naylor J, Delgado M, Roberts E et al (2010) Human IRGM regulates autophagy and cell-autonomous immunity functions through mitochondria. Nat Cell Biol 12(12):1154–1165

Small CL, Reid-Yu SA, McPhee JB, Coombes BK (2013) Persistent infection with Crohn's disease-associated adherent-invasive Escherichia coli leads to chronic inflammation and intestinal fibrosis. Nat Commun 4:1957

Snoeks L, Weber CR, Wasland K, Turner JR, Vainder C, Qi W et al (2009) Tumor suppressor FOXO3 participates in the regulation of intestinal inflammation. Lab Invest 89(9):1053–1062

Sokol H, Pigneur B, Watterlot L, Lakhdari O, Bermudez-Humaran LG, Gratadoux JJ et al (2008) *Faecalibacterium prausnitzii* is an anti-inflammatory commensal bacterium identified by gut microbiota analysis of Crohn disease patients. Proc Natl Acad Sci U S A 105(43):16731–16736

Solem CA, Loftus EV Jr, Tremaine WJ, Harmsen WS, Zinsmeister AR, Sandborn WJ (2005) Correlation of C-reactive protein with clinical, endoscopic, histologic, and radiographic activity in inflammatory bowel disease. Inflamm Bowel Dis 11(8):707–712

Song J, Medline A, Mason JB, Gallinger S, Kim YI (2000) Effects of dietary folate on intestinal tumorigenesis in the apcMin mouse. Cancer Res 60(19):5434–5440

Staley BK, Irvine KD (2010) Warts and Yorkie mediate intestinal regeneration by influencing stem cell proliferation. Curr Biol 20(17):1580–1587

Stecher B, Paesold G, Barthel M, Kremer M, Jantsch J, Stallmach T et al (2006) Chronic Salmonella enterica serovar Typhimurium-induced colitis and cholangitis in streptomycin-pretreated Nramp1 +/+ mice. Infect Immun 74(9):5047–5057

Stoll M, Corneliussen B, Costello CM, Waetzig GH, Mellgard B, Koch WA et al (2004) Genetic variation in DLG5 is associated with inflammatory bowel disease. Nat Genet 36(5):476–480

Sturlan S, Oberhuber G, Beinhauer BG, Tichy B, Kappel S, Wang J et al (2001) Interleukin-10-deficient mice and inflammatory bowel disease associated cancer development. Carcinogenesis 22(4):665–671

Su L, Shen L, Clayburgh DR, Nalle SC, Sullivan EA, Meddings JB et al (2009) Targeted epithelial tight junction dysfunction causes immune activation and contributes to development of experimental colitis. Gastroenterology 136(2):551–563

Szanto I, Rubbia-Brandt L, Kiss P, Steger K, Banfi B, Kovari E et al (2005) Expression of NOX1, a superoxide-generating NADPH oxidase, in colon cancer and inflammatory bowel disease. J Pathol 207(2):164–176

Szilagyi A, Leighton H, Burstein B, Xue X (2014) Latitude, sunshine, and human lactase phenotype distributions may contribute to geographic patterns of modern disease: the inflammatory bowel disease model. Clin Epidemiol 6:183–198

Takeda K, Clausen BE, Kaisho T, Tsujimura T, Terada N, Forster I et al (1999) Enhanced Th1 activity and development of chronic enterocolitis in mice devoid of Stat3 in macrophages and neutrophils. Immunity 10(1):39–49

Tanaka S, Kobayashi T, Songjinda P, Tateyama A, Tsubouchi M, Kiyohara C et al (2009) Influence of antibiotic exposure in the early postnatal period on the development of intestinal microbiota. FEMS Immunol Med Microbiol 56(1):80–87

Tanaka T, Kohno H, Suzuki R, Yamada Y, Sugie S, Mori H (2003) A novel inflammation-related mouse colon carcinogenesis model induced by azoxymethane and dextran sodium sulfate. Cancer Sci 94(11):965–973

Tao Y, Drabik KA, Waypa TS, Musch MW, Alverdy JC, Schneewind O et al (2006) Soluble factors from *Lactobacillus* GG activate MAPKs and induce cytoprotective heat shock proteins in intestinal epithelial cells. Am J Physiol Cell Physiol 290(4):C1018–C1030

Tarrant KM, Barclay ML, Frampton CM, Gearry RB (2008) Perianal disease predicts changes in Crohn's disease phenotype-results of a population-based study of inflammatory bowel disease phenotype. Am J Gastroenterol 103(12):3082–3093

Tlaskalova-Hogenova H, Stepankova R, Kozakova H, Hudcovic T, Vannucci L, Tuckova L et al (2011) The role of gut microbiota (commensal bacteria) and the mucosal barrier in the pathogenesis of inflammatory and autoimmune diseases and cancer: contribution of germ-free and gnotobiotic animal models of human diseases. Cell Mol Immunol 8(2):110–120

Todhunter CE, Sutherland-Craggs A, Bartram SA, Donaldson PT, Daly AK, Francis RM et al (2005) Influence of IL-6, COL1A1, and VDR gene polymorphisms on bone mineral density in Crohn's disease. Gut 54(11):1579–1584

Tokuyama H, Ueha S, Kurachi M, Matsushima K, Moriyasu F, Blumberg RS et al (2005) The simultaneous blockade of chemokine receptors CCR2, CCR5 and CXCR3 by a non-peptide chemokine receptor antagonist protects mice from dextran sodium sulfate-mediated colitis. Int Immunol 17(8):1023–1034

Toprak NU, Yagci A, Gulluoglu BM, Akin ML, Demirkalem P, Celenk T et al (2006) A possible role of *Bacteroides fragilis* enterotoxin in the aetiology of colorectal cancer. Clin Microbiol Infect 12(8):782–786

Toyoda H, Wang SJ, Yang HY, Redford A, Magalong D, Tyan D et al (1993) Distinct associations of HLA class II genes with inflammatory bowel disease. Gastroenterology 104(3):741–748

Turnbaugh PJ, Ridaura VK, Faith JJ, Rey FE, Knight R, Gordon JI (2009) The effect of diet on the human gut microbiome: a metagenomic analysis in humanized gnotobiotic mice. Sci Transl Med 1:6ra14

Tzelepis I, Kapsetaki SE, Panayidou S, Apidianakis Y (2013) *Drosophila melanogaster*: a first step and a stepping-stone to anti-infectives. Curr Opin Pharmacol 13(5):763–768

Uchida M, Mogami O, Matsueda K (2007) Characteristic of milk whey culture with *Propionibacterium freudenreichii* ET-3 and its application to the inflammatory bowel disease therapy. Inflammopharmacology 15(3):105–108

Uematsu S, Jang MH, Chevrier N, Guo Z, Kumagai Y, Yamamoto M et al (2006) Detection of pathogenic intestinal bacteria by Toll-like receptor 5 on intestinal CD11c + lamina propria cells. Nat Immunol 7(8):868–874

Ueno N, Fujiya M, Segawa S, Nata T, Moriichi K, Tanabe H et al (2011) Heat-killed body of lactobacillus brevis SBC8803 ameliorates intestinal injury in a murine model of colitis by enhancing the intestinal barrier function. Inflamm Bowel Dis 17(11):2235–2250

UK IBD Genetics Consortium, Barrett JC, Lee JC, Lees CW, Prescott NJ, Anderson CA et al (2009) Genome-wide association study of ulcerative colitis identifies three new susceptibility loci, including the HNF4A region. Nat Genet 41(12):1330–1334

Valentin-Vega YA, Okano H, Lozano G (2008) The intestinal epithelium compensates for p53-mediated cell death and guarantees organismal survival. Cell Death Differ 15(11):1772–1781

van Es JH, Clevers H (2005) Notch and Wnt inhibitors as potential new drugs for intestinal neoplastic disease. Trends Mol Med 11(11):496–502

van Nood E, Vrieze A, Nieuwdorp M, Fuentes S, Zoetendal EG, de Vos WM et al (2013) Duodenal infusion of donor feces for recurrent *Clostridium difficile*. N Engl J Med 368(5):407–415

Velcich A, Yang W, Heyer J, Fragale A, Nicholas C, Viani S et al (2002) Colorectal cancer in mice genetically deficient in the mucin Muc2. Science 295(5560):1726–1729

Vilaseca J, Salas A, Guarner F, Rodriguez R, Martinez M, Malagelada JR (1990) Dietary fish oil reduces progression of chronic inflammatory lesions in a rat model of granulomatous colitis. Gut 31(5):539–544

Vogelstein B, Fearon ER, Hamilton SR, Kern SE, Preisinger AC, Leppert M et al (1988) Genetic alterations during colorectal-tumor development. N Engl J Med 319(9):525–532

von Lampe B, Barthel B, Coupland SE, Riecken EO, Rosewicz S (2000) Differential expression of matrix metalloproteinases and their tissue inhibitors in colon mucosa of patients with inflammatory bowel disease. Gut 47(1):63–73

Wagner J, Catto-Smith AG, Cameron DJ, Kirkwood CD (2013) Pseudomonas infection in children with early-onset Crohn's disease: an association with a mutation close to PSMG1. Inflamm Bowel Dis 19(4):E58–E59
Walker DG, Bancil AS, Williams HR, Bunn C, Orchard TR (2011) How helpful are serological markers in differentiating crohn's disease from ulcerative colitis in indian asian inflammatory bowel disease patients? Gut 60(Suppl 1):A222–A223
Wang C, Zhao R, Huang P, Yang F, Quan Z, Xu N et al (2013) APC loss-induced intestinal tumorigenesis in *Drosophila*: Roles of Ras in Wnt signaling activation and tumor progression. Dev Biol 378(2):122–140
Wehkamp J, Harder J, Weichenthal M, Mueller O, Herrlinger KR, Fellermann K et al (2003) Inducible and constitutive beta-defensins are differentially expressed in Crohn's disease and ulcerative colitis. Inflamm Bowel Dis 9(4):215–223
Wehkamp J, Salzman NH, Porter E, Nuding S, Weichenthal M, Petras RE et al (2005) Reduced paneth cell alpha-defensins in ileal Crohn's disease. Proc Natl Acad Sci U S A 102(50):18129–18134
Wirtz S, Neurath MF (2007) Mouse models of inflammatory bowel disease. Adv Drug Deliv Rev 59(11):1073–1083
Wittkopf N, Billmeier U, Günther C, Wirtz SJ, Neurath MF, Becker C (2011) STAT3-mediated production of antimicrobial peptides by intestinal epithelial cells helps controlling C. rodentium induced infection. Gastroenterology 140(5, Supplement 1):S-325
Wong JM, de Souza R, Kendall CW, Emam A, Jenkins DJ (2006) Colonic health: fermentation and short chain fatty acids. J Clin Gastroenterol 40(3):235–243
Wu H, Wang MC, Bohmann D (2009a) JNK protects *Drosophila* from oxidative stress by trancriptionally activating autophagy. Mech Dev 126(8–9):624–637
Wu S, Rhee KJ, Albesiano E, Rabizadeh S, Wu X, Yen HR et al (2009b) A human colonic commensal promotes colon tumorigenesis via activation of T helper type 17 T cell responses. Nat Med 15(9):1016–1022
Wu Y, Antony S, Hewitt SM, Jiang G, Yang SX, Meitzler JL et al (2013) Functional activity and tumor-specific expression of dual oxidase 2 in pancreatic cancer cells and human malignancies characterized with a novel monoclonal antibody. Int J Oncol 42(4):1229–1238
Xavier RJ, Podolsky DK (2007) Unravelling the pathogenesis of inflammatory bowel disease. Nature 448(7152):427–434
Xu N, Wang SQ, Tan D, Gao Y, Lin G, Xi R (2011) EGFR, wingless and JAK/STAT signaling cooperatively maintain Drosophila intestinal stem cells. Dev Biol 354(1):31–43
Yan B, Wang H, Rabbani ZN, Zhao Y, Li W, Yuan Y et al (2006) Tumor Necrosis Factor-α is a potent endogenous Mutagen that promotes cellular transformation. Cancer Res 66(24):11565–11570
Zaidman-Remy A, Herve M, Poidevin M, Pili-Floury S, Kim MS, Blanot D et al (2006) The *Drosophila* amidase PGRP-LB modulates the immune response to bacterial infection. Immunity 24(4):463–473
Zaph C, Troy AE, Taylor BC, Berman-Booty LD, Guild KJ, Du Y et al (2007) Epithelial-cell-intrinsic IKK-beta expression regulates intestinal immune homeostasis. Nature 446(7135):552–556
Ze X, Duncan SH, Louis P, Flint HJ (2012) *Ruminococcus bromii* is a keystone species for the degradation of resistant starch in the human colon. ISME J 6(8):1535–1543
Zenewicz LA, Yancopoulos GD, Valenzuela DM, Murphy AJ, Stevens S, Flavell RA (2008) Innate and adaptive interleukin-22 protects mice from inflammatory bowel disease. Immunity 29(6):947–957
Zhernakova A, Festen EM, Franke L, Trynka G, van Diemen CC, Monsuur AJ et al (2008) Genetic analysis of innate immunity in Crohn's disease and ulcerative colitis identifies two susceptibility loci harboring CARD9 and IL18RAP. Am J Hum Genet 82(5):1202–1210
Zhou FX, Chen L, Liu XW, Ouyang CH, Wu XP, Wang XH et al (2012) *Lactobacillus crispatus* M206119 exacerbates murine DSS-colitis by interfering with inflammatory responses. World J Gastroenterol 18(19):2344–2356
Zindl CL, Lai J, Lee YK, Maynard CL, Harbour SN, Ouyang W et al (2013) IL-22–producing neutrophils contribute to antimicrobial defense and restitution of colonic epithelial integrity during colitis. Proc Natl Acad Sci 110:12768–12773

Chapter 8
Protein Quality Control in Brain Aging: Lessons from Protein Misfolding Disorders in *Drosophila*

Lorena de Mena, Pedro Fernandez-Funez and Diego E. Rincon-Limas

Abstract Protein quality control is an essential process for cellular survival. When protein damage occurs, a series of coordinated response mechanisms repair or degrade damaged proteins to avoid the accumulation of toxic protein aggregates and restore proteostasis. However, the amount of misfolded proteins increases during aging overwhelming the mechanisms responsible for protein quality control, thus leading to the development of several age-dependent neurodegenerative disorders. Interestingly, targeted expression of proteins causative of these diseases in flies reproduces the pathological behaviors seen in humans. This remarkable conservation provides a valuable experimental tool to elucidate the complex mechanisms associated with the maintenance of proteostasis. In this chapter, we summarize how *Drosophila* has contributed to understand the roles of the heat shock response, the unfolded protein response, autophagy and the ubiquitin proteasome system in brain aging and neurodegeneration associated with protein-misfolding disorders. In addition, we describe fundamental contributions of the fly system to the design of new therapeutic strategies for these devastating disorders.

Keywords Protein quality control · Proteostasis · Misfolded proteins · Neurodegenerative disorders · Heat shock response · Autophagy

---

L. de Mena · P. Fernandez-Funez · D.E. Rincon-Limas
Department of Neurology, McKnight Brain Institute, Florida, USA

P. Fernandez-Funez (✉) · D.E. Rincon-Limas (✉)
Department of Neuroscience, Genetics Institute, and Center for Translational Research in Neurodegenerative Disease, University of Florida, 1149 Newell Dr, Gainesville, FL 32611, USA
e-mail: pedro.fernandez@neurology.ufl.edu

D.E. Rincon-Limas
e-mail: diego.rincon@neurology.ufl.edu

A.M. Vaiserman et al. (eds.), *Life Extension*, Healthy Ageing and Longevity 3,
DOI 10.1007/978-3-319-18326-8_8

8.1 Introduction

Protein quality control or proteostasis refers to the cellular processes involved in the biogenesis, folding, maturation, distribution, and degradation of proteins in different cellular compartments. These activities are critical for proper cell function because they maintain active proteins and organelles, and also prevent the accumulation of toxic unfolded/misfolded proteins. When protein damage occurs, a series of coordinated response mechanisms repair or degrade damaged proteins to avoid the accumulation of toxic protein aggregates that can, ultimately, lead to cell death. Protein toxicity can be the result of different alterations in protein homeostasis associated with aberrant folding and aggregation, leading to the so-called proteinopathies or protein misfolding disorders. The most common causes of these disorders include missense mutations, abnormal post-translational modifications, protein overexpression, and exogenous stressors, including temperature and chemicals among others. Most of the cellular machinery implicated in protein quality control is essential for cell survival, highlighting the importance of protein homeostasis. This is particularly true for neurons, because of their weak capacity to regenerate and their long life span. It is, thus, obvious that maintaining protein homeostasis is critical for healthy brain aging and longevity.

A complex network of conserved cellular processes controls the quality of the proteome and restores proteostasis following the aberrant accumulation of misfolded proteins. This protective network can be grouped into mechanisms that detect and respond to protein unfolding/misfolding, which include the heat shock response (HSR) and the unfolded protein response (UPR), and mechanisms responsible for abnormal protein degradation, which comprises autophagy and the ubiquitin proteasome system (UPS). All four processes are highly integrated to prevent the cellular toxicity associated with the accumulation of protein aggregates (Fig. 8.1).

A particular group of human disorders has played a critical role in the characterization of protein quality control pathways in neuronal survival, the so-called protein-misfolding diseases or proteinopathies. A large number of human diseases are associated with protein misfolding, which can affect specific organs (brain, muscle, eye, skin) or have systemic effects (Valastyan and Lindquist 2014). Currently, more than 30 different human diseases are linked to aberrant protein accumulation, including the highly prevalent Alzheimer's disease (AD), Parkinson's disease (PD), and amyotrophic lateral sclerosis (ALS), which have a mostly sporadic etiology. A large number of rare genetic conditions also have protein aggregation as the main trigger, including Huntington's disease (HD) and several spinocerebellar ataxias. These disorders have distinct clinical presentations, affect different parts of the brain, and are caused by the misfolding and aggregation of unrelated proteins, which can be either wild type or mutant. A common thread to these diverse disorders, though, is that they all have late onset presentation. This link is revealing because in the genetic diseases the triggering protein is expressed throughout the lifetime, but several decades go by without clinical

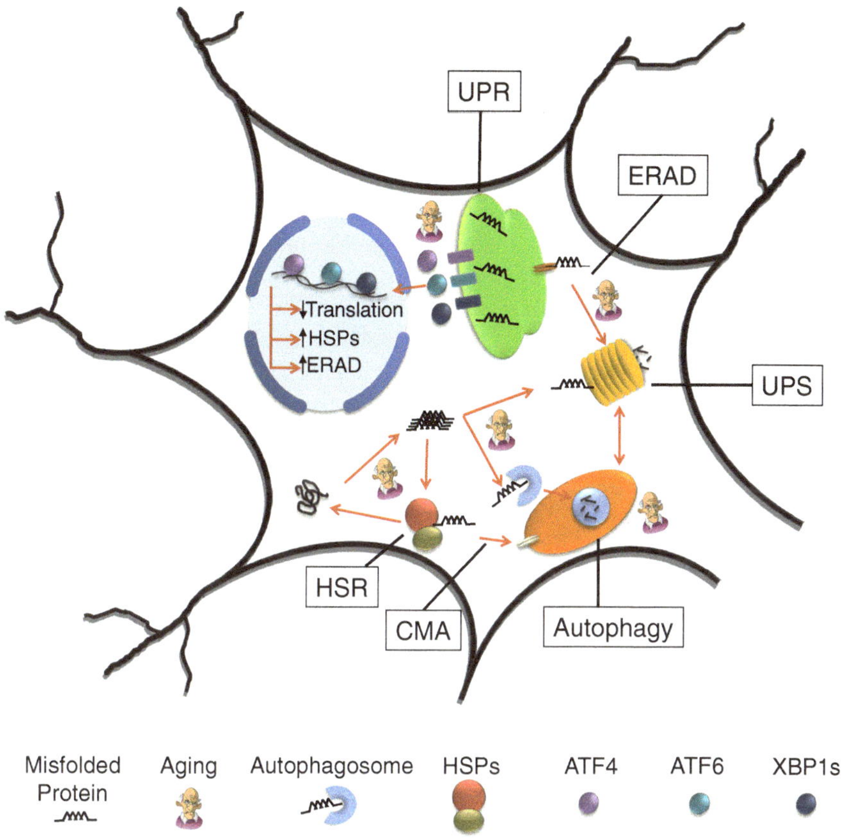

Fig. 8.1 Impact of aging and disease in cellular proteostasis. Under physiologic conditions, the heat shock response (HSR), the unfolded protein response (UPR), and two catalytic pathways (autophagy and UPS) cooperate to prevent the aggregation and toxicity of misfolded proteins. Heat shock proteins (HSPs) act as the first line of defense by binding misfolded substrates, promoting their refolding, and directing their degradation through chaperone-mediated autophagy (CMA). In the ER, accumulation of misfolded proteins activates the three UPR pathways that alter gene expression to restore proteostasis, including the up-regulation of ER associated degradation (ERAD). When misfolded proteins escape HSPs, they accumulate and aggregate in the cytosol, where become substrates for degradation by autophagy and the UPS. This complex, coordinated system maintains cellular proteostasis for decades and is critical for the function of post-mitotic neurons. However, protein misfolding disorders and aging place the proteostasis machinery under stress by overproducing misfolded proteins and reducing the cellular response to insults. Misfolded proteins saturate the response systems and prevent the regular turnover of old proteins and organelles. Aging is known to weaken gene expression of all stress-response pathways, thus dampening the ability to protect the cell. The combination of both stressors may be at the heart of multiple late-onset neurodegenerative disorders

signs. This observation suggests that the cellular machinery that controls protein homeostasis prevents protein aggregation and neuronal cell death for several decades. Then, in middle to old age, the same machinery seems to be overwhelmed by the accumulation of misfolded proteins and an age-related decrease in the expression of key factors of the quality control pathways. Overall, the tight connection of aging with protein-misfolding diseases enables the study of the cellular machinery that keeps protein quality in check during normal brain physiology and aging. In this chapter we summarize the proteostasis mechanisms, and how *Drosophila* has contributed to understand their roles in brain aging and neurodegeneration associated with protein-misfolding disorders.

Several model systems have been employed to investigate the pathways and protein interactions involved in proteostasis, including in vitro and cell culture paradigms. The small fruit fly *Drosophila melanogaster* has proven a useful tool for the study of multiple human disorders, protein misfolding diseases and aging. Around 75 % of human genes associated with human diseases have a homologous in *Drosophila*. Moreover, the low cost and easy handling make this small fly an excellent model organism. Another advantage of *Drosophila* is the large number of genetic tools available for gene manipulation (loss-of-function and gain-of-function alleles), tissue specific gene expression with the UAS/GAL4 system, temporal control of gene expression with Gal80TS, and the recent introduction of markers of neuronal architecture and function (del Valle Rodriguez et al. 2011).

Drosophila has also contributed significantly to understand human protein misfolding diseases. A large number of these diseases have been modeled in flies because the proteins causative of these diseases in human exhibit the same aberrant behavior in flies and, thus, misfold, aggregate, and induce toxicity in the *Drosophila* brain and eyes (Rincon-Limas et al. 2012). Interestingly, *Drosophila* models have played a critical role in unraveling the cellular mechanisms underlying disease pathogenesis. Indeed, unbiased genetic screens have revealed the involvement of unsuspected genes and pathways. Moreover, candidate approaches allowed the efficient testing of hypotheses in vivo, confirming the role of suspected pathways, including chaperones and autophagy. Finally, these models have also been used to examine the effectiveness of pharmacological modulators of suspected pathways, providing in vivo information that can shorten the path to the translation of effective therapies in humans.

8.2 Molecular Chaperones: The Cell First Line of Defense

Molecular chaperones are ubiquitous and highly conserved proteins with a main role in protein homeostasis. They recognize and bind unfolded and misfolded proteins to prevent aggregation and promote refolding. Chaperones were first identified through their transcriptional response to heat stress in *Drosophila*, but they can respond to several stresses such as cold, ischemia, hypoxia, oxidative stress, proteotoxic stress, chemicals, and heavy metals. Heat shock proteins (HSPs) are the major group of molecular chaperones inside the cell, and include several families with distinct localization in organelles and cellular compartments, and different functions (Kim et al. 2013).

HSPs contain a substrate-binding domain (SBD) that recognizes short hydrophobic stretches on unfolded/misfolded client proteins and this interaction is the first key step to prevent their aggregation. In fact, in the absence of chaperones, proteins with exposed hydrophobic stretches self-associate creating aggregates that can be unstructured or highly ordered into amyloid fibers (Douglas and Cyr 2009). Some HSPs contain ATPase domains and cooperate with ATP-independent co-chaperones to efficiently bind and refold substrates. HSPs are classified based on their molecular weight, ranging from small HSPs with less than 30 kDa to large proteins over 100 kDa. Each family comprises several members with different tissue and subcellular distribution (Voisine et al. 2010).

8.2.1 The Hsp70/Hsp40 System

The Hsp70 family is among the most conserved proteins in the animal world. Their main role is to promote the correct protein folding of both nascent and mature proteins by recognizing exposed hydrophobic residues on unfolded or misfolded substrates, and prevent their aggregation. Structurally, Hsp70 chaperones have two major functional domains: an N-terminal nucleotide binding domain (NBD) that binds ATP and hydrolyzes it to ADP and a C-terminal SBD that interacts with exposed linear hydrophobic segments. In its ATP-bound state, Hsp70 binds substrates with low affinity. ATP hydrolysis to ADP forces a conformational change in the SBD that increases the affinity for the substrate. Then, interaction with a nucleotide exchange factor restores ATP binding of Hsp70, which forces the release of the client protein, completing the Hsp70 cycle. Although Hsp70 exerts a potent refolding activity in vitro, the Hsp40 family regulates Hsp70 cycling in vivo. Hsp40s are ATP-independent chaperones characterized by a J-domain that interacts with the NBD of Hsp70 and stimulates ATP hydrolysis, thus strengthening the interaction with the client protein. In vitro studies suggest that Hsp40 first binds client proteins and presents them to the SBD of Hsp70 while stimulating ATP hydrolysis in the NDB domain, therefore enhancing Hsp70 refolding efficiency (Young 2010).

The characterization of several brain disorders linked to protein misfolding and deposition suggested that HSPs should be involved in the pathogenesis and could also play a role in therapy. Initially, a study in cell culture showed that overexpressed Hsp70 co-localized with mutant Ataxin1 (Atx1-82Q) in nuclear inclusions, which pointed to a stress response against pathogenic protein aggregation. To determine whether increased Hsp70 activity could harbor a protective role against protein aggregates, Warrick and cols. overexpressed HSPA1L, a human inducible Hsp70, in *Drosophila* showing aberrant eye morphology linked to mutant Ataxin3 (Atx3tr-78Q). Co-expression of HSPA1L completely suppressed the aberrant phenotype caused by Atx3tr-78Q, indicating the critical role of Hsp70 in the toxicity of this aggregate-prone polyglutamine allele (Warrick et al. 1999). Moreover, co-expression of Atx3tr-78Q and a constitutive *Drosophila* Hsp70

(Heat shock cognate 4, Hsc4) carrying the K71S substitution that inactivates the ATPase activity enhanced neurotoxicity, suggesting that the endogenous activity of Hsp70 mitigates the toxicity of Atx3tr-78Q. A prediction from these results was that Hsp70 overexpression would exhibit the same protective activity against other polyglutamine expansions, since the same mutation was responsible for the aberrant protein folding in a different protein. Spinal-bulbar muscular atrophy (SBMA) is linked to a polyglutamine expansion in the androgen receptor (AR) and expression of the mutant allele in flies induces neurotoxicity. As expected, co-expression of HSPA1L rescued the toxicity of AR-108Q, confirming the beneficial effect of Hsp70 against related misfolded substrates (Chan et al. 2000).

The next question in the field was whether Hsp70 could exert the same neuroprotective activity against other toxic amyloids unrelated to the polyglutamine expansions. The same lab addressed this question in a *Drosophila* model expressing α-synuclein, the most abundant protein in Lewy bodies, typically found aggregating in PD patients. HSPA1L not only co-localized with α-synuclein aggregates but also increased the survival rate of dopaminergic cells (Auluck et al. 2002). In contrast, co-expression of α-synuclein and dominant-negative Hsc4 resulted in increased dopaminergic cell death without changing the number of Lewy body-like aggregates (Auluck et al. 2002). Additionally, the prion protein (PrP), a membrane-anchored glycoprotein widely expressed in the brain, leads to aggressive vacuolar degeneration when misfolds. We asked whether Hsp70 could exert a protective activity against PrP misfolding in vivo either by indirectly maintaining intracellular homeostasis or directly by exiting the cell and interacting with PrP. To our surprise, we found that Hsp70 overexpression protected against PrP toxicity, reduced PrP levels, and inhibited PrP misfolding (Fernandez-Funez et al. 2009). Since we found that Hsp70 co-immunoprecipitated with PrP, Hsp70 seems to mediate its protective effects by binding directly to PrP at the membrane. This ability of Hsp70 and other chaperones to exit the cell has been described previously, but is associated with stressful conditions (Pittet et al. 2002; Fleshner and Johnson 2005). Thus, HSPA1L is beneficial against a wide variety of misfolded proteins that accumulate in the nucleus, cytosol, and the extracellular space.

With Hsp70 working as a potent suppressor of toxicity, the potential protective activity of other chaperones emerged over the next few years. Two independent screens for modifiers of Atx1-82Q and 65Q-only toxicity in flies identified the protective activity of dDnaJ-1/Hsp40 overexpression (Fernandez-Funez et al. 2000; Kazemi-Esfarjani and Benzer 2000). Interestingly, Hsp40 overexpression altered the distribution of nuclear inclusions of Atx1-82Q, which appeared to coalesce into a single compact aggregate per nucleus, suggesting a link between aggregate distribution and toxicity (Fernandez-Funez et al. 2000). Further prove of the protective activity of Hsp70 came from overexpression of the *Drosophila* orthologue of human Hdj-1/Hsp40 in fly models of polyglutamine toxicity. Flies overexpressing dHdj1 strongly suppressed Atx3tr-78Q toxicity, but only partially suppressed the toxicity of mutant Huntingtin (Htt-120Q). On the other hand, elimination of the J-domain or mutation of the SBD enhanced the toxicity of Atx3tr-78Q, suggesting that mutant dHdj1 behaves as a dominant-negative by blocking the protective activity of endogenous factors, i.e., Hsp70. However, expression of a second orthologue, dHdj2, partially rescued Atx3tr-78Q and did not rescue Htt-120Q, indicating that Hsp40s exhibit substrate selectivity. Moreover, a coordinated overexpression of both, dHdj1 and Hsp70 showed an even stronger protection (Chan et al. 2000; Bonini 2002). This different function supported the theory that chaperones from the same family can recognize different substrates and, in consequence, exert different functions (Bonini 2002). Interestingly, these studies failed to observe changes in nuclear inclusions, which suggested a decrease in protein neurotoxicity without reductions on protein aggregation (Chan et al. 2000).

Subsequent studies demonstrated that, besides Hsp70, other chaperones cooperate with Hsp40 to mitigate protein toxicity. Kuo and collaborators described a chaperone capable of enhancing the protective effect of DNAJ-1, Hsp70bc, an orthologue of human Hsp110. Despite the structural and functional conservation with Hsp70, Hsp70bc did not rescue polyQ toxicity on its own. However, Hsp70bc enhanced the protective activity of Hsp40, whereas inactivation of its ATPase domain led to a complete loss of such protective function (Kuo et al. 2013), demonstrating the cooperative activity of HSPs.

Another interesting member of the Hsp40 family of chaperones are the cysteine-string proteins (CSPs). CSPs contain the characteristic N-terminal J domain that recognize and bind to Hsc70/Hsp70 proteins promoting their ATPase activity (Miller et al. 2003; Zinsmaier 2010). Under physiological conditions, CSPs participate on the maintenance of synaptic function and structure by regulating the Soluble NSF Attachment Protein Receptor (SNARE) complex assembly and modulating presynaptic Ca^{2+} channels activity (Zinsmaier 2010). In addition, CSP chaperone activity could impede neurodegeneration by preventing toxic accumulation of misfolded synaptic proteins. However, when abnormal protein aggregation occurs in the neurons, sequestration of CSPs by the toxic aggregates compromise CSP function, reducing synapsis and triggering neurodegeneration (Miller et al. 2003; Fernandez-Chacon et al. 2004; Zinsmaier 2010). Recently, Wang et al. described that overexpression of Hip, a co-chaperone that enhanced and stabilized binding of Hsp70 to its substrates, promotes client protein's ubiquitination and poly Q-Androgen receptor clearance. In this regard, YM-1, a synthetic co-chaperone similar to Hip, led to similar results when supplied to a *Drosophila* model of SBMA. Thus, allosteric activators of Hsp70 can be used as new therapeutic approaches against the protein aggregation that occurs during age-depending neurodegenerative disorders (Wang et al. 2013).

8.2.2 Hsp90

Hsp90 is the most conserved chaperone involved in the folding, stability and maturation of the structural integrity of proteins. Hsp90 contains an N-terminal ATPase domain, a SBD in the middle, and a C-terminal domain for dimerization (Pearl and Prodromou 2006). Hsp90 assists in protein folding and stabilization in coordination with Hsp70 and Hsp40, but it seems to have a higher affinity for aberrant proteins (Waza et al. 2006; Luo et al. 2008; Koren et al. 2009). Hsp90 plays a key regulatory role in the heat shock response due to its binding to Heat shock factor 1 (HSF1), a transcription factor that regulates the expression of Hsp70 and other HSPs. Hsp70 and Hsp40 also bind HSF1, but do not inhibit its transcriptional activity. Under stress, Hsp90 releases HSF1, which forms trimers, relocates to the nucleus, and induces the expression of HSPs (Morimoto 2008; Luo 2010, 2013). Hsp90 inhibitors have been developed to reverse the permissive role of Hsp90 in tumorigenesis. Several geldanamycin derivatives show lower toxicity and improved activity, including 17-allylamino-17-demethoxygeldanamycin (17-AAG) and 17-(dimethylaminoethylamino)-17-demethoxygeldanamycin (17-DMAG). However, these Hsp90 inhibitors still have limitations for clinical use due to their toxicity.

Quite surprisingly, the potential benefit of Hsp90 refolding activity against amyloids has not been tested in animal models, so far. In fact, Hsp90 has been targeted for inhibition as a strategy to induce the protective activity of HSF1 and its downstream targets, including Hsp70. Administration of Hsp90 inhibitors in the media activates HSF-1 in *Drosophila*, which leads to suppression of Atx3tr-78Q, α-synuclein, and Htt-120Q (Auluck et al. 2002; Fujikake et al. 2008). This recovery correlates with an increase in Hsp70 and Hsp40 transcription, demonstrating for the first time the use of chaperones as therapeutic targets in protein-misfolding diseases. We had shown before the beneficial effect of Hsp70 overexpression against PrP neurotoxicity and wanted to find out whether the Hsp90 inhibitors also showed this protective activity. We found that the Hsp90 inhibitors geldanamycin and 17-DMAG had no effect on the accumulation of PrP. This was somehow expected because the protective effect of Hsp70 on PrP is weak due to their presence in different compartments. We argued that boosting Hsp70 induction with a second compound would enhance the activity of 17-DMAG and result in direct effects on PrP accumulation. We found that the glucocorticoid dexamethasone acted as an Hsp70 activator by increasing the transcriptional activity of HSF1 by stabilizing its binding to the transcriptional machinery. As predicted, the combination of 17-DMAG and dexamethasone significantly increased the levels of inducible Hsp70 and decreased PrP steady-state levels (Zhang et al. 2014). This treatment reduced final levels of pathogenic PrP, and improved locomotor activity without the potential deleterious effects observed when adding either drug in high doses.

8.2.3 Small Heat Shock Proteins (SHSPs)

Small HSPs (sHSPs) comprise all chaperones under 30 kDa with an 80 amino acid α-crystallin domain responsible for intra- and inter-molecular interactions. The sHSPs family includes 10 members in mammals (referred to as HSPB1 to 10) and four in *Drosophila* (Hsp22, Hsp23, Hsp26, and Hsp27) that exert different functions, although most inhibit protein aggregation and increase the clearance of abnormally folded proteins.

HSPBs dimerize through their α-crystallin domains and form oligomers that recognize and inhibit the aggregation of unfolded proteins (Carra et al. 2008, 2009). Although all HSPBs share structural similarities, they exert different activities in vivo. For instance, human HspB7 and HspB8 exhibited the strongest protective function against Atx3tr-78Q toxicity in flies. Cell culture assays showed that this protective activity was due to a high anti-aggregation function that does not require the refolding activity. Interestingly, the anti-aggregation activity associated with HSPB7 and HSPB8 is exerted through different mechanisms and is substrate-dependent (Vos et al. 2010). In contrast, HSPB1, HSPB4, and HSPB5 display milder protective effects against Atx3tr-78Q with little anti-aggregation activity, but a strong refolding effect. Regarding *Drosophila* sHsps, Hsp27 is the only one that improves paraquat-induced movement disorder and mild toxicity caused by a short polyQ tract (47 residues); however, it does not alleviate the severe toxicity caused by a long polyQ repeat (121 residues) (Liao et al. 2008). These data support the hypothesis that HSPBs have different functions and substrate selectivity to maintain protein homeostasis (Vos et al. 2010).

8.2.4 Heterologous Hsp104

The strong protective effect of classic HSPs against misfolded protein has sparked a renewed interest in new chaperones with therapeutic applications. Hsp104 is a yeast chaperone that has recently gain attention because of its ability to efficiently disaggregate amyloids, although it shows poor ability to prevent aggregation of unfolded substrates. Hsp104 forms a hexameric complex with two AAA^+-ATPase sites per monomer (NBD1 and 2) and a large central cavity for the interaction with aggregated substrates.

The interest on Hsp104 relies on its unusual ability to disaggregate large protein aggregates, which has been tested in different animal models including *Drosophila* and mice. Recently, Hsp104 demonstrated different activities against closely related substrates: Hsp104 suppressed the toxicity of Atx3tr-78Q without affecting its aggregation, but enhanced the toxicity of full length Atx3-78Q (Cushman-Nick et al. 2014). These opposite effects were explained by the interaction of Hsp104 with different domains that modulate Atx3 conformation. Another interesting finding in this study is that Hsp104, contrary to Hsp70, reduces Atx3tr-78Q aggregation after it has begun. By using the conditional Gene Switch system in *Drosophila* (Roman et al. 2001), the authors induced Hsp104 or Hsp70 at day 7 in flies that expressed Atx3tr-78Q constitutively in the eye. While Hsp70 had no effect, Hsp104 mitigated the degeneration associated with pre-existing Atx3tr-78Q aggregates (Cushman-Nick et al. 2014). These promising results suggest that this heterologous HSP could have therapeutic applications against proteinopathies after the onset of pathogenic protein-induced degeneration.

8.2.5 NMNAT

NAD synthase nicotinamide mononucleotide adenylyltransferase (NMNAT) is a protein that acts both as NAD synthase and chaperone in *Drosophila* (Zhai et al. 2006). NMNAT delays axonal degeneration and protects against protein toxicity in *Drosophila* models of spinocerebellar ataxia type 1 (SCA1), suggesting that NMNAT is required for neuronal maintenance and neuroprotection. Presumably, NMNAT reduces aggregation and promotes protein degradation through a proteasome-mediated pathway (Zhai et al. 2008). In addition, NMNAT promotes ubiquitination and clearance of toxic tau species in *Drosophila*, and when overexpressed suppresses tau-related phenotypes (Ali et al. 2012). Moreover, in fly brains overexpressing polyglutamine expanded proteins, NMNAT appears up-regulated and co-localizes with Hsp70 in protein aggregates; however, both proteins act independently and in an additive manner. These data suggest that NMNAT acts as a stress-response chaperone to maintain and protect neuronal cells (Zhai et al. 2008; Jaiswal et al. 2012).

8.2.6 Engineered Chaperones: Secreted Hsp70

One of the limitations of HSPs is that small amounts leak outside the cell under stressful conditions. However, several proteinopathies are characterized by protein aggregates in the extracellular space, where limited chaperone activity exists. We devised a new approach to increase the chaperone activity in the extracellular space by designing a secreted form of Hsp70 (secHsp70). This new chaperone consists in fusing the signal peptide of the human Immunoglobulin heavy chain V–III to HSPA1L. SecHsp70 can be detected in cell media and in the lumen of the eye imaginal disc, confirming the functionality of the signal peptide. Moreover, the activity of Hsp70 outside the cell neutralizes the toxicity of Aβ42 in a *Drosophila* model of AD, including life span extension. We confirmed that secHsp70 binds Aβ42, but this interaction has no effect on Aβ42 steady-state levels or aggregation. We propose that the protective activity of secHsp70 is mediated by masking key neurotoxic Aβ42 epitopes, thus preventing the interaction of Aβ42 with cellular substrates that mediate pathogenesis, including receptors and channels. This ability of secHsp70 to bind misfolded extracellular substrates may also prove useful to prevent the prion-like spread of different oligomeric assemblies proposed to mediate pathogenesis, including synucleinopathies, tauopathies, and TDP-43 proteinopathies (Fernandez-Funez et al. under review).

8.2.7 Chaperone Dynamics During Aging

Aggregation of damaged or unfolded proteins is associated with an impairment in the mechanisms involved in protein quality control associated with aging, and is a risk factor for the development of several age-dependent diseases. Chaperone proteins are the first cell defense to assure correct protein folding and ameliorate protein aggregation. However, during aging, protein aggregation is enhanced, which can be partly associated with a decrease in the expression and function of HSPs, or defects in capacity of HSPs (Soti and Csermely 2003). In *Drosophila*, expression of hsp22 appears up-regulated during aging, most likely as a response to stressors only present later in life; and, at normal temperatures, transient heat induced overexpression of Hsp70 increases life span (Tatar et al. 1997; King and Tower 1999). Furthermore, overexpression of *Drosophila* Hsp22, Hsp23 and Hsp26 resulted in a mean increase of life span in flies exposed to mild heat stress and paraquat-induced oxidative stress (Morrow and Tanguay 2003; Liao et al. 2008). However, as stated above, only HSP27 has the ability to ameliorate paraquat-induced movement disorder (Liao et al. 2008).

Interestingly, although expression of HSPs is positively-correlated with life span under mild stressors, under constant stress HSP expression can be an indicative of susceptibility and failure in homeostasis (Yang and Tower 2009).

8.3 ER Stress and Unfolded Protein Response

The endoplasmic reticulum (ER) is an essential organelle responsible for the translocation and folding of ER-resident proteins, membrane proteins, and secreted proteins. A complex protein network in the ER lumen regulates post-translational modifications to ensure correct protein function (Hetz et al. 2011). Alterations in ER homeostasis due to protein accumulation in the lumen results in a condition termed ER stress, which results in the activation of the unfolded protein response (UPR). Under normal circumstances, UPR controls abnormal protein aggregation by attenuating protein synthesis, promoting protein refolding, and enhancing abnormal protein degradation. However, chronic disruption in ER proteostasis can trigger cell death leading to neurodegeneration (Matus et al. 2011). Recent evidence indicates that chronic ER stress is involved in diverse diseases, including neurodegenerative conditions, cancer, and diabetes (Koumenis 2006; Lipson et al. 2006; Hetz and Glimcher 2009).

Three key ER stress sensors regulate the UPR: protein kinase RNA-like ER kinase (PERK), activating transcription factor 6 (ATF6), and inositol-requiring enzyme 1 (IRE1) (Hetz and Mollereau 2014). PERK is a type I ER transmembrane protein with a cytosolic kinase domain. When misfolded proteins accumulate in the ER, the cytosolic domain of PERK dimerizes and autophosphorylates, which leads to eIF2a phosphorylation and inactivation (Harding et al. 1999). EIF2a inactivation causes a decrease in the general protein synthesis rate and increases the translation of the transcription factor ATF4, which regulates expression of UPR genes involved in amino acid metabolism, autophagy, and resistance to oxidative stress (Hetz et al. 2011). The PERK signaling cascade is reversible once ER homeostasis is reestablished, recovering the normal translation rate.

ATF6 is a type II ER transmembrane protein that comprises a bZIP transcriptional motif in the cytosolic domain. Under normal conditions, the ATF6 is retained at the ER membrane. Under ER stress conditions, ATF6 migrates to the Golgi apparatus, where two protease cleavages release the cytosolic fragment. Cleaved ATF6, then, enters the nucleus to activate the transcription of several ER-chaperones and endoplasmic reticulum-associated protein degradation (ERAD)-associated genes among others. ATF6 response initiates early during the UPR and activates expression of unspliced X-box binding protein 1 (XBP1), which accumulates in the ER becoming available for the later IRE1 response.

IRE1 is the most conserved branch of the UPR, with homologues in plants, *C. elegans*, *Drosophila*, and mammals. IRE1 is a type I transmembrane protein with a luminal domain responsible for detecting ER stress and a cytosolic domain with protein kinase and endonuclease activities (Rasheva and Domingos 2009). Upon ER stress, IRE1 oligomerizes and autophosphorylates, activating an alternative splicing in XBP1 mRNA. Spliced XBP1 (XBP1s) is a potent transcription factor that regulates the expression of genes associated with protein folding, ERAD, protein translocation into the ER, and lipid synthesis among other processes. During the recovery phase of ER stress, the unspliced XBP1 acts as a negative regulator of XBP1s, which inhibits transcription of target genes (Hetz et al. 2011).

Under physiological circumstances, the UPR is activated as a protective mechanism against ER stress caused by abnormal protein aggregation associated with cellular differentiation and exposure to stressors. ER stress markers also appear up-regulated in several models of neurodegeneration as well as in postmortem brains of several neurodegenerative diseases suggesting a pathogenic role of UPR. While a transitory response can be beneficial for cell survival, a chronic UPR can lead to apoptotic mechanisms promoting neurodegeneration (Halliday and Mallucci 2014).

In recent years, *Drosophila* has contributed to elucidate the role of UPR against protein toxicity in vivo. Retinitis pigmentosa (RP) is a degenerative eye condition caused by an accumulation of misfolded mutant Rhodopsin 1 (Rh1*) in the ER. In flies, accumulation of misfolded Rh1* in the ER elevates the expression of XBP1s. Accordingly, reduction of XBP1 increases the retinal degeneration induced by Rh1*, which indicates that XBP1 and its target genes are protective (Ryoo and Steller 2007). Moreover, the authors describe that disruption of the ERAD pathway leads to an increase in Rh1 protein levels in *Drosophila*. Consistently, the ER stress caused by Rh1 is reduced when some subunits of the ERAD machinery are overexpressed (Kang and Ryoo 2009). Interestingly, UPR is also implicated in neurodegenerative diseases. In a fly model of tauopathy, UPR is increased due to phosphorylation of tau, and when levels of XBP1 are reduced, the ER stress response decreases promoting tau toxicity and neurodegeneration (Loewen and Feany 2010). In addition, we recently described that a fly model expressing Aβ42 displays ER stress-mediated neurodegeneration and that overexpression of XBP1s suppresses this phenotype (Casas-Tinto et al. 2013). Accordingly, XBP1 loss-of-function exacerbates Aβ42-induced toxicity. This protective effect of XBP1s was mediated by the down-regulation of ryanodine calcium channels in the ER, which prevents the release of pro-apoptotic levels of calcium in the cytoplasm (Casas-Tinto et al. 2013). Another study with *Drosophila* identified mild ER stress ("preconditioning") as a neuroprotective mechanism against human α-synuclein toxicity (Fouillet et al. 2012). These authors demonstrated that α-synuclein-expressing flies treated with a mild dose of tunicamycin, a chemical inducer of UPR, display activation of the IRE1-XBP1 pathway in the brain, improvement of locomotor function and protection of dopaminergic neurons. In addition, the authors found that this "preconditioned" mild ER stress mediates neuronal survival by blocking apoptosis in vivo. Interestingly, this ER-mediated protection was lost upon autophagy impairment suggesting that autophagic clearance contributes to the ER-mediated protection (Fouillet et al. 2012). These data agree with the hypothesis that UPR functions as neuroprotective mechanism when ER stress is not prolonged in time.

8.3.1 UPR Impairment During Aging

It is clear that a better resistance to stress leads to an increase in life span, which correlates with a better prevention against different harmful insults (Salminen and Kaarniranta 2010). However, the amount of misfolded proteins increases in the ER lumen during aging suggesting an involvement of ER stress in the aging process (Salminen and Kaarniranta 2010). Indeed, while the UPR is strong in young animals, it is compromised during aging and the expression of several ER proteins is reduced (Naidoo 2009). It is not surprising then, that some neurodegenerative disorders that involved accumulation of misfolded proteins present impaired UPR (Brown and Naidoo 2012). As it occurs with other protein quality control systems, duration and strength of the stressor have different consequences in the cell and life span of the organisms. Mild doses of a stressor can lead to a better stress tolerance, whereas a chronic stress exposure leads to cellular degeneration. This is especially true in the ER, where abnormal protein accumulation lead to a sustained UPR that triggers cell apoptosis.

8.4 Autophagy

Autophagy is a catabolic process for degrading and recycling organelles and misfolded proteins inside double membrane vesicles called autophagosomes. The fusion of the autophagosome with lysosomes allows the degradation of the contents of the vesicle, producing fresh building materials or energy. Autophagy occurs at basal levels in healthy cells but can be up-regulated by starvation, cellular stress, and other stimuli. This basic cellular process is critical for cell survival and renewal, but it is particularly important for quality control in neurons due to the limited regenerating capacity of the nervous system.

8.4.1 Macroautophagy and Protein Misfolding Diseases

In the last decade, several studies have shown presence of autophagy vesicles in neurodegenerative disorders such as HD or PD, pointing to a protective role of this mechanism against neurodegeneration. *Drosophila* has been a useful tool to study this process. Ravikumar and colleges showed that TOR (target of rapamycin, a key regulator of autophagy) is sequestered in polyQ aggregates and promotes autophagy (Ravikumar et al. 2004). These authors proposed that rapamycin, a negative regulator of TOR, would increase autophagy and protect against the toxicity of Htt-120Q. Flies expressing Htt-120Q in the eye fed with rapamycin experienced an increase in the number of photoreceptors (Ravikumar et al. 2004). In fact, rapamycin also reduces toxicity caused by wild type and mutant tau in *Drosophila* presumably because of autophagy degradation of insoluble tau (Ravikumar and Rubinsztein 2006). Moreover, overexpression of TSC1 and TSC2, which are negative regulators of TOR, inhibit neuronal degeneration caused by Htt-120Q (Wang et al. 2009a). Overexpression of the autophagy gene 1 (Atg1) also suppresses photoreceptor cell death in the Htt-120Q flies due to an induction of autophagy (Wang et al. 2009a).

In another study, inhibition of the small GTPase Rab5, which regulates endosome trafficking, enhances the toxicity of Htt-120Q, whereas overexpression enhances autophagosome synthesis, suppression of aggregation, and suppression of Htt-120Q toxicity (Ravikumar et al. 2008). These data suggest that besides the role of Rab5 in endocytosis, it also functions in the early stages of autophagosome formation (Ravikumar and Rubinsztein 2006; Ravikumar et al. 2008). On the other hand, a screen for modifiers of Atx3tr-78Q neurodegeneration identified genes that affect protein accumulation through autophagy confirming the protective effect of this process over protein deposition (Bilen and Bonini 2007).

However, not all aggregate-prone proteins are degraded by autophagy. P. Salvaterra's group found that expression of Aβ1-42 induces the formation of autophagic vesicles that compromise cell viability and led to neurological deficits, whereas Aβ1-40 did not show changes in autophagy (Ling et al. 2009). This suggests that, depending of the substrate, autophagy can begin as a protective mechanism but due to a decrease in its degradation function induces abnormal autophagy vesicles leading to neurodegeneration (Ling et al. 2009). In a posterior study, Ling and Salvaterra observed that this deterioration in the autophagy system shifts normal brain aging into pathological aging. The authors used temporal-dependent paradigms in flies and found an early protective effect of autophagy in life that becomes progressively deleterious during normal brain aging (Ling and Salvaterra 2010). Under these paradigms, expression of Aβ1-42 exacerbates the dysfunction of the autophagy system increasing the ratio of neurodegeneration (Ling and Salvaterra 2010). This observation agrees with the fact that age is a risk factor in the onset of AD.

8.4.2 Interaction Between Autophagy and Chaperones

Chaperone-mediated autophagy (CMA) is a process that selectively removes damaged or unfolded cytosolic proteins in order to maintain proteostasis (Cuervo and Wong 2014). Although CMA occurs under normal conditions, its activity is increased under stress conditions. In CMA, misfolded proteins are targeted on the cytosol by chaperones, which recognize a specific amino acid sequence, and, then, transported to the lysosomal membrane. Once there, the protein enters the lysosome by crossing the membrane through a translocation complex. CMA selective degradation increases over time in the presence of a stressor as a protective mechanism. In this regard, its specific substrate selection avoids the elimination of proteins or organelles that are essential for the cell survival, and preferentially targets non-essential proteins. Interestingly, CMA is impaired in several human diseases including neurodegenerative disorders as PD, AD or PolyQ diseases. Aberrant proteins such as alpha-synuclein or leucine-rich repeat kinase 2 (LRRK2) can interact with lysosome-associated membrane protein 2A (LAMP-2A), the receptor for chaperone-mediated autophagy, with high affinity but are unable to translocate to the lysosome lumen. This interaction not only fails to eliminate these aberrant or damaged proteins, but also affects the degradation of other CMA substrates by saturating the system (Cuervo et al. 2004; Orenstein et al. 2013). In the case of tauopathies, it seems that tau binds successfully LAMP-2A but undergoes a partial translocation to the lysosome lumen releasing a toxic tau fragment to the cytosol, therefore, increasing toxicity and cell death (Wang et al. 2009b; Cuervo and Wong 2014).

It is clear that chaperones play an important role against protein misfolding (Kim et al. 2013). However, some proteins cannot be refolded and need to be directed to degradation by an independent system. Some chaperones lack refolding activity and, instead, act as part of the autophagy system to target unfolded proteins to degradation by the lysosome. For instance, two members of HSPB family, HspB7 and HspB8, protect against polyQ toxicity in an autophagy-mediated manner (Vos et al. 2010). These findings suggest that autophagy acts downstream of the sHSP response. In support of this idea, overexpression of autophagy related 7 gene (Atg7) is sufficient to revert Hsp27 knockdown-mediated shortened life span in *Drosophila*. Conversely, knockdown of Atg7 blocks Hsp27-mediated long life span (Chen et al. 2012).

Altogether, these data suggests that the heat shock protein response functions upstream of autophagy and that the association between these pathways can be due to CMA or chaperone-assisted selective autophagy (CASA).

8.4.3 Autophagy and Aging

During normal aging, the expression of several autophagy genes is reduced in *Drosophila*. A decrease in autophagy is associated with an increase of neuronal damage and a reduced life span, however maintaining the expression of autophagy-related genes promotes longevity and prevents the age-dependent damage. For instance, when the *Atg8a* gene (*autophagy-related 8a*) is mutated in flies, life span is shorter and sensitivity to oxidative stress is higher (Simonsen et al. 2008). Interestingly, a raise in *Atg8a* expression in old fly brains results in an increase over 50 % in life span and promotes resistance to oxidative stress (Simonsen et al. 2008). Similarly, expression of Atg7, an essential factor for longevity of flies, has been consistently associaed with the regulation of aging (Hara et al. 2006; Chen et al. 2012). Furthermore, mutations in Atg5 and Atg7 genes, which are required for autophagosome formation, lead to increased neurodegeneration (Hara et al. 2006). These data support the hypothesis that autophagy plays an important role in regulation of aging and that its activation can be neuroprotective.

8.5 UPS and Interaction Between Protein Degradation Pathways

Protein quality control is a complex mechanism that needs a tight collaboration between several processes to avoid misfolded protein aggregation and cell death. When chaperones fail to unfold and refold abnormal proteins, degradation pathways are activated to prevent protein toxicity and ultimately cell death. The UPS is one of the pathways involved in the aberrant protein degradation (Hershko and Ciechanover 1998). In fact, a wide range of mutations in UPS-associated genes are associated with accumulation of proteasome substrates and, therefore, with neurodegeneration (Ciechanover and Brundin 2003). Interestingly, compromising the proteasome pathway in a *Drosophila* model of SBMA enhances degeneration and decreases poly(Q) protein solubility (Chan et al. 2002). Moreover, the protective activity of SCA3 observed against polyQ neurotoxicity in vivo requires the ubiquitin-associated activities of the protein and is dependent upon proteasome function (Warrick et al. 2005). In addition, reduction of the 26S proteasome activity in flies is associated with age-related accumulation of proteins and the duration of life span (Tonoki et al. 2009).

Recent evidences suggest a compensatory regulation between autophagy, UPR and UPS. For instance, in a fly model of SBMA, expression of histone deacetylase 6 (HDAC6), a microtubule-associated deacetylase that interacts with polyubiquitinated proteins, rescues degeneration associated with UPS impairment through autophagy (Pandey et al. 2007). Similarly, in a fly model for Gaucher's disease, activation of UPR leads to accumulation of *parkin* substrates triggering cell death most likely through an attenuation of normal autophagy (Maor et al. 2013). In addition, *Drosophila* proteasome mutants show increased autophagy, whereas autophagy mutants display increased ubiquinated protein agreggation (Chang and Neufeld 2010; Jaiswal et al. 2012). Moreover, flies treated with Bortezomib—a selective proteasome inhibitor—exhibit UPR induction that leads to autophagy activation to counterbalance UPS impairment (Velentzas et al. 2013). All these data illustrate the complexity of the regulatory mechanisms that coexist in the cell to maintain proteostasis during aging in an attempt to prevent protein aggregation and cell death.

8.6 Concluding Remarks

The functional similarities between *Drosophila* and mammalian proteomes make it possible to model human protein misfolding disorders with results that are relevant to human physiology. In fact, *Drosophila* has been behind many of the fundamental advances that have occurred in this field during the last 15 years. While fruit flies cannot replace the need for studying mammalian models, the studies outlined here illustrate their potential to untangle the molecular associations between protein aggregation, neurodegeneration and aging. In addition, these studies illustrate how *Drosophila* has contributed to understand a number of complex processes utilized by the cell to maintain protein homeostasis in the brain. It is clear now that promoting a correct refolding of misfolded proteins by overexpressing chaperones, enhancing the activity of the UPR, or promoting protein clearance with autophagy or the ubiquitin proteasome system reduces protein toxicity and improves life span. However, it is important to keep a balance among the pathways involved in proteostasis, otherwise a sustained stimulation can overwhelm the systems and contribute to potentiate cell death. Altogether, these important discoveries demonstrate the relevance of *Drosophila* as a model to study protein quality control in late-onset neurodegenerative diseases and its potential to design new therapeutic strategies for these devastating disorders. We anticipate, therefore, that the power of *Drosophila* genetics will further extend our understanding of neuronal proteostasis in new and unexpected directions in the years to come.

Acknowledgments We apologize to all authors whose work could not be cited due to space constraints. This work was supported in part by NIH grant R21NS081356 to DER-L, and by a McKnight Brain Institute Research Development Award and start-up funds from the UF Department of Neurology to PF-F and DER-L.

References

Ali YO, Ruan K, Zhai RG (2012) NMNAT suppresses tau-induced neurodegeneration by promoting clearance of hyperphosphorylated tau oligomers in a *Drosophila* model of tauopathy. Hum Mol Genet 21(2):237–250

Auluck PK, Chan HY, Trojanowski JQ, Lee VM, Bonini NM (2002) Chaperone suppression of alpha-synuclein toxicity in a *Drosophila* model for Parkinson's disease. Science 295(5556):865–868

Bilen J, Bonini NM (2007) Genome-wide screen for modifiers of ataxin-3 neurodegeneration in *Drosophila*. PLoS Genet 3(10):1950–1964

Bonini NM (2002) Chaperoning brain degeneration. Proc Natl Acad Sci USA 99(Suppl 4): 16407–16411

Brown MK, Naidoo N (2012) The endoplasmic reticulum stress response in aging and age-related diseases. Front Physiol 3:263

Carra S, Seguin SJ, Landry J (2008) HspB8 and Bag3: a new chaperone complex targeting misfolded proteins to macroautophagy. Autophagy 4(2):237–239

Carra S, Brunsting JF, Lambert H, Landry J, Kampinga HH (2009) HspB8 participates in protein quality control by a non-chaperone-like mechanism that requires eIF2{alpha} phosphorylation. J Biol Chem 284(9):5523–5532

Casas-Tinto S, Zhang Y, Sanchez-Garcia J, Gomez-Velazquez M, Rincon-Limas DE, Fernandez-Funez P (2013) The ER stress factor XBP1s prevents amyloid-beta neurotoxicity. Hum Mol Genet 20(11):2144–2160

Chan HY, Warrick JM, Gray-Board GL, Paulson HL, Bonini NM (2000) Mechanisms of chaperone suppression of polyglutamine disease: selectivity, synergy and modulation of protein solubility in *Drosophila*. Hum Mol Genet 9(19):2811–2820

Chan HY, Warrick JM, Andriola I, Merry D, Bonini NM (2002) Genetic modulation of polyglutamine toxicity by protein conjugation pathways in *Drosophila*. Hum Mol Genet 11(23):2895–2904

Chang YY, Neufeld TP (2010) Autophagy takes flight in *Drosophila*. FEBS Lett 584(7):1342–1349

Chen SF, Kang ML, Chen YC, Tang HW, Huang CW, Li WH, Lin CP, Wang CY, Wang PY, Chen GC, Wang HD (2012) Autophagy-related gene 7 is downstream of heat shock protein 27 in the regulation of eye morphology, polyglutamine toxicity, and lifespan in *Drosophila*. J Biomed Sci 19:52

Ciechanover A, Brundin P (2003) The ubiquitin proteasome system in neurodegenerative diseases: sometimes the chicken, sometimes the egg. Neuron 40(2):427–446

Cuervo AM, Wong E (2014) Chaperone-mediated autophagy: roles in disease and aging. Cell Res 24(1):92–104

Cuervo AM, Stefanis L, Fredenburg R, Lansbury PT, Sulzer D (2004) Impaired degradation of mutant alpha-synuclein by chaperone-mediated autophagy. Science 305(5688):1292–1295

Cushman-Nick M, Bonini NM, Shorter J (2014) Hsp104 suppresses polyglutamine-induced degeneration post onset in a *Drosophila* MJD/SCA3 model. PLoS Genet 9(9):e1003781

del Valle Rodriguez A, Didiano D, Desplan C (2011) Power tools for gene expression and clonal analysis in *Drosophila*. Nat Methods 9(1):47–55

Douglas PM, Cyr DM (2009) Interplay between protein homeostasis networks in protein aggregation and proteotoxicity. Biopolymers 93(3): 229–236

Fernandez-Chacon R, Wolfel M, Nishimune H, Tabares L, Schmitz F, Castellano-Munoz M, Rosenmund C, Montesinos ML, Sanes JR, Schneggenburger R, Sudhof TC (2004) The synaptic vesicle protein CSP alpha prevents presynaptic degeneration. Neuron 42(2):237–251

Fernandez-Funez P, Nino-Rosales ML, de Gouyon B, She WC, Luchak JM, Martinez P, Turiegano E, Benito J, Capovilla M, Skinner PJ, McCall A, Canal I, Orr HT, Zoghbi HY, Botas J (2000) Identification of genes that modify ataxin-1-induced neurodegeneration. Nature 408(6808):101–106

Fernandez-Funez P, Casas-Tinto S, Zhang Y, Gomez-Velazquez M, Morales-Garza MA, Cepeda-Nieto AC, Castilla J, Soto C, Rincon-Limas DE (2009) In vivo generation of neurotoxic prion protein: role for hsp70 in accumulation of misfolded isoforms. PLoS Genet 5(6):e1000507

Fleshner M, Johnson JD (2005) Endogenous extra-cellular heat shock protein 72: releasing signal(s) and function. Int J Hyperth 21(5):457–471

Fouillet A, Levet C, Virgone A, Robin M, Dourlen P, Rieusset J, Belaidi E, Ovize M, Touret M, Nataf S, Mollereau B (2012) ER stress inhibits neuronal death by promoting autophagy. Autophagy 8(6):915–926

Fujikake N, Nagai Y, Popiel HA, Okamoto Y, Yamaguchi M, Toda T (2008) Heat shock transcription factor 1-activating compounds suppress polyglutamine-induced neurodegeneration through induction of multiple molecular chaperones. J Biol Chem 283(38):26188–26197

Halliday M, Mallucci GR (2014) Targeting the unfolded protein response in neurodegeneration: a new approach to therapy. Neuropharmacology 76(Pt A):169–174

Hara T, Nakamura K, Matsui M, Yamamoto A, Nakahara Y, Suzuki-Migishima R, Yokoyama M, Mishima K, Saito I, Okano H, Mizushima N (2006) Suppression of basal autophagy in neural cells causes neurodegenerative disease in mice. Nature 441(7095):885–889

Harding HP, Zhang Y, Ron D (1999) Protein translation and folding are coupled by an endoplasmic-reticulum-resident kinase. Nature 397(6716):271–274

Hershko A, Ciechanover A (1998) The ubiquitin system. Annu Rev Biochem 67:425–479

Hetz C, Glimcher LH (2009) Fine-tuning of the unfolded protein response: assembling the IRE1alpha interactome. Mol Cell 35(5):551–561

Hetz C, Mollereau B (2014) Disturbance of endoplasmic reticulum proteostasis in neurodegenerative diseases. Nat Rev Neurosci 15(4):233–249

Hetz C, Martinon F, Rodriguez D, Glimcher LH (2011) The unfolded protein response: integrating stress signals through the stress sensor IRE1alpha. Physiol Rev 91(4):1219–1243

Jaiswal M, Sandoval H, Zhang K, Bayat V, Bellen HJ (2012) Probing mechanisms that underlie human neurodegenerative diseases in *Drosophila*. Annu Rev Genet 46:371–396

Kang MJ, Ryoo HD (2009) Suppression of retinal degeneration in *Drosophila* by stimulation of ER-associated degradation. Proc Natl Acad Sci USA 106(40):17043–17048

Kazemi-Esfarjani P, Benzer S (2000) Genetic suppression of polyglutamine toxicity in *Drosophila*. Science 287(5459):1837–1840

Kim YE, Hipp MS, Bracher A, Hayer-Hartl M, Hartl FU (2013) Molecular chaperone functions in protein folding and proteostasis. Annu Rev Biochem 82:323–355

King V, Tower J (1999) Aging-specific expression of *Drosophila* hsp22. Dev Biol 207(1):107–118

Koren J 3rd, Jinwal UK, Lee DC, Jones JR, Shults CL, Johnson AG, Anderson LJ, Dickey CA (2009) Chaperone signalling complexes in Alzheimer's disease. J Cell Mol Med 13(4):619–630

Koumenis C (2006) ER stress, hypoxia tolerance and tumor progression. Curr Mol Med 6(1):55–69

Kuo Y, Ren S, Lao U, Edgar BA, Wang T (2013) Suppression of polyglutamine protein toxicity by co-expression of a heat-shock protein 40 and a heat-shock protein 110. Cell Death Dis 4:e833

Liao PC, Lin HY, Yuh CH, Yu LK, Wang HD (2008) The effect of neuronal expression of heat shock proteins 26 and 27 on lifespan, neurodegeneration, and apoptosis in *Drosophila*. Biochem Biophys Res Commun 376(4):637–641

Ling D, Salvaterra PM (2010) Brain aging and Abeta(1)(-)(4)(2) neurotoxicity converge via deterioration in autophagy-lysosomal system: a conditional *Drosophila* model linking Alzheimer's neurodegeneration with aging. Acta Neuropathol 121(2):183–191

Ling D, Song HJ, Garza D, Neufeld TP, Salvaterra PM (2009) Abeta42-induced neurodegeneration via an age-dependent autophagic-lysosomal injury in *Drosophila*. PLoS ONE 4(1):e4201

Lipson KL, Fonseca SG, Urano F (2006) Endoplasmic reticulum stress-induced apoptosis and auto-immunity in diabetes. Curr Mol Med 6(1):71–77

Loewen CA, Feany MB (2010) The unfolded protein response protects from tau neurotoxicity in vivo. PLoS ONE 5(9):e13084

Luo W, Rodina A, Chiosis G (2008) Heat shock protein 90: translation from cancer to Alzheimer's disease treatment? BMC Neurosci 9(Suppl 2):S7

Luo W, Sun W, Taldone T, Rodina A, Chiosis G (2010) Heat shock protein 90 in neurodegenerative diseases. Mol Neurodegener 5:24

Luo D, Bu Y, Ma J, Rajput S, He Y, Cai G, Liao DF, Cao D (2013) Heat shock protein 90-alpha mediates aldo-keto reductase 1B10 (AKR1B10) protein secretion through secretory lysosomes. J Biol Chem 288(51):36733–36740

Maor G, Rencus-Lazar S, Filocamo M, Steller H, Segal D, Horowitz M (2013) Unfolded protein response in Gaucher disease: from human to *Drosophila*. Orphanet J Rare Dis 8:140

Matus S, Glimcher LH, Hetz C (2011) Protein folding stress in neurodegenerative diseases: a glimpse into the ER. Curr Opin Cell Biol 23(2):239–252

Miller LC, Swayne LA, Chen L, Feng ZP, Wacker JL, Muchowski PJ, Zamponi GW, Braun JE (2003) Cysteine string protein (CSP) inhibition of N-type calcium channels is blocked by mutant huntingtin. J Biol Chem 278(52):53072–53081

Morimoto RI (2008) Proteotoxic stress and inducible chaperone networks in neurodegenerative disease and aging. Gene Dev 22(11):1427–1438

Morrow G, Tanguay RM (2003) Heat shock proteins and aging in *Drosophila melanogaster*. Semin Cell Dev Biol 14(5):291–299

Naidoo N (2009) The endoplasmic reticulum stress response and aging. Rev Neurosci 20(1):23–37

Orenstein SJ, Kuo SH, Tasset I, Arias E, Koga H, Fernandez-Carasa I, Cortes E, Honig LS, Dauer W, Consiglio A, Raya A, Sulzer D, Cuervo AM (2013) Interplay of LRRK2 with chaperone-mediated autophagy. Nat Neurosci 16(4):394–406

Pandey UB, Nie Z, Batlevi Y, McCray BA, Ritson GP, Nedelsky NB, Schwartz SL, DiProspero NA, Knight MA, Schuldiner O, Padmanabhan R, Hild M, Berry DL, Garza D, Hubbert CC, Yao TP, Baehrecke EH, Taylor JP (2007) HDAC6 rescues neurodegeneration and provides an essential link between autophagy and the UPS. Nature 447(7146):859–863

Pearl LH, Prodromou C (2006) Structure and mechanism of the Hsp90 molecular chaperone machinery. Annu Rev Biochem 75:271–294

Pittet JF, Lee H, Morabito D, Howard MB, Welch WJ, Mackersie RC (2002) Serum levels of Hsp 72 measured early after trauma correlate with survival. J Trauma 52(4):611–617

Rasheva VI, Domingos PM (2009) Cellular responses to endoplasmic reticulum stress and apoptosis. Apoptosis 14(8):996–1007

Ravikumar B, Rubinsztein DC (2006) Role of autophagy in the clearance of mutant huntingtin: a step towards therapy? Mol Aspects Med 27(5–6):520–527

Ravikumar B, Vacher C, Berger Z, Davies JE, Luo S, Oroz LG, Scaravilli F, Easton DF, Duden R, O'Kane CJ, Rubinsztein DC (2004) Inhibition of mTOR induces autophagy and reduces toxicity of polyglutamine expansions in fly and mouse models of Huntington disease. Nat Genet 36(6):585–595

Ravikumar B, Imarisio S, Sarkar S, O'Kane CJ, Rubinsztein DC (2008) Rab5 modulates aggregation and toxicity of mutant huntingtin through macroautophagy in cell and fly models of Huntington disease. J Cell Sci 121(Pt 10):1649–1660

Rincon-Limas DE, Jensen K, Fernandez-Funez P (2012) *Drosophila* models of proteinopathies: the little fly that could. Curr Pharm Des 18(8):1108–1122

Roman G, Endo K, Zong L, Davis RL (2001) P[Switch], a system for spatial and temporal control of gene expression in *Drosophila melanogaster*. Proc Natl Acad Sci USA 98(22):12602–12607

Ryoo HD, Steller H (2007) Unfolded protein response in *Drosophila*: why another model can make it fly. Cell Cycle 6(7):830–835

Salminen A, Kaarniranta K (2010) ER stress and hormetic regulation of the aging process. Ageing Res Rev 9(3):211–217

Simonsen A, Cumming RC, Brech A, Isakson P, Schubert DR, Finley KD (2008) Promoting basal levels of autophagy in the nervous system enhances longevity and oxidant resistance in adult *Drosophila*. Autophagy 4(2):176–184

Soti C, Csermely P (2003) Aging and molecular chaperones. Exp Gerontol 38(10):1037–1040

Tatar M, Khazaeli AA, Curtsinger JW (1997) Chaperoning extended life. Nature 390(6655):30

Tonoki A, Kuranaga E, Tomioka T, Hamazaki J, Murata S, Tanaka K, Miura M (2009) Genetic evidence linking age-dependent attenuation of the 26S proteasome with the aging process. Mol Cell Biol 29(4):1095–1106

Valastyan JS, Lindquist S (2014) Mechanisms of protein-folding diseases at a glance. Dis Model Mech 7(1):9–14

Velentzas PD, Velentzas AD, Mpakou VE, Antonelou MH, Margaritis LH, Papassideri IS, Stravopodis DJ (2013) Detrimental effects of proteasome inhibition activity in *Drosophila melanogaster*: implication of ER stress, autophagy, and apoptosis. Cell Biol Toxicol 29(1):13–37

Voisine C, Pedersen JS, Morimoto RI (2010) Chaperone networks: tipping the balance in protein folding diseases. Neurobiol Dis 40(1):12–20

Vos MJ, Zijlstra MP, Kanon B, van Waarde-Verhagen MA, Brunt ER, Oosterveld-Hut HM, Carra S, Sibon OC, Kampinga HH (2010) HSPB7 is the most potent polyQ aggregation suppressor within the HSPB family of molecular chaperones. Hum Mol Genet 19(23):4677–4693

Wang T, Lao U, Edgar BA (2009a) TOR-mediated autophagy regulates cell death in *Drosophila* neurodegenerative disease. J Cell Biol 186(5):703–711

Wang Y, Martinez-Vicente M, Kruger U, Kaushik S, Wong E, Mandelkow EM, Cuervo AM, Mandelkow E (2009b) Tau fragmentation, aggregation and clearance: the dual role of lysosomal processing. Hum Mol Genet 18(21):4153–4170

Wang AM, Miyata Y, Klinedinst S, Peng HM, Chua JP, Komiyama T, Li X, Morishima Y, Merry DE, Pratt WB, Osawa Y, Collins CA, Gestwicki JE, Lieberman AP (2013) Activation of Hsp70 reduces neurotoxicity by promoting polyglutamine protein degradation. Nat Chem Biol 9(2):112–118

Warrick JM, Chan HY, Gray-Board GL, Chai Y, Paulson HL, Bonini NM (1999) Suppression of polyglutamine-mediated neurodegeneration in *Drosophila* by the molecular chaperone HSP70. Nat Genet 23(4):425–428

Warrick JM, Morabito LM, Bilen J, Gordesky-Gold B, Faust LZ, Paulson HL, Bonini NM (2005) Ataxin-3 suppresses polyglutamine neurodegeneration in *Drosophila* by a ubiquitin-associated mechanism. Mol Cell 18(1):37–48

Waza M, Adachi H, Katsuno M, Minamiyama M, Tanaka F, Sobue G (2006) Alleviating neurodegeneration by an anticancer agent: an Hsp90 inhibitor (17-AAG). Ann N Y Acad Sci 1086:21–34

Yang J, Tower J (2009) Expression of hsp22 and hsp70 transgenes is partially predictive of *Drosophila* survival under normal and stress conditions. J Gerontol A Biol Sci Med Sci 64(8):828–838

Young JC (2010) Mechanisms of the Hsp70 chaperone system. Biochem Cell Biol 88(2):291–300

Zhai RG, Cao Y, Hiesinger PR, Zhou Y, Mehta SQ, Schulze KL, Verstreken P, Bellen HJ (2006) *Drosophila* NMNAT maintains neural integrity independent of its NAD synthesis activity. PLoS Biol 4(12):e416

Zhai RG, Zhang F, Hiesinger PR, Cao Y, Haueter CM, Bellen HJ (2008) NAD synthase NMNAT acts as a chaperone to protect against neurodegeneration. Nature 452(7189):887–891

Zhang Y, Casas-Tinto S, Rincon-Limas DE, Fernandez-Funez P (2014) Combined pharmacological induction of Hsp70 suppresses prion protein neurotoxicity in *Drosophila*. PLoS ONE 9(2):e88522

Zinsmaier KE (2010) Cysteine-string protein's neuroprotective role. J Neurogenet 24(3):120–132

Chapter 9
Drosophila Models in Therapeutic Drug Discovery Related to Aging

Charles D. Nichols

Abstract *Drosophila melanogaster* is increasingly being used in drug discovery efforts. Although there are several other genetic model organisms including *C. elegans* that can and are being used to study aging, the fly has certain advantages. These include a complex brain capable of sophisticated behaviors, and several organ systems with certain degrees of structural and functional conservation with humans like a heart, that can be investigated with the powerful genetics afforded by the fly. With respect to aging and drug discovery, there are several opportunities to utilize the fly ranging from therapeutic discovery for age related diseases like Alzheimer's disease to drugs that extend life span to drugs that improve cognition.

Keywords Drug discovery • Drug screening • Life span • Neurodegeneration • Alzheimer's disease • *Drosophila melanogaster*

9.1 Introduction

Drosophila melanogaster is increasingly being used in drug discovery efforts. The exact role of the common fruit fly within the pipeline can vary depending on the nature of the disease process being studied. The traditional discovery pipeline for target-based drug discovery usually begins with brute force screening of up to several hundreds of thousands of chemically distinct small molecules from large compound libraries against cells in culture. The endpoints can vary, but most assays involve screening in 384-well cell culture plates that detect some measureable change in physiology of the cell produced by the drug. For example: cell growth

C.D. Nichols (✉)
Department of Pharmacology and Experimental Therapeutics,
Louisiana State University Health Sciences Center, 1901 Perdido St.,
New Orleans, LA 70112, USA
e-mail: cnich1@lsuhsc.edu

A.M. Vaiserman et al. (eds.), *Life Extension*, Healthy Ageing and Longevity 3,
DOI 10.1007/978-3-319-18326-8_9

or inhibition of growth, increases in intracellular calcium levels, or light produced from a reporter can all be measured by high throughput with the appropriate equipment. Unfortunately, most drugs fail for one reason or another at some point in the developmental process, and the cost of bringing a single new drug to market has recently been estimated to be over five billion dollars (Herper 2013). Incorporation of the model genetic organism *Drosophila melanogaster* into the development pipeline can significantly reduce time and costs by rapidly generating lead compounds of intrinsically higher quality prior to moving to expensive and time consuming preclinical rodent studies. Although there are several other genetic model organisms including *C. elegans* that can and are being used to study aging, the fly has certain advantages. These include a complex brain capable of sophisticated behaviors, and several organ systems with certain degrees of structural and functional conservation with humans like a heart, that can be investigated with the powerful genetics afforded by the fly. With respect to aging and drug discovery, there are several opportunities to utilize the fly ranging from therapeutic discovery for age related diseases like Alzheimer's disease to drugs that extend life span to drugs that improve cognition.

9.2 Traditional Drug Development

In traditional drug development, a high throughput assay is developed such that the function of a particular disease related enzyme or receptor can be measured by an appropriate outcome in an appropriate cell based system. Compounds from a chemical library are added to individual wells of a high density plate containing the cellular or enzymatic assay, the plates are incubated for the appropriate time, and the plates are then placed into a reader and the outcome variable measured. Several thousand chemicals can be screened in a day and several hundred thousand in weeks to months depending on the assay. Chemicals that produce the desired outcome (leads) are identified and evaluated for further study. A large screen may yield several hundred lead candidates. These are retested more vigorously in panels of cell based assays at different concentrations and multiple cell types relevant to the disease of question. A significant number of initial lead candidates will usually not demonstrate efficacy in these subsequent studies, and several more will likely prove to have unacceptable toxicity, or other negative property associated with them. After many more months of secondary screening, several hundred initial lead compounds may have been narrowed down to only a few. Candidates at this point are not usually suitable as drug candidates, however. Whereas they may have efficacy, it may be weak or the potency may be low, a certain degree of toxicity or off-target effects may still be present, or the nature of the structure confers low bioavailability to the compound. Medicinal chemistry is usually employed to modify the more promising structures to produce series of analogs that are then retested for the desired effects over the following months. Often, several rounds of testing and modification are performed that in sum can last up to several years in order to arrive at a small collection of promising leads to take further into animal models.

When moving to animals models, these have traditionally been rodent models using mice and/or rats. Rodent studies are expensive and time consuming, and can take up to several more years to demonstrate safety and efficacy. More often than not, a promising lead is found to have unexpected toxicity in the whole animal, or found to be ineffective in treating the disease condition it was developed for. After several years, and several hundred millions of dollars, a company is lucky to find one drug suitable for being moved to clinical trials in humans. Even then, most drugs today fail in clinical trials for one reason or another (DiMasi et al. 2010). In the past 20 years the number of successful drugs brought to market per year has declined sharply, while the cost of development per successful drug has sky-rocketed. On average, it now takes about 15 years to develop, test, and bring a new drug to market at a cost of over 5 billion dollars per drug (Herper 2012, 2013).

One theory that has been presented for the decreasing success in drug development is that 'all the low hanging fruit has been identified', and there are effective drugs on the market for most of the diseases that are caused by a defect in a single target (e.g. enzyme or receptor). It has been recently estimated that of the 30,000 genes encoded by the human genome, only 3000 of these genes are responsible for diseases when defective. It has also been estimated that there are about 3000 genes that encode for druggable targets (genes that encode for enzymes or receptors) (Hopkins and Groom 2002). Unfortunately, the overlap between these two sets of genes that represent the number of disease causing genes that also encode for druggable targets is estimated to only be 600–1500 genes (Hopkins and Groom 2002). Increasingly, drug discovery efforts are being aimed at diseases and conditions that are multicomponent (e.g. diabetes, heart disease), with several genes underlying their etiology. The traditional target-based brute force discovery pipeline is becoming obsolete, allowing for more sophisticated systems-based models to be used. In the systems-based approach, rather than the target consisting of a single enzyme/receptor, the target represents a heterogeneous system with several potential targets present that is believed to be more relevant to the disease as a whole.

9.3 The Role of *Drosophila Melanogaster* in Drug Development

The fly can have a valuable role in both realms, in traditional target-based as well as systems-based approaches. In the target-based pipeline, the process from going from the initial leads through rodent models can be lengthy and expensive, and the bulk of lead compounds will fail in rodents due to toxicity or ADME issues (adsorption, distribution, metabolism, and excretion). The fly can serve here as a valuable secondary screening platform. Once a fly model has been developed for the appropriate disease, several hundred initial leads can be screened in flies at a fraction of the time and cost as compared to rodents to produce a smaller and higher quality pool of candidates for further optimization or to directly take into rodents. Drugs that are toxic to flies will likely also be toxic to rodents, and ADME parameters are also correlative between fly and rodent. Importantly, drugs of several diverse classes where the target protein is conserved between fly and mammal usually demonstrate similar shared effects on both physiology and/or behaviors. For example, CNS active drugs like cocaine and methamphetamine stimulate activity and arousal through mediating increases in dopamine (Andretic et al. 2005; Kaun et al. 2012), and sulfonylurea drugs used to treat diabetes in humans influence glucose homeostasis in flies (Haselton and Fridell 2010; Kim and Rulifson 2004). Significantly, many disease models in flies now incorporate the transgenic expression of rodent or human gene counterparts relevant to the disease in the fly, increasing the probability of translation of therapeutic efficacy. For example, human alpha synuclein expressed in fly retina or brain has served as a discovery platform for Parkinson's disease (Whitworth 2011).

The fly can also be useful as a primary screening platform within the systems-based approach. In this scenario, a smaller collection of up to only a few thousand 'high quality' compounds are screened in the appropriate fly model. These higher quality libraries can represent sets of previously FDA approved medications, each with acceptable toxicity and ADME measures in humans. Given the difficulty of generating and identifying novel chemical entities to treat a condition, repurposing older medications for new indications has recently become very popular with some success (Padhy and Gupta 2011). Here, proceeding first with flies can narrow down several thousand compounds tested to only a few, but these few will have a greater chance of also achieving therapeutic success when tested in rodent models. Because these drugs are already human approved, in many instances, rodent trials may not even be necessary before using the drug in human patients directly. Alternatively, if the appropriate assay can be designed in flies, high throughput can be achieved that can test several tens of thousands of drugs within a few weeks to months to test any collection of small molecules in a primary screen to identify a small collection of leads with higher quality than would result from traditional cell culture screening.

The ability to generate sophisticated models of human diseases in flies has been steadily increasing as genetic tools and techniques for transgene expression have become more advanced. There are now informative models of diseases including cardiovascular disease, asthma, inflammatory bowel, and cancer (Pandey and Nichols 2011). With regard to CNS disorders, the use of flies to study neurodegenerative diseases like Alzheimer's disease (AD), Parkinson's disease (PD), and amyotrophic lateral sclerosis (ALS) have a long history and several drug discovery efforts have already been initiated with these models (Pandey and Nichols 2011). For normal CNS processes, the fly has also been used as a discovery platform to identify agents that modify sleep and cognitive processes like learning and memory.

9.4 Drug Discovery Using *Drosophila Melanogaster* Relevant to Age Related Diseases

The fly has been used to model several aspects of the aging process to understand the physiology of aging not only for the fly but applicable to higher organisms as well as humans, as discussed in the other chapters in this book. Specifically relevant to drug discovery, initial efforts were largely directed towards disorders associated with aging like progressive neurodegenerative disorders, which typically do not afflict an individual until middle or advanced age. Several models of Alzheimer's disease (AD), Parkinson's disease (PD), and Huntington's disease (HD) have been developed and used to understand relevant physiological processes associated with the diseases as well as discovery platforms for therapeutics.

Although AD, PD, and HD are different diseases with different causes, they have overlapping symptoms that include loss of motor coordination, and premature death in humans. In the case of AD, the precise etiology remains unknown, although there are several factors that contribute to neurodegeneration. The most popular theory is that abnormal processing of the amyloid precursor protein (APP) via β- and γ-secretase enzymes produce toxic aggregates of Aβ-40 and Aβ-42 peptides, and that accumulation these plaques within neurons leads to neurodegeneration (Querfurth and LaFerla 2010). Processing of APP by a different α-secretase enzyme produces different Aβ peptides that do not form aggregates and are not toxic (Querfurth and LaFerla 2010). Mutations within the APP can promote a higher degree of aggregation of toxic Aβ peptides. Within the fly, most components of this system are present, including APP-Like (APPL), and enzymes with γ-secretase activity, which is involved in the Notch signaling pathway (Prüßing et al. 2013). There are several published reports of developing flies that over express human APP, the γ-secretase presenilin, or the human version of another protein implicated in neurofibrilatory tangles associated with AD, Tau (Pandey and Nichols 2011; Prüßing et al. 2013). These flies can exhibit hallmark symptoms of AD that include loss of coordination, severe neurodegeneration, and reduced life span (Prüßing et al. 2013).

The vast majority of PD in humans results from unknown causes. There is, however, a small percentage (2–3 %) of patients that develop PD early in life (in their 20s or 30s). These early onset cases are for the most part the result of mutations in defined genes including *alpha synuclein*, *parkin*, *DJ-1*, and *PINK*. Mutations of these genes in humans leads to a rapid decline of dopaminergic neurons in the substantia nigra, and the development of PD symptoms (Dauer and Przedborski 2003). The hallmark of PD is the presence of Lewy bodies, comprised of aggregates of alpha synuclein and ubiquitin, among other proteins, in neurons (Dauer and Przedborski 2003). There have been several fly models of PD developed (Whitworth 2011). Notably, the fly does not express many of the known genes associated with PD like the one encoding for alpha synuclein. Nevertheless, overexpression of the human form of these genes in the fly has been reported by several groups to produce symptoms of loss of dopamine neurons, motor control, and premature death (Whitworth 2011).

Unlike AD or PD, the precise molecular nature of HD is known, and results from a polyglutamate expansion within the protein encoded by the *Hungtintin* gene (Walker 2007). There are several diseases that are due to polyglutamate expansions like spinal and bulbar muscular atrophy (androgen receptor gene; Kennedy's Disease) (Ross 2002). Disease models have been developed in the fly to express the human *Huntingtin/htt* gene with normal and expanded glutamine repeats of various lengths. In flies where the mutant protein is expressed, phenotypes of neurodegeneration, loss of motor control, and premature death correlate with the degree of polyglutamate expansion (Green and Giorgini 2012).

Even though these neurodegenerative diseases are caused by different factors and molecular lesions, the fly is used in each model in a similar fashion for drug discovery. The corresponding transgene (either the endogenous wild type or mutant fly or homologous human gene) is overexpressed in either the eye, all neurons, or the entire body, and a similar suite of behaviors and/or pathologies are examined. Why would the eye be such a powerful tool to screen for drug therapies against brain diseases like AD or PD? The eye is a very sophisticated organ that has been termed a neurocrystalline lattice (Ready et al. 1976), comprised of about 850 discreet units called ommatidium. Each ommatidium contains 8 photoreceptor neurons that form deeper connections in the lamina and medulla with interneurons that send projections to the brain. In addition, there are glial cells, and support cells like pigment cells and cone cells, which secrete the lens material covering the ommatidium. The high degree of structure present in the eye is the most attractive feature both literally and figuratively. Small perturbations to the development or function of a particular cell will often produce obvious defects in the eye structure that are easy to observe. The eye can appear rough, mis-shapen, smaller in size, lacking bristles, or not even be present. The degree of change often correlates with the degree of defect. Expression in the eye is easily achieved using the bipartite GAL4/UAS system (Brand and Perrimon 1993) where the driver strain is one such as GMR-GAL4 that produces the yeast transcription factor GAL4 in photoreceptor neurons of the eye, and the responder UAS strain has the binding and activation sites for the GAL4 protein upstream of the cDNA encoding for the human transgene of the disease-relevant proteins.

When the driver and responder strains are crossed, the human gene is expressed exclusively within photoreceptors of the eye. Over expression of the pathogenic forms of human proteins like alpha synuclein, Htt, or APP in photoreceptor neurons can lead to profound degeneration of the neurons that is easily visible and conducive to rapid screening by simple observation under a dissection microscope. Recently, a modification of the eye-based assay has been developed that incorporates the use of membrane targeted GFP, which may be more powerful to detect more mild neurodegeneration (Burr et al. 2014). Use of the eye is not limited to drug screening for neurodegenerative disease, but to many other diseases as well including cancer (Vidal and Cagan 2006). For drug screening using the eye, a fertilized egg can be placed into a 96 well plate (or other sized container) containing food or solid media to which an appropriate amount of the test molecule has been added. After hatching, the larva will ingest the drug throughout development until pupation. After the adult fly emerges (10 days between egg and adult at 25 °C), it can be collected, and the eye examined under a microscope for rescue of the degree of defect. Several thousand compounds can be tested by this technique within a few weeks to months. Compounds that rescue the eye defects of the model represent high quality leads to examine further. They are high quality because by virtue of the fly growing to adulthood and demonstrating rescue, the compound likely has little overall toxicity at the organismal level, and favorable pharmcodynamic properties. An example of the eye being used as a validation platform for potential chemicals identified by more traditional HTS that may be effective for enhancing autophagy and reducing toxicity in HD is that of Sarkar et al. (2007).

Rescue of locomotor activity is another powerful screening tool that can be used for discovery related to neurodegenerative disorders. Here, the pathological human disease gene is expressed exclusively in neurons using the GAL4/UAS system. Genes can be expressed in all neurons using the elav-GAL4 driver, or only in motor neurons using one of several drivers like OK6-GAL4. When proteins relevant to AD, PD, and HD are expressed in motor neurons they lead to a decrease in motor coordination and activity, with the severity of these effects correlating with the expression level of the transgene, or the severity of the allele. There are two basic assays to assess motor function in adults. The first is a simple assay that measures overall locomotor activity over time, and is accomplished by placing individual flies of the desired genotype within small glass capillary tubes in 32 tube arrays equipped with photobeams for each tube and measuring photobeam breaks with the Drosophila Activity Monitoring System (DAMS; Waltham, MA, USA). Multiple arrays can be used in parallel to examine several hundred flies simultaneously. Drug is included with food or media plugging one end of the tube, with cotton at the other (for ventilation), and the arrays placed in an incubator for several days and photobeam breaks monitored by computer. Drugs that rescue activity levels to control levels represent high quality leads to proceed with.

Another common adult motor assay that also takes coordination into account is the negative geotaxis assay (Nichols et al. 2012). Here, small groups of flies expressing pathological variants of human proteins in motor neurons are rapidly tapped down to the bottom of a vial, and the number of flies that climb to a certain height, or the average height climbed in a specific time interval, is determined. Normal control flies should be able to rapidly climb the interior. As coordination, and/or activity is disrupted fewer flies will have the ability to climb the interior wall, or will climb the wall slower. Whereas the eye-based screens and the DAMS method are relatively high throughput, and have the capacity to screen several hundred to thousand of drugs within a relatively short period of time, the negative geotaxis assay is comparatively lower in throughput because typically ~25 adult flies are needed per assay, larger amounts of food + drug are needed in larger spaces to grow the flies for treatment, and an investigator is usually necessary to tap the vials down. Therefore, negative geotaxis may be more appropriate for a secondary screen to cross validate results of leads identified from eye-based or photobeam beam-based assays, or even mammalian cell based HTS. An example of this type of screening was performed by McKoy and colleagues who used negative geotaxis as a secondary validation assay for lead compounds from a cell-based HTS that disrupt Aβ aggregates in a fly model of AD (McKoy et al. 2012).

Larval locomotion can also be used as a screening platform for rescue of locomotor deficits in neurodegenerative disease models. In this type of assay, larva are maintained on food substrate + drug, or larva are acutely fed drug for a period of time and placed on Petri dishes with a solid agarose substrate (Nichols et al. 2012). Locomotion is measured by either placing the dish over a grid and counting grids crossed in a defined time period, or by video tracking. Although this type screen is of somewhat low throughput, several hundred drugs can be screened in a few weeks. This type of screen was performed by Lawal et al. (2014) who identified potential new therapeutic entities for PD from a collection of 1000 known and FDA approved drugs.

9.5 Drug Discovery with *Drosophila Melanogaster* to Examine Longevity

Perhaps one of the easiest assays to perform in the fly, and the most relevant to the aging process, are longevity assays, which can also be adapted to drug screening. It has been estimated that there are several hundred genes involved in the *Drosophila* aging process (Pletcher et al. 2002). Many of these genes are known and have mammalian orthologs [reviewed in (Paaby and Schmidt 2009)]. A number of these genes, and likely many of those that remain to be characterized, represent druggable targets. In the longevity assay, newly emerged adult flies are maintained on food substrate with drug, and the number of days the flies survive measured, or flies are allowed to lay eggs onto food + drug, and the entire developmental timeline measured. This type of screen can be used as a primary screen or as a secondary validation platform. The life span of an adult fly in the laboratory can range from 25 to 60 days at 25 °C, depending on the strain, temperature, and humidity (Bhandari et al. 2007). Therefore, several factors such as the strain used and environmental conditions the flies are raised in need to be taken into account when using longevity assays for drug screening.

The choice of housing conditions is important. The life span is dependent on the temperature, with flies raised at lower temperatures (e.g. 18°) living about twice as long as flies raised at higher temperatures (e.g. 25 °C). The difference in longevity between high and low temperatures can be exacerbated in certain disease models, therefore the temperature most appropriate for the particular strain of fly (i.e. wild type or disease model) must be carefully chosen. Choice of food substrate is also a factor. Normal fly food usually contains yeast, and other microorganisms that can easily grow in the food. These microorganisms can sometimes metabolize test compounds and lower the effective concentration in the food over time. Whereas this may not be a factor in an assay like the locomotor assay that only lasts a few days, it may pose a significant issue over longer time periods. Alternatives are to use other food substrates like agarose supplemented with sucrose, or an instant food without yeast products.

The size of the vessel the fly is maintained in is also an important factor. When performing validation screens of only a few drugs, groups of flies can be maintained in larger vials or bottles with food + drug, and these can last up to several weeks before the flies will need to be placed on fresh food + drug. The disadvantage to this approach is that the larger containers will require a significant amount of drug. When performing more high throughput screens using fewer flies in smaller tubes (e.g. 5 ml culture tube with 0.5 ml food, and 5–10 adult flies), we have found that transferring flies to fresh food + drug every 3–4 days is necessary to prevent desiccation of the food, which will affect feeding. When feeding drugs to flies, they may have unpleasant tastes causing the flies to eat less. This can lead to dietary restriction, which in and of itself can promote longevity (Mair et al. 2003; Metaxakis and Partridge 2013), and confound data. If reduced food intake is a concern, or suspected of lead candidates resulting from a screen, feeding assays like the CAFE assay (Ja et al. 2007) can be performed incorporating the drug in question to examine the effects of drug on food intake.

Selection of the sex of fly to be tested is important. Mixing both sexes in a vial will result in extra energy utilization from procreation and egg laying, which will confound results of assays examining adult longevity, therefore examining only one sex per population of a vial is desired. Females are more sensitive to dietary restriction (Magwere et al. 2004; Mair et al. 2005) and are a popular choice. The use of only males, however, avoids potential confounding effects of egg laying (Spindler et al. 2012). Importantly, if a drug candidate is found to increase life span, it must then be tested in other assays to assess its influence on other processes like fecundity, metabolism, stress resistance, and for possible toxic effects that may be evident from examination of activity levels (Jafari 2010). An example of a small molecule screen for compounds that extend life span is that of Spindler and colleagues, who screened a small library of known kinase inhibitors to identify a small collection that increased life span by up to 35 % (Spindler et al. 2012).

9.6 *Drosophila Melanogaster* as a Discovery Platform for Therapeutics to Treat Age Related Declines in Cognitive Function

A significant effect of aging in humans is a decline in cognitive function in advanced years. The discovery of new chemical entities to enhance normal cognitive function and perhaps return cognitive abilities of the aged to their youth is an enticing prospect. Although there are a few drugs aimed at preserving cognitive function in dementia like donepizil (Doody et al. 2012), an acetylcholinesterase inhibitor, or wakefulness, arousal and vigilance in humans with CNS stimulants like modafinil (Minzenberg and Carter 2008) and caffeine (Griffiths and Woodson 1988), there are no medications approved by the FDA solely to enhance cognitive function of healthy individuals. Not surprisingly, many CNS stimulants like amphetamine, cocaine, modafinil, and caffeine have similar effects in flies as humans acting through conserved molecular mechanism (Nichols 2006). Further, also acting through conserved molecular mechanisms, many drugs that are CNS depressants in humans likc bcnzodiazepines have similar sedative effects in flies (Nichols 2006). Importantly, unlike lower model organisms, the fly demonstrates learning and memory behaviors with many conserved aspects to mammalian learning and memory (Davis 2005; Kahsai and Zars 2011). The fly is capable of short term memory, medium term memory, and long term memory comprised of acquisition, consolidation, and recall (Margulies et al. 2005). The fly can be trained and tested in operant paradigms and conditioned place preference assays. Learning and memory can be visual, olfactory, gustatory, or mediated by temperature. As in humans, learning and memory ability declines in the fly as it ages (Tonoki and Davis 2012). Because of each of these factors, the fly is an attractive platform to utilize learning and memory assays to discover therapeutics that may enhance normal cognitive performance.

Small molecule screening experiments can be designed for any of the available learning and memory assays. In general these assays are rather low throughput due to the need for an experimenter to test the populations of treated flies individually. The training component, at least for olfactory learning and memory, can be automated to some extent with the use of multichannel 'Robotrainers', which are capable of being programmed for a variety of training protocols and can be run in parallel. Nevertheless, using this method a single dedicated investigator should be able to perform a first pass screen of about 10–20 different drugs/week on long term olfactory learning and memory, and about 50 different drugs per week for effects on short term memory. Using the olfactory assay as an example (Krashes and Waddell 2011), the effects of a drug on different aspects of learning and memory can be examined. To examine short term learning and memory, populations of about 100 adult flies (2 days post eclosion for young healthy flies) are maintained on food + drug for 48 h, then trained by conditioned stimulus training where the unconditioned stimulus is an aversive odor, and the conditioned stimulus an electrical shock paired with a different aversive odor. The flies are immediately tested by transferring to the choice point of a T-maze where they are presented with both odors. The flies that go towards and away from the respective odors are counted and a performance index calculated (PI = [((# away from the paired odor)-(# toward the unpaired odor))/total # of flies tested]*100). A PI of zero indicates a 50/50 split, and no learning, and a PI of 100 indicates that all flies learned the shock pairing to the odor and went to the non-paired odor side of the maze. Eight half trials are usually performed, alternating the shock pairing, to give a final n = 4 for statistical analysis. A typical PI for robust short term learning and memory is between 70 and 90.

Different aspects of the learning and memory process can be assessed by increasing the time interval between training and testing. For example, long term memory (LTM) can be determined by testing the flies 24 h or more after training.

The most effective training for LTM is spaced training, where there are ten training sessions separated by 15 min. This training produces a robust response readily measured at 24 h, although the PI values are typically lower and in the range of 15–25 at 24 h. To measure the effects of a drug on acquisition or learning, drug is fed to flies for a specified time interval until training (e.g. 24–48 h). To measure the effects of a drug on the consolidation process, flies can be fed drug for 3–6 h immediately after the training session. To measure recall, flies can be fed drug for 2–3 h prior to testing. An example of screening several different drugs for their effects on both short term and long term olfactory learning and memory can be found in (Johnson et al. 2011).

Interestingly, the fly has been used to develop a cell-based screening platform for cognitive enhancers. It was discovered in 1995 that activation of the transcription factor CREB in flies improved olfactory memory performance (Yin et al. 1995), and later in another form of learning and memory, courtship suppression (Tubon et al. 2013). A mammalian cell-based HTS screening was developed based upon this premise that identified several enhancers of CREB function that were tested in flies for their ability to activate CREB and enhance aspects of memory (Scott et al. 2002). One of the more promising candidates identified through this pipeline was an inhibitor of PDE4 (Scott et al. 2002).

9.7 Summary

Drosophila has tremendous potential to facilitate drug discovery efforts. Not only in aging related diseases and processes, but several others as well. As discovery is moving away from target-based approaches to system-based approaches the fly offers a rapid and inexpensive platform for identifying high quality leads subsequent to more traditional HTS, or even as a primary screening platform. Compounds that rescue phenotypes in the relevant fly model and do not demonstrate toxicity are of high quality, and will have much more favorable ADME properties in subsequent whole animal rodent testing compared to compounds taken into rodents from the culture dish. The promises that the fly holds, when properly incorporated into the discovery pipeline, is to more rapidly bring more effective drugs to market at reduced development costs.

References

Andretic R, van Swinderen B, Greenspan RJ (2005) Dopaminergic modulation of arousal in *Drosophila*. Curr Biol CB 15:1165–1175

Bhandari P, Jones MA, Martin I, Grotewiel MS (2007) Dietary restriction alters demographic but not behavioral aging in *Drosophila*. Aging Cell 6:631–637

Brand A, Perrimon N (1993) Targeted gene expression as a means of altering cell fates and generating dominant phenotypes. Development 118:401–415

Burr AA, Tsou W-L, Ristic G, Todi SV (2014) Using membrane-targeted green fluorescent protein to monitor neurotoxic protein-dependent degeneration of *Drosophila* eyes. J Neurosci Res 92:1100–1109

Dauer W, Przedborski S (2003) Parkinson's disease. Neuron 39:889–909

Davis RL (2005) Olfactory memory formation in Drosophila: from molecular to systems neuroscience. Annu Rev Neurosci 28:275–302

DiMasi JA, Feldman L, Seckler A, Wilson A (2010) Trends in risks associated with new drug development: success rates for investigational drugs. Clin Pharmacol Ther 87:272–277

Doody RS, Cummings JL, Farlow MR (2012) Reviewing the role of donepezil in the treatment of Alzheimer's disease. Curr Alzheimer Res 9:773–781

Green EW, Giorgini F (2012) Choosing and using *Drosophila* models to characterize modifiers of Huntington's disease. Biochem Soc Trans 40:739–745

Griffiths RR, Woodson PP (1988) Reinforcing properties of caffeine: studies in humans and laboratory animals. Pharmacol Biochem Behav 29:419–427

Haselton AT, Fridell Y-WC (2010) Adult *Drosophila melanogaster* as a model for the study of glucose homeostasis. Aging (Albany NY) 2:523–526

Herper M (2012) The truly staggering cost of inventing new drugs. http://www.forbes.com/sites/matthewherper/2012/02/10/the-truly-staggering-cost-of-inventing-new-drugs. Forbes 2012

Herper M (2013) The cost of creating a new drug now $5 billion, pushing big pharma to change. http://www.forbes.com/sites/matthewherper/2013/08/11/how-the-staggering-cost-of-inventing-new-drugs-is-shaping-the-future-of-medicine. Forbes 2013

Hopkins AL, Groom CR (2002) The druggable genome. Nat Rev Drug Discov 1:727–730

Ja WW, Carvalho GB, Mak EM, La Rosa de NN, Fang AY, Liong JC, Brummel T, Benzer S (2007) Prandiology of *Drosophila* and the CAFE assay. Proc Nat Acad Sci USA 104:8253–8256

Jafari M (2010) *Drosophila melanogaster* as a model system for the evaluation of anti-aging compounds. Fly 4:253–257

Johnson O, Becnel J, Nichols CD (2011) Serotonin receptor activity is necessary for olfactory learning and memory in *Drosophila melanogaster*. Neuroscience 192:372–381

Kahsai L, Zars T (2011) Learning and memory in *Drosophila*: behavior, genetics, and neural systems. Int Rev Neurobiol 99:139–167

Kaun KR, Devineni AV, Heberlein U (2012) *Drosophila melanogaster* as a model to study drug addiction. Hum Genet 131:959–975

Kim SK, Rulifson EJ (2004) Conserved mechanisms of glucose sensing and regulation by Drosophila corpora cardiaca cells. Nature 431:316–320

Krashes MJ, Waddell S (2011) *Drosophila* aversive olfactory conditioning. Cold Spring Harb Protoc 2011(5) (pdb.prot5608)

Lawal HO, Terrell A, Lam HA, Djapri C, Jang J, Hadi R, Roberts L, Shahi V, Chou M-T, Biedermann T et al (2014) *Drosophila* modifier screens to identify novel neuropsychiatric drugs including aminergic agents for the possible treatment of Parkinson's disease and depression. Mol Psychiatry 19:235–242

Magwere T, Chapman T, Partridge L (2004) Sex differences in the effect of dietary restriction on life span and mortality rates in female and male *Drosophila melanogaster*. J Gerontol Ser A Biol Sci Med Sci 59:B3–B9

Mair W, Goymer P, Pletcher SD, Partridge L (2003) Demography of dietary restriction and death in *Drosophila*. Science (New York, NY) 301:1731–1733

Mair W, Piper MDW, Partridge L (2005) Calories do not explain extension of life span by dietary restriction in *Drosophila*. PLoS Biol 3:e223

Margulies C, Tully T, Dubnau J (2005) Deconstructing memory in *Drosophila*. Curr Biol CB 15:R700–R713

Metaxakis A, Partridge L (2013) Dietary restriction extends life span in wild-derived populations of *Drosophila melanogaster*. PLoS ONE 8:e74681

McKoy AF, Chen J, Schupbach T, Hecht MH (2012) A novel inhibitor of amyloid β (Aβ) peptide aggregation. J Biol Chem 287(46):38992–39000
Minzenberg MJ, Carter CS (2008) Modafinil: a review of neurochemical actions and effects on cognition. Neuropsychopharmacology: official publication of the American college. Neuropsychopharmacology 33:1477–1502
Nichols CD (2006) *Drosophila melanogaster* neurobiology, neuropharmacology, and how the fly can inform central nervous system drug discovery. Pharmacol Ther 112:677–700
Nichols CD, Becnel J, Pandey UB (2012) Methods to assay *Drosophila* behavior. J Visual Exp JoVE e3795
Paaby AB, Schmidt PS (2009) Dissecting the genetics of longevity in *Drosophila melanogaster*. Fly 3:29–38
Padhy BM, Gupta YK (2011) Drug repositioning: re-investigating existing drugs for new therapeutic indications. J Postgrad Med 57(2):153–160. doi:10.4103/0022-3859.81870
Pandey UB, Nichols CD (2011) Human disease models in *Drosophila melanogaster* and the role of the fly in therapeutic drug discovery. Pharmacol Rev 63:411–436
Pletcher SD, Macdonald SJ, Marguerie R, Certa U, Stearns SC, Goldstein DB, Partridge L (2002) Genome-wide transcript profiles in aging and calorically restricted *Drosophila melanogaster*. Curr Biol CB 12:712–723
Prüßing K, Voigt A, Schulz JB (2013) *Drosophila melanogaster* as a model organism for Alzheimer's disease. Mol Neurodegeneration 8:35
Querfurth HW, LaFerla FM (2010) Alzheimer's disease. N Engl J Med 362:329–344
Ready DF, Hanson TE, Benzer S (1976) Development of the *Drosophila* retina, a neurocrystalline lattice. Dev Biol 53:217–240
Ross CA (2002) Polyglutamine pathogenesis: emergence of unifying mechanisms for Huntington's disease and related disorders. Neuron 35:819–822
Sarkar S, Perlstein EO, Imarisio S, Pineau S, Cordenier A, Maglathlin RL, Webster JA, Lewis TA, O'Kane CJ, Schreiber SL et al (2007) Small molecules enhance autophagy and reduce toxicity in Huntington's disease models. Nat Chem Biol 3:331–338
Scott R, Bourtchuladze R, Gossweiler S, Dubnau J, Tully T (2002) CREB and the discovery of cognitive enhancers. J Mol Neurosci MN 19:171–177
Spindler SR, Li R, Dhahbi JM, Yamakawa A, Sauer F (2012) Novel protein kinase signaling systems regulating life span identified by small molecule library screening using *Drosophila*. PLoS ONE 7:e29782
Tonoki A, Davis RL (2012) Aging impairs intermediate-term behavioral memory by disrupting the dorsal paired medial neuron memory trace. Proc Natl Acad Sci USA 109(16):6319–6324
Tubon TC, Zhang J, Friedman EL, Jin H, Gonzales ED, Zhou H, Drier D, Gerstner JR, Paulson EA, Fropf R et al (2013) dCREB2-mediated enhancement of memory formation. J Neurosci Official J Soc Neurosci 33:7475–7487
Vidal M, Cagan R (2006) *Drosophila* models for cancer research. Curr Opin Genet Dev 16:10–16
Walker FO (2007) Huntington's disease. Lancet 369:218–228
Whitworth AJ (2011) *Drosophila* models of Parkinson's disease. Adv Genet 73:1–50
Yin JCP, Del Vecchio M, Zhou H, Tully T (1995) CREB as a memory modulator: induced expression of a dCREB2 activator isoform enhances long-term memory in *Drosophila*. Cell 81:107–115

Part III
Life-Extending Treatments

Chapter 10
Life-Extending Effect of Phytochemicals in *Drosophila*

Lee Shin-Hae and Min Kyung-Jin

Abstract Plant-derived compounds known as phytochemicals have attracted the attention of biologists as well as the general public for their ability to improve quality and quantity of life. Several phytochemicals have been found to exert beneficial effects in this regard, including inhibition of aging and extending the life spans of experimental animals such as yeast, worms, flies, and mice. *Drosophila melanogaster* is a particularly effective model system for evaluating anti-aging compounds. This insect is suitable for aging research since it has a rapid generation time and short life span. In this chapter, we review typical phytochemicals such as resveratrol, curcumin, and catechin that extend the life span of *Drosophila melanogaster* as well as discuss the molecular mechanisms underlying this effect.

Keywords *Drosophila* · Phytochemicals · Aging · Life extension · Resveratrol · Curcumin · Catechin

10.1 Introduction

Fruits, vegetables, and herbs have long been consumed to improve human health. Plant-derived products and their extracts have a variety of physiological effects, including anti-oxidant and anti-inflammatory properties, and are thus effective for protecting against numerous diseases such as cancer, diabetes, cardiovascular disease, and neurodegenerative disorders. For example, blueberries and green tea have anti-oxidant and anti-inflammatory activities that help prevent cancer, cardiovascular disease, memory impairment, obesity,

L. Shin-Hae · M. Kyung-Jin (✉)
Department of Biological Sciences, Inha University, Incheon 402-751, South Korea
e-mail: minkj@inha.ac.kr

A.M. Vaiserman et al. (eds.), *Life Extension*, Healthy Ageing and Longevity 3,
DOI 10.1007/978-3-319-18326-8_10

and cognitive impairment due to aging in experimental animals (Giacalone et al. 2011; Yang et al. 2014). In addition, pomegranate juice has been reported to reduce blood pressure via anti-oxidant activity (Aviram and Dornfeld 2001), whereas cinnamon is known to ameliorate diabetes by enhancing insulin signaling (Qin et al. 2003). The beneficial properties of these nutraceuticals can be attributed to their secondary metabolites, which are part of the biological systems by which plants resist various stresses. These secondary metabolites function as phytochemicals. Numerous phytochemicals have been identified, and the mechanisms underlying their protective effects against disease have been actively investigated. Recently, some phytochemicals have attracted attention due to their potential to retard the aging process and extend the life spans of various experimental animals.

To evaluate the beneficial properties of nutraceuticals and phytochemicals, efficient animal models with rapid generation times and short life spans are required. *Drosophila melanogaster* is an effective model for evaluating anti-aging compounds. The average life span of *Drosophila* is approximately 60–80 days depending on the strain. This short life span is useful for testing the anti-aging properties of phytochemicals using several replicates. Any effective model for aging research should also have organismal complexity comparable to that of humans. Although *D. melanogaster* is evolutionally distal to humans, the *Drosophila* genome contains 60 % of all human disease-related genes. Additionally, numerous signaling pathways and cellular processes such as, for example, the insulin signaling pathway, are conserved in both flies and mammals. Furthermore, well established databases and valuable genetic tools for modulating gene expression in these flies are available.

Over past decades, many phytochemicals have been identified as having anti-aging effects in *Drosophila*. However, some of these compounds need to be rigorously tested again to confirm their effects. Recently, Mahtab Jafari introduced an algorithm for evaluating anti-aging compounds using a *Drosophila* model system (Jafari 2010). Based on her study, compounds that increase life span and reduce mortality should be further tested to determine whether or not they affect other parameters such as food intake, reproduction rates, and fecundity. In addition, these compounds should not have adverse effect on health span that can be monitored by assessing fly locomotion. In this chapter, we discuss current studies on phytochemicals that have been shown to extend life span in a *Drosophila* model system.

10.2 Nutraceuticals and Phytochemicals

Several phytochemicals with known health beneficial effects are polyphenolic compounds. Polyphenols are characterized by hydroxylated phynyl moieties and classified as flavonoid polyphenols, such as catechine and fisetin, or non-flavonoid polyphenols, including resveratrol and curcumin, based on chemical structure.

10.2.1 Resveratrol

Resveratrol (3,5,4'-trihydroxysilbene), a non-flavonoid polyphenolic compound found in the skin of red grapes, is produced in response to fungal infection. This compound belongs to the stilbene group, which includes members that naturally exist as both *cis*- and *trans*-isomers. Resveratrol has several beneficial effects, including conferring protection against metabolic diseases such as diabetes, cardiovascular disease, and neurodegenerative disorders (Marchal et al. 2013). In addition, resveratrol is the most well characterized phytochemical for extending the life spans of various animal models, including yeast, worms, flies, short-lived fish *Nothobranchius furzeri*, and mice fed a high-fat diet (Howitz et al. 2003; Wood et al. 2004; Baur et al. 2006; Valenzano et al. 2006). In 2003, the effect of resveratrol on life span extension was first investigated using *Saccharomyces cerevisiae* (Howitz et al. 2003). It was found that resveratrol could increase the replicative life span of yeast via stimulation of sirtuin expression and activity. In 2004, the life span-extending effect of resveratrol was confirmed using *Caenorhabditis elegans* and *D. melanogaster* (Wood et al. 2004). Subsequently, Valenzano et al. reported the pro-longevity effect of resveratrol on short-lived fish *Nothobranchius furzeri* (Valenzano et al. 2006). Baur et al. also demonstrated that resveratrol increases survival of mice fed a high-calorie diet (Baur et al. 2006).

Several studies have shown the effect of resveratrol on life span extension in flies fed a normal diet, although these investigations have produced conflicting results. In 2004, Wood et al. and Bauer et al. independently reported that resveratrol (100 and 200 μM, respectively) could extend the life span of Canton-S wild-type *D. melanogaster* (Bauer et al. 2004; Wood et al. 2004). In contrast, Bass et al. showed in 2007 that the median life spans of Canton-S and Dahomey wild-type fruit flies were not extended by resveratrol supplementation at concentrations ranging from 1 to 1000 μM (Bass et al. 2007). In 2011, Xiang et al. confirmed the pro-longevity effect of resveratrol (50 μM) using a *yw* fruit fly strain (Xiang et al. 2011). Supplementation of larval diet with resveratrol was found to extend the life span of adult fruit flies (Chandrashekara and Shakarad 2011). More recently, the pro-longevity effect of resveratrol was found to be dependent on gender and dietary nutrient composition (Wang et al. 2013). Specifically, resveratrol (200 μM) extended the life span of females fed a low-sugar/high-protein diet while 400 μM resveratrol extended the life span of females fed a high-fat diet. These findings are in agreement with another report showing that resveratrol was able to extend the life span of mice fed a high-fat diet but not a standard diet (Baur et al. 2006). These results suggest that the ability of resveratrol to extend life span depends on the dietary conditions of the organism.

Several reports have suggested that the life span-extending effect of resveratrol is associated with activation of sirtuin. For instance, resveratrol was found to lack pro-longevity activity in *C. elegans* or *D. melanogaster* possessing mutated sirtuin genes (Wood et al. 2004). However, other studies have shown that resveratrol does not activate sirtuin (Borra et al. 2005, Kaeberlein et al. 2005), and different factors such as insulin-like growth factor-1 (IGF-1), AMP-activated protein kinase (AMPK), p53, and peroxisome proliferator-activated receptor-γ-coactivator-1α (PGC-1α) may instead be associated with the pro-longevity effect of resveratrol (Lagouge et al. 2006; Baur et al. 2006; Pirola and Frojdo 2008; Antosh et al. 2011).

10.2.2 Curcumin

Curcumin [1,7-bis-(4-hydroxy-3-methoxyphenyl)-1,6-heptadiene-3,5-dione], a non-flavonoid polyphenolic compound found in the rhizome of *Curcumma longa* (turmeric), has been wildly used as a spice, food additive, dye, and herbal medicine in Asia. Numerous studies have investigated the anti-oxidant activity of curcumin. However, this compound is also known to exert anti-inflammatory, anti-diabetic, anti-allergic, and anti-carcinogenic effects (Shen et al. 2013a).

The first ever demonstration of the life span-extending effect of curcumin using *Drosophila* was by Suckow and Suckow in (2006). In their study, supplementation with 1 mg/g of curcumin extended the life span of wild-type fruit flies. However, it remained unclear whether or not this pro-longevity effect is associated with secondary physiological confounding factors that extend life span, such as food consumption and fecundity. Furthermore, the population size used in their study was rather small. In 2010, Lee et al. investigated the life span-extending effect of curcumin using two independent strains of fruit flies, Canton-S and Ives (Lee et al. 2010). It was confirmed that the life span-extending effect of curcumin is not associated with changes in fecundity or food intake. They also observed that gene expression profiles were altered by curcumin. In 2011, tetrahydocurcumin (THC), an active metabolite of curcumin, was reported to have a pro-longevity effect on both male and female *yw* strain flies (Xiang et al. 2011). Additionally, THC supplementation increased survival rates of both of paraquat-treated and untreated wild-type Oregon-R fruit flies. In 2013, Shen et al. observed that 0.5 and 1 mg/g of curcumin could extend the mean and median life spans of both male and female Oregon-R flies as well as increase activity of superoxide dismutase (SOD) (Shen et al. 2013b). Further, Soh et al. showed that supplementation with 100 mM curcumin over the entire adult life span of fruit flies, or during periods in which their survival was less than 80 %, did not extend the median life span of the Ra strain (Soh et al. 2013). However, supplementation with 100 mM curcumin during the development stage, periods in which survival was greater than 90 %, or periods in which survival was from 80–90 % extended the median and maximum life spans of Ra flies. These findings suggest that the life span-extending effect of curcumin is associated with age-specific gene profiles.

The mechanism underlying the life span-extending effect of curcumin is being actively investigated. Curcumin supplementation affects many genes and signaling pathways, including the nuclear factor-κB (NF-κB), AMPK, and target ofrapamycin (TOR) signaling pathways, which are known to regulate organismal life span (Aggarwal 2010). The pro-longevity effect of THC is absent when dFoxo and dSir2 genes are mutated, indicating that activation of FOXO/4EBP and sirtuin plays a key role (Xiang et al. 2011). In addition, genome-wide analysis was recently performed using *Drosophila* to better understand the molecular alterations induced by curcumin supplementation (Zhang et al. 2013). Curcumin supplementation altered the transcription levels of five groups of genes, including those affecting the Notch signaling pathway, Wnt signaling pathway, cell cycle regulation, riboprotein synthesis, and p53 pathway.

10.2.3 Catechin

Catechin is a flavanol-type flavonoid (flavan-3-ol) that has two benzene rings as well as a dihydropyran heterocycle with a hydroxyl group at position 3 (Malaguti et al. 2013). Catechin is commonly found in foods and edible plants, such as green tea, cacao, and red wine. In tea, catechin mainly exists as epicatechin (EC), epicatechin-3-gallate (ECG), epigallocatechin (EGC), epigallocatechin-3-gallate (EGCG), and gallocatechin (GC) (Clifford et al. 2013). All of these possess various biological properties such as cardioprotective, anti-atherogenic, and anti-carcinogenic effects (Mak 2012). In particular, EGCG, the most abundant form of catechin in green tea, induces expression of anti-oxidant enzymes such as glutathione peroxidase, catalase, and glutathione S-transferase in mice (Na and Surh 2008). In addition, EGCG treatment has been reported to increase the expression of SOD and reduce ROS levels in *C. elegans* (Zhang et al. 2009). Furthermore, Huntington disease (HD)-related phenotypes such as photoreceptor degeneration and motor function were shown to improve when EGCG was fed to transgenic HD flies overexpressing a pathogenic htt exon 1 protein (Ehrnhoefer et al. 2006). Other studies have demonstrated the life span-extending effect of catechin using various animal models. In particular, EGCG was reported to extend the life span of *C. elegans* under stress (Zhang et al. 2009). In addition, Kitani et al. reported that 80 mg/L of green tea catechin could increase the mean life span of male mice (Kitani et al. 2004), whereas Si et al. showed that 0.25 % EC could increase the survival time of diabetic *db* mice (Si et al. 2011).

Using a *Drosophila* model system, several groups previously investigated the life span-extending effects of catechin. In 2007, Li et al. showed that 10 mg/mL of green tea catechin extract consisting of 62 % EGCG, 19 % EGC, 7 % EC, and 9 % ECG could extend the median and mean life spans of Oregon-R flies by 36 and 16 %, respectively (Li et al. 2007). In addition, they found that 10 mg/mL of green tea catechin could increase the survival time of paraquat-treated wild-type flies but not *SOD* or *catalase* mutants, and supplementation with green tea catechin increased SOD and catalase activities. These data suggest that the pro-longevity and anti-oxidant effects of green tea catechin are associated with SOD and catalase. Furthermore, the authors reported that supplementation with green tea catechin could reduce mortality and prolong the life span of flies fed a high-fat diet (Li et al. 2008). EC (0.01–8 m mol/L) extracted from cocoa has also been reported to increase the life span of *Drosophila* (Si et al. 2011), whereas black tea extract mixed with EC and theaflavins was found to increase SOD1 and catalase activities as well as extend the life span of *Drosophila* (Peng et al. 2009).

10.2.4 Others

Aside from the compounds described above, several other phytochemicals such as morphine, quercetin, and fisetin have also attracted attention from gerontologists as putative anti-aging and pro-longevity compounds. Previous studies have attempted to verify the life span-extending effects of these reagents using model animals such as *D. melanogaster*. Morphine, widely known as an narcotic analgesic extracted from opium poppies, reportedly extends the life span of *D. melanogaster* (Dubiley et al. 2011). Specifically, exposure to morphine hydrochloride at 0.01–0.25 mg/mL once per week increased the mean life span of Oregon-R male and female flies. In addition, morphine has been reported to reduce the expression of genes involved in metabolic functions (Loguinov et al. 2001), inhibit expression of NADPH oxidase, and modulate intracellular redox levels (Lee et al. 2004, Qian et al. 2007), suggesting that the pro-longevity effect of morphine hydrochloride is associated with a reduced metabolic rate and reactive oxygen species production. Furthermore, the pro-longevity effect of morphine is corroborated by several reports showing that morphine activates NF-κB, a factor involved in cellular senescence and age-related disease (Lin et al. 2007; Salminen et al. 2011b), as well as heat shockprotein 70 (HSP70), which helps regulate life span (Ammon-Treiber et al. 2004).

Quercetin (3,3′,4′,5,7-pentahydroxyflavone) is a flavonol-type flavonoid and one of the most important dietary flavonoids in foods (Peng et al. 2011). This compound is found in most fruits and vegetables, especially peels and seeds of apples, berries, onions, grapes, tea, and tomatoes as well as some medicinal plants. Quercetin has many beneficial effects, including anti-inflammatory, anti-oxidative, and anti-mutagenic activities, and has thus received much attention as a reagent to extend life span. The life span-extending effect of quercetin is mainly based on studies using a *C. elegans* model system. The pro-longevity activity of quercetin is supposedly mediated by insulin/IGF signaling components in *C. elegans* (Pietsch et al. 2009) but not by FOXO (Saul et al. 2008). Quercetin has no significant effect on the life span of yeast (Howitz et al. 2003) or long-lived F1 hybrid mice (Spindler et al. 2013). The effects of quercetin have not been examined in a *Drosophila* model and future research is needed to test it.

The flavonoid fisetin (3,7,3′,4′-tetrahydroxyflavone) has also been reported to extend the *Drosophila* life span (Wood et al. 2004). In 2004, the pro-longevity effect of fisetin as a sirtuin-activating compound (STAC) was analyzed. It was found that fisetin increased the median and mean life spans of *yw* male and female flies fed a normal diet but not flies given a restricted diet (Wood et al. 2004). This observation means that the life span-extending effect of fisetin is associated with caloric restriction. In addition, fisetin was reported to have neuroprotective, neurotrophic, and cognition-enhancing effects mediated by Ras/extracellular signal-regulated kinase (ERK) signaling in an animal model of Huntington's disease (Maher et al. 2011).

10.3 Plausible Underlying Mechanisms of Life Span Extension by Phytochemicals

10.3.1 Hormesis

The mechanisms underlying the beneficial health effects of dietary phytochemicals have only recently begun to be understood. One general mechanism of action involves hormesis, which activates the adaptive cellular stress response pathway. Hormesis is a biphasic dose-response with a low-dose beneficial effect and high-dose toxic effect. Mild stress stimulates a defense response at the organismal level, resulting in biologically beneficial effects. A typical example of hormesis extending life span can be observed with calorie restriction. Moderately reduced calorie intake acts as a mild stressor that stimulates several stress-induced signaling pathways along with gene expression. However, malnutrition acts as a severe stressor, harming organisms and shortening life span. Similar to the effects of calorie restriction, many phytochemicals extend life span when ingested at certain doses, whereas their life span-extending effects are abolished when ingested at high doses.

Hormesis is reportedly mediated by several cellular signaling pathways, such as sirtuins, nuclear factor-erythoroid 2-related factor (NRF2), and NF-κB. Genetic manipulation of these signaling pathways is known to regulate life span in animal models. Certain phytochemicals such as resveratrol activate the sirtuin-FOXO pathway while others activate NFκB and cAMP response element-binding protein (CREB) (Mattson et al. 2007; Salminen et al. 2008b). In addition, the expression of vitagenes encoding cytoprotective heat shock proteins, hemeoxygenase-1 (HO-1), and antioxidant enzymes is reportedly induced by phytochemicals (Calabrese et al. 2012).

10.3.2 NRF2

NRF2 is a redox-sensitive leucin zipper transcription factor. Under basal conditions, NRF2 resides in the cytoplasm bound to an inhibitory partner, Kelch-like ECH-associated protein (Keap1), which is anchored to the actin cytoskeleton and represses NRF2 activity. In the presence of electrophilic and oxidative stresses, cysteine residues of Keap1 become modified, resulting in disruption of the Keap1-NRF2 complex and translocation of NRF2 to the nucleus. Once inside the nucleus, NRF2 binds to an antioxidant response element (ARE) or electrophile response element (EpRE). Protective stress responses are subsequently generated via induced expression of several cytoprotective genes, including NADPH:quinone oidoreductase-1, heme oxygenase-1, glutamate cysteine ligase, glutathione S-transferase, glutathione peroxidase, and thioredoxin (Itoh et al. 1997). NRF2 nuclear translocation and activation are facilitated by upstream kinases such as mitogen-activated protein kinases (MAPKs), phosphatidylionositol-3-kinase (PI3 K), Akt, protein kinase C (PKC), and casein kinase-2. Furthermore, NRF2 interacts with other signaling pathways such as the NF-κB cascade during inflammation and aryl hydrocarbon receptor (AhR), which regulates xenobiotic detoxification (Kensler et al. 2007, Reddy et al. 2007).

The effect of NRF2 on longevity has been well established in invertebrate models, including *C. elegans* and *D. melanogaster* (Sykiotis et al. 2011). In *C. elegans*, SKN-1, the functional homologue of mammalian NRF2, is negatively regulated by the insulin/IGF-1 signaling pathway, which itself negatively regulates life span in many animals (Tullet et al. 2008). SKN-1 was found to accumulate in nuclei in worms having a mutated copy of *daf-2*, which is the insulin/IGF-1 analogue of *C. elegans*, and have a pro-longevity function mediated by p38 MAPK, AKT-1, and SGK-1. In addition, *skn-1* mutant worms do not respond to dietary restriction, indicating that the hormetic effect of dietary restriction on longevity is mediated by SKN-1 in *C. elegans* (Bishop and Guarente 2007). In *Drosophila*, homologues of Keap1 and NRF2 were reported to be activated by oxidants or increased tolerance to oxidative stress (Sykiotis and Bohmann 2008). Although over-activation of CncC, the *Drosophila* homologue of NRF2, was not found to have direct effects on life span extension, mutation of *Keap1* was observed to extend the life span of flies. These findings indicate that NRF2/Keap1 signaling plays a role in the regulation of longevity as a master mediator of anti-oxidant and detoxification responses (Sykiotis and Bohmann 2008).

Numerous phytochemicals such as curcumin, resveratrol, EGCG, sulphoraphane, and acetyl-L-carnitine are known to be potent antioxidants that activate NRF2 via several steps (Jeong et al. 2005, Wu et al. 2006, Surh et al. 2008). Some of these compounds such as sulforaphane, an isothiocyananate found in broccoli, oxidize or modify the cysteine thiol groups of Keap1, thereby stabilizing and activating NRF2 (Hong et al. 2005). In addition, many phytochemicals such as EGCG and curcumin can activate NRF2 signaling by stimulating upstream kinases that subsequently phosphorylate NRF2 protein (Chen et al. 2000; Rushworth et al. 2006).

10.3.3 Sirtuin

Silent information regulator 2 (Sir2 or sirtuin) is a highly conserved nicotinamide adenine dinucleotide (NAD+)-dependent deacetylase that targets histones (resulting in chromatin silencing) as well as non-histone proteins. Seven members of the sirtuin family, SIRT1 to SIRT7, have been identified in mammals. These factors are involved in different physiological processes, including metabolism, stress responses, cell survival, replicative senescence, inflammation, circadian rhythm, neurodegeneration, and cancer (Pallauf et al. 2013). In addition, mammalian SIRT1 is known to be a functional homologue of yeast *sir2* and was reported to interact with 34 known deacetylation targets, including FOXO, NF-κB, and p53 as well as six binding partners such as peroxisome proliferator-activated receptor γ (PPARγ) (Baur 2010).

Overexpression of *sir2* is known to extend the life spans of yeast, worms, and flies (Imai et al. 2000; Tissenbaum and Guarente 2001; Rogina and Helfand 2004). An extra copy of the *sir2* gene increases the replicative life span of yeast, whereas *sir*2 mutants have a shorter life span (Imai et al. 2000). In *C. elegans*, duplication of *sir-2.1*, the homologue of yeast *sir2*, extends life span (Tissenbaum and

Guarente 2001). In *Drosophila*, ubiquitous expression of *dSir2* using dSir2^{EP2300} and tubulin-GAL4 was shown to increase mean life span by 57 % (Rogina and Helfand 2004), whereas the life span of flies expressing *dSir2* in fat bodies using $S_1$106-GAL4 increased by 12 % (Hoffmann et al. 2013). Sirtuins are known to mediate the beneficial effects of calorie restriction, although their activity was found to be dependent on experimental conditions and genetic background in yeast (Lin et al. 2000; Tissenbaum and Guarente 2001).

Numerous compounds activate sirtuins, including chalcones, flavones, and stilbenes (Wood et al. 2004; Cho et al. 2013). The best known phytochemical that extends life span via sirtuin activation is resveratrol. Treatment with resveratrol increases expression of sirutins. Furthermore, the pro-longevity effect of resveratrol is absent from *C. elegans* and *D. melanogaster* with mutations in the sirtuin gene (Wood et al. 2004). These results indicate that the pro-longevity effect of resveratrol is dependent on sirtuin activation. In addition, EGCG is also known to extend life span and ameliorate age-related inflammation through activation of sirtuin (Niu et al. 2013).

10.3.4 NF-κB

NF-κB is a major regulator of the immune response. Under normal conditions, NF-κB is usually trapped in the cytoplasm after binding to inhibitory IκB proteins. When activating signals stimulate upstream kinases such as IκB kinase (IKK), calmodulin-dependent kinase II (CKII), and c-Jun N-terminal kinase (JNK), IκB becomes phosphorylated and degraded by ubiquitination-related proteasomes. The activated NF-κB complex then translocates into the nucleus where it activates transcription of several genes that control numerous cellular processes, including immune responses, proliferation, cancerous transformation, and cell survival, in response to stress and exposure to inflammatory cytokines (Baker et al. 2011).

Since increased proinflammatory status is known to stimulate the aging process, a condition known as inflamm-aging (Franceschi et al. 2000), many studies have shown that sustained activation of NF-κB signaling promotes cellular and organismal senescence (Rovillain et al. 2011; Salminen et al. 2011b). Over-activation of NF-κB is known to be associated with age-related diseases such as cancer, cardiovascular disease, type 2 diabetes, obesity, and neurodegenerative diseases (Adler et al. 2007; Prasad et al. 2010). Furthermore, it was reported that AMPK, sirtuins, and NRF2 (a target of phytochemicals) inhibit NF-κB signaling (Salminen et al. 2008a; Kim et al. 2010; Salminen et al. 2011a).

Various phytochemicals are known to regulate the NF-κB pathway (Aggarwal and Shishodia 2004). Flavonoids and terpenoids such as artemisinin, celastrol, kahweol, and lutein inhibit NF-κB activation along several points of the NF-κB signaling cascade. This may be accomplished through inhibition of IκB degradation or nuclear translocation of p65 (Salminen et al. 2011b). Additionally, resveratrol and many stilbenes activate sirtuin and subsequently inhibit NF-κB signaling (Chung et al. 2010). EGCG treatment was also recently reported to reduce the expression of NF-κB mRNA and protein in rat liver and kidney (Niu et al. 2013).

10.3.5 AMPK

AMPK, a serine/threonine kinase, is a crucial regulator of energy metabolic homeostasis. This factor is activated by elevated AMP levels and phosphorylation by upstream kinases such as liver kinase B1 (LKB1), CK, and transforming growth factor β-activated kinase 1 (Steinberg and Kemp 2009).

AMPK has been shown to have a significant effect on regulation of aging and age-related changes in lower organisms such as yeast and worms. Loss of SNF1p, a yeast homologue of AMPK, results in a 20 % increase in the life span of yeast cells (Ashrafi et al. 2000). In *C. elegans*, overexpression of *aak-2*, the worm homologue of AMPKα, extends life span and appears to be required for increased longevity associated with *daf-2*/insulin and deacetylase *sir-2.1* signaling (Curtis et al. 2006). In addition, AMPK was reported to mediate the pro-longevity effect of calorie restriction via phosphorylation of the transcription factor FOXO/DAF-16 (Greer et al. 2007). In *Drosophila*, inhibition of AMPKα expression using dAMPKα RNAi in muscle was reported to reduce survival of flies (Tohyama and Yamaguchi 2010). In addition, mild overexpression of LKB1 in whole bodies of female flies has been shown to stimulate activation of AMPK and life span extension (Funakoshi et al. 2011). Furthermore, overexpression of AMPK in adult abdominal fat bodies and muscle using $S_1$106-GAL4 and MHC-GAL4 could extend life span in flies (Stenesen et al. 2013). However, the role of AMPK in the regulation of aging in higher organisms is not clear. A recent report showed that long-term treatment with metformin can increase AMPK activity and extend the life span of male C57BL/6 mice (Martin-Montalvo et al. 2013). Nevertheless, there is currently no solid evidence indicating that AMPK has a direct effect on the life span of mammals.

Several pharmacological AMPK activators such as 5-aminoimidazole-4-carboxamide reboside (AICAR) and metformin have been identified. Additionally, many natural plant-derived compounds, including berberine, curcumin, quercetin, theaflavin, and ginsenoside Rh2, are also known to activate AMPK (Jeong et al. 2009). These compounds activate AMPK either by affecting upstream kinases or increasing local AMP concentrations. In particular, metformin, an oral anti-diabetic drug derived from French lilac, is a well-known potent activator of AMPK. Previous studies have shown that the life span-extending effect of metformin is mediated by AMPK activation in *C. elegans* (Onken and Driscoll 2010, De Haes et al. 2014). However, other investigations into life span extension in model systems have produced conflicting results (Anisimov et al. 2010, Smith et al. 2010, Slack et al. 2012). Curcumin is also known to activate AMPK similar to AICAR (Lee et al. 2009) as well as inhibit activation of the MAPK pathway, JNK, p38 MAPK, and ERK in human keratinocytes or mouse adipocytes (Cho et al. 2007; Ahn et al. 2010). Resveratrol increases AMPK and PGC-1α activity independent of Sir2 (Baur et al. 2006).

10.4 Conclusion

In this chapter, we summarized the results of current research on life span extension by phytochemicals in *Drosophila* and other organisms. Plant-derived chemicals and drugs have attracted a great deal of attention as reagents that can enhance quality of life and improve health. However, our understanding of the mechanisms underlying the effects of these phytochemicals is lacking. In addition, the impact of a wide range of plant-derived compounds on health and aging remains to be demonstrated along with the molecular mechanisms underlying their pro-longevity effects. Furthermore, the dose of phytochemicals required to exert beneficial effects on health and aging needs to be identified. Additional studies should be conducted to determine whether or not the effects of these compounds are similar among species. Finally, more systematic and integrated assays should be performed to identify and characterize phytochemicals that can extend life span as well as better understand how these reagents can improve quality of life.

Acknowledgments This work was supported by LG Yonam Foundation Professors' Oversea Research Program.

References

Adler AS, Sinha S, Kawahara TL, Zhang JY, Segal E, Chang HY (2007) Motif module map reveals enforcement of aging by continual NF-kappaB activity. Genes Dev 21(24):3244–3257

Aggarwal BB (2010) Targeting inflammation-induced obesity and metabolic diseases by curcumin and other nutraceuticals. Annu Rev Nutr 30:173–199

Aggarwal BB, Shishodia S (2004) Suppression of the nuclear factor-kappaB activation pathway by spice-derived phytochemicals: reasoning for seasoning. Ann N Y Acad Sci 1030:434–441

Ahn J, Lee H, Kim S, Ha T (2010) Curcumin-induced suppression of adipogenic differentiation is accompanied by activation of Wnt/beta-catenin signaling. Am J Physiol 298(6):C1510–C1516

Ammon-Treiber S, Grecksch G, Stumm R, Riechert U, Tischmeyer H, Reichenauer A, Hollt V (2004) Rapid, transient, and dose-dependent expression of hsp70 messenger RNA in the rat brain after morphine treatment. Cell Stress Chaperones 9(2):182–197

Anisimov VN, Piskunova TS, Popovich IG, Zabezhinski MA, Tyndyk ML, Egormin PA, Yurova MV, Rosenfeld SV, Semenchenko AV, Kovalenko IG, Poroshina TE, Berstein LM (2010) Gender differences in metformin effect on aging, life span and spontaneous tumorigenesis in 129/Sv mice. Aging 2(12):945–958

Antosh M, Whitaker R, Kroll A, Hosier S, Chang C, Bauer J, Cooper L, Neretti N, Helfand SL (2011) Comparative transcriptional pathway bioinformatic analysis of dietary restriction, Sir2, p53 and resveratrol life span extension in *Drosophila*. Cell cycle Georgetown, Tex 10(6):904–911

Ashrafi K, Lin SS, Manchester JK, Gordon JI (2000) Sip2p and its partner SNF1p kinase affect aging in S. cerevisiae. Genes Dev 14(15):1872–1885

Aviram M, Dornfeld L (2001) Pomegranate juice consumption inhibits serum angiotensin converting enzyme activity and reduces systolic blood pressure. Atherosclerosis 158(1):195–198

Baker RG, Hayden MS, Ghosh S (2011) NF-kappaB, inflammation, and metabolic disease. Cell Metab 13(1):11–22

Bass TM, Weinkove D, Houthoofd K, Gems D, Partridge L (2007) Effects of resveratrol on lifespan in drosophila melanogaster and caenorhabditis elegans. Mech Ageing Dev 128(10):546–552

Bauer JH, Goupil S, Garber GB, Helfand SL (2004) An accelerated assay for the identification of lifespan-extending interventions in *Drosophila melanogaster*. Proc Natl Acad Sci USA 101(35):12980–12985

Baur JA (2010) Biochemical effects of SIRT1 activators. Biochim Biophys Acta 1804(8):1626–1634

Baur JA, Pearson KJ, Price NL, Jamieson HA, Lerin C, Kalra A, Prabhu VV, Allard JS, Lopez-Lluch G, Lewis K, Pistell PJ, Poosala S, Becker KG, Boss O, Gwinn D, Wang M, Ramaswamy S, Fishbein KW, Spencer RG, Lakatta EG, Le Couteur D, Shaw RJ, Navas P, Puigserver P, Ingram DK, de Cabo R, Sinclair DA (2006) Resveratrol improves health and survival of mice on a high-calorie diet. Nature 444(7117):337–342

Bishop NA, Guarente L (2007) Two neurons mediate diet-restriction-induced longevity in *C. elegans*. Nature 447(7144):545–549

Borra MT, Smith BC, Denu JM (2005) Mechanism of human SIRT1 activation by resveratrol. J Biol Chem 280(17):17187–17195

Calabrese V, Cornelius C, Dinkova-Kostova AT, Iavicoli I, Di Paola R, Koverech A, Cuzzocrea S, Rizzarelli E, Calabrese EJ (2012) Cellular stress responses, hormetic phytochemicals and vitagenes in aging and longevity. Biochim Biophys Acta 1822(5):753–783

Chandrashekara KT, Shakarad MN (2011) Aloe vera or resveratrol supplementation in larval diet delays adult aging in the fruit fly *Drosophila melanogaster*. J Gerontol 66(9):965–971

Chen C, Yu R, Owuor ED, Kong AN (2000) Activation of antioxidant-response element (ARE), mitogen-activated protein kinases (MAPKs) and caspases by major green tea polyphenol components during cell survival and death. Arch Pharmacal Res 23(6):605–612

Cho JW, Lee KS, Kim CW (2007) Curcumin attenuates the expression of IL-1beta, IL-6, and TNF-alpha as well as cyclin E in TNF-alpha-treated HaCaT cells; NF-kappaB and MAPKs as potential upstream targets. Int J Mol Med 19(3):469–474

Cho SY, Cho M, Seo DB, Lee SJ, Suh Y (2013) Identification of a small molecule activator of SIRT1 gene expression. Aging 5(3):174–182

Chung S, Yao H, Caito S, Hwang JW, Arunachalam G, Rahman I (2010) Regulation of SIRT1 in cellular functions: role of polyphenols. Arch Biochem Biophys 501(1):79–90

Clifford MN, van der Hooft JJ, Crozier A (2013) Human studies on the absorption, distribution, metabolism, and excretion of tea polyphenols. The Am J Clin Nutr 98(6 Suppl):1619S–1630S

Curtis R, O'Connor G, DiStefano PS (2006) Aging networks in *Caenorhabditis elegans*: AMP-activated protein kinase (aak-2) links multiple aging and metabolism pathways. Aging Cell 5(2):119–126

De Haes W, Frooninckx L, Van Assche R, Smolders A, Depuydt G, Billen J, Braeckman BP, Schoofs L, Temmerman L (2014) Metformin promotes lifespan through mitohormesis via the peroxiredoxin PRDX-2. Proc Natl Acad Sci USA 111(24):E2501–E2509

Dubiley TA, Rushkevich YE, Koshel NM, Voitenko VP, Vaiserman AM (2011) Life span extension in Drosophila melanogaster induced by morphine. Biogerontology 12(3):179–184

Ehrnhoefer DE, Duennwald M, Markovic P, Wacker JL, Engemann S, Roark M, Legleiter J, Marsh JL, Thompson LM, Lindquist S, Muchowski PJ, Wanker EE (2006) Green tea (-)-epigallocatechin-gallate modulates early events in Huntington misfolding and reduces toxicity in Huntington's disease models. Hum Mol Genet 15(18):2743–2751

Franceschi C, Bonafe M, Valensin S, Olivieri F, De Luca M, Ottaviani E, De Benedictis G (2000) Inflamm-aging. An evolutionary perspective on immunosenescence. Ann N Y Acad Sci 908:244–254

Funakoshi M, Tsuda M, Muramatsu K, Hatsuda H, Morishita S, Aigaki T (2011) A gain-of-function screen identifies *WDB* and *LKB1* as lifespan-extending genes in *Drosophila*. Biochem Biophys Res Commun 405(4):667–672

Giacalone M, Di Sacco F, Traupe I, Topini R, Forfori F, Giunta F (2011) Antioxidant and neuroprotective properties of blueberry polyphenols: a critical review. Nutr Neurosci 14(3):119–125

Greer EL, Dowlatshahi D, Banko MR, Villen J, Hoang K, Blanchard D, Gygi SP, Brunet A (2007) An AMPK-FOXO pathway mediates longevity induced by a novel method of dietary restriction in *C. elegans*. Curr Biol 17(19):1646–1656

Hoffmann J, Romey R, Fink C, Yong L, Roeder T (2013) Overexpression of Sir2 in the adult fat body is sufficient to extend lifespan of male and female *Drosophila*. Aging 5(4):315–327

Hong F, Freeman ML, Liebler DC (2005) Identification of sensor cysteines in human Keap1 modified by the cancer chemopreventive agent sulforaphane. Chem Res Toxicol 18(12):1917–1926

Howitz KT, Bitterman KJ, Cohen HY, Lamming DW, Lavu S, Wood JG, Zipkin RE, Chung P, Kisielewski A, Zhang LL, Scherer B, Sinclair DA (2003) Small molecule activators of sirtuins extend *Saccharomyces cerevisiae* lifespan. Nature 425(6954):191–196

Imai S, Armstrong CM, Kaeberlein M, Guarente L (2000) Transcriptional silencing and longevity protein Sir2 is an NAD-dependent histone deacetylase. Nature 403(6771):795–800

Itoh K, Chiba T, Takahashi S, Ishii T, Igarashi K, Katoh Y, Oyake T, Hayashi N, Satoh K, Hatayama I, Yamamoto M, Nabeshima Y (1997) An NRF2/small MAF heterodimer mediates the induction of phase II detoxifying enzyme genes through antioxidant response elements. Biochem Biophy Res Commun 236(2):313–322

Jafari M (2010) *Drosophila melanogaster* as a model system for the evaluation of anti-aging compounds. Fly 4(3):253–257

Jeong HW, Hsu KC, Lee JW, Ham M, Huh JY, Shin HJ, Kim WS, Kim JB (2009) Berberine suppresses proinflammatory responses through AMPK activation in macrophages. Am J Physiol 296(4):E955–E964

Jeong WS, Keum YS, Chen C, Jain MR, Shen G, Kim JH, Li W, Kong AN (2005) Differential expression and stability of endogenous nuclear factor E2-related factor 2 (NRF2) by natural chemopreventive compounds in HepG2 human hepatoma cells. J Biochem Mol Biol 38(2):167–176

Kaeberlein M, McDonagh T, Heltweg B, Hixon J, Westman EA, Caldwell SD, Napper A, Curtis R, DiStefano PS, Fields S, Bedalov A, Kennedy BK (2005) Substrate-specific activation of sirtuins by resveratrol. J Biol Chem 280(17):17038–17045

Kensler TW, Wakabayashi N, Biswal S (2007) Cell survival responses to environmental stresses via the Keap1-NRF2-ARE pathway. Annu Rev Pharmacol Toxicol 47:89–116

Kim JE, You DJ, Lee C, Ahn C, Seong JY, Hwang JI (2010) Suppression of NF-kappaB signaling by KEAP1 regulation of IKKbeta activity through autophagic degradation and inhibition of phosphorylation. Cell Signal 22(11):1645–1654

Kitani K, Yokozawa T, Osawa T (2004) Interventions in aging and age-associated pathologies by means of nutritional approaches. Ann NY Acad Sci 1019:424–426

Lagouge M, Argmann C, Gerhart-Hines Z, Meziane H, Lerin C, Daussin F, Messadeq N, Milne J, Lambert P, Elliott P, Geny B, Laakso M, Puigserver P, Auwerx J (2006) Resveratrol improves mitochondrial function and protects against metabolic disease by activating SIRT1 and PGC-1alpha. Cell 127(6):1109–1122

Lee J, Kim MS, Park C, Jung EB, Choi DH, Kim TY, Moon SK, Park R (2004) Morphine prevents glutamate-induced death of primary rat neonatal astrocytes through modulation of intracellular redox. Immunopharmacol Immunotoxicol 26(1):17–28

Lee KS, Lee BS, Semnani S, Avanesian A, Um CY, Jeon HJ, Seong KM, Yu K, Min KJ, Jafari M (2010) Curcumin extends life span, improves health span, and modulates the expression of age-associated aging genes in *Drosophila melanogaster*. Rejuvenation Res 13(5):561–570

Lee YK, Lee WS, Hwang JT, Kwon DY, Surh YJ, Park OJ (2009) Curcumin exerts antidifferentiation effect through AMPKalpha-PPAR-gamma in 3T3-L1 adipocytes and antiproliferatory effect through AMPKalpha-COX-2 in cancer cells. J Agric Food Chem 57(1):305–310

Li YM, Chan HY, Huang Y, Chen ZY (2007) Green tea catechins upregulate superoxide dismutase and catalase in fruit flies. Mol Nutr Food Res 51(5):546–554

Li YM, Chan HY, Yao XQ, Huang Y, Chen ZY (2008) Green tea catechins and broccoli reduce fat-induced mortality in *Drosophila melanogaster*. J Nutr Biochem 19(6):376–383

Lin SJ, Defossez PA, Guarente L (2000) Requirement of NAD and SIR2 for life-span extension by calorie restriction in Saccharomyces cerevisiae. Science (New York, NY) 289(5487):2126–2128

Lin X, Li Q, Wang YJ, Ju YW, Chi ZQ, Wang MW, Liu JG (2007) Morphine inhibits doxorubicin-induced reactive oxygen species generation and nuclear factor kappaB transcriptional activation in neuroblastoma SH-SY5Y cells. Biochem J 406(2):215–221

Loguinov AV, Anderson LM, Crosby GJ, Yukhananov RY (2001) Gene expression following acute morphine administration. Physiol Genomics 6(3):169–181

Maher P, Dargusch R, Bodai L, Gerard PE, Purcell JM, Marsh JL (2011) ERK activation by the polyphenols fisetin and resveratrol provides neuroprotection in multiple models of Huntington's disease. Hum Mol Genet 20(2):261–270

Mak JC (2012) Potential role of green tea catechins in various disease therapies: progress and promise. Clin Exp Pharmacol Physiol 39(3):265–273

Malaguti M, Angeloni C, Hrelia S (2013) Polyphenols in exercise performance and prevention of exercise-induced muscle damage. Oxidative Med Cell Longevity 2013:825928

Marchal J, Pifferi F, Aujard F (2013) Resveratrol in mammals: effects on aging biomarkers, age-related diseases, and life span. Ann NY Acad Sci 1290:67–73

Martin-Montalvo A, Mercken EM, Mitchell SJ, Palacios HH, Mote PL, Scheibye-Knudsen M, Gomes AP, Ward TM, Minor RK, Blouin MJ, Schwab M, Pollak M, Zhang Y, Yu Y, Becker KG, Bohr VA, Ingram DK, Sinclair DA, Wolf NS, Spindler SR, Bernier M, de Cabo R (2013) Metformin improves healthspan and lifespan in mice. Nature Commun 4:2192

Mattson MP, Son TG, Camandola S (2007) Viewpoint: mechanisms of action and therapeutic potential of neurohormetic phytochemicals. Dose Response 5(3):174–186

Na HK, Surh YJ (2008) Modulation of NRF2-mediated antioxidant and detoxifying enzyme induction by the green tea polyphenol EGCG. Food Chem Toxicol 46(4):1271–1278

Niu Y, Na L, Feng R, Gong L, Zhao Y, Li Q, Li Y, Sun C (2013) The phytochemical, EGCG, extends lifespan by reducing liver and kidney function damage and improving age-associated inflammation and oxidative stress in healthy rats. Aging Cell 12(6):1041–1049

Onken B, Driscoll M (2010) Metformin induces a dietary restriction-like state and the oxidative stress response to extend *C. elegans* healthspan via AMPK, LKB1, and SKN-1. PLoS ONE 5(1):e8758

Pallauf K, Giller K, Huebbe P, Rimbach G (2013) Nutrition and healthy ageing: calorie restriction or polyphenol-rich "MediterrAsian" diet? Oxidative Med Cell Longevity 2013:707421

Peng C, Chan HY, Li YM, Huang Y, Chen ZY (2009) Black tea theaflavins extend the lifespan of fruit flies. Exp Gerontol 44(12):773–783

Peng C, Chan HY, Huang Y, Yu H, Chen ZY (2011) Apple polyphenols extend the mean lifespan of *Drosophila melanogaster*. J Agric Food Chem 59(5):2097–2106

Pietsch K, Saul N, Menzel R, Sturzenbaum SR, Steinberg CE (2009) Quercetin mediated lifespan extension in *Caenorhabditis elegans* is modulated by age-1, daf-2, sek-1 and unc-43. Biogerontology 10(5):565–578

Pirola L, Frojdo S (2008) Resveratrol: one molecule, many targets. IUBMB Life 60(5):323–332

Prasad S, Ravindran J, Aggarwal BB (2010) NF-kappaB and cancer: how intimate is this relationship. Mol Cell Biochem 336(1–2):25–37

Qian L, Tan KS, Wei SJ, Wu HM, Xu Z, Wilson B, Lu RB, Hong JS, Flood PM (2007) Microglia-mediated neurotoxicity is inhibited by morphine through an opioid receptor-independent reduction of NADPH oxidase activity. J Immunol 179(2):1198–1209

Qin B, Nagasaki M, Ren M, Bajotto G, Oshida Y, Sato Y (2003) Cinnamon extract (traditional herb) potentiates in vivo insulin-regulated glucose utilization via enhancing insulin signaling in rats. Diabetes Res Clin Pract 62(3):139–148

Reddy NM, Kleeberger SR, Yamamoto M, Kensler TW, Scollick C, Biswal S, Reddy SP (2007) Genetic dissection of the NRF2-dependent redox signaling-regulated transcriptional programs of cell proliferation and cytoprotection. Physiol Genomics 32(1):74–81

Rogina B, Helfand SL (2004) Sir2 mediates longevity in the fly through a pathway related to calorie restriction. Proc Natl Acad Sci USA 101(45):15998–16003

Rovillain E, Mansfield L, Caetano C, Alvarez-Fernandez M, Caballero OL, Medema RH, Hummerich H, Jat PS (2011) Activation of nuclear factor-kappa B signalling promotes cellular senescence. Oncogene 30(20):2356–2366

Rushworth SA, Ogborne RM, Charalambos CA, O'Connell MA (2006) Role of protein kinase C delta in curcumin-induced antioxidant response element-mediated gene expression in human monocytes. Biochem Biophys Res Commun 341(4):1007–1016

Salminen A, Hyttinen JM, Kaarniranta K (2011a) AMP-activated protein kinase inhibits NF-kappaB signaling and inflammation: impact on healthspan and lifespan. J Mole Med (Berlin, Germany) 89(7):667–676

Salminen A, Kauppinen A, Suuronen T, Kaarniranta K (2008a) SIRT1 longevity factor suppresses NF-kappaB -driven immune responses: regulation of aging via NF-kappaB acetylation? BioEssays 30(10):939–942

Salminen A, Ojala J, Huuskonen J, Kauppinen A, Suuronen T, Kaarniranta K (2008b) Interaction of aging-associated signaling cascades: inhibition of NF-kappaB signaling by longevity factors FoxOs and SIRT1. Cell Mol Life Sci 65(7–8):1049–1058

Salminen A, Kauppinen A, Kaarniranta K (2011b) Phytochemicals suppress nuclear factor-kappaB signaling: impact on health span and the aging process. Curr Opin Clin Nutr Metab Care 15(1):23–28

Saul N, Pietsch K, Menzel R, Steinberg CE (2008) Quercetin-mediated longevity in *Caenorhabditis elegans*: is DAF-16 involved? Mech Ageing Dev 129(10):611–613

Shen LR, Parnell LD, Ordovas JM, Lai CQ (2013a) Curcumin and aging. BioFactors (Oxford, England) 39(1):133–140

Shen LR, Xiao F, Yuan P, Chen Y, Gao QK, Parnell LD, Meydani M, Ordovas JM, Li D, Lai CQ (2013b) Curcumin-supplemented diets increase superoxide dismutase activity and mean lifespan in Drosophila. Age (Dordrecht, Netherlands) 35(4):1133–1142

Si H, Fu Z, Babu PV, Zhen W, Leroith T, Meaney MP, Voelker KA, Jia Z, Grange RW, Liu D (2011) Dietary epicatechin promotes survival of obese diabetic mice and *Drosophila melanogaster*. J Nutr 141(6):1095–1100

Slack C, Foley A, Partridge L (2012) Activation of AMPK by the putative dietary restriction mimetic metformin is insufficient to extend lifespan in *Drosophila*. PLoS ONE 7(10):e47699

Smith DL Jr, Elam CF Jr, Mattison JA, Lane MA, Roth GS, Ingram DK, Allison DB (2010) Metformin supplementation and life span in Fischer-344 rats. J Gerontol 65(5):468–474

Soh JW, Marowsky N, Nichols TJ, Rahman AM, Miah T, Sarao P, Khasawneh R, Unnikrishnan A, Heydari AR, Silver RB, Arking R (2013) Curcumin is an early-acting stage-specific inducer of extended functional longevity in *Drosophila*. Exp Gerontol 48(2):229–239

Spindler SR, Mote PL, Flegal JM, Teter B (2013) Influence on longevity of blueberry, cinnamon, green and black tea, pomegranate, sesame, curcumin, morin, pycnogenol, quercetin, and taxifolin fed iso-calorically to long-lived, F1 hybrid mice. Rejuvenation Res 16(2):143–151

Steinberg GR, Kemp BE (2009) AMPK in Health and Disease. Physiol Rev 89(3):1025–1078

Stenesen D, Suh JM, Seo J, Yu K, Lee KS, Kim JS, Min KJ, Graff JM (2013) Adenosine nucleotide biosynthesis and AMPK regulate adult life span and mediate the longevity benefit of caloric restriction in flies. Cell Metab 17(1):101–112

Suckow BK, Suckow MA (2006) Lifespan extension by the antioxidant curcumin in *Drosophila melanogaster*. Int J Biomed Sci 2(4):402–405

Surh YJ, Kundu JK, Na HK (2008) NRF2 as a master redox switch in turning on the cellular signaling involved in the induction of cytoprotective genes by some chemopreventive phytochemicals. Planta Med 74(13):1526–1539

Sykiotis GP, Bohmann D (2008) Keap1/NRF2 signaling regulates oxidative stress tolerance and lifespan in *Drosophila*. Dev Cell 14(1):76–85

Sykiotis GP, Habeos IG, Samuelson AV, Bohmann D (2011) The role of the antioxidant and longevity-promoting NRF2 pathway in metabolic regulation. Curr Opin Clin Nutr Metab Care 14(1):41–48

Tissenbaum HA, Guarente L (2001) Increased dosage of a sir-2 gene extends lifespan in Caenorhabditis elegans. Nature 410(6825):227–230

Tohyama D, Yamaguchi A (2010) A critical role of *SNF1A/dAMPKalpha* (*Drosophila AMP-activated protein kinase alpha*) in muscle on longevity and stress resistance in *Drosophila melanogaster*. Biochemical and biophysical research communications 394(1):112–118

Tullet JM, Hertweck M, An JH, Baker J, Hwang JY, Liu S, Oliveira RP, Baumeister R, Blackwell TK (2008) Direct inhibition of the longevity-promoting factor SKN-1 by insulin-like signaling in *C. elegans*. Cell 132(6):1025–1038

Valenzano DR, Terzibasi E, Genade T, Cattaneo A, Domenici L, Cellerino A (2006) Resveratrol prolongs lifespan and retards the onset of age-related markers in a short-lived vertebrate. Curr Biol 16(3):296–300

Wang C, Wheeler CT, Alberico T, Sun X, Seeberger J, Laslo M, Spangler E, Kern B, de Cabo R, Zou S (2013) The effect of resveratrol on lifespan depends on both gender and dietary nutrient composition in Drosophila melanogaster. Age (Dordrecht, Netherlands) 35(1):69–81

Wood JG, Rogina B, Lavu S, Howitz K, Helfand SL, Tatar M, Sinclair D (2004) Sirtuin activators mimic caloric restriction and delay ageing in metazoans. Nature 430(7000):686–689

Wu CC, Hsu MC, Hsieh CW, Lin JB, Lai PH, Wung BS (2006) Upregulation of heme oxygenase-1 by Epigallocatechin-3-gallate via the phosphatidylinositol 3-kinase/Akt and ERK pathways. Life Sci 78(25):2889–2897

Xiang L, Nakamura Y, Lim YM, Yamasaki Y, Kurokawa-Nose Y, Maruyama W, Osawa T, Matsuura A, Motoyama N, Tsuda L (2011) Tetrahydrocurcumin extends life span and inhibits the oxidative stress response by regulating the FOXO forkhead transcription factor. Aging 3(11):1098–1109

Yang CS, Chen G, Wu Q (2014) Recent scientific studies of a traditional chinese medicine, tea, on prevention of chronic diseases. Journal of traditional and complementary medicine 4(1):17–23

Zhang L, Jie G, Zhang J, Zhao B (2009) Significant longevity-extending effects of EGCG on Caenorhabditis elegans under stress. Free Radic Biol Med 46(3):414–421

Zhang ZG, Niu XY, Lu AP (2013) Xiao GG (2013) effect of curcumin on aged *Drosophila Melanogaster*: a pathway prediction analysis. Chin J Integr Med 23:0415–1672

Chapter 11
Life Extension in *Drosophila* by Histone Deacetylase Inhibitors

Alexander M. Vaiserman and Elena G. Pasyukova

Abstract In the last years, epigenetic regulatory mechanisms are increasingly appreciated as central to a diverse array of age-associated processes such as cellular and organismal senescence, genomic instability, and tumorigenesis. Recently, histone deacetylase inhibitors (HDACIs), a novel class of drugs targeting epigenetic pathways, have been proposed as a highly promising type of drugs with anti-aging effects. This chapter presents an overview of the anti-aging and life-extending effects of HDACIs such as phenilbutyrate, sodium butyrate, trichistatin A and suberoylanilide hydroxamic acid as well as their plausible mechanism(s) of action in *Drosophila melanogaster*. Data supporting the hypothesis that life span extension induced by HDACIs may be caused by generalized changes in epigenetic regulation of gene expression are discussed. Overall, findings reviewed in this chapter suggest that uncovering which genetic factors and signaling pathways contributing to healthy aging can be influenced by HDACIs in fruit fly may facilitate the development of new strategies for treating and preventing age-related human diseases and health span extension.

Keywords Histone deacetylase inhibitors · Epigenetic mechanisms · Aging · Life span extension · Fruit fly

A.M. Vaiserman (✉)
Laboratory of Epigenetics, Institute of Gerontology, Vyshgorodskaya St. 67, Kiev 04114, Ukraine
e-mail: vaiserman@geront.kiev.ua

E.G. Pasyukova
Laboratory of Genome Variation, Institute of Molecular Genetics of RAS, Kurchatov Sq. 2, Moscow, Russia
e-mail: egpas@rambler.ru

A.M. Vaiserman et al. (eds.), *Life Extension*, Healthy Ageing and Longevity 3,
DOI 10.1007/978-3-319-18326-8_11

11.1 Introduction

Age-related senescence is a process immanent to all living beings, and it is characterized by the gradual loss of physiological functions and is accompanied by decreasing fertility and increased risk of mortality with advancing age. Recent advances in biogerontology, however, give hope that senescence may be postponed and/or prevented by certain approaches. An increasing number of pharmacological and dietary interventions are being identified, suggested to have the anti-aging and life-extending effects (Dominguez et al. 2009; Kapoor et al. 2009).

In the last years, epigenetic regulatory mechanisms are increasingly appreciated as central to a diverse array of age-associated processes such as cellular and organismal senescence, genomic instability, and tumorigenesis (Muñoz-Najar and Sedivy 2011; Boyd-Kirkup et al. 2013; Wood and Helfand 2013). Epigenetic modifications refer to heritable but reversible changes in chromatin structure and gene function that occur without the change in the primary DNA sequence. The main epigenetic mechanisms include DNA methylation, modifications of histones that package the DNA, and microRNA regulatory pathways. The epigenetic processes normally are finely balanced throughout the life span. However, the epigenetic patterns can be significantly disrupted in abnormal cells, such as pre-cancerous or cancerous cells. The rate of aging may be influenced by a myriad of environmental cues that can be 'remembered' throughout the life span due to the changes in the epigenome (Boyd-Kirkup et al. 2013). The global DNA demethylation and promoter region-specific methylation of several specific genes including tumor suppressor genes are among the key age-associated epigenetic processes (Issa 1999). Epigenetic dysregulation has been shown to be implicated in a wide variety of age-related chronic diseases such as decline of immune function, atherosclerosis, type 2 diabetes, cancer, and neurodegenerative and psychiatric diseases (Berdasco and Esteller 2012).

It is noteworthy that, in contrast to genetic changes (mutations) that cannot be restored, epigenetic aberrations are reversible and can be relatively easily corrected (Wood and Helfand 2013; Tollefsbol 2014). The potential reversibility of epigenetic aberrations, which are induced by adverse environmental exposures, through nutritional or pharmacological interventions makes them attractive targets for therapeutic drug development. Recently, a novel class of drugs targeting epigenetic pathways ('epigenetic drugs') have been proposed as a highly promising class of drugs with anti-aging effects (Vaiserman and Pasyukova 2012). Among them, some members of superfamilies of histone deacetylases (HDACs) 1-11 and sirtuins (SIRTs) 1-7 are presently in the main focus of research activities.

At the present time, screening of agents with potential anti-aging properties is performed on invertebrate models (Lucanic et al. 2013). The fruit fly *Drosophila* has numerous advantages for such studies including its relatively short life span, ease of maintenance, a large variety of environmental and genetic manipulations that alter life span, availability of stocks containing altered genes, sequence of the full *Drosophila* genome, and clear distinction between developmental and adult stages (Helfand and Rogina 2003). The main focus of this chapter is a review of the literature describing the anti-aging and life-extending effects of synthetic inhibitors of HDAC activity and elucidating their mechanism(s) of action in *Drosophila melanogaster*.

11.2 Role of Histone Modification in Epigenetic Regulation in Drosophila

Two major epigenetic mechanisms influencing gene expression throughout the eukaryotic life cycle, including aging, are methylation of DNA cytosine residues and modification of histones (Huidobro et al. 2013). In some species, however the picture is not as straightforward. A question on the role of DNA methylation in fruit fly has been discussed controversially for many years. It has been assumed until recently that genomic DNA in *Drosophila* is completely unmethylated (Lyko et al. 2006). This assumption was based on the fact that most research has failed to detect methylated bases in the flies' genomes during their pupal to adult stages. The analysis of genomic DNA in early embryos, however, revealed low but significant levels of cytosine methylation. Thus, DNA methylation in *Drosophila*, in contrast to those in vertebrates, appears to be a transient epigenetic signal during early developmental stages, but not in adulthood. The obvious role of histone modifications in epigenetic regulation in *Drosophila*, in contrast, has been demonstrated in many studies (Boros 2012; Swaminathan et al. 2012).

The highly conserved core histones (H2A, H2B, H3, and H4) contain lysine-rich N-terminal tails. These residues are prime targets for a number of covalent post-translational modifications, such as acetylation, phosphorylation, ubiquitylation, biotinylation, sumoylation, and others (Kouzarides 2007). Modifications of residues in histone tails alter the histone–DNA interaction and create a "histone code" that coordinates the recruitment of transcription factors and polymerases, and regulates chromatin structure and gene expression. Among all known histone modifications, acetylation has the highest potential to induce chromatin unfolding, as it neutralizes the electrostatic interaction between the histone and the negatively charged DNA, making it more accessible to the transcriptional apparatus (Zentner and Henikoff 2013).

Both histone acetylation and deacetylation play a crucial role in chromatin remodeling and, thus, in gene expression. In normal, non-transformed cells, there is a fine balance between acetylation and deacetylation of histones (Davie and Spencer 1999). Histone acetylation is associated with an open chromatin and activation of gene expression, while histone deacetylation is associated with closed chromatin and repression of transcription. The enzymes catalyzing these modifications by addition or removal of acetyl groups to the tails of histone octamers are histone acetyltransferases (HATs) and HDACs, respectively (Kuo and Allis 1998). HATs catalyze the transfer of the acetyl moiety from acetyl coenzyme A to the ε-amino groups of histone lysine residues, thereby neutralizing the positive charge of the histone tails and reducing their affinity for DNA. This results in a more open chromatin state and greater access of DNA to transcription factors. HDACs, on the contrary, catalyze the removal of acetyl groups from lysine residues of histone tails, resulting in a more condensed, transcriptionally repressive chromatin conformation (Witt et al. 2009). HDACs balance the acetylation activities of HATs in chromatin remodeling and play a key role in cell proliferation, migration and apoptosis, as well as immune functions and angiogenesis (Zhou et al. 2011). There are four main classes of HDACs. In *Drosophila*, Class I HDACs include RPD3 and dHDAC3, Class II include dHDAC4 and dHDAC6, and Class III include dSIR2, dSIRT2, dSIRT4, dSIRT6 dSIRT7 (Chang and Min 2002). SIR2 protein shows NAD+-dependent HDAC activity (Imai et al. 2000). Finally, HDAC Class IV is represented by a single member, HDAC11 (Gao et al. 2002).

Many recent studies revealed a role of chromatin modification in both developmental and adult stages of life cycle, including aging. The heterochromatin loss model of aging proposed by Villeponteau (1997) suggests that heterochromatin domains are set up early in embryogenesis but then are gradually lost with aging, which results in aberrant gene expression associated with old age. An association between heterochromatic silencing and longevity has been revealed in various experimental models including yeast, *Caenorhabditis elegans*, mice and *Drosophila* (Wood and Helfand 2013). A dramatic reorganization of chromosomal regions with age in fruit flies was found in a whole genome study by Wood et al. (2010). An overall decline of the active chromatin marks, such as H3K4me3 and H3K36me3, as well as a significant decrease in the enrichment of the repressive heterochromatin H3K9me3 and heterochromatin protein 1 (HP1) marks at pericentric heterochromatin loci have been found with age. Such extensive alterations in repressive chromatin state were associated with age-related changes in gene expression. In a Larson et al. (2012) study conducted in transgenic *Drosophila* lines with genetically manipulated HP1 levels, the decreased heterochromatin levels were associated with dramatically shortened longevity, while increased heterochromatin levels resulted in extended life span compared to controls. These changes in the life span were associated with changes in muscle integrity. Specifically, HP1-overexpressing flies showed increased muscle function and structure with age, whereas flies with decreased HP1 levels demonstrated premature muscle degeneration.

11.3 Effects of HDAC Inhibitors on Life Span and Associated Life History Traits in *Drosophila Melanogaster*

Among the compounds affecting chromatin structure and gene expression through modulation of HDAC activity, HDAC inhibitors (HDACIs) seem the most promising agents for anti-aging therapies. A decrease in HDAC activity generally results in gene up-regulation. A decline in transcription of many genes, primarily metabolic and biosynthetic genes, is known to be observed in old age (Seroude et al. 2002). Therefore, there is hope that HDACIs will delay aging due to the preservation of the level of transcription of these genes, which is characteristic of the young, in aging individuals. HDAC inhibition can also result in an up-regulation of longevity-associated genes, such as inflammatory response and stress response genes. HDACIs, due to their anti-proliferative activity, are currently being investigated in human clinical trials as a new generation of anticancer therapeutics (Benedetti et al. 2014; Slingerland et al. 2014; West and Johnstone 2014). The antitumor effects of HDACIs are suggested to be attributed to both transcriptional repression of proto-oncogenes and transcriptional reactivation of silent tumor suppressor genes (Boumber and Issa 2011). HDACIs are considered to be very promising candidates in cancer treatments since these agents, targeting different malignant pathways, result in a preferential killing of the neoplastic cells, but they are relatively non-toxic to normal cells (Johnstone 2002).

Apart from their clinical application in oncology, HDACIs have recently been evaluated as potential therapeutics for many chronic pathological conditions including cardiovascular disorders (Baltan et al. 2013). Supportive evidence is obtained for their potent immunomodulatory and anti-inflammatory effects (Licciardi et al. 2013). HDACIs have been also identified as candidate drugs for the treatment of neurodegenerative disorders such as Parkinson's, Alzheimer's and Huntington's diseases (Abel and Zukin 2008; Hahnen et al. 2008).

High hopes were also placed on the potential therapeutic applications of modulators of NAD-dependent class III HDACs (also known as SIRTs) that play central roles in cell survival, inflammation, energy metabolism, and aging (Sinclair and Guarente 2014). SIRTs are key regulators of many important processes implicated in aging, such as DNA damage, genome stability, stress response, cell proliferation, differentiation and apoptosis, metabolism, energy homeostasis, and cancer (Satoh et al. 2011; Yuan et al. 2013). Since SIRTs are crucial to pathways that counter the age-associated health decline, pharmacological agents that modulate SIRT activity are expected to have clinical potential in preventing and/or treating many chronic conditions, including cardiovascular, metabolic and neurodegenerative diseases, arthritis, and cancer (Morris 2013).

In recent years, experimental research has emerged on the life-extending potential of synthetic HDACIs although several natural compounds, such as sulforaphane contained in broccoli, curcumin extracted from turmeric and garlic-derived diallyl disulfide seem to be very promising as well. The most commonly used HDACIs are listed in Table 11.1.

In *Drosophila*, each HDAC was shown to regulate transcription of a unique set of genes and to have a distinct pattern of temporal expression (Cho et al. 2005). Furthermore, a differential sensitivity of HDACs to HDACIs has been shown. The research findings supporting the anti-aging and life-extending properties of HDACIs in *Drosophila melanogaster* are reviewed in the sections below.

Table 11.1 Overview of most widely used HDACIs

Class	Compound name	HDAC specificity class
Short-chain fatty acids	Sodium Butyrate (SB)	I, II
	Phenylbutyrate (PBA)	I, II
	Valproic Acid (VPA)	I, II
Hydroxamic acids	Trichostatin A (TSA)	I, II, IV
	Vorinostat (suberoylanilide hydroxamic acid, SAHA)	I, II, IV
	Givinostat (ITF2357)	I, II
	Abexinostat (PCI-24781)	
	Belinostat (PXD101)	I, II, IV
	Panobinostat (LBH589)	I, II, IV
	Resminostat (4SC-201)	I, II, IV
	Quisinostat (JNJ-26481585)	I, II, IV
Cyclic peptides	Depsipeptide (romidepsin)	I
	Apicidin	I, II
Benzamides	Entinostat (MS-275)	I, II
	Mocetinostat (MGCD0103)	I

11.3.1 Phenylbutyrate

Sodium 4-phenylbutyrate (PBA) is a general HDAC inhibitor with potential antineoplastic activity. This chemical was shown to inhibit class I and II HDACs, which lead to elevated gene expression, reduced cellular proliferation, induction of apoptosis, and the enhanced cell differentiation in neoplastic cell populations (Iannitti and Palmieri 2011).

Kang et al. (2002) were the first reporting the life-extending potential of the sodium salt of PBA in *Drosophila melanogaster*. Feeding of flies with PBA resulted in a substantial extension of both mean and maximal life span by up to 30–50 % regardless of the fly's genetic background, without diminution of locomotor activity and resistance to stress. This result is not due to caloric restriction, known to extend life span in different model organisms, as it is evident from the similarity of weight and size between the flies fed with or without PBA. The effects obtained are also unlikely due to the decrease in reproductive activity, as is evident from the similar numbers of produced eggs and percentages of fertile eggs, and the offspring weight and size in flies fed with PBA and control animals. The treatment for a limited period, either early or late in adult life, has also been found to have potential to extend the flies' longevity, possibly by stimulating repair mechanisms and/or inhibiting the accumulation of damages (Kang et al. 2002). The life-extending effect of PBA was dose-dependent: 10 mM of PBA lead to extended longevity whereas lower concentrations of this drug have no effect and higher concentrations were toxic. Remarkably, this trend was highly consistent with those for the levels of acetylation of histones 3 and 4. The effects of PBA were also accompanied by marked changes in gene expression. DNA microarray-based global transcriptional analysis revealed that PBA treatment lead to either down- or up- regulation of several hundreds of genes.

A partial list of genes which were either repressed or induced by this treatment is presented in Table 11.2. Among genes which were significantly up-regulated by PBA treatment, several genes have previously been found to be involved in life span determination in *D. melanogaster*, including chaperones (Tatar et al. 1997), superoxide dismutase (SOD) (Orr and Sohal 1994; Sun et al. 2012), glutathione *S*-transferase (Toba and Aigaki 2000), cytochrome P450 (Doroszuk et al. 2012), and elongation factor 1 (Webster and Webster 1984). The level of transcription of SOD gene was most dramatically affected by PBA treatment—in flies fed PBA this level was 50 times higher than in control flies. It may be a reason why in this study the flies fed PBA were more resistant to the paraquat-induced oxidative stress than control ones. The general trends evident from this table are

Table 11.2 Partial list of genes induced or repressed by PBA, according to the Kang et al. (2002)

Function	Gene	Fold change
Detoxification	Superoxide dismutase	51.9
	Cytochrome P450-4d1	7.2
	Glutathione S-transferase	4.6
Chaperone	Hsc70	4.5
	Hsp60	6.7
	DnaJ like2	2.9
	DnaJ like1	−13.2
Translation	Translational elongation factor1 α	4.1
Neurotransmitter	Inebriated	26.8
Transcription factor	Daughterless	8.0
Signal transduction	Epididymal secretory protein	7.5
	Peroxisomal farnesylated protein	−1.8
Transporter	Mitochondrial phophate carrier protein	5.1
Growth factor	Imaginal disc growth factor1	5.3
Ligand binding	Transportin	8.0
	Calreticulin	−26.5
DNA binding	Osa	−5.0
Kinase	Cyclin-dependent kinase 9	−2.7
Ion channel	Porin	−1.4
Metabolism	Glyceraldehyde-3-phosphate dehydrogenase1	−3.7
	NADH:ubiquinone reductase 75-kD subunit precursor	−25.3
	Cytochrome c oxidase	−6.6
	Peptidyl gycine α hydroxylating monooxygenase	−1.4
	Fatty acid synthetase	−2.4
	Cytochrome c oxidase subunit VIb	−2.2
	Hexokinase	−5.0

up-regulation of the majority of genes involved in detoxification and chaperone activity, and down-regulation of genes involved in different metabolic pathways. These findings support the hypothesis that life span extension may be caused by overall generalized changes in epigenetic regulation (Vaiserman 2011).

11.3.2 Sodium Butyrate

In several studies, the life-extending capacity was also shown for sodium butyrate (SB), a short chain fatty acid having HDAC inhibition activity and known to markedly influence the processes of cell growth, differentiation and apoptosis in both normal and transformed cells (Buommino et al. 2000; Khan and Jena 2014). The epigenetic modifications potentially involved in the SB-induced life extension in *Drosophila* have been deeply studied in a series of experiments of the research group headed by Bai-Qu Huang in the Northeast Normal University, Changchun, China (see Table 11.3). The first evidence for the life-extending potential of SB in *Drosophila melanogaster* was obtained in the Zhao et al. (2005a) study. In the short-lived line, *iso4*, one-off treatment with SB in a dose of 10 mmol/l for 5 h in the larval stage resulted in an increase of both mean and maximum life span by 25.8 and 11.5 %, respectively. No obvious effect on longevity was observed in the long-lived line, *iso2*. The changes obtained were accompanied by the hyperacetylation of core histone H3 and elevated levels of expression of hsp22 and hsp70 genes. In the subsequent study by the same authors (Zhao et al. 2006), treatment with SB caused elevated acetylation levels at histone H3 located at both promoter and coding regions of the hsp70 gene, along with enhanced accessibility of heat-shock factor to target element and increased rate of RNA polymerase II-mediated transcription. The SB-induced hyperacetylation of histone H3 resulted in up-regulation of both basal and inducible hsp70 expression. In addition, SB affected the structure of chromatin at the site of cytogenetic location of the hsp70 gene on the polytene chromosome (Chen et al. 2002). Furthermore, in this study, SB lead to a markedly elevated level of transcription of the hsp70 gene, at an extent similar to that induced by heat shock. Zhao et al. (2005b) by using the chromatin immunoprecipitation, also located the regions of the SB-induced H3 hyperacetylation at both the promoter and the downstream of RNA polymerase II of the transcribing hsp22 gene. It has been found that acetylation of this histone stimulated the transcription initiation and promoted the transcription elongation, thereby up-regulating both basal and inducible expression of this gene. In the Zhao et al. (2007) study, the flies fed with SB during the third instar larval stage, demonstrated an elevated level of acetylation of histone H3, whereas the level of acetylation of histone H4 remained unchanged. The histone H3 acetylation levels were significantly increased at all the regulatory elements of the hsp26 gene promoter. The hyperacetylation of histone H3 was accompanied by a significantly decreased level of basal transcription of the hsp26 gene under the non-heat shock conditions, but increased level of inducible transcription under the heat shock. Specifically, when

Table 11.3 Epigenetic and phenotypic changes induced by SB treatment

Strain/model	Stage of treatment	Phenotypic modifications	Epigenetic changes	References
Short-lived *iso4* line	Third instar larvae	Increased MLS and MaxLS in a sex-pooled population	Hyperacetylation of core histone H3; elevated levels of expression of hsp22 and hsp70 genes	Zhao et al. (2005a)
	Larval and adult stages	No effect on lifespan	Hyperacetylation of core histone H3	
Long-lived *iso2* line	Third instar larvae	No effect on lifespan	Hyperacetylation of core histone H3; elevated levels of expression of hsp22 and hsp70 genes	
	Larval and adult stages	No effect on lifespan	Hyperacetylation of core histone H3	
Canton-S	Third instar larvae	ND	H3 hyperacetylation in the promoter and coding regions of hsp70 gene; up-regulation of basal and inducible hsp70 expression	Zhao et al. (2006)
NS	Third instar larvae	ND	Modification of chromatin structure at the site of cytogenetic location of hsp70 gene; elevated level of transcription of hsp70	Chen et al. (2002)
NS	Third instar larvae	ND	H3 hyperacetylation in the promoter and coding regions of hsp22 gene; up-regulation of basal and inducible expression of hsp22	Zhao et al. (2005b)

(continued)

Table 11.3 (continued)

Strain/model	Stage of treatment	Phenotypic modifications	Epigenetic changes	References
Canton-S	Third instar larvae	ND	H3 hyperacetylation in the promoter region of hsp26; decreased level of basal transcription and increased level of inducible transcription of hsp26	Zhao et al. (2007)
Canton-S, Sin3A^{lof}; Model of Parkinson's disease	First 5 days of adult life	Rescue of locomotor impairment and early mortality; elevated dopamine levels in the brain	Unchanged deficiency in the tyrosine hydroxylase mRNA level, but elevated dopamine levels in the fly brain	St Laurent et al. (2013)
Normal-lived *Ra* strain	Transition/senescent spans	Decreased mortality rate and increased MLS	ND	McDonald et al. (2013)
Long-lived *La* strain	Entire adult life span or healthspan	Decreased MLS	ND	
	Adult stage	Decreased MLS	ND	
Oregon-R	Larval and adult stages	Increased MLS in both sexes	ND	Vaiserman et al. (2012)
	Adult stage	Increased male MLS	ND	
Oregon-R	Larvae	Increased male MLS and MaxLS; increased female MaxLS	Up-regulation of inducible expression of *sir2* gene	Vaiserman et al. (2013)
Oregon-R	Adult stage	Increased male MLS; no effect on locomotion	ND	Symonenko et al. (2014)
w1118	Adult stage	Increased male MLS; increase in locomotion in 40 and 50 day old males	ND	Symonenko et al. (2014) and unpublished results

the larvae were fed with 10 mM SB for 6 h at 25 °C, the basal level of expression of the hsp26 gene was inhibited by approximate 60 %, whereas heat shock at 37 °C induced the transcription of hsp26, and if such heat activation was accompanied by the SB treatment, the level of inducible expression of this gene was elevated by about 30 %.

According to the authors, these findings suggest that the alterations in histone acetylation and, thereafter, the expression of chaperone genes, may be contributed to the life-extending effects of SB and other HDACIs in *Drosophila melanogaster* (Zhao et al. 2005a, b).

In the Tables 11.3 and 11.4: mean life span (MLS), maximum life span (MaxLS); not specified (NS); not determined (ND).

Other mechanisms, however, may also be contributing. In recent research by St Laurent et al. (2013), the treatment with 10 mM SB-supplemented food rescued the locomotor impairment and early mortality of the flies with the pesticide rotenone-induced Parkinson's disease. In this model, SB was selected as a therapeutic candidate because it is known to be able to correct the disrupted HDAC activity in Parkinson's disease and other neurodegenerative disorders. The SB-mediated rescue of rotenone-induced locomotor impairment was associated with elevated dopamine levels in the fly brain. At the same time, no significant differences in the serotonin content among the groups were observed. Additionally, treatment with SB did not improve significantly the deficiency in tyrosine hydroxylase (the rate-limiting enzyme for dopamine biosynthesis) mRNA levels and in SOD activity in rotenone-treated insects.

Phase separation in the adult life of fruit fly and other gradually aging organisms into a health span, a transition phase, and a senescent span was proposed by Arking et al. (2002). In analysis conducted in different model organisms, it has been shown that these life stages are characterized by different gene expression patterns. The health span is characterized by a tightly regulated gene expression pattern which leads to a maximized tissue function and to a minimized inflammatory and other damage response; the transition phase is characterized by a gradual decline of the cellular regulatory capacity, and the senescent span is characterized by a gradual deregulation of the gene expression pattern (Arking 2009). In a recent study by McDonald et al. (2013), a normal-lived Ra strain demonstrated a decreased mortality rate and an increased life span if it was administered with SB during transition or senescent spans, but a decreased life span when administered throughout the entire adult life span or health span only. In a long-lived La strain, however, treatment with SB resulted in mostly deleterious effects on longevity.

In our own study, both *Oregon-R* male and female flies fed 10 and 20 mmol/l SB during both pre-imaginal and imaginal stages showed significant increase in mean life span compared to control flies, whereas the treatment with 20 and 40 mmol/l SB throughout the adult stage only lead to a significant increase in male (but not female) longevity (Fig. 11.1). Dietary supplementation with SB in a high dose (160 mmol/l) during the adult stage resulted in significant decrease of both male and female life span (Vaiserman et al. 2012). The life-extending effects obtained were unlikely due to the decreased reproductive investment, because any

Table 11.4 Epigenetic and phenotypic changes induced by TSA treatment

Strain/model	Stage of treatment	Phenotypic modifications	Epigenetic changes	References
w^{1118}	Larvae	Increased lethality and delayed larval development	ND	Pile et al. (2001)
Short-lived *iso4* line	Third instar larvae	Increase of MLS by one-off treatment; increase of MLS and MaxLS by continuous treatment	Hyperacetylation of core histone H3; elevated levels of expression of hsp22 and hsp70 genes	Zhao et al. (2005a)
Long-lived *iso2* line		Increase of MLS by continuous treatment		
Canton-S	Third instar larvae	ND	H3 hyperacetylation in the promoter and coding regions of hsp70 gene; up-regulation of basal and inducible hsp70 expression	Zhao et al. (2006)
NS	Third instar larvae	ND	Modification of chromatin structure at the site of cytogenetic location of hsp70 gene; elevated level of transcription of hsp70 gene	Chen et al. (2002)
NS	Third instar larvae	ND	H3 hyperacetylation in the promoter and coding regions of hsp22 gene; up-regulation of basal and inducible expression of hsp22 gene	Zhao et al. (2005b)
Canton-S	Third instar larvae	ND	H3 hyperacetylation in the promoter region of *hsp26* gene; decreased level of basal transcription and increased level of inducible transcription of *hsp26* gene	Zhao et al. (2007)
Canton-S	Adult stage	Increased MLS and MaxLS in both sexes	Modified chromatin morphology at the locus of *hsp22* gene; promoted *hsp22* transcription	Tao et al. (2004)
Oregon-R	Adult stage	Increased male MLS; no effect on locomotion	ND	Symonenko et al. (2014)
w1118	Adult stage	Increased male MLS; increase in locomotion in 30, 40, and 50 day old males	ND	Symonenko et al. (2014 and unpublished results)

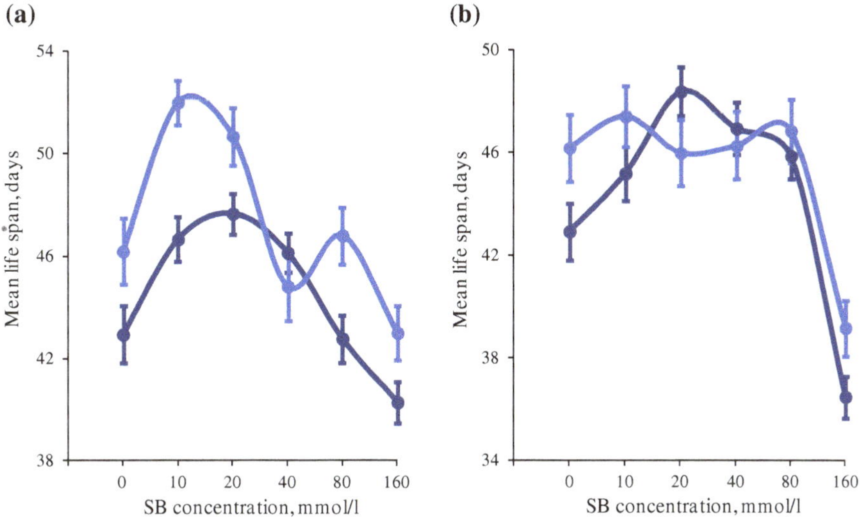

Fig. 11.1 Mean life spans (M ± m) of male (*dark line*) and female (*light line*) fruit flies supplemented with SB at both larval and imaginal stages (**a**) or at imaginal stage only (**b**) (adapted with changes from Vaiserman et al. 2012)

reduction in reproductive activity (fecundity) was revealed in SB-treated female flies. In our subsequent study (Vaiserman et al. 2013), supplementation with SB in a dose of 20 mmol/l led to a significant increase in the male mean life span; maximum life span was significantly increased in groups treated with SB in doses of 10, 20 and 40 mmol/l.

No changes in female life span were obtained after administration of SB, whereas maximum life span was significantly increased in group treated with the 10 mmol/l SB only. Increase in longevity of *Oregon-R* and w^{1118} males was further confirmed in our experiments in which SB (10 mM solution) was applied to the food surface during the adult stage (Symonenko et al. 2014). The effect was somewhat more pronounced and reproducible for w^{1118} line characterized by a lower life span, and, in old w^{1118} flies, was accompanied by a slight increase in locomotion which is often considered as a marker of aging (for review, see Ridgel and Ritzmann 2005). In addition, both basal and inducible levels of expression of *hsp70*, *sir2* and *InR* genes were determined in SB-treated flies and in control flies. Dietary supplementation with SB in a dose of 20 mmol/l at the larval stage had no effect on the basal levels of expression of these genes under normal (unstressed) conditions. The inducible levels of expression of *hsp70* and *InR* genes in the stressful conditions (heat shock and starvation, respectively) were also unchanged, whereas, under the starvation condition, 2.5-fold higher expression of *sir2* gene was detected in SB-treated group compared to control flies (Vaiserman et al. 2013). We also performed an RNA-seq analysis of transcriptomes of SB-treated and control w^{1118} males. According to our preliminary results, the following functional gene sets were associated with SB treatment: (i) defense response to bacteria; (ii) regulation of immune system; (iii) regulation of response to stress, including JNK signaling. Accordingly, up-regulation of *foxo*, *hep* and other genes involved in life span control was revealed.

To summarize, a wide variety of effects of SB on life span was observed. SB in concentrations varying from 10 to 40 mM demonstrated a potential to increase life span, whereas SB treatment in higher doses (more than 100 mM) decreased longevity. Stage and duration of SB supplementation seem to be important with respect to the SB effect on life span (Table 11.3). Treatment on the larval stage increased longevity in most experiments. This fact indicates that SB might indeed affect epigenetic mechanisms of life span extension based on modification of histones. Treatment with SB on adult stage had positive effect on longevity only in some cases, in particular, though not exclusively, when SB was supplied later in life and could prolong active transcription of genes essential for maintaining health span. Short/normal-lived strains were more sensitive to SB treatment than long-lived strains (Table 11.3). It must take into account that different authors used various genotypes, which also could contribute to the observed differences in SB effects. Contradictions between different experiments could, at least partially, be also explained by sex-specificity of SB effects (Table 11.3). In many experiments, life span was measured in a mixed population of males and females, and this could substantially bias the final result. However, despite the complexity and partial inconsistency of results, SB demonstrated a high potential as a life-extending agent.

11.3.3 Trichostatin A

Trichostatin A (TSA) is another widely used HDACI that demonstrates a broad spectrum of epigenetic activities, including inhibition of the cell cycle since the beginning of the growth stage and promotion of the expression of apoptosis-associated genes. TSA is recognized as a promising anticancer drug candidate. Possible mechanisms of action of this compound are induction of terminal differentiation, cell cycle arrest and apoptosis in different cancer cell lines and thereby inhibition of tumorigenesis (Vanhaecke et al. 2004).

In the study by Pile et al. (2001), it has been found that TSA may play an important role in normal developmental progression in *Drosophila melanogaster*. Supplementation with 5 microM TSA caused lethality and delayed larval development, and acted synergistically with the notched-wing mutations in the *rpd3* deacetylase gene. In the above-mentioned series of studies by Bai-Qu Huang and colleagues, the epigenetic and phenotypic effects of the TSA treatment were obtained that were very similar to those shown for the SB treatment (Table 11.4).

In the study by Zhao et al. (2005a), a treatment with TSA influenced the longevity of both short- and long-lived *Drosophila* lines, but to a different extent. In the short-lived *iso4* line, one-off treatment with TSA caused an increase in mean life span by 5 %, whereas continuous treatment resulted in increase of mean and maximum life span by 24.4 and 16.4 %, respectively. At the same time, in the long-lived *iso2* line, one-off treatment with TSA had no significant effects on both mean and maximum life span and continuous treatment with TSA led to increased mean life span by 15.6 %. These life-extending effects induced by the TSA treatment were accompanied by the hyperacetylation of core histone H3 in the promoter and coding regions of some chaperone genes, such as hsp22, hsp26 and hsp70, along with up-regulation, in most cases, of both basal and inducible expression of these genes. The life-extending potential of TSA was also demonstrated in the Tao et al. (2004) research. In this study, TSA treatment significantly extended the mean life span by 27.3 and 23.3 % for female and male flies, respectively. Maximum life span was also extended by 37.9 % for females and by 37.0 % for males. These phenotypic effects were accompanied by a significantly increased transcription level of the *hsp22* gene, and modified chromatin morphology at the locus of *hsp22* gene along the polytene chromosome. The authors suggest that the expression of chaperones can reduce the level of accumulation of damage, stimulate the repair mechanisms, and improve the cell stress resistance to create cellular and physiological environments that are favorable for longevity.

In our own study, both *Oregon-R* and w^{1118} males fed 10 mkM TSA throughout the adult stage showed significant increase in mean life span compared to control flies, whereas this effect was not observed in females. The effect was somewhat more pronounced for w^{1118} line characterized by a lower life span, and, in old w^{1118} flies, was accompanied by a slight increase in locomotion. We also performed an RNA-seq analysis of transcriptomes of TSA-treated and control w^{1118} males. According to our preliminary results, the following functional gene sets were associated with the differential expression in control and TSA treated flies: (i) DNA replication; (ii) cell fate determination, differentiation and development of various organ systems, (iii) mitochondria function, ATP synthesis. Surprisingly, up-regulation of many genes involved in development of the nervous system, heart and cuticula was revealed in TSA treated males in association with increased life span.

To summarize, effects of TSA on life span seem more consistent than effects of SB. TSA was shown to affect life span of both short/normal- and long-lived strains, and stage of TSA supplementation seems to be less important compared to SB treatment (Table 11.4). The life span-modulating effects of TSA were also found to be sex-dependent. Similarly to SB, TSA demonstrated a high potential as a life-extending agent. However, in our experiments, two HDACIs affected transcription of different sets of genes, with TSA treatment affecting transcription of much more genes and in a greater extent compared to SB treatment.

11.3.4 Suberoylanilide Hydroxamic Acid (SAHA)

One more HDACI that was shown to be able to extend life in fruit fly, is suberoylanilide hydroxamic acid (SAHA). In in vitro studies, SAHA was found to have similar effects as does SB although at much lower effective doses (Zhou et al. 2011). This compound is known to induce growth arrest in transformed cells

(Yin et al. 2007), and it was shown to be effective in preventing Huntington disease in various animal models including *Drosophila* (Steffan et al. 2001).

In the recent study by McDonald et al. (2013), the effects of administration with SAHA throughout *Drosophila* health span, transition phase, and senescent span have been studied. Treatment with SAHA during the transition or senescent spans resulted in decreased mortality rate and extended longevity compared to the control, while supplementation during the entire adult life span or during the health span only led to decreased longevity in the normal-lived Ra strain. At the same time, when the long-lived La strain was administered with SAHA by the same scheme, mostly deleterious effects were detected. All three SAHA doses used in research altered the late-life survivorship of the normal-lived Ra strain so that it resulted in the significant increase in both median and maximum longevity compared to the control flies. The analysis of mortality curves in all three experimental cohorts used in the study indicated that there were no significant effects of the SAHA administration until the age of ~50 days. After this time point, all cohorts showed an evident lowering of the mortality rates compared to the control flies. The supplementation with SAHA significantly influenced the mortality rate when applied to the transition or senescent phases of normal-lived strain but not of long-lived strain. Remarkably, the SAHA-treated normal-lived *Drosophila* strain showed the late-life extending effects similar to those seen in the same study for the other HDACI, SB. The fact that these two different HDACIs, SB and SAHA, had similar effects on mortality rate during the senescent span indicates the similarity of mechanisms that underlie beneficial effects for this class of HDACIs. The authors suggest that their findings demonstrate that the use of these HDACIs may significantly influence the mortality rate throughout the senescent phase by reducing the vulnerability of treated individuals, in a manner similar to that of dietary restriction. On the basis of their findings, the authors suggested that the HDACIs used may affect several pathways involved in regulating gene expression patterns associated with healthy aging. According to the authors, the obtained stage-specific ability of the studied HDACIs to adversely affect the survival of the both normal-lived Ra and long-lived La flies implies that the sensitivity of the organism to the preparation may be due to the presence of HDAC-dependent stage-specific patterns of gene expression. Disruption of these patterns can lead to life shortening. Conversely, the induction of these patterns of gene expression throughout senescence when they are not normally present may likely underlie the life-extending effects of HDACIs.

11.4 Conclusion

A number of environmental, life style and genetic interventions have clearly proven to be effective in prolonging life span in experimental animals (Jylhava 2014). Currently, an opinion is becoming common among researchers that life-extending interventions operate primarily to modify the epigenome (Vaiserman

2008; Rando and Chang 2012). A growing body of studies indicates that epigenetic changes which are associated with age may be beneficially affected by several lifestyle factors, such as diet or exercise. Several compounds, both chemical and natural, have been also proposed recently to positively affect aging and longevity including those presumably based on the fine-tuning of epigenetic regulation (Uchiumi et al. 2012). Among such epigenetically-targeted pro-longevity drugs, HDACIs are currently in focus of scientific interest. Importantly, a very accurate and precise fine-tuning is required for this purpose since any unbalancing in HDAC activity, similarly to unbalanced consumption of vitamins, antioxidants, or hormones, may result in disruption of finely-tuned mechanisms controlling homeostasis. However, as the regulation of transcriptional networks is mediated by central regulatory systems and thereby is a highly coordinated and orchestrated process, the mechanism of epigenetic regulation of gene expression appears a plausible candidate mechanism for the modulation of a highly integrated process such as aging (Vaiserman 2011; Bacalini et al. 2014). Thereby, nonspecific HDACIs, such as SB, TSA and SAHA, which may potentially influence the expression of thousands of genes including those which are involved in aging, can prove to be quite effective for further anti-aging treatments. The tissue-, stage-, and HDAC-specific inhibitors, however, may also be developed.

Drosophila, among other experimental models, is a very useful model in screening of anti-aging and pro-longevity drug candidates including HDACIs. In addition, determining which HDACIs may extend longevity in *Drosophila* and which genes are implicated in these effects, may provide important information about the genetic basis of aging. The critical issue is certainly whether the mechanisms of action of HDACIs are similar between invertebrates, including *Drosophila*, and mammals. Many of these pathways are likely substantially conserved in a wide variety of species, from invertebrates to humans (Lucanic et al. 2013). Thereby, uncovering which transcription factors and signaling pathways contributing to healthy aging can be influenced by HDACIs in fruit fly may facilitate the development of new strategies for treating and preventing age-related human diseases and health span extension.

Acknowledgments The authors would like to thank Oksana Zabuga for the assistance in preparing the manuscript. This work was supported by a joint grant of the Russian Foundation for Basic Research and the State Fund for Fundamental Research of Ukraine (project no. 11-04-90478-Ukr_f_a).

References

Abel T, Zukin RS (2008) Epigenetic targets of HDAC inhibition in neurodegenerative and psychiatric disorders. Curr Opin Pharmacol 8:57–64

Arking R (2009) Overview of the genomic architecture of longevity. In: Sell C, Lorenzini A, Brown-Borg HM (eds) Life Span extension: single cell organisms to man. Humana Press, Springer, Dordrecht, pp 59–73

Arking R, Novoseltseva J, Hwangbo DS et al (2002) Different age-specific demographic profiles are generated in the same normal-lived *Drosophila* strain by different longevity stimuli. J Gerontol A Biol Sci Med Sci 57:B390–B398

Bacalini MG, Friso S, Olivieri F et al (2014) Present and future of anti-ageing epigenetic diets. Mech Ageing Dev 136–137:101–115

Baltan S, Morrison RS, Murphy SP (2013) Novel protective effects of histone deacetylase inhibition on stroke and white matter ischemic injury. Neurotherapeutics 10:798–807

Benedetti R, Conte M, Altucci L (2014) Targeting HDACs in diseases: where are we? Antioxid Redox Signal Jan 1 (Epub ahead of print)

Berdasco M, Esteller M (2012) Hot topics in epigenetic mechanisms of aging: 2011. Aging Cell 11:181–186

Boros IM (2012) Histone modification in *Drosophila*. Brief Funct Genomics 11:319–331

Boumber Y, Issa JP (2011) Epigenetics in cancer: what's the future? Oncology (Williston Park) 25(220–226):228

Boyd-Kirkup JD, Green CD, Wu G, Wang D, Han JD (2013) Epigenomics and the regulation of aging. Epigenomics 5:205–227

Buommino E, Pasquali D, Sinisi AA et al (2000) Sodium butyrate/retinoic acid costimulation induces apoptosis-independent growth arrest and cell differentiation in normal and ras-transformed seminal vesicle epithelial cells unresponsive to retinoic acid. J Mol Endocrinol 24:83–94

Chang KT, Min KT (2002) Regulation of lifespan by histone deacetylase. Ageing Res Rev 1:313–326

Chen T, Sun H, Lu J et al (2002) Histone acetylation is involved in hsp70 gene transcription regulation in *Drosophila melanogaster*. Arch Biochem Biophys 408:171–176

Cho Y, Griswold A, Campbell C et al (2005) Individual histone deacetylases in *Drosophila* modulate transcription of distinct genes. Genomics 86:606–617

Davie JR, Spencer VA (1999) Control of histone modifications. J Cell Biochem Suppl 32–33:141–148

Dominguez LJ, Barbagallo M, Morley JE (2009) Anti-aging medicine: pitfalls and hopes. Aging Male 12:13–20

Doroszuk A, Jonker MJ, Pul N et al (2012) Transcriptome analysis of a long-lived natural *Drosophila* variant: a prominent role of stress- and reproduction-genes in lifespan extension. BMC Genomics 13:167

Gao L, Cueto MA, Asselbergs F et al (2002) Cloning and functional characterization of HDAC11, a novel member of the human histone deacetylase family. J Biol Chem 277:25748–25755

Hahnen E, Hauke J, Tränkle C et al (2008) Histone deacetylase inhibitors: possible implications for neurodegenerative disorders. Expert Opin Investig Drugs 17:169–184

Helfand SL, Rogina B (2003) Molecular genetics of aging in the fly: is this the end of the beginning? BioEssays 25:134–141

Huidobro C, Fernandez AF, Fraga MF (2013) Aging epigenetics: causes and consequences. Mol Aspects Med 34:765–781

Iannitti T, Palmieri B (2011) Clinical and experimental applications of sodium phenylbutyrate. Drugs R D 11:227–249

Imai S, Armstrong CM, Kaeberlein M et al (2000) The transcriptional silencing and longevity protein Sir2 is an NAD-dependent histone deacetylase. Nature 403:795–800

Issa JP (1999) Aging, DNA methylation and cancer. Crit Rev Oncol Hematol 32:31–43

Johnstone RW (2002) Histone-deacetylase inhibitors: novel drugs for the treatment of cancer. Nat Rev Drug Discov 1:287–299

Jylhava J (2014) Determinants of longevity: genetics, biomarkers and therapeutic approaches. Curr Pharm Des 20(38):6058–6070

Kang H-L, Benzer S, Min K-T (2002) Life extension in *Drosophila* by feeding a drug. Proc Natl Acad Sci USA 99:838–843

Kapoor VK, Dureja J, Chadha R (2009) Synthetic drugs with anti-ageing effects. Drug Discov Today 14:899–904
Khan S, Jena GB (2014) Protective role of sodium butyrate, a HDAC inhibitor on beta-cell proliferation, function and glucose homeostasis through modulation of p38/ERK MAPK and apoptotic pathways: study in juvenile diabetic rat. Chem Biol Interact 213:1–12
Kouzarides T (2007) Chromatin modifications and their function. Cell 128:693–705
Kuo MH, Allis CD (1998) Roles of histone acetyltransferases and deacetylases in gene regulation. Bioessays 20:615–626
Larson K, Yan S-J, Tsurumi A et al (2012) Heterochromatin formation promotes longevity and represses ribosomal RNA synthesis. PLoS Genet 8:e1002473
Licciardi PV, Ververis K, Tang ML et al (2013) Immunomodulatory effects of histone deacetylase inhibitors. Curr Mol Med 13:640–647
Lucanic M, Lithgow GJ, Alavez S (2013) Pharmacological lifespan extension of invertebrates. Ageing Res Rev 12:445–458
Lyko F, Beisel C, Marhold J et al (2006) Epigenetic regulation in *Drosophila*. Curr Top Microbiol Immunol 310:23–44
McDonald P, Maizi BM, Arking R (2013) Chemical regulation of mid- and late-life longevities in *Drosophila*. Exp Gerontol 48:240–249
Morris BJ (2013) Seven sirtuins for seven deadly diseases of aging. Free Radic Biol Med 56:133–171
Muñoz-Najar U, Sedivy JM (2011) Epigenetic control of aging. Antioxid Redox Signal 14:241–259
Orr WC, Sohal RS (1994) Extension of lifespan by overexpression of superoxide dismutase and catalase in *Drosophila melanogaster*. Science 263:1128–1130
Pile LA, Lee FW, Wassarman DA (2001) The histone deacetylase inhibitor trichostatin A influences the development of *Drosophila melanogaster*. Cell Mol Life Sci 11:1715–1718
Rando TA, Chang HY (2012) Aging, rejuvenation, and epigenetic reprogramming: resetting the aging clock. Cell 148:46–57
Ridgel AL, Ritzmann RE (2005) Insights into age-related locomotor declines from studies of insects. Ageing Res Rev 4:23–39
Satoh A, Stein L, Imai S (2011) The role of mammalian sirtuins in the regulation of metabolism, aging, and longevity. Handb Exp Pharmacol 206:125–162
Seroude L, Brummel T, Kapahi P et al (2002) Spatio-temporal analysis of gene expression during aging in *Drosophila melanogaster*. Aging Cell 1:47–56
Sinclair DA, Guarente L (2014) Small-molecule allosteric activators of sirtuins. Annu Rev Pharmacol Toxicol 54:363–380
Slingerland M, Guchelaar HJ, Gelderblom H (2014) Histone deacetylase inhibitors: an overview of the clinical studies in solid tumors. Anticancer Drugs 25:140–149
St Laurent R, O'Brien LM, Ahmad ST (2013) Sodium butyrate improves locomotor impairment and early mortality in a rotenone-induced *Drosophila* model of Parkinson's disease. Neuroscience 246:382–390
Steffan JS, Bodai L, Pallos J et al (2001) Histone deacetylase inhibitors arrest polyglutamine-dependent neurodegeneration in *Drosophila*. Nature 413:739–743
Sun X, Komatsu T, Lim J et al (2012) Nutrient-dependent requirement for SOD1 in lifespan extension by protein restriction in *Drosophila melanogaster*. Aging Cell 11:783–793
Swaminathan A, Gajan A, Pile LA (2012) Epigenetic regulation of transcription in *Drosophila*. Front Biosci 17:909–937
Symonenko AV, Roshina NV, Kolyada AK et al (2014) Changes in *Drosophila melanogaster* lifespan and gene expression profiling caused by the histone deacetylase inhibitors. In: Abstracts of the 3rd international conference "genetics of aging and longevity", Moscow, Russia, April 2014
Tao D, Lu J, Sun H et al (2004) Trichostatin A extends the lifespan of *Drosophila melanogaster* by elevating hsp22 expression. Acta Biochim Biophys Sinica 36:618–622

Tatar M, Khazaeli AA, Curtsinger JW (1997) Chaperoning extended life. Nature 390:30

Toba G, Aigaki T (2000) Disruption of the microsomal glutathione S-transferase-like gene reduces life span of *Drosophila melanogaster*. Gene 253:179–187

Tollefsbol TO (2014) Dietary epigenetics in cancer and aging. Cancer Treat Res 159:257–267

Uchiumi F, Oyama T, Ozaki K (2012) A new protocol to discover novel anti-aging compounds. Pharmaceut Anal Acta 3:7. doi:10.4172/2153-2435.1000166

Vaiserman AM (2008) Epigenetic engineering and its possible role in anti-aging intervention. Rejuvenation Res 11:39–42

Vaiserman A (2011) Hormesis and epigenetics: is there a link? Ageing Res Rev 10:413–421

Vaiserman AM, Pasyukova EG (2012) Epigenetic drugs: a novel anti-aging strategy? Front Genet 3:224

Vaiserman AM, Koliada AK, Koshel NM et al (2012) Effect of the histone deacetylase inhibitor sodium butyrate on the viability and life span in *Drosophila melanogaster*. Adv Gerontol 25:126–131 [In Russian]

Vaiserman AM, Koshel NM, Zabuga OG et al (2013) Determination of geroprotective potential of sodium butyrate in *Drosophila melanogaster*: long-term effects. Adv Gerontol 26:111–116 [In Russian]

Vanhaecke T, Papeleu P, Elaut G et al (2004) Trichostatin A-like hydroxamate histone deacetylase inhibitors as therapeutic agents: toxicological point of view. Curr Med Chem 11:1629–1643

Villeponteau B (1997) The heterochromatin loss model of aging. Exp Gerontol 32:383–394

Webster GC, Webster SL (1984) Specific disappearance of translatable messenger RNA for elongation factor one in aging *Drosophila melanogaster*. Mech Aging Dev 24:335–342

West AC, Johnstone RW (2014) New and emerging HDAC inhibitors for cancer treatment. J Clin Invest 124:30–39

Witt O, Deubzer HE, Milde T et al (2009) HDAC family: what are the cancer relevant targets? Cancer Lett 277:8–21

Wood JG, Helfand SL (2013) Chromatin structure and transposable elements in organismal aging. Front Genet 4:274

Wood JG, Hillenmeyer S, Lawrence C et al (2010) Chromatin remodeling in the aging genome of *Drosophila*. Aging Cell 9:971–978

Yin D, Ong JM, Hu J et al (2007) Suberoylanilide hydroxamic acid, a histone deacetylase inhibitor: effects on gene expression and growth of glioma cells in vitro and in vivo. Clin Cancer Res 3:1045–1052

Yuan H, Su L, Chen WY (2013) The emerging and diverse roles of sirtuins in cancer: a clinical perspective. Onco Targets Ther 6:1399–1416

Zentner GE, Henikoff S (2013) Regulation of nucleosome dynamics by histone modifications. Nat Struct Mol Biol 3:259–266

Zhao Y, Lu J, Sun H et al (2005a) Histone acetylation regulates both transcription initiation and elongation of hsp22 gene in *Drosophila*. Biochem Biophys Res Commun 326:811–816

Zhao Y, Sun H, Lu J et al (2005b) Lifespan extension and elevated hsp gene expression in *Drosophila* caused by histone deacetylase inhibitors. J Exp Biol 208(Pt 4):697–705

Zhao YM, Chen X, Sun H et al (2006) Effects of histone deacetylase inhibitors on transcriptional regulation of the hsp70 gene in *Drosophila*. Cell Res 16:566–576

Zhao Y, Lu J, Sun H et al (2007) Roles of histone acetylation modification in basal and inducible expression of hsp26 gene in *D. melanogaster*. Mol Cell Biochem 306:1–8

Zhou Q, Dalgard CL, Wynder C et al (2011) Histone deacetylase inhibitors SAHA and sodium butyrate block G1-to-S cell cycle progression in neurosphere formation by adult subventricular cells. BMC Neurosci 12:50. doi:10.1186/1471-2202-12-50

Chapter 12
An Evolutionary Analysis of Healthspan Extension Using Diet: Have We Come to the End of the Ponce de Leon Trail?

Grant A. Rutledge and Michael R. Rose

Abstract The search for compounds that enhance health span has been daunting. Many gerontological experiments on model organisms, including *Drosophila* species, have examined the effects of individual substances on life span solely. But it is now clear that effective alleviation of aging requires more than merely prolonged survival regardless of other functional effects. Monitoring other life-history characters is imperative. In addition, functional characters such as locomotor and cognitive capacities may be important too. Here we review the topic of healthspan extension using diet from the standpoint of evolutionary biology. We discuss proposed "rules" for evaluating candidate anti-aging substances. We point out the failings of some studies of anti-aging substances, such as resveratrol. We also critically review proposed anti-aging strategies that have been based on evolutionary reasoning, questioning some of our own earlier suggestions. Here we offer a new evolutionary strategy for dietary enhancement of healthspan, one that is as applicable to fruit flies as humans. However, our overall view is that the project of ameliorating aging using ingestible substances is without doubt challenging to a high degree.

Keywords Life span · Healthspan extension · Fruit fly · Anti-aging substances · Resveratrol

---

G.A. Rutledge (✉) · M.R. Rose
Department of Ecology and Evolutionary Biology, Francisco J. Ayala School of Biological Sciences, University of California, Irvine, California 92697-2525, USA
e-mail: grutledg@uci.edu

M.R. Rose
e-mail: mrrose@uci.edu

A.M. Vaiserman et al. (eds.), *Life Extension*, Healthy Ageing and Longevity 3,
DOI 10.1007/978-3-319-18326-8_12

12.1 The Ponce de Leon Trail Problem: Looking for Anti-aging Substances

12.1.1 The Ponce de Leon Trail is Very Old

One of the more universal features of the historical record of biological research is the search for a substance that can postpone or reverse the effects of aging on people. This is a ubiquitous topic in Taoist writings (vid. Needham's *Science and Civilization in China* books 1954–2008), and a commonplace theme of Traditional Chinese Medicine, which grew out of Taoist traditions. In the West, the topic was of interest in ancient civilizations, as illustrated by the legend of Gilgamesh from Sumerian civilization.

After Western civilization recovered from the Dark Ages and Middle Ages millennium of hostility to biological research, the topic resurfaced in the work of the Renaissance alchemists, such as Paracelsus. Perhaps the most famous early-modern Western example of this search for a restorative substance is the possibly apocryphal story of Ponce de Leon looking for a fountain of youth in Florida, after the voyages of Columbus to the New World (Haycock 2009). Of greater significance for academic biology, the polymath and founding figure of Western science Francis Bacon devoted an entire book to the topic of aging and how it can be influenced, *Historia Vitae et Mortis* (1637). But because Francis Bacon was by inclination and prescription skeptical, it is more appropriate to refer to the project of controlling aging by means of substances as the Ponce de Leon Trail (cf. Moment et al. 1978).

For a *very* long time by the standard of contemporary biology, the genus *Drosophila* has been a model system of choice for the control of aging. For example, Loeb and Northrop (1917) used temperature to control rates of demographic aging in laboratory fruit flies almost a century ago, a practice that has continued ever since (e.g. McArthur and Sohal 1982). This fruit fly aging literature has increased explosively, with hundreds of publications claiming to demonstrate the experimental manipulation of *Drosophila* aging using ingested substances and other interventions. This is a literature too vast to be enumeratively reviewed. Instead, what we offer here is a critique from the standpoint of evolutionary biology.

12.1.2 We Need to Study Healthspan

A central point for us, to begin with, is the importance of what is sometimes called "healthspan." Crudely speaking, this can be thought of as a combination of the capacity to survive together with a capacity, or capacities, to function. That is to say, to take an extreme example we do not regard the prolongation of human life in a medically-induced coma as a notable achievement of anti-aging. Effective mitigation of aging should be more than merely prolonged survival.

Fortunately, in the context of evolutionary theory there are well defined quantitative measures of healthspan that can be used for the purpose of objective experimentation. One such measure would be R_0, which is the summation over all ages of the products of survival probability to a particular age with the fecundity at that age (vid. Charlesworth and Charlesworth 1973). [Similar measures would include the survival probability *at* a particular age multiplied by the fecundity at that age, summed over all ages.] The point of such measures is that the central function of living things, from the standpoint of Darwinian theory, is reproduction. In effect, everything else about life-history is subservient to that end, with the appropriate modifications for inclusive fitness when there are significant transfers of resources between individuals, such as occur with parental care (vid. Hamilton 1964; Lee 2003). Thus it is more than just appropriate to use such summations of survival probabilities and fecundities to calculate total healthspan. We would argue that such indices provide correct scientific measures of healthspan. However, for the present purpose we need only argue that some measure of healthspan of this kind is necessary for the measurement of net effects on aging, properly considered as a whole.

12.1.3 Rules for Studying Anti-aging Candidate Substances

A useful starting point for the study of the healthspan effects of ingested substances was supplied by Jafari and Rose (2006), which proposed a set of rules for the design of model organism tests of candidate anti-aging substances. One of their starting points, which we share, is the demographic partitioning of life-history into three phases: development, aging, and late life (cf. Mueller et al. 2011). Though McCay's classic experiments on dietary restriction in rodents incorporated life span extension arising from either protracted developed or prolonged adult survival (vid. McCay et al. 1939), almost all gerontologists since then have agreed on the point that useful anti-aging trials should focus on adult life, after the completion of development.

What is still controversial is the status of late life in anti-aging experiments. Late life is a distinctive phase of life first well-characterized from human demographic data by Greenwood and Irwin (1939) as a plateauing in mortality rates at very late ages, after the age of 90 years in their data. However, the phenomenon of late-life mortality rate plateaus was not generally credited as a significant biological phenomenon until the publications of Carey et al. (1992) and Curtsinger et al. (1992), which used laboratory medflies and *Drosophila melanogaster*, respectively. Rauser et al. (2003, 2005, 2006) subsequently demonstrated a comparable, though not synchronous, plateau in later-life fecundity. Within evolutionary genetic theory, the selective pressures that characterize aging and late life phases are qualitatively different (e.g. Mueller and Rose 1996; Charlesworth 2001; Mueller et al. 2011), which is how evolutionary biologists like ourselves explain the distinctly different demographic patterns of these two parts of adult life-history.

Jafari and Rose (2006) suggested that experimental trials using *Drosophila* should study the effects of candidate anti-aging substances on mortality rate during the aging phase of life only. This is an elegant solution to the quantitative complexity of the full adult life-cycle. However, if we are successful at slowing human aging demographically, many more people will survive to reach late life than were found to do so by Greenwood and Irwin (1939). This makes the impact of candidate anti-aging substances on the post-aging late-life phase also of interest. However, what is indubitable is that there is potential for significant confusion about the impact of a candidate anti-aging substance *if* the demographic analysis of its effects docs *not* take into account the existence of post-aging adults in an experimental cohort of fruit flies. Unlike the human case at present, some *Drosophila* laboratory cohorts have many individuals surviving into late life (e.g. Shahrestani et al. 2012), much as found by Carey et al. (1992) for medflies. Overall then, we are more agnostic than Jafari and Rose (2006) about the advisability of confining the study of the effects of candidate anti-aging substances to the aging demographic phase only.

A classic concern of pharmacologists like Jafari (vid. Jafari et al. 2007a, b) is that one cannot be sure that the candidate substance, rather than some artifact, is having the inferred anti-aging effect unless there is a dose-dependent pattern to the response. That is, the healthspan effects of a candidate anti-aging substance should scale with the dose. Again, we have some qualifications that we will apply to this stricture from Jafari and Rose (2006), particularly where qualitative changes in diet are concerned.

Jafari and Rose (2006) further contended that experimental *Drosophila* that are being used in a test of a candidate anti-aging substance should not be hypometabolic. As humans are homeotherms with fairly stable metabolic rates, drugs and other interventions that act via gross lowering of metabolic rates in poikilotherms like fruit flies, producing a state of hypometabolism, are not appropriate candidates for anti-aging interventions among human subjects. This was patently the case in the work of Loeb and Northrop (1917), and it is a well-known phenomenon in experimental physiology. In Djawdan et al. (1996), no differences in metabolic rate were observed between the experimentally evolved longer-lived and shorter-lived flies of Rose (1984), and that was a material point in the case for the value of those *Drosophila* for aging research (vid. Rose et al. 2004). Therefore, a drug that increases life span at the expense of a decrease in metabolism is not an ideal candidate anti-aging substance for adoption by humans.

In the same vein, Jafari and Rose (2006) argue that candidate anti-aging substances should not curtail fecundity. It has been well established that lowering fecundity in fruit flies can dramatically increase longevity, for example by dietary restriction (e.g. Chippindale et al. 1993), but also when fecundity is depressed by other means (e.g. Maynard Smith 1958). Compounds that substantially lower fecundity may increase longevity from reduced 'cost of reproduction' effects alone. Again, a key point is that gross depression of total fecundity is not associated with evolutionarily postponed aging (Rose 1984; Leroi et al. 1994; Rose et al. 2004). However, in the framework that we are developing, measures like R_0 naturally take depressed fecundity into account, so this problem in effect washes out in the quantitative measures that we recommend.

Following the same line of argument, Jafari and Rose (2006) emphasize that experimental model organisms should not have general nervous system impairment as a result of a candidate substance. A typical example of such an effect can be achieved by a general-purpose tranquilizing substance. But again, appropriate healthspan measurement should directly obviate this problem, in that heavily tranquilized fruit flies are not going to be mating, feeding, or reproducing at a normal rate.

12.2 Elixirs of Life? Single Substances Have a Problematic Record in the *Drosophila* Aging Literature

12.2.1 Why Healthspan Studies Must Consider Reproduction

Fruit flies are increasingly being used to test candidate pharmaceuticals for long-term health benefits. There are many anti-aging studies of *Drosophila* supplementation with a wide variety of substances, from antioxidants such as resveratrol and lipoic acid to histone deacetylase inhibitors like phenyl butyrate (e.g. Bauer 2004; Kang 2002). However Matsagas et al. (2009) demonstrate that some single substances have ostensibly beneficial effects when only longevity or mortality rates are monitored, effects that might be an artifact of functional impairment of reproductive characters.

An example of this problem is provided by the study of Bahadorani et al. (2008). Vitamins A, C, and E are each thought to play an important role in mitigating oxidative stress. Accordingly, each was administered to *Drosophila* cohorts under oxidative stress conditions. Under chronic oxidative stress conditions, some of these supplements increased life span and some decreased life span. However, only life span was measured in this study. Bonilla et al. (2002) is another example of pharmaceutical supplementation research studying only fruit fly life span effects. Melatonin, a hormone that is thought to prevent oxidative damage to fly tissues, was added to nutritional medium. Life span was significantly increased from 61.2 days in the controls to 81.5 days in the melatonin-fed flies. Once again only life span was observed in this study.

As a general rule, effects on reproduction and other functional characters are often not measured in fruit fly drug studies that measure survival rates or longevity. Yet decades of genetic and manipulative *Drosophila* research have shown that longevity is just one part of the spectrum of life-history characters that jointly respond when fruit fly longevity is impacted significantly. Thus average longevity on its own may be a poor measure of the full spectrum of effects of administered substances. In other words, most *Drosophila* studies of the effects of dietary substances fail to adequately document the range of healthspan effects.

Although not intentionally achieved by supplementation with pharmaceutical substances, dietary restriction (DR) in model organisms like *Drosophila* is well-known in animal cohorts to increase average life span in conjunction with reduced fertility (e.g. Chippindale et al. 1993, 1997). Figure 12.1 demonstrates hypothetical results of DR fly studies that monitor *both* survivorship and fecundity. Chippindale et al. (1993) performed a series of experiments in which the amount

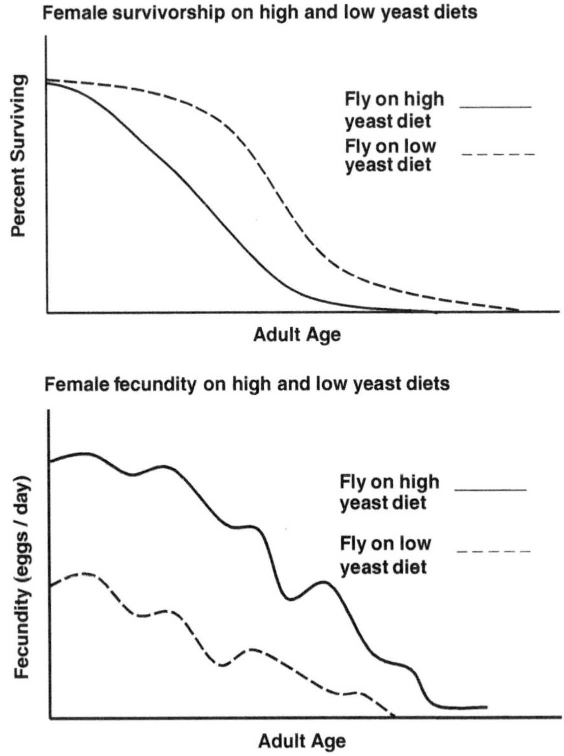

Fig. 12.1 Hypothetical effects of dietary restriction on *Drosophila* using high and low levels of yeast inoculate

of live yeast inoculate applied to the substrate was varied. Lower yeast levels, which significantly reduce fecundity, enhanced longevity. However it is also important to note that an overly-extreme reduction of food levels will lead to a reduction in life span and fecundity.

Chronic exposure of an experimental cohort to a pharmaceutical drug could have a superficially beneficial effect if it reduces nutritional intake due to the flies' perceived noxiousness of the drug for the model organism. Also an animal may be sickened to the point of lethargy by a substance, even if its feeding rate is not reduced—such as would be the case with an addictive opiate analog, leading to reduced reproduction.

Research with urea supplementation in adult *Drosophila* provides a clear example of toxicity-induced increase in life-span. Joshi et al. (1996) and Santos et al. (in prep.) demonstrate that when adult *D. melanogaster* are maintained on food supplemented with urea, longevity of both males and females is significantly increased. In addition, female flies maintained on urea-supplemented food exhibit a consistent decline in fecundity over time, relative to those maintained on regular food (Fig. 12.2 left) (Joshi et al. 1996; Santos et al. in prep.). The toxicity of urea is apparent when you expose larva to it: there is a significant decrease in mean adult longevity and an increase in age-specific mortality. Female flies exhibit a dramatic

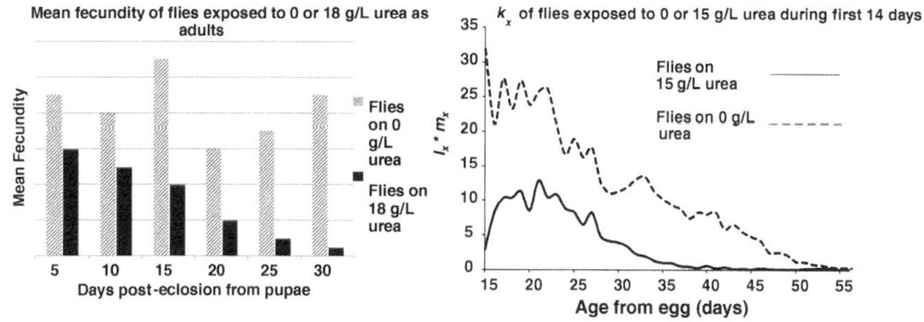

Fig. 12.2 The effect of urea on *D. melanogaster*. *Left* Data plotted from Joshi et al. (1996) showing mean fecundity of adult flies maintained on 0 or 18 g/L of urea. *Right* Data from Rutledge et al. (in prep.) showing female k_x when exposed to 0 or 15 g/L of urea only in larval stage and first few days of adult stage (prior to day 14). After day 14, flies were given food with no urea

decrease in fecundity as a result of exposure to urea as larvae too. The right panel of Fig. 12.2 graphically instantiates this using a healthspan measure known as "k_x" [product of female survivorship (l_x) and eggs per surviving female (m_x)] (Rutledge et al. in preparation).

12.2.2 Single-Substance Failures of Replication

A key strategy in the publication of "successful" single-substance interventions is to (1) avoid collecting adverse healthspan information, (2) avoid detecting adverse healthspan side-effects by using inadequate replication or technique, or (3) suppress/fail to publish any such adverse results if they have been obtained. Usually tactic (3) is not necessary, because biologists can be expert at avoiding the collection of data that would impinge on the "story" that they want to tell. We have ourselves been involved in collaborations where our (now former) colleagues have suppressed results that were adverse to their favored thesis.

As an example of practices that are at least less than ideal, we have the polyphenol resveratrol, a natural compound found in commonly-consumed plants and notoriously present in red wine. Resveratrol has received much attention in scientific studies (Howitz 2003; Baur 2006; Lagouge 2006; Valenzano 2006; Morselli 2010; Miller 2011). However, the life span results have been variable. Resveratrol is thought to be a sirtunin2-activating antioxidant compound (Bauer 2004; Wood 2004). The authors of some studies have suggested that resveratrol acts as a caloric restriction mimetic due to general sirtuin activation (Howitz et al. 2003; Wood et al. 2004). However, Kaeberlein et al. (2005) found that resveratrol has no detectable effect on Sir2 activity in vivo or on life span in yeast. On the other hand, resveratrol has been shown to increase life span in *Drosophila* studies with little or no obvious effects on fecundity as shown in the left panel of Fig. 12.3 (Wood et al. 2004).

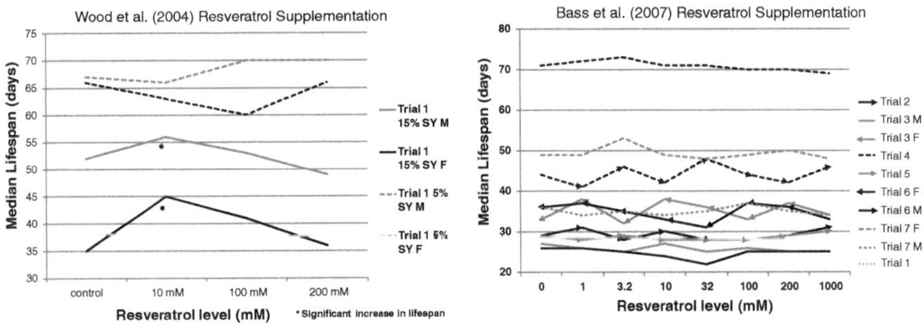

Fig. 12.3 Two studies of the dose-dependent effects of resveratrol on the lifespan of *D. melanogaster. Left* Data from Wood et al. (2004) showing the effect of various levels of resveratrol on median lifespan. Trials using the wild type strain Canton-S under a 15 % sugar-yeast (SY) diet (*solid lines*) and 5 % SY diet (*dotted lines*) are graphed. *Right* Data from Bass et al. (2007) showing the effect of various levels of resveratrol on Median lifespan. Some trials are split into two trend lines, when males and females were tabulated separately

Yet Bass et al. (2007) found no significant effects of resveratrol on life span in seven independent trails, three of which used the same strain as was used in Wood et al. (2004) (Fig. 12.3 right). Bass et al. (2007) was not able to reproduce the longevity increasing property of resveratrol on the *D. melanogaster* strain Canton-S in three of their trials, and they also did not obtain positive results using the strain Dahomey. One possible explanation for such inconsistent results is that the effect of a candidate anti-aging substance can be dependent on the genetic ancestry of the cohort(s) undergoing pharmacological trials. It is a well-established principle of epistasis that some genetic backgrounds will respond differently to the introduction of the same mutation. In the case of anti-aging drug trials, it is possible that a particular compound might increase life span in a stock that has accidentally fixed a particular gene, or set of genes, yet the same compound given to a different stock of fruit flies might have no effect on healthspan. In such cases, it would be fair to say that the impact of the candidate anti-aging compound depends critically on the genetics of *D. melanogaster*, making its general value dubious.

Kang et al. (2002) reported that feeding *Drosophila* with 4-phenylbutyrate (PBA) can significantly increase life span without a reduction in other healthspan characteristics like reproductive ability. However Jafari et al. (2006) pointed out that 10 mM of PBA resulted in life span extension in a *white* mutant strain while the wild type strain only required 5 mM of PBA for life span extension.

Another possible cause of ambiguous or inconsistent results from single-substance trials could be effects on metabolic rate and locomotion. Avanesian et al. (2010) tested the common anticonvulsant Lamotrigine. It was hypothesized that this chemical increased life span at the expense of decreasing healthspan. They found that lamotrigine did in fact increase life span, however a reduction in locomotor activity and metabolic rate depression were also observed. Matsagas et al. (2009) performed experiments testing the effect of sedatives on life span and healthspan of *Drosophila*. Lithium, a commonly used sedative, slightly elevated mean longevity at the two lowest doses. However, there was a significant negative impact on fecundity and male mating success even at those doses.

Another possible cause of difficulties with single-substance trials arises when the effect of a medication is highly sensitive to the fly culture environment. Among other things, it is possible that recondite environmental effects on longevity are fairly minor, but the effects on fecundity or mating behavior are much greater. Under these conditions, some anti-aging medications may have a beneficial direct effect on adult survival, but inconsistent deleterious side-effects which are difficult to control or to resolve, especially if no data are collected on the effects of the substance on the other life-history characters. It is at least conceivable that the inconsistent results observed with resveratrol are due to poorly controlled recondite effects on other life-history characters, such as female fecundity. Wood et al. (2004) tested the effect of resveratrol in a low calorie environment (Fig. 12.3 left dotted lines) and determined that no significant increase in life span was observed. They concluded that the lack of a response in the DR environment suggested that resveratrol must extend life span through some mechanism that is related to caloric restriction. We would suggest that, given the difficulty of reproducing an anti-aging effect of resveratrol, variable secondary life-historical effects could be obscuring its general impact on healthspan. However, pharmacological anti-aging effects that are not robust over trials that have a range of culture conditions suggest that such drug treatments may not have the consistency that would warrant their further study for the purpose of medical applications. Table 12.1 summarizes various longevity-increasing compounds and their possible adverse effects on healthspan.

12.3 Hamiltonian and Genomic Approaches to Healthspan Manipulation

12.3.1 Hamiltonian Gerontology Versus Aging as Cumulative Damage

The common assumption among many gerontologists, particularly those that do not study aging from an evolutionary perspective, is that aging is a process of accumulating damage. With age, it is supposed that organisms accumulate damage through oxidation, free radicals and the like. It is doubtful that significant progress will be made in the manipulation of aging with these presuppositions.

Bluntly put, the falsity of conventional *damage* theories of aging is well demonstrated by the following facts. (1) There are fissile organisms that show no detectable aging, both unicellular and multicellular (Finch 1990; Rose 1991). (2) Non-fissile species with ovigerous reproduction nonetheless are sustained by unbroken cell lineages that are hundreds of millions of years old, whether these lineages engage in sex or not. (3) Aging in some laboratory cohorts of sufficient size comes to a halt at later ages, as discussed previously. Together these findings

Table 12.1 Longevity-increasing compounds and their potential adverse effects on healthspan

Longevity stimulating substance	Reduces food intake	Possibly sedates	Reduces male mating success	Decreases viability	Decreases metabolism	Depends on genetic background	Overly dependent on environment	Reference
Urea	X							Joshi et al. (1996)
Lithium		X	X					Matsagas et al. (2009)
Resveratrol						X	X	Bass et al. (2007), Wood et al. (2004)
Penylbutyrate (PBA)						X		Kang et al. (2002)
Lamotrigine		X			X			Avanesian et al. (2010)
Deuterium				X				Hammel et al. (2013)

falsify any theory of aging that is based on a universal process of cumulative damage akin the Second Law of Thermodynamics.

Aging is instead due to the declines in the Hamilton's Forces of natural selection which occur at the start of adult life in species with clear separation of the products of reproduction from the adult soma (Hamilton 1966; Charlesworth 1980; Rose 1991; Rose et al. 2007). Because natural selection produces adaptation, as the power of natural selection declines, a decline in adaptation with age is expected. Evolutionary biologists have further been able to readily and substantially postpone fruit fly aging by manipulating Hamilton's Forces (Rose and Charlesworth 1980; Luckinbill et al. 1984; Rose 1991; Rose et al. 2004), a track record that is unmatched by attempts to manipulate aging based on non-evolutionary gerontological theories such as those based on cumulative damage. This leads us to conclude that Hamiltonian gerontology, as outlined in Rose (1991) and developed further in Mueller et al. (2011), delivers the best scientific foundation on which to design or evaluate attempts to intervene in aging.

Rose et al. (2010) addresses at length the question of how to develop Hamiltonian strategies with which to ameliorate human aging. The strategies that they discuss are based on starting with organisms that have had their aging slowed by manipulating Hamilton's forces of natural selection and then reverse engineering the biology of those longer-lived organisms to discover interventions that can be used to ameliorate aging in other organisms, including humans. The so-called "Methuselah Flies" that have evolved slower or delayed aging (Rose et al. 2004) are readily available sources of physiological and genomic information with which to find candidate substances that might ameliorate healthspan. In particular, these flies have also been shown to have greater (i) stress resistance, (ii) total reproductive output, and (iii) athletic capacity (Rose et al. 2004). Thus these are not flies that have achieved greater life span as a result of reduced overall reproductive output; rather, they have massively extended healthspan.

12.3.2 *Finding Which Genes to Target with Pharmaceuticals or Nutritional Substances*

As explained in Rose et al. (2010), in 2006 Rose and colleagues compared whole-genome gene-expression patterns in Methuselah Flies with their matched controls. They found about 1000 genes showing statistically consistent differences in expression. These genes are presumptive indicators of the genetic changes that underlie the substantially ameliorated aging achieved using Hamiltonian methods in fruit flies. Seven hundred of these genes had matching orthologous loci in the human genome and about 100 of the 700 human genes were considered candidate pathways to target to slow aging in *both* fruit flies and humans based on parallel findings from genomic analysis in the two species.

But that early gene-expression analysis was only a first step toward the genomic analysis of the genetic foundations of Hamiltonian healthspan extension.

Re-sequencing studies in *Drosophila* have shown that experimentally evolved differences in aging involve SNP frequency changes at hundreds of locations across the fruit fly genome (Burke et al. 2010; Rose et al. 2011). Because of this, it will be very difficult, if not impossible, to find effective anti-aging pharmaceutical agents that increase healthspan by targeting a single pathway.

12.3.3 Can We Use Multiple Supplements to Slow Aging?

If aging is due to just a few "master regulatory genes" (vid. Guarente and Kenyon 2000) or a small number of types of accumulating damage (e.g. de Grey and Rae 2007), then we can suppose that massively effective "anti-aging" supplements containing just a few substances might be discovered. Radically successful anti-aging formulations would then only have to target those few genes or stop a few pathways of accumulating damage. But all the experimental evidence on this point suggests instead that aging is rarely, and perhaps never, due to just a few master regulatory genes.

From a Hamiltonian perspective, it is clear that in order to slow aging with substances, we will need to retune hundreds of genetically defined mechanisms of aging. Natural selection can do this for us as Methuselah Flies demonstrates. It will be very hard to get a small number of powerful pharmaceuticals to do this, but numerous substances of individually small effect conceivably might. Thus Rose et al. (2010) propose that the best strategy to emulate the effects of natural selection in extending life span might be nutritional supplementation with many supplements that individually have physiological effects of small magnitude.

But this does not necessarily mean ingesting the hundreds of supplements that many modern-day molecular biologists and physicians recommend. [In fact, we predict a failure of such supplementation to produce extended human healthspans, for reasons we will discuss later.] The Hamiltonian perspective suggests using nutritional supplements in the same manner as evolution often uses genetic variants of small effect. Rose et al. (2010) proposed screening candidate "nutrigenomic agents" for small to moderate benefits, just as natural selection screens new genetic variants for their beneficial effects. In the case of genomically-informed substance testing, the experimenter can choose candidate substances based on biochemical information about the effects of candidate substances on the specific pathways that genomic analysis of healthspan extension has identified. This seems like a plausible strategy. However, we will suggest here that it may face potentially fatal challenges.

12.4 The Poisoned Chalice Problem: Do Animals Perceive Too Many Novel Substances as Poison?

There is a widespread belief that supplementing our diets with large amounts of isolated nutrients or vitamins will enhance healthspan, a belief that motivates many thousands of people to take a plethora of supplements that have one or another claimed or merely conjectured health benefit (vid. Kurzweil and Grossman 2004, 2009). Perhaps because of the universal failure to find a single Ponce de Leon substance that provides everlasting youth in fruit flies, mice, or humans, the present-day hope is to combine hundreds of substances for a net enhancement of healthspan. The Hamiltonian strategy of Rose et al. (2010) is no exception to this general ambition. It seems plausible that if you have many single-substance successes, one could combine these substances and in sum propitiously compound life-extending properties.

Unfortunately, studies of both humans and mice have not found that combined supplementation is more successful than supplementation with single substances. For example, Macpherson et al. (2013) performed a meta-analysis of randomized controlled studies in humans. Across all studies, no effect of multivitamin treatment on all-cause mortality was seen. Furthermore, cohort studies of human multivitamin use and mortality have found no benefit (Watkins 2000; Park et al. 2011). Among the diverse studies of multifold supplementation, it is clear that those of Spindler (see, e.g., Spindler 2012; Spindler et al. 2014) have achieved the highest standards of design and replication. Spindler et al. (2014) performed isocaloric studies in mice to test the hypothesis that complex mixtures of dietary supplements including vitamins, phytochemicals, and other nutraceuticals could increase the longevity of initially healthy mammals. In addition, nutraceutical, vitamin, or mineral combinations that have had success in previous studies were tested again. Spindler et al. (2014) found that there was no significant increase in rodent life span for any supplement mixture *including combinations that had been reported to increase life span in previous experiments.* Also, some of the more complex mixtures tested significantly decreased life span.

We have an evolutionary hypothesis that we would like to offer to explain these experimental results. We also suggest that this hypothesis provides a cautionary note even for the Hamiltonian and genomic strategies advocated by Rose et al. (2010). We call this the "poisoned chalice" hypothesis.

Metazoa are not Erlenmeyer flasks. That is, our bodies are not inert vessels in which numerous parallel biochemical reactions occur independently of each other. Instead, natural selection has created enormous "kluged" networks of physiology that collectively enhance our Darwinian fitness, often by incorporating bits and pieces of molecular machinery that act both summatively and sometimes in antagonism with each other. Furthermore, this complex large-scale interacting network has feedback circuits that respond to features of the environment, much like control-theory designed stabilizing components of complex electronics function to sustain circuit signaling integrity and to prevent destructive overload of circuits.

Thus, in the case of *Drosophila,* we know that flies actively modify their physiological functioning in the event of elevated temperatures, the so-called "heat shock" response. Likewise, moderate dietary restriction abruptly modifies reproductive activity (Chippindale et al. 1993), which is a response that is implicated in the physiological machinery underlying the extension of longevity in conjunction with the decrease in reproduction observed in dietary restriction. Likewise, exposure to urea elicits an abrupt reduction in reproduction, which may at least partly explain the resulting extension in life span, as we have discussed here.

Perhaps the provision of many novel substances to humans, mice, or fruit flies elicits the same kind of physiological responses as those elicited by urea exposure or moderate dietary restriction? That is, in the specific case of *Drosophila,* when fruit flies are exposed to culture medium that this is so novel that their physiology reacts as if they are in an environment which is suboptimal for reproduction, they may shut down functional components of their aggregate physiology. Such "shut downs" may reduce reproduction, or they may curtail activity through sedation, or they may indeed curtail cellular repair processes vital to organismal survival. Likewise, we would suggest, assaulting human physiology with numerous substances that our digestive machinery and other pieces of our metabolic machinery react to as low-grade poisons may trigger toxicity reactions. The net effect of too many of these toxicity reactions may be to reduce overall healthspan, not increase it.

In effect, we suggest, the provision of multiple, purified, novel substances of individually small effect may result in a supplementation cocktail that is a poisoned chalice for the kind of metabolic machinery that differentiated multicellular animals possess. In a phrase, almost all such complex supplementation regimens may amount to a poisoned chalice for healthspan, not some elixir of life.

As an alternative, we suggest, what is needed is dietary intervention that our physiologies unequivocally "accept" as healthy food, forestalling any adverse poisoned-chalice reaction. The question then becomes, what would such an optimal food be like? How can we find such an ideal dietary regime?

12.5 Is There a Hamiltonian Holy Grail for Human Healthspan Extension? Going Backward in Evolutionary Time as You Go Forward in Biological Time

Recently we had another idea of some relevance for the discovery of better diets based on evolutionary biology. We developed this idea from considering the evolution of a population that has undergone a substantial change of diet in recent evolutionary time. The evidence we have from experimental evolution suggests rapid adaptation to a novel environment (e.g. Matos et al. 2000), particularly for early components of fitness such as developmental speed and initial fecundity. But Hamilton's forces of natural selection fall with adult age in almost all cases, which should produce weaker adaptation at later ages in the first generations after dietary change (Mueller et al. 2011). Explicit simulations of this have the expected effect: lack of adaptation to the new environment at later ages (Rutledge et al. in preparation).

We have experimentally tested this idea in our *D. melanogaster* lab populations, because they have undergone a major change in diet since their introduction to the laboratory in 1975, approximately 1000 generations ago. As expected from this age-dependent effect on adaptation, at later ages our flies are better adapted to

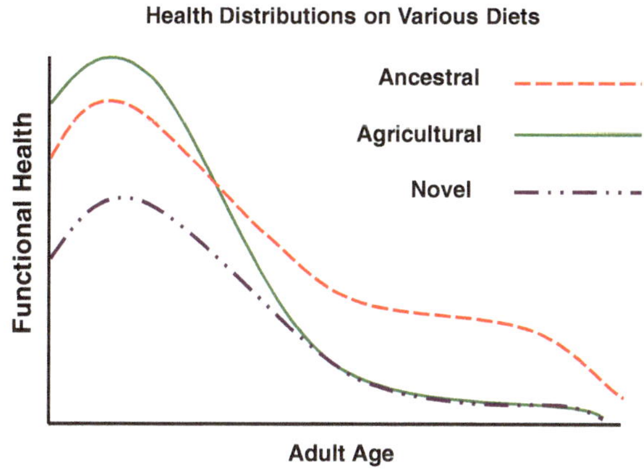

Fig. 12.4 The functional health measure in our research with *Drosophila* is usually measured as the product of an individual's probability of survival to a specific adult age and fertility at that age. The graph summarizes recent data of ours in which individuals cope as well or better with an evolutionarily recent "agricultural" diet as on their ancestral diet, but *only* at early ages. At later ages, the Hamiltonian diet hypothesis infers that older individuals should fare better on an ancestral diet, when Hamilton's forces of natural selection have weakened enough to have short-changed adaptation to agricultural food. Individuals raised on evolutionarily novel "industrial" foods fare considerably worse than those raised on ancestral foods at all ages

a crude approximation of their ancestral diet in the wild (Rutledge et al. in prep.) (Fig. 12.4). So the idea works in explicit theory and in careful laboratory experiments. Its practical application is that older humans might be able to improve their healthspans by switching to diets that resemble those of our Paleolithic ancestors, given that our adoption of the agricultural Neolithic diet is relatively recent in evolutionary terms, about 200–400 generations ago.

But the more general principle that this line of research suggests is that there indeed *are* complex dietary changes that can be made which will enhance our healthspans. In the particular case of our fruit flies given their ancestral diet from more than 1000 generations ago, we have stumbled in the direction of such a dietary change based on our knowledge of their evolutionary history, at least over the last few hundred years. This supports the general ambition to provide improved diets for human healthspans based on evolutionary insights.

What remains an entirely unanswered question is whether or not we can ever do better, for fruit flies or humans, than the adoption of the diet that evolution long ago tuned our physiologies to exploit efficiently. We can of course more effectively home in on what evolution has already achieved, by learning more about the details of fruit fly or human evolutionary histories. But can we do even better than to exploit what evolution has already accomplished, with respect to the creation of a maximal healthspan? That remains a tantalizing question for which we have no answer at present.

References

Avanesian A, Khodayari B, Felgner JS, Jafari M (2010) Lamotrigine extends lifespan but compromises health span in *Drosophila melanogaster*. Biogerontology 11(1):45–52. doi:10.1007/s10522-009-9227-1

Bacon F (1637) Francisci Baconis … historia vitae et mortis. Maire

Bahadorani S, Bahadorani P, Phillips JP, Hilliker AJ (2008) The effects of vitamin supplementation on *Drosophila* life span under normoxia and under oxidative stress. J Gerontol Ser A Biol Sci Med Sci 63(1):35–42

Bass TM, Weinkove D, Houthoofd K, Gems D, Partridge L (2007) Effects of resveratrol on lifespan in *Drosophila melanogaster* and *Caenorhabditis elegans*. Mech Ageing Dev 128(10):546–552. doi:10.1016/j.mad.2007.07.007

Bauer JH, Goupil S, Garber GB, Helfand SL (2004) An accelerated assay for the identification of lifespan-extending interventions in *Drosophila melanogaster*. Proc Natl Acad Sci USA 101(35):12980–12985. doi:10.1073/pnas.0403493101

Baur JA, Pearson KJ, Price NL, Jamieson HA, Lerin C, Kalra A, Prabhu VV, Allard JS, Lopez-Lluch G, Lewis K, Pistell PJ, Poosala S, Becker KG, Boss O, Gwinn D, Wang M, Ramaswamy S, Fishbein KW, Spencer RG, Lakatta EG, Le Couteur D, Shaw RJ, Navas P, Puigserver P, Ingram DK, de Cabo R, Sinclair DA (2006) Resveratrol improves health and survival of mice on a high-calorie diet. Nature 444(7117):337–342. doi:10.1038/nature05354

Bonilla E, Medina-Leendertz S, Dias S (2002) Extension of life span and stress resistence of Drosophila melanogaster by long-term supplementation with melatonin. Exp Gerontol 37(5):629–638. doi:10.1016/S0531-5565(01)00229-7

Burke MK, Dunham JP, Shahrestani P, Thornton KR, Rose MR, Long AD (2010) Genome-wide analysis of a long-term evolution experiment with *Drosophila*. Nature 467(7315):587–590. doi:http://www.nature.com/nature/journal/v467/n7315/abs/nature09352.html-supplementary-information

Carey JR, Liedo P, Orozco D, Vaupel JW (1992) Slowing of mortality rates at older ages in large medfly cohorts. Science 258(5081):457–461

Charlesworth B (1980) Evolution in age-structured populations. Cambridge University Press, Cambridge

Charlesworth B (2001) Patterns of age-specific means and genetic variances of mortality rates predicted by the mutation-accumulation theory of ageing. J Theor Biol 210(1):47–65. doi:http://dx.doi.org/10.1006/jtbi.2001.2296

Charlesworth B, Charlesworth D (1973) The measurement of fitness and mutation rate in human populations. Ann Hum Genet 37(2):175–187

Chippindale AK, Leroi AM, Kim SB, Rose MR (1993) Phenotypic plasticity and selection in *Drosophila* life-history evolution. I. Nutrition and the cost of reproduction. J Evol Biol 6(2):171–193. doi:10.1046/j.1420-9101.1993.6020171.x

Chippindale AK, Leroi AM, Saing H, Borash DJ, Rose MR (1997) Phenotypic plasticity and selection in *Drosophila* life history evolution. 2. Diet, mates and the cost of reproduction. J Evol Biol 10(3):269–293. doi:10.1046/j.1420-9101.1997.10030269.x

Curtsinger JW, Fukui HH, Townsend DR, Vaupel JW (1992) Demography of genotypes: failure of the limited life-span paradigm in *Drosophila melanogaster*. Science 258(5081):461–463

De Grey A, Rae M (2007) Ending aging: the rejuvenation breakthroughs that could reverse human aging in our lifetime. St. Martin's Print, New York

Djawdan M, Sugiyama TT, Schlaeger LK, Bradley TJ, Rose MR (1996) Metabolic aspects of the trade-off between fecundity and longevity in *Drosophila melanogaster*. Physiol Zool 69(5):1176–1195. doi:10.2307/30164252

Finch CE (1990) Longevity, senescence, and the genome. The University of Chicago Press, Chicago, London

Greenwood M, Irwin JO (1939) The biostatistics of senility. Hum Biol 11(1):1–23. doi:10.2307/41447403

Guarente L, Kenyon C (2000) Genetic pathways that regulate ageing in model organisms. Nature 408(6809):255–262. doi:10.1038/35041700

Hamilton WD (1964) The genetical evolution of social behaviour. I. J Theor Biol 7(1):1–16. doi:http://dx.doi.org/10.1016/0022-5193(64)90038-4

Hamilton WD (1966) The moulding of senescence by natural selection. J Theor Biol 12(1):12–45. doi:http://dx.doi.org/10.1016/0022-5193(66)90184-6

Hammel SC, East K, Shaka AJ, Rose MR, Shahrestani P (2013) Brief early-life non-specific incorporation of deuterium extends mean life span in *Drosophila melanogaster* without affecting fecundity. Rejuvenation Res 16(2):98–104. doi:10.1089/rej.2012.1368

Haycock DB (2009) Mortal coil: a short history of living longer. Yale University Press, New Haven

Howitz KT, Bitterman KJ, Cohen HY, Lamming DW, Lavu S, Wood JG, Zipkin RE, Chung P, Kisielewski A, Zhang LL, Scherer B, Sinclair DA (2003) Small molecule activators of sirtuins extend *Saccharomyces cerevisiae* lifespan. Nature 425(6954):191–196. doi:10.1038/nature01960

Jafari M, Long AD, Mueller LD, Rose MR (2006) The pharmacology of aging in *Drosophila*. Curr Drug Targets 7(11):1479–1483. doi:10.2174/138945011060701 1479

Jafari M, Felgner JS, Bussel II, Hutchili T, Khodayari B, Rose MR, Vince-Cruz C, Mueller LD (2007a) Rhodiola: a promising anti-aging Chinese herb. Rejuvenation Res 10(4):587–602. doi:10.1089/rej.2007.0560

Jafari M, Khodayari B, Felgner J, Bussel II, Rose MR, Mueller LD (2007b) Pioglitazone: an anti-diabetic compound with anti-aging properties. Biogerontology 8(6):639–651. doi:10.1007/s10522-007-9105-7

Jafari M, Rose MR (2006) Rules for the use of model organisms in anti-aging pharmacology. Aging Cell 5(1):17–22. doi:10.1111/j.1474-9726.2006.00195.x

Joshi A, Shiotsugu J, Mueller LD (1996) Phenotypic enhancement of longevity by environmental urea in *Drosophila melanogaster*. Exp Gerontol 31(4):533–544

Kaeberlein M, McDonagh T, Heltweg B, Hixon J, Westman EA, Caldwell SD, Napper A, Curtis R, DiStefano PS, Fields S, Bedalov A, Kennedy BK (2005) Substrate-specific activation of sirtuins by resveratrol. J Biol Chem 280(17):17038–17045. doi:10.1074/jbc.M500655200

Kang H-L, Benzer S, Min K-T (2002) Life extension in *Drosophila* by feeding a drug. Proc Natl Acad Sci 99(2):838–843. doi:10.1073/pnas.022631999

Kurzweil R, Grossman T (2004) Fantastic voyage: live long enough to live forever. Rodale, Emmaus, PA

Kurzweil R, Grossman T (2009) Transcend: nine steps to living well forever. Rodale, Emmaus, PA

Lagouge M, Argmann C, Gerhart-Hines Z, Meziane H, Lerin C, Daussin F, Messadeq N, Milne J, Lambert P, Elliott P, Geny B, Laakso M, Puigserver P, Auwerx J (2006) Resveratrol improves mitochondrial function and protects against metabolic disease by activating SIRT1 and PGC-1alpha. Cell 127(6):1109–1122. doi:10.1016/j.cell.2006.11.013

Lee RD (2003) Rethinking the evolutionary theory of aging: transfers, not births, shape senescence in social species. Proc Natl Acad Sci 100(16):9637–9642. doi:10.1073/pnas.1530303100

Leroi AM, Kim SB, Rose MR (1994) The evolution of phenotypic life-history trade-offs: an experimental study using *Drosophila melanogaster*. Am Nat 144(4):661–676. doi:10.2307/2462943

Loeb J, Northrop JH (1917) What determines the duration of life in metazoa? Proc Natl Acad Sci USA 3(5):382–386

Luckinbill LS, Arking R, Clare MJ, Cirocco WC, Buck SA (1984) Selection for delayed senescence in *Drosophila melanogaster*. Evolution 38(5):996–1003. doi:10.2307/2408433

Macpherson H, Pipingas A, Pase MP (2013) Multivitamin-multimineral supplementation and mortality: a meta-analysis of randomized controlled trials. Am J Clin Nutr 97(2):437–444. doi:10.3945/ajcn.112.049304

Maynard Smith J (1958) The effects of temperature and of egg-laying on the longevity of *Drosophila subobscura*. J Exp Biol 35:832–842

Matos M, Rose MR, Pité MTR, Rego C, Avelar T (2000) Adaptation to the laboratory environment in *Drosophila subobscura*. J Evol Biol 13(1):9–19. doi:10.1046/j.1420-9101.2000.00116.x

Matsagas K, Lim DB, Horwitz M, Rizza CL, Mueller LD, Villeponteau B, Rose MR (2009) Long-term functional side-effects of stimulants and sedatives in *Drosophila melanogaster*. PLoS One 4(8):e6578. doi:10.1371/journal.pone.0006578

McArthur MC, Sohal RS (1982) Relationship between metabolic rate, aging, lipid peroxidation, and fluorescent age pigment in milkweed bug, *Oncopeltus fasciatus* (Hemiptera). J Gerontol 37(3):268–274

McCay CM, Maynard LA, Sperling G, Barnes LL (1939) Retarded growth, lifespan, ultimate body size and age changes in the albino rat after feeding diets restricted in calories. Nutr Rev 33(8):241–243. doi:10.1111/j.1753-4887.1975.tb05227.x

Miller RA, Harrison DE, Astle CM, Baur JA, Boyd AR, de Cabo R, Fernandez E, Flurkey K, Javors MA, Nelson JF, Orihuela CJ, Pletcher S, Sharp ZD, Sinclair D, Starnes JW, Wilkinson JE, Nadon NL, Strong R (2011) Rapamycin, but not resveratrol or simvastatin, extends life span of genetically heterogeneous mice. J Gerontol Ser A Biol Sci Med Sci 66A(2):191–201. doi:10.1093/gerona/glq178

Moment G, Behnke J, Finch C (1978) The Ponce de Leon Trail Today. The Biology of Aging. Springer, US, pp 1–17

Morselli E, Maiuri MC, Markaki M, Megalou E, Pasparaki A, Palikaras K, Criollo A, Galluzzi L, Malik SA, Vitale I, Michaud M, Madeo F, Tavernarakis N, Kroemer G (2010) Caloric restriction and resveratrol promote longevity through the Sirtuin-1-dependent induction of autophagy. Cell Death Dis 1:e10. doi:10.1038/cddis.2009.8

Mueller LD, Rauser CL, Rose MR (2011) Does aging stop? Oxford University Press, Oxford

Mueller LD, Rose MR (1996) Evolutionary theory predicts late-life mortality plateaus. Proc Natl Acad Sci 93(26):15249–15253

Needham J (1954–2008) Science and civilization in China. 7 vols. Cambridge University Press, United Kingdom

Park S-Y, Murphy SP, Wilkens LR, Henderson BE, Kolonel LN (2011) Multivitamin use and the risk of mortality and cancer incidence: the multiethnic cohort study. Am J Epidemiol. doi:10.1093/aje/kwq447

Rauser CL, Abdel-Aal Y, Shieh JA, Suen CW, Mueller LD, Rose MR (2005) Lifelong heterogeneity in fecundity is insufficient to explain late-life fecundity plateaus in *Drosophila melanogaster*. Exp Gerontol 40(8–9):660–670. doi:10.1016/j.exger.2005.06.006

Rauser CL, Mueller LD, Rose MR (2003) Aging, fertility, and immortality. Exp Gerontol 38(1–2):27–33

Rauser CL, Tierney JJ, Gunion SM, Covarrubias GM, Mueller LD, Rose MR (2006) Evolution of late-life fecundity in *Drosophila melanogaster*. J Evol Biol 19(1):289–301. doi:10.1111/j.1420-9101.2005.00966.x

Rose MR (1984) Laboratory evolution of postponed senescence in *Drosophila melanogaster*. Evolution 38(5):1004–1010. doi:10.2307/2408434

Rose MR (1991) Evolutionary biology of aging. Oxford University Press, New York

Rose MR, Long DA, Mueller LD, Rizza CL, Matsagas KC, Greer LF, Villeponteau B (2010) Evolutionary nutrigenomics, The future of aging: pathways to human life extension. Springer, Dordrecht

Rose MR, Mueller LD, Burke MK (2011) New experiments for an undivided genetics. Genetics 188(1):1–10. doi:10.1534/genetics.111.128900

Rose MR, Passananti HB, Matos M (2004) Methuselah flies: a case study in the evolution of aging. World Scientific Publishing, Singapore

Rose MR, Rauser CL, Benford G, Matos M, Mueller LD (2007) Hamilton's forces of natural selection after forty years. Evolution 61(6):1265–1276. doi:10.1111/j.1558-5646.2007.00120.x

Shahrestani P, Quach J, Mueller LD, Rose MR (2012) Paradoxical physiological transitions from aging to late life in *Drosophila*. Rejuvenation Res 15(1):49–58. doi:10.1089/rej.2011.1201

Spindler SR (2012) Review of the literature and suggestions for the design of rodent survival studies for the identification of compounds that increase health and life span. Age 34(1):111–120. doi:10.1007/s11357-011-9224-6

Spindler SR, Mote PL, Flegal JM (2014) Lifespan effects of simple and complex nutraceutical combinations fed isocalorically to mice. Age 36(2):705–718. doi:10.1007/s11357-013-9609-9

Valenzano DR, Terzibasi E, Genade T, Cattaneo A, Domenici L, Cellerino A (2006) Resveratrol prolongs lifespan and retards the onset of age-related markers in a short-lived vertebrate. Curr Biol 16(3):296–300. doi:http://dx.doi.org/10.1016/j.cub.2005.12.038

Watkins ML, Erickson JD, Thun MJ, Mulinare J, Heath CW (2000) Multivitamin use and mortality in a large prospective study. Am J Epidemiol 152(2):149–162. doi:10.1093/aje/152.2.149

Wood JG, Rogina B, Lavu S, Howitz K, Helfand SL, Tatar M, Sinclair D (2004) Sirtuin activators mimic caloric restriction and delay ageing in metazoans. Nature 430(7000):686–689. doi:10.1038/nature02789

Chapter 13
Atmosphere, Metabolism and Longevity

Khatchik Muradian

Omne vivum ex ovo in hypoxia et hypercapnia

Abstract *Are the current atmosphere gaseous composition and unlimited oxygen consumption mode optimal for the health and longevity?* Food and oxygen are the two tightly related substances critical for the life support. Recommendations to restrict food consumption below the ad libitum level are recognized and have many followers worldwide. No analogous suggestions concerning the O_2 consumption are known. Living beings originated and most part of their evolution occurred in atmospheres with extraordinarily high CO_2 and low O_2. In contrast, O_2 content in the modern atmosphere is exceeding CO_2 more than 500 fold. Such dramatic changes should provoke conflicting situations. According to the proposed 'nostalgia' concept, living systems somehow 'remember' and are striving to return to the less conflicting primordial environments. Maintenance of *Drosophila* in hypoxic atmospheres (5, 10 and 15 % of O_2) started from the 20 day extended their mean but not maximum life span. Optimal hypoxia was lower for the older flies (15 % O_2 started from the 40 days and 18 % for the 50 days). Data accumulated do not exclude that modified atmospheres and diets could have additive positive effects on longevity. People in the developed countries are already living in artificial atmospheres with optimized physical parameters—the air is conditioned, filtered, ozonized, ionized, humidified, deodorized etc. However, the same air composition could hardly be optimal for everyone and in all situations. Supplementation of the air conditioners with additional gadgets could ensure optimization of the atmosphere gaseous composition, as well. Despite the importance and technical availability, little is known about efficiency of the individually and situationally optimized atmospheres in human aging and longevity.

K. Muradian (✉)
State Institute of Gerontology of National Academy of Medical Sciences of Ukraine,
67, Vyshgorodskaya Street, 04114 Kiev, Ukraine
e-mail: kkm@geront.kiev.ua

A.M. Vaiserman et al. (eds.), *Life Extension*, Healthy Ageing and Longevity 3,
DOI 10.1007/978-3-319-18326-8_13

Keywords Aging · Evolution · Oxygen · Metabolism · *Drosophila*

13.1 Introduction

Living beings originated and most part of their evolution occurred in atmospheres with extraordinarily high CO_2 and low O_2, whereas in the modern atmosphere the O_2 content exceeds CO_2 more than 500 fold (Walker 1985; Nunn 1998; Berner 1999; Holland 2006; Kurbel 2014; Mills et al. 2014; Planavsky et al. 2014). Such dramatic changes might induce principal alterations in the intracellular and external environment, e.g., shifting the cellular redox from reducing to an oxidizing milieu or modifying the biogeochemical cycles of the carbon, nitrogen, oxygen, sulfur etc. Moreover, the atmosphere composition and temperature underwent substantial fluctuations causing mass extinction of populations and species (Huey and Ward 2005; Goldblatt et al. 2006; Harrison et al. 2006; Sessions et al. 2009; Taylor and McElwain 2010). The survived species apparently had to reshape their adaptability resources. However, redistribution of the initially limited adaptive potential may occur at expanse of declined functionality of the other life supporting systems. Therefore, it is pertinent asking whether the current atmosphere gaseous composition is optimal for aging and longevity.

13.2 'Nostalgia' Concept of Longevity

The basic life supporting systems are highly conservative. Analyses of the microbiota, stromatolites and other fossils from the Archaean and Proterozoic eons support the contention that the basic features of the living beings remained almost the same throughout the billions of years passed from the times of the universal common ancestor (Tyler and Barghoorn 1954; Schopf 1994; Scorpf et al. 2002; Sharma and Shulka 2009). But why the relatively small repertoire, e.g., few nucleotides and around 20 amino acids were selected for the genome and proteins among the thousands of other possible candidates? A simplest answer could be: because they were optimal at those early life conditions. Later, the conditions changed dramatically but the basic life-supporting elements remained practically the same inevitably provoking conflicting situations. Such conflicts were apparently resolved by the evolutionary trade-offs and compromises primarily for the benefit of reproduction and other situational short-term interests. Longevity could hardly be among them. Modeling long-passed environmental conditions could at least partly neutralize such conflicts in the modern species. According to the 'nostalgia concept', living systems somehow 'remember' and are striving to return to the less conflicting conditions. It may explain why modeling evolutionary earlier atmospheres often resulted in positive effects, as it was in our experiments with hypoxic and hypercapnic atmospheres (Muradian 2008).

13.3 Co-evolution of the Atmosphere and Living Beings

The emergence of photosynthetic microorganisms around 3 billion years ago marked the beginning of co-evolution of the earth atmosphere, flora and fauna (Beerling and Berner 2005; Taylor and McElwain 2010; Lyons et al. 2014). Using the atmospheric CO_2 and the sun ultraviolet, than plenty on the earth surface, the microorganisms synthesized glucose and other necessary compounds and released O_2 as a by-product. The mission appeared so successful that in cooperation with plants they managed decreasing the CO_2 content to the current 0.04 % reciprocally increasing O_2 up to the 20.95 %. It is remarkable that in the body of most modern animal species the O_2 and CO_2 concentrations are comparable, as it possibly could be in the atmosphere at the times of the multicellular organisms' origin. In our sampling of mammalian species (data collected from the publications available in the PubMed and in the AnAge database) the blood partial pressure of O_2 (Po_2) was almost twice higher than Pco_2 ($Po_2/Pco_2 = 1.89 \pm 0.88$). However, it is known that another half of CO_2 produced in the cells is 'buffered' in biological liquids and cells in the form of HCO_3^- which could be converted back to CO_2 by the carbonic anhydrase. The latter is known as an extraordinary efficient enzymatic system encoded by many dozens of multifunctional and evolutionary conserved genes (Gilmour and Perry 2009; Everaert et al. 2011; Imtaiyaz et al. 2013). The necessity of such powerful enzymatic network highlights the importance of the balanced Pco_2 for the viability.

The basic energy generating processes could simply be described as: $R_1 + O_2 \leftrightarrow R_2 + CO_2$; where R_1 and R_2 are bioorganic compounds. According to the universal laws of chemical kinetics, increased Po_2 or decreased Pco_2 should shift the equilibrium to the right and increase the rate of energy generation and vice versa. In mammals, the non-parametric Spearman rank order coefficient revealed negative correlations of Pco_2 and $[HCO_3^-]$ with the relative metabolic rate (RMR), supporting the idea that elevated CO_2 content could suppress the metabolic rate (Fig. 13.1).

As for the expected positive correlation between the RMR and Po_2, the data available at this moment seems ambivalent. After excluding the three outlier rodent species with higher metabolic rates (mouse, rat and hamster with logRMR > 0.6), the correlation between the blood Po_2 and RMR was positive ($R = 0.601$, $P < 0.04$). However, when all currently available species were taken into consideration, the correlation remained positive but far from being significant (Fig. 13.1c). It is not excluded that rodents developed mechanisms to stimulate their energy generation bypassing the Po_2 pathway. It is understood that the presented correlations should be regarded as preliminary results to be checked in larger mammalian and non-mammalian samplings. Nevertheless, it is intriguing whether archaic species could have lower metabolic rate and longer life span due to the higher CO_2 and lower O_2 in the atmosphere.

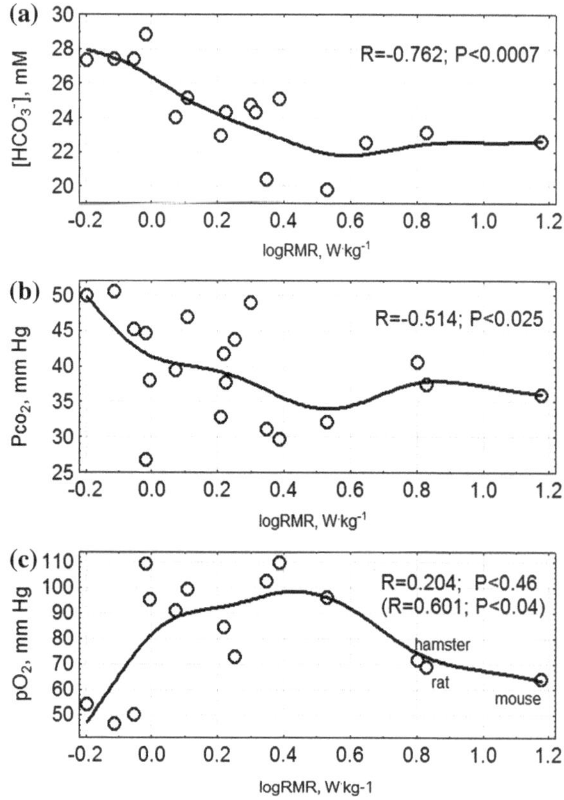

Fig. 13.1 Correlation (non-parametric Spearman) between the mammalian relative metabolic rate (RMR) and the blood concentration of HCO_3^- (**a**) and partial pressures of CO_2 (**b**) and O_2 (**c**)

13.4 Metabolic Rate and Life Span

From the very childhood, we comprehend a simple truth: intensively working objects wear and broke sooner. Biological systems could hardly be an exception from this universal rule. It may explain why the inverse relationship between the rate of metabolism and longevity, first shown by Rubner in 1907, remains a popular gerontological concept since then. In fact, the negative relationship between the metabolic and aging rates is compatible with and could be considered as a more generalized concept of many known hypotheses of aging and longevity primarily associated with the metabolic rate, e.g., the cellular garbage accumulation, generation of the reactive oxygen species, DNA mutations etc. Universality and plasticity of the relationship between the metabolism and longevity was proved on numerous phylogenetic and ontogenetic models (Frolkis and Muradian 1991). Nevertheless, it was surprising to find out that the essence of idea could be known from the ancient times, as it follows from the manuscript of Aristotle written around 350 year BS and entitled "On Longevity and Shortness of Life" (Aristotle 2007).

An important assumption of the Rubner's low is that slower metabolism could be a cause or, at least, a predictor of extended life span. Shown by Leob and Northrop as early as in 1917 (Loeb and Northrop 1917), *Drosophila* incubated at different temperatures has become a popular object for study of such relationship. In our experiments, life span of *Drosophila* incubated within the range of temperatures typical for their natural habitat (15–30 °C) changed in the scales well outmatching other known means of life span modulation. In the studied range of temperatures, the four fold increase of the mean life span (Fig. 13.2a) was associated with around the same degree decrease of the O_2 consumption (Fig. 13.2b) and CO_2 production (Fig. 13.2c) in young and old imagoes. The observed alterations of the metabolism could partly be intermediated by the locomotor activity which changed over 20 fold in the same range of temperatures (Fig. 13.2d).

According ANOVA, significance of the temperature influence on the determinants of longevity was extremely high ($P < 10^{-30}$), much exceeding the effects of aging (usually $P < 0.05$). It is worth mentioning, that temperature-dependency of the *Drosophila* metabolism and survival is a unique model allowing manipulations of not only the aging rate. By a relatively simple change of the temperature, it is possible to enforce an organism to age following any given algorithm—gradual acceleration or deceleration etc. No other known means allow so elegantly modulate the generally complex and rigid processes of metabolism and aging.

Comparative study of mammals could be another impressive demonstration of the universality of the negative relation between the metabolism and longevity (Fig. 13.3). Significance of the non-parametric correlation between the mammalian maximum life span (MLS) and RMR reached the same extraordinary high levels ($P < 10^{-30}$) as at the incubation of *Drosophila* in different temperatures It is understood that the negative relationship between the longevity and metabolism should have certain limits. Life span primarily depends on the genetic background and low predictable environmental variables. To reveal statistical significance, the background variation should adequately exceed unpredictable effects (the 'noise'). Therefore, the absence of statistically significant correlation between the metabolic rate and life span did not obligatorily contradict the Rubner's low but may be explained by the insufficient variation of life span in the analyzed samples compared with the 'noise', as it could be when comparing individuals from the same population, laboratory strains of the same species or conducting comparative analysis within small taxonomic groups. Perhaps that explains the low efficiency of attempts to extend life span by drug-induced inhibition of the metabolic processes or thermoregulation. With few exceptions, such attempts resulted in only marginal effects, as it was in our experiments with application of various inhibitors of the nuclear and mitochondrial transcription, translation and uncoupling into the feeding medium of *Drosophila* or drinking water of mice (Frolkis and Muradian 1991). Although short-term application of such compounds could be a necessary life-saving treatment in certain critical situations, in the life-long chronic experiments on healthy objects this kind of approaches appeared low efficient primarily because of possible negative side-effects.

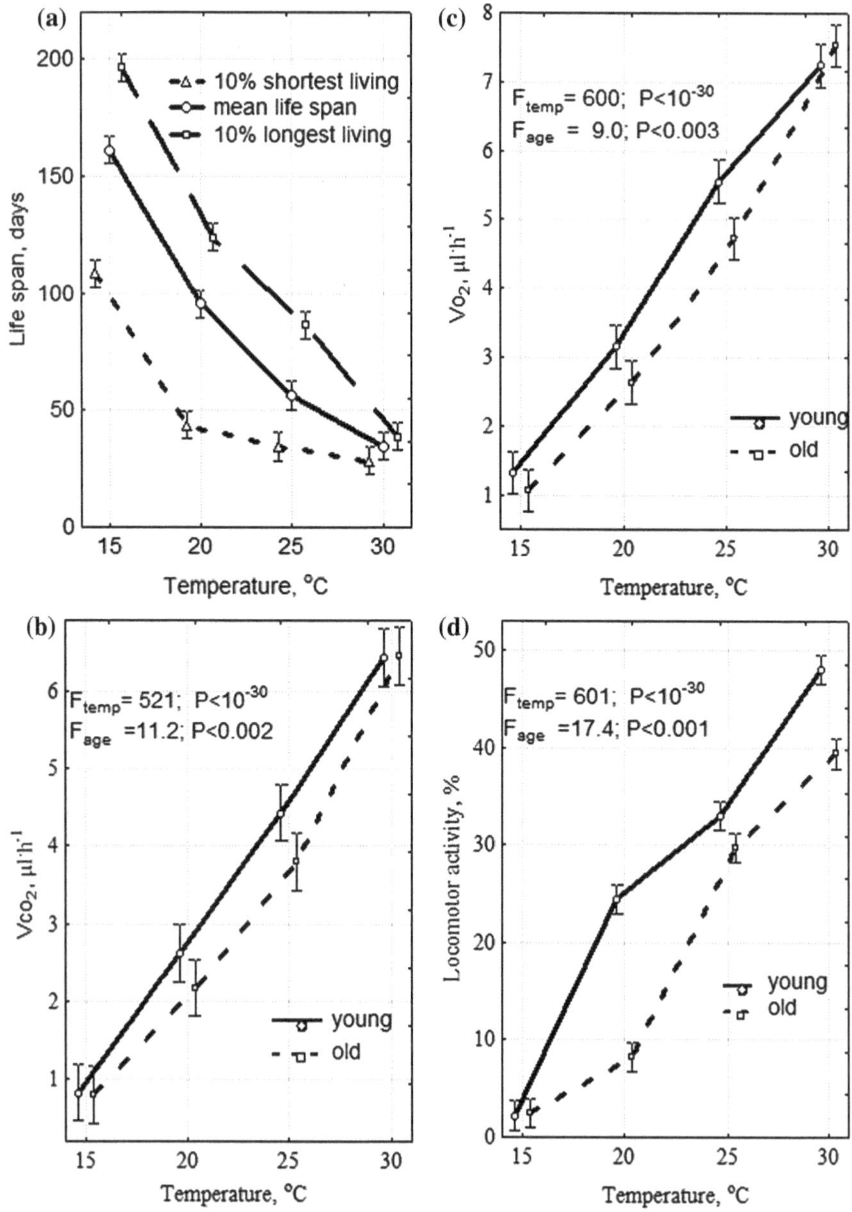

Fig. 13.2 Life span (**a**), rates of the O_2 consumption (**b**), CO_2 production (**c**) and locomotor activity (**d**) of young (10–15 days) and old (45–55 days) *Drosophila melanogaster* incubated at different temperatures. Statistical significance was calculated by the ANOVA

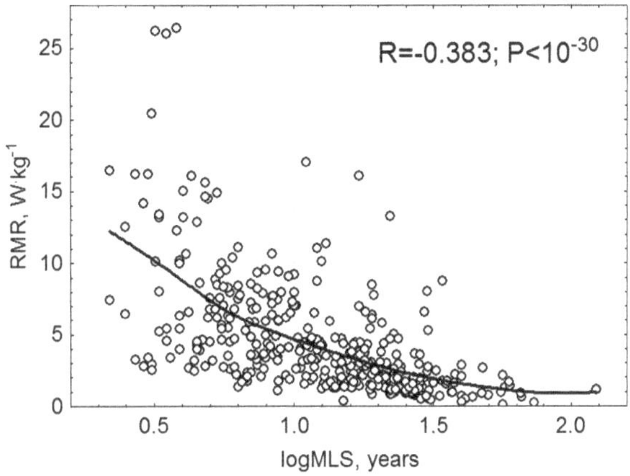

Fig. 13.3 Correlation (non-parametric Spearman) between the mammalian maximum life span (MLS) and relative metabolic rate (RMR)

13.5 'Pull and Push Back' Concept of Longevity

In search of less harmful physiological means of slowing down the metabolic rate, maintenance of biological objects in atmospheres with modified gaseous composition seems more attractive because it is associated with lesser external invasions. Additionally, it could switch on numerous forward and backward loops of the physiological self-regulation. To slow down the rate of metabolic processes it is obviously necessary either decreasing the cellular Po_2 or increasing Pco_2 or the both. The decreased O_2 content in the atmosphere could result in hypoxia and kind of 'pulling' the $O_2 \rightarrow CO_2$ stream back, whereas the increased CO_2 content may counteract the CO_2 escape from the organism, thus 'pushing' the stream back. Both types of manipulations should decline the metabolic rate and hopefully extend life span (Muradian 2013). When checking predictions of the concept on mammals of different species it was found that the blood Pco_2 positively correlated with species maximum life span ($P < 0.02$). Thc positive correlation was even stronger ($P < 0.01$) when relation between the species longevity and blood [HCO_3^-] were analyzed (data not shown). However, we failed to demonstrate the expected negative correlation between the maximum life span and Po_2 (Fig. 13.4).

The stronger correlation of CO_2 with the rates of metabolism and life span compared with the O_2 (Figs. 13.3 and 13.4) could, at least, partly be explained by the differences of their evolutionary history and possible targets in the energy generating processes. As a metabolism regulator, CO_2 obviously appeared much earlier and occupies more basic position than O_2. Pco_2 may regulate the rates of the three reactions of pyruvate decarboxylation in the citric acid cycle occurring according the scheme: R–COOH $\leftrightarrow$ R + CO_2. Higher CO_2 concentrations will obviously move the reaction equilibrium to the left thus slowing the process of energy generation. Unlike the CO_2, the molecular O_2 does not directly participate in the citric acid cycle. Its regulatory role could be comparatively 'passive',

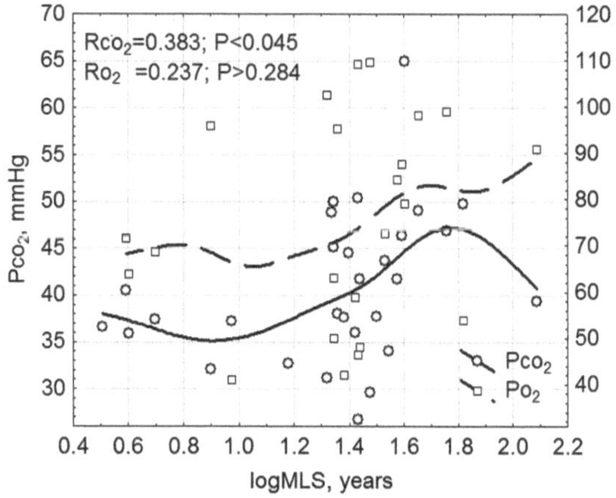

Fig. 13.4 Correlation (non-parametric Spearman) between the mammalian maximum life span (MLS) and the blood partial pressures of CO_2 and O_2 (Pco_2 and Po_2)

because its participation in the energy generation is limited by oxidation of the electrons which reached the very bottom of the mitochondrial electron transport chain.

13.6 Hypoxia

Hypoxia is often regarded as a precondition of malfunction and pathology development. Nevertheless, positive effects of moderate hypoxia, e.g., in vasodilation, ameliorations of the lipid metabolism, hypertension, insulin resistance and glucose tolerance etc. are well established (Kayser 1992; Netzer and Breitenbach 2010; Pollock et al. 2014). Moderate inhibition of the mitochondrial respiration could slow down the metabolic rate and extend life span of the short-lived laboratory animal species (Honda et al. 1993; Feng et al. 2001; Cheung et al. 2008; Copeland et al. 2009; Klok and Harrison 2009; Forgan and Forster 2010; Lee et al. 2010; Hwang and Lee 2011). Hypoxia significantly extends replicative life span of the human cells in culture (Bell et al. 2007; Poulios et al. 2007). Many types of stem cells live in a hypoxic microenvironment, and the hypoxia inducible factor (HIF 1α) is essential for their self-renewal, telomere length maintenance and proliferation (Davy and Allsopp 2011). Moreover, mild hypoxia is well known for all multicellular organisms from the very beginning of their life cycle. During the embryogenesis, all organisms are practically defenseless and should be protected from the environmental hazards by the egg shell in the egg-laying species or by the mother womb in the placental animals. The protecting 'shields', however, become a barrier for the gaseous exchange creating a hypoxic and hypercapnic environment (Muradian 2008, 2013). Thus, the well known Latin proverb "Omne vivum ex ovo" might be extended as "*Omne vivum ex ovo in hypoxia et hypercapnia*" (All living beings originate from an egg in hypoxia and hypercapnia). Crocodiles are few vertebrates which survived the evolutionary atmospheric fluctuations with little morphologic changes. Incubation of American alligator eggs in hypoxia (12 % O_2), normoxia (21 % O_2) and hyperoxia (30 % O_2) showed that hypoxic animals were smaller at hatching. Their post-hatching growth rate, O_2 consumption and food utilization capacity were lower (Owerkowicz et al. 2009). It is remarkable that hypoxia and restricted diets could have additive life-prolonging effects. As it was found in experiments on *Caenorhabditis elegans*, worms treated by combined action of the dietary restriction and hypoxia lived almost twice longer than the controls, outliving the animals kept on the hypoxia or restricted diet alone (Leiser et al. 2013). In the same kind preliminary experiments performed on *Drosophila*, we failed to show additive effects of the hypoxia and diets because of the known difficulties of diet effects on life span extension in flies (data not shown).

Drosophila has obvious advantages as an experimental model specifically for studying the effects of modified atmospheres because it possesses a superior ventilation and O_2 delivery system. Their tracheae branches into smaller tracheoles up to 1 μm in diameter piercing and ensuring gaseous exchange directly in the cells. Apparently due to the more efficient gaseous exchange system, *Drosophila* can survive and may possibly live longer in atmospheres with O_2 content up to 1 %, significantly outperforming mammalian species (Strehler 1962; Harrison et al. 2006). To study the effects of artificial atmospheres on the metabolic and aging rates, we have developed a simple system for modeling atmospheres with given gaseous composition. In experiments with hypoxia, vials with flies were kept in hermetically closed 100 ml syringes containing mixtures of the air with N_2. In the life-extending experiments, the usage of long-lived populations is a critical issue. Otherwise, possible extended life span of the initially short-lived control populations could be interpreted as a 'neutralization' of the negative life-shortening factors. In these experiments, we used a long-lived population of *Drosophila* with mean life span around 60–70 days. *Drosophila* imagoes were tested at three ages for exploring possible 'age-thresholds' of the hypoxia effects on aging. In the first group, the hypoxic treatments were started in the middle of the initial plateau of the survival curve (20 days) when the mortality rate is minimal. Flies of the second group (40 days old) corresponded to the transient period between the plateau and the steeper decline of the survival curve. It obviously is associated with the intriguing age when hidden age-damages are abundant but are not yet manifested in the form of accelerated mortality. The third group (50 days) corresponded to the conditions when aging effects were fully deployed.

13.6.1 Hypoxia Started from the 20 Days

The life-long hypoxic treatment (15, 10 and 5 % O_2) modeled by addition of corresponding amounts of N_2 to the air and started from the 20 days-old imagoes induced a moderate though significant increase of the mean life span (12, 8 and 15 % correspondingly) (Fig. 13.5a).

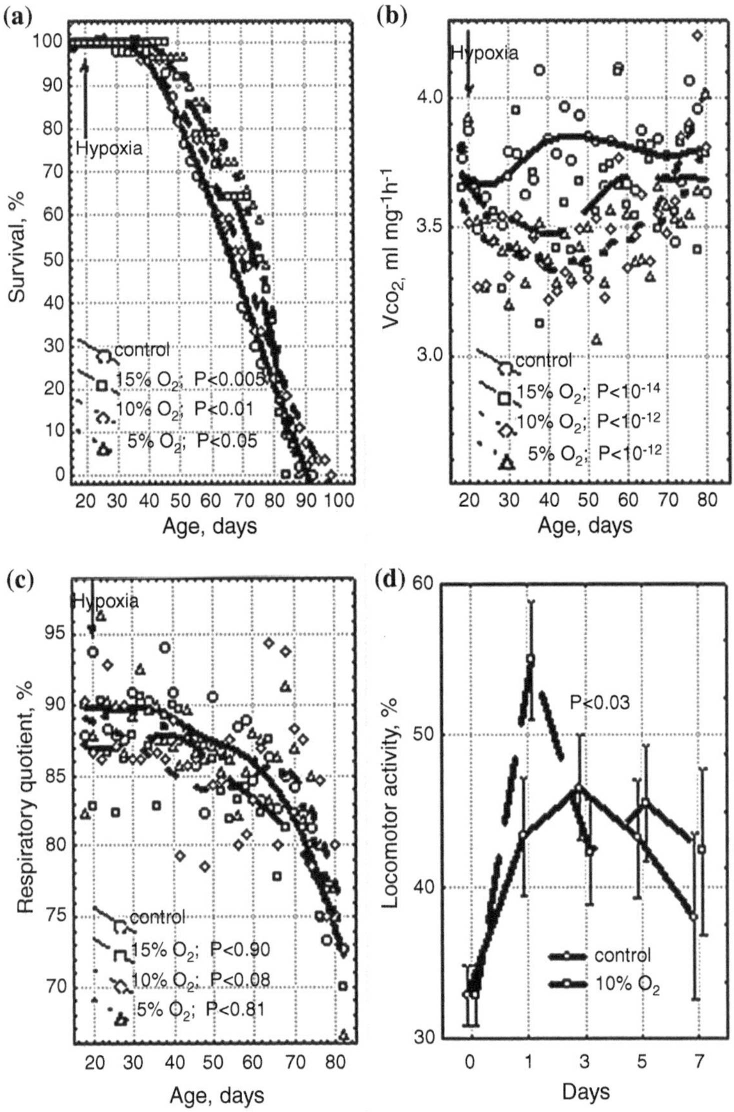

Fig. 13.5 Survival (**a**), rate of CO_2 production (**b**), respiratory quotient (**c**) and locomotor activity (**d**) of *Drosophila melanogaster* incubated in hypoxic atmospheres with 5, 10 and 15 % O_2 starting from the 20 days of life. Statistical significance between the hypoxic and control groups (kept in the syringes with air) was assessed by the non-parametric Wilcoxon's criteria

Two control groups were used: flies kept either in the syringes with 100 % air or in the open air. The two groups did not significantly differ by the age-dynamics of survival or the rate of gaseous exchange. Therefore, only data for flies living in the syringes with air was presented. Although according the non-parametric

Wilcoxon matched pair test the observed effects of life span extension at hypoxia were highly significant ($P < 0.0001$), it is necessary to recognize that the changes of life span were moderate and remained within the boundary typical for the variations of life span of the control animals in different series of experiments. Moreover, the maximal life span practically did not change at hypoxia, resulting in a steeper decline of the survival curve at the advanced ages (Fig. 13.5a). There was certain analogy with the lower mortality of populations living at the moderate highlands. Lower frequency of the cardiovascular diseases, stroke and some types of cancer were often referred as evidences of the beneficial effects of the mild hypoxia on longevity. However, the lower rates of pathology occurrence at the higher attitudes were paradoxically associated with elevated rates of the disease progress and mortality when the pathological processes were already initiated (Burtscher 2013). Such relations should apparently result in steeper decline of the survival curve, as it was in our experiments with *Drosophila*.

In the hypoxia groups, the rate of CO_2 production (Vco_2) decreased in the middle ages (30–50 days) but normalized at the advanced periods of life (Fig. 13.5b). It is remarkable that hypoxia could decrease oxygen consumption via modulation of the basic regulatory mechanisms, e.g., by enhancing degradation of the main physiological stimulators of the metabolism—thyroxin and tri-iodothyronine (Raguso and Luthy 2011). Apparently, *Drosophila* can easily compensate the O_2 deficit in the moderately hypoxic atmosphere by activation of the ventilation apparatus. Nevertheless, as it is typical for many other physiological systems, the compensation was obviously partial. Interesting age-dynamics were found for the respiratory quotient ($RQ = 100 * Vco_2/Vo_2$). It is known that RQ varies within 75–100 % depending on the substrate nature: for oxidation of the carbohydrates RQ is equal 100 %, whereas for the fatty acids and proteins it decreases up to 75 %. For young flies kept on a routine sucrose-yeast feeding medium, RQ is usually stabilized at the level around 90 %. Although in the groups of flies kept in the hypoxia RQ was generally lower, it did not significantly differ from the control groups, at least, partly because of the high variation of this index. However, RQ negatively correlated with the age revealing an especially steeper decline after the 60–70 days of age ($P < 0.02$). If very old flies have problems with the food digestion, as it is typical for other animal species, it should obviously activate oxidation of the stored fatty acids and proteins which will decrease the RQ (Fig. 13.5c). Flies locomotor activity was assessed during the first week of hypoxia at 10 % O_2. Within the first few days, the control and experimental flies demonstrated an elevated pattern of spontaneous motor activity. Among other reasons, it could be explained by the well known 'curiosity' of flies and other animal species placed into a new environment (the syringes). In the hypoxia, additional stimuli for anxiety and elevated movement could appear. However, at the end of the week the motor activity was practically normalized in the both control and hypoxia groups (Fig. 13.5d).

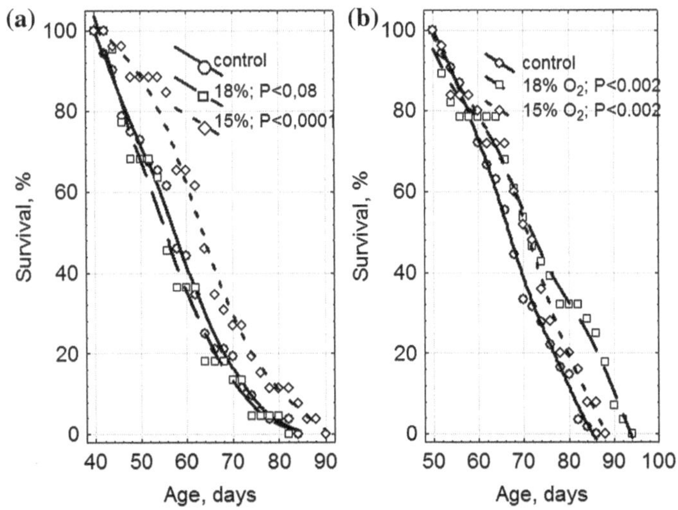

Fig. 13.6 Survival of *Drosophila melanogaster* incubated in hypoxic atmospheres with 15 and 18 % O_2 starting from the 40 days (**a**) and 50 days (**b**). Statistical significance between the hypoxic and control groups (kept in the syringes with air) was assessed by the non-parametric Wilcoxon's criteria

13.6.2 Hypoxia Started from the 40 and 50 Days

Because of the declined adaptability of old animals, the flies in these experiments were kept in milder hypoxia (15 and 18 % O_2). When hypoxic treatments were started on the 40-days-old imagoes, a moderate life span extending effect was observed for the 15 % O_2 (Fig. 13.6a). At the 18 % O_2 conditions, life span did not significantly change for the group started from 40-days age, but increased in the experiments started from 50 days of age (Fig. 13.6b). Apparently, the older the flies, the milder optimal hypoxia should be.

13.7 Concluding Remarks

In the developed countries, people are actually living in artificial atmospheres with optimized physical parameters. At office and home, the air is conditioned, filtered, ozonized, ionized, humidified, deodorized etc. Disproportionally little has been done in a more important medico-biological issue—optimizing the air gaseous composition. The accumulated phylogenetic and ontogenetic data infer that a moderate hypoxia and hypercapnia could be beneficial for longevity. However, it should be recognized that the results of the direct experiments are not much optimistic. Firstly, the life span extending effects were shown only for the short-lived laboratory species; secondly, the scale of extension was moderate and did not exceed the species specific maximum life span; thirdly, older animals were especially prone to the negative effects of hypoxia. There is an obvious similarity in the effects of O_2 and food consumptions. Until recently, most people believed, and some do even now, that 'the body knows better what and how much to eat'. Nowadays, however, it is apparent that this knowledge can in fact be counterproductive. More than often we have to avoid what the body wants to eat. The same could be true with the gaseous exchange and especially in combination of the modified food and O_2 consumptions. Unfortunately, currently we know too little to discuss possible effects of the modified atmospheres on the health and longevity. Nevertheless, there are reasons to believe that atmospheres with individually and occupationally optimized compositions could be necessary not only for the deep space travelers but in the everyday life because these technically available manipulations show obvious potentials for the metabolic rate and life span modulation.

Acknowledgments The author is grateful to Prof. Vadim Fraifeld for his valuable suggestions and contribution in collecting the mammalian data.

References

Aristotle (2007) On longevity and shortness of life. The University of Adelaide Library. eBooks@Adelaide

Beerling DJ, Berner RA (2005) Feedbacks and the coevolution of plants and atmospheric CO_2. Proc Natl Acad Sci USA 102:1302–1305

Bell EL, Klimova TA, Eisenbart J, Schumacker PT, Chandel NS (2007) Mitochondrial reactive oxygen species trigger hypoxia-inducible factor-dependent extension of the replicative life span during hypoxia. Mol Cell Biol 27:5737–5745

Berner RA (1999) Atmospheric oxygen over Phanerozoic time. Proc Natl Acad Sci USA 96:10955–10957

Burtscher M (2013) Effects of living at higher altitudes on mortality: a narrative review. Aging Dis 5:274–280

Cheung SG, Chan HY, Liu CC, Shin PK (2008) Effect of prolonged hypoxia on food consumption, respiration, growth and reproduction in marine scavenging gastropod Nassarius festivus. Mar Pollut Bull 57:280–286

Copeland JM, Cho J, Lo T Jr, Hur JH, Bahadorani S, Arabyan T et al (2009) Extension of *Drosophila* life span by RNAi of the mitochondrial respiratory chain. Curr Biol 13:1591–1598

Davy P, Allsopp R (2011) Hypoxia: are stem cells in it for the long run? Cell Cycle 10:206–211

Everaert N, Willemsen H, Willems E, Franssens L, Decuypere E (2011) Acid-base regulation during embryonic development in amniotes, with particular reference to birds. Respir Physiol Neurobiol 178:118–128

Feng J, Bussière F, Hekimi S (2001) Mitochondrial electron transport is a key determinant of life span in Caenorhabditis elegans. Dev Cell 1:633–644

Forgan LG, Forster ME (2010) Oxygen-dependence of metabolic rate in the muscles of craniates. J Comp Physiol B 180:715–729

Frolkis VV, Muradian KhK (1991) Life span prolongation. CRC Press, Boca Raton

Gilmour KM, Perry SF (2009) Carbonic anhydrase and acid-base regulation in fish. J Exp Biol 212:1647–1661

Goldblatt C, Lenton TM, Watson AJ (2006) Biostability of atmospheric oxygen and the great oxidation. Nature 443:643–645
Harrison J, Frazier MR, Henry JR, Kaiser A, Klok CJ, Rascón B (2006) Responses of terrestrial insects to hypoxia or hyperoxia. Respir Physiol Neurobiol 154:4–17
Holland HD (2006) The oxygenation of the earth and oceans. Phil Trans R Soc B 361:903–915
Honda S, Ishii N, Suzuki K, Matsuo M (1993) Oxygen-dependent perturbation of life span and aging rate in the nematode. J Gerontol 48:B57–B61
Huey RB, Ward PD (2005) Hypoxia, global warming, and terrestrial late Permian extinctions. Science 308:398–401
Hwang AB, Lee SJ (2011) Regulation of life span by mitochondrial respiration: the HIF-1 and ROS connection. Aging (Albany NY) 3:304–310
Imtaiyaz HM, Shajee B, Waheed A, Ahmad F, Sly WS (2013) Structure, function and applications of carbonic anhydrase isozymes. Bioorg Med Chem 21:1570–1582
Kayser B (1992) Nutrition and high altitude exposure. Int J Sports Med 13:129–132
Klok CJ, Harrison JF (2009) Atmospheric hypoxia limits selection for large body size in insects. PLoS ONE 4:e3876
Kurbel S (2014) Animal evolution and atmospheric po2: is there a link between gradual animal adaptation to terrain elevation due to ural orogeny and survival of subsequent hypoxic periods? Theor Biol Med Model 11:47
Lee SJ, Hwang AB, Kenyon C (2010) Inhibition of respiration extends *C. elegans* life span via reactive oxygen species that increase HIF-1 activity. Curr Biol 20:2131–2136
Leiser SF, Fletcher M, Begun A, Kaeberlein M (2013) Life-span extension from hypoxia in *Caenorhabditis elegans* requires both HIF-1 and DAF-16 and is antagonized by SKN-1. J Gerontol A Biol Sci Med Sci 68:1135–1144
Loeb J, Northrop JH (1917) What determines the duration of life in metazoa? Proc Natl Acad Sci U S A 3:382–386
Lyons TW, Reinhard CT, Planavsky NJ (2014) The rise of oxygen in Earth's early ocean and atmosphere. Nature 506:307–315
Mills DB, Ward LM, Jones C, Sweeten B, Forth M, Treusch AH et al (2014) Oxygen requirements of the earliest animals. Proc Natl Acad Sci U S A 111:4168–4172
Muradian KK (2008) Artificial atmosphere, rejuvenation and longevity. Probl Aging Longevity 17:457–477 (in Russian)
Muradian K (2013) "Pull and push back" concepts of longevity and life span extension. Biogerontology 14:687–691
Netzer NC, Breitenbach M (2010) Metabolic changes through hypoxia in humans and in yeast as a comparable cell model. Sleep Breath 14:221–225
Nunn JF (1998) Evolution of the atmosphere. Proc Geol Assoc 109:1–13
Owerkowicz T, Elsey RM, Hicks JW (2009) Atmospheric oxygen level affects growth trajectory, cardiopulmonary allometry and metabolic rate in the American alligator (*Alligator mississippiensis*). J Exp Biol 212:1237–1247
Planavsky NJ, Reinhard CT, Wang X, Thomson D, McGoldrick P, Rainbird RH et al (2014) Earth history. Low mid-Proterozoic atmospheric oxygen levels and the delayed rise of animals. Science 346:635–638
Pollock JP, Patel HM, Randolph BJ, Heffernan MJ, Leuenberger UA, Muller MD (2014) Ascorbic acid does not enhance hypoxia-induced vasodilation in healthy older men. Physiol Rep 2(7):e12091
Poulios E, Trougakos IP, Chondrogianni N, Gonos ES (2007) Exposure of human diploid fibroblasts to hypoxia extends proliferative life span. Ann N Y Acad Sci 1119:9–19
Raguso CA, Luthy C (2011) Nutritional status in chronic obstructive pulmonary disease: role of hypoxia. Nutrition 27:138–143
Schopf JW (1994) Disparate rates, differing fates: tempo and mode of evolution changed from the Precambrian to the Phanerozoic. Proc Natl Acad Sci U S A 1991:6735–6742

Schopf JW, Kudryavtsev AB, Agresti DG, Wdowiak TJ, Czaja AD (2002) Laser-Raman imagery of Earth's earliest fossils. Nature 416:73–76
Sessions AL, Doughty DM, Welander PV, Summons RE, Newman DK (2009) The continuing puzzle of the great oxidation event. Curr Biol 19:R567–R574
Sharma M, Shukla Y (2009) The evolution and distribution of life in the Precambrian eon-global perspective and the Indian record. J Biosci 34:765–776
Strehler B (1962) The distribution of cellular aging. Time, cells, and aging. Academic Press, New York, pp 33–85
Taylor CT, McElwain JC (2010) Ancient atmospheres and the evolution of oxygen sensing via the hypoxia-inducible factor in metazoans. Physiology (Bethesda) 25:272–279
Tyler SA, Barghoorn ES (1954) The occurrence of structurally preserved plants in Precambrian rocks of the Canadian Shield. Science 119:606–608
Walker JC (1985) Carbon dioxide on the early earth. Orig Life Evol Bioph 16:117–127

Chapter 14
Effects of Mild Stresses Applied in Adults on Aging and Longevity

Éric Le Bourg

Abstract It is now known that some mild stresses can have positive effects on longevity, aging, and resistance to severe stresses in several species, and particularly in *Drosophila melanogaster* flies. This chapter describes the effects of some mild stresses (heat, cold, hypergravity, oxidative stress, fasting, and so on) in flies and the possible mechanisms of these mild stresses (antioxidant enzymes, heat shock proteins, NF-κB).

Keywords *Drosophila melanogaster* · Mild stress · Hormesis · Aging · Longevity · Severe stresses · Heat shock proteins · Antioxidant enzymes · NF-κB

14.1 Introduction

The idea that a stress could be of positive value to an organism does not appear at a first sight to be particularly sound. However, there is now a large corpus of data showing it is indeed the case in *Drosophila melanogaster* flies, as well as in other organisms such as the tiny worm *Caenorhabditis elegans* (e.g. Cypser and Johnson 2002), but also in mammals such as mice or rats (review in Le Bourg 2009), pigs (e.g. Lavitrano et al. 2004), and human beings (reviews in Rattan and Le Bourg 2014).

Obviously, stress has no positive effect if too intense: positive effects are only observed when mild stresses are used. Therefore, the first condition to envisage a stress could have positive effects is to accept that the effect of any treatment is dependent on the strength of this treatment. This idea is not really a new one, as Paracelsus stated five centuries ago that anything can be a poison or not, depending

É. Le Bourg (✉)
Centre de Recherches sur la Cognition Animale, UMR CNRS 5169,
Université Paul-Sabatier, 31062 Toulouse cedex 9, France
e-mail: eric.le-bourg@univ-tlse3.fr

A.M. Vaiserman et al. (eds.), *Life Extension*, Healthy Ageing and Longevity 3,
DOI 10.1007/978-3-319-18326-8_14

on the dose (Borzelleca 2000). This idea thus opened the way to the idea that a poison (or a stress) can have positive effects if used at a low dose. However, with the rise of toxicology in the last century, this idea was not seriously considered, when not disregarded or even clearly discarded, and the threshold dose-response or the linear no-threshold models were preferred (Calabrese 2014). Favoring these models in toxicology has the obvious consequence that any dose is considered as neutral or toxic (threshold dose-response model) or always toxic (linear no-threshold model). Obviously, it is easier and cheaper to draw a dose-response curve by studying some high doses with highly toxic effects and extrapolating the dose-response curve up to the origin than to study the real effect of low doses, which can require a large sample to bring to the fore a significant effect. Adopting the threshold dose-response or linear no-threshold model allows to protect the population and workers against toxicity of chemicals or radioactive isotopes and it is easy to manage the risk. For instance, dosimeters worn by workers in nuclear power plants cumulate the doses and the annual dose has to remain below a threshold. However, a less desired consequence of adopting this view is that if low doses have positive effects, and thus could be of therapeutic value, this could remain ignored. It thus can be understood there is a hard debate on the need to include the positive effects of low doses (hormetic effects) in environmental safety regulations. For instance, Calabrese (2014) is of the view that "the hormetic dose response model should replace the threshold and linear dose response models and become the default dose response model" whereas Thayer et al. (2005) stressed that "even if certain low-dose effects were sometimes considered beneficial, this should not influence regulatory decisions to allow increased environmental exposures to toxic and carcinogenic agents, given factors such as inter-individual differences in susceptibility and multiplicity in exposures".

It is the opinion of the author that Thayer et al. (2005) are right when stressing the risks of adopting the hormetic dose-response model in risk assessment but it is not to say that hormesis should not be considered in therapy and biogerontology. However, it happened that the esteemed gerontologist Sacher (1977) considered that hormetic effects were to be observed only in animals living in suboptimal conditions and therefore that "hormesis is in one sense an obstacle in the path of gerobiological research, and efforts to understand and annul it would be well justified. A first step is to breed vigorous animal genotypes and a second step is to develop living environments that are optimal for their behavioral, physiological and immunological health". Because this view was held by a famous gerontologist, it is easy to understand that, after this article was published, any scientist observing a hormetic effect was prone to conclude it was due to suboptimal rearing conditions, which obviously precludes any publication. In sharp contrast, the idea that a mild stress could be used to improve aging was considered as valuable in the 1970s in the former Soviet Union (Frolkis 1982).

Despite the Sacher's article (1977), some authors however reported hormetic effects on resistance to severe stress, aging and longevity and the seminal review of Minois (2000) revitalized the interest for hormetic effects on aging and longevity. After that, the journal *Human and Experimental Toxicology* organized a debate among experts on the use of hormesis in aging research (Volume 20, issue 6, 2001) and Le Bourg and Rattan (2008) edited the first book on hormesis and aging. Results in mammals and human beings have been reviewed elsewhere (e.g. Volume 7, issue 1 of *Ageing Research Reviews*, Le Bourg 2009; Rattan and Le Bourg 2014) and the present chapter is concerned with the effects of several mild stresses on aging and longevity in *D. melanogaster*.

The privileged hypothesis explaining the existence of hormetic effects is that, when confronted with a mild stress, animals implement strategies to cope with it by inducing an overcompensation response to the disruption of homeostasis but, in case of chemicals, the hormetic response can also be based on specific receptors (see lists in Calabrese 2008, 2013). As summarized by Calabrese (2008), "a low dose of a toxic agent induces stress and/or damage; following the damage a repair response is initiated which leads to a slight overcompensation". This explanation allows to predict that various mild stresses could have rather similar positive effects because these effects are not mediated by specific receptors. Therefore, it is of interest to study the effects of various mild stresses to know whether they provoke similarly positive or contrasted responses of the organism. The most studied stresses in *D. melanogaster* have been heat, cold, hypergravity, and irradiation. However, the effects of irradiation will not be reviewed here because the usual procedure is to irradiate eggs, and not adults (for reviews see Vaiserman 2008; Le Bourg 2009), but see Antosh et al. (2014) for recent results showing no positive effect of irradiation on longevity of adults. Some other studied stresses are fasting, oxidative stress, desiccation, or hydrogen sulfide but only a few studies have been done. The effects of these mild stresses are described in the following, as well as some studies on their possible mechanisms.

14.2 Heat Stress

14.2.1 Effects on Longevity

Several experiments have studied the effects of heat shocks on longevity. Table 14.1 shows the results of some studies reporting mean life spans: except for the results of Hercus et al. (2003), the effects are not of a large magnitude, when they indeed exist. Furthermore, these results can be fragile because Le Bourg et al. (2009) failed to confirm the results of Le Bourg et al. (2001) reported in the table. Other studies are reported in Le Bourg (2009). More recently, Defays et al. (2011) used recombinant-inbred lines from parental lines selected for low (D48 line: 32 lines) or high (SH2 line: 20 lines) resistance to heat to study the effect of 35 min at 35.5 °C at 4 and 7 days of age. In each sex, this heat shock increased longevity in ca 25 % of the D48 recombinant lines (usually by ca 5 days), decreased it in 25 % of them and had no effect in the second half of lines, these figures being respectively ca 5, 45 and 50 % in the SH2 lines. On the whole, it seems possible to conclude that the effect of heat shock is modest and a meta-analysis concluded that heat shocks do not increase longevity in *D. melanogaster* (Lagisz et al. 2013) but, after this article was published, Sarup et al. (2014) still reported that 3 h at 34 °C at the ages of 3, 6 and 9 days increased life span by 12 %, which shows that some procedures are nevertheless efficient.

Table 14.1 Summary of some studies on the effect of heat shock on life span

T (°C)	Duration	Strain	Sex and mating status	Age at heat shock (days)	Life span of controls (days)	Effect of heat (days)	Reference
36	70 min	Inbred line 1	Virgin females	4	29.6	1.3	Khazaeli et al. (1997)
			Virgin males		35.3	3.9	
			Mated females		27.1	0.9	
			Mated males		37.8	2.1	
		Inbred line 2	Virgin females		23.9	1.3	
			Virgin males		28.6	2.8	
			Mated females		23.2	−0.2	
			Mated males		27.1	2.7	
37	100 min	Ra line	Mated males and females	5–7	36.7	1.5	Kuether and Arking (1999)
				27–29	38.7	0.3	
		La line	Mated males and females	5–7	58.8	−5.9	
				27–29	58.4	−1.2	
				62–64	74.3	−4.2	
37	5 min/day for 5 days	Wild-type	Virgin males	5, 6, 7, 8, 9	42.1	1.9	Le Bourg et al. (2001)
			Virgin females		45.5	2.0	
34	3 h at 4 ages	Wild-type	Mated females in pairs	3, 6, 9, 12	54.7	8.6	Hercus et al. (2003)
					55.2	6.9	
			Mated females in groups		63.3	6.6	
					64.6	10.0	

In most of experiments life span was recorded from the day after heat shocks and thus is not life expectancy at emergence

14.2.2 Effects on Behavioral Aging

On the one hand, heat shock had no effect on climbing activity recorded throughout life, i.e. the ability to climb up on the vertical side of a vial, in a wild-type strain (Le Bourg et al. 2001), in the inbred strain w^{1118}, in transgenic flies overexpressing the *hsp70* gene, and in their control strain with no extra-copies of this gene (Minois et al. 2001). On the other hand, heat shock had, respectively, no effect or increased spontaneous locomotor activity observed throughout life in these two studies. Therefore, heat shocks do not seem to clearly delay behavioral aging.

14.2.3 Effects on Resistance to Stress

A pretreatment with heat often increased resistance to a strong heat stress (e.g. Krebs and Loeschcke 1994a, b; Khazaeli et al. 1997; Dahlgaard et al. 1998; Le Bourg et al. 2001; Hercus et al. 2003; Sejerkilde et al. 2003; Sørensen et al. 2007, but see Kuether and Arking 1999). Heat shocks have also been reported to increase resistance to a cold shock (10–15 min at 20 °C) or to the oxidant paraquat in the inbred strain w^{1118}, in transgenic flies overexpressing the *hsp70* gene, and in their control strain with no extra-copies of this gene (Minois 2001). However, heat shocks slightly decreased starvation resistance in the same strains (Minois 2001) and did not protect, in both sexes of a wild-type strain, against hydrogen peroxide, another oxidant (Le Bourg 2008), or fungal infection (Le Bourg et al. 2009). A pretreatment with heat (1 h at 36 °C) increased resistance to heat (38.5 °C) of females but also decreased resistance to desiccation and had no significant effect on resistance to cold (time to recovery after 3 h at 0 °C) or to starvation (Bubliy et al. 2012). All the previous experiments were done with adults but, in larvae, a heat stress can also protect against a severe cold stress (Burton et al. 1988), confer neuroprotection at a high temperature (Karunanithi et al. 1999), or preserve locomotor competence (Klose et al. 2005).

14.2.4 Conclusion

Heat has no clear positive effect on life span, behavioral aging, and some severe stresses, but it can increase resistance to other severe stresses. Thus, hormetic effects of a heat pretreatment do exist but are not very important.

14.3 Cold Stress

14.3.1 Effects on Longevity

Five daily cold stresses (1 h at 0 °C) either during the first or second week of life (from 5 to 9 or from 12 to 17 days of age) had no effect. However, the same cold stress applied during these two periods (weeks 1 + 2: no stress at days 10 and 11) increased longevity of males by 10 % or more (5–10 days; Le Bourg 2007a). The effects were more inconstant in females, because cold either increased (Le

Bourg 2007a; Le Bourg et al. 2009), decreased (Le Bourg 2010a, 2012), or had no effect on longevity (Le Bourg 2007a, 2011a). Applying the cold stress at various ages (weeks 1 + 2, or 2 + 3, or 3 + 4, or 4 + 5) significantly (weeks 1 + 2 or 3 + 4) or marginally (weeks 2 + 3) increased male longevity, no positive effect being observed when the stress was applied at the oldest ages (Le Bourg 2011a). Figure 14.1 shows however that the cold stress had both negative and positive effects on mortality if applied at rather old ages, whereas nearly no negative effect was observed when used at the youngest age.

In another experiment, flies were subjected to the weeks 1 + 2 cold pretreatment and thereafter, from 19 days of age, fed on saccharose only, a poorly

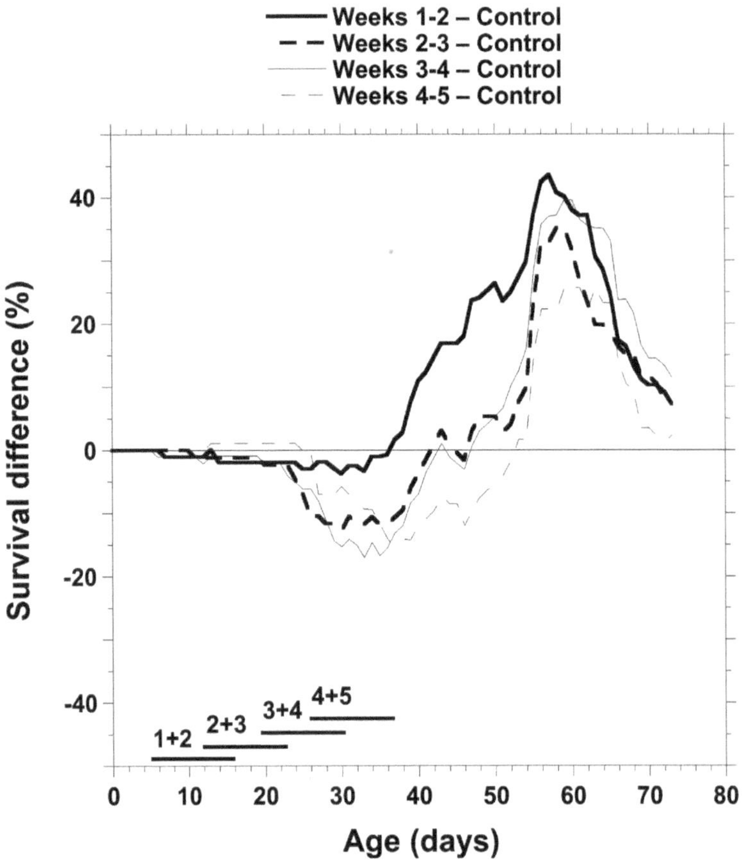

Fig. 14.1 Survival difference (%) between the cold-pretreated groups and the control one. If the percentage of survivors at, say 40 days of age, is 60 % in an experimental group and 50 % in the control one, the survival difference is 10 %. By contrast, if the control group has a higher survival than the experimental one the survival difference is negative. The bars at the bottom indicate the time of pretreatment by cold for each experimental group (1 h at 0 °C during two periods of 5 days separated by 2 days with no cold shock). For instance, "1 + 2" stands for a cold pretreatment during the two first weeks of life (redrawn from Le Bourg 2011a)

nutritious medium known to decrease longevity. The cold pretreatment increased remaining longevity by 2 days in both sexes (21 vs. 19 days; Le Bourg 2008).

Severe cold stresses are detrimental, because 1, 2, or 3 h daily at −3 °C from the third day of life decreased longevity of both sexes by ca 2 weeks in the Oregon strain and in the vestigial mutant (*vg*; Ayar et al. 2010). One may wonder whether milder (e.g. 0 °C) or shorter stresses (e.g. for 2 weeks rather than daily throughout life) could have increased longevity in these strains.

14.3.2 Effects on Behavioral Aging

The weeks 1 + 2 cold pretreatment described above increased climbing scores, particularly at 4–6 weeks of age in males, and at 3–4 weeks of age in females (Le Bourg 2007a; see also Le Bourg et al. 2009; Le Bourg 2012).

14.3.3 Effects on Resistance to Stress

The weeks 1 + 2 pretreatment increased survival time at 37 °C throughout life (Le Bourg 2007a, 2012). In addition, cold stresses applied at later ages increased survival time at 37 °C at 6 weeks of age (Le Bourg 2011a). Positive effects of the weeks 1 + 2 pretreatment were also observed in flies subjected to a proxy for "summer heatwave", i.e. a total of 4 heat shocks (30 or 45 min at 37 °C, twice a week) from 4 weeks of age during a two-week period. These repeated heat shocks do not kill flies but decrease longevity by 50 %: cold-pretreated flies lived 15 % longer (+2 days) than control ones after the first 30-min heat shock, but no positive effect was observed with 45-min shocks (Le Bourg 2007a).

The weeks 1 + 2 cold pretreatment increased percent of survival to a long cold shock (16 h at 0 °C) in both sexes at 3, 4, 5 and 6 weeks of age, but not at 7 weeks of age (Le Bourg 2007a, 2012). These results are in accordance with many other results showing that pretreatments with cold increase resistance to later cold shocks (e.g. Czajka and Lee 1990; Kelty and Lee 1999, 2001; Sejerkilde et al. 2003; Rako and Hoffmann 2006; Jensen et al. 2007; see also Marshall and Sinclair 2012).

The weeks 1 + 2 cold pretreatment also increased survival time to fungal infection by *Beauveria bassiana*, mainly in males and throughout life (Le Bourg et al. 2009; see also Le Bourg 2012). Applying the cold stresses at later ages also increased survival of males at 6 weeks of age, except when they were applied during weeks 4 and 5 (Le Bourg 2011a).

By contrast, the weeks 1 + 2 cold pretreatment hardly increased resistance to oxidative stress (hydrogen peroxide) in both sexes (Le Bourg 2008, 2012). Finally, this pretreatment decreased survival time in starvation conditions, mainly in females (Le Bourg 2007a, 2012).

Another study showed that a pretreatment with cold (5 days at 11 °C) increased resistance of females to cold (time to recovery after 3 h at 0 °C), but decreased that to starvation and had no effect on resistance to heat and desiccation (Bubliy et al. 2012).

14.3.4 Conclusion

In summary, cold pretreatments can increase longevity, delay behavioral aging, and improve resistance to various severe stresses, but not to all stresses.

14.4 Hypergravity Stress

14.4.1 Effects on Longevity

Hypergravity (HG) is a gravity level higher than the Earth gravity level (1 g). HG is hardly encountered by organisms living in the wild, not to say never, but people can be subjected to short stays in HG. For instance, a mild HG episode is observed in high-speed elevators, in cars during the braking phase, or in aircrafts at take-off, but there is an intense HG phase at launch of rockets or during acrobatic flights of aircrafts. Furthermore, short HG and microgravity episodes can be obtained during the parabolic flights of space agencies aircrafts (an example in Le Bourg et al. 1995).

The weight is increased during HG, e.g. thrice at 3 g. Therefore, HG probably induces a stress because the organism has to cope with this increased weight with, in rats acclimated to HG, the consequence of an increased resting energy expenditure (Wade et al. 2002). In rats, living continuously in HG provokes a deformation of the vertebra, impairs growth, suppresses fat (Oyama 1982), non-significantly decreases life span (Economos et al. 1982), impairs body mass gain (Pitts and Oyama 1979; Kita et al. 2006), or decreases the number of pups (Megory and Oyama 1984). A short stay in HG (2, 3, or 4 g for 21 days at 9 weeks of age) has also negative effects in mice (Bojados and Jamon 2014; see also Jamon and Serradj 2009). Thus, in mammals, HG is probably too strong a stress for expecting favorable effects of exposure to HG (other results have been gathered for example in hamsters, e.g. Sondag et al. 1997). Nevertheless, HG is a suitable procedure to increase metabolism in insects without the consequences of increased temperatures in poikilotherms (speeding chemical reactions, decreased longevity, and so on) and without the problems linked to a large weight, as in mammals (the ability to tolerate HG decreases with body size increase).

Living in HG for 2 weeks (up to 5 or 7.4 g) can increase longevity of males (10–20 %), but has no effect or decreases longevity in females (e.g. Le Bourg and

Minois 1997; Le Bourg et al. 2000). On the one hand, 3 weeks in HG can still increase life span of males (Le Bourg et al. 2000) but longer or lifetime exposures can decrease it (e.g. Lints et al. 1993). On the other hand, a one-week exposure has no effect, as well as 3 periods of 4 days in HG followed by 3 days at 1 g (i.e. 12 days in HG). Thus, 3 weeks in HG is the limit between a mild stress with hormetic effects and a strong stress with no positive effects, and a continuous exposure to HG is needed for positive effects to be observed.

Using conditions known to increase or decrease longevity can modify the limit between positive and absence or negative effects of HG. Males kept in individual vials (which increases life span) after having lived 25 days in HG live longer than those always kept at 1 g, whereas no HG effect is observed if they live in groups of 15 males throughout life, the usual living condition (Le Bourg et al. 2000). In contrast, the positive effect of 2 weeks in HG is suppressed in males continuously kept with females or in males transferred at 28 or 30 °C after having lived in HG at 25 °C, probably because mating and high temperature decrease longevity (Le Bourg et al. 2004). Therefore, HG can have hormetic effects or not, depending on rearing conditions. Conditions that decrease longevity (mating, high temperature) can erase the hormetic effect of HG, whereas those that increase longevity favor hormetic effects to be observed (individual rearing). This is exactly the opposite of the Sacher's reasoning (1977, see the introduction).

14.4.2 Effects on Behavioral Aging

HG has been shown to delay the age-linked decrease of climbing scores, mainly in males (Le Bourg and Minois 1999; Le Bourg et al. 2002), but this effect was not always observed (Le Bourg 2012). HG has however no positive effect on spontaneous locomotor activity and patterns of movement, i.e. the shape of paths of flies walking in a circular arena (Le Bourg and Minois 1999).

14.4.3 Effects on Resistance to Stress

A 2-week exposure (but also 1- and 4-week exposures) to HG increased survival time at 37 °C in both sexes (Le Bourg and Minois 1997; Minois and Le Bourg 1999; Minois et al. 1999; Le Bourg 2012) or to simulated "summer heatwave" (see above), in males only (Le Bourg et al. 2004; Le Bourg 2005). HG had either no effect (Minois and Le Bourg 1999) or increased resistance to cold (Le Bourg 2012) in experiments using however different procedures. HG did not confer protection against fungal infection (Le Bourg et al. 2009; Le Bourg 2012) or oxidative stress (Le Bourg 2008, 2012) and, finally, decreased resistance to desiccation or starvation (Minois and Le Bourg 1999, Le Bourg 2012).

14.4.4 Conclusion

In summary, HG can increase longevity of males and may have positive effects on behavioral aging and resistance to heat, but it has no clear effects on other severe stresses.

14.5 Fasting Stress

Dietary restriction (DR), i.e. a reduction of available food without malnutrition applied throughout adult life or at least for a long period, is often considered to increase longevity and improve healthspan in most of species (reviews in Everitt et al. 2010), but this conclusion is probably wrong (reviews in Le Bourg 2010b; Nakagawa et al. 2012; Swindell 2012). DR is probably not a mild stress because, whereas mild stresses can increase longevity and severe ones decrease it, DR is more efficient in rodents as its duration and the reduction of food increase (Bertrand et al. 1999), except if the malnutrition threshold is reached (Fig. 17.4 in Speakman and Mitchell 2011). Furthermore, DR increases mean (up to +50 % in rodents) and maximal longevities whereas mild stress only increases mean longevity (+20 % at a maximum, Fig. 17.1 in Minois 2000). Therefore, the features and effects of DR and mild stress are different (Le Bourg 2009, but see the debate among experts on hormesis and DR in *Belle Newsletter* Volume 8, issue 3, 2000).

In the following, we are thus not concerned with DR but with fasting, i.e. the complete absence of food for a limited period with water available. Fasting could be a signal for impaired environmental conditions, preparing for worse living conditions by increasing resistance to severe stresses, which often co-occur with food limitation in the wild (e.g. heat or cold shocks in summer and winter). A short starvation could thus be a stimulus disturbing homeostasis without inducing severe damages, and provoking an adaptive response enhancing the ability to resist severe stresses, i.e. fasting could be a mild stress with hormetic effects.

14.5.1 Effects on Resistance to Stress

The very few studies of fasting in *D. melanogaster* have focused on stress resistance. Vigne et al. (2009) fed young flies on a highly diluted medium for 2 days before an anoxia followed by reoxygenation (this is similar to a cardiac ischemia-reperfusion insult in mammals) and this strongly increased survival to this stress. A caveat is that these authors did not fast flies but simply diluted the rearing medium. Via nitric oxide release, which is active against gram-negative bacteria (Foley and O'Farrell 2003), starvation for 24 h protected *relish* flies against gram-negative bacteria despite the fact that the Imd innate immunity pathway protecting flies against these bacteria is deficient in *relish* flies (Brown et al. 2009). Bubliy et al. (2012) showed that three periods of 18 h of fasting followed by 20 h of ad libitum feeding increased survival time of females to starvation (+20 %) and the number of survivors to desiccation (+21 %), but decreased longevity and resistance to heat (38.5 °C) or cold (time to recovery after 3 h at 0 °C). Le Bourg (2013) starved flies for 24 h and observed resistance to heat (37 °C), cold (0 °C), oxidative stress (hydrogen peroxide), and fungal infection (*B. bassiana*). Fasted flies better resisted to cold (up to 48 h at 0 °C) at young and middle age, results being rather inconsistent at old age, but fasting had no effect or decreased resistance to the other stresses.

14.5.2 Conclusion

Therefore, it is somewhat difficult to conclude on the effects of fasting on flies, due to a lack of studies. Nevertheless, for the time being, it seems that the positive effects are limited.

14.6 Oxidative Stress

14.6.1 Effects on Longevity

Le Bourg (2007b) transferred 2 or 4 week-old flies on a solution of saccharose, a poor medium known to decrease life span (−50 %). For one half of the flies, the solution contained hydrogen peroxide at various concentrations (60 mM to one M). Hydrogen peroxide decreased longevity, but the lowest concentration increased mean life span of middle-aged males (mean life span at 4 weeks of age, respectively, of males living on the usual corn medium, on saccharose, and on saccharose with 60 mM hydrogen peroxide: 20, 11, 16 days). Thus, hydrogen peroxide reduced the negative effect of the poor medium. However, another study showed that daily, weekly, or twice-weekly repeated 8 h exposures throughout life to a lower dose (0.1 mM) decreased life span of females feeding on the usual corn medium, in a linear manner with increased exposures (Pickering et al. 2013).

14.6.2 Effects on Behavioral Aging

Le Bourg (2007b) did not observe any positive effect on climbing scores of the hydrogen peroxide dose that reduced the negative effect on life span of a poor medium.

14.6.3 Effects on Resistance to Stress

Similarly, Le Bourg (2007b) did not observe any positive effect on resistance to heat or to starvation of the hydrogen peroxide dose that reduced the negative effect on life span of a poor medium. In contrast, subjecting female flies at 7–9 days of age to a single 6 h exposure to a 0.1 mM hydrogen peroxide solution increased survival time, 16 h later, to a 4.4 M solution (ca 63 vs. 49 h; Pickering et al. 2013). The effect of this pretreatment was lost if flies were pretreated at 13–15 or 25–27 days of age and 2 or 3 exposures, daily, weekly, or twice-weekly, at any age, also nullified the positive effect.

14.6.4 Conclusion

In summary, the positive effects of oxidative stress are modest and the previous two studies show that the window to observe such effects is narrow.

14.7 Miscellaneous Stresses

14.7.1 Desiccation

Bubliy et al. (2012) kept female flies in vials containing the usual medium and a desiccant. Probably due to the presence of the medium, the relative humidity level only fell to 7–10 % in ca 11 h. However, this pretreatment decreased life span (−3 days). The same pretreatment increased resistance to desiccation (15 h at 30–17 % relative humidity) and to heat (38.5 °C), but decreased that to cold (3 h at 0 °C) and had no effect on survival time to starvation. The positive effects of desiccation appear to be modest.

14.7.2 Hydrogen Sulfide

Hydrogen sulfide (H_2S) in air has been shown to have no effect on survival to starvation in females but to increase survival time in desiccation conditions (e.g. 29 vs. 23 h, Zhong et al. 2010).

14.7.3 Dead Spores of a Fungus

Covering flies with dead spores of the entomopathogenic fungus *Metarhizium robertsii* two days before a heat shock (38 °C for 45 min) increases survival, but 3 or 5 exposures have no effect or even slightly decrease survival. In the same way, a single exposure to dead spores increased life span (+3 days or ca +9 %, both sexes pooled) and fecundity, but decreased survival time to live spores (ca −0.8 day or −10 %). Thus, there is a balance between positive and negative effects of the mild stress (McClure et al. 2014).

14.8 Mechanisms of Mild Stress-Induced Hormetic Effects

The previous results have shown that various mild stresses can have similar beneficial effects. For instance, they can increase life span (cold, HG), delay the age-linked decline of climbing scores (cold, HG), increase resistance to heat (heat, cold, HG, desiccation), cold (heat?, cold, HG?, fasting), desiccation (fasting, desiccation, hydrogen sulfide), or to oxidative stress (heat?, cold?, oxidative stress?). However, resistance to fungal infection is increased only by cold and that to starvation by fasting. Mild stresses can also have no effect on resistance to severe stresses, or can even decrease it. For instance, the resistance to heat or cold is decreased by fasting and that to desiccation is lowered by HG or heat.

As most of these previous stresses are environmental changes, their effects are probably explained by an overcompensation response to the disruption of homeostasis, i.e. a general response of the organism, and not by a physical link of a molecule with a specific receptor, for instance at the surface of cells. This link could be observed if the mild stress were a specific molecule such as hydrogen sulfide, which is naturally produced by flies (Zhong et al. 2010).

Various stimuli can provoke the translocation in the nucleus of transcription factors and this translocation can induce not a single gene but many genes with various effects. For instance, the translocation of DAF-16, the homologue of FOXO in *C. elegans*, is induced by flavonoids (Kampfkötter et al. 2007), and DAF-16 is necessary to observe a longevity increase after heat shocks, but not an increased thermotolerance (Cypser and Johnson 2003). dFOXO, the homologue of FOXO in flies, can be activated by various stresses and directs the expression of several defense mechanisms (apoptosis, antioxidant, immune, or metabolic responses: review in Puig and Mattila 2011). Therefore, transcription factors are candidates to explain the effects of mild stress. Enzymes known to counteract oxidative challenge, such as the antioxidant enzymes superoxide dismutase (SOD) and catalase (CAT), could also be appropriate candidates.

14.8.1 Antioxidant Enzymes

The activity of SOD (converts O_2^- to H_2O_2) and CAT (converts H_2O_2 to H_2O) has been measured at 2, 4, or 6 weeks of age in individual flies kept in HG for the first two weeks of adult life and in control ones always living at 1 g. These two weeks in HG, known to increase longevity of males and thermotolerance in both sexes (see above), had no effect on SOD and CAT at any age and in either sex (Le Bourg and Fournier 2004).

Therefore, the increased longevity and thermotolerance observed after exposure to HG are probably not explained by these enzymes and it may be understood why HG does not increase resistance to oxidative stress (e.g. Le Bourg 2008). However, no study of SOD and CAT has been performed in flies subjected to another stress than HG.

14.8.2 The NF-κB like Factor DIF

The DIF transcription factor, a homolog of NF-κB in flies, is involved in resistance to gram-positive bacteria and fungal infection. The activation of the Toll pathway by fungi directs the expression of the *drosomycin* and *metchnikowin* antimicrobial peptide genes by inducing the translocation into the nucleus of the DIF transcription factor (review in e.g. Ganesan et al. 2011).

Dif mutants and their control strain pretreated with cold (1 h at 0 °C, daily, from 5 to 9 and from 12 to 17 days of age) better survived to a severe cold stress (16 or 18 h at 0 °C) than flies not subjected to the pretreatment and, in each of these groups, the *Dif* mutation had no effect on the percentage of survivors. By contrast, the cold pretreatment increased survival time to heat (37 °C) in the control strain but not in *Dif* flies. These results thus show that DIF probably does not mediate the effect of the pretreatment on resistance to cold but can, at least partly, explain the resistance to heat. The effects of DIF and of the pretreatment on resistance to fungal infection were complex, because this pretreatment had no effect in the control strain and decreased resistance of *Dif* flies (Le Bourg et al. 2012), while it is known to increase resistance to fungi in wild-type flies (e.g. Le Bourg et al. 2009). The genetic background of the control strain (*cn bw* mutant) could explain this absence of a positive effect in these rather frail flies (low longevity and viability of the eggs, see Le Bourg 2011b), while the absence of DIF in this genetic background could explain the negative effect. However, testing the effect of the cold pretreatment in wild-type flies subjected to a pharmacological inhibition of NF-κB (Moskalev and Shaposhnikov 2011) could clarify the issue.

It seems possible to conclude that the DIF transcription factor can explain some effects of mild stress.

14.8.3 Heat-Shock Proteins

Heat-shock proteins (HSP) mediate protein refolding or degradation after a heat shock via the translocation in the nucleus of the heat shock transcription factor HSF (reviews in Morrow and Tanguay 2003; Tower 2011).

Flies subjected to HG do not synthesize the 70 kDa HSP (HSP70) and, thus, HSP70 cannot explain the increased longevity of males living in HG for the two first weeks of adult life. However, if flies of both sexes are heat-shocked, those having lived in HG synthesize more HSP70 than flies that have always lived at

1 g (Minois et al. 1999; Le Bourg et al. 2002). This better resistance could thus be explained by an increased HSP70 synthesis. As a consequence, flies overexpressing the *hsp70* gene should take advantage of an increased HSP70 synthesis after having been subjected to HG, but it happened that the positive effects of HG on survival time at 37 °C were similar in a transgenic strain and in a control strain with no extra-copies of the *hsp70* gene (Le Bourg et al. 2002). Therefore, the synthesis of HSP70 can explain the better thermotolerance of flies that have lived in HG, but it is useless to increase this synthesis beyond the level observed in wild-type flies.

In males, 2 h at 34 °C at 3, 6 and 9 days of age provoked the up-regulation of heat shock proteins genes 10 and 26 days after the last heat stress (Sarup et al. 2014). In females, 3 h at 34 °C at 2, 4, and 6 days of age induced a higher induction of HSP70 after a severe heat stress at 32 days of age (Kristensen et al. 2003). A similar result has been observed in flies, mainly males, harboring a heat shock transcription factor inactivated at 30 °C and rescued by a functional *hsf* gene. These flies also better resisted to a severe heat shock after a pretreatment with heat (3 h at 34 °C at 3, 6, and 9 days of age), whereas no such effects were observed in flies not rescued by a *hsf* gene (Sørensen et al. 2007). In contrast, wild-type flies of both sexes subjected to heat shocks at young age, which increased survival time at 37 °C, did not synthesize more HSP70 at any age after a 37 °C heat shock than control flies (Le Bourg et al. 2001). For the time being, the effect of other stresses on the synthesis of HSPs has not been studied, even if it is known that various HSPs can be induced during the recovery phase (8 h at 25 °C) after a cold shock (9 h at 0 °C, Colinet et al. 2010a) and that knocking down *hsp22* or *hsp23* genes impairs recovery from cold (Colinet et al. 2010b).

This whole set of results indicates that HSP70 explains, at least partly, the higher thermotolerance observed after a mild stress.

14.9 General Conclusions to the Chapter

It is now a well established fact that mild stresses can have positive effects on longevity, aging and resistance to severe stress in *D. melanogaster*. A better resistance to severe stress can be observed at old age even if a mild stress is applied at young age, and the mild stress can be efficient even if applied at rather old ages (Le Bourg 2011a). In addition, the positive effects of combining two mild stresses with positive effects in the same flies are additive. For instance, HG and cold can have additive positive effects on resistance to heat or to cold stresses and on longevity of males, but a caveat is that when one of the mild stresses has negative effects the result of the combination of the two pretreatments can be negative (Le Bourg 2011a). The mechanisms of the positive effects of mild stresses are still largely unknown but it can be safely concluded that transcription factors are at play, as they are in mammals subjected to a mild stress (e.g. Calvert et al. 2009).

Being able to resist strong stresses after having been subjected to a mild stress can be of a high interest in the wild and thus of selective value. For instance, a common threat in the wild is excessive heat and a brief heat shock could be a signal for forthcoming strong heat shocks, drought, and/or starvation. Animals able to induce a hormetic response after such a brief heat shock (a very hot day, for instance) would probably be more able to resist summer heatwave than those not displaying this hormetic response: indeed, a very hot summer can also be a dry summer with less food available. This would confer a selective advantage in the wild and these animals previously subjected to a brief heat shock would have a higher chance to survive and to reproduce.

The duration of the protection offered by hormetic responses in flies is in the range of a few weeks, at a maximum. This range is the same as their life span and one could wonder whether, in mammals for instance, the hormetic response would last for a few weeks (i.e. the absolute duration would be the same in short- and long-lived species), or for a significant part of their life span, e.g. for months in rodents or years in humans. If the latter result were observed, it would mean that a mild stress can really have effects at an old age, even if applied a long time before, like a vaccine in a way. On the contrary, results in flies would be more easily explained by the features of the hormetic response than by real effects on the aging process.

Performing more experiments in mammals, and particularly in rodents, could allow to clarify the issue because, for the time being, such studies are scarce (see e.g. Calvert et al. 2009; see also in pigs for instance Lavitrano et al. 2004). These studies could help to close the gap between studies in invertebrates, mainly done with flies, and those on humans (reviews in Rattan and Le Bourg 2014), and to know to what extent mild stresses can be of therapeutic value, particularly at old age.

Yet, one can guess that new studies on flies will be done in the future but, obviously, flies are not humans, and this would be a fantasy to imagine subjecting young people to a mild stress such as HG in the hope to increase their resistance to winter infections or summer heatwave at old age. Nevertheless, mild stresses fitted to human physiology can be used. For instance, sauna (60 °C, 15 min daily for 2 weeks) can decrease blood pressure and fasting plasma glucose in patients suffering from hypertension or hypercholesterolemia or can decrease body weight and fat in obese patients (Biro et al. 2003). Indeed, there is now a corpus of results showing that mild stress can be of therapeutic value (reviews in Rattan and Le Bourg 2014). In such conditions, studies of mild stress in *D. melanogaster*, beyond the attempt to establish the most efficient stresses and their mechanisms in this fly, could be like suggestion-boxes for clinic research and help to find out new therapeutic tools to improve daily life of elderly people.

References

Antosh M, Fox D, Hasselbacher T, Lanou R, Neretti N, Cooper LN (2014) *Drosophila melanogaster* show a threshold effect in response to radiation. Dose-Response 12(4):551–581

Ayar A, Uysal H, Altun D (2010) The effects of cold shock on the longevity in *Oregon R* wild and *vestigial* mutant of *Drosophila melanogaster* (Diptera: Drosophilidae). Ekoloji 74:38–44

Bertrand HA, Herlihy JT, Ikeno Y, Yu BP (1999) Dietary restriction. In: Yu BP (ed) Methods in aging research. CRC, Boca Raton, pp. 271–300

Biro S, Masuda A, Kihara T, Tei C (2003) Clinical implications of thermal therapy in lifestyle-related diseases. Exp Biol Med 228:1245–1249

Bojados M, Jamon M (2014) The long-term consequences of the exposure to increasing gravity levels on the muscular, vestibular and cognitive functions in adult mice. Behav Brain Res 264:64–73

Borzelecca JF (2000) Paracelsus: herald of modern toxicology. Toxicol Sci 53:2–4

Brown AE, Baumbach J, Cook PE, Ligoxygakis P (2009) Short-term starvation of immune deficient *Drosophila* improves survival to gram-negative bacterial infections. PLoS ONE 4(2):e4490

Bubliy OA, Kristensen TN, Kellermann V, Loeschcke V (2012) Plastic responses to four environmental stresses and cross-resistance in a laboratory population of *Drosophila melanogaster*. Funct Ecol 26:245–253

Burton V, Mitchell HK, Young P, Petersen NS (1988) Heat shock protection against cold stress of *Drosophila melanogaster*. Mol Cell Biol 8:3550–3552

Calabrese EJ (2008) What is hormesis? In: Le Bourg E, Rattan SIS (eds) Mild stress and healthy aging. Applying hormesis in aging research and therapy. Springer, Dordrecht, pp 5–19

Calabrese EJ (2013) Hormetic mechanisms. Crit Rev Toxicol 43:580–606

Calabrese EJ (2014) Hormesis and risk assessment. In: Rattan SIS, Le Bourg E (eds) Hormesis in health and disease. CRC Press, Boca Raton, pp 339–355

Calvert JW, Jha S, Gundewar S, Elrod JW, Ramachandran A, Pattillo CB, Kevil CG, Lefer DJ (2009) Hydrogen sulfide mediates cardioprotection through Nrf2 signaling. Circ Res 105:365–374

Colinet H, Lee SF, Hoffmann A (2010a) Temporal expression of heat-shock genes during cold stress and recovery from chill coma in adult *Drosophila melanogaster*. FEBS J 277:174–185

Colinet H, Lee SF, Hoffmann A (2010b) Knocking down expression of Hsp22 and Hsp23 by RNA interference affects recovery from chill coma in *Drosophila melanogaster*. J Exp Biol 213:4146–4150

Cypser JR, Johnson TE (2002) Multiple stressors in *Caenorhabditis elegans* induce stress hormesis and extended longevity. J Gerontol Biol Sci 57A:B109–B114

Cypser JR, Johnson TE (2003) Hormesis in *Caenorhabditis elegans* dauer-defective mutants. Biogerontology 4:203–214

Czajka MC, Lee RE (1990) A rapid cold-hardening response protecting against cold shock injury in *Drosophila melanogaster*. J Exp Biol 148:245–254

Dahlgaard J, Loeschcke V, Michalak P, Justesen J (1998) Induced thermotolerance and associated expression of the heat-shock protein Hsp70 in adult *Drosophila melanogaster*. Funct Ecol 12:786–793

Defays R, Gómez FH, Sambucetti P, Scannapieco AC, Loeschcke V, Norry FM (2011) Quantitative trait loci for longevity in heat-stressed *Drosophila melanogaster*. Exp Geront 46:819–826

Economos AC, Miquel J, Ballard RC, Blemden M, Lindseth KA, Fleming J, Philpott DE, Oyama J (1982) Effects of simulated increased gravity on the rate of aging of rats. Implications for the rate of living theory of aging. Arch Gerontol Geriatr 1:349–363

Everitt AV, Rattan SIS, Le Couteur DG, de Cabo R (eds) (2010) Calorie restriction, aging and longevity. Springer, Dordrecht

Foley E, O'Farrell PH (2003) Nitric oxide contributes to induction of innate immune responses to gram-negative bacteria in *Drosophila*. Genes Dev 17:115–125

Frolkis VV (1982) Aging and life-prolonging processes. Springer, Heidelberg

Ganesan S, Aggarwal K, Paquette N, Silverman N (2011) NF-κB/Rel proteins and the humoral immune responses of *Drosophila melanogaster*. Cur Top Microbiol 349:25–60

Hercus MJ, Loeschcke V, Rattan SIS (2003) Lifespan extension of *Drosophila melanogaster* through hormesis by repeated mild heat stress. Biogerontology 4:149–156

Jamon M, Serradj N (2009) Ground-based researches on the effects of altered gravity on mice development. Microgravity Sci Technol 21:327–337

Jensen D, Overgaard J, Sørensen JG (2007) The influence of developmental stage on cold shock resistance and ability to cold-harden in *Drosophila melanogaster*. J Insect Physiol 53:179–186
Kampkötter A, Nkwonkam CG, Zurawski RF, Timpel C, Chovolou Y, Wätjen W, Kahl R (2007) Effects of the flavonoids kaempferol and fisetin on thermotolerance, oxidative stress and FoxO transcription factor DAF-16 in the model organism *Caenorhabditis elegans*. Arch Toxicol 81:849–858
Karunanithi S, Barclay JW, Robertson RM, Brown IR, Atwood HL (1999) Neuroprotection at *Drosophila* synapses conferred by prior heat shock. J Neurosci 19:4360–4369
Kelty JD, Lee RE (1999) Induction of rapid cold hardening by cooling at ecologically relevant rates in *Drosophila melanogaster*. J Insect Physiol 45:719–726
Kelty JD, Lee RE (2001) Rapid cold-hardening of *Drosophila melanogaster* (diptera: drosophilidae) during ecologically based thermoperiodic cycles. J Exp Biol 204:1659–1666
Khazaeli AA, Tatar M, Pletcher SD, Curtsinger JW (1997) Heat-induced longevity extension in *Drosophila*. I. Heat treatment, mortality, and thermotolerance. J Gerontol Biol Sci 52A:B48–B52
Kita S, Shibata S, Kim H, Otsubo A, Ito M, Iwasaki KI (2006) Dose-dependent effects of hypergravity on body mass in mature rats. Aviat Space Environ Med 77:842–845
Klose MK, Chu D, Xiao C, Seroude L, Robertson RM (2005) Heat-shock-mediated thermoprotection of larval locomotion compromised by ubiquitous overexpression of Hsp70 in *Drosophila melanogaster*. J Neurophysiol 94:3563–3572
Krebs RA, Loeschcke V (1994a) Effects of exposure to short-term heat stress on fitness components in *Drosophila melanogaster*. J Evol Biol 7:39–49
Krebs RA, Loeschcke V (1994b) Costs and benefits of activation of the heat-shock response in *Drosophila melanogaster*. Funct Ecol 8:730–737
Kristensen TN, Sørensen JG, Loeschcke V (2003) Mild heat stress at a young age in *Drosophila melanogaster* leads to increased Hsp70 synthesis after stress exposure later in life. J Genet 82:89–94
Kuether K, Arking R (1999) *Drosophila* selected for extended longevity are more sensitive to heat shock. Age 22:175–180
Lagisz K, Hector L, Nakagawa S (2013) Life extension after heat shock exposure: assessing meta- analytic evidence for hormesis. Ageing Res Rev 12:653–660
Lavitrano M, Smolenski RT, Musumeci A, Maccherini M, Slominska E, Di Florio E, Bracco A, Mancini A, Stassi G, Patti M, Giovannoni R, Froio A, Simeone F, Forni M, Bacci ML, D'Alise G, Cozzi E, Otterbein LE, Yacoub MH, Bach FH, Calise F (2004) Carbon monoxide improves cardiac energetics and safeguards the heart during reperfusion after cardiopulmonary bypass in pigs. FASEB J 18:1093–1095
Le Bourg E (2005) Hormetic protection of *Drosophila melanogaster* middle-aged male flies from heat stress by mildly stressing them at young age. Naturwiss 92:293–296
Le Bourg E (2007a) Hormetic effects of repeated exposures to cold at young age on longevity, aging and resistance to heat or cold shocks in *Drosophila melanogaster*. Biogerontology 8:431–444
Le Bourg E (2007b) Hormetic effects on longevity of hydrogen peroxide in *Drosophila melanogaster* flies living on a poorly nutritious medium. Biogerontology 8:327–344
Le Bourg E (2008) Three mild stresses known to increase longevity in *Drosophila melanogaster* flies do not increase resistance to oxidative stress. Am J Pharmacol Toxicol 3:134–140
Le Bourg E (2009) Hormesis, aging and longevity. Biochim Biophys Acta 1790:1030–1039
Le Bourg E (2010a) Combined effects of suppressing live yeast and of a cold pretreatment on longevity, aging and resistance to several stresses in *Drosophila melanogaster*. Biogerontology 11:245–254
Le Bourg E (2010b) Predicting whether dietary restriction would increase longevity in species not tested so far. Ageing Res Rev 9:289–297

Le Bourg E (2011a) A cold stress applied at various ages can increase resistance to heat and fungal infection in aged *Drosophila melanogaster* flies. Biogerontology 12:185–193

Le Bourg E (2011b) The NF-kB like factor DIF has weaker effects on *Drosophila melanogaster* immune defenses than previously thought. J Comp Physiol B 181:741–750

Le Bourg E (2012) Combined effects of two mild stresses (cold and hypergravity) on longevity, behavioral aging, and resistance to severe stresses in *Drosophila melanogaster*. Biogerontology 13:313–328

Le Bourg E (2013) Fasting can protect young and middle-aged *Drosophila melanogaster* flies against a severe cold stress. Biogerontology 14:513–529

Le Bourg E, Fournier D (2004) Is lifespan extension accompanied by improved antioxidant defences? A study of superoxide dismutase and catalase in *Drosophila melanogaster* flies that lived in hypergravity at young age. Biogerontology 5:261–266

Le Bourg E, Grimal A, Fresquet N, Lints FA (1995) Spontaneous locomotor activity of *Drosophila melanogaster* flies at various gravity levels (0 g, 1 g, 1.8 g) during parabolic flights. Behav Proc 34:175–184

Le Bourg E, Malod K, Massou I (2012) The NF-κB-like factor DIF could explain some positive effects of a mild stress on longevity, behavioral aging, and resistance to strong stresses in *Drosophila melanogaster*. Biogerontology 13:455–465

Le Bourg E, Massou I, Gobert V (2009) Cold stress increases resistance to fungal infection throughout life in *Drosophila melanogaster*. Biogerontology 10:613–625

Le Bourg E, Minois N (1997) Increased longevity and resistance to heat shock in *Drosophila melanogaster* flies exposed to hypergravity. C R Acad Sci Paris 320:215–221

Le Bourg E, Minois N (1999) A mild stress, hypergravity exposure, postpones behavioral aging in *Drosophila melanogaster*. Exp Geront 34:157–172

Le Bourg E, Minois N, Bullens P, Baret P (2000) A mild stress due to hypergravity exposure at young age increases longevity in *Drosophila melanogaster* males. Biogerontology 1:145–155

Le Bourg E, Rattan SIS (eds) (2008) Mild stress and healthy aging. Applying hormesis in aging research and therapy, Springer, Dordrecht

Le Bourg E, Toffin E, Massé A (2004) Male *Drosophila melanogaster* flies exposed to hypergravity at young age are protected against a non-lethal heat shock at middle age but not against behavioral impairments due to this shock. Biogerontology 5:431–443

Le Bourg E, Valenti P, Lucchetta P, Payre F (2001) Effects of mild heat shocks at young age on aging and longevity in *Drosophila melanogaster*. Biogerontology 2:155–164

Le Bourg E, Valenti P, Payre F (2002) Lack of hypergravity-associated longevity extension in *Drosophila melanogaster* flies overexpressing *hsp70*. Biogerontology 3:355–364

Lints FA, Bullens P, Le Bourg E (1993) Hypergravity and aging in *Drosophila melanogaster*: 7. New longevity data. Exp Geront 28:611–615

Marshall KE, Sinclair BJ (2012) The impacts of repeated cold exposure on insects. J Exp Biol 215:1607–1613

McClure CD, Zhong W, Hunt VL, Chapman FM, Hill FV, Priest NK (2014) Hormesis results in trade-offs with immunity. Evolution 68:2225–2233

Megory E, Oyama J (1984) Hypergravity effects on litter size, nursing activity, prolactin, TSH, T3, and T4 in the rat. Aviat Space Environ Med 55:1129–1135

Minois N (2000) Longevity and aging: beneficial effects of exposure to mild stress. Biogeront 1:15–29

Minois N (2001) Resistance to stress as a function of age in transgenic *Drosophila melanogaster* overexpressing *hsp70*. J Insect Physiol 47:1007–1012

Minois N, Guinaudy MJ, Payre F, Le Bourg E (1999) HSP70 induction may explain the long-lasting resistance to heat of *Drosophila melanogaster* having lived in hypergravity. Mech Ageing Dev 109:65–77

Minois N, Khazaeli AA, Curtsinger JW (2001) Locomotor activity as a function of age and life span in *Drosophila melanogaster* overexpressing *hsp70*. Exp Geront 36:1137–1153

Minois N, Le Bourg E (1999) Resistance to stress as a function of age in *Drosophila melanogaster* living in hypergravity. Mech Ageing Dev 109:53–64

Morrow G, Tanguay RM (2003) Heat shock proteins and aging in *Drosophila melanogaster*. Seminars Cell Develop Biol 14:291–299

Moskalev A, Shaposhnikov M (2011) Pharmacological inhibition of NF-κB prolongs lifespan of *Drosophila melanogaster*. Aging (Albany NY) 3:391–394

Nakagawa S, Lagisz M, Hector KL, Spencer HG (2012) Comparative and meta-analytic insights into life extension via dietary restriction. Aging Cell 11:401–409

Oyama J (1982) Metabolic effects of hypergravity on experimental animals. In: Miquel J, Economos AC (eds) Space gerontology, vol 2248. Nasa Conference Publication, pp 37–51

Pickering AM, Vojtovich L, Tower J, Davies KJA (2013) Oxidative stress adaptation with acute, chronic, and repeated stress. Free Rad Biol Med 55:109–118

Pitts GC, Oyama J (1979) Rat growth during chronic centrifugation. In: Holmquist R (ed) COSPAR life sciences and space research, vol 17, pp 225–229

Puig O, Mattila J (2011) Understanding Forkhead box class O function: lessons from *Drosophila melanogaster*. Antioxid Redox Signal 14:635–647

Rako L, Hoffmann AA (2006) Complexity of the cold acclimation response in *Drosophila melanogaster*. J Insect Physiol 52:94–104

Rattan SIS, Le Bourg E (eds) (2014) Hormesis in health and disease. CRC Press, Boca Raton

Sacher GA (1977) Life table modification and life prolongation. In: Finch CE, Hayflick L (eds) Handbook of the Biology of Aging. Van Nostrand Reinhold Company, New York, pp 582–638

Sarup P, Sørensen JG, Loeschcke V (2014) The long-term effects of a life-prolonging heat treatment on the *Drosophila melanogaster* transcriptome suggest that heat shock proteins extend lifespan. Exp Geront 50:34–39

Sejerkilde M, Sørensen JG, Loeschcke V (2003) Effects of cold- and heat hardening on thermal resistance in *Drosophila melanogaster*. J Insect Physiol 49:719–726

Sondag HNPM, de Jong HAA, Oosterveld WJ (1997) Altered behaviour in hamsters conceived and born in hypergravity. Brain Res Bull 43:289–294

Sørensen JG, Kristensen TN, Kristensen KV, Loeschcke V (2007) Sex specific effects of heat induced hormesis in *Hsf*-deficient *Drosophila melanogaster*. Exp Geront 42:1123–1129

Speakman JR, Mitchell SE (2011) Caloric restriction. Mol Aspects Med 32:159–221

Swindell WR (2012) Dietary restriction in rats and mice: a meta-analysis and review of the evidence for genotype-dependent effects on lifespan. Ageing Res Rev 11:254–270

Thayer KA, Melnick R, Burns K, Davis D, Huff J (2005) Fundamental flaws of hormesis for public health decisions. Environ Health Perspect 113:1271–1276

Tower J (2011) Heat shock proteins and *Drosophila* aging. Exp Geront 46:355–362

Vaiserman AM (2008) Irradiation and hormesis. In: Le Bourg E, Rattan SIS (eds) Mild stress and healthy aging. Applying hormesis in aging research and therapy. Springer, Dordrecht, pp 21–41

Vigne P, Tauc M, Frelin C (2009) Strong dietary restrictions protect *Drosophila* against anoxia/reoxygenation injuries. PLoS ONE 4:e5422

Wade CE, Moran MM, Oyama J (2002) Resting energy expenditure of rats acclimated to hypergravity. Aviat Space Environ Med 73:859–864

Zhong JF, Wang SP, Shi XQ, Mu LL, Li GQ (2010) Hydrogen sulfide exposure increases desiccation tolerance in *Drosophila melanogaster*. J Insect Physiol 56:1777–1782

Chapter 15
Strategies for Stage-Specific Extension of Longevity

Robert Arking

Abstract It is generally recognized that the adult life span of *Drosophila* is not a continuous process from fertilization to death, but rather consists of four demographically defined phases: the developmental, health, transition, and senescent spans. These are marked by differences in gene expression patterns, physiology, homeostasis, and mortality rates. I review our empirical data showing that drugs known to significantly extend longevity in either early or late life do so via stage-specific effects such that beneficial effects are observed at one stage but neutral or detrimental effects at another stage. Whole-life feeding of these drugs leads to failure to detect their pro-longevity effect, suggesting that the gene-based drug targets are not necessarily present in all four stages.

Keywords *Drosophila* • Life span • Developmental phase • Healthspan • Transition span • Senescent span

15.1 Introduction

Current drug screening protocols are based on the paradigm that whole life feeding of candidate drugs will not yield any false negative data. Our data indicates that this paradigm is flawed. It might be useful to perform the initial whole-life and stage-specific testing of candidate pro-longevity drugs on flies prior to any testing on mammals. The lower cost fly experiments would allow the characterization of any stage-specific effects of the candidate drug on conserved pathways, and would inform the experimental design of subsequent mouse studies. One testable prediction of the current data is that drugs with beneficial late life effects are likely to be enriched for molecules capable of inhibiting or countering age-related

R. Arking (✉)
Department of Biological Sciences, Wayne State University, Detroit 48098, MI, USA
e-mail: aa2210@wayne.edu

A.M. Vaiserman et al. (eds.), *Life Extension*, Healthy Ageing and Longevity 3,
DOI 10.1007/978-3-319-18326-8_15

diseases, while drugs with early life beneficial effects are likely to be enriched for healthspan extension effects.

15.1.1 Demographic Data Supporting Different Life Span Phases

The lives of animals can be lengthened in three ways: increasing both their mean and maximum life span; increasing their mean but not their maximum life span; or increasing their maximum but not their mean life span (Fig. 15.1a–c)

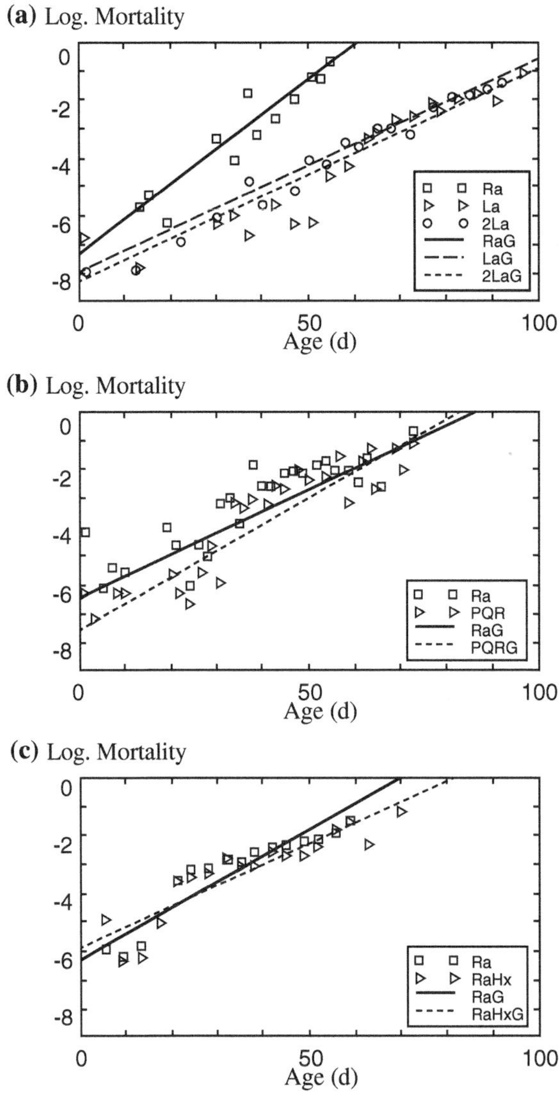

◀ **Fig. 15.1** Three types of survival curves for extended longevity. **a** Survival curves of the normal-lived Ra strain and of two long-lived strains (La and 2La) derived from it by a direct selection for delayed female fecundity. The data points represent the observed survival data and are based on the age-specific values obtained from two or three replicate cohorts consisting for at least 250 mixed sex individuals each. The Ra, La, and 2La curves are significantly different (log-rank test, $\chi^2 = 530.16$, 2 df, $p < 0.0001$). The continuous lines are the Weibull approximations of the empirical data (see Arking 1987 for details). **b** Survival curves of the normal-lived Ra strain and the PQR strain selected from it by direct selection for paraquat resistance. The data points represent the observed survival data and are based on the age-specific values obtained from mixed sex cohorts of 250–450 animal each. The two curves are significantly different (log-rank test, $\chi^2 = 24.76$, 1 df, $p = 0.0005$). The continuous lines are the Weibull approximations of the empirical data (see Vettraino et al. 2001 for details). **c** Survival curves of the normal-lived Ra control strain and the longer-lived Ra heat-treated strain. The latter animals were subjected to a non-lethal heat shock (37 °C for 90 min) early in life at days 5–7 after eclosion. They were then maintained under controlled optimal conditions and their survival monitored. The two curves are significantly different (log-rank test, $\chi^2 = 17.84$, 1 df, $p < 0.0005$). The continuous lines are the Weibull approximations of the empirical data (see Kuether and Arking 1999 for details)

(Arking et al. 2004). Although each of the three treatments resulted in a significant alteration of the survival curves, only the first treatment resulted in a significant slowing of the age-specific mortality rate (Fig 15.2a–c) (Arking et al. 2004). The latter two approaches transiently decreased the mortality rate only at the beginning or end of life, and neither case represents an optimal increase of healthy life span. Only the first approach caused the organism to enhance its existing repair and maintenance capabilities so as to actually slow the aging rate and thereby delay the onset of senescence.

The age-specific mortality rate is defined as the probability of dying at a certain age among those individuals of the birth cohort who have survived to that age. Gompertz (1825) found that the age -specific death rate increases as an exponential function of age:

$$q_x = (q_0)^{ex} \tag{15.1}$$

where q_0 = further expectation of life at the time of birth, or the y-intercept; q_x = further expectation of life at the beginning of an age interval, x; and x = the slope constant.

The Eq. 15.1 can also be written in linear form as: $\ln q_x = \ln q_0 + \mathrm{x}$, which is, of course, a particular form of the general equation for a straight line: $y = b + mx$. Non-linear calculations of the age-specific mortality rate can be made using multiple likelihood statistics as described by Pletcher and Geyer (1999).

The three longevity types are probably general, in the sense that they exhaust the logical permutations of variations in the linear mortality curves. There is no reason to believe that this one strain is special and every reason to believe that many if not all wild-type strains in this and other species are capable of mounting the same set of responses. The presence of three alternative longevity phenotypes in the same organism means that the genetic mechanisms regulating longevity are more complex than is often assumed. Presumably this flexibility evolved as to allow the organism to reproduce effectively under different environmental conditions.

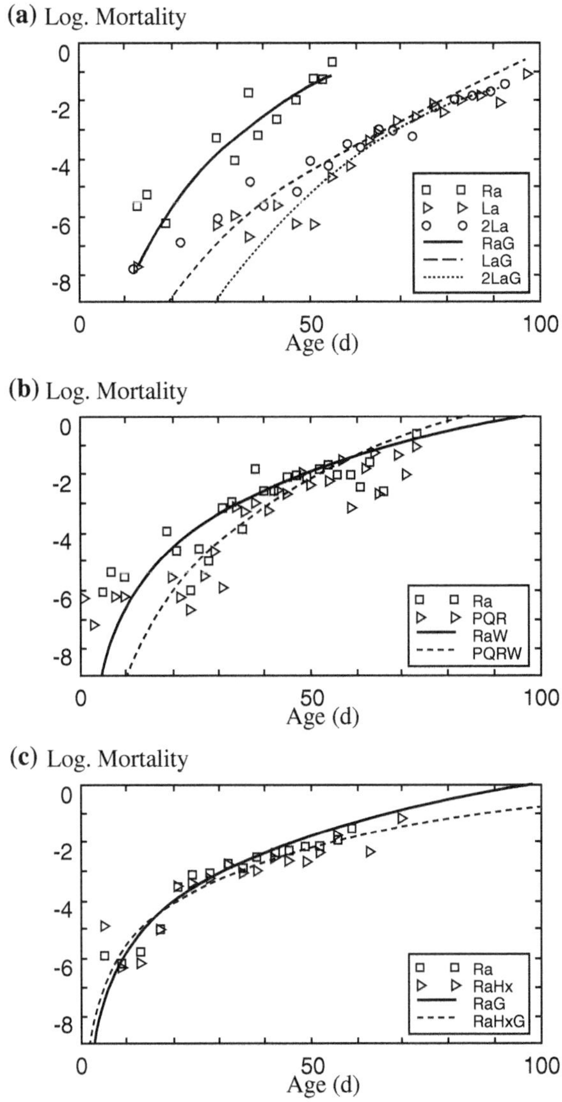

Fig. 15.2 Three patterns of age-specific mortality rates and their effect on longevity. **a** Age-specific mortality curves of the normal-lived Ra strain and of two long-lived strains (La and 2La). The Ra and La curves, and the Ra and 2La curves, are significantly different from each other, whereas the La and 2La curves are not statistically different from each other. **b** Age-specific mortality curves of the normal-lived Ra strain and the PQR strain. The data points and their Gompertz approximations are shown (Ra-G and PQR-G). Statisical analysis shows that the curves are significantly different ($p < 0.0036$); the intercepts differ significantly whereas the slopes of the two curves have no significant difference. **c** Age-specific mortality curves of the normal-lived Ra strain and the longer-lived heat-treated strain (RaHX). The experimental points and the Gompertz approximations (Ra-G and Ra-HXG) are shown. There is no significant difference between the two curves ($p < 0.35$). The intercepts do not differ significantly (from Arking et al. 2004)

The extended healthspan phenotype of Figs. 15.1a and 15.2a would clearly be the desired outcome for an eventual longevity intervention in humans. This desirability arises from the fact that the delay of the inflection point in the survival curve results in an extended healthspan while having minimal effect on the length of the senescent span (Fig. 15.2). The extra longevity is added on to the healthspan and not to the senescent span; this observation suggests that the adult life span is based on a multiphasic genetic architecture. The existence of a complex architecture implies the possibility of multiple points at which one might intervene in the life span and further raises the possibility that the several phases may be lightly connected and thus susceptible to different stimuli.

The analysis of the data from our laboratory as well as of that from the literature leads us to conclude that the life span is not just one long process of erosion but rather a complex process in which different stimuli can give rise to alternate patterns of gene expression that give rise via differential mortality to the three longevity phenotypes of Fig. 15.1. Our long-lived La strain of *Drosophila* lives longer because its healthspan has been genetically altered, not because its senescent span is different in any way. In this view, the life span consists of four phases: namely, developmental span, healthspan, and senescent span; with the latter two being separated by a variable transition span. Actually, all four phases are variable in their length of time however expressed, and in their observed effects on the organism and it's age-specific mortality rate. A brief summary of these four spans follows.

15.1.1.1 The Developmental Phase

The developmental phase in Drosophila covers the period of time from conception to the eclosion of the adult. Embryonic and larval development is clearly understood to be a program in the sense that development consists of a series of gene expression modules in which the output of one module is the input for the next (Davidson et al. 2002; Davidson and Erwin 2006; Tu et al. 2014). Developmental gene expression patterns for any species are highly complex yet very predictable in that they depend almost entirely on the nature of the *cis*-acting regulatory elements of the genes involved (Segal et al. 2008). They may be viewed as leading to a stage of highest optimal functionality in the young adult, as judged by their minimum values of the age-specific mortality rate. Although development is mostly internally driven, chance may still play an important role in individuals (Finch and Kirkwood 2000, Finch 2007). These developmental expression patterns are susceptible to certain environmental influences affecting their basic regulatory mechanisms and thus altering their future life span (Barker 1995; Danese et al. 2007). For this reason, the developmental phase of the life span cannot be excluded when considering the overall genetic architecture of longevity.

15.1.1.2 The Health Span

The healthspan begins when the development period ends. It is defined by a low age-specific mortality rate and a high survival rate. The adult cohort starts out with 100 % survival and this high level is maintained for some time in most healthy cohorts. The healthspan ends when the survival rate drops below 90 %, or when there is a negative inflection in thc survival curvc, whichever comes first.

15.1.1.3 The Transition Span

The Transition Span is defined as that portion of the life span encompassed between the adult survival rates of 89 % (e.g., immediately following the end of the healthspan) and 80 % (e.g., immediately preceding the beginning of the senescent span). The transition span is variable, being obvious in *Drosophila* strains raised under conditions that allow a robust healthspan. Some strains and/or cohorts display an almost continuous decrease in the survival curve, suggesting that these animals are somehow weakened such that they have a relatively high age-specific mortality rate from eclosion onwards.

15.1.1.4 The Senescent Span

The Senescent Span is characterized by the obvious loss of function implicit in the decreasing survival as well as by the changing physiology of the older adults. It is this portion of the life span that is commonly referred to by the word aging. Humans and the common laboratory model organisms used to study senescence are all species with determinate growth that undergo a gradual senescence. The canonical patterns of senescence in these species were elucidated by Finch (2007) as encompassing most or all of the following: mortality acceleration, reproductive decline, slowed movements, cardiovascular dysfunctions, abnormal growths, oxidative damage, and neuron loss. The senescent span ends with the death of the organisms and the cohort.

15.1.1.5 Architecture of the Life Span

A diagrammatic summary of these demographic divisions of the life span, as well as brief summaries of their genetic and physiological traits, is presented in Fig. 15.3. Much discussion is available about aging mechanisms, but this term lumps together processes involved in our healthspan with those mostly involved in our senescent span. The data presented in Figs. 15.1 and 15.2 make it likely that the healthspan can be regulated independently of the senescent span. Our more recent data, summarized in this chapter, confirms that the mechanisms operative in one phase are not necessarily present in another phase (McDonald et al. 2013;

LIFE SPAN =

DEVELOPMENTAL SPAN	HEALTH SPAN	TRANSITION PHASE	SENESCENT SPAN
gene network innately differentiates, but can be epigenetically perturbed by maternal or developmental effects	***gene-dependent longevity assurance mechanisms operative***	***event-& history- dependent loss of optimal function***	***stochastic effects on unique networked genome leads to individualized loss of function***
\Gene/Protein Interaction Networks differentiate via developmental program	Gene/Protein Interaction Networks are at optimal levels but may have been affected by epigenetic perturbation	Loss of inter-or intra-cellular signals alters balance of cell defenses. Gene interaction networks slip into lower state of function which may be reversible. Damage accumulates.	Degradation of Gene/Protein Interaction Networks brought on by spoke gene perturbation of gene network dynamic equilibrium
Cell replication and body growth dominant activities	*Longevity Determinant Mechanisms Operative; Cells are in growth/reproduction mode but still have high levels of repair and maintenance Homeostatic Ability Sensitive & Reliable*	*Timing & nature of transition may be influenced by cell network states induced during development Repression of age-related disease genes weakens. Abnormal proteins aggregate, exceed chaperone capacity.*	*Cell's Regulatory Ability Decreases due to sub-optimal expression within or between cells Age-related Diseases Systemic Deleterious Effects on Organism. Increased Number of Senescent Cells in Tissues accelerate Loss of Tissue Function.*
Cells in some systems (neural, metabolic) may be perturbed into other long term equilibrium states by pre- and/or post-natal events	Cells have sufficient reserve capacity to deal with various stressors	Damage Accumulates and Initiates a Positive Feedback of Damage Induction. Cell Function Decreases. Damaged Cells Survive.	Positive Feedback Damage Cascades Lessen Cell's Homeostatic Ability; Tissue function suffers.
		←------ *Possible Reversals*	
Minimum age-specific mortality rate reached just prior to sexual maturation.	Age-specific mortality rate may be flat or show only a slight increase starting from a low base level.	Age-specific mortality rate may show sharp increase over prior decades.	Critical Thresholds Passed; Cell Function Ceases
pre-conception to ~20 yrs	20 -- ~55 yrs	~55 -- 65 yrs	~ 65 -- 122 yrs

Fig. 15.3 A diagrammatic view of the architecture of the life span overview of the changes in physiology, gene expression, cell functions, mortality, and resilience during the developmental, health, and senescent phases of the life span as well as during the transition phase between the latter two spans. See text for discussion (modified after Arking 2009)

Soh et al. 2013). The data summarized in Fig. 15.3 demonstrates that the four phases of the life span are quite different at multiple functional levels. They should be separated for clarity of thought, which is essential if we wish to be able to understand and manipulate the processes regulating longevity and aging.

The reader will notice that I list the possibility of reversing aging from the transition or senescent spans to the healthspan. This has now been shown to take place in several precisely defined conditions. The work of Chung and colleagues (see Adler et al. 2007, 2008) demonstrated the reversal of skin aging in a mouse genetically engineered so as to inhibit in late life the activity of NFkB, a transcription factor intimately involved in chronic ROS production. Once the pro-inflammatory signals were inhibited, the skin histology changed to resemble that of a younger mouse. The work of Conboy and colleagues (Conboy et al. 2005; Carlson et al. 2014) showed that age-related loss of mice and humans to undergo muscle wound healing could be restored by providing or enhancing the MAPK/Notch signal molecules needed to maintain that process. Both projects show that cell and tissue function is dependent on the presence of appropriate signals and so either the presence of a pro-inflammatory signal or the absence of a necessary signal may induce the aging process. The basic validity of these experiments is supported by the description of genetic processes in other systems which either repress the onset of senescent span gene expression or activate genes capable of protecting against age-related senescent diseases. Both of these processes also result in an effective extension of the healthspan (Liu et al. 2011; Gentilini et al. 2013; Lu et al. 2014). The important point about these several processes is that they either alter the gene expression patterns characteristic of the senescent span into patterns characteristic of the healthspan, or else maintain the healthspan gene expression patterns by repressing genes which would otherwise shift the organism into the senescent span. The fact that these processes act so as to maintain the healthspan against insults suggest that we can view the healthspan as being the default state of the adult organism. It also suggests that the demographic divisions of Fig. 15.3 actually represent functionally distinct compartments at the genetic level.

15.1.2 Goal of this Chapter

The question posed by this chapter is: What is the empirical evidence supporting the concept that the demographic divisions of the life span as shown in Fig. 15.3 actually do reflect distinct phases of gene expression patterns, with all that is implied by that statement? The most direct way to fully answer the question is to do an analysis of stage-specific gene expression patterns for each of the four life stages, and compare them. We recently began such an analysis and those data will be reported elsewhere at a later date. Our present data indicate that there are at least ~1000 genes which exhibit significantly different expression differences in the long-lived La females relative to the normal-lived Ra females. Our analysis is in progress and so we cannot now use a direct approach to answer our question. in the meantime, we will use indirect measure of stage-specific gene expression effects to provide an interim answer to our question.

15.2 Empirical Data Supporting the Demographic Divisions

15.2.1 Background Information

Our motivation in undertaking the series of experiments discussed in this chapter was to determine if there existed any small molecule compounds (i.e., drugs) which might induce an extended longevity in the normal-lived Ra strains similar to that observed in the long-lived La strains. We decided to use both whole-life as well as stage-specific patterns of exposure to various drugs. Experimental pro-longevity drugs are usually administered over the entire adult life span of the test organism. Given that different life stages have different patterns of gene expression, and given that molecules regulating longevity may have a gene-specific target pattern, then there is reason to believe that whole-life interventions will not always be the most effective intervention. Such compounds may have stage-specific positive effects in one part of the life span but neutral or negative effects in another part. Given the potential of whole life experiments to yield false negative conclusions, then temporally targeted investigations may yield better insights into drug efficacy. The various reagents and techniques involved in these experiments were fully discussed in Soh et al. (2013) and McDonald et al. (2013), and should be consulted for those details as well as for the data mentioned but not shown below. Animals were raised on yeast-sucrose-agar food with a standard caloric (protein/carbohydrate) content/ml (AL food = ~3.1 cal/10 ml) or a 50 % reduction (DR food). The survival statistical data cited herein are each based on the log-rank test as specified in the GraphPad Prism 5.04 software; the relevant parameters are in the figure legends or text. Mortality kinetics were derived using multiple likelihood analysis; the calculations are available in the supplementary data of the original publications.

15.2.2 Summary Information

We now report that curcumin and certain HDAC inhibitors have beneficial effects in the normal-lived Ra. However, the two drugs do this in strikingly different manners. Curcumin increases longevity when administered in the developmental or healthspan stages, but has a negative effect when administered over the entire adult life span or over the senescent stage only. The two HDAC inhibitors tested, on the other hand, increased longevity during the transition and senescent spans but had negative effects on longevity when given during the developmental or healthspans. Table 15.1 briefly summarizes the pattern of these effects in the Ra and La strains. Note the clear difference between the sensitive periods of the normal-lived Ra animals to these two drugs. Note also that the long-lived La animals are not affected by curcumin at all at any stage, but are negatively affected by sodium butyrate (SB) in early life but not in late life.

Table 15.1 Summary of the stage-specific effects of the tested drugs on the Ra and La strains

Drugs	Development span		Healthspan		Transition span		Senescent span		Whole life	
Strain	Ra	La	Ra	La	Ra	La	Ra	La	Ra	La
Curcumin	+	0	+	0	–	0	–	0	–	0
SB	–	–	–	–	+	–	+	–	–	–
SAHA	nd	nd	nd	nd	+	–	+	–	nd	nd

Ra normal-lived strain, *La* long-lived strain, + positive effect of drug on longevity when fed at the indicated stage, 0 neutral effect of drug on longevity when fed at the indicated stage, – negative effect of drug on longevity when fed at the indicated stage, *nd* no data

In the remainder of this chapter, I will present the detailed data supporting Table 15.1 and then conclude with comments, based on these data, regarding how to improve the current protocols for testing pro-longevity drug candidates.

15.3 Results of Stage-Specific Curcumin Treatment on Longevity and Gene Expression

15.3.1 Developmental Span

15.3.1.1 Effect of Curcumin Feeding

A dose-response test was done on both the Ra and La strains by testing the effect of 0, 10, 100 and 200 mM curcumin food fed to larvae only, and assaying the subsequent longevity of the adult flies raised on the AL food. The optimal extended longevity response was displayed by the Ra adults fed 100 mM as larvae (days −10 to −5) which displayed an 80 % increase in the length of their healthspan relative to the control (Fig. 15.4). and the resulting adults expressed decreased mRNA levels for a number of known longevity loci. The ten genes studied and their relative responses to larval curcumin are shown in Table 15.2, while Fig. 15.5 visually details the response of *Tor* (which is a nutritionally sensitive regulator of RNA and protein synthesis). The results are expressed as the mean ±SEM of fold change in curcumin relative to control expression at each age, as determined by RTq-PCR.

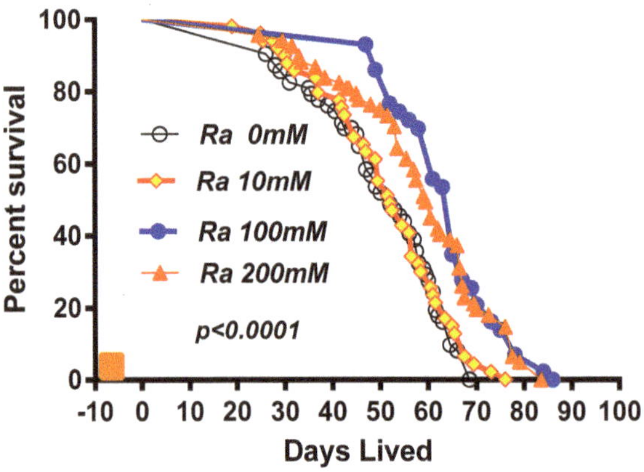

Fig. 15.4 Larval curcumin Dose-Response studies. Eggs were laid on curcumin-containing media and exposed to the chemical from −10 to −5 days as indicated by the solid box on the X-axis. Resulting female adults were raised on control AL food. The 150 normal-lived Ra animals show a maximum response at 100 mM (e.g., a ~ 85 % delay in the age of onset of senescence relative to the 150 controls (48 days vs. 26 days), and significant increases in median and maximum life spans relative to controls (log rank test, $\chi^2 = 26.79$, 1 df, $p < 0.0001$) (from Soh et al. 2013)

Table 15.2 qRT-PCR-based expression values of candidate genes affected by curcumin treatment (from Soh et al. 2013)

	Curcumin/Control relative gene expression					
Stage	Larval	day 1	day 5	day 15	day 30	day 45
GENE						
Tor	0.65	2.22	0.67	0.78	0.93	1.14
4eBP	0.41	3.45	0.34	1.21	0.64	0.99
S6 k	0.75	2.61	0.54	1.11	0.95	1.13
Sir2	0.44	2.58	0.3	0.85	1.06	1.27
hsp22	0.54	1.6	0.16	2.37	0.94	1.38
hsp27	0.37	1.98	0.18	1.42	0.99	1.04
hsp70	0.05	0.86	0.1	1.03	0.67	1.11
Cat	0.77	2.5	0.53	0.75	0.95	1.49
Foxo	1.03	1.84	0.6	0.776	0.99	1.11
Sod	2.09	1.46	0.75	1.02	0.91	1.02

The day 1 results are anomalous in all cases and likely reflect the stresses of eclosion on the young adult. Most of the genes tested are repressed in late larval life and at day 5 but return to normal levels by day 15 or 30. Both *4e-BP* and *S6k* are downstream target branches from *Tor*, while *Foxo* (a transcription factor intimately involved in stress resistance responses) is indirectly downstream on another branch from *Tor*. *Sir2* is a major longevity gene indirectly affecting *Tor*. The *hsps* are downstream chaperones genes and only indirectly involved in longevity. *Cat* and

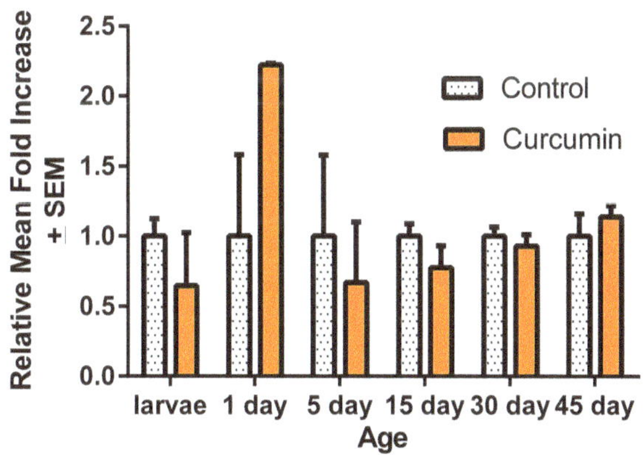

Fig. 15.5 Curcumin/control relative tor gene expression values at each age. These preliminary qRT-PCR data were obtained from one experiment consisting of three dependent replicates of ~50 males/time point. Wild-type parents laid eggs on control or curcumin-containing food, and the F1 were flash-frozen in liquid N_2 at the indicated stages (larval = late non-feeding 3rd instar) or male adult ages. qRT-PCR was done using standard techniques. Results are expressed as the mean ±SEM of fold change in curcumin relative to control expression at each age. Table 15.2 shows the relative expression values of nine other genes assayed during the same experiment (from Soh et al. 2013)

Sod are also downstream genes involved in reactive oxygen species scavenging and indirectly involved in longevity. The significance of these gene expression patterns is not yet understood. We are presently involved in the analysis of microarray data which should increase our understanding of the processes influenced by curcumin treatment. Chandrashekara et al. (2014) have data suggesting that curcumin affects both Akt activity and mitochondrial energy efficiency, among other effects. Akt affects many different signaling pathways and may be responsible for the multiple effects of curcumin noted in the research literature.

The long-lived La animals showed no significant response to curcumin, possibly because their selection regime involved alterations in the same longevity mechanisms affected by curcumin in the normal-lived Ra animals. The La animals do not respond to a DR diet either, supporting the idea that the mechanism(s) involved in both the DR and the curcumin responses may be similar to one another. Curcumin thus appears to be genome specific in that it does not extend the longevity of at least one long-lived strain. The developmental span must be considered as an integral part of the life span.

15.3.1.2 Similarity of Curcumin Effect to the Crowding Effect

When first isolated and tested, the La animals expressed a strong gene-environment effect such that the extended longevity phenotype was not expressed

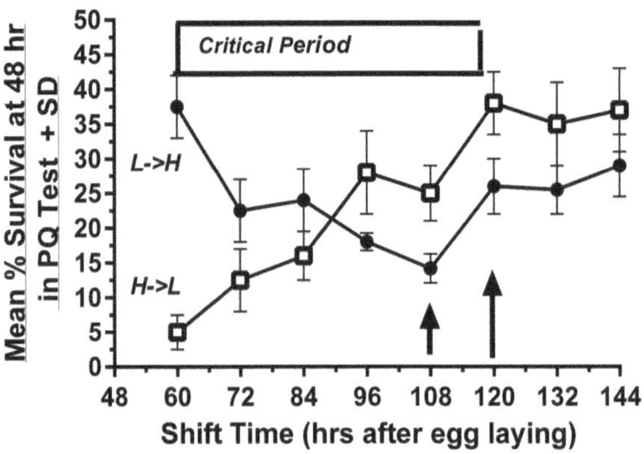

Fig. 15.6 Timing of the critical larval period determining the adult expression of the long-lived phenotype. Determination of the critical period during which time the larvae must be exposed to high larval densitiy (HD; 50–60 eggs/vial) conditions if they are to exhibit the extended longevity phenotype, and conversely, which time the larvae must be exposed to low larval density (LD; 10–12 eggs/vial) conditions if they are not to express the extended longevity phenotype. Timed La strain larvae were started at one density condition, shifted to the other at the indicated time, and assayed with the paraquat test at 5 days of aging to see if they expressed the extended longevity phenotype. The *points* represent the mean ±SEM of three independent replicate experiments. The *numbers* indicated the total number of vials, each containing 10 animals, used for each indicated shift. The entire experiment involved ~2550 animals exclusive of controls (not shown). *Short arrow* Estimated time the larvae stop feeding. *Long arrow* Approximate observed time of pupation. *Closed circles* LD to HD shifts. Open squares HD to LD shifts (from Buck et al. 1993). The critical period likely ends at 108 hrs, when the larvae stop feeding and presumably receive no further signals

unless the La larvae were raised under conditions of high larval density (i.e., >50 eggs/vial) (Arking 1987). Controlled density shift experiments delineated a critical period during larval life for expression of the adult extended longevity phenotype ranging from 60 to 96 h after egg laying (Fig. 15.6). Failure to be exposed to high density during this time period results in the La animal expressing only a normal longevity (Buck et al. 1993). This density-dependent phenotype was lost when the animals were inadvertently raised on a very low pH food, the effect of which was to convert the extended longevity of the La strain from an environmentally inducible phenotype to a constitutively expressed phenotype. All of our work since then has been done using the density-insensitive strain which survived the low pH food. I long thought that the original Ra strain, since it was the progenitor of the La strain, likely contained some sort of environmentally dependent longevity phenotype. I suggest that the curcumin-dependent ability of the Ra strain to express an extended longevity is likely to operate through the same mechanism which gave the original La strain its density-dependent inducible longevity phenotype. It may also have been one of the biological processes acted on by our selection protocol, unbeknownst to us (Luckinbill et al. 1984).

15.3.1.3 Transgenerational Effects of Curcumin

The F_1 female offspring of larvae fed on curcumin food show a beneficial transgenerational effect on their longevity (Fig. 15.7). The offspring had no direct contact themselves with curcumin food. The females showed a larger significant increase in longevity than did the males. Our data doesn't allow us to decide if the

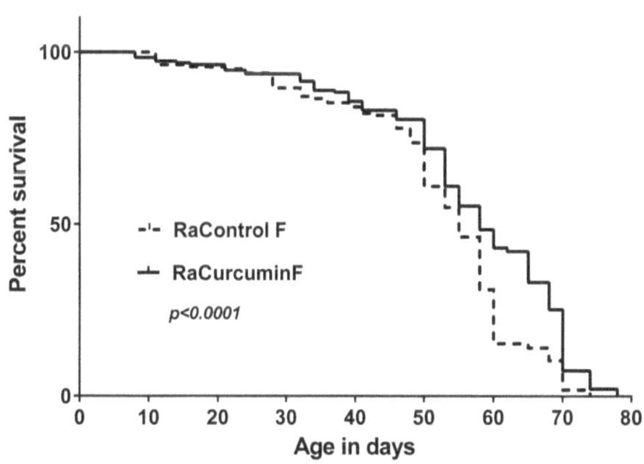

Fig. 15.7 Transgenerational effects of curcumin. Ra eggs were laid on food containing 100 mM curcumin. The resultant adults were raised on AL food without curcumin and allowed to mate. These young adults laid eggs on AL food without curcumin. The F1 offspring were then raised on this standard food and their adult longevity measured. The F1 females of curcumin-fed parents live longer than F1 females from control parents not fed on curcumin (log-rank test, 1 df, $\chi^2 = 17.17$, $p < 0.001$). F1 males show a similar statistically significant ($p = 0.0001$) extension of longevity (data not shown) (Arking unpublished data)

transgenerational effect is due to epigenetic effects or simply due to an improved health of the offspring of curcumin fed mothers. Chandrashekara et al. (2014) have data showing that the transgenerational effect extends to the F2 generation. Their finding deserves attention since it may provide insight into alternative methods of increasing the healthspan of the progeny of treated mothers.

15.3.1.4 Effects of Larval Curcumin Feeding on the Adult Age-Specific Mortality Rate

Changes in longevity must flow from changes in mortality. Curcumin-fed Ra larvae yield adults of both sexes with significantly altered mortality kinetics (Fig. 15.8). The age-specific mortality values of the curcumin treated females and males are significantly different ($p < 0.0001$, F-test) than the controls, and are best described by different curves. Curcumin treatment changed the Gompertz-Makeham type curve of the control animals into the Gompertz curve of the experimental cohort. The Gompertz curves of the untreated long-lived La animals are shown for comparison. Note that the decreased slope and reduced early mortality of the curcumin-treated Ra females approximate that of the Gompertz curve for the long-lived La female control. Curcumin significantly

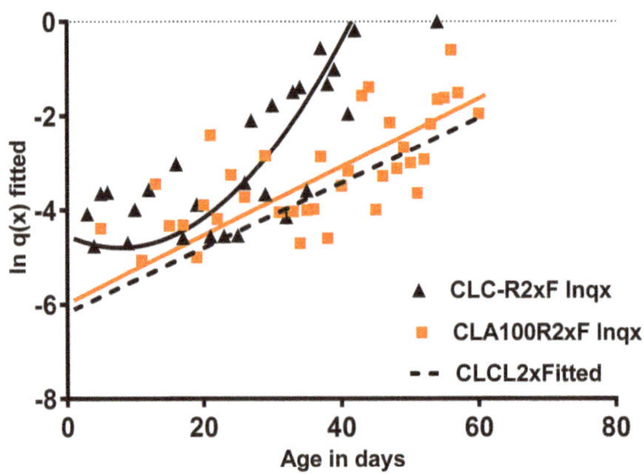

Fig. 15.8 Effect of curcumin on mortality rate kinetics. Curcumin affects both Gompertz parameters and increases mortality rate doubling time in both sexes (male data not shown). The age-specific mortality rate (lnqx) was calculated from the survival curve of each cohort, plotted, and the best fit mortality curve fitted to the data using maximum likelihood analysis (Winmodest), calculated and graphed. The *data symbols* represent the actual lnqx values. The *solid lines* represent the best fit curves: a Gompertz-Makeham plot for the control females (CLC-R2xF, N = 120) and a Gompertz plot for the experimental female (CLA100R2xF, N = 87) cohort. The *dotted line* represents the Gompertz curves of the long-lived La females (N = 188) on standard food without curcumin. Note that the fitted Gompertz curve of the curcumin treated Ra females approximates the comparable curve of the La female (from Soh et al. 2013, which should also be consulted for the male data)

alters the male mortality response ($P < 0.0001$, F-test) such that they are also best described by a Gompertz curve without additional mortality terms (data not shown). These data unequivocally show that altering the normal developmental by feeding the Ra larvae on a high dose (100 mM) of curcumin (Fig. 15.5) alters the normal gene expression patterns such that the Ra adult longevity phenotype now mimics that of the long-lived La adult. These data reinforce the presumption that the larval crowding-dependent mechanism (see 15.3.1.2) is closely related in some way to the larval curcumin-dependent mechanism. In these strains at least, adult longevity is the outcome of an environmentally-dependent development and genetic process.

15.3.1.5 Effects of Larval Curcumin Feeding on the Adult Trophic Behaviors

Curcumin feeding during the larval span significantly improved the adult females climbing ability in a standard negative geotactic test relative to controls (Fig. 15.9). The enhanced climbing response lasts for the first four weeks of the adult life span. The enhanced negative geotropic response suggests that the extended longevity of the treated animals is based on a more robust or efficient metabolism. The La animals are known to be very active as well as long-lived.

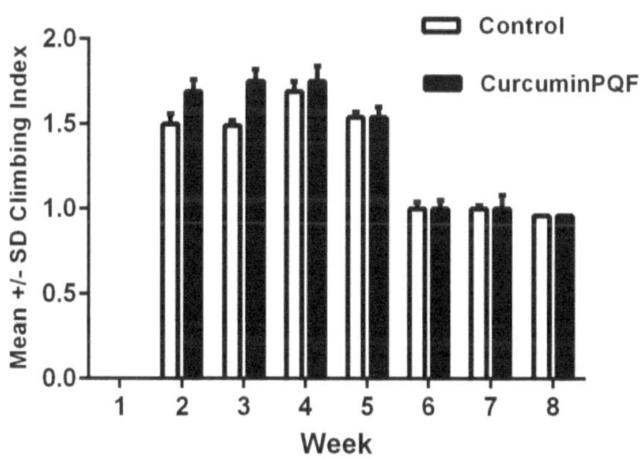

Fig. 15.9 Curcumin enhances locomotor activity. Negative geotactic climbing behavior is enhanced by curcumin regardless of adult diet. Each column represents the mean ±SD of the climbing index compiled from 6 cohorts of 25 flies each which were tested twice consecutively (i.e., replicated tests of 150 animals/cohort/time point) and used to calculate the mean climbing index value for each cohort. Females raised on AL food with curcumin had a significantly higher climbing index over weeks 2–4 relative to controls fed AL food only (t-test, $t = 4.574$, $df = 2$, $p = 0.0441$). Flies raised on dietary restriction (dr) food with curcumin had a significantly higher climbing index than controls fed DR food only over weeks 2–8 (t-test, $t = 3.342$, $df = 6$, $p = 0.0156$; data not shown). There is no significant difference between the AL control and the dietary restriction (DR) control (t-test = 0.4053, $df = 6$, $p = 0.6993$; data not shown). Similar results were found in males (from Soh et al. 2013, which should also be consulted for the male data)

The fact that larval curcumin treatment increases the Ra animals' activity as well as decreases its age-specific mortality to a level coincident with that of the long-lived La strain indicates that the curcumin effect is not limited to longevity alone, or to climbing alone, but rather to a complex of activities that make the treated Ra animals closely resemble the selected La animal. If this hypothesis is confirmed by our ongoing analysis of our microarray data, then it may well be the case that curcumin is activating a pathway(s) responsible for a coordinated alteration of multiple traits into an integrated extended longevity phenotype.

15.3.2 Entire Adult Life Span

We investigated the effects of feeding curcumin over the entire adult life span but not in the developmental span. Feeding curcumin during the entire adult life span (from days 5–65) decreases median life span, but not maximum life span, in both males and females (Fig. 15.10). This decrease is borderline significant in the Ra males but is highly significant in the Ra females.

In both cases, the negative effects of lifetime feeding manifest themselves in an earlier age of the end of the healthspan and an earlier age of onset of senescence relative to either controls or to animals fed curcumin only during the healthspan (see below).

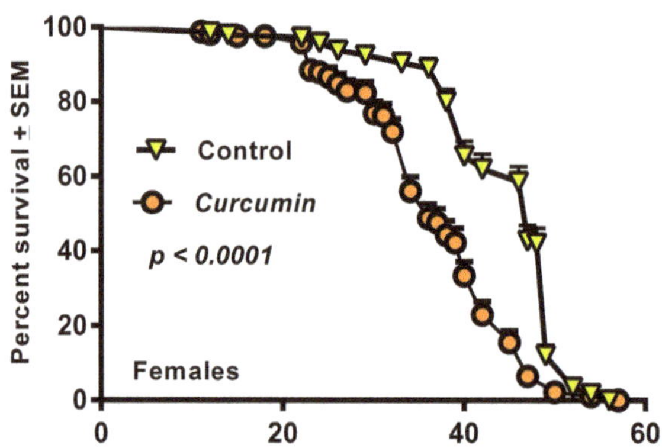

Fig. 15.10 Effect of curcumin feeding over the entire adult life span. Curcumin has an inhibitory effect if fed thoughout the adult life span in either sex. Ra females were fed 100 mM curcumin only during the entire adult life span. The curcumin fed females (N = 156, median life span = 36 days) respond with significant 30 % decrease in median but not maximum life span relative to their controls (N = 230, median = 47 days). The negative effects of lifetime feeding of curcumin manifest themselves in an earlier age of the end of the healthspan and an earlier age of onset of the senescent span relative to either control or to animals fed curcumin only during the healthspan. Males (data not shown) showed a similar response (from Soh et al. 2013, which should also be consulted for the male data)

15.3.3 *Effects of Curcumin Feeding in the Healthspan on Longevity*

When Ra males were fed an AL diet with curcumin only during their healthspan (days 5–27; see 15.1.1.2) and then transferred to an AL diet without curcumin for the rest of their lives, they displayed a delayed onset of senescence which resulted in a significant increase in their median (49 %) and maximum longevities (49 %) (see Fig. 15.11a). Ra females treated in the same manner also showed a significant increase in longevity (see Fig. 15.11d).

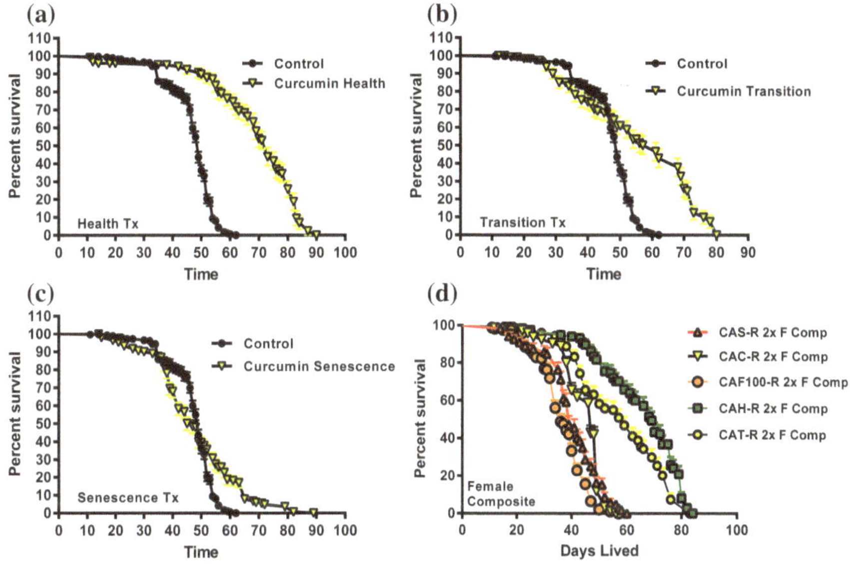

Fig. 15.11 Effects of curcumin feeding during specific phases of adult life span. Adult curcumin intervention has an early life stage-specific effect in Ra males and females. Panel **a** compares effect of feeding curcumin during the Healthspan (days 5–27; orange, N = 125, median life span = 73 days) relative to AL fed only male controls (*black*, N = 290, median LS = 49; log-rank test $\chi^2 = 211.0$, 1 df, $p < 0.0001$). Panel **b** compares the effect of feeding curcumin (N = 121, median = 57) during the Transition Span (days 28–40) to male controls (N = 290, median = 49) (log rank test, $\chi^2 = 63.38$, 1 df, $p < 0.0001$). Panel **c** shows the effect of feeding curcumin (N = 181, median = 45) during the Senescent Span (days 38–89) relative to male controls (N = 290, median = 49) (log-rank test, $\chi^2 = 12.91$, 1 df, $p = 0.0003$). Panel **d** is a composite graph of female longevities following curcumin feeding during the entire life span (N = 147, median = 36 days), Healthspan (N = 126, median = 69, $\chi^2 = 127.7$, $p < 0.0001$); Transition Span (N = 126, median = 59, $\chi^2 = 59.4$, $p < 0.0001$); or Senescent Span (N = 125, median = 39, $\chi^2 = 7.567$, $p = 0.0059$) spans relative to controls (N = 230, median = 47). Note that both sexes show the same stage-specific response as well as the same decreased longevity after whole-life feeding (from Soh et al. 2013)

15.3.4 *Effects of Curcumin Feeding in the Transition Span on Longevity*

When Ra males were fed an AL diet with curcumin only during their transition span (days 27–40; see 15.1.1.3) and fed with an AL diet without curcumin before and after that period, for the rest of their lives, they displayed a lesser increase in their median and minimum longevities of 25 % (see Fig. 15.11b). Ra females treated in the same manner also showed a significant in their median longevity relative to animals treated only in the healthspan (see Fig. 15.11d).

15.3.5 *Effects of Curcumin Feeding in the Senescent Span on Longevity*

When Ra males were fed an AL diet with curcumin only during their senescent span (days 38–89, see 15.1.1.4) and fed an AL diet with no curcumin prior to day 38, they displayed a 4 % decrease in median longevity and an ~11 % decrease in maximum longevity (see Fig. 15.11c). Ra females treated in the same manner also showed a significant decrease in their median longevity relative to controls (see Fig. 15.11d).

15.4 Effects of HDAC Inhibitors on Longevity and Gene Expression of Ra and La Strains

15.4.1 *Background Information*

We were curious as to whether other drugs would have stage-specific effects on the longevity of our Ra or La strains. Kang et al. (2002) showed that phenylbutyrate fed to flies resulted in an extension of longevity, particularly when fed during the later phase of the healthspan and the transition span. This latter result was the first to present proof of principle that a drug could significantly increase the longevity of an organism. The suggestion of that data that SB might be active at a stage when curcumin was not led us to test its stage-specific effects to determine if it did in fact have a different pattern of stage-specific effects on longevity. We tested the effects of SB and suberoylanilide hydroxamic acid (SAHA) on the longevity of our Ra and La strains, using the same stage-specific strategy as described above with curcumin. SB was chosen because it is a broad spectrum, if relatively insensitive, histone deacetylase (HDAC) inhibitor representative of the Class I, II and IV zinc-binding enzymes (Witt et al. 2009). SAHA has similar effects on cells as does SB but at much lower doses (Zhou et al. 2011). The experimental design employed in these SB/SAHA studies is described in McDonald et al. (2013) and is similar to that described above for the curcumin tests.

15.4.2 Effects of Continuous Larval + Adult Exposure to SB on Ra and La Animals

Ra animals raised on AL + SB food from the time of egg-laying until death showed significant dose-dependent longevity effects (Fig. 15.12a). Both treatments showed a short delay (~6 days) in the onset of senescence relative to controls; they differed however in their effects on the senescent span itself. In the low dose group, the treatment also induced a significantly longer life span ($p < 0.0001$) which involves an alteration of the survival pattern such that there is a 9 day increase in the median life span. In the high dose group, the treatment induced a significant decreased longevity which involved a 10 day decrease in the median life span. Neither dose affected the maximum longevity. These data suggest that the low dose slowed the mortality rate during the transition and senescent stages while the high dose accelerated it during these same stages.

Exposure of the La strain to the drug continuously during its entire life time leads to a set of survival curves which are statistically different from the controls, but in different directions (Fig. 15.12b). The high dose group

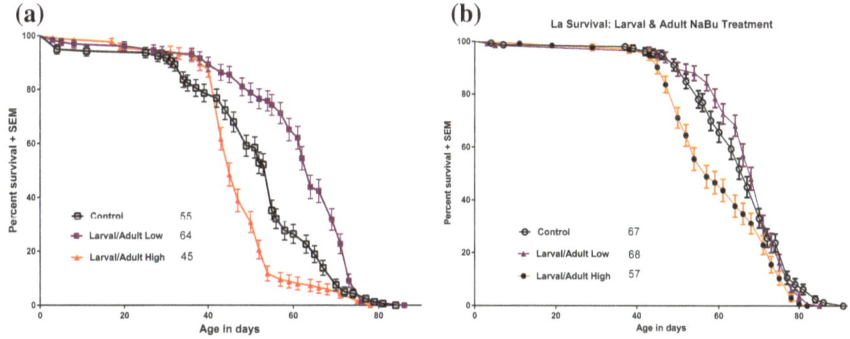

Fig. 15.12 Effects of whole-life feeding of SB on longevity of normal-lived and long-lived females. Feeding of SB during the entire life span has dose-dependent effects on longevity of Ra and La adults. SB fed at either dose over the entire life span has different effects on the Ra (Panel **a**) or La (Panel **b**) strains relative to controls. In the Ra strains, the low dose results in a significant extension of median, but not maximum, age relative to controls (log-rank test, $\chi^2 = 26.81$, 1 df, $p < 0.0001$) while the high dose results in a significant decrease of median but not maximum longevity ($\chi^2 = 20.21$, 1 df, $p < 0.0001$). In the La strains, the low dose has no effect ($p = 0.9415$) while the high dose significantly decreases the median but not maximum life span ($\chi^2 = 13.04$, 1 df, $p = 0.0003$). The median life spans of the several cohorts are listed in the legends of Panels **a** and **b**. The N for each Panel **a** cohort is: *control* 169; *low dose* 136; and *high dose* 132. The N for each Panel **b** cohort is: *control* 160; *low dose* 135; and *high dose* 135 (from McDonald et al. 2013)

shows a significant decrease in longevity relative to controls ($p = 0.0244$), which seems to begin at the end of the healthspan. The low dose group is not statistically different from the control. These data suggest that the low dose had no effect on the La mortality rate, while the high dose accelerated its increase relative to controls.

15.4.2.1 Effects of Continuous Adult Exposure Only to SB on Ra and La Animals

The high dose of SB during the adult phase of the Ra strain significantly lowers the longevity relative to controls ($p < 0.0001$) (see Fig. 15.13a). This appears to occur at a decreased length of late life (e.g., post 45 day longevity). The low dose of SB has an obvious but borderline shortening of the later life span ($p = 0.0589$). We conclude that the Ra animals are negatively affected by this treatment.

The response of the La animals to continuous adult exposure to SB was similar to that of the Ra animals in that the high dose induced a significant decrease ($p = 0.0244$) in its longevity but the low dose had no statistically significant ($p = 0.2869$) effect. We conclude that the La animals are negatively affected by this treatment (see Fig. 15.13b).

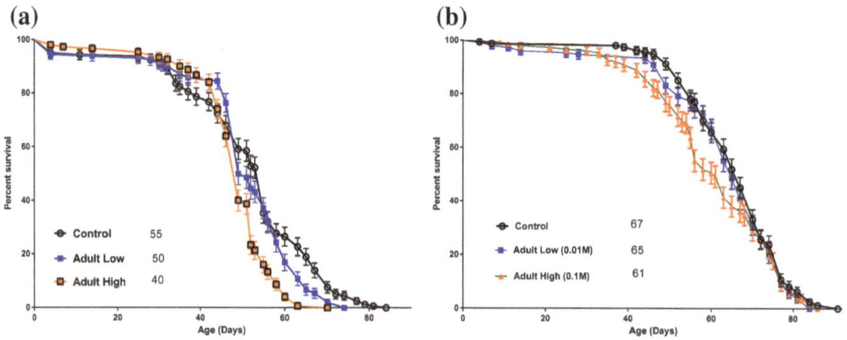

Fig. 15.13 Effect of continuous SB feeding during the entire adult life span on Ra and La females. Feeding of SB only during the adult stage decreases longevity of Ra (Panel **a**) and La (Panel **b**) adults. The low dose (10 mM) has no statistically significant effect relative to controls (log-rank test, $p = 0.0589$ (Ra) or $p = 0.2869$ (La). The high dose (100 mM) yields signficant decreases relative to controls in Ra ($p < 0.0001$) and La ($p = 0.0244$). The median life spans of the several cohorts are listed in the legends of Panels **a** and **b**. The N for each Panel **a** cohort is: *control* 159; *low dose* 148; and *high dose* 150. The N for each Panel **b** cohort is: *Control* 160, *low dose* 154, and *high dose* 152 (from McDonald et al. 2013)

15.4.3 *Effects of SB Feeding Only During the Health or Transition or Senescent Span*

Figure 15.14 summarizes the results of exposure to the low dose of SB to Ra animals (Fig. 15.14a) or to La animals (Fig. 15.14b). SB decreased longevity of Ra animals at both low and high (data not shown) doses when applied during the healthspan only (1–21 days). However treatment with the low dose during the transition span (21–42 days) yielded a highly significant ($p < 0.0001$) 12.4 % increase in the median longevity, even though the drug was not applied until the mid-life stage of the control cohort's life span. Treatment of Ra animals with SB during the senescent span (43–64 days) also yielded a highly significant increase in longevity. These data have been replicated (Supp. Figure 1 of McDonald et al. 2013). We conclude that SB is harmful to Ra adults if administered during the healthspan but is beneficial if

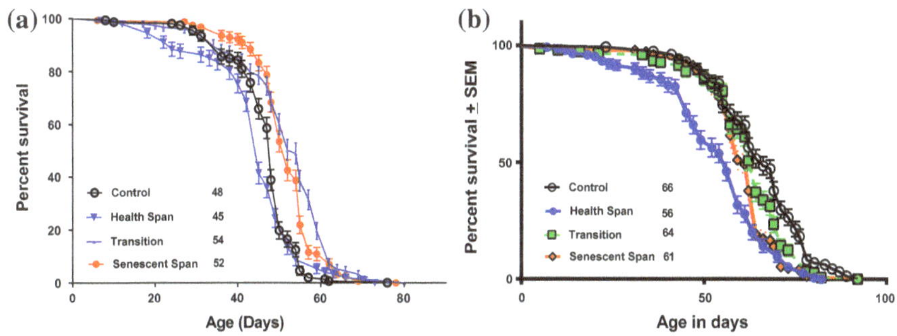

Fig. 15.14 Effect of SB feeding only during the healthspan, or transition span, or senescent span on longevity of Ra or La females. Feeding SB to Ra adults only during their health, transition or senescent spans leads to significant stage-specific effects on longevity. The *dotted lines* in the figure indicate the approximate boundaries of these three spans. Panel **a** reports the Ra results. At a low dose, there is a non-significant decrease in median longevity relative to controls (log-rank test, $p = 0.2781$) for intervention in the healthspan. However a significant increase in the median and late-life longevities is observed following feeding of SB during the transition ($\chi^2 = 52.98$, 1 df, $p < 0.0001$) or senescent ($\chi^2 = 38.52$, e df, $p < 0.0001$) spans. Note that SB feeding added 12.4 % to the median life span of the control even thought the treatment did not start until 54 % of the control longevity was already over at the treatment start. The N for each Panel **a** cohort is: control = 152; health = 147; transition = 133; and senescent = 155. Panel **b** reports the La results. Feeding SB at low doses to La adults only during their health (log-rank test, $\chi^2 = 49.83$, 1 df, $p < 0.0001$), transition ($\chi^2 = 6.72$, 1 df, $p = 0.0095$) or senescnet ($\chi^2 = 25.91$, 1 df, $p < 0.0001$) spans lead to significant decreases relative to controls in their median but not maximum life spans. Feeding SB at high doses also leads to similar decreases in longevity (data not shown). The N for each Panel **b** cohort is: *control* 159, *health* 129; *transition* 158, and *senescent* 152 (from McDonald et al. 2013)

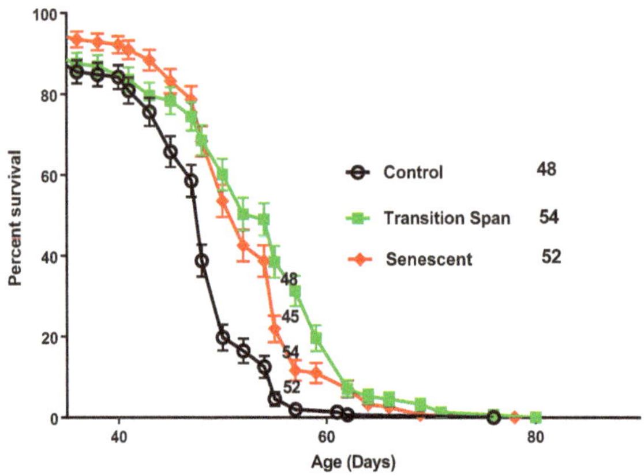

Fig. 15.15 Expanded view of SB-induced extended survival in late life. The Ra survival from Fig. 15.14a were rescaled so as to examine the details of extended survival during the senescent span (days 43 and following). It is clear that low dose SB treatment during the transition span has a greater positive effect on survival and longevity in the senescent span than does intervention during the senescent span itself (log-rank test, $\chi^2 = 10.31$, 1df, $p = 0.0013$). Other N and p values are as noted in Fig. 15.14 (from McDonald et al. 2013)

administered during the transition or senescent span. Figure 15.15 illustrates the significant extension of the senescent span that is brought about by this late-life acting drug.

15.4.3.1 Effects of SAHA Feeding

SAHA has a similar effect as SB on the late life survival of the Ra females, as shown in Fig. 15.16. There is no statistical difference between the three different doses of SAHA, the 1 μM dose being equivalent to the 20 μM dose.

15.4.4 Effects of Late-Life SB or SAHA Feeding on Mortality Kinetics of the Ra and La Animals

SB feeding, at either dose, significantly reduces the age-specific mortality rate of the late-life (i.e., >42 days) Ra females relative to that of the untreated control (Fig. 15.17). The analysis was done using the Winmodest program (Pletcher et al. 2000); a Gompertz-logistic curve is the best fit to the observed mortality data derived from Fig. 15.15. Note that there seems to be a difference in the manner by which each does brings about a similar phenotype. In the case of the La animals, each of the two doses significantly increase the age-specific mortality during the senescent span, which phenomenon underlies the shortened longevity of the La animals in every experimental situation involving the long-lived animals (data not shown).

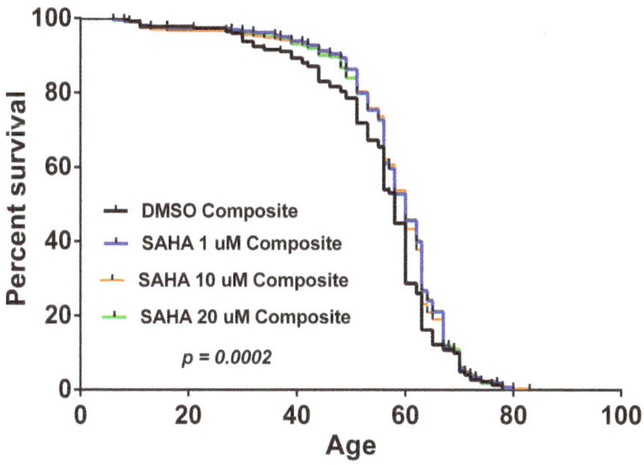

Fig. 15.16 Effect of SAHA on late-life survival of Ra females. Late life treatment with SAHA increases late life survival. Sister cohorts of Ra females were treated with the vehicle (dimethyl sulfoxide; DMSO) only (N = 225), or 1 μM SAHA in DMSO (N = 301), or 10 μM SAHA in DMSO (N = 125), or 20 μM SAHA in DMSO (N = 288). The three experimental cohorts are highly significantly different from the controls (log-rank test, $\chi^2 = 19.50$, 1 df, $p = 0.0002$), but do not statistically differ between themselves (log-rank test, $\chi^2 = 1.233$, 1 df, $p = 0.5400$). The vertical arrow approximates the age (42 days) at which the experimental cohorts were introduced to SAHA (from McDonald et al. 2013)

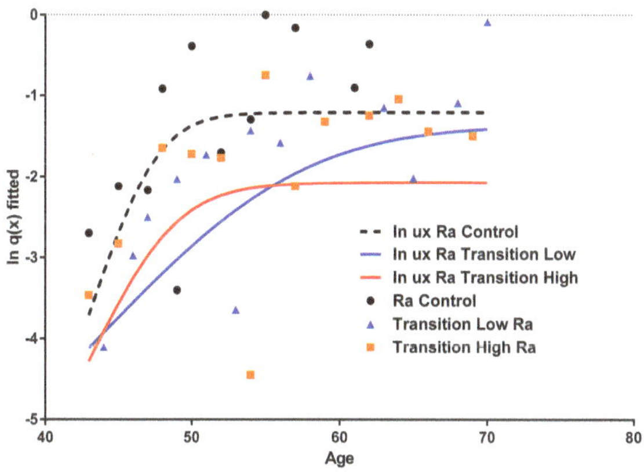

Fig. 15.17 Analysis of SB effects on mortality kinetics during the senescent span. Mortality rates were calculated from the survival data of Fig. 15.15 as ux = ln(px), where px is the probability of survival from age x to age x + 1, and are presented on the logarithmic scale. The Winmodest program was used to determine that a logistic-Gompertz model was the best fit to the observed data (*solid symbols* in graph). Both low and high doses or SB started in the mid-life period yields significant decreases in the mortality rates of Ra females (*solid lines*) during the senescent span relative to controls (*dotted line*) (from McDonald et al. 2013)

A similar analysis was made of the effects of SAHA on the mortality kinetics of late-life Ra females. A Gompertz-logistic curve also fits the survival data of Fig. 15.16. Note that there is very little difference in the mortality curves of the three doses tested.

15.5 Conclusions

15.5.1 Existence of Stage-specific Domains

The detailed data presented above supports the concept of stage-specific domains. It is likely, but not yet proven, that these domains are defined by stage-specific gene expression patterns, and by stage-specific gene targets unique to particular drugs. Once the stage-specific gene expression patterns have been assayed and catalogued, then the identification of potential stage-specific gene product targets for curcumin and SB might be made easier. Awareness of presumably critical stage- and pathway-specific gene products available in each phase of the life span might also lead to the reverse engineering of potential 'designer drugs' targeted for the alleviation of particular conditions.

15.5.2 Curcumin has Stage-specific Longevity Effects

The results of whole life feeding of curcumin to mice led to the conclusion that curcumin has no effect on longevity (Strong et al. 2013). This conclusion contradicted prior preliminary studies on mice which showed that early to mid-life feeding of tetrahydrocurcumin led to an increased longevity (Kitani et al. 2007). In our hands, whole life feeding of high doses of curcumin to flies led to significant decreases in median longevity (Fig. 15.10), while feeding curcumin to flies at low doses led to no or minimal increases in their healthspan or median longevity (Fig. 15.5; Lee et al. 2010). The possibility exists that the failure of curcumin to induce a longevity effect in mice might be due to problems with the dosage and/or with feeding it over the whole life span. It seems as if some investigators may have used curcumin dosages derived from cell-based studies and so may have failed to create an effective bioavailability of the drug in an intact animal. Our use of high doses (100 mM) yields repeatable and verifiable effects.

There is a possibility that current drug screening protocols are based on a faulty paradigm of drug-gene interactions; namely that whole life feeding of candidate drugs will not yield any false negative data if the drug in question actually has a stage-specific effect. Given the expense of drug screens on mammals, and given the real possibility that whole life feeding may yield false negative data due to its different stage-specific effects on the organism, then it might be useful to perform the initial whole-life and stage-specific testing of candidate pro-longevity drugs on worms or flies prior to any testing on mice or other mammals. The lower cost fly experiments would allow the characterization of any stage-specific effects of the candidate drug on conserved pathways, and would inform the experimental design of subsequent mouse studies.

15.5.3 Testable Implications of the Stage-specific Hypothesis

One testable prediction of the current data is that drugs with beneficial late life effects are likely to be enriched for molecules capable of inhibiting or countering age-related diseases, while drugs with early life beneficial effects are likely to be ineffective against late life age-related diseases. The use of stage-specific testing protocols might enrich the candidate drug pool of molecules capable of inhibiting the expression of late-life age-related diseases. Future research on life extension should focus on identifying and characterizing those active gene repression/ protection mechanisms that prevent the expression of age-related disease genes. The 'escaper' subset of centenarians is known to delay the onset of any major age-related disease and increase the length of their healthspan by as much as 40 years due to their ability to inhibit disease gene expression in mid- and late-life (Andersen et al. 2012; Sebastiani et al. 2012, 2013; Sebastiani 2013). Given the effectiveness of stage-specific drug effects in extending the healthspan (e.g., curcumin) or in inducing a healthier senescent span (e.g., SB or SAHA) then identifying and characterizing more small molecules with stage-specific effects would be an effective way of enabling those of us who are not centenarians to also extend our healthspans or reduce the frailty of our senescent spans. Extending the healthspan and/or decreasing the morbidity of the senescent span may do much to increase the quality and productivity of human life. The data generated by a continued investigation of stage-specific mechanisms may help guide us to that desired future.

References

Adler AS, Sinha S, Kawahara TLA, Zhang JY, Segal E, Chang HY (2007) Motif module map reveals enforcement of aging by continual NFkB activity. Genes Develop 21:3244–3257

Adler AS, Kawahara TLA, Segal E, Chang HY (2008) Reversal of aging by NFkB blockade. Cell Cycle 7:556–559

Andersen SL, Sebastiani P, Dworkis DA, Feldman L, Perls TT (2012) Health span approximates life span among many supercentenarians: compression of morbidity at the approximate limit of life span. J Gerontol A Biol Sci Med Sci 67:395–405

Arking R (1987) Successful selection for increased longevity in Drosophila: analysis of the survival data and presentation of a hypothesis on the genetic regulation of longevity. Exp Gerontol 22:199–220

Arking R (2009) Overview of the Genomic Architecture of Longevity. In: Sell C, Lorenzini A, Brown-Borg HM (eds) Life span extension: single cell organisms to man. Humana Press Springer, Dordrecht, pp 59–73

Arking R, Novoseltsev V, Novoseltseva J (2004) The human life span is not that limited: effects of multiple longevity phenotypes. J Gerontology Biol Sci 59A:697–704

Barker DJP (1995) Intrauterine programming of adult disease. Mol Med Today 1:418–423

Buck S, Nicholson M, Dudas S, Wells R, Force A, Baker GT, Arking R (1993) Larval regulation of adult longevity in a genetically-selected long-lived line of Drosophila. Heredity 71:21–32

Carlson ME, Suette C, Conboy MJ, Aagea P, Mackey A, Kjaar M, Conboy I (2014) Molecular aging and rejuvenation of human muscle stem cells. EMBO Mol Medicine 1:381–391

Chandrashekara KT, Popli S, Shakarad MN (2014) Curcumin enhances parental reproductive lifespan and progeny viability in Drosophila melanogaster. Age 36:9702

Conboy IM, Conboy MJ, Wagers AJ, Girma ER, Weissman IL, Rando TA (2005) Rejuvenation of aged progenitor cells by exposure to a young systemic environment. Nature 433:760–764

Danese A, Pariante CM, Caspi A, Taylor A, Poultan R (2007) Childhood maltreatment predicts adult inflammation in a life course study. Proc Natl Acad Sci USA 104:1319–1324

Davidson EH, Erwin DH (2006) Gene regulatory networks and the evolution of animal body plans. Science 311:796–800

Davidson EH, Rast JP, Oliveri P (2002) A provisional regulatory gene network for specification of endomesoderm in the sea urchin embryo. Develop Biol 246:162–190

Finch CE (2007) The biology of human longevity: inflammation, nutrition, and aging in the evolution of life spans. Academic Press-Elsevier, Burlington

Finch CE, Kirkwood TBL (2000) Chance, Development, and Aging. Oxford University Press, New York

Gentilini D, Mari D, Castaldi D, Remondini D, Ogliari G, Ostan R, Bucci L, Sirchia SM, Tabano S, Cavagnini F, Monti D, Franceschi C, Di Blasio AM, Vitale G (2013) Role of epigenetics in human aging and longevity: genome-wide DNA methylation profile in centenarians and centenarians' offspring. Age (Dodr) 35:1961–1975

Gompertz B (1825) On the nature of the function expressive of the law of human mortality, and on a new mode of determining the value of life contingencies. Phil Trans R Soc Lond 115:513–583

Kang HL, BenzerBenzer S, Min KT (2002) Life extension in *Drosophila* by feeding a drug. Proc Natl Acad Sci U S A 99:838–843

Kitani K, OsawaOsawa Toshihiko, Yokozaw Takako (2007) The effects of tetrahydrocurcumin and green tea polyphenol on the survival of male C57BL/6. Biogerontology 8:567–573

Kuether K, Arking R (1999) *Drosophila* selected for extended longevity are more sensitive to heat shock. AGE 22:175–180

Lee K-S, Lee BS, Semnani S, Avanesian A, Um C-Y, Jeon H-J, Seong K-M, Yu K, Min K-J, Jafari M (2010) Curcumin extends life span, improves health span, and modulates the expression of age-associated aging genes in *Drosophila melanogaster*. Rejuvenation Res 13:561–570

Liu L, van Green T, Kadish I et al (2011) Insufficient DNA methylation affects healthy aging and promotes age-related health problems. Clin Epigenetics 2:349–360

Lu T, Aron L, Zullo J et al (2014) REST and stress resistance in ageing and Alzheimer's disease. Nature 507:448–454

Luckinbill LS, Arking R, Clare M, Cirocco W, Buck S (1984) Selection for delayed senescence in *Drosophila melanogaster*. Evolution 38:996–1003

McDonald P, Maiza BM, Arking R (2013) Chemical regulation of mid- and late-life longevity in *Drosophila*. Exp Gerontol 48:240–249

Pletcher SD, Geyer CJ (1999) The genetic analysis of age-dependent traits: Modeling the character process. Genetics 153:825–835

Pletcher SD, Khazelli AA, Curtsinger JW (2000) Why do lifespans differ? Partitioning mean longevity differences in terms of age-specific mortality parameters. J Gerontol A Biol Sci Med Sci 55:B381–389

Sebastiani P (2013) Families enriched for exceptional longevity also have increased health-span: findings from the long life family study. Front Pub Health 1:38

Sebastiani P, Solovieff N, DeWan AT, Walsh KM, Puca A, Hartley SW, Melista E, Andersen S, Dworkis DA, Wilk JB, Myers RH, Steinberg MH, Montano M, Aldwin CT, Hoh J, Perls TT (2012) Genetic signatures of exceptional longevity in humans. PLoS ONE 7(1):e29848

Sebastiani P, Bae H, Sun FX, Andersen SL, Daw EW, Malovini A, Kojima T, Hirose N, Schupf N, Puca A, Perls TT (2013) Meta-analysis of genetic variants associated with human exceptional longevity. Aging (Albany NY) 5:653–661

Segal E, Raveh-Sadka T, Schroeder M, Unnerstall U, Gaul U (2008) Predicting expression patterns from regulatory sequence in *Drosophila* segmentation. Nature 451:535–540

Soh J-W, Marowsky N, Nichols TJ, Rahman AB, Miah T, Sarao P, Khasawneh R, Unnikrishnan A, Heydari AR, Silver RB, Arking R (2013) Curcumin is an early-acting stage-spewcific inducer of extended functional longevity in *Drosophila*. Exp Geron 48:229–239

Strong R, Miller RA, Astle CM, Baur JA, de Cabo R, Fernandez E, Guo W, Javors M, Kirkland JL, Nelson JF, Sinclair DA, Teter B, Williams D, Zaveri N, Nadon NL, Harrison DE (2013) Evaluation of resveratrol, green tea extract, curcumin, oxaloacetic acid, and medium-chain triglyceride oil on life span of genetically heterogeneous mice. J Gerontol A Biol Sci Med Sci 68:6–16

Tu Q, Cameron RA, Davidson EH (2014) Quantitative developmental transcriptomes of the sea urchin *Strongylocentrotus purpurtas*. Dev Biol 385:160–167

Vettraino J, Buck S, Arking R (2001) Direct selection for paraquat resistance in *Drosophila* results in a different extended longevity phenotype. J Gerontol Biol Sce 56A:B415–B425

Witt O, Deubzer HE, Milde T, Oehme I (2009) HDAC family: What are the cancer relevant targets? Cancer Lett 277:8–21

Zhou Q, Dalgard CL, Wynder C, Doughty ML (2011) Histone deacetylase inhibitors SAHA and sodium butyrate block G1-to-S cell cycle progression in neurosphere formation by adult subventricular cells. BMC Neuroscience 12:50

Erratum to: Parallels Between Mammals and Flies in Inflammatory Bowel Disease

Theodoulakis Christofi and Yiorgos Apidianakis

Erratum to:
Chapter 7 in: A.M. Vaiserman et al. (eds.), *Life Extension*, Healthy Ageing and Longevity 3, DOI 10.1007/978-3-319-18326-8_7

The original version of this book was inadvertently published with an incorrect author name for chapter 7. The correct author name should be Theodoulakis Christofi.

Space has been inserted in the word 'inflammationof' in the sentence Inflammatory bowel disease (IBD) is a condition caused by 'inflammationof the' and the word now correctly read as 'inflammation of'.

The online version of the original chapter can be found under
DOI 10.1007/978-3-319-18326-8_7

T. Christofi · Y. Apidianakis (✉)
Department of Biological Sciences, University of Cyprus, Nicosia, Cyprus
e-mail: apidiana@ucy.ac.cy

T. Christofi
e-mail: christofi.theodoulakis@ucy.ac.cy

A.M. Vaiserman et al. (eds.), *Life Extension*, Healthy Ageing and Longevity 3,
DOI 10.1007/978-3-319-18326-8_16

Index

A.M. Vaiserman et al. (eds.), *Life Extension*, Healthy Ageing and Longevity 3,
DOI 10.1007/978-3-319-18326-8

FSC
www.fsc.org
MIX
Papier | Fördert gute Waldnutzung
FSC® C083411

Kolibri 360 GmbH
Ferdinand-Jühlke-Straße 7
99095 Erfurt, Deutschland
produktsicherheit@kolibri360.de

Yasuyuki Suzuki   Kálmán Varga

Stochastic Variational Approach to Quantum-Mechanical Few-Body Problems

Springer

Authors

Yasuyuki Suzuki
Department of Physics
Niigata University
Ikarashi, Niigata 959-2181, Japan
suzuki@nt.sc.niigata-u.ac.jp

Kálmán Varga
Physics Division
Argonne National Laboratory
9700 South Cass Avenue
Argonne, IL 60439, USA
varga@theory.phy.anl.gov

Cataloging-in-Publication Data applied for.

Die Deutsche Bibliothek - CIP-Einheitsaufnahme

Suzuki, Yasuyuki:
Stochastic variational approach to quantum mechanical few-body problems / Yasuyuki Suzuki ; Kalman Varga. - Berlin ; Heidelberg ; New York ; Barcelona ; Hong Kong ; London ; Milan ; Paris ; Singapore ; Tokyo : Springer, 1998
(Lecture notes in physics)
ISBN 3-540-65152-7

ISSN 0940-7677 (Lecture Notes in Physics. Monographs)
ISBN 3-540-65152-7 Springer-Verlag Berlin Heidelberg New York

Typesetting: Camera-ready by authors
Cover design: *design & production* GmbH, Heidelberg
SPIN: 10552619 55/3144-543210 - Printed on acid-free paper

Preface

The variational method is one of the most widely used techniques to solve the eigenvalue problems of quantum-mechanical Hamiltonians. Its popularity is due to its simplicity and flexibility. The most crucial point in the variational approach is the choice of a variational trial function. One usually attempts to construct the trial function from some adequate basis functions which contain a number of nonlinear parameters. The direct method of diagonalizing the Hamiltonian matrix on such a basis set may not be feasible, except for simple systems, because of the large number of degrees of freedom involved in specifying the system. One thus faces a problem of selecting the most suitable basis set. It is by the stochastic variational method, that is, by a trial and error procedure with an admittance test that we give an answer to this problem. The stochastic variational method has been developed through the search for precise solutions of nuclear few-body problems. In this method we set up the basis element one after the other because it enables us to test many parameters as fast as possible and moreover to monitor the energy convergence.

The aim of this book is to give a unified and reasonably simple recipe for solutions of few-body bound-state problems with the use of the stochastic variational method and to present its application to various few-body problems which one encounters in atomic, molecular, nuclear, subnuclear and solid state physics.

Though a unified approach to the diverse quantum systems is in general extremely difficult and challenging, great advances have been made in recent years, especially for systems of a few particles and it has become possible to obtain accurate solutions for the eigenvalue problem of various quantum-mechanical Hamiltonians. The main interest in the few-body problems lies in, e.g., finding an accurate solution for the system so as to understand the dynamics of its constituents, testing the equation of motion and the conservation laws and symmetries, or looking for unknown interactions governing the system.

Quantum mechanics plays a fundamental role in atomic and subatomic physics. It is via quantum mechanics that one can understand the binding mechanism of atoms, molecules and atomic nuclei, that is, the structure of the building blocks of matter. The interaction between the particles depends on the system: For example, the long-range Coulomb interaction dominates in atoms and molecules but the very different mass ratio of the electrons and the atomic nucleus plays a key role as well. In contrast, the protons and neutrons in nuclei have almost equal masses and the interaction between them is short-ranged.

The variational foundation for the time-independent Schrödinger equation provides a solid and arbitrarily improvable framework for the solution of diverse bound-state problems. As mentioned above, the most crucial point in the variational approach is the choice of the trial function. There are two widely applied strategies for this choice: One is to use the most appropriate functional form to describe the short-range as well as the long-range correlations among the particles. Such calculations, however, are fairly complex for systems of more than three particles, and the integration involved is performed by the Monte Carlo method. Another way is to approximate the solution as a combination of a number of simple basis states which facilitate the analytical calculation of matrix elements. We follow the latter course in this book and show that the stochastic variational method selects the most important basis set without any bias, keeps the dimension of the basis low and, most importantly, provides a very accurate solution.

The book is conceptually divided into two parts. The first seven chapters present the basic concepts of the variational method and the formulation using Gaussian basis functions. The latter four chapters of the book cover applications of the formulation to various quantum-mechanical few-body bound-state problems. In Chap. 2 a general formulation is developed to express the physical operators which are needed to specify the Hamiltonian in terms of an arbitrary set of independent relative coordinates. The linear transformation of the relative coordinates induced by the permutation of identical particles is also established in this chapter. In Chap. 3 we review the basic principles of the variational method with particular emphasis on the case where the variational trial function is given as a linear combination of nonorthogonal basis functions. We introduce in Chap. 4 a key algorithm used in this book, the stochastic variational method, and show that its trial and error search procedure makes it possible to select the most important basis functions without any bias in the function space spanned by the basis functions. Some other methods to solve few-body problems are briefly introduced in Chap. 5. Chapter 6 defines the type of variational trial functions used extensively in the book, the correlated Gaussians and the correlated Gaussian-type geminals. They are chosen because they enable us to evaluate matrix elements analytically and because they provide us with precise solutions for most problems of real interest. A simple but powerful angular function is introduced to describe orbital motion with nonzero orbital angular momentum. To facilitate the systematic and unified evaluation of matrix elements, it is shown that the above Gaussian basis functions are all obtained from a generating function. In Chap. 7 we show that the generating function plays a vital role in deriving the matrix elements of the Gaussian basis functions for an N-body system of essentially any interaction. Explicit formulas are given in this chapter for the simplest possible Gaussian basis functions, because they are already found to be very useful. The matrix elements for a general case are detailed in the appendix. We show also in this chapter that the method can be extended to evaluate the matrix elements of nonlocal potentials and the semirelativistic kinetic energy as well. Chapters 8–11 present application of the stochastic variational method to various systems: small atoms and molecules (Chap. 8), baryon spectroscopy (Chap. 9), excitonic complexes and quantum dots in solid state physics (Chap. 10), and nuclear few-body problems (Chap. 11).

We hope that this book will be found useful by students who want to understand and make use of the variational approach to quantum-mechanical few-body problems, while it may also be of interest to researchers who are familiar with the subjects. It will be our pleasure if this book serves to bridge the gap between graduate lectures and the literature in scientific journals, as well as to give impetus to further development in the deeper understanding of quantum-mechanical few-body systems. We assume that readers have taken courses on quantum mechanics and mathematical physics at an undergraduate level. No special knowledge is assumed of, e.g., atomic physics or nuclear physics. To help readers to understand the text, we have attempted to make the book self-contained, put as much emphasis as possible on clarity, and given several Complements of an explanatory nature. The Complements are intended to further develop or to reinforce the arguments and ideas presented in the text. We have collected the derivation of formulas that may possibly be difficult for readers as exercises with solutions at the end of the chapters.

Depending on their interest, readers may adopt several reading strategies. A thorough-going reader is advised to read all of the text including the Complements. A reader who wants to understand only the basic formulation can omit the Complements. Anyone who is familiar with the variational method may skip Chaps. 2 and 3. Experts, or those readers who are only interested in the performance of the stochastic variational approach or the physical consequences of the results, may skip Chaps. 2–7.

The Gaussian basis has long been used in many areas of physics. The correlated Gaussians were first introduced in quantum chemistry by S.F. Boys and K. Singer. The application of the Gaussian basis is one of the key elements of the success of the ab initio calculations in quantum chemistry. The stochastic variational method is actually very similar to the so-called "random tempering", that has been used to find the optimal parameters of the basis in quantum chemistry. In the random tempering pseudo-random parameters are generated, and the best basis functions are selected by sorting out the states which improve the energy. There exists another method which is also similar to the stochastic variational method, called the stochastic diagonalization. This method, originally developed in solid state physics, attempts to find the lowest eigenvalue of huge eigenvalue problems by randomly testing the contributions of different states. The random selection of the parameters of a Gaussian basis, named the stochastic variational method, was first used by V.I. Kukulin and V.M. Krasnopol'sky in nuclear physics.

We are grateful to Prof. V.I. Kukulin for his interest and encouragement during this work. We are indebted to our collaborators, K. Arai, Y. Fujiwara, L.Ya. Glozman, R.G. Lovas, J. Mitroy, C. Nakamoto, H. Nemura, Y. Ohbayasi, Z. Papp, W. Plessas, G.G. Ryzhikh, N. Tanaka, J. Usukura, and R.F. Wagenbrunn for useful discussions and for many of the calculations which are included in this book. We wish to express our hearty thanks to Prof. K.T. Hecht, Prof. D. Baye, Prof. M. Kamimura and Prof. R.G. Lovas for their careful reading of the manuscript and for much advice and many comments. Despite all these efforts mistakes may still remain. Needless to say, these are our own responsibility. Suggestions and criticisms from our readers would be welcomed. We are grateful for the use of the computer VPP500 of the Institute of Physical and Chemical Research (RIKEN). One of the authors (K.V.) is grateful for the hospitality of the Physics Department of Niigata University and the Linac Laboratory of RIKEN. We would like to thank the funding agencies, Grants-in-Aid for Scientific Research of the Ministry of Education, Science and Culture (Japan), OTKA Grants (Hungary), the Japan Society for the Promotion of Science (JSPS), the Science and Technology Agency (STA) of Japan, the U. S. Department of Energy, Nuclear Physics Division, and the Hungarian Academy of Sciences (HAS), for their support which has been vital for the completion of the work. Finally, without the support and patience of our families, this book would not have been possible.

Niigata Y. Suzuki
Argonne K. Varga
September 1998

Table of Contents

1. Introduction

There are a countless number of examples of quantum-mechanical few-body systems: constituent quarks in subnuclear physics, few-nucleon or few-cluster systems in nuclear physics, small atoms and molecules in atomic physics or few-electron quantum dots in solid state physics, etc. The intricate feature of the few-body systems is that they develop individual characters depending on the number of constituent particles. The mesons and baryons, the alpha-particle and the ^{6}Li nucleus, or the He atom and the Be atom have very different physical properties. The most important causes of these differences are the correlated motion and the Pauli principle. This individuality requires specific methods for the solution of the few-body Schrödinger equation. Approximate solutions which assume restricted model spaces, mean field, etc. fail to describe the behavior of the few-body systems.

The goal of this book is to show how to find the energy and the wave function of any few-particle system in a simple, unified approach. The system will normally be in the minimum energy quantum state. As forewarned, however, to find this state, the ground state, is a complicated matter. The present stage of the development of computer technology, however, makes a very simple approach possible: Searching for the ground state by "gambling". Without any a priori information on the true ground state, completely random states are generated. Provided that the random states are general enough, after a series of trials one finds the ground state in a good approximation. The reader may find this a little suspicious but there are indeed a number of fine tricks in the trial and error procedure which makes the whole idea really practicable.

Before bombarding the reader with sophisticated details, let us demonstrate the random search with an example. Let us try to determine the energy of a Coulombic three-body system: a positron and two electrons, the positronium negative ion, Ps^{-}. The example is simple enough so that a graphic illustration of the wave function can be given, but its solution is far from being trivial. The ground state of this particular system can be calculated by different methods accurately. In the rest of this chapter we will give just the outline of the gambling method without paying much attention to the details.

To look for the wave function of the ground state of the system, we generate random functions. The functions are random in the sense that we assume some functional form which depends on some discrete or continuous parameters or "quantum numbers" and these parameters are randomly chosen. For example, we assume a simple Gaussian form

$$\Psi(\alpha, \beta, x, y) = \mathrm{e}^{-\alpha x^2 - \beta y^2}, \tag{1.1}$$

where $x = |\boldsymbol{r}_2 - \boldsymbol{r}_3|$ denotes the distance between the two electrons and $y = |(\boldsymbol{r}_2 + \boldsymbol{r}_3)/2 - \boldsymbol{r}_1|$ denotes the distance between the center-of-mass of the electrons and the positron.

A certain number of parameter sets α and β are randomly chosen and then the energy expectation values of the generated states are calculated and compared. The one among them giving the lowest value is selected to be a successful parameter set. Figures 1.1(a)–(d) show examples of four random states and their energy expectation values which are obtained with such successful parameters. Not surprisingly, the energies of the "configurations" appreciably depend on the shape of the functions, that is, on the random parameters. The lowest energy, -0.11 in atomic units (a.u.) in Fig. 1.1(d), is higher than that of the exact ground state (-0.262). Here in the atomic units m_e (the electron mass), e (the electron charge magnitude), and $\hbar$ are chosen to be basic quantities of units, so the unit of length is Bohr radius $a_0 = \hbar^2/(m_e e^2) = 5.29 \times 10^{-11}$ m, the unit of energy is $m_e e^4/\hbar^2 = 27.2$ eV, and the unit of time is $\hbar^3/(m_e e^4) = 2.42 \times 10^{-17}$ s. (Actually the minimum energy calculated analytically with one term of Eq. (1.1) is -0.177, but it may not be known for a general case.) That means that the trial function is not general enough to describe the ground state.

To improve the trial function, let us take a linear combination of two of the above functions of Eq. (1.1), where one of the two is fixed as the one already selected and another is newly selected after a number of random trials. Figures 1.2(a)–(d) show examples for random wave functions which are calculated by the two terms. The energy improved because the model space increased but we still miss a substantial amount of binding energy. Increasing the model space further by adding more and more functions to the linear combination one by one, we reach the exact energy and the wave function (Fig. 1.3) with

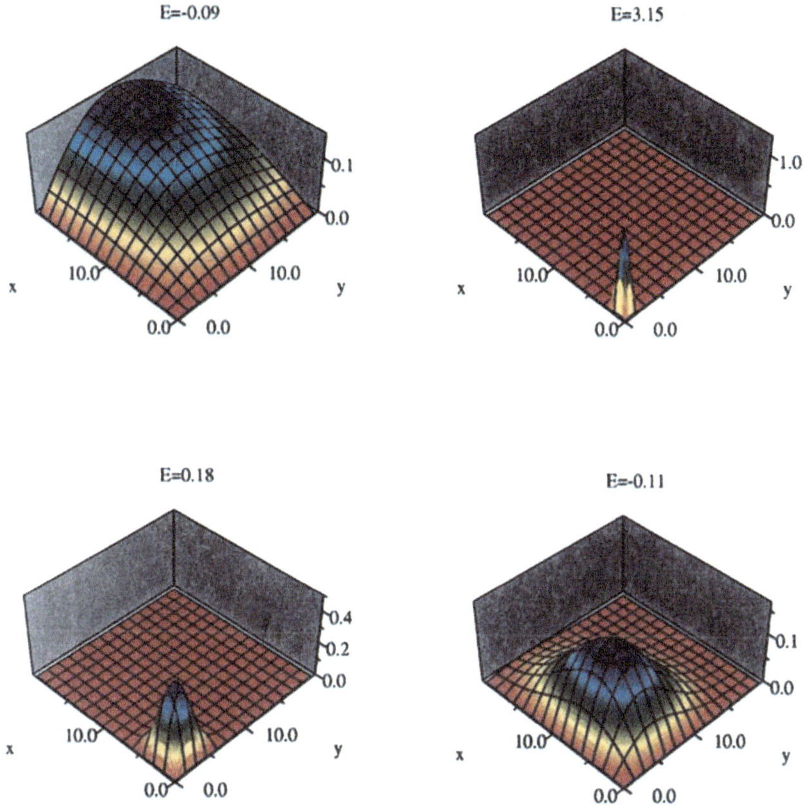

Fig. 1.1. Examples of the energy expectation value and the (normalized) wave function $xy\Psi(\alpha, \beta, x, y)$ of Ps^- for one random basis state. Figure 1(a) is in the upper left, 1(b) in the upper right, 1(c) in the lower left, and 1(d) in the lower right. In each section the x axis denotes the distance between the two electrons and the y axis denotes the distance between the center-of-mass of the two electrons and the positron. Atomic units are used.

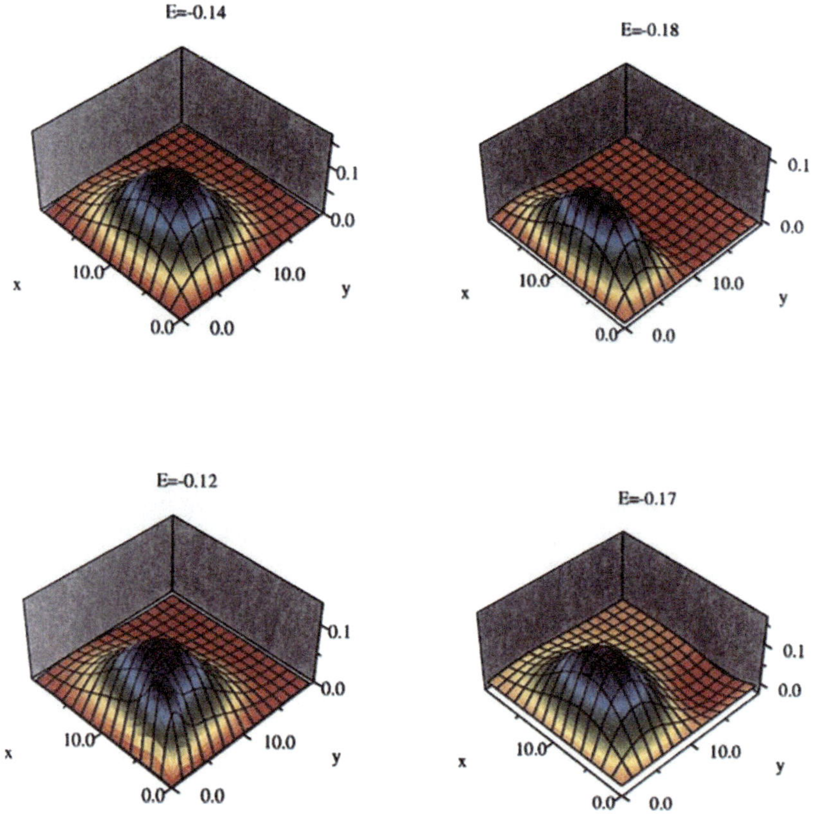

Fig. 1.2. Examples of the energy expectation value and the wave function, $xy\{c_1\Psi(\alpha_1,\beta_1,x,y)+c_2\Psi(\alpha_2,\beta_2,x,y)\}$, of Ps^- for combinations of two random basis states. Figure 2(a) is in the upper left, 2(b) in the upper right, 2(c) in the lower left, and 2(d) in the lower right. See also the caption of Fig. 1.1.

a combination of about 150 functions of the form (1.1). The convergence of the energy versus the number of the functions in the linear combination is shown in Fig. 1.4. The energy gain is large in the first few steps and then it slowly approaches the exact energy.

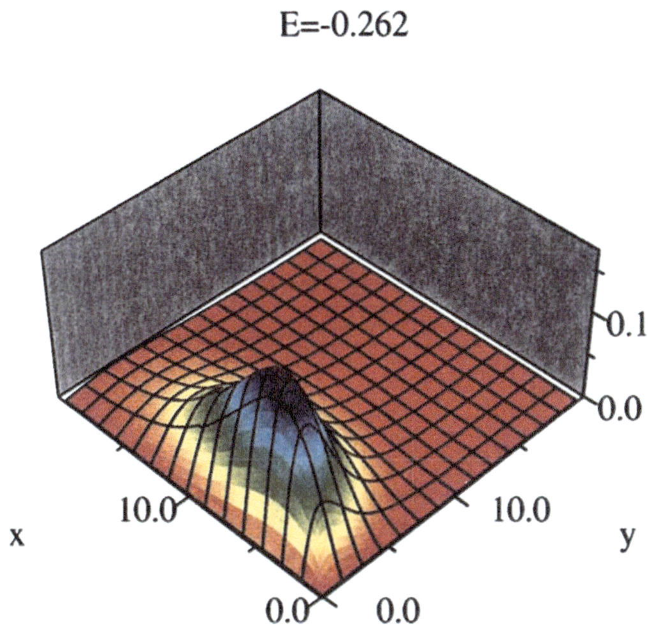

Fig. 1.3. The energy and the wave function of Ps^- obtained by the trial and error selection of 150 basis states

The alert reader may question the importance of the steps that are taken to increase the model space and can ask why not start with, say, a linear combination of 150 functions and then, by random trials of all the parameters, one may find the solution. The reason for increasing the number of functions in the combinations is to control the "convergence" of the energy. By comparing the energy gain in the successive steps one can guess how far the exact ground-state energy is. There is no guarantee that the ad hoc 150 functions would be

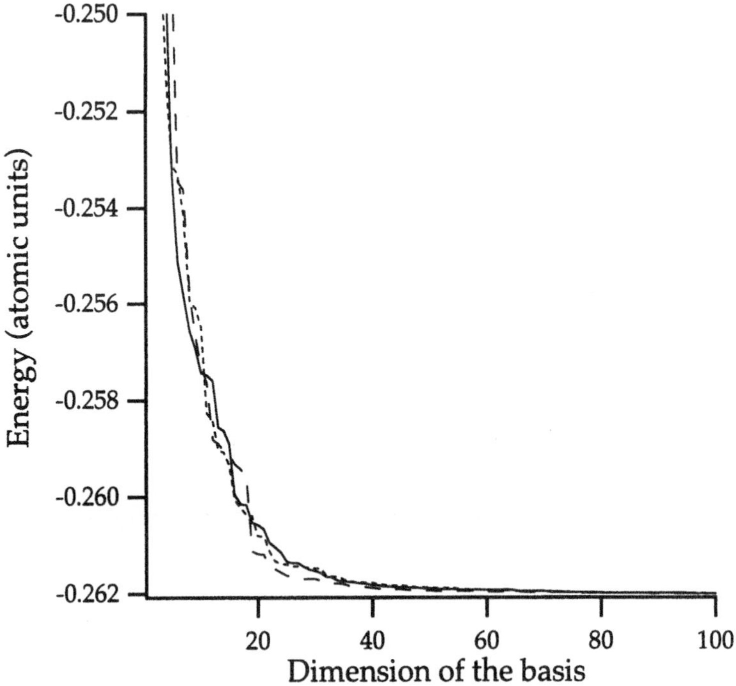

Fig. 1.4. The energy convergence of Ps^- as a function of dimension of basis states that are increased one by one with the trial and error selection. The solid, dashed and dotted curves correspond to three different random paths.

sufficient to reach the solution in cases where the exact energy is not known.

A skeptic may say that one should, instead of the above gambling method, try a deterministic parameter search such as that furnished by the Newton or conjugate gradient method. While this may be true for small systems, the random trials are more successful in most of the cases by avoiding being trapped in the omnipresent local minima. Moreover, one may have discrete parameters or quantum numbers where the deterministic search may not be suitable.

One may also wonder if the solution we reach is really (close to) the ground state energy. These and other questions will be answered in the succeeding chapters.

2. Quantum-mechanical few-body problems

As a first step toward the goal of giving a unified and reasonably simple recipe for solutions of few-body bound-state problems the basic notations and concepts are introduced here.

The motion of the few-body system is governed by the Hamilton operator (Hamiltonian) and described by the eigenfunction of the Hamiltonian. This wave function depends on the positions and other degrees of freedom of the particles. See, for example, [1, 2] for textbooks on quantum mechanics. One can define the positions of the particles in several different ways by using single-particle or relative coordinates. The single-particle coordinates are useful if the particles move almost independently of each other. The relative coordinates are advantageous to emphasize correlations between the particles and to separate the center-of-mass motion. One can relate the different coordinate systems to each other by linear transformations. The definitions and the properties of these transformations are elaborated in this chapter.

One often deals with indistinguishable particles. To comply with the Pauli principle the wave function has to be properly symmetrized. In the spatial part of the wave function the symmetrization induces permutations of the coordinates. It is shown that these permutations can be imposed by linear transformations in the relative coordinate space.

2.1 Hamiltonian

Let us consider an isolated system of N particles with masses $m_1, \cdots, m_N$ and let $\boldsymbol{r}_1, \cdots, \boldsymbol{r}_N$ denote the position vectors of the particles. All of the particles may be identical or different, or some group(s) of them may be identical. The Hamiltonian of the system, with the center-of-mass kinetic energy $T_{\rm cm}$ being subtracted, reads as

$$H = \sum_{i=1}^{N} \frac{\boldsymbol{p}_i^2}{2m_i} - T_{\rm cm} + \sum_{j>i=1}^{N} V_{ij}. \tag{2.1}$$

As the interaction V_{ij} remains unchanged by the displacement of the origin, the above Hamiltonian is translationally invariant and depends on the internal degrees of freedom only.

One may encounter few-body problems where the system is subjected to some external field or the particles move in a single-particle potential. The Hamiltonian then reads as

$$H = \sum_{i=1}^{N} \frac{\boldsymbol{p}_i^2}{2m_i} + \sum_{i=1}^{N} U_i + \sum_{j>i=1}^{N} V_{ij}. \tag{2.2}$$

Even with the presence of the one-body potential U_i it may happen that one wants to remove the contribution from the center-of-mass motion. This will be considered later in evaluating the matrix elements of various operators. An extension of the nonrelativistic kinetic energy to the relativistic kinematics will be discussed in Sect. 7.6. Three-body potentials can be treated in principle but are suppressed for the sake of simplicity.

The two-body potentials are assumed to be local (but nonlocal potentials will also be considered in Sect. 7.5) and can in general be expressed as

$$\begin{aligned} V_{ij} &= \sum_{p} V^p(\boldsymbol{r}_i - \boldsymbol{r}_j)\, O_{ij}^p \\ &= \sum_{p} \int V^p(\boldsymbol{r})\delta(\boldsymbol{r}_i - \boldsymbol{r}_j - \boldsymbol{r})\, O_{ij}^p \, d\boldsymbol{r}. \end{aligned} \tag{2.3}$$

Here p is the short-hand notation to specify the component of the potential which is characterized by O_{ij}^p (e.g., central, spin-orbit, etc.) and $V^p(\boldsymbol{r})$ is the corresponding form factor. To specify the particle motion, several degrees of freedom are in general needed besides the spatial one. For example, the nucleon has spin and isospin degrees of freedom, and the quark has spin, flavor and color degrees of freedom. The specific potentials related to nuclear, subnuclear and other systems will be discussed in later sections.

2.2 Relative coordinates

The separation of the center-of-mass motion is important to describe intrinsic excitations of the system. The center-of-mass motion is taken care of most conveniently by introducing relative and the center-of-mass coordinates $\tilde{\boldsymbol{x}} = (\boldsymbol{x}_1, \boldsymbol{x}_2, ..., \boldsymbol{x}_N)$. The symbol ~ stands for a transpose of a matrix. In particular, $\boldsymbol{x}$ is often used to stand for an $N \times 1$ one-column matrix whose ith element is $\boldsymbol{x}_i$, whereas $\tilde{\boldsymbol{x}}$ a $1 \times N$ one-row matrix. Here $\boldsymbol{x}_N$ is chosen to be the center-of-mass coordinate of the system and the rest of the coordinates $\{\boldsymbol{x}_1, ..., \boldsymbol{x}_{N-1}\}$ is a set of independent relative coordinates. They are in general related to the single-particle coordinates $\tilde{\boldsymbol{r}} = (\boldsymbol{r}_1, ..., \boldsymbol{r}_N)$ by a linear transformation:

$$\boldsymbol{x}_i = \sum_{j=1}^{N} U_{ij}\boldsymbol{r}_j, \qquad \boldsymbol{r}_i = \sum_{j=1}^{N} (U^{-1})_{ij}\boldsymbol{x}_j \quad (i = 1, ..., N). \tag{2.4}$$

We show two examples of relative coordinates $\boldsymbol{x}$. See Fig. 2.1. One is the *Jacobi coordinate* set, which is defined by the matrix

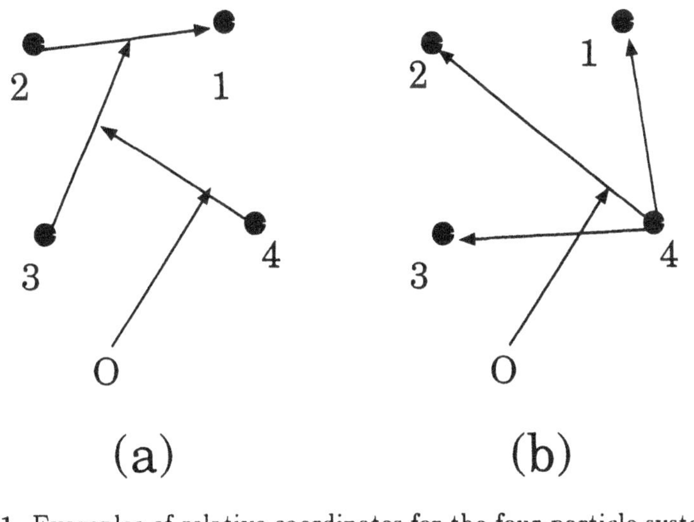

Fig. 2.1. Examples of relative coordinates for the four-particle system. (a) the Jacobi coordinate set, $\boldsymbol{x}_1 = \boldsymbol{r}_1 - \boldsymbol{r}_2$, $\boldsymbol{x}_2 = (m_1\boldsymbol{r}_1 + m_2\boldsymbol{r}_2)/(m_1 + m_2) - \boldsymbol{r}_3$, $\boldsymbol{x}_3 = (m_1\boldsymbol{r}_1 + m_2\boldsymbol{r}_2 + m_3\boldsymbol{r}_3)/(m_1 + m_2 + m_3) - \boldsymbol{r}_4$, and (b) the heavy-particle center coordinate set, $\boldsymbol{x}_1 = \boldsymbol{r}_1 - \boldsymbol{r}_4$, $\boldsymbol{x}_2 = \boldsymbol{r}_2 - \boldsymbol{r}_4$, $\boldsymbol{x}_3 = \boldsymbol{r}_3 - \boldsymbol{r}_4$.

$$U_J = \begin{pmatrix} 1 & -1 & 0 & \cdots & 0 \\ \frac{m_1}{m_{12}} & \frac{m_2}{m_{12}} & -1 & \cdots & 0 \\ \vdots & & & & \vdots \\ \frac{m_1}{m_{12\cdots N-1}} & \frac{m_2}{m_{12\cdots N-1}} & \cdots & \cdots & -1 \\ \frac{m_1}{m_{12\cdots N}} & \frac{m_2}{m_{12\cdots N}} & \cdots & \cdots & \frac{m_N}{m_{12\cdots N}} \end{pmatrix}. \tag{2.5}$$

Its inverse matrix is given by

$$(U_J)^{-1} = \begin{pmatrix} \frac{m_2}{m_{12}} & \frac{m_3}{m_{123}} & \cdots & \frac{m_N}{m_{12\cdots N}} & 1 \\ -\frac{m_1}{m_{12}} & \frac{m_3}{m_{123}} & \cdots & \frac{m_N}{m_{12\cdots N}} & 1 \\ 0 & -\frac{m_{12}}{m_{123}} & & \vdots & \vdots \\ \vdots & \vdots & & \vdots & \vdots \\ 0 & 0 & \cdots & -\frac{m_{12\cdots N-1}}{m_{12\cdots N}} & 1 \end{pmatrix}, \tag{2.6}$$

where $m_{12\cdots i} = m_1+m_2+\cdots+m_i$ and, especially, $m_{12\cdots N}$ is the total mass of the system. Another is the *"heavy-particle center" coordinate* set defined by

$$U_C = \begin{pmatrix} 1 & 0 & 0 & \cdots & -1 \\ 0 & 1 & 0 & \cdots & -1 \\ \vdots & & & & \vdots \\ 0 & 0 & \cdots & \cdots & -1 \\ \frac{m_1}{m_{12\cdots N}} & \frac{m_2}{m_{12\cdots N}} & \cdots & \cdots & \frac{m_N}{m_{12\cdots N}} \end{pmatrix}, \tag{2.7}$$

and its inverse matrix is given by

$$(U_C)^{-1} = \begin{pmatrix} 1-\frac{m_1}{m_{12\cdots N}} & -\frac{m_2}{m_{12\cdots N}} & \cdots & -\frac{m_{N-1}}{m_{12\cdots N}} & 1 \\ -\frac{m_1}{m_{12\cdots N}} & 1-\frac{m_2}{m_{12\cdots N}} & \cdots & -\frac{m_{N-1}}{m_{12\cdots N}} & 1 \\ \vdots & -\frac{m_2}{m_{12\cdots N}} & & \vdots & \vdots \\ \vdots & \vdots & & \vdots & \vdots \\ -\frac{m_1}{m_{12\cdots N}} & -\frac{m_2}{m_{12\cdots N}} & \cdots & -\frac{m_{N-1}}{m_{12\cdots N}} & 1 \end{pmatrix}. \tag{2.8}$$

The choice of the heavy-particle center coordinates may be natural when the mass of the Nth particle is heavier than the masses of the other particles. See Complement 2.1. Note, however, that it is always possible to transform from one coordinate set to another. The Jacobian corresponding to the coordinate transformation (2.4) is unity for both U matrices.

Corresponding to the transformation (2.4), the momentum $\boldsymbol{p}_i$ is expressed in terms of the momentum $\boldsymbol{\pi}_j = -i\hbar\frac{\partial}{\partial \boldsymbol{x}_j}$ conjugate to the coordinate $\boldsymbol{x}_j$:

$$\boldsymbol{p}_i = \sum_{j=1}^{N} U_{ji}\boldsymbol{\pi}_j, \qquad \boldsymbol{\pi}_i = \sum_{j=1}^{N}(U^{-1})_{ji}\boldsymbol{p}_j \quad (i = 1, ..., N). \tag{2.9}$$

Note that $\boldsymbol{\pi}_N = \sum_{j=1}^{N} \boldsymbol{p}_j$ is the total momentum. The center-of-mass kinetic energy is given by $T_{\text{cm}} = \boldsymbol{\pi}_N^2/(2m_{12\cdots N})$. The kinetic energy operator with the center-of-mass kinetic energy subtracted is then expressed as

$$\begin{aligned}\sum_{i=1}^{N}\frac{\boldsymbol{p}_i^2}{2m_i} - T_{\text{cm}} &= \sum_{i=1}^{N}\sum_{j=1}^{N}\Big(\frac{1}{2m_i}\delta_{ij} - \frac{1}{2m_{12\cdots N}}\Big)\boldsymbol{p}_i\cdot\boldsymbol{p}_j \\ &= \frac{1}{2}\sum_{i=1}^{N-1}\sum_{j=1}^{N-1}\Lambda_{ij}\boldsymbol{\pi}_i\cdot\boldsymbol{\pi}_j, \end{aligned} \tag{2.10}$$

where

$$\Lambda_{ij} = \sum_{k=1}^{N} U_{ik}U_{jk}\frac{1}{m_k} \qquad (i, j = 1, ..., N-1). \tag{2.11}$$

To evaluate the potential energy matrix elements for the wave functions that contain no dependence on the center-of-mass coordinate, it is convenient to express the single-particle coordinate relative to the center-of-mass, $\boldsymbol{r}_i - \boldsymbol{x}_N$, the interparticle-distance vector, $\boldsymbol{r}_i - \boldsymbol{r}_j$, and the momentum difference, $\boldsymbol{p}_i - \boldsymbol{p}_j$, in terms of $\boldsymbol{x}$ and $\boldsymbol{\pi}$, respectively. This is easily done by using Eqs. (2.4) and (2.9):

$$\boldsymbol{r}_i - \boldsymbol{x}_N = \sum_{k=1}^{N-1}(U^{-1})_{ik}\boldsymbol{x}_k \equiv \widetilde{w^{(i)}}\boldsymbol{x}, \tag{2.12}$$

$$\boldsymbol{r}_i - \boldsymbol{r}_j = \sum_{k=1}^{N-1}\Big((U^{-1})_{ik} - (U^{-1})_{jk}\Big)\boldsymbol{x}_k \equiv \widetilde{w^{(ij)}}\boldsymbol{x}, \tag{2.13}$$

$$\frac{1}{2}(\boldsymbol{p}_i - \boldsymbol{p}_j) = \frac{1}{2}\sum_{k=1}^{N-1}(U_{ki} - U_{kj})\boldsymbol{\pi}_k \equiv \widetilde{\zeta^{(ij)}}\boldsymbol{\pi}. \tag{2.14}$$

Note that the contribution from the term proportional to the total momentum, $(m_i - m_j)/(2m_{12\cdots N})\boldsymbol{\pi}_N$, is omitted in Eq. (2.14). This is legitimate when we are in the center-of-mass system ($\boldsymbol{\pi}_N = 0$) and interested in the intrinsic motion of the system. The Hamiltonian of type (2.1) is now expressible in terms of independent relative coordinates alone. In what follows $\boldsymbol{x}$ and $\boldsymbol{\pi}$ are meant to represent only the relative coordinates unless otherwise stated, so that they are $(N-1)\times 1$ one-column matrices. In Eqs. (2.12)–(2.14), $(N-1)\times 1$ one-column matrices (i.e. $(N-1)$-dimensional vectors) $w^{(i)}$, $w^{(ij)}$, and $\zeta^{(ij)}$ are introduced to simplify the notation. E.g., $\widetilde{w^{(i)}}\boldsymbol{x}$ is a Cartesian vector which is formed by the usual matrix multiplication of $\widetilde{w^{(i)}}$ and $\boldsymbol{x}$, i.e., $\widetilde{w^{(i)}}\boldsymbol{x} = \sum_{k=1}^{N-1} w_k^{(i)}\boldsymbol{x}_k$. Note that $w^{(ij)}$ and $\zeta^{(ij)}$ satisfy the following equalities

$$\sum_{k=1}^{N-1} w_k^{(ij)}\zeta_k^{(ij)} = \widetilde{w^{(ij)}}\zeta^{(ij)} = \widetilde{\zeta^{(ij)}}w^{(ij)} = 1 - \delta_{ij},$$

$$\sum_{j>i=1}^{N} w_k^{(ij)}\zeta_l^{(ij)} = \left(\sum_{j>i=1}^{N} w^{(ij)}\widetilde{\zeta^{(ij)}}\right)_{kl}$$

$$= \frac{N}{2}\delta_{kl} \quad (k, l = 1, ..., N-1), \tag{2.15}$$

where the relations $U_{Ni} = m_i/m_{12\cdots N}$, $(U^{-1})_{iN} = 1$, $\sum_{i=1}^{N} U_{ki} = 0$ for $k \neq N$ are used.

For an $(N-1)\times(N-1)$ symmetric matrix A, let the quadratic form, $\tilde{\boldsymbol{x}}A\boldsymbol{x}$, represent scalar products of the Cartesian vectors:

$$\tilde{\boldsymbol{x}}A\boldsymbol{x} = \sum_{i=1}^{N-1}\boldsymbol{x}_i\cdot(A\boldsymbol{x})_i = \sum_{i=1}^{N-1}\sum_{j=1}^{N-1}A_{ij}\boldsymbol{x}_i\cdot\boldsymbol{x}_j. \tag{2.16}$$

It is easy to show, with this convention, that

$$\sum_{i=1}^{N}\alpha_i(\boldsymbol{r}_i - \boldsymbol{x}_N)^2 = \tilde{\boldsymbol{x}}\left(\sum_{i=1}^{N}\alpha_i w^{(i)}\widetilde{w^{(i)}}\right)\boldsymbol{x},$$

$$\sum_{j>i=1}^{N} \alpha_{ij}(\boldsymbol{r}_i - \boldsymbol{r}_j)^2 = \tilde{\boldsymbol{x}}\left(\sum_{j>i=1}^{N} \alpha_{ij} w^{(ij)} \widetilde{w^{(ij)}}\right)\boldsymbol{x}. \tag{2.17}$$

As a special case of the above relations, we have the following equalities for the moment of inertia around the center-of-mass:

$$\sum_{i=1}^{N} m_i(\boldsymbol{r}_i - \boldsymbol{x}_N)^2 = \sum_{j>i=1}^{N} \frac{m_i m_j}{m_{12\cdots N}}(\boldsymbol{r}_i - \boldsymbol{r}_j)^2 = \sum_{i=1}^{N-1} \mu_i \boldsymbol{x}_i^2, \tag{2.18}$$

where the reduced mass μ_i corresponding to the Jacobi coordinate $\boldsymbol{x}_i$ is defined by

$$\mu_i = \frac{m_{i+1} m_{12\cdots i}}{m_{12\cdots i+1}} \qquad (i = 1, ..., N-1). \tag{2.19}$$

We also obtain

$$\sum_{j>i=1}^{N} (\boldsymbol{r}_i - \boldsymbol{r}_j)^2 = N\sum_{i=1}^{N}(\boldsymbol{r}_i - \boldsymbol{x}_N)^2 - \left(\sum_{i=1}^{N}(\boldsymbol{r}_i - \boldsymbol{x}_N)\right)^2. \tag{2.20}$$

The term $\sum_{i=1}^{N}(\boldsymbol{r}_i - \boldsymbol{x}_N)$ vanishes when the masses of all the particles are the same.

The total orbital angular momentum operator $\boldsymbol{L}$ in the center-of-mass system ($\boldsymbol{\pi}_N = 0$) can be expressed in terms of the operators relevant to the relative coordinates:

$$\hbar \boldsymbol{L} \equiv \sum_{i=1}^{N}(\boldsymbol{r}_i \times \boldsymbol{p}_i) - (\boldsymbol{x}_N \times \boldsymbol{\pi}_N) = \sum_{i=1}^{N-1}(\boldsymbol{x}_i \times \boldsymbol{\pi}_i), \tag{2.21}$$

$$\begin{aligned}
&\sum_{j>i=1}^{N}\left((\boldsymbol{r}_i - \boldsymbol{r}_j) \times \frac{1}{2}(\boldsymbol{p}_i - \boldsymbol{p}_j)\right) \\
&\quad = \sum_{j>i=1}^{N}\left((\boldsymbol{r}_i - \boldsymbol{r}_j) \times \widetilde{\zeta^{(ij)}}\boldsymbol{\pi}\right) \\
&\quad + \sum_{j>i=1}^{N}\left((\boldsymbol{r}_i - \boldsymbol{r}_j) \times \frac{m_i - m_j}{2m_{12\cdots N}}\boldsymbol{\pi}_N\right) \\
&\quad = \sum_{j>i=1}^{N}\left(\widetilde{w^{(ij)}}\boldsymbol{x} \times \widetilde{\zeta^{(ij)}}\boldsymbol{\pi}\right) = \frac{N}{2}\hbar \boldsymbol{L}.
\end{aligned} \tag{2.22}$$

In Eq. (2.22) use is made of Eq. (2.15) in the last step. Thus the equality of Eq. (2.22) holds in the case one considers the orbital angular momentum in the system of $\boldsymbol{\pi}_N = 0$ or treats a system of particles with equal masses.

In atomic physics the trial function of Hylleraas type or correlated exponential type is often used with success. This function contains the exponential form expressed in terms of the *interparticle-distance coordinates.* Instead of the exponential function let us consider its Gaussian analogue

$$\Psi = \exp\Big\{ -\frac{1}{2} \sum_{j>i=1}^{N} \alpha_{ij} (\boldsymbol{r}_i - \boldsymbol{r}_j)^2 \Big\}. \tag{2.23}$$

Here $N(N-1)/2$ parameters α_{ij} determine the falloff of the interparticle functions and may be considered variational parameters. Since not all of the $N(N-1)/2$ interparticle-distance vectors, $\boldsymbol{r}_i - \boldsymbol{r}_j$, are independent, e.g., $\boldsymbol{r}_1 - \boldsymbol{r}_3 = (\boldsymbol{r}_1 - \boldsymbol{r}_2) + (\boldsymbol{r}_2 - \boldsymbol{r}_3)$, it is convenient to rewrite Ψ in terms of the independent coordinates $\boldsymbol{x}$. This is needed to calculate, e.g., the norm of Ψ. Equation (2.17) enables one to rewrite Eq. (2.23) as

$$\Psi = \exp\Big(-\frac{1}{2} \tilde{\boldsymbol{x}} A \boldsymbol{x} \Big), \tag{2.24}$$

where the $(N-1) \times (N-1)$ symmetric matrix A and the parameters α_{ij} are related by

$$A_{kl} = \sum_{j>i=1}^{N} \alpha_{ij} w_k^{(ij)} w_l^{(ij)} = \sum_{j>i=1}^{N} \alpha_{ij} \Big(w^{(ij)} \widetilde{w^{(ij)}} \Big)_{kl},$$

$$\alpha_{ij} = -\sum_{k=1}^{N-1} \sum_{l=1}^{N-1} U_{ki} A_{kl} U_{lj} \equiv -(\tilde{U} A U)_{ij} \quad (i<j). \tag{2.25}$$

By substituting A_{kl} of the first relation to the second relation one can check the validity of the second relation. The above two forms, (2.23) and (2.24), are thus equivalent. We introduce both because in some applications the first form or in some others the second form is more advantageous. See Complements 6.4 and 7.1. The norm $\langle \Psi | \Psi \rangle$ becomes finite if and only if A is positive-definite and is then given by $((2\pi)^{N-1}/\mathrm{det}A)^{\frac{3}{2}}$ with the use of Eq. (6.32) or Exercise 6.2.

A necessary and sufficient condition of the positive-definiteness of a symmetric matrix A is that the matrix A is expressible as $A =$

$\tilde{G}DG$, where G is an $(N-1)\times(N-1)$ orthogonal matrix containing $(N-1)(N-2)/2$ parameters and D is a diagonal matrix, $D_{ij}=d_i\delta_{ij}$, including $(N-1)$ positive parameters d_i.

Before closing this section, we remark that the matrix U_J of Eq. (2.5) is not orthogonal in general. In some cases it is convenient to use a coordinate system that is obtained from the single-particle coordinates by an orthogonal matrix. We show such an example in Complement 2.2.

2.3 Symmetrization

The wave function of the system must have a proper symmetry for the interchange of identical particles. This is achieved by operating on the basis function with a suitable operator $\mathcal{P}$, which is the symmetrizer for bosons or the antisymmetrizer for fermions. The operator $\mathcal{P}$ can in general be given by a combination of the permutations P with suitable phases. The permutation $P=\begin{pmatrix}1 & 2 & \cdots & N\\ p_1 & p_2 & \cdots & p_N\end{pmatrix}$ of particle indices transforms the single-particle coordinates as $\boldsymbol{r}_i\to\boldsymbol{r}_{p_i}$:

$$P\boldsymbol{r}=T_P\boldsymbol{r}, \tag{2.26}$$

where the matrix T_P is

$$(T_P)_{ij}=\delta_{j\,p_i}\quad (i,j=1,...,N). \tag{2.27}$$

E.g., the matrices T_P are given as follows in the case of three particles:

$$T_{\left(\begin{smallmatrix}123\\123\end{smallmatrix}\right)}=\begin{pmatrix}1&0&0\\0&1&0\\0&0&1\end{pmatrix},\quad T_{\left(\begin{smallmatrix}123\\213\end{smallmatrix}\right)}=\begin{pmatrix}0&1&0\\1&0&0\\0&0&1\end{pmatrix},$$
$$T_{\left(\begin{smallmatrix}123\\321\end{smallmatrix}\right)}=\begin{pmatrix}0&0&1\\0&1&0\\1&0&0\end{pmatrix},\quad T_{\left(\begin{smallmatrix}123\\132\end{smallmatrix}\right)}=\begin{pmatrix}1&0&0\\0&0&1\\0&1&0\end{pmatrix},$$
$$T_{\left(\begin{smallmatrix}123\\231\end{smallmatrix}\right)}=\begin{pmatrix}0&1&0\\0&0&1\\1&0&0\end{pmatrix},\quad T_{\left(\begin{smallmatrix}123\\312\end{smallmatrix}\right)}=\begin{pmatrix}0&0&1\\1&0&0\\0&1&0\end{pmatrix}. \tag{2.28}$$

The operation of P on the wave function is thus not a problem when it is written in terms of the single-particle coordinates.

When the wave function is expressed in terms of the relative coordinates, we have to note that the permutation P induces a linear transformation of the relative coordinates as follows:

$$P\boldsymbol{x} = T_P\boldsymbol{x}, \tag{2.29}$$

where by using Eqs. (2.4), (2.26) and (2.27) the matrix T_P is now given by

$$(T_P)_{ij} = \sum_{k=1}^{N} U_{ik}(U^{-1})_{p_k j} \quad (i, j = 1, ..., N-1). \tag{2.30}$$

Though we use the same notation of T_P, note that the size of the matrix is now $(N-1)\times(N-1)$ because the center-of-mass coordinate, which is unchanged under the permutation, is eliminated. We will show in Chap. 6 that the properties of Eqs. (2.26)–(2.30) can be used to advantage for Gaussian functions. See Eqs. (6.28) and (6.29).

2.4 Permutation of the Jacobi coordinates

One can generate different sets of Jacobi coordinates by interchanging the particle labels by permutations. Figure 2.2 shows an example of three different sets, $\boldsymbol{x}^{(1)}$, $\boldsymbol{x}^{(2)}$, and $\boldsymbol{x}^{(3)}$, of the Jacobi coordinates for a three-body system. These sets are completely equivalent from the point of view of the dynamical description of the system, but they can be advantageously used in the calculation of the matrix elements.

In the Jacobi coordinate set, defined by Eq. (2.5), the first Jacobi coordinate is equal to $\boldsymbol{r}_1 - \boldsymbol{r}_2$. By permutations one can create other Jacobi coordinate sets $\boldsymbol{x}^{(k)}$ where the first Jacobi coordinate takes the form of $\boldsymbol{r}_i - \boldsymbol{r}_j$. The index k refers to the kth permutation and serves to distinguish the permuted Jacobi coordinate set. As the two-body potential depends on this coordinate, the set created by the permutation will prove to be useful. See Appendix A.5 for an example of using the permuted Jacobi coordinate set.

Another peculiarity of the Jacobi coordinates is that if we are in the center-of-mass system ($\boldsymbol{\pi}_N = 0$) then by using Eqs. (2.5) and (2.9) $\boldsymbol{p}_N = -\boldsymbol{\pi}_{N-1}$, that is, one of the single-particle momenta is equal to one of the relative momenta. By permutations we can express any of the single-particle momenta in terms of a certain momentum of the relative momenta. This simple expression can be exploited in several ways, because instead of evaluating matrix elements of operators

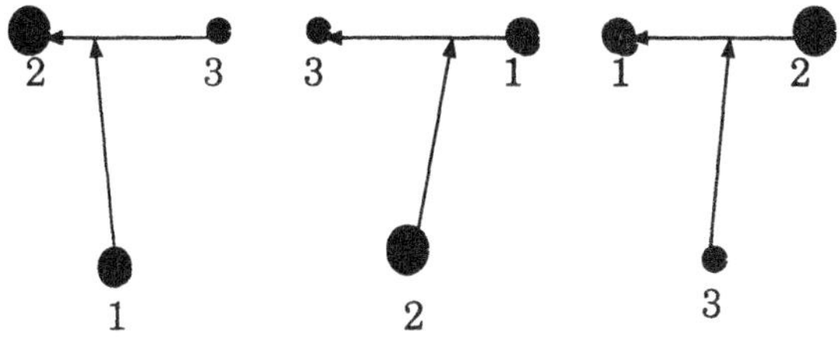

Fig. 2.2. Different sets of the Jacobi coordinates for the three-particle system. From left to right: $\boldsymbol{x}^{(1)}$, $\boldsymbol{x}^{(2)}$, and $\boldsymbol{x}^{(3)}$, respectively.

containing single-particle momentum we can do so with operators of relative momentum and vice versa. See Sect. 7.6 for more details and an application to the calculation of matrix elements for the semirelativistic kinetic energy.

The permutation of the Jacobi coordinates is defined by the permutation of the column vectors of U_J and the corresponding exchange of the masses. For example, for the cyclic permutation $(1,2,...,N) = \begin{pmatrix} 1 & 2 & \dots & N \\ 2 & 3 & \dots & 1 \end{pmatrix}$ the transformation matrix between the single-particle coordinates $\boldsymbol{r}$ and the Jacobi coordinates $\boldsymbol{x}^{(k)}$ takes the form

$$U_J^{(k)} = \begin{pmatrix} 0 & 1 & -1 & \cdots & 0 \\ 0 & \frac{m_2}{m_{23}} & \frac{m_3}{m_{23}} & \cdots & 0 \\ \vdots & & & & \vdots \\ -1 & \frac{m_2}{m_{23\cdots N}} & \frac{m_3}{m_{23\cdots N}} & \cdots & \frac{m_N}{m_{23\cdots N}} \\ \frac{m_1}{m_{12\cdots N}} & \frac{m_2}{m_{12\cdots N}} & \frac{m_3}{m_{12\cdots N}} & \cdots & \frac{m_N}{m_{12\cdots N}} \end{pmatrix}. \tag{2.31}$$

The relative coordinates in the different Jacobi coordinate sets can be expressed by each other through the transformation:

$$\boldsymbol{x}^{(k)} = T^{(k)}\boldsymbol{x}, \qquad T^{(k)} = U_J^{(k)}U^{-1}. \tag{2.32}$$

The symmetrization in Sect. 2.3 and the permutation of the Jacobi coordinates are formally very similar. The main difference is that while the symmetrization is assumed for identical (and therefore equal mass) particles, in the latter case we allow for particles with different masses.

Complements

2.1 *An N-particle Hamiltonian in the heavy-particle center coordinate set*

Suppose that the mass of one of the particles, say the Nth particle, is the heaviest among all the particles. In this case the choice of the heavy-particle center coordinate may be natural particularly when the position of the Nth particle is close to the center-of-mass of the system, like in an atom, where the Nth particle is a nucleus. The Hamiltonian (2.1) can be reduced to the form which contains no explicit dependence on the variables of the Nth particle.

By using Eqs. (2.7), (2.10) and (2.11) the kinetic energy operator is expressed in the heavy-particle center coordinates as

$$\sum_{i=1}^{N} \frac{\boldsymbol{p}_i^2}{2m_i} - T_{\text{cm}} = \sum_{i=1}^{N-1} \sum_{j=1}^{N-1} \left(\frac{1}{2m_i} \delta_{ij} + \frac{1}{2m_N} \right) \boldsymbol{\pi}_i \cdot \boldsymbol{\pi}_j$$

$$= \sum_{i=1}^{N-1} \frac{\boldsymbol{\pi}_i^2}{2\mu_{iN}} + \frac{1}{m_N} \sum_{j>i=1}^{N-1} \boldsymbol{\pi}_i \cdot \boldsymbol{\pi}_j, \tag{2.33}$$

where $\mu_{iN} = m_i m_N/(m_i + m_N)$ is the reduced mass for the ith and Nth particles. The potential energy V_{iN} $(i = 1, ..., N-1)$ is a function of $\boldsymbol{r}_i - \boldsymbol{r}_N = \boldsymbol{x}_i$ and can be considered the one-body potential U_i acting on the ith particle. On the other hand the potential energy V_{ij} acting between the lighter particles is a function of $\boldsymbol{r}_i - \boldsymbol{r}_j = \boldsymbol{x}_i - \boldsymbol{x}_j$. The Hamiltonian can thus be reduced to the type of Eq. (2.2)

$$H = \sum_{i=1}^{N-1} \frac{\boldsymbol{\pi}_i^2}{2\mu_{iN}} + \sum_{i=1}^{N-1} U_i + \sum_{j>i=1}^{N-1} \left(V_{ij} + \frac{1}{m_N} \boldsymbol{\pi}_i \cdot \boldsymbol{\pi}_j \right). \tag{2.34}$$

The problem is thus reduced to that of an $(N-1)$-particle system with an external field U_i and an additional separable "two-body potential" $\boldsymbol{\pi}_i \cdot \boldsymbol{\pi}_j$. In the limit of $m_N \to \infty$, μ_{iN} goes to m_i and the $\boldsymbol{\pi}_i \cdot \boldsymbol{\pi}_j$ term disappears. The effect of the finite nuclear mass in the helium atom was discussed in [3].

The above formulation can also be applied to a system of identical particles when some, or most, of them form an inert core. As an example let us take up the nuclear shell model, where the nucleons are divided into two groups, the passive or inactive nucleons and a relatively small number of active nucleons. The passive nucleons form an inert core by filling the lowest possible single-particle orbits and exert a potential field U_i on each of the active nucleons. The active nucleons may occupy several single-particle orbits belonging to several major shells outside the inert core. In this approximation it is desirable to describe the motion of the active nucleons without inclusion of any excitations arising from the center-of-mass motion. It is clear that such excitations can be excluded by taking the full Hamiltonian to have the form (2.34) with the core as the heavy particle.

2.2 *Canonical Jacobi coordinates*

As was noted in Sect. 2.2, the Jacobi coordinates $\boldsymbol{x}$ have the property that the transformation matrix U_J of Eq. (2.5) is not orthogonal, that is, neither $U_J\widetilde{U_J}$ nor $\widetilde{U_J}U_J$ is equal to the unit matrix. Instead, the following relations are fulfilled:

$$\widetilde{U_J}\Lambda^{-1}U_J = L^{-1}, \qquad U_J L\widetilde{U_J} = \Lambda, \tag{2.35}$$

with

$$L_{ij} = \frac{1}{m_i}\delta_{ij}, \qquad \Lambda_{ij} = \frac{1}{\mu_i}\delta_{ij}, \tag{2.36}$$

where μ_i $(i = 1, ..., N-1)$ is the reduced mass belonging to the ith Jacobi coordinate and is defined in Eq. (2.19), while μ_N is equal to the total mass $m_{12\cdots N}$. By using the transformation between the single-particle and the Jacobi coordinate systems, $\boldsymbol{x} = U_J\boldsymbol{r}$ and $\boldsymbol{p} = \widetilde{U_J}\boldsymbol{\pi}$, of Eqs. (2.4) and (2.9), the above relations lead to the following identities

$$\sum_{i=1}^{N} m_i\boldsymbol{r}_i^2 = \sum_{i=1}^{N}\mu_i\boldsymbol{x}_i^2, \qquad \sum_{i=1}^{N}\frac{\boldsymbol{p}_i^2}{m_i} = \sum_{i=1}^{N}\frac{\boldsymbol{\pi}_i^2}{\mu_i}. \tag{2.37}$$

Alternatively, one can introduce a "canonical" set of Jacobi coordinates:

$$\boldsymbol{\xi}_i = \sum_{j=1}^{N} V_{ij}\boldsymbol{r}_j \qquad \text{with} \qquad V = \Lambda^{-\frac{1}{2}}U_J L^{\frac{1}{2}}. \tag{2.38}$$

The square root matrices of the diagonal matrices L and Λ are simply given by the square root of the elements. This system of Jacobi coordinates belongs to an orthogonal transformation:

$$V\tilde{V} = \tilde{V}V = 1, \tag{2.39}$$

and therefore

$$\boldsymbol{r}_i = \sum_{j=1}^{N} V_{ji}\boldsymbol{\xi}_j, \qquad \text{and} \qquad \sum_{i=1}^{N} \boldsymbol{\xi}_i^2 = \sum_{i=1}^{N} \boldsymbol{r}_i^2. \tag{2.40}$$

Let the momentum canonically conjugate to $\boldsymbol{\xi}_i$ be denoted $\boldsymbol{\eta}_i$. Similarly to the transformation of coordinates, we obtain the transformation of momenta as follows:

$$\boldsymbol{\eta}_i = \sum_{j=1}^{N} V_{ij}\boldsymbol{p}_j, \qquad \boldsymbol{p}_i = \sum_{j=1}^{N} V_{ji}\boldsymbol{\eta}_j. \tag{2.41}$$

The total kinetic energy does not take a diagonal form with respect to $\boldsymbol{\eta}$ and can be expressed as follows:

$$\sum_{i=1}^{N} \frac{\boldsymbol{p}_i^2}{m_i} = \sum_{i=1}^{N}\sum_{j=1}^{N} \left(VL\tilde{V}\right)_{ij}\boldsymbol{\eta}_i \cdot \boldsymbol{\eta}_j = \tilde{\boldsymbol{\eta}} VL\tilde{V}\boldsymbol{\eta}. \tag{2.42}$$

The two systems of the Jacobi coordinates $\boldsymbol{x} = U_J\boldsymbol{r}$ and the canonical Jacobi coordinates $\boldsymbol{\xi} = V\boldsymbol{r}$ can easily be related by

$$\boldsymbol{x} = U_J\tilde{V}\boldsymbol{\xi}, \qquad \boldsymbol{\xi} = V(U_J)^{-1}\boldsymbol{x}. \tag{2.43}$$

For a three-particle system the transformation matrix V reads as

$$V = \begin{pmatrix} \sqrt{\frac{m_2}{m_{12}}} & -\sqrt{\frac{m_1}{m_{12}}} & 0 \\ \sqrt{\frac{m_1 m_3}{m_{12}m_{123}}} & \sqrt{\frac{m_2 m_3}{m_{12}m_{123}}} & -\sqrt{\frac{m_{12}}{m_{123}}} \\ \sqrt{\frac{m_1}{m_{123}}} & \sqrt{\frac{m_2}{m_{123}}} & \sqrt{\frac{m_3}{m_{123}}} \end{pmatrix}, \tag{2.44}$$

and for particles with equal mass

$$V = \begin{pmatrix} \frac{1}{\sqrt{2}} & -\frac{1}{\sqrt{2}} & 0 \\ \frac{1}{\sqrt{6}} & \frac{1}{\sqrt{6}} & -\frac{2}{\sqrt{6}} \\ \frac{1}{\sqrt{3}} & \frac{1}{\sqrt{3}} & \frac{1}{\sqrt{3}} \end{pmatrix}. \tag{2.45}$$

The disadvantage of the canonical Jacobi coordinates $\boldsymbol{\xi}$ is that if the masses m_i are not equal then the center-of-mass motion is not separated easily ($\boldsymbol{\xi}_N$ is not the center-of-mass coordinate).

3. Introduction to variational methods

The variational method is one of the most popular approaches to tackle quantum-mechanical few-body problems. Though it gives only an approximate solution except for some special cases (the Ritz variational method, for example, gives only an upper bound of the energy), one can get a virtually exact solution with an appropriately chosen function space. The function space is defined by basis states and the wave function of the system is expanded in that basis. In this chapter we briefly introduce the theorems requisite for obtaining a variational solution.

3.1 Variational principles

Let us consider a physical system whose Hamiltonian H is self-adjoint (Hermitian), bounded from below and time-independent. We are interested in finding the discrete eigenvalues of H and its corresponding (normalized) eigenstates:

$$H\Phi_n = E_n\Phi_n, \quad n = 1, 2, ... \tag{3.1}$$

The energies E_n are real and are ordered such that $E_1 \leq E_2 \leq ....$ We assume that the ground state is non-degenerate.

Although H is known, this does not mean that E_n and Φ_n are known. In general, it is difficult to solve the eigenvalue equation (3.1). When we do not know how to diagonalize H exactly, the variational method becomes useful for any type of Hamiltonian.

Theorem 3.1 (Ritz Theorem). *For an arbitrary function Ψ of the state space the expectation value (mean value) of H in the state Ψ is such that*

$$E \equiv \frac{\langle\Psi|H|\Psi\rangle}{\langle\Psi|\Psi\rangle} \geq E_1, \tag{3.2}$$

where the equality holds if and only if Ψ is an eigenstate of H with the eigenvalue E_1.

Proof. The proof of this theorem is elementary and can be found in textbooks of quantum mechanics. See for example [1, 2]. The function Ψ may be expanded in terms of the energy eigenstates

$$\Psi = \sum_{i=1}^{\infty} a_i \Phi_i. \tag{3.3}$$

In this expansion the eigenstates with continuous eigenvalues are included and the sum must be extended appropriately to include an integration over them. Then we can show that for an integer n

$$\frac{\langle\Psi|H^n|\Psi\rangle}{\langle\Psi|\Psi\rangle} - E_1^n = \frac{\sum_{i=2}^{\infty}(E_i^n - E_1^n)|a_i|^2}{\sum_{i=1}^{\infty}|a_i|^2}. \tag{3.4}$$

Clearly the right-hand side of Eq. (3.4) is non-negative for $n = 1$ because $E_i - E_1 \geq 0$ for $i \geq 2$. It vanishes if and only if $a_i = 0$ for $i \geq 2$, that is, Ψ is an eigenstate of H with eigenvalue E_1. This proves the Ritz theorem. Only the case of $n = 1$ was used to prove the Ritz theorem but other cases will be needed later to derive Temple's bound.

The Ritz theorem is the basis for a method of approximate determination of the ground-state energy E_1, namely it is exploited as the minimization principle of the mean value of H. Suppose that we choose a family of functions $\Psi(\alpha)$ which are characterized by a finite number of parameters denoted α. In the case of Eq. (1.1), there are two parameters α and β. We calculate the mean value $E(\alpha)$ of the Hamiltonian H in this trial function, and minimize $E(\alpha)$ with respect to α. The minimal value obtained in this way is an approximation to the ground-state energy E_1 of the system. Clearly, whether this variational method produces a satisfactory result or not substantially depends on the choice of the trial functions $\Psi(\alpha)$.

The Ritz theorem can be generalized to excited states as well:

Theorem 3.2 (Generalized Ritz Theorem). *The expectation value of the Hamiltonian H is stationary in the neighborhood of its discrete eigenvalues.*

Proof. Let us calculate an increment δE of the mean value using $E\langle\Psi|\Psi\rangle = \langle\Psi|H|\Psi\rangle$ when Ψ is changed to $\Psi + \delta\Psi$, where $\delta\Psi$ is assumed to be infinitely small. Neglecting terms of higher order than a linear term in $\delta\Psi$, we obtain

$$\langle\Psi|\Psi\rangle\delta E = \delta(\langle\Psi|H|\Psi\rangle) - E\delta(\langle\Psi|\Psi\rangle)$$
$$= \langle\delta\Psi|H - E|\Psi\rangle + \langle\Psi|H - E|\delta\Psi\rangle. \tag{3.5}$$

The mean value E is stationary if $\delta E = 0$ for any infinitesimal $\delta\Psi$, that is, if

$$\langle\delta\Psi|H - E|\Psi\rangle + \langle\Psi|H - E|\delta\Psi\rangle = 0. \tag{3.6}$$

If $\delta\Psi$ is chosen to be $\varepsilon(H - E)\Psi$, where ε is an infinitesimal real number, the above equation implies that the norm of the function $(H - E)\Psi$ is zero, and thus the function $(H - E)\Psi$ itself must be a null function, namely $H\Psi = E\Psi$. Therefore the mean value E is stationary if and only if the state Ψ from which E is calculated is an eigenstate of H, and the stationary value E is the corresponding eigenvalue of the Hamiltonian.

This generalized Ritz theorem allows an approximate determination of the eigenvalues of the Hamiltonian. If the $E(\alpha)$ has several extrema, they are the approximate values of some of the energies E_n.

In most cases for practical applications the trial function is given as a linear combination of a finite number of independent functions $\Psi(\alpha)$:

$$\Psi = \sum_{i=1}^{K} c_i \Psi(\alpha_i). \tag{3.7}$$

The linear independence of the functions will be discussed soon later. We do not always assume that $\Psi(\alpha_1), ..., \Psi(\alpha_K)$ are mutually orthogonal because the use of nonorthogonal functions is in fact quite useful. They can, however, always be made orthogonal if necessary, e.g. by a Gram-Schmidt orthogonalization procedure. The variational method then reduces to the eigenvalue problem of the Hamiltonian inside the state space $\mathcal{V}_K$ spanned by the set $\{\Psi(\alpha_1), ..., \Psi(\alpha_K)\}$, that is, the space containing all linear combinations of $\Psi(\alpha_1), ..., \Psi(\alpha_K)$. The mean value E is given by

$$E = \frac{c^\dagger \mathcal{H} c}{c^\dagger \mathcal{B} c}, \tag{3.8}$$

where c is a K-dimensional column vector whose ith element is c_i and $c^\dagger$ is the Hermitian conjugate of c. The $K \times K$ Hamiltonian and overlap matrices $\mathcal{H}$ and $\mathcal{B}$ are defined by

$$\mathcal{H}_{ij} = \langle \Psi(\alpha_i)|H|\Psi(\alpha_j)\rangle, \qquad \mathcal{B}_{ij} = \langle \Psi(\alpha_i)|\Psi(\alpha_j)\rangle. \tag{3.9}$$

The linear parameter c_i can be determined by the generalized Ritz theorem. The condition that E is stationary with respect to an arbitrary, infinitesimal change of c_i leads to the generalized eigenvalue problem

$$\mathcal{H}c = E\mathcal{B}c, \quad \text{i.e.,} \quad \sum_{j=1}^{K}(\mathcal{H}_{ij} - E\mathcal{B}_{ij})c_j = 0 \quad (i = 1, ..., K). \tag{3.10}$$

The restriction of the eigenvalue problem of H to the subspace $\mathcal{V}_K$ can thus simplify the solution.

We discuss the linear independence of the functions $\Psi(\alpha_1)$,..., $\Psi(\alpha_K)$. The linear independence of the functions means that no vector c except for $c = 0$ can make a linear combination, $\sum_{i=1}^{K} c_i\Psi(\alpha_i)$, a null function. In other words, the combination becomes a null function if and only if $c = 0$. This is equivalent to saying that the equation $\mathcal{B}c = 0$ for c has a unique solution of $c = 0$. (To understand this, we show that $\sum_{i=1}^{K} c_i\Psi(\alpha_i) = 0$ is equivalent to $\mathcal{B}c = 0$. When $\sum_{i=1}^{K} c_i\Psi(\alpha_i) = 0$, we obviously have $\sum_{i=1}^{K} c_i\langle\Psi(\alpha_j)|\Psi(\alpha_i)\rangle = 0$ for $j = 1, ..., K$, which is nothing but $(\mathcal{B}c)_j = 0$ for $j = 1, ..., K$. Conversely, when $(\mathcal{B}c)_j = 0$ for $j = 1, ..., K$, by multiplying c_j^* (the complex conjugate of c_j) and summing over j, we obtain $c^\dagger\mathcal{B}c = \langle\sum_{i=1}^{K} c_i\Psi(\alpha_i)|\sum_{i=1}^{K} c_i\Psi(\alpha_i)\rangle = 0$, which leads us to $\sum_{i=1}^{K} c_i\Psi(\alpha_i) = 0$.) As the existence of a unique solution $c = 0$ for the equation $\mathcal{B}c = 0$ is possible only when $\det\mathcal{B} \neq 0$, the linear independence of the functions is assured by the condition $\det\mathcal{B} \neq 0$. Because the overlap matrix $\mathcal{B}$ is at least positive-semidefinite, that is $\langle\Psi|\Psi\rangle = c^\dagger\mathcal{B}c \geq 0$ for any vector c, all eigenvalues μ of $\mathcal{B}$ become real, positive when the functions $\Psi(\alpha_1), ..., \Psi(\alpha_K)$ are linearly independent. The positiveness of μ is understood as follows: For the eigenvector $c \neq 0$ corresponding to the eigenvalue μ, we have the relation $\mu c^\dagger c = \langle\Psi|\Psi\rangle$. As the basis functions $\Psi(\alpha_i)$ are linearly independent, Ψ is not identically zero because otherwise we can make $\Psi = \sum_{i=1}^{K} c_i\Psi(\alpha_i)$ vanish identically with $c \neq 0$, which contradicts the assumption of the linear independence of the basis functions. Thus $\langle\Psi|\Psi\rangle$ is positive and of course $c^\dagger c$ is positive, so the eigenvalue μ has to be positive. We can also state that if there exists a vanishing eigenvalue of $\mathcal{B}$ then the basis set $\{\Psi(\alpha_1), ..., \Psi(\alpha_K)\}$ is linearly dependent. The solution of the eigenvalue problem, $\mathcal{B}c^{(\mu)} = \mu c^{(\mu)}$, gives us an orthonormal set:

$$\phi_\mu = \frac{1}{\sqrt{\mu}} \sum_{i=1}^{K} c_i^{(\mu)} \Psi(\alpha_i), \tag{3.11}$$

where $c^{(\mu)}$ is assumed to be normalized to $c^{(\mu)\dagger} c^{(\mu)} = 1$. In many practical problems it may happen that $\mathcal{B}$ has one or several very small eigenvalues. Then the eigenstate corresponding to the small eigenvalue has very large expansion coefficients $c_i^{(\mu)}/\sqrt{\mu}$. If this occurs then a small error in the matrix elements of $\mathcal{H}$ or $\mathcal{B}$ can lead to a larger error in the solution of Eq. (3.10).

When the ill condition mentioned above does not occur, the generalized eigenvalue problem (3.10) can be solved safely. The eigenvalues ϵ_i $(i = 1, ..., K)$ are arranged in increasing order $\epsilon_1 \leq \epsilon_2 \leq ...$ The lowest eigenvalue ϵ_1 may be a good approximation to the ground-state energy E_1 if the state space $\mathcal{V}_K$ is chosen to include the physically most important configurations. Two functions Ψ and Ψ' belonging to the eigenvalues ϵ_i and ϵ_j, respectively, have the overlap

$$\langle \Psi | \Psi' \rangle = c^\dagger \mathcal{B} c', \tag{3.12}$$

where c and c' satisfy the following equation

$$\mathcal{H} c = \epsilon_i \mathcal{B} c, \quad \text{and} \quad \mathcal{H} c' = \epsilon_j \mathcal{B} c'. \tag{3.13}$$

To normalize Ψ, c has to be normalized to $c^\dagger \mathcal{B} c = 1$ and likewise $c'^\dagger \mathcal{B} c' = 1$. Using the Hermiticity of $\mathcal{H}$ and $\mathcal{B}$ and the reality of the eigenvalues ϵ_i in Eq. (3.13) leads one to $(\epsilon_i - \epsilon_j) c^\dagger \mathcal{B} c' = 0$. Thus the two eigenstates belonging to different eigenvalues ϵ_i and ϵ_j are orthogonal. In the case of degeneracy, they can be made orthogonal by an appropriate procedure.

The relationship between the eigenvalues ϵ_i of the truncated problem and the eigenvalues E_n of the full Hamiltonian is elucidated by the Mini-Max theorem.

Theorem 3.3 (Mini-Max Theorem). *Let H be a Hermitian operator with discrete eigenvalues $E_1 \leq E_2 \leq ....$ Let $\epsilon_1 \leq \epsilon_2 \leq ... \leq \epsilon_K$ be the eigenvalues of H restricted to the subspace $\mathcal{V}_K$ of a linearly independent set of K functions $\Psi(\alpha_1), ..., \Psi(\alpha_K)$. Then*

$$E_1 \leq \epsilon_1, \quad E_2 \leq \epsilon_2, \quad ..., \quad E_K \leq \epsilon_K. \tag{3.14}$$

Proof. Let $\mathcal{W}_K$ be the subspace spanned by the orthonormal eigenstates $\Phi_1, \Phi_2, ..., \Phi_K$ of the operator H. We will first show that there is at least one normalized function in $\mathcal{V}_K$ with the property

$$\langle\Psi|H|\Psi\rangle \geq E_K. \tag{3.15}$$

To show this we consider the projected function $P\Psi$ for any normalized function Ψ in $\mathcal{V}_K$, where P is the projection operator that projects onto $\mathcal{W}_K$; $P = \sum_{i=1}^{K} |\Phi_i\rangle\langle\Phi_i|$. There are two possibilities:

(i) there exists a function Ψ_0 in $\mathcal{V}_K$ such that $P\Psi_0 = 0$.
(ii) $P\Psi \neq 0$ for all functions Ψ in $\mathcal{V}_K$.

In case (i) Ψ_0 is a linear combination of $\Phi_{K+1}, \Phi_{K+2}, ...,$ and hence the normalized Ψ_0 has the mean value $\langle\Psi_0|H|\Psi_0\rangle \geq E_{K+1} \geq E_K$. In case (ii) any two different normalized functions Ψ_1 and Ψ_2 in $\mathcal{V}_K$ are projected to different functions $P\Psi_1$ and $P\Psi_2$ in $\mathcal{W}_K$ because otherwise $P(\Psi_1 - \Psi_2) = 0$, which contradicts the assumption. Namely, any two different normalized functions in $\mathcal{V}_K$ are projected by P to different functions in $\mathcal{W}_K$. As both $\mathcal{V}_K$ and $\mathcal{W}_K$ have the same dimension, it follows that there must exist a normalized function Ψ in $\mathcal{V}_K$ with the property $P\Psi = a\Phi_K$ $(a \neq 0)$. Then Ψ can be expressed as $a\Phi_K + b\phi$, where ϕ is a normalized function given as a linear combination of $\Phi_{K+1}, \Phi_{K+2}, ...; |a|^2 + |b|^2 = 1$. This function has the mean value: $\langle\Psi|H|\Psi\rangle = |a|^2 E_K + |b|^2\langle\phi|H|\phi\rangle = E_K + |b|^2(\langle\phi|H|\phi\rangle - E_K) \geq E_K$. Because ϵ_K is the largest eigenvalue of H restricted to $\mathcal{V}_K$, we have

$$\epsilon_K \geq \langle\Psi|H|\Psi\rangle \geq E_K. \tag{3.16}$$

Next we define the $(K-1)$-dimensional state space $\mathcal{V}_{K-1}$ to be the set of all those functions in $\mathcal{V}_K$ which are orthogonal to the eigenvector belonging to ϵ_K and by induction can show $\epsilon_{K-1} \geq E_{K-1}$ and so on.

It is easy to derive the following theorem from the Mini-Max theorem.

Theorem 3.4. *Let $E_1 \leq E_2 \leq ... \leq E_K$ be a number K of the lowest eigenvalues of a Hermitian operator H. Let $\Psi(\alpha_1)$, $\Psi(\alpha_2), ..., \Psi(\alpha_K)$ be a set of linearly independent functions. Then*

$$\sum_{i=1}^{K} E_i \leq \mathrm{Tr}(\mathcal{B}^{-1}\mathcal{H}), \tag{3.17}$$

where the matrices $\mathcal{B}$ and $\mathcal{H}$ are defined by Eq. (3.9).

Proof. Let $\epsilon_1 \leq \epsilon_2 \leq ... \leq \epsilon_K$ be the eigenvalues of H restricted to the subspace $\mathcal{V}_K$ of the K functions $\Psi(\alpha_1), ..., \Psi(\alpha_K)$ and let

$\phi_1, \phi_2, ..., \phi_K$ be the orthonormal eigenstates corresponding to the eigenvalues $\epsilon_1, \epsilon_2, ..., \epsilon_K$. Then ϕ_i is given by

$$\phi_i = \sum_{j=1}^{K} U_{ji} \Psi(\alpha_j) \tag{3.18}$$

with U satisfying the orthonormality condition $U^\dagger \mathcal{B} U = 1$. From Theorem 3.3 we obtain

$$\begin{aligned}\sum_{i=1}^{K} E_i \le \sum_{i=1}^{K} \epsilon_i &= \sum_{i=1}^{K} \langle \phi_i | H | \phi_i \rangle = \mathrm{Tr}(U^\dagger \mathcal{H} U) = \mathrm{Tr}((\mathcal{B}U)^{-1} \mathcal{H} U) \\ &= \mathrm{Tr}(U^{-1} \mathcal{B}^{-1} \mathcal{H} U) = \mathrm{Tr}(\mathcal{B}^{-1} \mathcal{H}). \end{aligned} \tag{3.19}$$

It is noted that the right-hand side of Eq. (3.17) is determined only by the subspace spanned by the functions and not by the particular choice of the basis because any non-singular linear transformation of the basis does not change the trace (3.17). Namely, when a new basis set $\chi_1, ..., \chi_K$ is given by a transformation $\chi_i = \sum_{j=1}^{K} T_{ij} \Psi(\alpha_j)$ $(i = 1, ..., K)$ with a non-singular matrix T $(\det T \neq 0)$, we can show that $\mathrm{Tr}(\mathcal{B}'^{-1} \mathcal{H}') = \mathrm{Tr}(\mathcal{B}^{-1} \mathcal{H})$, where $\mathcal{B}'_{ij} = \langle \chi_i | \chi_j \rangle$ and $\mathcal{H}'_{ij} = \langle \chi_i | H | \chi_j \rangle$. This leads to the density-functional formalism where the density can be used as a basic variable for calculating the energies of excited states as well as the ground state. See [4, 5] for this interesting idea.

The next theorem is fundamental to answer a question of how the eigenvalues obtained in a restricted subspace change as the basis dimension is increased.

Theorem 3.5. *Let $\epsilon_1 \le \epsilon_2 \le ... \le \epsilon_K$ be the eigenvalues of a Hermitian operator H restricted to the subspace $\mathcal{V}_K$ of linear combinations of independent functions $\Psi(\alpha_1)$, $\Psi(\alpha_2), ..., \Psi(\alpha_K)$. Let $\epsilon'_1 \le \epsilon'_2 \le ... \le \epsilon'_{K+1}$ be the eigenvalues of H restricted to the subspace $\mathcal{V}_{K+1}$ of linear combinations of independent functions $\Psi(\alpha_1), \Psi(\alpha_2), ..., \Psi(\alpha_K)$, and $\Psi(\alpha_{K+1})$. Then*

$$\epsilon'_1 \le \epsilon_1 \le \epsilon'_2 \le \epsilon_2 \le ... \le \epsilon'_K \le \epsilon_K \le \epsilon'_{K+1}. \tag{3.20}$$

Proof. Let $\phi_1, \phi_2, ..., \phi_K$ be the orthonormal eigenstates corresponding to the eigenvalues $\epsilon_1, \epsilon_2, ..., \epsilon_K$. Any function Ψ in $\mathcal{V}_{K+1}$ can be expressed as

$$\Psi = \sum_{i=1}^{K+1} c_i \phi_i, \tag{3.21}$$

where the normalized function ϕ_{K+1} is constructed from $\Psi(\alpha_{K+1})$ to make it orthogonal to any function in $\mathcal{V}_K$:

$$\phi_{K+1} = \left(\langle \Psi(\alpha_{K+1}) | \Psi(\alpha_{K+1}) \rangle - \sum_{i=1}^{K} |\langle \phi_i | \Psi(\alpha_{K+1}) \rangle|^2 \right)^{-\frac{1}{2}}$$
$$\times \left\{ \Psi(\alpha_{K+1}) - \sum_{i=1}^{K} \phi_i \langle \phi_i | \Psi(\alpha_{K+1}) \rangle \right\}. \tag{3.22}$$

The eigenvalue problem using the basis set $\{\phi_1, \phi_2, ..., \phi_{K+1}\}$ takes the simple form

$$\begin{pmatrix} \epsilon_1 & 0 & \cdots & 0 & h_1 \\ 0 & \epsilon_2 & \cdots & 0 & h_2 \\ \vdots & \vdots & & \vdots & \vdots \\ 0 & 0 & \cdots & \epsilon_K & h_K \\ h_1^* & h_2^* & \cdots & h_K^* & h_{K+1} \end{pmatrix} \begin{pmatrix} c_1 \\ c_2 \\ \vdots \\ c_K \\ c_{K+1} \end{pmatrix} = E \begin{pmatrix} c_1 \\ c_2 \\ \vdots \\ c_K \\ c_{K+1} \end{pmatrix}, \tag{3.23}$$

where $h_j = \langle \phi_j | H | \phi_{K+1} \rangle$ and h_j^* is the complex conjugate of h_j. The characteristic function $D(E)$ to determine the eigenvalues reads as

$$D(E) \equiv (h_{K+1} - E) \prod_{i=1}^{K} (\epsilon_i - E) - \sum_{i=1}^{K} \left(|h_i|^2 \prod_{\substack{j=1 \\ j \neq i}}^{K} (\epsilon_j - E) \right)$$
$$= \left(\prod_{i=1}^{K} (\epsilon_i - E) \right) \left(h_{K+1} - E - \sum_{j=1}^{K} \frac{|h_j|^2}{\epsilon_j - E} \right) = 0. \tag{3.24}$$

Here it is assumed that all h_i $(i = 1, ..., K)$ are nonzero. The $(K+1)$ eigenvalues are obtained by finding the roots of the equation

$$E - h_{K+1} = \sum_{i=1}^{K} \frac{|h_i|^2}{E - \epsilon_i}. \tag{3.25}$$

The right-hand side of the above equation becomes negative for $E < \epsilon_1$ and positive for $E > \epsilon_K$. A graphic solution of Eq. (3.25) is displayed in Fig. 3.1. It is clear that the cross points of curves, namely the new eigenvalues ϵ_i', satisfy the relation (3.20).

When there are n vanishing h_i's, the corresponding ϵ_i's become the solutions of Eq. (3.23), i.e., $\epsilon_i' = \epsilon_i$. The remaining $(K+1-n)$ new

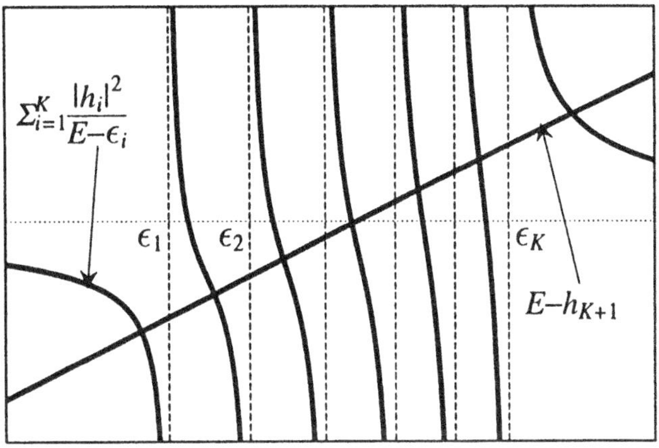

Fig. 3.1. Graphic solution of Eq. (3.25) to obtain the eigenvalues of Eq. (3.23)

eigenvalues are obtained through the eigenvalue equation, which has the same structure as Eq. (3.23). The relation (3.20) trivially holds also in this case.

The significance of this theorem is that, by including a further term in the basis, the K lowest eigenvalues cannot become worse. Therefore, this theorem implies the Mini-Max theorem because, in the limiting case that the subspace approaches the full Hilbert space as the basis dimension increases, the K lowest eigenvalues ϵ_i converge to the exact eigenvalues of the Hamiltonian.

Suppose that one wants to see how the lowest eigenvalue changes when one varies only the $(K+1)$th function $\Psi(\alpha_{K+1})$ while keeping the rest of all the K functions unchanged. In this case one does not need to solve the generalized eigenvalue problem of type (3.10) but only needs to find the smallest root of Eq. (3.25). Of course the latter is much simpler and faster than diagonalizing the matrix. This advantage is used to sample more random trials in selecting a suitable basis function in the stochastic variational method, as explained in Sects. 4.2.5 and 4.2.6.

3.2 The variance of local energy

It would be nice if there were a method to judge the accuracy of the solution obtained variationally. The expectation value E is the upper bound of the ground-state energy E_1. If we can calculate a lower bound, the difference between the two bounds gives an estimate for how close E may be to E_1. This makes sense if and only if the lower bound can be calculated as closely as possible to the ground-state energy. In order to discuss the lower bound, we define the variance of the energy expectation value, σ^2, by the norm of the residue function $(H-E)\Psi/\langle\Psi|\Psi\rangle^{\frac{1}{2}}$

$$\sigma^2 \equiv \frac{\langle\Psi|(H-E)^2|\Psi\rangle}{\langle\Psi|\Psi\rangle} = \frac{\langle\Psi|H^2|\Psi\rangle}{\langle\Psi|\Psi\rangle} - E^2. \tag{3.26}$$

According to Weinstein [6, 7] the following theorem can be proved.

Theorem 3.6. *There is at least one exact eigenvalue in the interval* $[E-\sigma, E+\sigma]$.

Proof. By using Eq. (3.3), the variance can be expressed as

$$\sigma^2 = \frac{\sum_{i=1}^{\infty}(E_i - E)^2|a_i|^2}{\sum_{i=1}^{\infty}|a_i|^2}. \tag{3.27}$$

Suppose that E_k is the eigenvalue which is closest to E. Then we have

$$\sum_{i=1}^{\infty}(E_i - E)^2|a_i|^2 \geq (E_k - E)^2\sum_{i=1}^{\infty}|a_i|^2. \tag{3.28}$$

Thus we have $\sigma^2 \geq (E_k - E)^2$, which proves the theorem.

The Weinstein criterion guarantees that there is an eigenstate whose energy is in the interval $[E-\sigma, E+\sigma]$ but does not indicate which one. In case E is sufficiently close to the ground-state energy, the theorem gives the lower bound as

$$E - \sigma \leq E_1. \tag{3.29}$$

Another lower bound called Temple's bound [8] is also expressed in terms of the variance of the energy as

$$E - \frac{\sigma^2}{\varepsilon - E} \leq E_1. \tag{3.30}$$

Here ε is an arbitrary energy such that $E < \varepsilon \le E_2$, where E_2 is the energy of the first excited state which has the same conserved quantum numbers as the ground state. The best bound is clearly obtained by setting ε equal to E_2. Temple's bound is easily derived as follows:

$$\begin{aligned} E_1 - \left(E - \frac{\sigma^2}{\varepsilon - E}\right) &= \frac{1}{\varepsilon - E}\left[-(\varepsilon + E_1)(E - E_1) + \frac{\langle\Psi|H^2|\Psi\rangle}{\langle\Psi|\Psi\rangle} - E_1^2\right] \\ &= \frac{1}{\varepsilon - E}\,\frac{\sum_{i=2}^{\infty}(E_i - \varepsilon)(E_i - E_1)|a_i|^2}{\sum_{i=1}^{\infty}|a_i|^2} \ge 0, \end{aligned} \tag{3.31}$$

where use is made of Eq. (3.4) for $E - E_1$ (with $n = 1$) and for $\langle\Psi|H^2|\Psi\rangle / \langle\Psi|\Psi\rangle - E_1^2$ (with $n = 2$).

The two bounds, (3.29) and (3.30), indicate that the trial function that gives the smallest value of σ^2 is best among various trial functions giving the same expectation value. Temple's bound often gives better lower bounds than the Weinstein bound. See [9] for the extension of variational bounds.

The variance of the expectation value can also be defined through the local energy

$$E_{\rm loc}(\boldsymbol{R}) = \frac{H\Psi}{\Psi}, \tag{3.32}$$

where $\boldsymbol{R}$ stands for the "configuration" of the particles. More precisely, if the coordinates of the ith particle are denoted by $\boldsymbol{r}_i$ and ρ_i, where ρ_i stands for the coordinates other than the position coordinate $\boldsymbol{r}_i$, then $\boldsymbol{R}$ stands for $\{\boldsymbol{r}_1, \rho_1, \boldsymbol{r}_2, \rho_2, \ldots\}$. The variance defined by Eq. (3.26) is equal to the variance of the local energy

$$\sigma^2 = \frac{\langle\Psi||E_{\rm loc}(\boldsymbol{R}) - E|^2|\Psi\rangle}{\langle\Psi|\Psi\rangle} = \int P(\boldsymbol{R})|E_{\rm loc}(\boldsymbol{R}) - E|^2 d\boldsymbol{R} \tag{3.33}$$

with

$$P(\boldsymbol{R}) = \frac{|\Psi(\boldsymbol{R})|^2}{\langle\Psi|\Psi\rangle}, \tag{3.34}$$

where E, the mean energy of H in the state Ψ, can also be rewritten as an average of the local energy:

$$E = \frac{\langle\Psi|E_{\rm loc}(\boldsymbol{R})|\Psi\rangle}{\langle\Psi|\Psi\rangle} = \int P(\boldsymbol{R})E_{\rm loc}(\boldsymbol{R})d\boldsymbol{R}. \tag{3.35}$$

Here $P(\boldsymbol{R})$, the probability density of finding the system at the configuration point $\boldsymbol{R}$, is non-negative and satisfies the relation $\int P(\boldsymbol{R})d\boldsymbol{R} = 1$.

If the trial function Ψ is the exact ground state of the Hamiltonian, the local energy becomes $\boldsymbol{R}$-independent and equals the ground-state energy E_1. A similar statement holds for excited states as well. If Ψ is the exact eigenstate with energy E_n, the local energy would be equal to E_n. This opens up another possibility of the variational method instead of the minimization of the mean energy: The minimization of the variance of the local energy [10]. The adjustable parameters of the trial function may be varied to minimize the variance. The calculation of the variance is in general much more difficult than that of the mean energy. An advantage of minimizing the variance of the local energy is, however, that the quantity to be minimized has a known lower bound, namely zero. Since in any eigenstate the variance of the local energy is zero, this minimization principle applies to obtain excited states as well as the ground state.

The stationarity condition for σ^2 is obtained by calculating an increment $\delta\sigma^2$ in the same way as was done for the mean energy in Theorem 3.2. By multiplying both sides of Eq. (3.26) by $\langle\Psi|\Psi\rangle$ and calculating the increment to a linear change in $\delta\Psi$, we obtain the following equation

$$\begin{aligned}\langle\Psi|\Psi\rangle\delta\sigma^2 &= \delta(\langle\Psi|H^2|\Psi\rangle) - \delta(E^2\langle\Psi|\Psi\rangle) - \sigma^2\delta(\langle\Psi|\Psi\rangle)\\ &= \langle\delta\Psi|H^2 - E^2 - \sigma^2|\Psi\rangle + \langle\Psi|H^2 - E^2 - \sigma^2|\delta\Psi\rangle - 2E\langle\Psi|\Psi\rangle\delta E\\ &= \langle\delta\Psi|H^2 - 2EH + E^2 - \sigma^2|\Psi\rangle\\ &\quad + \langle\Psi|H^2 - 2EH + E^2 - \sigma^2|\delta\Psi\rangle. \end{aligned} \tag{3.36}$$

Here use is made of Eq. (3.5) in the last step. The stationarity condition is equivalent to requiring $(H^2 - 2EH + E^2)\Psi = \sigma^2\Psi$. When the trial function is given as a sum of a finite number of functions $\Psi(\alpha)$ as in Eq. (3.7), this condition reads as the following equation which can be used to determine the linear parameters c_i

$$(\mathcal{Q} - 2E\mathcal{H} + E^2\mathcal{B})c = \sigma^2\mathcal{B}c, \qquad \mathcal{Q}_{ij} = \langle\Psi(\alpha_i)|H^2|\Psi(\alpha_j)\rangle. \tag{3.37}$$

Since E defined by Eq. (3.8) depends on c, the matrix in the round bracket of the above equation also depends on c. Therefore the parameter c must be determined self-consistently: One assumes an initial value for c, calculates the energy E, and solves Eq. (3.37) to determine c and σ^2. As the value c obtained in this way may in general not be equal to the initial c value, one now uses the solution c as the initial value and repeats this cycle until both values of c become the same. The minimization of σ^2 becomes cumbersome even though the calculation of Q_{ij} is feasible.

The variational Monte Carlo (VMC) calculations employ Eq. (3.35) to determine the mean energy. The VMC method [11], or the method of the Amalgamation of Two-body correlations into Multiple Scattering (ATMS) [12] similar to it, usually chooses a trial function that attempts to incorporate both the short distance behavior and the asymptotic behavior of the correct wave function, leading to complicated functional forms. Thus the analytic evaluation of the matrix elements is in general hopeless. Instead, the VMC requires only the evaluation of the wave function and its Laplacian to determine the local energy. If the interaction between the particles contains the spin-orbit force, the first derivatives of the wave function are required as well. These derivatives are obtained numerically by appropriate difference formulas. The mean energy is then estimated by a Monte Carlo integration. The sampling of a set of configuration points $\boldsymbol{R}$ is made by, e.g. the Metropolis algorithm [13]. The accuracy of the VMC or ATMS calculation is limited by the choice of the trial function as well as by the statistical error inherent in the Monte Carlo integration. To reduce the statistical error one has to sample a great number of points, which is computer time consuming. To optimize the variational parameters of the trial wave function one has to repeat the mean energy calculation using a different set of parameters. This is a hard job considering that the VMC or ATMS has an inherent statistical error particularly in the case one has a number of parameters.

The variational method is a very flexible approximation method which can be adapted to diverse problems. The variational method is particularly valuable when physical intuitions lead us to an idea of the qualitative form of the solution. It easily gives good values for the energy. The variational solution may, however, present unpredictable erroneous values for other physical observables. It is unfortunately very difficult to evaluate their error.

3.3 The virial theoren

In this section we summarize two theorems, the Hellmann-Feynman theorem and the virial theorem, which are valid for the exact solution. They can be used to check the quality of the wave function.

Theorem 3.7 (Hellmann-Feynman Theorem). *Let $H(\lambda)$ be a Hermitian operator which depends on a real parameter λ, and $\Phi(\lambda)$ a normalized eigenstate of $H(\lambda)$ of eigenvalue $E(\lambda)$. Then the theorem states that*

$$\frac{d}{d\lambda}E(\lambda) = \langle\Phi(\lambda)|\frac{\partial}{\partial\lambda}H(\lambda)|\Phi(\lambda)\rangle. \tag{3.38}$$

Proof. Differentiating, with respect to λ, the relation

$$E(\lambda) = \langle\Phi(\lambda)|H(\lambda)|\Phi(\lambda)\rangle, \tag{3.39}$$

we have

$$\frac{d}{d\lambda}E(\lambda) = \langle\Phi(\lambda)|\frac{\partial}{\partial\lambda}H(\lambda)|\Phi(\lambda)\rangle + \langle\frac{\partial}{\partial\lambda}\Phi(\lambda)|H(\lambda)|\Phi(\lambda)\rangle + \langle\Phi(\lambda)|H(\lambda)|\frac{\partial}{\partial\lambda}\Phi(\lambda)\rangle. \tag{3.40}$$

The last two terms on the right-hand side of Eq. (3.40) vanish because

$$\begin{aligned}
&\langle\frac{\partial}{\partial\lambda}\Phi(\lambda)|H(\lambda)|\Phi(\lambda)\rangle + \langle\Phi(\lambda)|H(\lambda)|\frac{\partial}{\partial\lambda}\Phi(\lambda)\rangle \\
&= E(\lambda)\Big[\langle\frac{\partial}{\partial\lambda}\Phi(\lambda)|\Phi(\lambda)\rangle + \langle\Phi(\lambda)|\frac{\partial}{\partial\lambda}\Phi(\lambda)\rangle\Big] \\
&= E(\lambda)\frac{d}{d\lambda}\langle\Phi(\lambda)|\Phi(\lambda)\rangle = 0,
\end{aligned} \tag{3.41}$$

where use is made of the conditions that $H(\lambda)\Phi(\lambda) = E(\lambda)\Phi(\lambda)$ and $\langle\Phi(\lambda)|\Phi(\lambda)\rangle = 1$.

When the evaluation of the matrix element of $\frac{\partial}{\partial\lambda}H(\lambda)$ does not pose any difficulty, the Hellmann-Feynman theorem can be used to check the consistency of the variational solution. If $H(\lambda) = H_0 + \lambda H_1$, then the matrix elements needed are just the terms of the energy expectation value. It is straightforward to generalize the Hellmann-Feynman theorem to the case that the Hamiltonian contains a number of parameters $\lambda_1, \lambda_2, \ldots$.

We note that the Hellmann-Feynman theorem can be stated in an integral form

$$E(\lambda_2) - E(\lambda_1) = \frac{\langle\Phi(\lambda_2)|H(\lambda_2) - H(\lambda_1)|\Phi(\lambda_1)\rangle}{\langle\Phi(\lambda_2)|\Phi(\lambda_1)\rangle}$$

$$= \int_{\lambda_1}^{\lambda_2} \langle \Phi(\lambda)|\frac{\partial}{\partial \lambda} H(\lambda)|\Phi(\lambda)\rangle d\lambda. \tag{3.42}$$

The first equality is trivial for the exact solution and has nothing to do with the Hellmann-Feynman theorem. The equality holds, however, only approximately for the variational solution and may be used to check the quality of the solution.

The following theorem plays a key role in the derivation of the virial theorem.

Theorem 3.8. *Let Φ be an eigenstate of a Hermitian operator H. For any operator A we obtain*

$$\langle \Phi|[H, A]|\Phi\rangle = 0. \tag{3.43}$$

The proof is easily obtained by noting $H\Phi = E\Phi$ with eigenvalue E. To test the accuracy of the variational solution, one often uses a special operator of the form

$$A = \frac{i}{\hbar} \sum_{i=1}^{N} \boldsymbol{r}_i \cdot \boldsymbol{p}_i = \sum_{i=1}^{N} \boldsymbol{r}_i \cdot \frac{\partial}{\partial \boldsymbol{r}_i}, \tag{3.44}$$

because the commutator $[H, A]$ is easily calculated. We note that the operator A is a generator of the dilation operator, e^{aA}, which has the property, in a single-variable case,

$$\exp\left(ax\frac{d}{dx}\right) f(x) = f(\mathrm{e}^a x). \tag{3.45}$$

This is easily shown by expanding $f(x)$ in power series and using $(x\frac{d}{dx})^k x^n = n^k x^n$, valid for any integer value of n.

Assume that the Hamiltonian for an N-particle system takes the form

$$H = T + W = \sum_{i=1}^{N} \frac{\boldsymbol{p}_i^2}{2m_i} - T_{\mathrm{cm}} + W(\boldsymbol{r}_1, \boldsymbol{r}_2, ..., \boldsymbol{r}_N). \tag{3.46}$$

Whether or not the center-of-mass kinetic energy is subtracted from the Hamiltonian is irrelevant to the discussion on the virial theorem. The potential W is assumed to contain no momentum operators. We can then show that

$$\left[T,\ \boldsymbol{r}_i \cdot \frac{\partial}{\partial \boldsymbol{r}_i}\right] = 2\left(\frac{\boldsymbol{p}_i^2}{2m_i} - \frac{1}{2m_{12\cdots N}} \boldsymbol{p}_i \cdot \boldsymbol{\pi}_N\right),$$

$$\left[W,\, \boldsymbol{r}_i\cdot\frac{\partial}{\partial \boldsymbol{r}_i}\right] = -\boldsymbol{r}_i\cdot\frac{\partial W}{\partial \boldsymbol{r}_i}, \tag{3.47}$$

where $m_{12\cdots N} = \sum_{i=1}^{N} m_i$ is the total mass of the system and $\boldsymbol{\pi}_N = \sum_{i=1}^{N} \boldsymbol{p}_i$ is the total momentum. Using Eqs. (3.44), (3.46) and (3.47) we obtain

$$[H, A] = 2T - W_A \qquad \text{with} \qquad W_A = \sum_{i=1}^{N} \boldsymbol{r}_i\cdot\frac{\partial W}{\partial \boldsymbol{r}_i}. \tag{3.48}$$

Substituting this into Eq. (3.43) leads to the following theorem.

Theorem 3.9 (The virial Theorem). *Let Φ be an eigenstate of the Hamiltonian (3.46). For the operator A of Eq. (3.44) we obtain*

$$2\langle\Phi|T|\Phi\rangle - \langle\Phi|W_A|\Phi\rangle = 0. \tag{3.49}$$

Therefore a quantity η defined by

$$\eta = \left|\frac{\langle\Phi|W_A|\Phi\rangle}{2\langle\Phi|T|\Phi\rangle} - 1\right| \tag{3.50}$$

vanishes for the exact solution and can be used to check the quality of the solution.

One has to know W_A to make use of the virial theorem. It becomes particularly simple if the potential is a homogeneous function of degree s, namely

$$\begin{aligned} W(\lambda\boldsymbol{r}_1, \lambda\boldsymbol{r}_2, ..., \lambda\boldsymbol{r}_N) &\equiv W(\lambda x_1, \lambda y_1, \lambda z_1, ..., \lambda x_N, \lambda y_N, \lambda z_N) \\ &= \lambda^s W(\boldsymbol{r}_1, \boldsymbol{r}_2, ..., \boldsymbol{r}_N), \end{aligned} \tag{3.51}$$

because then we obtain by the use of Euler's theorem or by differentiating, with respect to λ, both sides of Eq. (3.51) and setting λ equal to $\lambda = 1$

$$W_A = sW. \tag{3.52}$$

In this special case Eq. (3.49), together with $\langle\Phi|T|\Phi\rangle + \langle\Phi|W|\Phi\rangle = E$, yields the well-known relation

$$\langle\Phi|T|\Phi\rangle = \frac{s}{s+2}E, \quad \langle\Phi|W|\Phi\rangle = \frac{2}{s+2}E \quad (s \neq -2). \tag{3.53}$$

Because the homogeneity condition is fulfilled for the Coulomb potential ($s = -1$), the quantity η is conveniently used to check the quality of the solution for a system of particles interacting via Coulomb potentials.

For a general non-homogeneous potential function of the form

$$W(\boldsymbol{r}_1, \boldsymbol{r}_2, ..., \boldsymbol{r}_N) = \sum_{i=1}^{N} U_i(\boldsymbol{r}_i) + \sum_{j>i=1}^{N} V_{ij}(\boldsymbol{r}_i - \boldsymbol{r}_j), \tag{3.54}$$

Eq. (3.52) must be replaced by

$$W_\Lambda = \sum_{i=1}^{N} \left(\boldsymbol{r} \cdot \frac{\partial U_i(\boldsymbol{r})}{\partial \boldsymbol{r}} \right)_{\boldsymbol{r}=\boldsymbol{r}_i} + \sum_{j>i=1}^{N} \left(\boldsymbol{r} \cdot \frac{\partial V_{ij}(\boldsymbol{r})}{\partial \boldsymbol{r}} \right)_{\boldsymbol{r}=\boldsymbol{r}_i - \boldsymbol{r}_j}. \tag{3.55}$$

For a spherically symmetric potential the operator $\boldsymbol{r} \cdot \frac{\partial}{\partial \boldsymbol{r}}$ reduces simply to $r \frac{d}{dr}$. If the calculation of matrix elements of the operator W_Λ is not difficult, the virial theorem can be used as in the Coulombic case.

When the potential W depends on a parameter λ, we obtain

$$\langle \Phi | W_\Lambda | \Phi \rangle = \langle \Phi | W_\Lambda + \lambda \frac{\partial}{\partial \lambda} W | \Phi \rangle - \lambda \frac{d}{d\lambda} E(\lambda). \tag{3.56}$$

Here the Hellmann-Feynman theorem is used to express the expectation value of $\lambda \frac{\partial}{\partial \lambda} W$ in terms of the energy eigenvalue $E(\lambda)$. In a molecular system λ may be a set of parameters which stand for the positions of nuclei and if W consists of only the Coulomb potentials $W_\Lambda + \lambda \frac{\partial}{\partial \lambda} W$ simply reduces to $-W$. See Complement 8.2 for an application of this relation.

4. Stochastic variational method

The most direct approach to the variational solution of quantum-mechanical bound-state problems is to diagonalize the Hamiltonian in a state space spanned by some appropriate functions $\Psi(\alpha_1)$, $\Psi(\alpha_2)$,..., $\Psi(\alpha_K)$. The applicability of this approach is, however, very limited, because the diagonalization may not be feasible if the dimension K of the state space is very large. This is typical in many-particle problems such as the Hubbard model or the nuclear shell model. This method can be called a "direct approach", because one sets up a basis in a well-defined way, e.g., by using a complete set of states that contains no parameters, and then obtains the energy by a diagonalization.

Another possibility is basis optimization, which is actually designed to avoid the problem of the huge basis dimension in the direct approach. In this case one specifically selects the basis states that are really essential to get the energy and the wave function of the system to a certain accuracy. It is obvious that this selection may be state-dependent: Some functions might be adequate to describe the ground state, but some others would be more appropriate to approximate an excited state, particularly when the ground state and the excited state have different spatial extensions. The stochastic variational method uses this second route by selecting the most appropriate basis functions in a trial and error procedure.

4.1 Basis optimization

It goes without saying that the quality of the variational approximation crucially depends on the choice of the basis functions. Our primary aim here is to solve N-particle problems. For this we will choose basis functions that meet the following requirements:

1. They can be easily generalized for an N-body system.

2. Their matrix elements are analytically calculable.
3. They are easily adaptable to the permutational symmetry of the system.
4. They are flexible enough to approximate even rapidly changing functions.

The important condition (3) is non-trivial because the permutational symmetry looks very complicated when expressed in terms of the relative coordinates to be used.

A possible choice for the basis functions fulfilling the above conditions is the correlated Gaussian of Eqs. (2.23)–(2.25):

$$\exp\Big(-\frac{1}{2}\tilde{\boldsymbol{x}}A\boldsymbol{x}\Big) \equiv \exp\Big(-\frac{1}{2}\sum_{i=1}^{N-1}\sum_{j=1}^{N-1} A_{ij}\boldsymbol{x}_i\cdot\boldsymbol{x}_j\Big)$$

$$= \exp\Big\{-\frac{1}{2}\sum_{j>i=1}^{N} \alpha_{ij}(\boldsymbol{r}_i-\boldsymbol{r}_j)^2\Big\}. \tag{4.1}$$

The matrix elements A_{ij} or, equivalently, α_{ij} are nonlinear parameters of the basis. These functions are spherically symmetric. To take into account non-spherical states, the basis function has to be multiplied by an appropriate orbital angular function. In addition to the spatial degree of freedom, the particles may have other degrees of freedom, such as spin and flavor, and thus the function has also to be multiplied by suitable trial functions in these additional spaces. These functions may bring other parameters or sets of quantum numbers. These sets of quantum numbers (e.g., total and intermediate spins, orbital angular momenta, etc.) can be considered as discrete parameters, and will often be referred to as channels in the following. (More details on the trial function will be given in Chap. 6.)

The actual form of the basis function is not very important at this stage. The above discussion serves to draw the reader's attention to the fact that the basis function depends on many linear, nonlinear and discrete parameters. The parameters define the shape of the basis function and determine how well the variational function space contains the true eigenfunction. To find the best possible solution, one has to optimize the parameters. To have a crude guess of how much work the optimization amounts to, let us consider an N-particle system with the simplest basis function (4.1). This correlated Gaussian has $N(N-1)/2$ parameters, and, by assuming that we need a linear combination of K functions, we face an optimization problem of

$K(N(N-1)/2)$ parameters. This number increases quadratically with the number of particles. By taking $N = 4$ and $K = 200$ as a typical case, we end up with 1200 parameters.

The main problem of the minimization of a function is the omnipresence of local minima. A local minimum is the point where the function reaches a minimum in a finite interval of variables and the number of such minima tends to increase exponentially with the size of the problem. There are plenty of different methods for the optimization of a function, and these optimizations can be divided into two categories: the deterministic and the stochastic optimizations.

A deterministic optimization moves downwards on the slope of the function according to a certain well-defined strategy. There exist many elaborated algorithms (conjugate gradient, Powell (direction set), etc. [14]) and they are deterministic in the sense that, starting from a given point, they always reach the same (local or global) minimum. The drawback of these techniques is that they are time consuming and tend to converge to whichever local minimum they first encounter. The solution in these cases may not be the global minimum but a local minimum. These methods are sensitive to the starting point and are unable to search further for a better solution after a local minimum is reached.

Stochastic optimizations address the problem of finding the global minimum in the presence of a large number of undesired local minima by making random steps [15, 16]. One can, for example, start deterministic searches from several random starting points and then pick up the minimum of these. Numerous strategies have been developed in the last few years, e.g., simulated annealing [17], genetic algorithm [18] etc. The simplest (and actually not very economical) stochastic optimization is a random search where one picks up random points and tries to find the minimum. This may not sound very sophisticated, but the optimization with a random trial and error procedure seems to be the most efficient one. In some cases it would be nearly impossible to apply other strategies than simple random trials.

To reduce the load of optimization, an alternative is to shift the burden of minimization from a large number of parameters to a smaller number of more sensitive, "tempering", parameters. Several such possibilities have been explored, e.g., geometric progression [19, 20], random tempering [21, 22, 23], and Chebyshev grid [24]. The common property of these methods is that they use a "grid" in the parameter space. The grid is given by some ad hoc rule and each point of the grids defines a basis function. The grid can be defined by some simple functions which may depend on some additional parameters to be optimized. The number of parameters contained in the functions which define the grid is chosen to be much smaller than that of the original function to be optimized. For example, in the case of a three-particle system, each of the basis functions has three parameters: A_{11}, $A_{12} = A_{21}$ and A_{22} or α_{12}, α_{13} and α_{23}. Some convenient choices for basis parameters are shown in Table 4.1.

Table 4.1. Choice of parameters. In the random tempering p is an index to denote a prime number that is used to generate pseudo-random numbers. See Eq. (4.8) for the notation $< i, j >$.

Geometric progression	
$A_{11} = \left(a_1 q_1^{k-1}\right)^{-2}$	$(k = 1, \ldots, m_1)$
$A_{22} = \left(a_2 q_2^{k-1}\right)^{-2}$	$(k = 1, \ldots, m_2)$
$A_{12} = 0$	
Random tempering	
$\alpha_{12} = \exp\big(d_1 < k,p > + d_2 < k, p+1 >\big)$	$(k = 1, \ldots, m_1)$
$\alpha_{13} = \exp\big(d_3 < k,p+2 > + d_4 < k, p+3 >\big)$	$(k = 1, \ldots, m_2)$
$\alpha_{23} = \exp\big(d_5 < k,p+4 > + d_6 < k, p+5 >\big)$	$(k = 1, \ldots, m_3)$
Chebyshev grid	
$A_{11} = a_1 \tan\left(\frac{\pi}{2} \frac{2k-1}{2m_1}\right)$	$(k = 1, \ldots, m_1)$
$A_{22} = a_2 \tan\left(\frac{\pi}{2} \frac{2k-1}{2m_2}\right)$	$(k = 1, \ldots, m_2)$
$A_{12} = 0$	

The basic problem of the grid methods is that they distribute a large number of parameters over the physically realistic interval irrespective of whether these points really contribute to the solution or not. Therefore, they lead to a prohibitively large basis dimension as the number of particles increases.

The behavior of the wave function can be quite different from place to place in the configuration space. It can be very smooth in some places (and that part might be described by relatively few functions) and it may change rapidly in other regions (where densely packed basis functions might be necessary). This feature is not a priori known and it cannot be followed by grid-type bases. In the optimal case, one would distribute a certain number of basis functions to approximate the bulk part of the wave function and then one might increase the density of the basis elements in regions where finer resolution is needed.

4.2 A practical example

To illustrate the point of the previous section, let us consider an example for basis optimization. The example will at the same time provide insight into some other aspects of the methods as well. We consider the system of a positron and two electrons, Ps^-, used in Chap. 1. The energy of this system has been very accurately calculated by various approaches and it has been found to be −0.262005 in atomic units (a.u.). (Energy and length are given in a.u. throughout this chapter unless otherwise mentioned.) The quality of the different optimization strategies used in this section can be judged by comparing their results with this value. Let the positron be labelled 1 and the electrons labelled 2 and 3. With the electron spins coupled to $S = 0$, the antisymmetry requires that the orbital part of the trial function ought to be symmetric with respect to the interchange of the electron coordinates. The wave function is thus expanded as

$$\Psi = \sum_{k=1}^{K} c_k \, (1 + P_{23}) \exp\Bigl(- \frac{1}{2} \tilde{\boldsymbol{x}} A_k \boldsymbol{x} \Bigr), \tag{4.2}$$

where the exchange operator P_{23} is introduced to assure the symmetry requirement. The coordinates are $\boldsymbol{x}_1 = \boldsymbol{r}_1 - \boldsymbol{r}_2$ and $\boldsymbol{x}_2 = (\boldsymbol{r}_1 + \boldsymbol{r}_2)/2 - \boldsymbol{r}_3$, and A_k is a 2×2 symmetric matrix with three nonlinear parameters A_{11}, A_{12} and A_{22}. The wave function Ψ has $3K$ parameters to be optimized. The linear parameter c_k is determined by solving the generalized eigenvalue problem (3.10). To reach the accuracy required in atomic physics, the minimum basis size of a three-body problem is at least $K = 100$, as will be shown later. That means that even in this simple example we would have 300 parameters. This is already almost beyond the capability of most of the computer codes for optimization and this "full" optimization is out of the question for larger systems. The fact that the optimization of a large number of parameters is not feasible is just a small part of the problem. To have an efficient optimization one needs fast function evaluation as many times as required. There are two steps which are necessary to calculate the energy expectation value: The calculation of the overlap and Hamiltonian matrix elements and the diagonalization. In a full optimization, one has to recalculate all matrix elements (the required time increases as K^2) and to rediagonalize the Hamiltonian (a K^3-process) at each calculation of the energy. This is a considerably heavy computational load even for a small system.

Instead, one can try a partial optimization. One can, for example, fix the parameters of all basis states but one. In this partial optimization only one row (column) of the Hamiltonian and the overlap matrix has to be recalculated (the required time is of order K). Let us assume that the Hamiltonian has been diagonalized over the fixed basis states. Then, in the successive step, only one row (column) is changed. As has been shown in Theorem 3.5, after the $N \times N$ diagonalization there is no need for an extra diagonalization to solve the eigenvalue problem on the $(N+1)$-dimensional basis. Consequently, the computational time required by the partial optimization is only a small fraction of that of the full optimization, and, moreover, only a small number of parameters (three in the present example) has to be optimized.

4.2.1 Geometric progression

First let us try to use a grid defined by a geometric progression. One can define three different sets of Jacobi coordinates, depending on which two particles are connected first: (23)1, (31)2, and (12)3. See Sects. 2.4 and 7.6 and Fig. 2.2. Let us denote the Jacobi coordinates in these systems by $\boldsymbol{x}^{(1)}$, $\boldsymbol{x}^{(2)}$ and $\boldsymbol{x}^{(3)}$. The trial function is assumed to be given in the form

$$\Psi = \sum_{p=1}^{3} \sum_{k=1}^{K} \sum_{l} c_{kl}^{(p)} \exp\Big(-\frac{1}{2} \widetilde{\boldsymbol{x}^{(p)}} A_k^{(p)} \boldsymbol{x}^{(p)} \Big) \theta_{(ll)00}(\boldsymbol{x}^{(p)}), \tag{4.3}$$

with the angular function

$$\begin{aligned} \theta_{(ll)00}(\boldsymbol{x}) &= [\mathcal{Y}_l(\boldsymbol{x}_1) \times \mathcal{Y}_l(\boldsymbol{x}_2)]_{00} \\ &= \sum_{m=-l}^{l} \frac{(-1)^{l-m}}{\sqrt{2l+1}} \mathcal{Y}_{lm}(\boldsymbol{x}_1) \mathcal{Y}_{l\,-m}(\boldsymbol{x}_2). \end{aligned} \tag{4.4}$$

Here p denotes the arrangement channel and $\mathcal{Y}_{lm}(\boldsymbol{r})$ is a solid spherical harmonic

$$\mathcal{Y}_{lm}(\boldsymbol{r}) = r^l Y_{lm}(\hat{\boldsymbol{r}}), \tag{4.5}$$

where $Y_{lm}(\hat{\boldsymbol{r}})$ is a spherical harmonic. The solid spherical harmonic is a polynomial of degree l in the Cartesian coordinate (see Complement 6.2). The matrix $A_k^{(p)}$ is always chosen to be diagonal. Therefore, the spherical part of Eq. (4.3) is considered a special case of Eq. (4.1). (See also Complement 8.4.) The trial function of type (4.3) is used in the so-called Coupled Rearrangement Channel Variational Method [20, 25, 26]. The basic idea of this approach is that, by taking into account the different Jacobi sets, one can introduce various correlations. This method has been used with great success especially for Coulombic three-body problems [20].

Because of the symmetry requirement imposed on the orbital part of the trial function, we may assume that

$$c_{kl}^{(2)} = c_{kl}^{(3)}, \qquad A_k^{(2)} = A_k^{(3)}. \tag{4.6}$$

The 2×2 diagonal matrix $A_k^{(p)}$ $(k = 1, ..., K)$ has two nonlinear parameters and they are taken as a geometric progression

$$\left(A_k^{(p)}\right)_{ii} = \left(\frac{1}{a_i^{(p)}\left(q_i^{(p)}\right)^{k-1}}\right)^2 \qquad (i = 1,\, 2). \tag{4.7}$$

In principle one can use different geometric progression parameters, $a_i^{(p)}$ and $q_i^{(p)}$, corresponding to the different arrangement channels and the Jacobi coordinates, that is, they would depend on p and i in order to get better convergence. They might even depend on the angular momentum l. For simplicity, in this example we use the same parameters, a and q, for all arrangement channels and all sets of Jacobi coordinates. The main reason for using the geometric progression as parametrization is clear: The number of parameters of the basis is reduced to just 2 (a and q), so the optimization is simple. Another reason for the choice of this specific parametrization is that the overlap integral of the basis functions can be easily controlled by a choice of q and thus the danger of the linear dependence of the basis functions can be avoided. (See Sect. 3 for the danger arising from the linear dependence of the basis functions.)

The disadvantage of the use of these diagonal matrices is that, without having some polynomials like the scalar product of Eq. (4.4), the energy does not converge to its accurate value. As is shown in Table 4.2, one reproduces only the first two figures ($E = -0.2618530$) of the ground-state energy without the polynomial part, i.e., by taking only $l = 0$ in the expansion.

Table 4.2. The energy of Ps^- in different arrangement channels. "Exact" energy is −0.262005.

Channel	Partial wave	E (a.u.)
(23)1	$l = 0$	−0.2068096
(23)1	$l = 0, 2$	−0.2392726
(31)2+(12)3	$l = 0$	−0.2609626
(31)2+(12)3	$l = 0, 1$	−0.2619622
(31)2+(12)3	$l = 0, 1, 2$	−0.2619717
(23)1+(31)2+(12)3	$l = 0$	−0.2618530
(23)1+(31)2+(12)3	$l = 0, 1$	−0.2619804
(23)1+(31)2+(12)3	$l = 0, 1, 2$	−0.2619816

At this moment a comment is in order. As the spherical harmonics form a complete set for angular functions, one may think that the partial-wave expansion using only one of the Jacobi coordinate sets would be sufficient to represent the wave function. The convergence as a function of l is, however, very slow if only one set of the Jacobi coordinates is used. (The slow convergence in the partial-wave expansion will also be studied in detail in Complement 11.2.) E.g., the energy calculated with the (23)1 channel using the partial waves of up to $l = 2$ is just −0.2392726, which is much higher than the energy of the three-channel calculation ($E = -0.2618530$) using only $l = 0$. Since the lowest threshold of Ps^- is the $Ps+e^-$ channel at the energy of −0.25, this result indicates that this single-channel calculation cannot bind the system. Moreover, the expansion with l has for practical purposes to be truncated to low values because the calculation of the matrix elements of high l waves generally becomes more time consuming. Table 4.2 shows that, by combining the expansion in the Jacobi coordinate sets, the energy converges faster and only the first few terms of the partial-wave expansion are needed. By using 10 grid points for $A_k^{(p)}$ we have found the optimal values of the parameters to be $a = 0.06$ and $q = 2.6$. The variational energy obtained with this $600\,(= 2 \times 10 \times 10 \times 3)$-dimensional calculation is −0.2619816. The computational load with one particular basis set is the calculation of $180300\,(= 600 \times 601/2)$ matrix elements and the solution of a generalized eigenvalue problem of dimension 600. For the optimization this has to be repeated a few dozen times. By increasing the basis size one can get closer and closer to the exact energy.

4.2.2 Random tempering

Another popular tempering method is random tempering [21, 22, 23]. This approach involves the generation of (pseudo-) random numbers with the basis parameters defined by the following prescription:

$$\alpha_k = \exp\left(\sum_{j=1}^{I} d_j < k, j > \right) \qquad (k = 1, ..., K), \tag{4.8}$$

where $< k, j >$ is a pseudo-random numer, the fractional part of $(k(k+1)/2)\sqrt{P(j)}$ with $P(j)$ being a prime number in the sequence 2, 3, 5, 7,..., that is, $P(1) = 2$, $P(2) = 3, ...$ By using this tempering formula, one optimizes a number I of parameters, d_j, instead of the original $3K$ parameters. A possible application of the formula for a three-particle case is given in Table 4.1.

The origin of this formula is the following. One may assume that the ground-state wave function is an integral transform of some known function with some weight function. The known function in our case is the correlated Gaussian and its nonlinear parameters are the integration variables. The simplest way to carry out the integral transformation is to use the Monte Carlo method. The quadrature points required in this integration can be generated by the above formula. As this formula provides "good lattice points" for the integration, these nonlinear parameters can be thought to be adequate to represent the wave function in a variational approach.

While the random tempering works in many examples in a superb way, one has to note that it has a serious problem: It often leads to (almost) linearly dependent bases. In our actual example, the best energy of Ps^- with random tempering is -0.261872, and further improvement of this value was difficult due to the linear dependence. The parameters of the random tempering used are, in the notation of Table 4.1, $m_1 = m_2 = 7$, $m_3 = 5$, $d_1 = 4$, $d_2 = -8$, $d_3 = 4$, $d_4 = -7$, $d_5 = -1$, $d_6 = -11$, $p = 1$, and the basis dimension is $K = 245$. The parameters d_i are nearly optimized.

4.2.3 Random basis

The above calculations have suggested that not all of the grid points are equally important but the different grids can give nearly the same energy. The explanation is simple: The basis functions are nonorthogonal to each other, none of them is indispensable; they are dense, that is, any of them can be omitted because some others will compensate for the loss. This property of the basis functions and the success of the random tempering suggest the idea of a completely random distribution of the parameters. To illustrate this possibility, let us generate $K = 100$ sets of basis parameters in the expansion (4.2). The elements of A_k are randomly chosen from a "physical" interval:

$$0 \leq \frac{1}{\sqrt{\alpha_{12}}} \leq 10, \qquad 0 \leq \frac{1}{\sqrt{\alpha_{13}}} \leq 10, \qquad 0 \leq \frac{1}{\sqrt{\alpha_{23}}} \leq 20. \qquad (4.9)$$

The advantage of using the parametrization by α_{ij} instead of A_k is that $1/\sqrt{\alpha_{ij}}$ corresponds to the "distance" between the particles. (Note that, for a Gaussian function $e^{-\frac{1}{2}\alpha r^2}$, the expectation value relevant to the distance is, e.g., $\langle r \rangle = \sqrt{4/(\pi\alpha)}$, or $\sqrt{\langle r^2 \rangle} = \sqrt{3/(2\alpha)}$.) The parameters inside this interval describe the $(e^+e^-e^-)$ system, where the average distance between the positron and the electrons is expected to be less than 10 a.u. and the distance between the electrons is expected to be less than 20 a.u. This limitation serves practical purpose, and it is based on the physical intuition that the particles in a bound system cannot move very far away from each other.

The difference between the random tempering and the random distribution of the parameters is that in the former there are several parameters which generate the parameters of the basis functions and these generating parameters are optimized, whereas in the latter the parameters of the basis functions are randomly chosen.

The energies of different random bases are shown in Fig. 4.1. By increasing the size of the basis, the energy goes very close to the exact value. See Table 4.3. The energy is better than that obtained with the basis set of geometric progression type, but the comparison is difficult. In the case of the geometric progression, the matrix of nonlinear parameters is diagonal and the partial-wave expansion is introduced to enable the trial function to span the full model space. This partial-wave expansion increases the basis size significantly. When we use the full A matrix and randomly distribute α_{ij}, this partial-wave expansion is not necessary (see Complement 11.2). One may try to choose α_{ij} in a geometric progression, but our experience is that both the geometric progression and the random tempering of α_{ij} tend to lead to linear dependence.

To avoid the linear dependence in the basis (which does not occur frequently with the fully random basis), the overlap of the random basis functions should be smaller than a prescribed limit. Any random points which do not satisfy this condition are to be omitted and

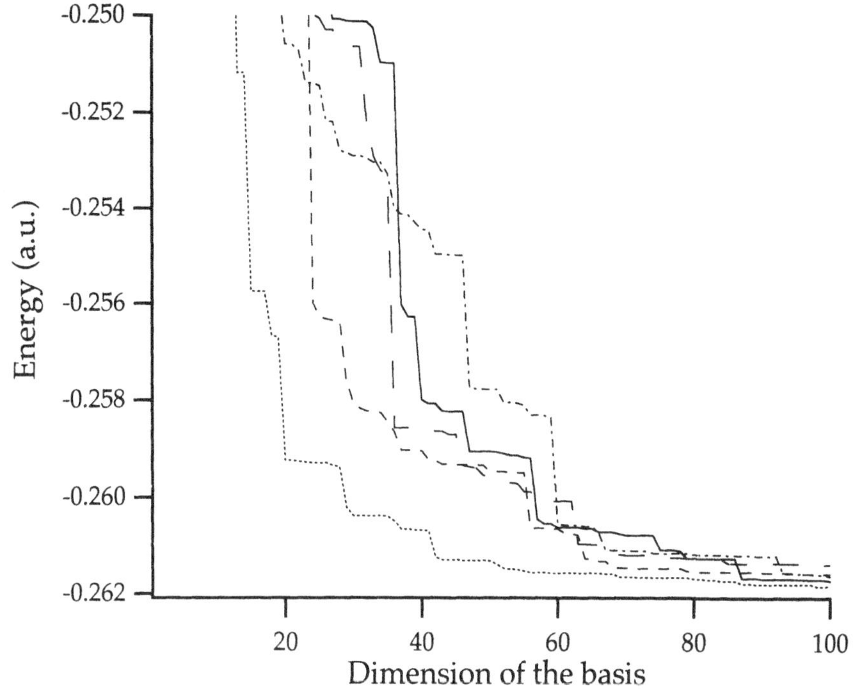

Fig. 4.1. Energies of Ps^- for different basis sets. The parameters of the basis states are completely randomly chosen and no preselection is made.

Table 4.3. The energy of Ps^- (in a.u.) for K basis states that are selected randomly. "Exact" energy is −0.262005. The energies Ei are obtained by starting from different random points. The energy of "Best 100" is the one calculated by sorting the best 100 basis states from 400 basis states, where 50 random trials are probed at each step of the basis selection. N_{eval} is the number of matrix elements evaluated during the optimization.

	$K = 100$	$K = 200$	$K = 400$	Best 100
E1	−0.2617619	−0.2619798	−0.2620032	−0.2619995
E2	−0.2617281	−0.2619793	−0.2620026	−0.2619978
E3	−0.2617918	−0.2619669	−0.2620016	−0.2619956
E4	−0.2618826	−0.2619675	−0.2620008	−0.2619953
E5	−0.2616285	−0.2619824	−0.2620024	−0.2619987
N_{eval}	5050	20100	80200	80200

replaced by a new trial. Note that this cure cannot be applied to the procedures of the random tempering and of the geometric progression. In these procedures there is no such prescription. One can just omit some grid points depending on the actual parameters.

The energies listed in Table 4.3 are surprisingly good, but the size of the basis seems larger than absolutely necessary. For the solution of a three-body problem a basis dimension of $K = 400$ is excessive. We would like to decrease the basis size by selecting a minimal, indispensable set of basis functions. This will be described in the following subsections.

4.2.4 Sorting

One may wonder: Are all the basis states equally important? What happens if we omit a few states from these bases? To answer this question, we reordered the basis states by the following random selection process:

1. A number n of basis states were picked up randomly and the one giving the lowest energy was selected to be the first element of a restricted basis.
2. Then n basis states, again randomly chosen from the remaining pool, were tested in a 2-dimensional calculation with the first state. Again the state that produced the lowest energy was selected.
3. This reordering is continued until the last basis state.

One can choose a minimal set of L basis states from the original set in such a way that the energy is very close to the energy obtained with the full basis set. The improvement one can reach by this sorting is illustrated in Table 4.3. The best $L = 100$ basis states selected from among $n = 50$ candidates give almost as good an energy as the full $K = 400$-dimensional basis. This improvement also occurs in the basis set determined by the random tempering or the geometric progression.

4.2.5 Trial and error search

The randomly chosen nonlinear parameters give good results provided the basis dimension is big enough. For an N-particle problem, however, the basis dimension might become prohibitively large to use this simple procedure. One can obviously improve the convergence by selecting the most important states and by not admitting all random trials in the basis set. This increase of basis dimension by searching the best among many random trials is a key of the stochastic variational method (SVM). The original procedure of the SVM, proposed in [27], has recently been developed further [28] and successfully applied to multicluster descriptions of light exotic nuclei [29, 30]. Learning from these applications to nuclear few-body problems, we have generalized and refined the method further to encompass diverse quantum-mechanical few-body systems emerging in nuclear and atomic physics [31, 32, 33].

The sorting method has shown that from a large number of random basis states one can select a smaller set without a significant loss of accuracy. Moreover, the accuracy can be improved by increasing the basis size. This experience motivates one to apply the following trial and error procedure, combined with an admittance test, in order to set up the most important basis set: Let A_k be the parameter set defining the kth basis function, and let us assume that the sets $A_1, ..., A_{k-1}$ have already been selected, and the $(k-1)$-dimensional eigenvalue problem has already been solved. The next step is the following:

Competitive selection

s1. A number n of different sets of $(A_k^1, ..., A_k^n)$ are generated randomly.
s2. By solving the n eigenvalue problems of k-dimension, the corresponding energies $(E_k^1, ..., E_k^n)$ are determined.
s3. The parameter set A_k^m that produces the lowest energy from among the set $(E_k^1, ..., E_k^n)$ is selected to be the kth parameter set.
s4. Increase k to $k+1$.

The essential reason motivating this strategy is the need to sample different parameter sets as fast as possible. The advantage of this procedure is that it is not necessary to recompute the whole Hamiltonian matrix nor is it necessary to perform a new diagonalization at each time when a new parameter A_k is generated. See Theorem 3.5.

The convergence of the energy with the basis selected in this way is shown in Fig. 4.2. The convergence is much faster than in the previous case in Fig. 4.1, and it can be made even faster by increasing the number n of trials. The energies found by using $K = 10$- and $K = 100$-dimensional basis sets are shown in Tables 4.4 and 4.5, respectively.

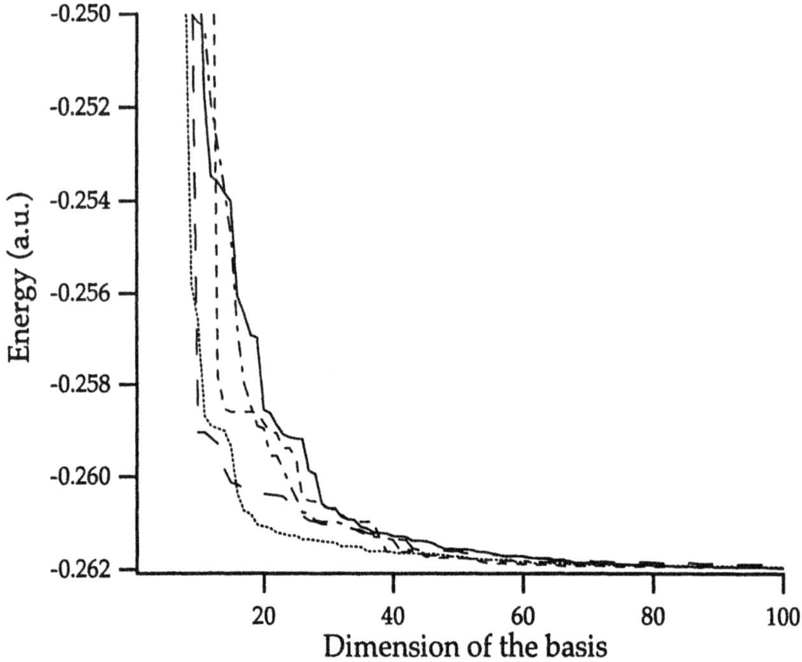

Fig. 4.2. Energies of Ps^- for different basis sets that are selected by the trial and error procedure

Table 4.4. Energy of Ps^- (in a.u.) by different optimization strategies. "Exact" energy is -0.262005. The basis size is set to $K = 10$. The energies Ei are obtained by starting from five different random points. The number n denotes the random candidates probed in the trial and error procedure of SVM. N_{eval} is the number of matrix elements evaluated during the optimization. In the case of the full optimization by the Powell algorithm 3900 diagonalizations were also required. The time in units of seconds is on a Digital Alpha 2100 (250MHz) workstation. The refining cycle is repeated 10 times.

Method	Powell	SVM($n = 1$)	SVM($n = 10$)	Refining ($n = 10$)
E1	−0.261251	−0.244364	−0.251083	−0.261180
E2	−0.261398	−0.234532	−0.250729	−0.261165
E3	−0.261283	−0.236207	−0.252659	−0.261040
E4	−0.261259	−0.243323	−0.252956	−0.261175
E5	−0.261193	−0.226697	−0.251260	−0.261039
N_{eval}	214500	55	550	88055
Time	60	0.8	1.0	11

Table 4.5. Energy of Ps^- (in a.u.) by different optimization strategies. The basis size is set to $K = 100$. In the case of the full optimization by the Powell algorithm 7200 diagonalizations were required. The refining cycle is repeated only once. See also the caption of Table 4.4.

Method	Powell	SVM($n = 1$)	SVM($n = 10$)	Refining ($n = 10$)
E1	−0.26200016	−0.26176191	−0.26199427	−0.26200231
E2	−0.26199947	−0.26172812	−0.26199382	−0.26200351
E3	−0.26200164	−0.26179182	−0.26199733	−0.26200312
E4	−0.26200135	−0.26188261	−0.26199778	−0.26200301
E5	−0.26200193	−0.26162811	−0.26199836	−0.26200271
N_{eval}	35466150	11615	86150	805050
Time	7200	27	43	195

These tables also contain a comparison with the performance of a deterministic procedure. One can conclude that one can easily reach energy convergence with this simple selection procedure. Moreover, the energy converges to the exact value.

4.2.6 Refining

It is obvious that the basis size cannot be increased forever. Moreover, when the Kth basis state is selected, the previous states are kept fixed. This means that we try to find the optimal state with respect to previously selected basis states, but actually some of the basis states selected earlier might not be so important anymore because the succeeding states took over their role. So one may include a refining procedure where the previous states are probed again as described by the following steps:

Refinement cycle

r1. $(A_i^1, ..., A_i^n)$ random parameter sets are generated.
r2. The parameters of the ith basis state are replaced by the new candidates and the energies $E_K^1, ... E_K^n$ are calculated.
r3. If the best of the new energies is better than the original one, then replace the old parameters with the new ones, otherwise keep the original ones.
r4. Cycle this procedure through the basis states from $i = 1$ to K.

Note that, again, there is no need for diagonalization because of Theorem 3.5.

The energy improvement reached by the refinement cycle is shown in Table 4.6. See also energy improvement by the refining as shown in Tables 4.4 and 4.5. The refinement increases the accuracy to a great extent. The energy improvement gets smaller after the repetition of several cycles and finally it gets saturated and no further improvement can be reached. To improve the energy further, one has to increase the basis size. With the enlarged basis, the refining process can be repeated. For example, as will be shown in Chap. 8, to reproduce the energy of Ps^- up to 10 digits, one has to increase the basis size up to $K = 600$, and the refinement has to be repeated many times. The fact that we can reproduce the most accurate energies obtained by other methods justifies the usage of the trial and error selection and the refinement: It shows the power of the SVM.

Table 4.6. Improvement of the energy of Ps^- in the refining steps. The basis size is $K = 100$ and 5 random states are probed. Atomic units are used.

Refinement cycle	E (a.u.)
(Starting value)	−0.261992767
1st	−0.262002742
2nd	−0.262003677
3rd	−0.262004063
4th	−0.262004178
5th	−0.262004255
6th	−0.262004314
7th	−0.262004339
8th	−0.262004372
10th	−0.262004398
"Exact"	−0.262005

4.2.7 Comparison of different optimizing strategies

To summarize what we have done so far, we compare the performances of different procedures of basis optimization for basis sizes of $K = 10$ and $K = 100$. The numbers of basis parameters are 30 and 300, respectively. The first approach we consider is an attempt of full optimization by Powell's method [14]. This method is quite time-consuming and we have terminated the calculation after a given computer time (60 units for $K = 10$ and 7200 units for $K = 100$). Letting the program run for a longer time, we might get somewhat better energies, but the available computer time is limited in any case. Moreover, the other methods reach at least the same energy in a fraction of this time as we have seen. Tables 4.4 and 4.5 show that the full optimization seems to converge to different points depending on the starting point, which indicates that the procedure leads to different local minima. This is confirmed by explicit inspections of the parameters.

In the third column of Tables 4.4 and 4.5 the energies obtained by using different sets of completely random parameters are given. The next column shows the results of a trial and error selection and the last one gives the energy after refinement cycles. We have carried out 10 refining cycles (each basis state has been cyclically probed 10 times) in the 10-dimensional, and one refinement cycle in the 100-dimensional case.

In summary, one may conclude that the stochastic selection of basis parameters leads to energy convergence independently of different random paths and, moreover, to very accurate results in a fraction of time and by much less computational load than the full optimization. And while the latter is certainly superior in principle (one may need fewer basis states to reach a given accuracy), the former suits more practical applications. Moreover, in more complicated problems where, in addition to the nonlinear parameters, one has to find the most adequate channels (quantum numbers, etc.) to describe the system, the random selection seems to be the only viable method. Since the energy convergence attainable is of course not precise in a mathematical sense, one should check the quality of convergence of the wave function as well.

In the competitive selection the relatively best candidate found in the steps is always admitted to be a basis state even though its contribution to the energy may sometimes happen to be very small. This indicates that the newly selected basis has very strong overlap with (linear combinations of) the previous basis states. To avoid this an alternative approach called a utility selection [28] can be used to select the basis states. The steps go as follows:

Utility selection

s1. Generate randomly a parameter set A_k.
s2. Determine the energy $\epsilon_1(k)$ by solving the k-dimensional eigenvalue problem.

s3. Admit the parameter set A_k to be the kth element if the energy gained by including it is larger than a preset value $\delta\epsilon$, namely

$$\epsilon_1(k) < \epsilon_1(k-1) - \delta\epsilon.$$

Otherwise return to the step (s1) for the next attempt. If no parameter set was found to pass the utility test out of n consecutive attempts, reduce $\delta\epsilon$ to, say half of its original value, and return to (s1).

s4. Increase k to $k+1$.

When competitive selection is used, convergence is inferred by the energy curve's flattening out, while when utility test is used, convergence is signalled by an insistent failure to find further elements that pass the test.

Instead of the random trial strategy described in (s1)-(s2)-(s3)-(s4) and (r1)-(r2)-(r3)-(r4), one may think of more efficient and sophisticated approaches, like simulated annealing or genetic algorithms. The latter approaches may give faster convergence, although the random selection strategy may have a slight advantage: It is easily implementable in a parallel way because the random trials are absolutely independent of each other.

One should note that the basis elements selected by the above admittance tests are in general never optimal, not even locally. A better candidate which will gain more energy could be found in the neighborhood of the basis element admitted. The incorporation of a fine tuning, that is, an additional search for even better parameters in the vicinity of the element that passed the admittance test would clearly accelerate the energy convergence. When the evaluation of the matrix elements does not require heavy computational loads, this fine tuning is recommended to reach faster convergence. Or even a deterministic selection such as Powell's method [34, 35] might be used to find out the locally best basis element.

4.3 Optimization for excited states

The Mini-Max theorem (see Sect. 3.1) shows that by diagonalizing the Hamiltonian in a K-dimensional basis one gets an upper bound not only for the ground state but also for the excited states as well. It may happen that the basis set found for the ground state is fairly good to predict the energies of the excited states with the same conserved quantum numbers (angular momentum, parity etc.) as the ground state. By optimizing only the ground state, however, the energies of the excited states will not necessarily converge (see Fig. 4.3). The energies certainly decrease because we increase the basis size, but except for the energy of the second excited state, the upper bounds are not very accurate. (See Chap. 8 for the detail of the basis function used in the present section.)

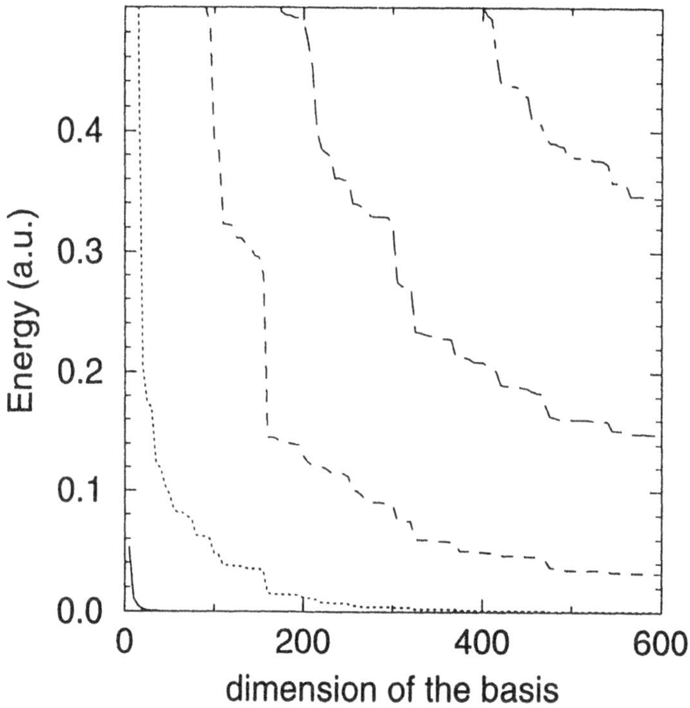

Fig. 4.3. Convergence of the energies of the first five ^1S states of the Helium atom when only the ground-state wave function is optimized. The energy difference $E_i - E_i^{\mathrm{exact}}$ is shown in the figure. See Table 4.7 for the exact energies.

As the ith eigenvalue E_i obtained by diagonalization is the upper bound of the energy of the ith excited state, one may try to optimize this upper bound to get an accurate estimate for the energy of the excited state. In practice we repeat the same procedure as before, but now the basis selection is governed by the requirement that the ith eigenvalue, not the ground-state energy E_1, should be improved. To get the ith eigenvalue we need at least an i-dimensional basis to start with, but practical numerical considerations suggest that it is better to start with a basis in which all the lower eigenvalues ($k = 1, ..., i-1$) are already "stable". This means that we need a first guess for the lower eigenvalues, otherwise it may happen that when improving the first excited state, for example, we pick up such components that lower the ground state.

As the energies of the excited states of the Helium atom are known to a high accuracy, we will test our strategy in this case. First we optimize the ground state of the He atom on a $K = 100$-dimensional basis. The energy of the ground state is very accurate and even the energy of the first excited state is acceptable. Starting from this $K = 100$-dimensional basis, we increase the basis size one by one, picking up basis states which improve the energy of the first excited state. This procedure quickly improves the energy of the first excited state, while the energy of the ground state also improves a little bit (the important thing to note is that it does not get worse) because the basis size is increased. After reaching the basis size of $K = 200$ we switch to the second excited state and so on. The convergence of the result is shown in Fig. 4.4. One can see that the energy quickly converges to the exact value after its turn of optimization started.

The above procedure gives us a basis where all the excited states are accurate up to a certain digit. We can of course create bases which give an accurate energy for an individual state, while the energy of the other states might be poor. In that case we simply carry out the stochastic search for a given state starting from a first guess basis (like the $K = 100$-dimensional basis in the previous case). This optimization may include refinement cycles as well, if necessary. The results we obtained in this way are compared to the "exact" (i.e., the best calculation in the literature) values [36, 37] in Table 4.7. One can thus get as accurate energies for the excited states as for the ground state. Examples of caculation of the energies of excited states will be given for the $td\mu$ molecule in Complement 8.4, the baryon spectroscopy in Chap. 9 and the four-nucleon system in Sect. 11.2.

Before closing this section, we remark that the SVM assumes that the numerical accuracy of the energy calculation in each step is high. If the energy is calculated by, say the Monte Carlo method and has an inherent statistical uncertainty, then the admittance test of the SVM will never work. Even when the energy is calculated by a deterministic method, such as a diagonalization, the accuracy of the numerical calculations has to be good enough to guarantee the required precision

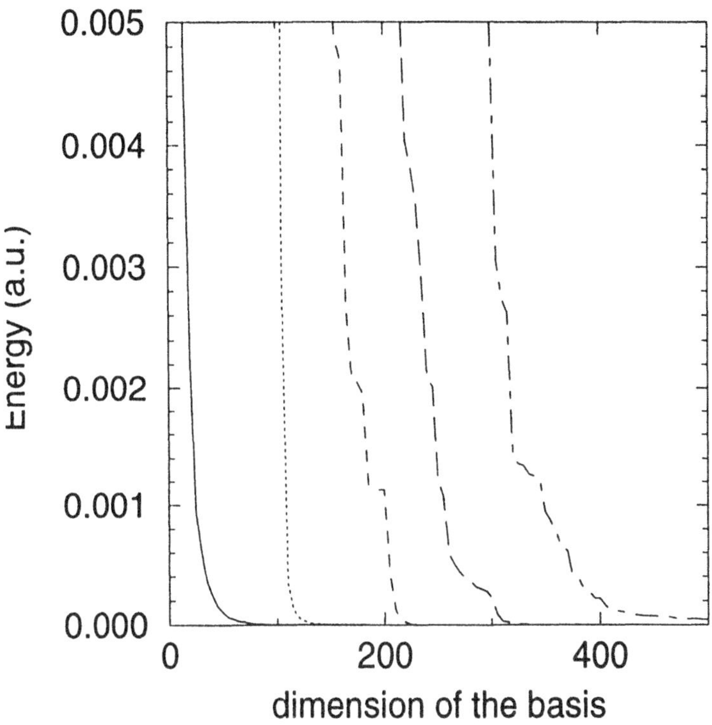

Fig. 4.4. Convergence of the energies of the first five ^1S states of the Helium atom. The first one hundred basis states are selected to optimize the ground state, the next one hundred (from 101 till 200) basis states are selected to optimize the first excited state, and so on. The energy difference $E_i - E_i^{\text{exact}}$ is shown in the figure. See Table 4.7 for the exact energies.

Table 4.7. Energies in atomic units of the ground state and the first four ^1S excited states of the Helium atom. In column A the basis is optimized successively for all the states as described in the text ($K = 500$). In column B the basis is optimized separately for each state, leading to five different bases ($K = 600$) tailored for the respective states. The "exact" values are taken from [36, 37].

State	A	B	"Exact"
E_1	−2.9037243758	−2.9037243769	−2.90372437698
E_2	−2.1459737740	−2.1459740452	−2.14597404605
E_3	−2.0612718887	−2.0612719880	−2.06127198974
E_4	−2.0335865085	−2.0335866779	−2.03358671702
E_5	−2.0211312479	−2.0211768312	−2.02117685157

of the energy. An analytic evaluation of the matrix elements is thus an essential prerequisite in the SVM.

Recently an importance sampling algorithm called the stochastic diagonalization method [38] has been presented to compute the smallest eigenvalue and the corresponding eigenvector of extremely large matrices. It is interesting to note that though the algorithm used there has been developed independently of the SVM, it includes procedures analogous to those of SVM. The stochastic diagonalization was applied to matrices of order up to $10^{35} \times 10^{35}$.

All the above procedures are based on the minimization of the energy expectation value. This is, however, not the only possibility to find accurate energies and wave functions. As described in Sect. 3.2, the minimization of the variance of the local energy is another possibility in case the variance can be calculated analytically. This will be studied in a simple example of the following Complement.

Complements

4.1 *Minimization of energy versus variance*
Here we compare the two ways of minimization by treating a solvable problem, an S-wave motion of a particle of mass m in the potential of $V(r) = -V_0 \exp(-r/a)$ ($V_0 > 0$). By expressing the wave function Ψ as $\chi/\sqrt{4\pi}r$, the Schrödinger equation for this problem takes the form

$$-\frac{\hbar^2}{2m}\frac{d^2\chi}{dr^2} - V_0\exp\Big(-\frac{r}{a}\Big)\chi = E_1\chi. \tag{4.10}$$

To simplify the notation, we introduce

$$\xi = \sqrt{\frac{8ma^2V_0}{\hbar^2}}, \quad \text{and} \quad \nu = \sqrt{\frac{-8ma^2E_1}{\hbar^2}}. \tag{4.11}$$

The change of variable $z = \xi\exp(-r/2a)$ reduces Eq. (4.10) to the differential equation of the Bessel functions

$$\frac{d^2\chi}{dz^2} + \frac{1}{z}\frac{d\chi}{dz} + \Big(1 - \frac{\nu^2}{z^2}\Big)\chi = 0. \tag{4.12}$$

The boundary conditions for bound states, $\chi(r) = 0$ at $r = 0$ and ∞, lead to the solution

$$\chi(r) = cJ_\nu\Big(\xi\exp(-\frac{r}{2a})\Big) \quad \text{with} \quad c^{-2} = \int_0^\infty J_\nu\Big(\xi\exp(-\frac{r}{2a})\Big)^2 dr, \tag{4.13}$$

where the value of ν is determined by the condition

$$J_\nu(\xi) = 0, \tag{4.14}$$

and the smallest ν value gives the ground-state energy E_1 through Eq. (4.11). In order to have a bound state, the potential strength V_0 must be such that ξ is larger than about 2.405.

To estimate the ground-state energy by the variational method, let us assume a simple basis function for χ, $r\exp(-\alpha r/a)$, where α is a variational parameter. There is no minimum of σ^2 for a single basis function. Table 4.8 compares the energy and the variance obtained by using the minimization principle for E or σ^2. The parameter values of α are determined by the SVM. For a deeper potential of $\xi = 5.0$, a combination of two basis functions already reproduces the energy reasonably well. For a very shallow potential of $\xi = 2.6$, however, at least

Table 4.8. Comparison of variational solutions obtained by E-minimization and σ^2-minimization for an exponential well. Units of E and σ^2 are $\hbar^2/(8ma^2)$ and $(\hbar^2/(8ma^2))^2$, respectively. The mean values, $\langle r\rangle$ and $\langle r^2\rangle^{1/2}$, are in unit of a. The overlap integral of the variational solution with the exact one is also given.

ξ	K		E	σ^2	$\langle r\rangle$	$\langle r^2\rangle^{\frac{1}{2}}$	Overlap
		E-min.	−3.5840	0.3236	1.4362	1.6169	0.99997
5.0	2	σ^2-min.	−3.5824	0.05437	1.4301	1.6044	0.99983
		Exact	−3.5848	0	1.4376	1.6213	1
		E-min.	−0.016446	1.275×10^{-4}	9.4174	12.236	0.99999
2.6	5	σ^2-min.	−0.016415	3.491×10^{-6}	9.1025	11.619	0.99940
		Exact	−0.016447	0	9.4331	12.285	1

five basis functions are needed to obtain the energy to better accuracy than three digits. The mean values of $\langle r\rangle$ and $\langle r^2\rangle$ are calculated to learn the sensitivity of physical quantities other than the energy to the wave function. One has to use more basis functions to reproduce the mean values of $\langle r\rangle$ and $\langle r^2\rangle$ to better accuracy.

As Fig. 4.5 displays, the local energy curve obtained by the minimization of σ^2 has larger deviation from the exact energy than the one obtained by the minimization of E. This unexpected result can be understood from the difference in contribution to the variance at short distances.

The basis function adopted above may cope with correct asymptotic behavior at large distances, but does not show correct behavior near $r = 0$. To make the local energy constant, we should demand that the trial function satisfy $\chi''/\chi = $ const. at $r = 0$ as the potential is smooth at the origin. However, as the ratio χ''/χ of the trial function has a $1/r$ behavior near the origin, therefore the minimization of σ^2 tends to decrease the deviation of the local energy at short distances. This is probably the reason why the σ^2 minimization does not lead to as good results as the energy minimization. This indicates that a use of the trial function which satisfies all the boundary conditions is important to make the minimization principle of σ^2 an accurate and substantial tool [39]. Unfortunately, it is not easy to construct a simple trial function which meets this requirement even in the present simple example.

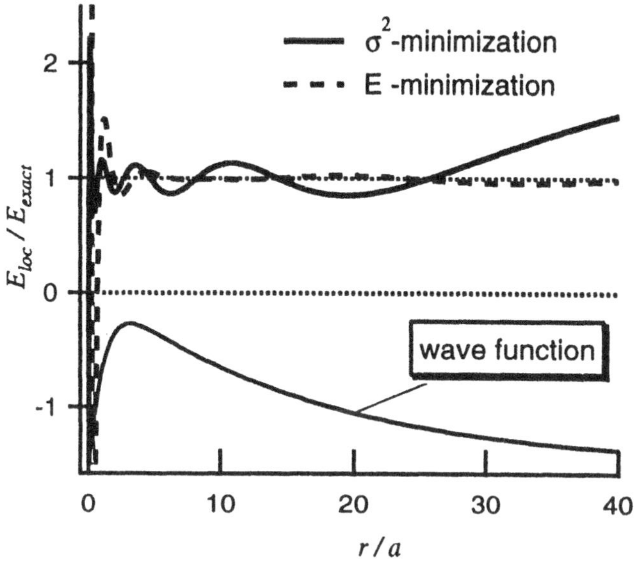

Fig. 4.5. The local energy curve for the shallow potential: $\xi = 2.6$. The exact wave function is also shown in an arbitrary unit.

5. Other methods to solve few-body problems

In this chapter we briefly show the essential points of other approaches to few-body bound-state problems. There are many different approaches and it is far beyond the scope of this book to discuss all of them. Included here are either only those methods which have some sort of connection with our approach or their results are frequently compared with that of SVM, and therefore a short explanation might be useful.

5.1 Quantum Monte Carlo method: The imaginary-time evolution of a system

The evolution of a quantum-mechanical system is governed by the time-dependent Schrödinger equation. When the time t is replaced by an imaginary time, $-i\hbar\alpha$ $(\alpha \geq 0)$, the Schrödinger equation transforms into a diffusion-like equation

$$\frac{\partial \Psi(\alpha)}{\partial \alpha} = -H\Psi(\alpha), \tag{5.1}$$

which has the formal solution

$$\Psi(\alpha) = \mathrm{e}^{-H\alpha}\Psi, \tag{5.2}$$

provided that the Hamiltonian H is time-independent. Here Ψ is an initial wave function $\Psi(\alpha = 0)$. The imaginary-time evolution of the wave function can be used to estimate the ground-state energy. The basis of this idea is

$$\lim_{\alpha\to\infty} \frac{\Psi(\alpha)}{\langle\Psi(\alpha)|\Psi(\alpha)\rangle^{\frac{1}{2}}} = \frac{a_1}{|a_1|}\Phi_1, \tag{5.3}$$

that is, the wave function $\Psi(\alpha)$ approaches the exact ground-state wave function at $\alpha \to \infty$. To prove this, we only need to use the expansion (3.3) for Ψ in Eq. (5.2) and assume that $a_1 \neq 0$, namely the initial wave function has a non-vanishing overlap with the ground-state wave function. In what follows we assume that this condition is fulfilled.

If we consider $\Psi(\alpha)$ a variational trial function depending on the imaginary time α, the mean value $E(\alpha)$ of H in $\Psi(\alpha)$ gives an upper bound to the ground-state energy E_1. In fact it is easy to show that

$$\begin{aligned} E(\alpha) &\equiv \frac{\langle\Psi(\alpha)|H|\Psi(\alpha)\rangle}{\langle\Psi(\alpha)|\Psi(\alpha)\rangle} \\ &= \frac{E_1 + \sum_{i=2}^{\infty} E_i \mathrm{e}^{-2\alpha(E_i-E_1)}|\frac{a_i}{a_1}|^2}{1+\sum_{i=2}^{\infty} \mathrm{e}^{-2\alpha(E_i-E_1)}|\frac{a_i}{a_1}|^2} \\ &= E_1 + \frac{\sum_{i=2}^{\infty}(E_i - E_1)\mathrm{e}^{-2\alpha(E_i-E_1)}|\frac{a_i}{a_1}|^2}{1+\sum_{i=2}^{\infty} \mathrm{e}^{-2\alpha(E_i-E_1)}|\frac{a_i}{a_1}|^2} \geq E_1. \end{aligned} \tag{5.4}$$

Clearly $E(\alpha)$ goes to E_1 when α goes to infinity.

Another quantity called the asymptotic energy estimator is defined by

$$E(\alpha, \tau) = \frac{1}{2\tau} \ln \frac{\langle\Psi(\alpha)|\Psi(\alpha)\rangle}{\langle\Psi(\alpha+\tau)|\Psi(\alpha+\tau)\rangle}. \tag{5.5}$$

This has the following properties:

$$\lim_{\tau\to 0} E(\alpha, \tau) = E(\alpha), \tag{5.6}$$

$$E(\alpha, \tau) = E_1 + \frac{1}{2\tau} \ln \frac{1+\sum_{i=2}^{\infty} \mathrm{e}^{-2\alpha(E_i-E_1)}|\frac{a_i}{a_1}|^2}{1+\sum_{i=2}^{\infty} \mathrm{e}^{-2(\alpha+\tau)(E_i-E_1)}|\frac{a_i}{a_1}|^2}. \tag{5.7}$$

Equation (5.6) can be derived by noting that the left-hand side is equal to $(-1/2)\frac{d}{d\alpha}\ln\langle\Psi(\alpha)|\Psi(\alpha)\rangle$ and that the derivative can be easily calculated by Eq. (5.1). It is clear that, for $\tau > 0$, $E(\alpha, \tau) \geq E_1$ and $E(\alpha, \tau)$ converges to E_1 as α goes to infinity.

In fact, the quantum Monte Carlo (QMC) method or the Green's function Monte Carlo (GFMC) method [40–44] uses $\mathrm{e}^{-H\alpha}$ as a vehicle to project out the ground state from the initial wave function. The point here is not a construction of a good state space which approximates the solution but a calculation of the imaginary-time evolution of $\mathrm{e}^{-H\alpha}$. In general, one cannot compute $\mathrm{e}^{-H\alpha} = \mathrm{e}^{-(H_0+H_1)\alpha}$ because H_0 and H_1 do not commute, but by dividing the time α into many small steps $\Delta\alpha = \alpha/n$, we can approximate $G(\boldsymbol{R}, \boldsymbol{R}') \equiv \langle \boldsymbol{R}|\mathrm{e}^{-H\Delta\alpha}|\boldsymbol{R}'\rangle$ by $\langle \boldsymbol{R}|\mathrm{e}^{-H_0\Delta\alpha}\mathrm{e}^{-H_1\Delta\alpha}|\boldsymbol{R}'\rangle$ in a factorized form. Then the full propagator

$$\langle \boldsymbol{R}|\mathrm{e}^{-H\alpha}|\boldsymbol{R}'\rangle = \int\int\cdots\int G(\boldsymbol{R}, \boldsymbol{R}_{n-1})G(\boldsymbol{R}_{n-1}, \boldsymbol{R}_{n-2})\cdots \times G(\boldsymbol{R}_1, \boldsymbol{R}')d\boldsymbol{R}_{n-1}d\boldsymbol{R}_{n-2}\cdots d\boldsymbol{R}_1 \tag{5.8}$$

can be evaluated by the Monte Carlo method. In practice, one must use several time steps $\Delta\alpha$ and extrapolate to $\Delta\alpha = 0$ in order to eliminate time step errors arising from the non-commuting nature of the kinetic energy and potential energy operators.

When we use the propagator (5.8) to calculate $E(\alpha)$ in Eq. (5.4), both the numerator and denominator of Eq. (5.4) are subject to the statistical errors of Monte Carlo integration. Then the nice feature of having an upper bound to the ground-state energy may be lost in the QMC method. One of the most serious problems in the QMC is the so-called minus-sign problem [43, 45]. The propagator $G(\boldsymbol{R}, \boldsymbol{R}')$ is not always positive. This makes it difficult to use importance sampling techniques based on the classical stochastic process such as a Markov process or a Molecular Dynamics technique in computing the propagator by the Monte Carlo method.

5.2 Hyperspherical harmonics expansion method

We briefly introduce an approach which is based on the expansion of the trial function in terms of hyperspherical harmonic (HH) functions. Just as the spherical harmonics are useful to expand the angular function of a single-particle motion, the basic idea of the HH method is to generalize this simplicity to a system of particles by introducing a global length ρ called the hyperradius and a set of angles Ω.

Let us suppose that all of the particles oscillate harmonically with angular frequency ω

$$H_{\mathrm{HO}} = \sum_{i=1}^{N} \frac{\boldsymbol{p}_i^2}{2m_i} + \sum_{i=1}^{N} \frac{1}{2}m_i\omega^2\boldsymbol{r}_i^2. \tag{5.9}$$

Using the Jacobi coordinate set, one can separate the Hamiltonian into the intrinsic and center-of-mass parts

$$H_{\mathrm{HO}} = \sum_{i=1}^{N-1} \frac{\boldsymbol{\pi}_i^2}{2\mu_i} + \sum_{i=1}^{N-1} \frac{1}{2}\mu_i\omega^2\boldsymbol{x}_i^2 + \frac{\boldsymbol{\pi}_N^2}{2m_{12\cdots N}} + \frac{1}{2}m_{12\cdots N}\omega\boldsymbol{x}_N^2, \tag{5.10}$$

where μ_i is defined by Eq. (2.19).

Among the $3(N-1)$ degrees of freedom for the intrinsic motion, the hyperradius ρ is defined by

$$\rho^2 = \frac{1}{\mu}\sum_{i=1}^{N-1} \mu_i \boldsymbol{x}_i^2. \tag{5.11}$$

The hyperradius contains only intrinsic coordinates, but it can also be expressed as $\sum_{i=1}^{N} m_i(\boldsymbol{r}_i - \boldsymbol{x}_N)^2/\mu$, which is proportional to the moment of inertia of the system (see Eq. (2.18)). The choice of μ can be arbitrary. The hyperradius is apparently symmetric with respect to the interchange of particle labels. The potential energy of the intrinsic part is simply expressed as $\frac{1}{2}\mu\omega^2\rho^2$. Just as the single-particle kinetic energy operator is separated into radial and angular parts, the intrinsic kinetic energy operator is reduced as follows

$$\sum_{i=1}^{N-1} \frac{\boldsymbol{\pi}_i^2}{2\mu_i} = -\frac{\hbar^2}{2\mu}\left[\frac{\partial^2}{\partial\rho^2} + \frac{3N-4}{\rho}\frac{\partial}{\partial\rho} - \frac{\boldsymbol{L}^2}{\rho^2}\right], \tag{5.12}$$

where $\boldsymbol{L}$ is called the grand angular momentum, which is expressible in terms of $3N-4$ angles Ω.

Although there is no unique way to choose the angles, one possibility is to choose $2(N-1)$ angles ϑ_i and ϕ_i, the polar coordinate of $\boldsymbol{x}_i$ $(i = 1, ..., N-1)$ and the other $N-2$ angles γ_k $(0 \le \gamma_k \le \frac{\pi}{2},\ k = 2, 3, ..., N-1)$ called the hyperangles as follows

$$\begin{aligned}
&\sqrt{\mu_{N-1}}\, x_{N-1} = \sqrt{\mu}\,\rho\cos\gamma_{N-1}, \\
&\sqrt{\mu_k}\, x_k = \sqrt{\mu}\,\rho\sin\gamma_{N-1}\ldots\sin\gamma_{k+1}\cos\gamma_k \qquad (k = 2, ..., N-2), \\
&\sqrt{\mu_1}\, x_1 = \sqrt{\mu}\,\rho\sin\gamma_{N-1}\ldots\sin\gamma_3\sin\gamma_2.
\end{aligned} \tag{5.13}$$

Then $\boldsymbol{L}$ is related to the usual angular momentum $\boldsymbol{l}_k = \frac{1}{\hbar}\boldsymbol{x}_k \times \boldsymbol{\pi}_k$ as follows

$$\begin{aligned}
\boldsymbol{L}^2 &= \boldsymbol{L}_{N-1}^2, \\
\boldsymbol{L}_k^2 &= -\frac{\partial^2}{\partial\gamma_k^2} - \left\{3(k-2)\cot\gamma_k + 4\cot 2\gamma_k\right\}\frac{\partial}{\partial\gamma_k} \\
&\quad + \frac{\boldsymbol{l}_k^2}{\cos^2\gamma_k} + \frac{\boldsymbol{L}_{k-1}^2}{\sin^2\gamma_k} \qquad (k = 2, ..., N-2),
\end{aligned}$$

$$\boldsymbol{L}_1^2 = \boldsymbol{l}_1^2. \tag{5.14}$$

The eigenfunctions of $\boldsymbol{L}^2$, called the HH functions,

$$\boldsymbol{L}^2 \mathcal{Y}_{\mathcal{K}}(\Omega) = K(K + 3N - 5)\mathcal{Y}_{\mathcal{K}}(\Omega) \tag{5.15}$$

are characterized by the quantum number $\mathcal{K}$ which comprises K ($K = 0, 1, ...$) and $3N - 5$ eigenvalues of the angular operators.

Expressing the intrinsic wave function as a sum, over $\mathcal{K}$, of products of radial functions $\mathcal{R}_{\mathcal{K}}(\rho)$ and the HH functions $\mathcal{Y}_{\mathcal{K}}(\Omega)$,

$$\Psi = \sum_{\mathcal{K}} \rho^{-\frac{3N-4}{2}} \mathcal{R}_{\mathcal{K}}(\rho)\mathcal{Y}_{\mathcal{K}}(\Omega), \tag{5.16}$$

the N-body Schrödinger equation, after eliminating the center-of-mass motion, is reduced to the matrix equation for the radial functions

$$\left(-\frac{\hbar^2}{2\mu}\left[\frac{d^2}{d\rho^2} - \frac{\mathcal{L}(\mathcal{L}+1)}{\rho^2}\right] - E\right)\mathcal{R}_{\mathcal{K}}(\rho)$$
$$+ \sum_{\mathcal{K}'} \langle \mathcal{Y}_{\mathcal{K}} | \sum_{j>i=1}^{N} V_{ij} | \mathcal{Y}_{\mathcal{K}'} \rangle \mathcal{R}_{\mathcal{K}'}(\rho) = 0, \tag{5.17}$$

with

$$\mathcal{L} = K + \frac{3}{2}(N - 2). \tag{5.18}$$

The volume element is given by

$$\prod_{i=1}^{N-1} d\boldsymbol{x}_i = \left(\frac{\mu^{N-1}}{\mu_1 \mu_2 \cdots \mu_{N-1}}\right)^{\frac{3}{2}} \rho^{3N-4} d\rho d\Omega_1$$
$$\times \prod_{k=2}^{N-1} d\Omega_k \sin^{3k-4}\gamma_k \cos^2\gamma_k d\gamma_k, \tag{5.19}$$

where $d\Omega_k = \sin\vartheta_k d\vartheta_k d\phi_k$ is the usual surface element. Because $\mathcal{L}$ is positive for $N \geq 3$-particle systems, it turns out that the "centrifugal barrier" is always present in $N \geq 3$-particle systems even for zero total orbital angular momentum states.

For the HH expansion method to be practical, of course, the evaluation of the potential energy matrix elements must be feasible. Moreover one has to have a guide for the truncation of the HH basis, even though the expansion converges with a few K values, because the degeneracy of the HH functions is high. It is also necessary to construct the HH functions of proper symmetry for a system of identical particles. This is a difficult problem, and it limits the application of the HH expansion method to systems of rather small numbers of particles ($N \leq 5$).

A sophisticated version of the HH expansion method [46] employs a correlation factor $F = \prod_{j>i=1}^{N} f_{ij}(\boldsymbol{r}_i - \boldsymbol{r}_j)$ [47], replacing Ψ in Eq. (5.16) with $F\Psi$. In [36, 48] the correlation factor is chosen to satisfy a special boundary condition like the cusp condition for the Coulombic system. (See Complement 8.1 for the cusp condition.) The use of the correlation factor accelerates the convergence and leads to precise solutions.

The HH expansion method was introduced in 1935 by Zernike and Brinkman and reintroduced 25 years later by Delves and Smith. See for example [49, 50] for details and [51] for recent developments.

5.3 Faddeev method

The Faddeev method [52] transcribes the content of the Schrödinger equation for a three-body system into a set of coupled equations in such a unique manner that eliminates undesirable disconnected diagrams in the Lippmann-Schwinger equation and accurate solutions based on the Faddeev equations are considered the best for a three-body problem.

The Faddeev method and its applications have been discussed extensively. See, for example, [53, 54]. The wave function Ψ of the three-body system is given as a sum of three Faddeev components

$$\Psi = \psi_1 + \psi_2 + \psi_3, \tag{5.20}$$

where for the bound state each component is related to Ψ by

$$\psi_i = G_0 V_i \Psi, \qquad \text{or} \qquad (E - H_0)\psi_i = V_i \Psi, \tag{5.21}$$

where the free three-body propagator is given by

$$G_0 = \frac{1}{E - H_0}. \tag{5.22}$$

Here H_0 is the three-body kinetic energy with the center-of-mass kinetic energy being subtracted. The potential V_i is expressed in terms of "odd man out" notation, that is, $V_1 = V_{23}$, $V_2 = V_{31}$, $V_3 = V_{12}$. It is easy to see that Eqs. (5.20)–(5.22) is equivalent to the Schrödinger equation $(E - H)\Psi = 0$ with $H = H_0 + V_1 + V_2 + V_3$.

We show two forms of the Faddeev equations which are widely used for numerical solutions. One is the differential form which results by rewriting the second form in Eq. (5.21):

$$
\begin{aligned}
(E - H_0 - V_1)\psi_1 &= V_1(\psi_2 + \psi_3), \\
(E - H_0 - V_2)\psi_2 &= V_2(\psi_3 + \psi_1), \\
(E - H_0 - V_3)\psi_3 &= V_3(\psi_1 + \psi_2).
\end{aligned} \tag{5.23}
$$

Another is the integral form of the above equation

$$
\psi_i = G_i V_i(\psi_j + \psi_k) \qquad (j \neq i,\ k \neq i,\ j \neq k), \tag{5.24}
$$

with

$$
G_i = \frac{1}{E - H_0 - V_i}. \tag{5.25}
$$

The integral form can be simply expressed in a matrix form

$$
\begin{pmatrix} \psi_1 \\ \psi_2 \\ \psi_3 \end{pmatrix} = \begin{pmatrix} G_1 & 0 & 0 \\ 0 & G_2 & 0 \\ 0 & 0 & G_3 \end{pmatrix} \begin{pmatrix} 0 & V_1 & V_1 \\ V_2 & 0 & V_2 \\ V_3 & V_3 & 0 \end{pmatrix} \begin{pmatrix} \psi_1 \\ \psi_2 \\ \psi_3 \end{pmatrix}. \tag{5.26}
$$

The bound-state solution has to satisfy the boundary condition that all of the components ψ_i vanish at large distances. This condition is, however, met only for a particular value of energy E, which is nothing but the energy eigenvalue of the three-body system.

The Faddeev method is thus a direct method to solve the eigenvalue problem. It does not rely on the variational principle. There are three possible choices for the Jacobi coordinates $\boldsymbol{x}^{(1)}$, $\boldsymbol{x}^{(2)}$, $\boldsymbol{x}^{(3)}$. See Fig. 2.2. It is natural to express the Faddeev component ψ_1 in terms of $\boldsymbol{x}^{(1)}$, and likewise ψ_2 by $\boldsymbol{x}^{(2)}$ and ψ_3 by $\boldsymbol{x}^{(3)}$. The most difficult part in numerical works of the Faddeev method then comes from the evaluation of matrix elements between the functions expressed in terms of the different Jacobi coordinate sets. Readers are referred to [55, 56] for some details for nuclear and Coulombic three-body bound-state problems.

In Chap. 11 we will show that a careful optimization of the correlated Gaussians with the use of the SVM gives solutions that are in excellent agreement with the results of the Faddeev method for nuclear three-body problems.

A real power of the Faddeev method is probably not only in bound-state problems but in applications to scattering and continuum problems. See [54] for the recent developments on this subject.

It is possible to extend the Faddeev method to four-particle system. The Faddeev equations are then called Faddeev-Yakubovsky equations which consist of seven components, four for $1+3$-partition of four particles and three for $2+2$-partition. Numerical solutions become increasingly difficult.

5.4 The generator coordinate method

There is a theory called the generator coordinate method (GCM), which is a natural extension of Eq. (3.7) to the case of continuous superpositions. The GCM is a powerful method and has a variety of applications such as the restoration of symmetries or good quantum numbers (e.g., angular momentum and particle number) in many-body wave functions, the microscopic description of reaction dynamics between two composite particles and collective motions of many-body systems.

The GCM assumes that the trial function is a continuous superposition of the basis function $\Psi(\alpha)$, which is often called the generating function:

$$\Psi = \int f(\alpha)\Psi(\alpha)d\alpha. \tag{5.27}$$

The function $f(\alpha)$ is called a weight function. A real or complex parameter α specifies the generating function. It is called the generator coordinate to distinguish it from a physical coordinate. The symbol α may represent more than one parameter: $\alpha = (\alpha_1, \alpha_2, ...)$. The integral in Eq. (5.27) is then a multidimensional integration over all the parameters. The choice of the generating functions $\Psi(\alpha)$ is most important in the GCM.

The variational principle is used to determine $f(\alpha)$ and leads to the equation known as the Hill-Wheeler equation [57]

$$\int \mathcal{H}(\alpha,\alpha')f(\alpha')d\alpha' = E\int \mathcal{B}(\alpha,\alpha')f(\alpha')d\alpha', \tag{5.28}$$

with

$$\mathcal{H}(\alpha,\alpha') = \langle\Psi(\alpha)|H|\Psi(\alpha')\rangle, \qquad \mathcal{B}(\alpha,\alpha') = \langle\Psi(\alpha)|\Psi(\alpha')\rangle. \quad (5.29)$$

Here $\mathcal{H}(\alpha,\alpha')$ and $\mathcal{B}(\alpha,\alpha')$ are linear integral operators and called the Hamiltonian kernel and the overlap (or norm) kernel, respectively. Formally the GCM looks very similar to the diagonalization of the Hamiltonian in the basis states $\Psi(\alpha)$. This is true if the parameter α is discretized and if only a finite number of functions $\Psi(\alpha_1),...,\Psi(\alpha_K)$ are used. Then the Hill-Wheeler equation reduces to the generalized eigenvalue problem (3.10). For further details see [57–60]. Examples of using discretized generator coordinates are given in Sects. 8.4 and 10.4.

6. Variational trial functions

Due care of the correlation between the particles is important for a precise variational solution. The variational trial functions used in this text, correlated Gaussians and correlated Gaussian-type geminals, are formulated by using a generating function g. Orbital functions with arbitrary angular momenta or Cartesian polynomials around arbitrary centers are constructed from g by using a simple, well-defined prescription. It is also shown that the generating function can be related to the product of single-particle Gaussian wave-packets through an integral transformation. This 'uncorrelated' form is useful to extend to a many-body system of identical particles. The spin function for an N-fermion system is briefly discussed.

6.1 Correlated Gaussians and correlated Gaussian-type geminals

Most crucial for the variational approach is the choice of the trial function. To solve an N-particle problem, it is of prime importance to describe the correlation between the particles properly. The correlation between the particles can be described by functions of appropriate relative coordinates. The correlation is then conveniently represented by a correlation factor, $F = \prod_{j>i=1}^{N} f_{ij}(\boldsymbol{r}_i - \boldsymbol{r}_j)$. There are two widely applied strategies: (1) One is to use this form of F directly by selecting the most appropriate functional form to describe the short-range as well as long-range correlations. Such calculations are, however, fairly involved for systems of more than three particles and the integrations involved require the Monte Carlo technique. (2) An alternative way to incorporate the correlation is to approximate f_{ij} by a number, possibly a large number, of simple terms which facilitate the analytical calculation of the matrix elements. We follow the second course by using an expansion over a correlated Gaussian "basis".

It is important that two basic requirements are satisfied in the second course. First, in the limit of large dimensions the basis functions should become *complete* so that the results obtained by means of some systematic increase of the number of basis functions could converge to the exact eigenvalue. Second, for the approach to be practicable, computational effectiveness is required for the basis functions. The matrix elements have to be evaluated with ease; otherwise calculations with combinations of a great number of basis functions would be extremely difficult. In our opinion only the correlated Gaussian basis meets these requirements. The usefulness of Gaussians was already suggested in 1960 independently by Boys [61] and Singer [62], and since then exploited by many authors [63, 22, 64, 65, 66]. Though a mathematically sound proof may not be available for the completeness of the Gaussian functions, a heuristic argument for it is possible as follows: As is discussed in Complement 6.1, any square-integrable, well-behaved function with angular momentum lm can be approximated, to any desired accuracy, by a linear combination of nodeless harmonic-oscillator functions (Gaussians) of a continuous size parameter a (Eq. (6.37)). By generalizing this to the N-particle system, the N-particle basis function then contains a product of these Gaussians: $F = \prod_{j>i=1}^{N} \exp\{-\frac{1}{2}\alpha_{ij}(\boldsymbol{r}_i - \boldsymbol{r}_j)^2\} = \exp\{-\frac{1}{2}\sum_{j>i=1}^{N} \alpha_{ij}(\boldsymbol{r}_i - \boldsymbol{r}_j)^2\}$. This simple correlated Gaussian is actually widely used in variational calculations. For a discussion on the completeness of Gaussians for specific cases, see for example [67, 35].

Of course, we have to keep in mind that the Gaussian is not economical in describing the asymptotic behavior of the wave function at large distances (see Fig. 6.2). Moreover, it does not predict a correct value for some specific quantities such as the cusp ratio. See Complement 8.1. Both the asymptotics and the cusp ratio could be well described with exponential functions. For example, we know that the Hylleraas-type functions give very accurate results for Coulombic few-body systems (see, e.g. [68, 69]). The correlated exponential functions are not, however, amenable to analytic evaluation of matrix elements for a system of more than three particles. This makes it difficult to use the exponential functions as a variational trial function for a general N-particle system.

We extend the above argument further to define the correlated Gaussians. For this we consider separately two cases for two types of Hamiltonians. First in case of Eq. (2.1) the Hamiltonian can be expressed in terms of a set of independent relative coordinates

$\tilde{\boldsymbol{x}} = (\boldsymbol{x}_1, ..., \boldsymbol{x}_{N-1})$. As was shown in Eqs. (2.23)–(2.25), it is then convenient to express F in terms of $\boldsymbol{x}$, instead of $N(N-1)/2$ interparticle-distance vectors, $\boldsymbol{r}_i - \boldsymbol{r}_j$. An N-particle basis function, a so-called correlated Gaussian, then looks like

$$\text{Type I}: \quad \Psi = \exp\Big(-\frac{1}{2}\tilde{\boldsymbol{x}}A\boldsymbol{x}\Big)\theta(\boldsymbol{x}), \tag{6.1}$$

where A is an $(N-1) \times (N-1)$ positive-definite, symmetric matrix of nonlinear parameters, specific to each basis element. As mentioned in Sect. 2.2, the matrix A with these properties can in general be written as $\tilde{G}DG$ with the use of an orthogonal matrix G and a diagonal matrix D with all positive diagonal elements. The function $\theta(\boldsymbol{x})$ is a generalization of the solid spherical harmonics $\mathcal{Y}$ of Eq. (6.56) to the many-particle case. More details will be given below.

The second case is suited to the Hamiltonian (2.2). The motion of each particle is governed by both the single-particle potential and the two-body interaction. It is thus useful to extend Eq. (6.1) to include an independent motion of the ith particle around some point $\boldsymbol{R}_i$. Here $\boldsymbol{R}_i$ is not a dynamical coordinate but just a parameter vector. In this case we do not need to use the relative and center-of-mass coordinates but can use the single-particle coordinates. Therefore we are led to the following type of correlated Gaussians which are often called correlated Gaussian-type geminals :

$$\Psi = \exp\Big\{-\frac{1}{2}\sum_{j>i=1}^{N} \alpha_{ij}(\boldsymbol{r}_i - \boldsymbol{r}_j)^2\Big\}$$

$$\times \exp\Big\{-\frac{1}{2}\sum_{i=1}^{N} \beta_i(\boldsymbol{r}_i - \boldsymbol{R}_i)^2\Big\}\theta(\boldsymbol{r} - \boldsymbol{R}).$$

Here $\boldsymbol{r} - \boldsymbol{R}$ stands for a set of vectors $\{\boldsymbol{r}_1 - \boldsymbol{R}_1, ..., \boldsymbol{r}_N - \boldsymbol{R}_N\}$. As $\sum_{j>i=1}^{N} \alpha_{ij}(\boldsymbol{r}_i - \boldsymbol{r}_j)^2$ can be written compactly as $\tilde{\boldsymbol{r}}A\boldsymbol{r}$ with a matrix A defined through $A_{ij} = A_{ji} = -\alpha_{ij}$ $(i < j)$, $A_{ii} = \sum_{j=1}^{i-1} \alpha_{ji} + \sum_{j=i+1}^{N} \alpha_{ij}$, the correlated Gaussian-type geminals can in general be expressed as

$$\text{Type II}: \quad \Psi = \exp\Big\{-\frac{1}{2}\tilde{\boldsymbol{r}}A\boldsymbol{r} - \frac{1}{2}(\widetilde{\boldsymbol{r} - \boldsymbol{R}})B(\boldsymbol{r} - \boldsymbol{R})\Big\}$$

$$\times\, \theta(\boldsymbol{r} - \boldsymbol{R}), \tag{6.2}$$

where B is a diagonal matrix whose ith element is β_i. As it satisfies the relation $\sum_{i=1}^{N} A_{ij} = \sum_{j=1}^{N} A_{ij} = 0$, the matrix A is not positive-definite. (To see this it is enough to show $\det A = 0$; this is true because the equation for c_j, $\sum_{j=1}^{N} A_{ij}c_j = 0$, has at least two different solutions, $c_1 = \cdots = c_N = 0$ and $c_1 = \cdots = c_N = 1$.) The exponent of Eq. (6.2) can be expressed simply as $\exp\{-\frac{1}{2}(\boldsymbol{r} - \boldsymbol{S})C(\boldsymbol{r} - \boldsymbol{S})\}$, except for a constant factor, where $C = A + B$, and $\boldsymbol{S} = (A + B)^{-1}B\boldsymbol{R}$. Therefore the matrix $A + B$ needs to be positive-definite in order for Eq. (6.2) to be square integrable.

We emphasize the usefulness of Eq. (6.2) even for the translation-invariant Hamiltonian (2.1). In fact the basis function (6.2) is quite useful in molecular physics and is often used to calculate the energy of the electronic system in the Born-Oppenheimer approximation, where $\boldsymbol{r}$ stands for the single-particle coordinates of the electrons and $\boldsymbol{R}$ are position parameters which can be varied to obtain the ground-state energy. As the nuclear and electronic motions can be separated to a good approximation in the molecule and the relative motion between the nuclei shows a pronounced peak at a fixed separation distance, it is natural to apply the wave function (6.2) to this case. However, even in the molecular case it would be more convenient to use the convention that the vector $\boldsymbol{r}$ in Eq. (6.2) be understood to be the relative coordinates $\boldsymbol{x}$ of the whole molecular system excluding only the center-of-mass coordinate $\boldsymbol{x}_N$, because one can then calculate the total energy including the effect of the nuclear motion. With this slight change of notation, one obtains the energy which can be directly compared to experiment. We will show this example in Chap. 8. To clarify the formulation, we assume here that the variables $\boldsymbol{r}$ stand for the single-particle coordinates but note that it is straightforward to switch to the case where they represent the relative coordinates.

As will be shown later, the correlated Gaussians satisfy the second requirement, that is their Hamiltonian matrix elements can be calculated analytically in a unified framework, thus enabling one to avoid the formidable calculation involving the correlation factor F. The fact that the correlated Gaussians allow us to evaluate the matrix elements analytically is an important point to be emphasized especially in the application of the SVM. The usefulness of correlated Gaussians has long been underestimated since they were claimed to have much slower convergence than functions containing powers of $|\boldsymbol{r}_i - \boldsymbol{r}_j|$. It will be shown, however, in Chap. 8 that this is not the case; a careful optimization of nonlinear parameters allows us to obtain high-quality solutions with expansions containing not too many terms.

The Gaussian of type I is a special case of type II. To unify the description for both cases, we change the notation in what follows throughout this chapter according to the following conventions: As it is cumbersome to use $\boldsymbol{x}$ or $\boldsymbol{r}$ depending on the type of problems, we will use $\boldsymbol{x}$ even in the case of the presence of an external field. To have N as the number of $\boldsymbol{x}_i$ coordinates in both cases, we consider $(N+1)$ particles for the basis of type I and N particles for the basis of type II, respectively. Table 6.1 summarizes our conventions for the notations.

Table 6.1. Convention of the coordinates $\tilde{\boldsymbol{x}} = (\boldsymbol{x}_1, \boldsymbol{x}_2, ..., \boldsymbol{x}_N)$ in the correlated Gaussians and the correlated Gaussian-type geminals. SP and CM are abbreviations of 'single-particle' and 'center-of-mass', respectively.

	Correlated Gaussians (Type I)	Correlated Gaussian-type geminals (Type II)
Number of particles	$N+1$	N
Meaning of $\boldsymbol{x}$	Relative coordinates	SP coordinates
Transformation matrix from SP coordinates $\boldsymbol{r}$	U: $\boldsymbol{x} = U\boldsymbol{r}$	$U = 1$: $\boldsymbol{x} = \boldsymbol{r}$
CM coordinate: $\boldsymbol{x}_{\rm CM}$	$\boldsymbol{x}_{N+1}$	$\frac{1}{m_{12\cdots N}} \sum_{i=1}^{N} m_i \boldsymbol{x}_i$
SP coordinates relative to CM: $\boldsymbol{r}_i - \boldsymbol{x}_{\rm CM}$	$(U^{-1}\boldsymbol{x})_i$	$\boldsymbol{x}_i - \boldsymbol{x}_{\rm CM}$
Relative distance vector $\boldsymbol{r}_i - \boldsymbol{r}_j$	$(U^{-1}\boldsymbol{x})_i - (U^{-1}\boldsymbol{x})_j$	$\boldsymbol{x}_i - \boldsymbol{x}_j$

Before discussing the function $\theta(\boldsymbol{x})$ in Eq. (6.1) we distinguish two types of variational calculations: One is a type of calculation called a *variation before projection.* Here the trial function used is not necessarily an eigenstate of some of the conserved quantities belonging to the symmetries of the Hamiltonian such as parity and angular momentum. After the variational calculation, the conserved quantities are projected out from the trial function. For example, we know that an eigenstate of a rotationally invariant Hamiltonian should have a good angular momentum. Yet one often uses a variational trial function that does not have the proper rotational symmetry but, after the variational calculation, the symmetry is restored by angular momentum projection. The restoration of good angular momentum is ensured by a superposition of the rotated variational solutions and the weight function involved in the superposition can be determined, e.g., by the generator coordinate method of Sect. 5.4. Another type of calculation is what is called *variation after projection*, where the trial function is constructed so as not to mix different quantum numbers of the conserved quantities. Clearly the variation after projection is superior to the variation before projection because the variation in the former case is carried out only in the state space which has the same symmetry as the exact solution, while in the latter the variational solution tends to reach a minimum in the space including the basis states with different quantum numbers. Even a calculation of variation before projection type, if thoroughly done, would reach a solution that has the proper symmetry in the limit that the state space is complete. See Sect. 10.4. Of course calculations of the variation after projection type become more challenging because it is in general not easy to calculate matrix elements with trial functions that are eigenstates of the conserved observables. Our objective is, however, to use trial functions with the proper symmetry.

The function $\theta(\boldsymbol{x})$ in Eq. (6.1) describes non-spherical motion. To describe the orbital motion with the orbital angular momentum L and its projection M, a direct generalization of solid spherical harmonics $\mathcal{Y}$ (4.5) (see Complement 6.2) to a many-particle case is a vector-coupled product of the solid spherical harmonics of the relative coordinates:

$$\theta_{LM}(\boldsymbol{x}) = \Big[[[\mathcal{Y}_{l_1}(\boldsymbol{x}_1) \times \mathcal{Y}_{l_2}(\boldsymbol{x}_2)]_{L_{12}} \times \mathcal{Y}_{l_3}(\boldsymbol{x}_3)]_{L_{123}} \times \cdots \times \mathcal{Y}_{l_N}(\boldsymbol{x}_N)\Big]_{LM}$$

$$= \sum_{\kappa=\{m_1,m_2,\ldots,m_N\}} c_\kappa \prod_{i=1}^{N} \mathcal{Y}_{l_i m_i}(\boldsymbol{x}_i), \tag{6.3}$$

where c_κ is a product, $\langle l_1 m_1 l_2 m_2 | L_{12}\, m_1{+}m_2\rangle \langle L_{12}\, m_1{+}m_2\, l_3 m_3 | L_{123}\, m_1{+}m_2{+}m_3\rangle \cdots \langle L_{12\cdots N-1}\, m_1{+}m_2{+}\cdots{+}\, m_{N-1}\, l_N m_N | LM\rangle$, of the Clebsch-Gordan coefficients needed to couple the orbital angular momenta to the specified quantum numbers. Here each relative motion has a definite angular momentum. See, e.g., [2, 70, 71, 72] for the details of angular momentum algebra. Angular momentum recoupling coefficients used in this book will be defined in Complement 6.3.

Since the angular momentum of the relative motion is not a conserved quantity, it may be important for a realistic description to include several sets of angular momenta $(l_1, l_2, ..., l_N; L_{12}, L_{123}, ...)$. This is the case especially in nuclear few-body problems [32, 25] because the non-central components of the nuclear potential necessitate higher partial waves. For example the tensor force couples S waves to D waves. It is also noted that a faster convergence is in general obtained by allowing the use of different sets of relative coordinates together with suitable sets of angular momenta [30, 20, 25], because a particular type of correlations can be best described in the coordinate set conforming to the type of correlation. From the fact that $\theta_{LM}(\boldsymbol{x})$ can be expressed by different partial-wave decompositions in different relative coordinate systems, one can conclude that the use of partial waves may not be so important after all. Besides, the various possible partial wave contributions increase the basis dimension. Moreover, the calculation of matrix elements for this choice of $\theta_{LM}(\boldsymbol{x})$ sooner or later becomes too complicated. This choice is therefore obviously inconvenient especially as the number of particles increases and/or different sets of relative coordinates are employed.

This difficulty can be avoided by adopting a different generalization of $\mathcal{Y}$ for $\theta_{LM}(\boldsymbol{x})$ [33, 31]:

$$\begin{aligned}\theta_{LM}(\boldsymbol{x}) &= v^{2K}\mathcal{Y}_{LM}(\boldsymbol{v}) \\ &= v^{2K+L}Y_{LM}(\hat{\boldsymbol{v}}) \qquad \text{with} \quad \boldsymbol{v} = \sum_{i=1}^{N} u_i\boldsymbol{x}_i = \tilde{u}\boldsymbol{x}. \qquad (6.4)\end{aligned}$$

Only the total orbital angular momentum, which is a good quantum number in most cases (at least approximately), appears in this expression. The real vector $\tilde{u} = (u_1, ..., u_N)$ defines a global vector, $\boldsymbol{v}$, a linear combination of the relative coordinates, and the wave function of the system is expanded in terms of its angle $\hat{\boldsymbol{v}}$. The vector u may be considered a variational parameter and one may try to minimize the energy functional with respect to it. The energy minimization then amounts to finding the most suitable angle or a linear combination of angles. The continuity of the parameter u can be more advantageous in a variational calculation than the discrete nature of the set of the angular momenta $(l_1, l_2, ..., l_N; L_{12}, L_{123}, ...)$ because the change of the energy functional can be continuously seen in the former case. The factor of v^{2K+L} plays an important role in improving the short-range behavior of the basis function. A remarkable advantage of this form of $\theta_{LM}(\boldsymbol{x})$ is that the calculation of matrix elements becomes much simpler than in the former case because the coupling of N angular momenta is completely avoided. See the appendix for details.

For the function $\theta(\boldsymbol{r} - \boldsymbol{R})$ in Eq. (6.2) the construction of the trial function with good orbital angular momentum may not be an immediate concern, because the vector $\boldsymbol{R}$ is intended to determine a specific shape of the system. The function $\theta(\boldsymbol{r} - \boldsymbol{R})$ is chosen in the spherical basis as

$$\theta(\boldsymbol{x} - \boldsymbol{R}) = \prod_{i=1}^{N} |\boldsymbol{x}_i - \boldsymbol{R}_i|^{2k_i} \mathcal{Y}_{l_i m_i}(\boldsymbol{x}_i - \boldsymbol{R}_i), \tag{6.5}$$

or in the Cartesian basis as

$$\theta(\boldsymbol{x} - \boldsymbol{R}) = \prod_{i=1}^{N} \prod_{p=1}^{3} (x_{ip} - R_{ip})^{n_{ip}}, \tag{6.6}$$

where the index $p = 1, 2, 3$ stands for x, y, z components of the vectors $\boldsymbol{x}_i$ and $\boldsymbol{R}_i$, and n_{ip} is a non-negative integer for the p-direction of the ith particle. Because there is a simple transformation for the polynomial parts between the spherical basis and the Cartesian basis, the above two representations are actually equivalent. See Complement 6.2. In the following we assume that θ is given by Eqs. (6.3) or (6.4) for the correlated Gaussian of type I and by Eq. (6.5) or Eq. (6.6) for the correlated Gaussian-type geminal of type II.

The Gaussians of type I have definite parity of either $(-1)^{l_1+\cdots+l_N}$ or $(-1)^L$, depending on the choice of $\theta_{LM}(\boldsymbol{x})$, while the Gaussians of type II are not eigenstates of the parity operator. To project out good parity from the trial function, one has to take a combination of two Gaussians with the centers at $\boldsymbol{R}$ and $-\boldsymbol{R}$.

6.2 Orbital functions with arbitrary angular momentum

The construction of a trial function with good angular momentum is very important for obtaining solutions of a rotationally invariant Hamiltonian. In the previous section we introduced the forms of $\theta_{LM}(\boldsymbol{x})$ to describe the orbital function with good angular momentum. Since Eq. (6.3) appears to be better established but Eq. (6.4) takes a particularly simple form and makes the calculation of matrix elements very simple, it is useful to understand the relationship between the two angular functions.

Theorem 6.1. *Any functions of type*

$$v^{2K}\mathcal{Y}_{LM}(\boldsymbol{v}) \quad \text{with } \boldsymbol{v}=\sum_{i=1}^{N} u_i \boldsymbol{x}_i$$

can be expressed in terms of a linear combination of terms

$$x_1^{2k_1} x_2^{2k_2} \cdots x_N^{2k_N} \Big[[[\mathcal{Y}_{l_1}(\boldsymbol{x}_1)\times\mathcal{Y}_{l_2}(\boldsymbol{x}_2)]_{L_{12}}\times\mathcal{Y}_{l_3}(\boldsymbol{x}_3)]_{L_{123}} \times\cdots\times\mathcal{Y}_{l_N}(\boldsymbol{x}_N)\Big]_{LM},$$

where k_i and l_i are all non-negative and satisfy $2k_1+l_1+2k_2+l_2+\cdots+2k_N+l_N=2K+L$.

Proof. We prove the theorem by induction. For a two-variable vector $\boldsymbol{v}=u_1\boldsymbol{x}_1+u_2\boldsymbol{x}_2$, the assertion is true because we know the following equality [73, 74, 75] (see Exercise 6.1 for a simple proof):

$$v^{2K}\mathcal{Y}_{LM}(\boldsymbol{v}) = \sum_{2k_1+l_1+2k_2+l_2=2K+L} D^{KL}_{k_1l_1k_2l_2} u_1^{2k_1+l_1} u_2^{2k_2+l_2} \times x_1^{2k_1} x_2^{2k_2}\Big[\mathcal{Y}_{l_1}(\boldsymbol{x}_1)\times\mathcal{Y}_{l_2}(\boldsymbol{x}_2)\Big]_{LM}, \tag{6.7}$$

where $D^{KL}_{k_1l_1k_2l_2}$ is given by

$$D^{KL}_{k_1l_1k_2l_2} = \frac{(2K+L)!}{(2k_1+l_1)!\,(2k_2+l_2)!}\,\frac{B_{k_1l_1}B_{k_2l_2}}{B_{KL}}\,C(l_1l_2;L). \tag{6.8}$$

Here B_{kl} is defined by

$$B_{kl} = \frac{4\pi(2k+l)!}{2^k k!(2k+2l+1)!!}, \tag{6.9}$$

and C is the coefficient needed to couple two spherical harmonics with the same argument [70]:

$$\big[Y_l(\hat{\boldsymbol{r}})\times Y_{l'}(\hat{\boldsymbol{r}})\big]_{LM} = C(ll';L)\,Y_{LM}(\hat{\boldsymbol{r}}), \tag{6.10}$$

and reads as

$$C(ll';L) = \sqrt{\frac{(2l+1)(2l'+1)}{4\pi(2L+1)}}\,\langle l0l'0|L0\rangle. \tag{6.11}$$

The coefficient C vanishes unless $l + l' + L$ is even.

For a general case of the vector $\boldsymbol{v} = u_1\boldsymbol{x}_1 + \cdots + u_N\boldsymbol{x}_N$, we put $\boldsymbol{v} = \boldsymbol{v}_{N-1} + u_N\boldsymbol{x}_N$ with $\boldsymbol{v}_{N-1} = u_1\boldsymbol{x}_1 + \cdots + u_{N-1}\boldsymbol{x}_{N-1}$ and use the above formula to show that $v^{2K}\mathcal{Y}_{LM}(\boldsymbol{v})$ can be expressed in terms of the vector-coupled products of the terms of $v_1^{2K_{12\cdots N-1}}\mathcal{Y}_{L_{12\cdots N-1}}(\boldsymbol{v}_{N-1})$ and $x_N^{2k_N}\mathcal{Y}_{l_N}(\boldsymbol{x}_N)$ with $2K_{12\cdots N-1} + L_{12\cdots N-1} = 2K + L - 2k_N - l_N$. By further decomposing $\boldsymbol{v}_{N-1}$ to $\boldsymbol{v}_{N-2} + u_{N-1}\boldsymbol{x}_{N-1}$, one can apply the same argument to $v_{N-1}^{2K_{12\cdots N-1}}\mathcal{Y}_{L_{12\cdots N-1}}(\boldsymbol{v}_{N-1})$ and use this process repeatedly. Then it is clear that we can show that the statement is true.

Theorem 6.2. *A vector-coupled function of two solid spherical harmonics*

$$\Big[\mathcal{Y}_{l_1}(\boldsymbol{x}_1) \times \mathcal{Y}_{l_2}(\boldsymbol{x}_2)\Big]_{LM} \qquad \text{with} \quad (-1)^{l_1+l_2} = (-1)^L$$

can be expressed in terms of a linear combination of terms

$$x_1^{2k_1}x_2^{2k_2}v^{2q}\mathcal{Y}_{LM}(\boldsymbol{v}), \tag{6.12}$$

where $2k_1 + 2k_2 + 2q = l_1 + l_2 - L$ *and the vector* $\boldsymbol{v}$ *of each term has the form of* $\boldsymbol{v} = u_1\boldsymbol{x}_1 + u_2\boldsymbol{x}_2$ *with appropriate coefficients* u_1 *and* u_2.

Proof. For a given L value, $l_1 + l_2 - L$ is even and non-negative, and thus $l_1 + l_2$ may be set equal to $2k + L$ with a non-negative integer k. We use induction with respect to k. First we show that the statement is true for $k = 0\,(l_1 + l_2 = L)$. As Eq. (6.7) shows, for the $K = 0$ case there are only terms with $k_1 = k_2 = 0$ (see also Eq. (6.108)), and $\mathcal{Y}_{LM}(\boldsymbol{v})$ with $\boldsymbol{v} = u_1\boldsymbol{x}_1 + u_2\boldsymbol{x}_2$ consists of $(L+1)$ terms of $[\mathcal{Y}_l(\boldsymbol{x}_1) \times \mathcal{Y}_{L-l}(\boldsymbol{x}_2)]_{LM}$ $(l = 0, ..., L)$, each multiplied by $u_1^l u_2^{L-l}$. By taking $(L+1)$ different vectors, $\boldsymbol{v}_i = a_i\boldsymbol{x}_1 + \boldsymbol{x}_2$ $(i = 1, ..., L+1)$ with $a_i \neq 0$, we obtain

$$\mathcal{Y}_{LM}(\boldsymbol{v}_i) = \sum_{l=0}^{L}(a_i)^l\sqrt{\frac{4\pi(2L+1)!}{(2l+1)!(2L-2l+1)!}} \times [\mathcal{Y}_l(\boldsymbol{x}_1) \times \mathcal{Y}_{L-l}(\boldsymbol{x}_2)]_{LM}. \tag{6.13}$$

One can view the above equation as a system of simultaneous linear equations for $[\mathcal{Y}_l(\boldsymbol{x}_1) \times \mathcal{Y}_{L-l}(\boldsymbol{x}_2)]_{LM}$ (multiplied by the square root factor). The determinant of $(L+1)\times(L+1)$ coefficient matrix (a_i^l) is nothing but Vandermonde's determinant and it becomes nonzero for different a_i's. Therefore the solution of the linear equation exists, and hence it is possible to express $[\mathcal{Y}_l(\boldsymbol{x}_1)\times\mathcal{Y}_{L-l}(\boldsymbol{x}_2)]_{LM}$ as a combination of terms of $\mathcal{Y}_{LM}(\boldsymbol{v}_i)$. Thus our assertion has been proved.

Next we assume that the statement holds for $k \le K-1$, and will show that it also holds for $k = K$. To prove this, we note that a general $[\mathcal{Y}_{l_1}(\boldsymbol{x}_1)\times\mathcal{Y}_{l_2}(\boldsymbol{x}_2)]_{LM}$ with $l_1+l_2 = 2k+L = 2K+L$ which satisfies the triangular relation $|l_1-l_2| \le L \le l_1+l_2$ takes the form $[\mathcal{Y}_{K+l}(\boldsymbol{x}_1)\times \mathcal{Y}_{K+L-l}(\boldsymbol{x}_2)]_{LM}$ $(l=0,...,L)$. What we have to show is that the term $[\mathcal{Y}_{K+l}(\boldsymbol{x}_1)\times\mathcal{Y}_{K+L-l}(\boldsymbol{x}_2)]_{LM}$ is expressible as a combination of terms of the form of Eq. (6.12). Now, looking at Eq. (6.7) in its full generality, the expansion of $v^{2K}\mathcal{Y}_{LM}(\boldsymbol{v})$ contains all of these $(L+1)$ functions, each multiplied by $u_1^{K+l}u_2^{K+L-l}$ and some further terms. These have the form const.$\times u_1^{2k_1+l_1}u_2^{2k_2+l_2}x_1^{2k_1}x_2^{2k_2}[\mathcal{Y}_{l_1}(\boldsymbol{x}_1)\times\mathcal{Y}_{l_2}(\boldsymbol{x}_2)]_{LM}$, where at least one of k_1 and k_2 is nonzero and l_1+l_2 is equal to $2(K-k_1-k_2)+L$ with $k = K-k_1-k_2 \le K-1$. Thus, by the assumption that statement of the theorem holds for $k \le K-1$, all factors $[\mathcal{Y}_{l_1}(\boldsymbol{x}_1)\times \mathcal{Y}_{l_2}(\boldsymbol{x}_2)]_{LM}$ in these terms are expressible in the form of Eq. (6.12) with appropriate $\boldsymbol{v}$ vectors. Equation (6.7) can now be rewritten as

$$v^{2K}\mathcal{Y}_{LM}(\boldsymbol{v}) = \sum_{l=0}^{L} u_1^{K+l}u_2^{K+L-l}c_{KLl}[\mathcal{Y}_{K+l}(\boldsymbol{x}_1)\times\mathcal{Y}_{K+L-l}(\boldsymbol{x}_2)]_{LM} + \cdots, \tag{6.14}$$

where c_{KLl} is a suitable constant factor and the symbol $\cdots$ indicates the terms that are already expressed in the form of Eq. (6.12) as stated above. By taking $(L+1)$ different vectors $\boldsymbol{v}_i$ in $v^{2K}\mathcal{Y}_{LM}(\boldsymbol{v})$ in exactly the same manner as before, Eq. (6.14) can be viewed as a system of linear equations for $[\mathcal{Y}_{K+l}(\boldsymbol{x}_1)\times\mathcal{Y}_{K+L-l}(\boldsymbol{x}_2)]_{LM}$, which is also solvable. The solution yields $[\mathcal{Y}_{K+l}(\boldsymbol{x}_1)\times\mathcal{Y}_{K+L-l}(\boldsymbol{x}_2)]_{LM}$ as combinations of terms of Eq. (6.12). This completes the proof. As is clear from the proof, note that the vector $\boldsymbol{v}$ in Eq. (6.12) is not uniquely determined.

It is easy to see that repeated application of Theorem 6.2 leads to the following result.

Theorem 6.3. *A vector-coupled product of solid spherical harmonics*

$$[[[\mathcal{Y}_{l_1}(\boldsymbol{x}_1)\times\mathcal{Y}_{l_2}(\boldsymbol{x}_2)]_{L_{12}}\times\mathcal{Y}_{l_3}(\boldsymbol{x}_3)]_{L_{123}}\times\cdots\times\mathcal{Y}_{l_N}(\boldsymbol{x}_N)]_{LM}$$

with $(-1)^{L_{12}} = (-1)^{l_1+l_2}$, $(-1)^{L_{123}} = (-1)^{L_{12}+l_3}$, ..., *and* $(-1)^L = (-1)^{L_{12\cdots N-1}+l_N}$ *can be expressed in terms of a linear combination of terms*

$$\left\{ \prod_{j\geq i=1}^{N} (\boldsymbol{x}_i \cdot \boldsymbol{x}_j)^{k_{ij}} \right\} v^{2q} \mathcal{Y}_{LM}(\boldsymbol{v})$$

with $2\sum_{i=1}^{N} k_{ii} + 2\sum_{j>i=1}^{N} k_{ij} + 2q = \sum_{i=1}^{N} l_i - L$, *where the vector* $\boldsymbol{v}$ *of each term is given by* $\boldsymbol{v} = \sum_{i=1}^{N} u_i \boldsymbol{x}_i$ *with appropriate coefficients* u_i*'s.*

Since the factors x_i^2 and $(\boldsymbol{x}_i \cdot \boldsymbol{x}_j)$ are scalar, they play no active roles in describing the rotational motion. Theorems 6.1 and 6.3 establish the equivalence between the angular functions of Eqs. (6.3) and (6.4) under the condition that the intermediate angular momenta are restricted as stated in Theorem 6.3 and that the parity of the basis function is given by $(-1)^L$ through the angular momentum L. The basis function whose parity is given by $(-1)^L$ is called to have a natural parity.

The construction of a general angular function with unnatural parity must be based on the vector-coupled form of Eq. (6.3). Unfortunately there is no simple function analogous to Eq. (6.4) for the unnatural parity case. One way to construct the angular function with unnatural parity is for $L \geq 1$

$$\theta_{LM}(\boldsymbol{x}) = v^{2K} [\mathcal{Y}_{L-1}(\boldsymbol{v}) \times \boldsymbol{w}]_{LM} \tag{6.15}$$

with

$$\boldsymbol{v} = \tilde{u}\boldsymbol{x} \quad \text{and} \quad \boldsymbol{w} = \sum_{ij} S_{ij} (\boldsymbol{x}_i \times \boldsymbol{x}_j), \tag{6.16}$$

where S is an $N \times N$ skew-symmetric matrix which satisfies $S_{ij} = -S_{ji}$. For the special case of $L^\pi = 0^-$, $\mathcal{Y}_{L-1}(\boldsymbol{v})$ must be replaced with $\mathcal{Y}_1(\boldsymbol{v})$. A slightly simpler angular function would be possible by introducing another vector $\boldsymbol{v}'$ as follows:

$$\theta_{LM}(\boldsymbol{x}) = v^{2K} [\mathcal{Y}_L(\boldsymbol{v}) \times \boldsymbol{v}']_{LM}, \quad \boldsymbol{v}' = \tilde{u'}\boldsymbol{x}. \tag{6.17}$$

We note in Eq. (6.7) that in the case of $K = 0$ both k_1 and k_2 are limited to zero and only the stretched coupling, namely $l_1 + l_2 = L$, is allowed. See Exercise 6.1. With an increasing K value the possible values of partial waves l_1 and l_2 increase including the case of non-stretched coupling. This applies to the case of many variables as well.

To increase K is thus one way to include higher partial waves in the calculation. The matrix A of the correlated Gaussian is often assumed to be diagonal in order to reduce the number of nonlinear parameters. In this case one has to increase K when the contribution of high partial waves is expected to be important. However, in the case where A is not diagonal, additional and important partial-wave contributions come from the cross terms of the exponential part of the correlated Gaussian even with $K = 0$. E.g., the term $\exp(-A_{ij}\boldsymbol{x}_i\cdot\boldsymbol{x}_j)$, when expanded into power series, contains many terms of the form $(\boldsymbol{x}_i\cdot\boldsymbol{x}_j)^n$, which can describe high partial waves associated with the coordinates $\boldsymbol{x}_i$ and $\boldsymbol{x}_j$. This is easily understood by noting the following relation for arbitrary vectors $\boldsymbol{a}$ and $\boldsymbol{b}$:

$$
\begin{aligned}
(\boldsymbol{a}\cdot\boldsymbol{r})^n &= \sum_{2k+l=n} B_{kl}a^{2k}r^{2k}\sum_{m=-l}^{l}\mathcal{Y}_{lm}(\boldsymbol{a})^*\mathcal{Y}_{lm}(\boldsymbol{r}) \\
&= \sum_{2k+l=n} B_{kl}a^{2k}r^{2k}(-1)^l\sqrt{2l+1}\,[\mathcal{Y}_l(\boldsymbol{a})\times\mathcal{Y}_l(\boldsymbol{r})]_{00}, \qquad (6.18)
\end{aligned}
$$

which results from Eq. (6.45) and the addition theorem (6.54) for spherical harmonics. This implies that even the basis function with $K = 0$ is expected to be useful if a general matrix A is used in the variational calculation. The calculation for the $dt\mu$ molecule given in Complement 8.4 will clarify the point discussed in this paragraph.

6.3 Generating function

The calculation of the matrix elements becomes simpler if one uses a generating function for the correlated Gaussian. In fact, the following function g, which contains an auxiliary "vector" $\tilde{\boldsymbol{s}} = (\boldsymbol{s}_1, ..., \boldsymbol{s}_N)$, is found to generate the correlated Gaussians of both type I and type II conveniently:

$$
g(\boldsymbol{s}; A, \boldsymbol{x}) = \exp\Big(-\frac{1}{2}\tilde{\boldsymbol{x}}A\boldsymbol{x} + \tilde{\boldsymbol{s}}\boldsymbol{x}\Big). \qquad (6.19)
$$

To relate the generating function g to the correlated Gaussian we use the following formula

$$
B_{kl}a^{2k+l}r^{2k}\mathcal{Y}_{lm}(\boldsymbol{r}) = \int Y_{lm}(\hat{\boldsymbol{a}})(\boldsymbol{a}\cdot\boldsymbol{r})^{2k+l}d\hat{\boldsymbol{a}}
$$

$$= \int Y_{lm}(\hat{\boldsymbol{a}}) \left(\frac{\partial^{2k+l}}{\partial \lambda^{2k+l}} \mathrm{e}^{\lambda \boldsymbol{a}\cdot\boldsymbol{r}} \right)_{\lambda=0} d\hat{\boldsymbol{a}}, \tag{6.20}$$

which is easily proved by using Eq. (6.18). Then the vector-coupled product (6.3) can be generated from the factor $\mathrm{e}^{\tilde{s}\boldsymbol{x}}$ of the function g as follows: By choosing $\boldsymbol{s}_i = \alpha_i \boldsymbol{t}_i$ with a unit vector $\boldsymbol{t}_i$ we obtain

$$\exp\Big(-\frac{1}{2}\tilde{\boldsymbol{x}}A\boldsymbol{x}\Big) \prod_{i=1}^{N} \mathcal{Y}_{l_i m_i}(\boldsymbol{x}_i)$$
$$= \left(\left\{ \prod_{i=1}^{N} \frac{1}{B_{0l_i}} \int d\hat{\boldsymbol{t}}_i Y_{l_i m_i}(\hat{\boldsymbol{t}}_i) \frac{\partial^{l_i}}{\partial {\alpha_i}^{l_i}} \right\} g(\alpha|\boldsymbol{t}; A, \boldsymbol{x}) \right)_{\substack{\alpha_1=0,\ldots,\alpha_N=0\\ |\boldsymbol{t}_1|=1,\ldots,|\boldsymbol{t}_N|=1}}. \tag{6.21}$$

By a symbol $\alpha|\boldsymbol{t}$ we mean a "vector" such that each component is given by $\alpha_i \boldsymbol{t}_i$, where $\tilde{\alpha} = (\alpha_1, ..., \alpha_N)$ with α_i a real number and $\tilde{\boldsymbol{t}} = (\boldsymbol{t}_1, ..., \boldsymbol{t}_N)$ is a $1 \times N$ one-row matrix of Cartesian vectors $\boldsymbol{t}_i$.

The key point in the above equation is Eq. (6.20) which relates the solid spherical harmonics to $\mathrm{e}^{\lambda \boldsymbol{a}\cdot\boldsymbol{r}}$. Since Eq. (6.20) yields a more general term $r^{2k}\mathcal{Y}_{lm}(\boldsymbol{r})$ than just the solid spherical harmonics, it may also be useful to use a simpler relation, which generates just the solid spherical harmonics. The relation (6.58) serves for this purpose. By choosing $\boldsymbol{t}_j = (1-{\tau_j}^2, i(1+{\tau_j}^2), -2\tau_j)$ $(j = 1, ..., N)$, we obtain

$$\exp\Big(-\frac{1}{2}\tilde{\boldsymbol{x}}A\boldsymbol{x}\Big) \prod_{i=1}^{N} \mathcal{Y}_{l_i m_i}(\boldsymbol{x}_i)$$
$$= \left(\left\{ \prod_{i=1}^{N} \frac{1}{C_{l_i m_i}} \frac{\partial^{l_i}}{\partial {\alpha_i}^{l_i}} \frac{\partial^{l_i - m_i}}{\partial {\tau_i}^{l_i - m_i}} \right\} \right.$$
$$\left. \times\, g(\alpha|\boldsymbol{t}; A, \boldsymbol{x}) \Big|_{\substack{\boldsymbol{t}_j=(1-{\tau_j}^2, i(1+{\tau_j}^2), -2\tau_j)\\ (j=1,\ldots,N)}} \right)_{\substack{\alpha_1=0,\ldots,\alpha_N=0\\ \tau_1=0,\ldots,\tau_N=0}}, \tag{6.22}$$

where

$$C_{lm} = (-2)^l\, l! \sqrt{\frac{4\pi(l-m)!}{(2l+1)(l+m)!}}. \tag{6.23}$$

Note that the vector $\boldsymbol{t}_i$ satisfies $\boldsymbol{t}_i{\cdot}\boldsymbol{t}_j = -2(\tau_i - \tau_j)^2$, particularly ${\boldsymbol{t}_i}^2 = 0$. This formula requires only differentiation in contrast to Eq. (6.21), where both differentiation and integration are needed. To couple the solid spherical harmonics to the desired function in Eq. (6.3), one has to multiply Eq. (6.21) or (6.22) by c_κ and sum over κ.

To construct the correlated Gaussian-type geminals from g we note

$$\exp\Big(-\frac{1}{2}\tilde{\boldsymbol{R}}B\boldsymbol{R}-\tilde{\boldsymbol{s}}\boldsymbol{R}\Big)g(\boldsymbol{s}+B\boldsymbol{R};A+B,\boldsymbol{x})$$

$$=\exp\Big\{-\frac{1}{2}\tilde{\boldsymbol{x}}A\boldsymbol{x}-\frac{1}{2}(\widetilde{\boldsymbol{x}-\boldsymbol{R}})B(\boldsymbol{x}-\boldsymbol{R})+\tilde{\boldsymbol{s}}(\boldsymbol{x}-\boldsymbol{R})\Big\}. \qquad (6.24)$$

The factor $e^{\tilde{\boldsymbol{s}}(\boldsymbol{x}-\boldsymbol{R})}$ serves to generate the function $\theta(\boldsymbol{x}-\boldsymbol{R})$ of Eq. (6.5) or Eq. (6.6). It is easy to show that

$$\exp\Big\{-\frac{1}{2}\tilde{\boldsymbol{x}}A\boldsymbol{x}-\frac{1}{2}(\widetilde{\boldsymbol{x}-\boldsymbol{R}})B(\boldsymbol{x}-\boldsymbol{R})\Big\}$$

$$\times\prod_{i=1}^{N}|\boldsymbol{x}_i-\boldsymbol{R}_i|^{2k_i}\mathcal{Y}_{l_im_i}(\boldsymbol{x}_i-\boldsymbol{R}_i)$$

$$=\Bigg(\Bigg\{\prod_{i=1}^{N}\frac{1}{B_{k_il_i}}\int d\hat{\boldsymbol{t}}_iY_{l_im_i}(\hat{\boldsymbol{t}}_i)\frac{\partial^{2k_i+l_i}}{\partial{\alpha_i}^{2k_i+l_i}}\Bigg\}$$

$$\times\exp\Big(-\frac{1}{2}\tilde{\boldsymbol{R}}B\boldsymbol{R}-\widetilde{\alpha|\boldsymbol{t}}\boldsymbol{R}\Big)$$

$$\times\, g(\alpha|\boldsymbol{t}+B\boldsymbol{R};A+B,\boldsymbol{x})\Bigg)_{\substack{\alpha_1=0,\ldots,\alpha_N=0\\|\boldsymbol{t}_1|=1,\ldots,|\boldsymbol{t}_N|=1}}, \qquad (6.25)$$

or

$$f_n(A,B,\boldsymbol{R},\boldsymbol{x})$$

$$\equiv\exp\Big\{-\frac{1}{2}\tilde{\boldsymbol{x}}A\boldsymbol{x}-\frac{1}{2}(\widetilde{\boldsymbol{x}-\boldsymbol{R}})B(\boldsymbol{x}-\boldsymbol{R})\Big\}\prod_{i=1}^{N}\prod_{p=1}^{3}(x_{ip}-R_{ip})^{n_{ip}}$$

$$=\Bigg(\Bigg\{\prod_{i=1}^{N}\prod_{p=1}^{3}\frac{\partial^{n_{ip}}}{\partial{t_{ip}}^{n_{ip}}}\Bigg\}\exp\Big(-\frac{1}{2}\tilde{\boldsymbol{R}}B\boldsymbol{R}-\tilde{\boldsymbol{t}}\boldsymbol{R}\Big)$$

$$\times\, g(\boldsymbol{t}+B\boldsymbol{R};A+B,\boldsymbol{x})\Bigg)_{\boldsymbol{t}_1=0,\ldots,\boldsymbol{t}_N=0}, \qquad (6.26)$$

where $\boldsymbol{n}$ stands for the set of $\{n_{11}, n_{12}, n_{13}, ..., n_{N1}, n_{N2}, n_{N3}\}$. In the Cartesian representation the parameter α plays no active role but the x, y, and z components of each vector $\boldsymbol{t}_i$, $(t_{i_1}, t_{i_2}, t_{i_3})$, serve to construct the function $\theta(\boldsymbol{x} - \boldsymbol{R})$.

To construct the vector-coupled product $\theta_{LM}(\boldsymbol{x})$ of Eq. (6.3), one has to sum over m_i's with appropriate Clebsch-Gordan coefficients in Eqs. (6.21) or (6.22). Apparently this is a very tedious task particularly when the number of particles is large.

The choice of $\theta_{LM}(\boldsymbol{x})$ of Eq. (6.4) leads us to the following very simple equation which relates the correlated Gaussian to g. By choosing $\boldsymbol{t}_1 = \boldsymbol{t}_2 = \cdots = \boldsymbol{t}_N = \lambda \boldsymbol{e}$ with a unit vector $\boldsymbol{e}$, we obtain for $\boldsymbol{v} = \tilde{u}\boldsymbol{x}$

$$f_{KLM}(u, A, \boldsymbol{x}) \equiv \exp\left(-\frac{1}{2}\tilde{\boldsymbol{x}}A\boldsymbol{x}\right) v^{2K} \mathcal{Y}_{LM}(\boldsymbol{v})$$

$$= \frac{1}{B_{KL}} \int Y_{LM}(\hat{\boldsymbol{e}}) \left(\frac{d^{2K+L}}{d\lambda^{2K+L}} g(\lambda \boldsymbol{e} u; A, \boldsymbol{x}) \right)_{\lambda=0, e=|\boldsymbol{e}|=1} d\hat{\boldsymbol{e}}. \tag{6.27}$$

We can see from Eqs. (6.21), (6.22), and (6.25)–(6.27) that the correlated Gaussians are explicitly constructed from the generating function g. Depending on the choice of the vector $\boldsymbol{s}$, g leads to different forms of the correlated Gaussians, when followed by suitable operations acting on $\boldsymbol{s}$. These are summarized in Table 6.2. The construction of $f_{KLM}(u, A, \boldsymbol{x})$ is simplest among others and it has a wide range of applications as will be shown in later chapters.

The correlated Gaussian of Eq. (6.27) contains only the relative coordinates and the center-of-mass motion is dropped from the outset. Thus there is no problem arising from the coupling between the intrinsic motion and the center-of-mass motion. If one uses the single-particle coordinates $\boldsymbol{r}$ instead of $\boldsymbol{x}$ in Eq. (6.27), the coupling between them occurs in general and has to be taken care of appropriately in order to calculate the energy of the intrinsic motion. A suitable choice of A and u will, however, lead to the result that the center-of-mass motion can be separated from the intrinsic motion. This will be discussed in more detail in Complement 6.4.

In Chap. 2 we discussed the linear transformation of the coordinates $\boldsymbol{x}$ induced by the permutation P. It is important to know the

Table 6.2. Relationship of the two types of the correlated Gaussians to the generating function g of Eq. (6.19). The symbol $\alpha|\boldsymbol{t}$ indicates a set of vectors $\{\alpha_1\boldsymbol{t}_1,\dots,\alpha_N\boldsymbol{t}_N\}$. B_{kl} and C_{lm} are defined in Eqs. (6.9) and (6.23), respectively.

Correlated Gaussians

$$\exp\Big(-\tfrac{1}{2}\tilde{\boldsymbol{x}}A\boldsymbol{x}\Big)\prod_{i=1}^{N}\mathcal{Y}_{l_im_i}(\boldsymbol{x}_i) = \Big(\Big\{\prod_{i=1}^{N}\frac{1}{B_{0l_i}}\int d\widehat{\boldsymbol{t}}_iY_{l_im_i}(\widehat{\boldsymbol{t}}_i)\frac{\partial^{l_i}}{\partial{\alpha_i}^{l_i}}\Big\}g(\alpha|\boldsymbol{t};A,\boldsymbol{x})\Big)_{\substack{\alpha_i=0,|\boldsymbol{t}_i|=1\\(i=1,\dots,N)}}$$

$$\exp\Big(-\tfrac{1}{2}\tilde{\boldsymbol{x}}A\boldsymbol{x}\Big)\prod_{i=1}^{N}\mathcal{Y}_{l_im_i}(\boldsymbol{x}_i) = \Big(\Big\{\prod_{i=1}^{N}\frac{1}{C_{l_im_i}}\frac{\partial^{l_i}}{\partial{\alpha_i}^{l_i}}\frac{\partial^{l_i-m_i}}{\partial{\tau_i}^{l_i-m_i}}\Big\} \times g(\alpha|\boldsymbol{t};A,\boldsymbol{x})\Big|_{\substack{\boldsymbol{t}_j=(1-{\tau_j}^2,i(1+{\tau_j}^2),-2\tau_j)\\(j=1,\dots,N)}}\Big)_{\substack{\alpha_i=0,\tau_i=0\\(i=1,\dots,N)}}$$

$$\exp\Big(-\tfrac{1}{2}\tilde{\boldsymbol{x}}A\boldsymbol{x}\Big)|\boldsymbol{v}|^{2K}\mathcal{Y}_{LM}(\boldsymbol{v})\quad \Big(\boldsymbol{v}=\sum_{i=1}^{N}u_i\boldsymbol{x}_i\Big) = \frac{1}{B_{KL}}\int d\hat{\boldsymbol{e}}Y_{LM}(\hat{\boldsymbol{e}})\Big(\frac{d^{2K+L}}{d\lambda^{2K+L}}g(\lambda\boldsymbol{e}u;A,\boldsymbol{x})\Big)_{\lambda=0,e=|\boldsymbol{e}|=1}$$

Correlated Gaussian-type geminals

$$\exp\Big\{-\tfrac{1}{2}\tilde{\boldsymbol{x}}A\boldsymbol{x}-\tfrac{1}{2}(\widetilde{\boldsymbol{x}-\boldsymbol{R}})B(\boldsymbol{x}-\boldsymbol{R})\Big\}\prod_{i=1}^{N}|\boldsymbol{x}_i-\boldsymbol{R}_i|^{2k_i}\mathcal{Y}_{l_im_i}(\boldsymbol{x}_i-\boldsymbol{R}_i) = \Big(\Big\{\prod_{i=1}^{N}\frac{1}{B_{k_il_i}}\int d\widehat{\boldsymbol{t}}_iY_{l_im_i}(\hat{\boldsymbol{t}}_i)\frac{\partial^{2k_i+l_i}}{\partial{\alpha_i}^{2k_i+l_i}}\Big\} \times\exp\Big(-\tfrac{1}{2}\widetilde{\boldsymbol{R}}B\boldsymbol{R}-\widetilde{\alpha|\boldsymbol{t}}\boldsymbol{R}\Big)g(\alpha|\boldsymbol{t}+B\boldsymbol{R};A+B,\boldsymbol{x})\Big)_{\substack{\alpha_i=0,|\boldsymbol{t}_i|=1\\(i=1,\dots,N)}}$$

$$\exp\Big\{-\tfrac{1}{2}\tilde{\boldsymbol{x}}A\boldsymbol{x}-\tfrac{1}{2}(\widetilde{\boldsymbol{x}-\boldsymbol{R}})B(\boldsymbol{x}-\boldsymbol{R})\Big\}\prod_{i=1}^{N}\prod_{p=1}^{3}(x_{ip}-R_{ip})^{n_{ip}} = \Big(\Big\{\prod_{i=1}^{N}\prod_{p=1}^{3}\frac{\partial^{n_{ip}}}{\partial{t_{ip}}^{n_{ip}}}\Big\}\exp\Big(-\tfrac{1}{2}\widetilde{\boldsymbol{R}}B\boldsymbol{R}-\tilde{\boldsymbol{t}}\boldsymbol{R}\Big) \times g(\boldsymbol{t}+B\boldsymbol{R};A+B,\boldsymbol{x})\Big)_{\substack{\boldsymbol{t}_i=0\\(i=1,\dots,N)}}$$

effect of P on the correlated Gaussians. Since the correlated Gaussian is generated from the generating function, it suffices to examine the transformation property of g due to P. By using the relation $P\boldsymbol{x} = T_P\boldsymbol{x}$ (see Eqs. (2.26) and (2.29)), we obtain a very simple result

$$Pg(\boldsymbol{s}; A, \boldsymbol{x}) = g(\widetilde{T_P}\boldsymbol{s}; \widetilde{T_P}AT_P, \boldsymbol{x}). \tag{6.28}$$

One only needs to change the matrix A and the vector $\boldsymbol{s}$ appropriately. An important fact is that the generating function preserves its functional form under P. This is also true for a more general linear transformation $\mathcal{T}$ of the coordinates $\boldsymbol{x}$, e.g., a transformation from one set of coordinates to another. Combining this fact with Eq. (6.27), we obtain a very useful property of the correlated Gaussian f_{KLM}. Namely for the transformation of $\mathcal{T}\boldsymbol{x} = T\boldsymbol{x}$, we have

$$\mathcal{T} f_{KLM}(u, A, \boldsymbol{x}) = f_{KLM}(\widetilde{T}u, \widetilde{T}AT, \boldsymbol{x}). \tag{6.29}$$

Thanks to this nice property one only needs to redefine the parameters A and u of the basis function to construct the transformed wave function. This property plays an important role in evaluating the matrix elements.

The generating function (6.19) plays a key role in generating the correlated Gaussians and, moreover, facilitates the evaluation of the matrix elements of physical operators. It is therefore desirable that one can use the formulation based on the generating function in a many-body system as well. In extending the results to larger systems of identical particles we need to cope with the symmetry adaptation of the wave function. In such a case it would sometimes be useful if the generating function were expressed in an "uncorrelated" form of the coordinates $\boldsymbol{x}$, because the symmetry adaptation can then be simplified by using the technique of Slater determinants or permanents.

In fact we can show that the generating function can be related to the product of the Gaussian wave-packets centered around $\widetilde{\boldsymbol{R}} = (\boldsymbol{R}_1, ..., \boldsymbol{R}_N)$ through an integral transformation. Using the definition of Eq. (6.51) of Complement 6.1, we can express the product of the Gaussian wave-packets as

$$\prod_{i=1}^{N} \varphi_{\boldsymbol{R}_i}^{\gamma_i}(\boldsymbol{x}_i) = \left(\frac{\det\Gamma}{\pi^N}\right)^{\frac{3}{4}} \exp\Big(-\frac{1}{2}\tilde{\boldsymbol{x}}\Gamma\boldsymbol{x} + \widetilde{\boldsymbol{R}}\Gamma\boldsymbol{x} - \frac{1}{2}\widetilde{\boldsymbol{R}}\Gamma\boldsymbol{R}\Big) \tag{6.30}$$

with an $N \times N$ diagonal matrix

$$\Gamma = \begin{pmatrix} \gamma_1 & 0 & \cdots & 0 \\ 0 & \gamma_2 & \cdots & \vdots \\ \vdots & & & \vdots \\ 0 & \cdots & \cdots & \gamma_N \end{pmatrix}. \tag{6.31}$$

A direct calculation using

$$\int \exp\Big(-\frac{1}{2}\tilde{\boldsymbol{x}}A\boldsymbol{x} + \tilde{\boldsymbol{s}}\boldsymbol{x}\Big) d\boldsymbol{x} = \Big(\frac{(2\pi)^N}{\det A}\Big)^{\frac{3}{2}} \exp\Big(\frac{1}{2}\tilde{\boldsymbol{s}}A^{-1}\boldsymbol{s}\Big) \tag{6.32}$$

proves the following equation which relates g of Eq. (6.19) to the product of the Gaussians of Eq. (6.30):

$$\begin{aligned} g(\boldsymbol{s}; A, \boldsymbol{x}) \\ &= \Big(\frac{(\det\Gamma)^3}{(4\pi)^N(\det(\Gamma - A))^2}\Big)^{\frac{3}{4}} \exp\Big\{ -\frac{1}{2}\tilde{\boldsymbol{s}}(\Gamma - A)^{-1}\boldsymbol{s}\Big\} \\ &\times \int g\Big(\Gamma(\Gamma - A)^{-1}\boldsymbol{s}; A(\Gamma - A)^{-1}\Gamma, \boldsymbol{R}\Big) \Big(\prod_{i=1}^{N} \varphi_{\boldsymbol{R}_i}^{\gamma_i}(\boldsymbol{x}_i)\Big) d\boldsymbol{R}. \end{aligned} \tag{6.33}$$

Note that $A(\Gamma - A)^{-1}\Gamma = \Gamma(\Gamma - A)^{-1}\Gamma - \Gamma$ is a symmetric matrix. The function g in the right-hand side of the above equation depends on the integration variable $\boldsymbol{R}$ and serves as just a weight which is needed to convert the product of the Gaussian wave-packets to the generating function. Equation (6.32) is proved in Exercise 6.2.

We have seen that the correlated Gaussians are all generated from the same generating function as summarized in Table 6.2. Furthermore, the latter can be obtained by the integral transformation of Eq. (6.33) involving the product of single-particle Gaussian wave-packets. From this we see that the calculations can be reduced to the matrix elements of N-particle wave functions involving the product of the Gaussian wave-packets. The width parameters γ_i of the wave packets can be chosen arbitrarily. To choose uniform width parameter for all of them is most convenient if $\boldsymbol{x}_i$ indicates the single-particle coordinate of the identical particles, because then the permutational symmetry of the wave function is simply reduced to that of the "generator coordinates" $\boldsymbol{R}$. Even when $\boldsymbol{x}$ denotes the set of relative coordinates, this nice property can be used by including the center-of-mass coordinate as shown in [31]. The integral representation of the generating function has been successfully employed in accurate solutions of few-body problems [31] as well as in microscopic descriptions of nuclear systems in multicluster models [29, 30].

6.4 The spin function

The spin function of the N-fermion system cannot be uniquely specified by the total spin S alone. One elementary way to construct it is to use a successive coupling of the spins of the particles. The intermediate spin quantum numbers S_{12}, S_{123}, etc., appearing in the successive coupling serve to specify the spin function. They do not, however, constitute good quantum numbers in general and thus several sets of the spin quantum numbers may be needed to obtain a good wave function. The order of the spin coupling in the successive couplings is fixed and in general does not naturally follow the choice of the relative coordinates. In addition the discrete nature of the set of intermediate spins may not be convenient from the point of view of variational calculations.

For example, in the case of three electrons, the spin function with $S = 1/2$ has two independent states, labelled by $S_{12} = 0$ or 1. Therefore, the general spin function in this case can be expressed as

$$\chi_{\frac{1}{2}M} = \cos\vartheta \left|\tfrac{1}{2}\tfrac{1}{2}(0), \tfrac{1}{2}; \tfrac{1}{2}M\right\rangle + \sin\vartheta \left|\tfrac{1}{2}\tfrac{1}{2}(1), \tfrac{1}{2}; \tfrac{1}{2}M\right\rangle. \tag{6.34}$$

Alternatively, we may set up the spin function with a continuous parameter such as that of Eq. (6.34) by an elementary method instead of using the successive coupling. Its merit is that the construction of the spin function is simple, that the evaluation of spin matrix elements is easy, and moreover that one can use continuous parameters such as ϑ in Eq. (6.34), in the variational calculation.

The construction is done as follows. The Young diagram for the spin function with spin S for the N-fermion system is $[n_+ \; n_-] = [(N/2) + S \; (N/2) - S]$. The maximum weight function with $M = S$, χ_{SS}, must have n_+ spin-up functions and n_- spin-down functions. The number of terms distributing n_+ spin-up functions among the N particles is $N_S = \binom{N}{n_+}$. Therefore χ_{SS} is expressed as a sum over these terms:

$$\chi_{SS}(\lambda) = \sum_{i=1}^{N_S} \lambda_i |i\rangle. \tag{6.35}$$

The coefficients $(\lambda_1, .., \lambda_{N_S}) \equiv \tilde{\lambda}$, characterizing the spin function, are not independent of each other but must satisfy the condition, $S_+ \chi_{SS}(\lambda) = 0$, where S_+ is the spin-raising operator. Acting with S_+ on $\chi_{SS}(\lambda)$ yields $n_+ + 1$ spin-up functions and $n_- - 1$ spin-down functions in each term and thus leads to $\binom{N}{n_+ + 1}$ independent terms. Since the coefficient of each term must vanish, one has

$$\binom{N}{n_+} - \binom{N}{n_+ + 1} - 1 = \frac{(2S+1)N!}{(\frac{1}{2}N + S + 1)!\,(\frac{1}{2}N - S)!} - 1 \tag{6.36}$$

independent parameters to specify the vector λ completely. Here the last minus one of Eq. (6.36) comes from the normalization of the spin function. χ_{SM} is easily obtained from $\chi_{SS}(\lambda)$ with the use of the spin-lowering operator and thus denoted as $\chi_{SM}(\lambda)$. The overlap of two spin functions is independent of M and simply given by $\langle \chi_{SM}(\lambda) | \chi_{SM}(\lambda') \rangle = \langle \chi_{SS}(\lambda) | \chi_{SS}(\lambda') \rangle = \tilde{\lambda}\lambda'$. The independent parameters needed to specify λ can be varied continuously in the variational calculation.

The isospin function is also constructed in exactly the same manner. The action of the permutation P on the spin and isospin parts simply produces a reordering of the terms in Eq. (6.35), leading to a linear transformation of the vector λ to another vector denoted $\lambda(P)$.

Complements

6.1 *Nodeless harmonic-oscillator functions as a basis*
In variational calculations for bound states it is very important to have a set of basis functions which can approximate square-integrable functions to any desired accuracy. For a single-variable function one may use the well-known complete, orthonormal eigenfunctions of, e.g., the harmonic-oscillator (HO) Hamiltonian. Or one may try to use nonorthogonal functions which may not be complete mathematically but are able to cope with many problems flexibly and, from a practical point of view, accurately. The purpose of this Complement is to examine the possibility of nodeless HO functions as a set of such "basis" functions. We will see the performance of the nodeless HO functions by studying how well they approximate a given function $f(\boldsymbol{r})$.

The nodeless HO functions with angular momentum lm have a continuous parameter a:

$$\Gamma_{alm}(\boldsymbol{r}) = N_{al}\left(\frac{\nu^3}{\pi}\right)^{\frac{1}{4}} \exp\Big(-\frac{a}{2}\nu r^2\Big)\mathcal{Y}_{lm}(\sqrt{\nu}\boldsymbol{r}) \tag{6.37}$$

with a solid spherical harmonic (see the following Complement). where the normalization constant N_{al} is given by

$$N_{al} = \left(\frac{2^{l+2}a^{l+\frac{3}{2}}}{(2l+1)!!}\right)^{\frac{1}{2}}. \tag{6.38}$$

Here ν is introduced to scale the length. The Gaussians employed in Complement 8.1 to solve the ground state of the hydrogen atom are special cases with $l = 0$. The overlap of two nodeless HO functions is simply given by

$$\langle \Gamma_{alm}|\Gamma_{a'lm}\rangle = \frac{N_{al}N_{a'l}}{\left(N_{\frac{a+a'}{2}l}\right)^2} = \left(\frac{aa'}{(\frac{a+a'}{2})^2}\right)^{\frac{1}{2}(l+\frac{3}{2})}. \tag{6.39}$$

We attempt to approximate a normalized function $f_{lm}(\boldsymbol{r})$ with angular momentum lm in terms of combinations of $\Gamma_{alm}(\boldsymbol{r})$

$$f_{lm}(\boldsymbol{r}) \sim \sum_{i=1}^{K} c_i\Gamma_{a_ilm}(\boldsymbol{r}). \tag{6.40}$$

The error of the approximation is estimated by calculating the variance

$$\sigma^2(f) = \int \left| f_{lm}(\boldsymbol{r}) - \sum_{i=1}^{K} c_i \Gamma_{a_i lm}(\boldsymbol{r}) \right|^2 d\boldsymbol{r}. \tag{6.41}$$

One may choose a set of parameters $\{a_1, a_2, ..., a_K\}$ by some appropriate way or may set them up by the trial and error procedure of the SVM. Once they are selected, minimizing σ^2 leads to a linear equation to determine the c_i's

$$\sum_{j=1}^{K} \langle \Gamma_{a_i lm} | \Gamma_{a_j lm} \rangle c_j = \langle \Gamma_{a_i lm} | f_{lm} \rangle \qquad (i = 1, ..., K), \tag{6.42}$$

and then σ^2 is given by $1 - \sum_{i,j=1}^{K} c_i^* c_j \langle \Gamma_{a_i lm} | \Gamma_{a_j lm} \rangle$.

As a test function $f_{lm}(\boldsymbol{r})$ we first take up the HO wave functions $\psi_{nlm}(\boldsymbol{r})$ with the angular frequency $\omega = \hbar\nu/m$, where n is the number of radial nodes. We employ a phase convention such that the radial HO wave functions are all positive for r greater than the outermost nodal point. The overlap of the functions $\Gamma_{alm}(\boldsymbol{r})$ and $\psi_{nlm}(\boldsymbol{r})$ is needed to calculate σ^2. It can be obtained by using a generating function of the HO functions

$$\begin{aligned} A(\boldsymbol{k}, \boldsymbol{r}) &\equiv \left(\frac{\nu}{\pi}\right)^{\frac{3}{4}} \exp\left\{ -\frac{1}{2}\boldsymbol{k}^2 + \sqrt{2\nu}\boldsymbol{k}\cdot\boldsymbol{r} - \frac{1}{2}\nu\boldsymbol{r}^2 \right\} \\ &= \sum_{nlm} \psi_{nlm}(\boldsymbol{r})^* P_{nlm}(\boldsymbol{k}), \end{aligned} \tag{6.43}$$

where $P_{nlm}(\boldsymbol{k})$, a polynomial of the complex Bargmann space variable $\boldsymbol{k}$, is the Bargmann transform [76, 77] of the HO wave functions in a spherical basis and its explicit form is given in terms of the solid spherical harmonics of (4.5) [78]

$$P_{nlm}(\boldsymbol{k}) = \sqrt{\frac{B_{nl}}{(2n+l)!}} (\boldsymbol{k}\cdot\boldsymbol{k})^n \mathcal{Y}_{lm}(\boldsymbol{k}), \tag{6.44}$$

where B_{nl}, defined in Eq. (6.9), is the coefficient needed to express x^m in terms of the Legendre polynomials $P_l(x)$:

$$x^m = \sum_{2n+l=m} \frac{2l+1}{4\pi} B_{nl} P_l(x). \tag{6.45}$$

Expanding the overlap $\langle \Gamma_{alm}(\boldsymbol{r}) | A(\boldsymbol{k}, \boldsymbol{r}) \rangle$ in terms of the polynomials (6.44), we obtain

$$\langle \Gamma_{alm}|\psi_{nlm}\rangle = \sqrt{\frac{(2n+2l+1)!!}{(2n)!!\,(2l+1)!!}}\left(\frac{1-a}{1+a}\right)^n \left(\frac{2\sqrt{a}}{1+a}\right)^{l+\frac{3}{2}} . \tag{6.46}$$

Obviously, the approximation in Eq. (6.40) becomes less trivial as the number of radial nodes of f increases. For $\psi_{nlm}(\boldsymbol{r})$ with $n = 1$ or 2, the combination of merely a few (≤ 3) terms leads to very good approximations, whereas for $n = 5$ it is hard to make σ^2 less than 10^{-10} with a double precision calculation. This is so because all a_i's tend to take a value of unity, reflecting the fact that such a HO function can be expressed by combinations of higher order derivatives of $\Gamma_{alm}(\boldsymbol{r})$ with respect to a at $a = 1$, and the optimization procedure tends to construct such derivatives out of Γ_{alm} of almost equal parameters. One thus has to deal with an almost singular overlap matrix $(\langle \Gamma_{a_ilm}|\Gamma_{a_jlm}\rangle)$ to obtain the solution of c_i's.

Instead of approximating each of the $\psi_{nlm}(\boldsymbol{r})$ with different sets of a_i's, we minimized the average value, $\langle\sigma^2\rangle_N = (\sum_{n=0}^{N}\sigma^2(\psi_{nlm}))/(N+1)$, with a single set of a_i's. The values of the a_i's were assumed to follow a geometric progression and its first term and the ratio of successive terms were determined by Powell's method [14]. A combination of ten nodeless HO functions yielded $\langle\sigma^2\rangle_8 = 0.6\times 10^{-10}$ for $l = 0$.

The next test example is the shifted Gaussian defined by

$$\psi_{slm}(\boldsymbol{r}) = F_{sl}\left(\frac{\nu^3}{\pi}\right)^{\frac{1}{4}} \exp\Big(-\frac{1}{2}\nu(r^2+s^2)\Big) i_l(\nu s r) Y_{lm}(\hat{\boldsymbol{r}}), \tag{6.47}$$

where the function $i_l(x)$ is the modified spherical Bessel function of the first kind

$$i_l(x) = \sqrt{\frac{\pi}{2x}} I_{l+\frac{1}{2}}(x) = \sum_{k=0}^{\infty} \frac{x^{2k+l}}{(2k)!!\,(2k+2l+1)!!}, \tag{6.48}$$

and where the normalization constant F_{sl} is

$$F_{sl} = \left(\frac{4\mathrm{e}^{\varsigma}}{i_l(\varsigma)}\right)^{\frac{1}{2}} \qquad \text{with} \qquad \varsigma = \frac{1}{2}\nu s^2. \tag{6.49}$$

To get the normalization constant, we have used the formula

$$\int_0^\infty x\,\mathrm{e}^{-ax^2} I_\nu(bx) I_\nu(cx)dx = \frac{1}{2a}\exp\Big(\frac{b^2+c^2}{4a}\Big) I_\nu\Big(\frac{bc}{2a}\Big), \tag{6.50}$$

which is valid for $\mathrm{Re}\,a > 0$, $\mathrm{Re}\,\nu > -1$. The function (6.47) is named shifted Gaussian because it is closely related to the single-particle Gaussian wave-packet centered around $\boldsymbol{s}$

$$\varphi_s^\nu(\boldsymbol{r}) = \left(\frac{\nu}{\pi}\right)^{\frac{3}{4}} \exp\left(-\frac{1}{2}\nu(\boldsymbol{r}-\boldsymbol{s})^2\right) \tag{6.51}$$

through the relation

$$\varphi_s^\nu(\boldsymbol{r}) = \sum_{lm} \frac{4\sqrt{\pi}}{F_{sl}} \psi_{slm}(\boldsymbol{r}) Y_{lm}(\hat{\boldsymbol{s}})^*. \tag{6.52}$$

Here use is made of the equation

$$\mathrm{e}^{\nu \boldsymbol{r}\cdot\boldsymbol{s}} = \mathrm{e}^{\nu r s \cos\vartheta} = \sum_l (2l+1) i_l(\nu r s) P_l(\cos\vartheta) \tag{6.53}$$

with the angle ϑ between $\boldsymbol{r}$ and $\boldsymbol{s}$, and the well-known addition theorem for the spherical harmonics

$$\begin{aligned} P_l(\cos\vartheta) &= \frac{4\pi}{2l+1} \sum_{m=-l}^{l} Y_{lm}(\hat{\boldsymbol{r}}) Y_{lm}(\hat{\boldsymbol{s}})^* \\ &= 4\pi \frac{(-1)^l}{\sqrt{2l+1}} [Y_l(\hat{\boldsymbol{r}}) \times Y_l(\hat{\boldsymbol{s}})]_{00}. \end{aligned} \tag{6.54}$$

The radial part of the shifted Gaussian is hence peaked at $r \sim s = \sqrt{2\varsigma/\nu}$. It would therefore be challenging to approximate the shifted Gaussian with large s in terms of the combinations of Γ_{alm}. Note that $\varphi_s^\nu(\boldsymbol{r})$ is expressible in terms of A of Eq. (6.43) as $\mathrm{e}^{-\frac{1}{4}\nu s^2} A(\sqrt{\frac{\nu}{2}}\boldsymbol{s}, \boldsymbol{r})$. The overlap between the functions $\Gamma_{alm}(\boldsymbol{r})$ and $\psi_{slm}(\boldsymbol{r})$ is

$$\langle \Gamma_{alm} | \psi_{slm} \rangle = \frac{1}{\sqrt{2}} N_{al} F_{sl} \frac{(2\varsigma)^{\frac{1}{2}l}}{(1+a)^{l+\frac{3}{2}}} \exp\left(-\frac{a}{1+a}\varsigma\right). \tag{6.55}$$

Assuming again that the a_i's follow a geometric progression, σ^2 was minimized by Powell's method for a given value of ς. For the case of $l = 0$ it is possible to make $\sigma^2 \leq 10^{-10}$ with ten nodeless HO functions for the shifted Gaussians of up to $\varsigma \leq 10$. All a_i then tend to take values close to each other for large ς, and this requires a very precise solution of the linear equation (6.42). Increasing the number of nodeless HO functions gives us even smaller σ^2 values and widens the range of maximum ς value in which the shifted Gaussian can be approximated well.

The above examples strongly suggest that the nodeless HO functions can approximate square-integrable functions to any desired accuracy, though the number of nodeless HO functions needed depends on the shape of the test function. A remarkable example given in [66] shows that the linear function r itself, defined in a finite range, can be approximated to high accuracy in terms of a combination of Gaussians $(1 - e^{-ar^2})$.

We stress two remarkable points of the nodeless HO functions. One is that a combination of a few nodeless HO functions can approximate the shifted Gaussian even with a large value of ς. For example, it is easy to make σ^2 less than 10^{-10} with a combination of only 15 nodeless HO functions. In an expansion in terms of HO functions, one would need many more terms to obtain the same accuracy. This indicates the flexibility of a nonorthogonal Gaussian "basis" compared to the orthogonal basis such as the HO functions. Another point is that there are many, possibly an infinite number, of sets $\{a_1, ..., a_K\}$ which approximate a given function equally well, even though the number K is fixed. This is what we have experienced in the above examples.

We display graphic illustrations of Gaussian expansions for different functions with $l = 0$, $m = 0$. The most appropriate nonlinear parameters a_i are determined by optimization, while the linear ones c_i are given by the solution of the least square equation (6.42). To point out the dependence on the number of Gaussians, we use different numbers of terms in the expansion ($K = 5$, 10 and 20).

In the first example we approximate the HO function $\psi_{500}(r)$ by Gaussians. This function is smooth but oscillates (it has five nodes) and asymptotically falls off like a Gaussian function. As shown in Fig. 6.1, $K = 10$ and $K = 20$ Gaussians give a perfect fit to the exact function, so these curves are practically indistinguishable.

In the second case we attempt to fit an exponential function $f(r) = e^{-r}$, the wave function of the ground state of the hydrogen atom. The asymptotics of this function is quite different from that of a Gaussian. To approximate the asymptotic part of this function one needs many terms of Gaussians as is illustrated in Fig. 6.2. By increasing the number of Gaussians one has better and better agreement in the asymptotics. After a certain distance, the Gaussians fall off much more rapidly than the exponential function. In many practical applications, especially for bound states, however, one can always use enough Gaussians to reach the required accuracy. We note that the Gaussian fit gives a poor value (zero) for the derivative at the origin (the exact value is -1) as will be discussed in Complement 8.1.

In the next example we try to approximate the absolute value function $f(r) = |2.5 - r|$. The Gaussian expansion, again, does a pretty

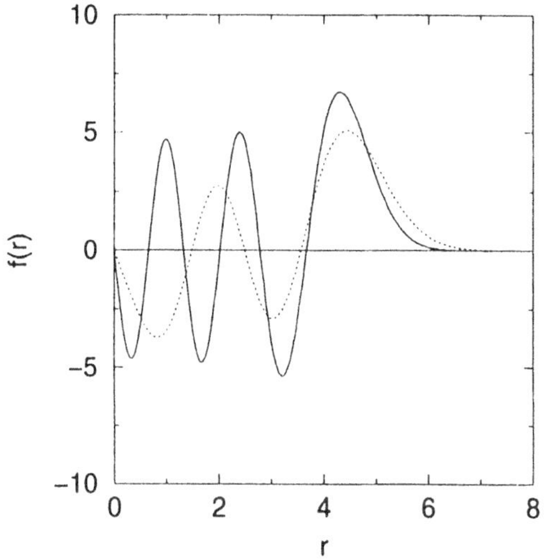

Fig. 6.1. The harmonic-oscillator function ($n = 5$, $l = 0$, $m = 0$) and its approximations by Gaussian expansions. The solid curve is the (unnormalized) harmonic-oscillator function $f(r) \sim \psi_{500}$, and the dotted, dashed and long-dashed curves are the $K = 5$-, 10- and 20-term Gaussian expansions. The 10- and 20-term expansions are practically indistinguishable from the exact curve.

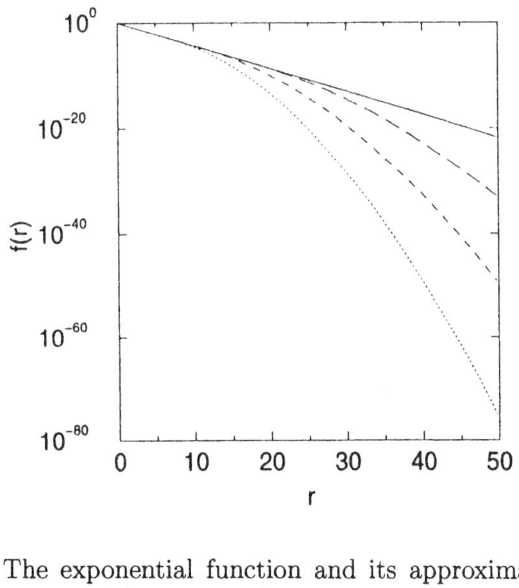

Fig. 6.2. The exponential function and its approximations by Gaussian expansions. The solid curve is the exponential function $f(r) = \exp(-r)$, and the dotted, dashed and long-dashed curves are the $K = 5$-, 10- and 20-term Gaussian expansions.

good job (Fig. 6.3), and by the inclusion of further Gaussians one would expect an even better fit.

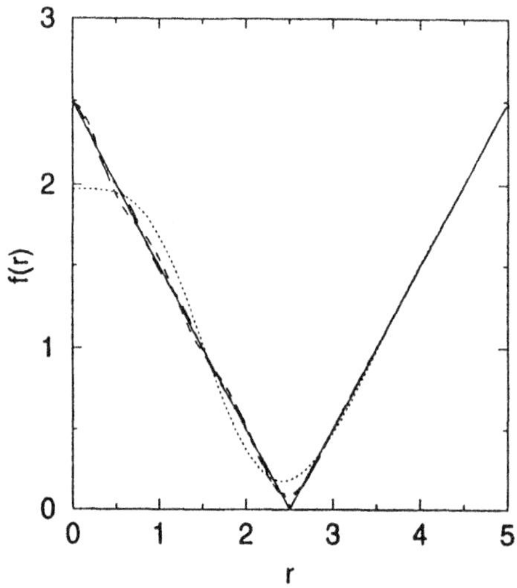

Fig. 6.3. The absolute value function and its approximations by Gaussian expansions. The solid line is the absolute value function $f(r) = |2.5 - r|$, and the dotted, dashed and long-dashed curves are the K = 5-, 10- and 20-term Gaussian expansions.

The approximation for the step function, $f(r) = 1$ if $r \leq 2.5$, $f(r) = 2$ if $r > 2.5$, is less impressive (Fig. 6.4), but it goes without saying that it is not trivial to fit that function. One can improve the fit by increasing the number of Gaussians, but the convergence might be pretty slow.

The last example shows that a Gaussian expansion may be inadequate in some cases. Let us try to approximate $f(r) = r^{12}\mathrm{e}^{-r^2}$ with Gaussians. This function is practically zero near the origin, and it simulates the behavior of a wave function of a system with very strong repulsive core. The Lenard-Jones or other hard-core potentials may produce such a function. Figure 6.5 shows that the Gaussians generally give a good fit to this function. By scrutinizing the inner part of the approximation (see Fig. 6.6) one can see, however, that in reality the Gaussians do not produce exact zero near the origin, but

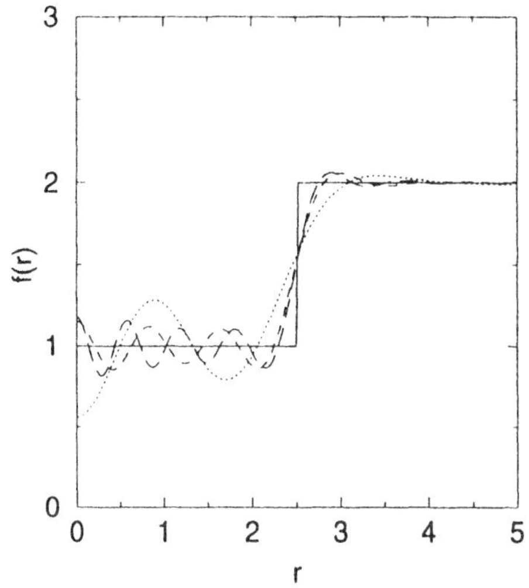

Fig. 6.4. The step function and its approximations by Gaussian expansions. The solid line is the step function $f(r) = 1$ if $r \leq 2.5$, $f(r) = 2$ if $r > 2.5$, and the dotted, dashed and long-dashed curves are the K = 5-, 10- and 20-term Gaussian expansions.

the approximate function oscillates around the exact one even for 20 Gaussians. This oscillation may be very unpleasent: To tame the hard core potential, we need a wave function which is effectively zero near the origin. The oscillating function leads to numerical problems, which makes it rather difficult to obtain the solution of problems where the interaction has a very strong hard-core part.

It is important to know the completeness of the basis functions in L^2 (square integrable functions), H^1 and H^2 (first and second Sobolev) spaces because the bound state in quantum mechanics is traditionally formulated in L^2, and the mathematical solution of the Schrödinger equation is formulated in H^1 and H^2 spaces. (The space H^m is a set of functions whose derivatives of up to the mth order are all square integrable. Note that the kinetic energy operator requires the second order derivative of a basis function.) The convergence properties of the Ritz variational method are demonstrated in H^1 and H^2 spaces, for example in [79], and the completeness in L^2 is not sufficient to guarantee the convergence of the Ritz method. The proof of the completeness in the space of L^2 and in the spaces of H^1 and H^2 is presented in [80]. This work proves that any function can be approximated to any

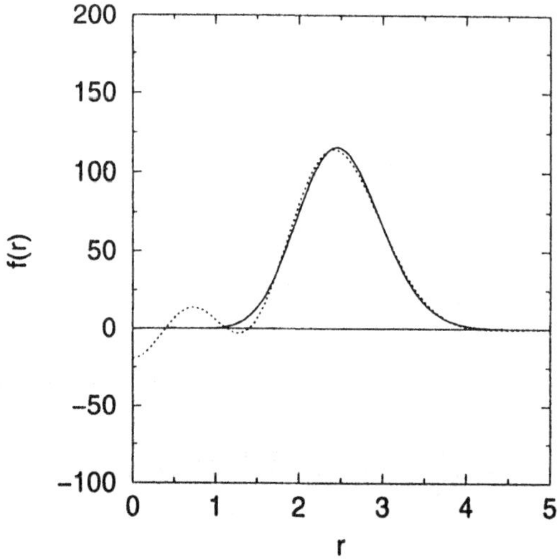

Fig. 6.5. The function $f(r) = r^{12}e^{-r^2}$ and its approximations by Gaussian expansions. The solid curve is the function and the dotted, dashed and long-dashed curves are the K = 5-, 10- and 20-term Gaussian expansions. The 10 and 20 term expansions are practically indistinguishable from the exact curve.

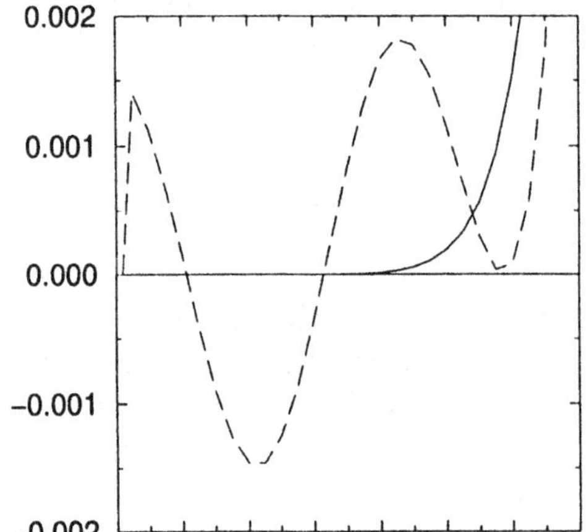

Fig. 6.6. The same as in Fig. 6.5, but the inner part is magnified. The curves with the K = 5- and 10-term Gaussian expansions are out of the scope and not drawn here.

prescribed accuracy with a linear combination of Gaussians, provided that the number of Gaussians in the expansion is sufficiently large and the parameters of the Gaussians are appropriately chosen.

6.2 *Solid spherical harmonics*

The solid spherical harmonic $\mathcal{Y}_{lm}(\boldsymbol{r}) = r^l Y_{lm}(\hat{\boldsymbol{r}})$ of Eq. (4.5) is a solution, regular at $r = 0$, of the Laplace equation, $\nabla^2 f(\boldsymbol{r}) = 0$. The irregular solution is given by $r^{-l-1} Y_{lm}(\hat{\boldsymbol{r}})$. The solid spherical harmonic is a homogeneous polynomial of degree l in the Cartesian coordinate:

$$r^l Y_{lm}(\hat{\boldsymbol{r}}) = \sqrt{\frac{2l+1}{4\pi}(l+m)!\,(l-m)!} \sum_{pq} \frac{1}{p!\,q!\,(l-p-q)!} \times \left(-\frac{x+iy}{2}\right)^p \left(\frac{x-iy}{2}\right)^q z^{l-p-q}, \tag{6.56}$$

where p and q are positive integers which satisfy the conditions $p+q \leq l$, $p-q=m$. The complex conjugation leads to $Y_{lm}(\hat{\boldsymbol{r}})^* = (-1)^m Y_{l\,-m}(\hat{\boldsymbol{r}})$.

It is easy to show that with a special complex vector $\boldsymbol{t} = (1-\tau^2, i(1+\tau^2), -2\tau)$ we obtain

$$(\boldsymbol{t}\cdot\boldsymbol{r})^l = \Big(x+iy-\tau^2(x-iy)-2\tau z\Big)^l = \sum_{m=-l}^{l} \frac{1}{(l-m)!} C_{lm} \tau^{l-m} \mathcal{Y}_{lm}(\boldsymbol{r}), \tag{6.57}$$

where C_{lm} is given in Eq. (6.23). For a given l the power of τ must satisfy the condition $0 \leq l-m \leq 2l$, which guarantees the range of m as it should. Equation (6.57) leads to the equality

$$\left(\frac{\partial^l}{\partial\alpha^l}\frac{\partial^{l-m}}{\partial\tau^{l-m}} e^{\alpha \boldsymbol{t}\cdot\boldsymbol{r}}\right)_{\substack{\alpha=0\\ \tau=0}} = C_{lm}\mathcal{Y}_{lm}(\boldsymbol{r}). \tag{6.58}$$

This simple relation was used in Sect. 6.3 to generate the angular part of variational trial functions.

By expressing the Laplacian as $\nabla^2 = \frac{1}{r^2}\frac{\partial}{\partial r}\left(r^2\frac{\partial}{\partial r}\right) - \frac{\boldsymbol{l}^2}{r^2}$, we see that the solutions of the equation

$$\left[\nabla^2 - \frac{(n-l)(n+l+1)}{r^2}\right] f(\boldsymbol{r}) = 0 \tag{6.59}$$

can be expressed in polar coordinates as $r^n Y_{lm}(\boldsymbol{r})$ or $r^{-n-1}Y_{lm}(\boldsymbol{r})$. When $n = l$ or $n = -l-1$, the equation reduces to the Laplace equation.

The inverse relation of Eq. (6.56) is expressed as follows:

$$x^p y^q z^{n-p-q} = r^n \sum_l \sum_{m=-l}^{l} B_{lm}^{npq}\, Y_{lm}(\hat{\boldsymbol{r}}), \tag{6.60}$$

where l takes the values $n, n-2, ..., 0$ or 1. A hint to calculate B_{lm}^{npq} is given in Exercise 6.3.

6.3 *Angular momentum recoupling*

The purpose of this Complement is to define the angular momentum recoupling coefficients, $6j$ symbol or Racah coefficient and the $9j$ symbol in unitary form, as used in this book. For states involving three commuting angular momentum operators, they may be specified by a basis

$$|j_1 m_1 j_2 m_2 j_3 m_3\rangle, \tag{6.61}$$

which diagonalizes six commuting operators $\boldsymbol{j}_1^2, j_{1_z}, \boldsymbol{j}_2^2, j_{2_z}, \boldsymbol{j}_3^2, j_{3_z}$. It will be advantageous to construct a basis with a good total angular momentum $\boldsymbol{J}$ and its z component J_z. To specify that basis completely, the set of the five commuting operators $\boldsymbol{j}_1^2, \boldsymbol{j}_2^2, \boldsymbol{j}_3^2, \boldsymbol{J}^2, J_z$ are not sufficient, but one more operator is needed. Noting that we can couple the total angular momentum $\boldsymbol{J}$ in two ways

$$\begin{aligned} \boldsymbol{J} &= (\boldsymbol{j}_1 + \boldsymbol{j}_2) + \boldsymbol{j}_3 = \boldsymbol{J}_{12} + \boldsymbol{j}_3 \\ &= \boldsymbol{j}_1 + (\boldsymbol{j}_2 + \boldsymbol{j}_3) = \boldsymbol{j}_1 + \boldsymbol{J}_{23}, \end{aligned} \tag{6.62}$$

it is possible to choose a basis in which $\boldsymbol{j}_1^2, \boldsymbol{j}_2^2, \boldsymbol{j}_3^2, \boldsymbol{J}_{12}^2, \boldsymbol{J}^2, J_z$ are diagonal

$$|j_1 j_2 (J_{12}), j_3; JM\rangle \tag{6.63}$$

or a basis in which $\boldsymbol{j}_1^2, \boldsymbol{j}_2^2, \boldsymbol{j}_3^2, \boldsymbol{J}_{23}^2, \boldsymbol{J}^2, J_z$ are diagonal

$$|j_1, j_2 j_3 (J_{23}); JM\rangle. \tag{6.64}$$

The transformation from the one basis to the other is unitary and it is expressed in terms of a U-coefficient as

$$|j_1 j_2 (J_{12}), j_3; JM\rangle = \sum_{J_{23}} U(j_1 j_2 J j_3; J_{12} J_{23})\, |j_1, j_2 j_3 (J_{23}); JM\rangle.$$

(6.65)

The U-coefficient is often called a unitary Racah coefficient. It is given by the overlap of the two basis states

$$U(j_1j_2Jj_3;J_{12}J_{23}) = \langle j_1, j_2j_3(J_{23}); JM | j_1j_2(J_{12}), j_3; JM\rangle, \tag{6.66}$$

and is in fact independent of the magnetic quantum number M.

By using Clebsch-Gordan coefficients, the unitary Racah coefficient can be expressed as

$$U(j_1j_2Jj_3;J_{12}J_{23}) = \sum_{m_1m_2m_3M_{12}M_{23}} \langle j_1m_1j_2m_2|J_{12}M_{12}\rangle \\ \times \langle J_{12}M_{12}j_3m_3|JM\rangle\langle j_2m_2j_3m_3|J_{23}M_{23}\rangle\langle j_1m_1J_{23}M_{23}|JM\rangle. \tag{6.67}$$

Note that in the above equation M is fixed to a certain value in the range of $-J \le M \le J$, so that actually only two m values are independent in the summation. As the Racah coefficient is real and unitary, the inverse transformation of Eq. (6.65) is simply given by ($U^{-1} = U^\dagger = \tilde{U}$)

$$|j_1, j_2j_3(J_{23}); JM\rangle = \sum_{J_{12}} U(j_1j_2Jj_3;J_{12}J_{23})\, |j_1j_2(J_{12}), j_3; JM\rangle. \tag{6.68}$$

If the basis of Eq. (6.63) is to be transformed to a basis in which the angular momenta $\boldsymbol{j}_1$ and $\boldsymbol{j}_3$ are first coupled to $\boldsymbol{J}_{13}$ and then it is coupled with $\boldsymbol{j}_2$ to the total angular momentum $\boldsymbol{J}$, then the transformation coefficient is given by

$$\begin{aligned} |j_1j_2(J_{12}), j_3; JM\rangle &= (-1)^{j_1+j_2-J_{12}}|j_2j_1(J_{12}), j_3; JM\rangle \\ &= (-1)^{j_1+j_2-J_{12}}\sum_{J_{13}} U(j_2j_1Jj_3;J_{12}J_{13})\,|j_2, j_1j_3(J_{13}); JM\rangle \\ &= \sum_{J_{13}} (-1)^{j_1+J-J_{12}-J_{13}} U(j_2j_1Jj_3;J_{12}J_{13})\,|j_1j_3(J_{13}), j_2; JM\rangle. \end{aligned} \tag{6.69}$$

The unitary Racah coefficients are transformation coefficients between complete sets of states, so that they obey orthonormality and completeness relations. The orthonormality condition $\langle j_1, j_2j_3(J_{23}'); JM|j_1, j_2j_3(J_{23}); JM\rangle = \delta_{J_{23}J_{23}'}$ reads, with the use of Eq. (6.68), as

$$\sum_{J_{12}} U(j_1j_2Jj_3; J_{12}J_{23})\, U(j_1j_2Jj_3; J_{12}J_{23}') = \delta_{J_{23}J_{23}'}. \tag{6.70}$$

The completeness relation may be expressed as

$$\sum_{J_{12}} \langle j_1j_3(J_{13}), j_2; JM|j_1j_2(J_{12}), j_3; JM\rangle \times \langle j_1j_2(J_{12}), j_3; JM|j_1, j_2j_3(J_{23}); JM\rangle = \langle j_1j_3(J_{13}), j_2; JM|j_1, j_2j_3(J_{23}); JM\rangle. \tag{6.71}$$

This equation can be transcribed in terms of the Racah coefficients to the following relation

$$\sum_{J_{12}} (-1)^{j_1+J-J_{12}-J_{13}} U(j_2j_1Jj_3; J_{12}J_{13})\, U(j_1j_2Jj_3; J_{12}J_{23}) = (-1)^{j_2+j_3-J_{23}} U(j_1j_3Jj_2; J_{13}J_{23}). \tag{6.72}$$

There are symmetry relations with respect to the interchange among the six angular momentum labels. These relations can be best seen by introducing the so-called $6j$ symbol [70, 72, 75] written by curly brackets

$$U(j_1j_2Jj_3; J_{12}J_{23}) = (-1)^{j_1+j_2+J+j_3}\sqrt{(2J_{12}+1)(2J_{23}+1)} \times \begin{Bmatrix} j_1 & j_2 & J_{12} \\ j_3 & J & J_{23} \end{Bmatrix}, \tag{6.73}$$

because the symmetry relation of the $6j$ symbol is rather simple. E.g., one can show that

$$U(j_1j_2Jj_3; J_{12}J_{23}) = U(j_2j_1j_3J; J_{12}J_{23}) = U(Jj_3j_1j_2; J_{12}J_{23}) = U(j_3Jj_2j_1; J_{12}J_{23}) = U(j_1Jj_2j_3; J_{23}J_{12}) = (-1)^{j_2+J-J_{12}-J_{23}}\sqrt{\frac{(2J_{12}+1)(2J_{23}+1)}{(2j_2+1)(2J+1)}} \times U(J_{12}j_1j_3J_{23}; j_2J)$$

$$= (-1)^{j_1+j_3-J_{12}-J_{23}} \sqrt{\frac{(2J_{12}+1)(2J_{23}+1)}{(2j_1+1)(2j_3+1)}}$$

$$\times U(j_2 J_{12} J_{23} J; j_1 j_3). \tag{6.74}$$

If one of the labels, j_1, j_2, j_3, or J, is zero, the unitary Racah coefficient has the value +1. If J_{12} or J_{23} is zero, then we have

$$U(j_1 j_2 j_1 j_2; J_{12} 0) = U(j_1 j_1 j_2 j_2; 0 J_{12})$$

$$= (-1)^{j_1+j_2-J_{12}} \sqrt{\frac{2J_{12}+1}{(2j_1+1)(2j_2+1)}}. \tag{6.75}$$

The above discussion is extended to the case involving four commuting angular momentum operators, $\boldsymbol{j}_1, \boldsymbol{j}_2, \boldsymbol{j}_3, \boldsymbol{j}_4$. A basis with a good total angular momentum $\boldsymbol{J}$ and its z component J_z is constructed as, for example,

$$|j_1 j_2(J_{12}), j_3 j_4(J_{34}); JM\rangle \quad \text{or} \quad |j_1 j_3(J_{13}), j_2 j_4(J_{24}); JM\rangle. \tag{6.76}$$

The transformation from the basis in which J_{12} and J_{34} are good quantum numbers to the basis in which J_{13} and J_{24} are good quantum numbers is again real and unitary:

$$|j_1 j_2(J_{12}), j_3 j_4(J_{34}); JM\rangle$$

$$= \sum_{J_{13} J_{24}} \begin{bmatrix} j_1 & j_2 & J_{12} \\ j_3 & j_4 & J_{34} \\ J_{13} & J_{24} & J \end{bmatrix} |j_1 j_3(J_{13}), j_2 j_4(J_{24}); JM\rangle. \tag{6.77}$$

The transformation coefficient is a recoupling coefficient between four angular momenta called a $9j$ symbol in unitary form. It is independent of M and given by $\langle j_1 j_3(J_{13}), j_2 j_4(J_{24}); JM | j_1 j_2(J_{12}), j_3 j_4(J_{34}); JM\rangle$, that is

$$\begin{bmatrix} j_1 & j_2 & J_{12} \\ j_3 & j_4 & J_{34} \\ J_{13} & J_{24} & J \end{bmatrix} = \sum_{m_1 m_2 m_3 m_4 M_{12} M_{34} M_{13} M_{24}}$$

$$\times \langle j_1 m_1 j_2 m_2 | J_{12} M_{12}\rangle \langle j_3 m_3 j_4 m_4 | J_{34} M_{34}\rangle \langle J_{12} M_{12} J_{34} M_{34} | JM\rangle$$

$$\times \langle j_1 m_1 j_3 m_3 | J_{13} M_{13}\rangle \langle j_2 m_2 j_4 m_4 | J_{24} M_{24}\rangle \langle J_{13} M_{13} J_{24} M_{24} | JM\rangle. \tag{6.78}$$

Note again that M is fixed to a certain value in the range of $-J \leq M \leq J$, so that only three m values are independent in the above summation. The inverse transformation is given by

$$|j_1 j_3(J_{13}), j_2 j_4(J_{24}); JM\rangle = \sum_{J_{12}J_{34}} \begin{bmatrix} j_1 & j_2 & J_{12} \\ j_3 & j_4 & J_{34} \\ J_{13} & J_{24} & J \end{bmatrix} |j_1 j_2(J_{12}), j_3 j_4(J_{34}); JM\rangle. \tag{6.79}$$

The orthonormality and completeness relations for the unitary $9j$ coefficients can be derived as before:

$$\sum_{J_{13}J_{24}} \begin{bmatrix} j_1 & j_2 & J_{12} \\ j_3 & j_4 & J_{34} \\ J_{13} & J_{24} & J \end{bmatrix} \begin{bmatrix} j_1 & j_2 & J_{12}' \\ j_3 & j_4 & J_{34}' \\ J_{13} & J_{24} & J \end{bmatrix} = \delta_{J_{12}J_{12}'}\delta_{J_{34}J_{34}'}.$$

$$\sum_{J_{12}J_{34}} (-1)^{j_3+j_4-J_{34}} \begin{bmatrix} j_1 & j_2 & J_{12} \\ j_3 & j_4 & J_{34} \\ J_{13} & J_{24} & J \end{bmatrix} \begin{bmatrix} j_1 & j_2 & J_{12} \\ j_4 & j_3 & J_{34} \\ J_{14} & J_{23} & J \end{bmatrix} = (-1)^{j_3-j_4-J_{23}+J_{24}} \begin{bmatrix} j_1 & j_3 & J_{13} \\ j_4 & j_2 & J_{24} \\ J_{14} & J_{23} & J \end{bmatrix}. \tag{6.80}$$

There are many symmetry properties of the unitary $9j$ coefficient. These will have their simplest form in terms of the familiar $9j$ symbol [70, 72, 75] written in curly bracket

$$\begin{bmatrix} j_1 & j_2 & J_{12} \\ j_3 & j_4 & J_{34} \\ J_{13} & J_{24} & J \end{bmatrix} = \sqrt{(2J_{12}+1)(2J_{34}+1)(2J_{13}+1)(2J_{24}+1)} \times \begin{Bmatrix} j_1 & j_2 & J_{12} \\ j_3 & j_4 & J_{34} \\ J_{13} & J_{24} & J \end{Bmatrix}. \tag{6.81}$$

Using the symmetry properties of the $9j$ symbol, we have for example

$$\begin{bmatrix} j_1 & j_2 & J_{12} \\ j_3 & j_4 & J_{34} \\ J_{13} & J_{24} & J \end{bmatrix} = \begin{bmatrix} j_1 & j_3 & J_{13} \\ j_2 & j_4 & J_{24} \\ J_{12} & J_{34} & J \end{bmatrix}$$

$$= \sqrt{\frac{(2J_{12}+1)(2J_{34}+1)(2J_{13}+1)}{(2j_1+1)(2j_3+1)(2J+1)}} \begin{bmatrix} j_2 & J_{12} & j_1 \\ j_4 & J_{34} & j_3 \\ J_{24} & J & J_{13} \end{bmatrix}$$

$$= \sqrt{\frac{(2J_{13}+1)(2J_{24}+1)(2J_{34}+1)}{(2j_3+1)(2j_4+1)(2J+1)}} \begin{bmatrix} J_{13} & J_{24} & J \\ j_1 & j_2 & J_{12} \\ j_3 & j_4 & J_{34} \end{bmatrix}$$

$$= (-1)^{j_1+j_2+j_3+j_4+J_{12}+J_{34}+J_{13}+J_{24}+J} \begin{bmatrix} j_3 & j_4 & J_{34} \\ j_1 & j_2 & J_{12} \\ J_{13} & J_{24} & J \end{bmatrix}. \tag{6.82}$$

One may consider other basis states than those given in Eq. (6.76). For example, the basis in which J_{12} and J_{34} are diagonal can be transformed to a basis in which the angular momentum is successively coupled as follows:

$$|j_1 j_2(J_{12}), j_3 j_4(J_{34}); JM\rangle$$
$$= \sum_{J_{123}} U(J_{12} j_3 J j_4; J_{123} J_{34}) \, |j_1 j_2(J_{12}), j_3(J_{123}), j_4; JM\rangle. \tag{6.83}$$

The $9j$ coefficient can be expressed in terms of products of unitary Racah coefficients. The use of Eqs. (6.69) and (6.68) enables one to rewrite the state in Eq. (6.83), which is obtained by the successive coupling, as follows:

$$|j_1 j_2(J_{12}), j_3(J_{123}), j_4; JM\rangle$$
$$= \sum_{J_{13}} (-1)^{j_1+J_{123}-J_{12}-J_{13}} U(j_2 j_1 J_{123} j_3; J_{12} J_{13})$$
$$\times |j_1 j_3(J_{13}), j_2(J_{123}), j_4; JM\rangle$$
$$= \sum_{J_{13} J_{24}} (-1)^{j_1+J_{123}-J_{12}-J_{13}} U(j_2 j_1 J_{123} j_3; J_{12} J_{13})$$
$$\times U(J_{13} j_2 J j_4; J_{123} J_{24}) \, |j_1 j_3(J_{13}), j_2 j_4(J_{24}); JM\rangle. \tag{6.84}$$

Substituting this result into Eq. (6.83) and comparing with Eq. (6.77) leads to a useful identity

$$\begin{bmatrix} j_1 & j_2 & J_{12} \\ j_3 & j_4 & J_{34} \\ J_{13} & J_{24} & J \end{bmatrix}$$

$$= \sum_{J_{123}} (-1)^{j_1+J_{123}-J_{12}-J_{13}} U(J_{12}j_3Jj_4; J_{123}J_{34})$$

$$\times U(j_2j_1J_{123}j_3; J_{12}J_{13})\, U(J_{13}j_2Jj_4; J_{123}J_{24}). \qquad (6.85)$$

As an application of the above equation, we note that if one of nine angular momenta is zero the $9j$ coefficient is expressible in terms of Racah coefficients. E.g., we have

$$\begin{bmatrix} j_1 & 0 & j_1 \\ j_3 & j_4 & J_{34} \\ J_{13} & j_4 & J \end{bmatrix} = U(j_1j_3Jj_4; J_{13}J_{34}),$$

$$\begin{bmatrix} j_1 & j_1 & 0 \\ j_3 & j_4 & J \\ J_{13} & J_{24} & J \end{bmatrix} = \sqrt{\frac{2J_{13}+1}{(2j_1+1)(2j_3+1)}}\, U(J_{13}j_1Jj_4; j_3J_{24}),$$

$$\begin{bmatrix} 0 & j_2 & j_2 \\ j_3 & j_4 & J_{34} \\ j_3 & J_{24} & J \end{bmatrix} = (-1)^{j_4+J-J_{24}-J_{34}} U(j_2j_4Jj_3; J_{24}J_{34}). \quad (6.86)$$

6.4 *Separation of the center-of-mass motion from correlated Gaussians*

The correlated Gaussian (6.27) in the global vector representation is, among the several possibilities of the correlated Gaussians, most easily constructed from the generating function. Thanks to this simplicity it has a wide range of applications. The relative coordinate $\tilde{\boldsymbol{x}} = (\boldsymbol{x}_1, ..., \boldsymbol{x}_{N-1})$ is used to represent the correlated Gaussian, so that the center-of-mass motion is dropped from the beginning. The aim of this Complement is to show that, even when the correlated Gaussian is expressed in terms of the single-particle coordinates $\boldsymbol{r}$ as

$$f_{KLM}(u, A, \boldsymbol{r}) = \exp\left(-\frac{1}{2}\tilde{\boldsymbol{r}}A\boldsymbol{r}\right) v^{2K} \mathcal{Y}_{LM}(\boldsymbol{v})$$

$$v = \sum_{i=1}^{N} u_i \boldsymbol{r}_i, \tag{6.87}$$

it is possible to separate the intrinsic motion, expressed in terms of the relative coordinates, from the center-of-mass motion by requiring some special conditions on the parameters A and u. Here A is an $N \times N$ symmetric matrix and u is an $N \times 1$ one-column matrix. Therefore, in the correlated Gaussian basis the use of the relative coordinates is not always absolutely necessary but the single-particle coordinates can be used as well. We shall show the conditions in the following.

By using Eq. (2.4), the correlated Gaussian of Eq. (6.87) can be expressed by the relative coordinates $\boldsymbol{x}$ and the center-of-mass coordinate $\boldsymbol{x}_N$, which are again denoted as $\boldsymbol{x}$:

$$f_{KLM}(u, A, \boldsymbol{r}) = f_{KLM}(u', A', \boldsymbol{x}) \tag{6.88}$$

with

$$A' = \widetilde{U^{-1}} A U^{-1} \qquad \text{and} \qquad u' = \widetilde{U^{-1}} u. \tag{6.89}$$

The matrix elements $A'_{Ni} = A'_{iN}$ $(i = 1, ..., N-1)$ connect the relative and center-of-mass coordinates and give rise to a center-of-mass dependence of the function. The element A'_{NN} is a parameter describing the center-of-mass motion which has Gaussian form. Thus the separation between the intrinsic motion and the center-of-mass motion is made possible by requiring that

$$A'_{Ni} = 0 \quad (i = 1, ..., N-1) \qquad \text{and} \qquad A'_{NN} = c,$$

where c is an arbitrary positive constant. As $(U^{-1})_{iN}$=1 $(i = 1, ..., N)$, the above conditions are rewritten as follows:

$$\sum_{j=1}^{N} \sum_{k=1}^{N} A_{jk} (U^{-1})_{ki} = 0 \quad (i = 1, ..., N-1) \qquad \text{and}$$

$$\sum_{j=1}^{N} \sum_{k=1}^{N} A_{jk} = c. \tag{6.90}$$

The center-of-mass contamination in the angular function v^{2K} $\mathcal{Y}_{LM}(\boldsymbol{v})$ can be removed by requiring that $u'_N = 0$, that is,

$$\sum_{i=1}^{N} u_i = 0. \tag{6.91}$$

The conditions (6.90) and (6.91) ensure that the correlated Gaussian (6.87) is free from any contaminations due to the center-of-mass motion. The intrinsic motion is described by the correlated Gaussian of type (6.27) and the center-of-mass motion is given by $\exp(-c\boldsymbol{x}_N^2/2)$. The number of parameters contained in the function (6.87) is $N(N+1)/2+N$, whereas since there are $(N+1)$ conditions we have $(N-1)(N+2)/2$ free parameters. This number is of course equal to the one which the correlated Gaussian (6.27) has: $(N-1)N/2+(N-1)=(N-1)(N+2)/2$.

It is instructive to see how Eq. (6.90) leads to the separation of the center-of-mass motion. For this purpose we note that the exponential part of Eq. (6.87) can be rewritten as

$$\exp\left(-\frac{1}{2}\tilde{\boldsymbol{r}}A\boldsymbol{r}\right)=\exp\left(-\frac{1}{2}\sum_{j>i=1}^{N}\alpha_{ij}(\boldsymbol{r}_i-\boldsymbol{r}_j)^2-\frac{1}{2}\sum_{i=1}^{N}\beta_i\boldsymbol{r}_i^2\right), \tag{6.92}$$

where α_{ij} and β_i are related to the matrix A by

$$\alpha_{ij}=-A_{ij}\quad(i\neq j),\qquad \beta_i=\sum_{k=1}^{N}A_{ki}. \tag{6.93}$$

The coupling between the intrinsic motion and the center-of-mass motion arises from the second term in the exponent. As $\boldsymbol{r}_i$ is expressed as

$$\boldsymbol{r}_i=\sum_{j=1}^{N-1}(U^{-1})_{ij}\boldsymbol{x}_j+\boldsymbol{x}_N, \tag{6.94}$$

we have

$$\sum_{i=1}^{N}\beta_i\boldsymbol{r}_i^2=\left(\sum_{i=1}^{N}\beta_i\right)\boldsymbol{x}_N^2+2\sum_{j=1}^{N-1}\left\{\sum_{i=1}^{N}\beta_i(U^{-1})_{ij}\right\}\boldsymbol{x}_j\cdot\boldsymbol{x}_N+\cdots. \tag{6.95}$$

Here the symbol $\cdots$ denotes the terms that are quadratic in the relative coordinates and have no dependence on $\boldsymbol{x}_N$. Substituting β_i of Eq. (6.93) and using the condition (6.90), we obtain that $\sum_{i=1}^{N}\beta_i=c$ and $\sum_{i=1}^{N}\beta_i(U^{-1})_{ij}=\sum_{k=1}^{N}\sum_{i=1}^{N}A_{ki}(U^{-1})_{ij}=0$.

6.5 *Three electrons with* $S = 1/2$

Let us construct the spin function with $S = 1/2$. The basic terms for $M = 1/2$ are

$$|1\rangle = |\uparrow\uparrow\downarrow\rangle, \quad |2\rangle = |\uparrow\downarrow\uparrow\rangle, \quad |3\rangle = |\downarrow\uparrow\uparrow\rangle, \tag{6.96}$$

where the particle indices are assumed to be in increasing order, e.g. in the state $|1\rangle$ the first and second electrons are in a spin-up state while the third electron is in a spin-down state. The requirement that $S_+\chi(\lambda)$ must vanish leads to the condition of $\lambda_1+\lambda_2+\lambda_3 = 0$. By taking account of the normalization the spin function can be parametrized by a single variable $\vartheta(-\pi/2 \le \vartheta \le \pi/2)$ as follows:

$$\chi_{\frac{1}{2}\frac{1}{2}}(\lambda) = \sqrt{\frac{2}{3}}\sin\vartheta\,|\uparrow\uparrow\downarrow\rangle + \left(\sqrt{\frac{1}{2}}\cos\vartheta - \sqrt{\frac{1}{6}}\sin\vartheta\right)|\uparrow\downarrow\uparrow\rangle$$
$$- \left(\sqrt{\frac{1}{2}}\cos\vartheta + \sqrt{\frac{1}{6}}\sin\vartheta\right)|\downarrow\uparrow\uparrow\rangle. \tag{6.97}$$

The ϑ value is chosen such that $\vartheta = 0$ corresponds to the spin function with the intermediate spin $S_{12} = 0$ whereas $\vartheta = \pm\pi/2$ to the one with $S_{12} = 1$.

By acting with S_- on Eq. (6.97), we obtain

$$\chi_{\frac{1}{2}\,-\frac{1}{2}}(\lambda) = -\sqrt{\frac{2}{3}}\sin\vartheta\,|\downarrow\downarrow\uparrow\rangle - \left(\sqrt{\frac{1}{2}}\cos\vartheta - \sqrt{\frac{1}{6}}\sin\vartheta\right)|\downarrow\uparrow\downarrow\rangle$$
$$+ \left(\sqrt{\frac{1}{2}}\cos\vartheta + \sqrt{\frac{1}{6}}\sin\vartheta\right)|\uparrow\downarrow\downarrow\rangle. \tag{6.98}$$

These spin functions can be used for the spin part of three-quark baryon wave functions. See Chap. 9.

6.6 *Four electrons in an arbitrary spin arrangement*

Arithmetic similar to that of Complement 6.5 gives us the following result for the general spin function of the four-electron system. For $S = 0$:

$$\chi_{00}(\lambda) = \sqrt{\frac{1}{3}}\sin\vartheta\,|\uparrow\uparrow\downarrow\downarrow\rangle + \left(\frac{1}{2}\cos\vartheta - \sqrt{\frac{1}{12}}\sin\vartheta\right)|\uparrow\downarrow\uparrow\downarrow\rangle$$
$$- \left(\frac{1}{2}\cos\vartheta + \sqrt{\frac{1}{12}}\sin\vartheta\right)|\uparrow\downarrow\downarrow\uparrow\rangle - \left(\frac{1}{2}\cos\vartheta + \sqrt{\frac{1}{12}}\sin\vartheta\right)|\downarrow\uparrow\uparrow\downarrow\rangle$$

$$+\left(\frac{1}{2}\cos\vartheta-\sqrt{\frac{1}{12}}\sin\vartheta\right)|\downarrow\uparrow\downarrow\uparrow\rangle+\sqrt{\frac{1}{3}}\sin\vartheta\,|\downarrow\downarrow\uparrow\uparrow\rangle, \tag{6.99}$$

where the parameter ϑ satisfies $-\pi/2 \le \vartheta \le \pi/2$ and is chosen such that $\vartheta = 0$ corresponds to the spin function with $S_{12} = 0$, $S_{123} = 1/2$; whereas $\vartheta = \pm\pi/2$ corresponds to the one with $S_{12} = 1$, $S_{123} = 1/2$. In the positronium molecule consisting of two electrons and two positrons, the most important spin function is such that the spins of the identical particles are coupled to zero, which corresponds to $\vartheta = 0$. See Sect. 8.3. For $S = 1$:

$$\begin{aligned}\chi_{11}(\lambda) &= \sqrt{\frac{3}{4}}\sin\vartheta\sin\varphi\,|\uparrow\uparrow\uparrow\downarrow\rangle\\ &+\left(\sqrt{\frac{2}{3}}\sin\vartheta\cos\varphi-\sqrt{\frac{1}{12}}\sin\vartheta\sin\varphi\right)|\uparrow\uparrow\downarrow\uparrow\rangle\\ &+\left(\sqrt{\frac{1}{2}}\cos\vartheta-\sqrt{\frac{1}{6}}\sin\vartheta\cos\varphi-\sqrt{\frac{1}{12}}\sin\vartheta\sin\varphi\right)|\uparrow\downarrow\uparrow\uparrow\rangle\\ &-\left(\sqrt{\frac{1}{2}}\cos\vartheta+\sqrt{\frac{1}{6}}\sin\vartheta\cos\varphi+\sqrt{\frac{1}{12}}\sin\vartheta\sin\varphi\right)|\downarrow\uparrow\uparrow\uparrow\rangle,\end{aligned} \tag{6.100}$$

where the parameters ϑ and φ satisfy $-\pi/2 \le \vartheta \le \pi/2$ and $-\pi/2 \le \varphi \le \pi/2$. The three independent spin functions with definite S_{12} and S_{123} values correspond to the ϑ and φ values as follows: for $S_{12} = 0$ and $S_{123} = 1/2$, $\vartheta = 0$; for $S_{12} = 1$ and $S_{123} = 1/2$, $\vartheta = \pm\pi/2$ and $\varphi = 0$; and for $S_{12} = 1$ and $S_{123} = 3/2$, $\vartheta = \pm\pi/2$ and $\varphi = \pm\pi/2$.

6.7 *Six electrons with* $S = 0$

The spin functions may be constructed from those of a smaller group of electrons through angular momentum couplings. As an example, let us construct the spin function of the six-electron system with $S = 0$, $\chi_{00}(\lambda)$, from two groups of three electrons. Clearly $\chi_{00}(\lambda)$ is given as a combination of two terms with $S_{123} (= S_{456}) = 1/2$ or $3/2$:

$$\chi_{00}(\lambda) = \cos\vartheta\left|\tfrac{1}{2}\tfrac{1}{2};00\right\rangle + \sin\vartheta\left|\tfrac{3}{2}\tfrac{3}{2};00\right\rangle. \tag{6.101}$$

As was already mentioned, the spin function with $S_{123} = 1/2$ has two independent states with $S_{12} = 0$ or 1, and similarly the function with $S_{456} = 1/2$ has two independent states with $S_{45} = 0$ or 1.

Therefore the first term of the right-hand side of Eq. (6.101) can be parametrized as follows:

$$
\begin{aligned}
\left|\tfrac{1}{2}\tfrac{1}{2};00\right\rangle &= \cos\vartheta_1 \left|(S_{12}=0)\tfrac{1}{2},\ (S_{45}=0)\tfrac{1}{2};00\right\rangle \\
&\quad + \sin\vartheta_1 \cos\vartheta_2 \left|(S_{12}=0)\tfrac{1}{2},\ (S_{45}=1)\tfrac{1}{2};00\right\rangle \\
&\quad + \sin\vartheta_1 \sin\vartheta_2 \cos\vartheta_3 \left|(S_{12}=1)\tfrac{1}{2},\ (S_{45}=0)\tfrac{1}{2};00\right\rangle \\
&\quad + \sin\vartheta_1 \sin\vartheta_2 \sin\vartheta_3 \left|(S_{12}=1)\tfrac{1}{2},\ (S_{45}=1)\tfrac{1}{2};00\right\rangle. \qquad (6.102)
\end{aligned}
$$

On the other hand, the spin function with $S_{123} = 3/2$ or $S_{456} = 3/2$ has just one independent state and needs no angle parameters. Each term of the above equation and the second term of the right-hand side of Eq. (6.101) can be easily written down in the form (6.35), e.g.,

$$
\begin{aligned}
&\left|(S_{12}=0)\tfrac{1}{2},\ (S_{45}=0)\tfrac{1}{2};00\right\rangle \\
&\quad = -\frac{1}{2\sqrt{2}}\Big(|\uparrow\downarrow\uparrow\downarrow\uparrow\downarrow\rangle - |\uparrow\downarrow\uparrow\uparrow\downarrow\downarrow\rangle - |\downarrow\uparrow\uparrow\downarrow\uparrow\downarrow\rangle + |\downarrow\uparrow\uparrow\uparrow\downarrow\downarrow\rangle \\
&\qquad - |\downarrow\uparrow\downarrow\uparrow\downarrow\uparrow\rangle + |\downarrow\uparrow\downarrow\downarrow\uparrow\uparrow\rangle + |\uparrow\downarrow\downarrow\uparrow\downarrow\uparrow\rangle - |\uparrow\downarrow\downarrow\downarrow\uparrow\uparrow\rangle \Big). \qquad (6.103)
\end{aligned}
$$

The four parameters ϑ, ϑ_1, ϑ_2, ϑ_3 determine the relative weight of the five independent spin states for the six-electron system with $S = 0$.

It is of course possible to start with other groupings such as two groups of four electrons and two electrons or three groups of two electrons.

Exercises

6.1. Prove Eq. (6.7).

Solution. Let us consider the quantity

$$\mathcal{I} \equiv \int Y_{LM}(\hat{\boldsymbol{a}})(\boldsymbol{a}\cdot\boldsymbol{v})^{2K+L} d\hat{\boldsymbol{a}} \qquad \text{with } \boldsymbol{v} = \boldsymbol{x}_1 + \boldsymbol{x}_2. \tag{6.104}$$

Of course $\mathcal{I}$ is equal to $B_{KL}a^{2K+L}v^{2K}\mathcal{Y}_{LM}(\boldsymbol{v})$ from Eq. (6.18). To calculate $\mathcal{I}$ in another way, we use the relation

$$(\boldsymbol{a}\cdot\boldsymbol{v})^{2K+L} = \sum_{p+q=2K+L} \frac{(2K+L)!}{p!q!}(\boldsymbol{a}\cdot\boldsymbol{x}_1)^p(\boldsymbol{a}\cdot\boldsymbol{x}_2)^q. \tag{6.105}$$

By using the expansion (6.18) for the terms $(\boldsymbol{a}\cdot\boldsymbol{x}_1)^p$ and $(\boldsymbol{a}\cdot\boldsymbol{x}_2)^q$, we see that $(\boldsymbol{a}\cdot\boldsymbol{x}_1)^p(\boldsymbol{a}\cdot\boldsymbol{x}_2)^q$ involves terms of the form $[\mathcal{Y}_{l_1}(\boldsymbol{a}) \times \mathcal{Y}_{l_1}(\boldsymbol{x}_1)]_{00}[\mathcal{Y}_{l_2}(\boldsymbol{a}) \times \mathcal{Y}_{l_2}(\boldsymbol{x}_2)]_{00}$. To carry out the integration, it is convenient to arrange the four angular momenta given in the coupling order of l_1, l_1, l_2, and l_2 to those in the order of l_1, l_2, l_1, and l_2. See Complement 6.3. This can be done as follows:

$$\begin{aligned}
&[\mathcal{Y}_{l_1}(\boldsymbol{a}) \times \mathcal{Y}_{l_1}(\boldsymbol{x}_1)]_{00}[\mathcal{Y}_{l_2}(\boldsymbol{a}) \times \mathcal{Y}_{l_2}(\boldsymbol{x}_2)]_{00} \\
&\quad = \sum_{\lambda} \begin{bmatrix} l_1 & l_1 & 0 \\ l_2 & l_2 & 0 \\ \lambda & \lambda & 0 \end{bmatrix} \\
&\qquad \times [[\mathcal{Y}_{l_1}(\boldsymbol{a}) \times \mathcal{Y}_{l_2}(\boldsymbol{a})]_{\lambda} \times [\mathcal{Y}_{l_1}(\boldsymbol{x}_1) \times \mathcal{Y}_{l_2}(\boldsymbol{x}_2)]_{\lambda}]_{00} \\
&\quad = \sum_{\lambda} \sqrt{\frac{2\lambda+1}{(2l_1+1)(2l_2+1)}}\, a^{l_1+l_2} C(l_1\, l_2; \lambda) \\
&\qquad \times [Y_{\lambda}(\hat{\boldsymbol{a}}) \times [\mathcal{Y}_{l_1}(\boldsymbol{x}_1) \times \mathcal{Y}_{l_2}(\boldsymbol{x}_2)]_{\lambda}]_{00}.
\end{aligned} \tag{6.106}$$

The coupled form $[\mathcal{Y}_{l_1}(\boldsymbol{a}) \times \mathcal{Y}_{l_2}(\boldsymbol{a})]_{\lambda\mu}$ is reduced to $a^{l_1+l_2}C(l_1\, l_2; \lambda)$ $Y_{\lambda\mu}(\hat{\boldsymbol{a}})$ by using Eqs. (6.10) and (6.11).

When using Eq. (6.106) in Eq. (6.104) and integrating over $\hat{\boldsymbol{a}}$, we see that λ is restricted to L, because otherwise the integral vanishes due to the orthogonality of the spherical harmonics. We thus obtain

$$\int Y_{LM}(\hat{\boldsymbol{a}})[Y_{\lambda}(\hat{\boldsymbol{a}}) \times [\mathcal{Y}_{l_1}(\boldsymbol{x}_1) \times \mathcal{Y}_{l_2}(\boldsymbol{x}_2)]_{\lambda}]_{00} d\hat{\boldsymbol{a}}$$

$$= \delta_{\lambda L} \frac{(-1)^L}{\sqrt{2L+1}} [\mathcal{Y}_{l_1}(\boldsymbol{x}_1) \times \mathcal{Y}_{l_2}(\boldsymbol{x}_2)]_{LM}. \tag{6.107}$$

Combining these results leads to Eq. (6.7).

Consider Eq. (6.105) for the special case of $K = 0$. Then $p + q$ is equal to L. Since l_1 and l_2 have to satisfy the condition $L \leq l_1 + l_2 \leq p + q$, $l_1 + l_2$ must also be equal to L. Equation (6.7) then simplifies to the well-known formula

$$\begin{aligned}
\mathcal{Y}_{LM}(\boldsymbol{x}_1 + \boldsymbol{x}_2) &\equiv |\boldsymbol{x}_1 + \boldsymbol{x}_2|^L Y_{LM}(\widehat{\boldsymbol{x}_1 + \boldsymbol{x}_2}) \\
&= \sum_{l=0}^{L} D^{0L}_{0l0\, L-l} [\mathcal{Y}_l(\boldsymbol{x}_1) \times \mathcal{Y}_{L-l}(\boldsymbol{x}_2)]_{LM} \\
&= \sum_{l=0}^{L} \sqrt{\frac{4\pi(2L+1)!}{(2l+1)!\,(2L-2l+1)!}} [\mathcal{Y}_l(\boldsymbol{x}_1) \times \mathcal{Y}_{L-l}(\boldsymbol{x}_2)]_{LM}.
\end{aligned} \tag{6.108}$$

6.2. Prove Eq. (6.32).

Solution. A positive-definite, symmetric matrix A can be diagonalized by a suitable orthogonal matrix T, namely $\tilde{T}AT$ becomes a diagonal matrix D with its diagonal element $D_{ii} = d_i$ being positive. With the use of a transformation $\boldsymbol{x} \to \boldsymbol{y}$ ($\boldsymbol{x} = T\boldsymbol{y}$), the integral $\mathcal{I}$ of Eq. (6.32) is evaluated as follows:

$$\mathcal{I} = \int \exp\Big(-\frac{1}{2}\tilde{\boldsymbol{y}}D\boldsymbol{y} + \widetilde{\tilde{T}\boldsymbol{s}}\boldsymbol{y}\Big)(\det T)^3 d\boldsymbol{y}, \tag{6.109}$$

where $(\det T)^3$ is the Jacobian (functional determinant) $\det(\partial\boldsymbol{x}/\partial\boldsymbol{y})$. Though $\det T$ is ± 1 in general, we may assume it to be $+1$ in order to preserve the volume element, $d\boldsymbol{x} = d\boldsymbol{y}$.

Since D is diagonal, the integration can now be performed separately in each component of $\boldsymbol{y}$, which reads as

$$\int \exp\Big(-\frac{1}{2}d_i\boldsymbol{y}_i^2 + (\tilde{T}\boldsymbol{s})_i\cdot\boldsymbol{y}_i\Big)d\boldsymbol{y}_i = \left(\frac{2\pi}{d_i}\right)^{\frac{3}{2}} e^{\frac{1}{2d_i}(\tilde{T}\boldsymbol{s})_i^2}. \tag{6.110}$$

Therefore we obtain

$$\mathcal{I} = \prod_{i=1}^{N} \left(\frac{2\pi}{d_i}\right)^{\frac{3}{2}} \exp\Big(\frac{1}{2d_i}(\tilde{T}\boldsymbol{s})_i^2\Big). \tag{6.111}$$

It is easy to note that $\prod_{i=1}^{N} d_i$ is equal to $\det D = \det(\tilde{T}AT) = \det A$, and $\sum_{i=1}^{N}(\tilde{T}\boldsymbol{s})_i^2/d_i$ can be expressed as $\sum_{i=1}^{N}(D^{-1})_{ii}(\tilde{T}\boldsymbol{s})_i\cdot(\tilde{T}\boldsymbol{s})_i = \widetilde{\tilde{T}\boldsymbol{s}}D^{-1}\tilde{T}\boldsymbol{s} = \tilde{\boldsymbol{s}}T(\tilde{T}A^{-1}T)\tilde{T}\boldsymbol{s} = \tilde{\boldsymbol{s}}A^{-1}\boldsymbol{s}$. Using these results in Eq. (6.111) leads to the proof.

It is useful to generalize Eq. (6.32) for the case where all the vectors $\boldsymbol{x}_i$ and $\boldsymbol{s}_i$ are d-dimensional. Following the above derivation we get

$$\int \exp\Big(-\frac{1}{2}\tilde{\boldsymbol{x}}A\boldsymbol{x} + \tilde{\boldsymbol{s}}\boldsymbol{x}\Big)d\boldsymbol{x} = \left(\frac{(2\pi)^N}{\det A}\right)^{\frac{d}{2}} \exp\Big(\frac{1}{2}\tilde{\boldsymbol{s}}A^{-1}\boldsymbol{s}\Big), \quad (6.112)$$

where the scalar (inner) product $(\boldsymbol{a}\cdot\boldsymbol{b})$ for d-dimensional vectors $\tilde{\boldsymbol{a}} = (a_1, a_2, ..., a_d)$ and $\tilde{\boldsymbol{b}} = (b_1, b_2, ..., b_d)$ is to be understood as $(\boldsymbol{a}\cdot\boldsymbol{b}) = \sum_{m=1}^{d} a_m b_m$.

Some useful formulas related to this integral are collected below. By differentiating both sides of the above equation with respect to the mth component of the vector $\boldsymbol{s}_i$, $(\boldsymbol{s}_i)_m$, we obtain

$$\begin{aligned}&\int (\boldsymbol{x}_i)_m \exp\Big(-\frac{1}{2}\tilde{\boldsymbol{x}}A\boldsymbol{x} + \tilde{\boldsymbol{s}}\boldsymbol{x}\Big)d\boldsymbol{x}\\ &\quad = ((A^{-1}\boldsymbol{s})_i)_m \left(\frac{(2\pi)^N}{\det A}\right)^{\frac{d}{2}} \exp\Big(\frac{1}{2}\tilde{\boldsymbol{s}}A^{-1}\boldsymbol{s}\Big). \end{aligned} \quad (6.113)$$

Further differentiation with respect to $(\boldsymbol{s}_j)_n$ leads us to

$$\begin{aligned}&\int (\boldsymbol{x}_i)_m(\boldsymbol{x}_j)_n \exp\Big(-\frac{1}{2}\tilde{\boldsymbol{x}}A\boldsymbol{x} + \tilde{\boldsymbol{s}}\boldsymbol{x}\Big)d\boldsymbol{x}\\ &\quad = \Big\{(A^{-1})_{ij}\delta_{mn} + ((A^{-1}\boldsymbol{s})_i)_m((A^{-1}\boldsymbol{s})_j)_n\Big\}\\ &\quad \times \left(\frac{(2\pi)^N}{\det A}\right)^{\frac{d}{2}} \exp\Big(\frac{1}{2}\tilde{\boldsymbol{s}}A^{-1}\boldsymbol{s}\Big). \end{aligned} \quad (6.114)$$

Setting $m = n$ and summing over m leads to

$$\begin{aligned}&\int (\boldsymbol{x}_i\cdot\boldsymbol{x}_j) \exp\Big(-\frac{1}{2}\tilde{\boldsymbol{x}}A\boldsymbol{x} + \tilde{\boldsymbol{s}}\boldsymbol{x}\Big)d\boldsymbol{x}\\ &\quad = \Big\{d(A^{-1})_{ij} + \Big((A^{-1}\boldsymbol{s})_i\cdot(A^{-1}\boldsymbol{s})_j\Big)\Big\}\\ &\quad \times \left(\frac{(2\pi)^N}{\det A}\right)^{\frac{d}{2}} \exp\Big(\frac{1}{2}\tilde{\boldsymbol{s}}A^{-1}\boldsymbol{s}\Big). \end{aligned} \quad (6.115)$$

In the case of $d = 3$, let us define a tensor product, $[\boldsymbol{a} \times \boldsymbol{b}]_{\kappa\mu} = \sum_{mm'} \langle 1m\, 1m' | \kappa\mu \rangle a_m b_{m'}$ with $a_m = \sqrt{4\pi/3}\, a Y_{1m}(\hat{\boldsymbol{a}})$ $(m = 1, 0, -1)$ etc. for three-dimensional vectors $\boldsymbol{a}$ and $\boldsymbol{b}$, that is, $a_1 = -\frac{1}{\sqrt{2}}(a_x + i a_y)$, $a_0 = a_z$, $a_{-1} = \frac{1}{\sqrt{2}}(a_x - i a_y)$. Here $\kappa = 0$ or 1 or 2 and $-\kappa \le \mu \le \kappa$. The scalar product is a special case of the tensor product, that is, $(\boldsymbol{a} \cdot \boldsymbol{b}) = -\sqrt{3}[\boldsymbol{a} \times \boldsymbol{b}]_{00} = \sum_{m=-1}^{1} (-1)^m a_m b_{-m}$, whereas $\kappa = 1$ and $\kappa = 2$ give a vector (outer) product and a second rank tensor. See also Eq. (11.2). The use of Eq. (6.114) leads to the following result

$$\int [\boldsymbol{x}_i \times \boldsymbol{x}_j]_{\kappa\mu} \exp\Big(-\frac{1}{2}\tilde{\boldsymbol{x}} A \boldsymbol{x} + \tilde{\boldsymbol{s}} \boldsymbol{x} \Big) d\boldsymbol{x}$$

$$= \Big\{ -\sqrt{3}\, \delta_{\kappa 0} \delta_{\mu 0} (A^{-1})_{ij} + \Big[(A^{-1}\boldsymbol{s})_i \times (A^{-1}\boldsymbol{s})_j \Big]_{\kappa\mu} \Big\}$$

$$\times \left(\frac{(2\pi)^N}{\det A} \right)^{\frac{3}{2}} \exp\Big(\frac{1}{2} \tilde{\boldsymbol{s}} A^{-1} \boldsymbol{s} \Big). \tag{6.116}$$

6.3. Derive B_{lm}^{npq} in Eq. (6.60).

Solution. The explicit formula for B_{lm}^{npq} is a bit lengthy, and here we give a hint for its derivation. First we recall that

$$x = \sqrt{\frac{2\pi}{3}}\, r \big(Y_{1\,-1}(\hat{\boldsymbol{r}}) - Y_{11}(\hat{\boldsymbol{r}}) \big), \quad y = i \sqrt{\frac{2\pi}{3}}\, r \big(Y_{1\,-1}(\hat{\boldsymbol{r}}) + Y_{11}(\hat{\boldsymbol{r}}) \big),$$

$$z = \sqrt{\frac{4\pi}{3}}\, r Y_{10}(\hat{\boldsymbol{r}}). \tag{6.117}$$

Thus the power $x^p y^q z^{n-p-q}$ can be expressed as

$$x^p y^q z^{n-p-q} = r^n \left(\frac{\pi}{3} \right)^{\frac{n}{2}} 2^{n - \frac{p+q}{2}} i^q \sum_{\mu=0}^{p} \binom{p}{\mu} \sum_{\nu=0}^{q} \binom{q}{\nu} (-1)^\mu$$

$$\times \big(Y_{11}(\hat{\boldsymbol{r}}) \big)^{\mu+\nu} \big(Y_{1\,-1}(\hat{\boldsymbol{r}}) \big)^{p+q-\mu-\nu} \big(Y_{10}(\hat{\boldsymbol{r}}) \big)^{n-p-q}. \tag{6.118}$$

The power of the spherical harmonics can be expressed as

$$\big(Y_{1m}(\hat{\boldsymbol{r}}) \big)^k = \sum_L D_L^{(k,m)} Y_{L\,km}(\hat{\boldsymbol{r}}), \tag{6.119}$$

where $D_L^{(k,m)}$ can be derived by using Eq. (6.10) successively and it is calculated from the equation

$$D_L^{(k,m)} = \sqrt{4\pi} \sum_{L_1 L_2 \ldots L_{k-1}} \times \prod_{i=1}^{k} \Big(C(L_{i-1} 1; L_i) \, \langle L_{i-1} \, (i-1)m \; 1 \, m | L_i \, im \rangle \Big), \qquad (6.120)$$

where $L_0 = 0$ and $L_k = L$. In the case of $m = \pm 1$, L_i is uniquely determined as $L_i = i$, but for $m = 0$ there are several possibilities for the value of L_i. Thus the product of the spherical harmonics in Eq. (6.118) can be reduced to the following:

$$\big(Y_{11}(\hat{\boldsymbol{r}})\big)^{\mu+\nu} \big(Y_{1\,-1}(\hat{\boldsymbol{r}})\big)^{p+q-\mu-\nu} \big(Y_{10}(\hat{\boldsymbol{r}})\big)^{n-p-q} = D_{\mu+\nu}^{(\mu+\nu,1)} D_{p+q-\mu-\nu}^{(p+q-\mu-\nu,-1)} \sum_L D_L^{(n-p-q,0)} \times Y_{\mu+\nu \;\, \mu+\nu}(\hat{\boldsymbol{r}}) \, Y_{p+q-\mu-\nu \;\, -p-q+\mu+\nu}(\hat{\boldsymbol{r}}) \, Y_{L0}(\hat{\boldsymbol{r}}). \qquad (6.121)$$

The product of the three spherical harmonics is coupled to a single spherical harmonic by using Eq. (6.10):

$$Y_{\mu+\nu \;\, \mu+\nu}(\hat{\boldsymbol{r}}) \, Y_{p+q-\mu-\nu \;\, -p-q+\mu+\nu}(\hat{\boldsymbol{r}}) \, Y_{L0}(\hat{\boldsymbol{r}}) = \sum_{L'l} C(\mu+\nu \; p+q-\mu-\nu; L') \, C(L' \, L; l) \times \langle \mu+\nu \;\, \mu+\nu \;\, p+q-\mu-\nu \;\, -p-q+\mu+\nu \,|\, L' \; -p-q+2\mu+2\nu \rangle \times \langle L' \; -p-q+2\mu+2\nu \; L \, 0 \,|\, l \; -p-q+2\mu+2\nu \rangle \, Y_{l \; -p-q+2\mu+2\nu}(\hat{\boldsymbol{r}}). \qquad (6.122)$$

The above formulas provide us with the ingredients needed to derive B_{lm}^{npq}.

7. Matrix elements for spherical Gaussians

In this chapter we show that the generating function introduced in the previous chapter can be used to advantage to derive the Hamiltonian matrix elements of an N-particle system for both the correlated Gaussians and the correlated Gaussian-type geminals. Matrix elements of various physical operators between the generating functions will be tabulated in a compact form. Examples of matrix elements for spherical or primitive Gaussians are given in this chapter and more general cases will be treated in the appendix. We can thus calculate matrix elements of essentially any interaction for a many-body system. We also show that the matrix elements involving the correlated Gaussians become particularly simple in two-dimensional problems. An extension to the matrix element for nonlocal potentials is also presented. We discuss the kinetic energy operator for the relativistic kinematics and calculate its matrix element for the correlated Gaussians as an example.

7.1 Matrix elements of the generating function

As the correlated Gaussian basis function can be constructed from the generating function g (6.19), it is convenient to derive the matrix element between the basis functions from that between the functions g. This is true even though we use more than one set of coordinates because the transformation of the coordinates simply leads to a change of the parameters A and $\boldsymbol{s}$, as shown in Eq. (6.28).

It is convenient to use g not only because thereby all types of correlated Gaussians can be constructed in a simplified manner but also because the matrix elements between functions g can be evaluated with ease. Table 7.1 lists formulas for the matrix elements of some basic operators $\mathcal{O}$

Table 7.1. Matrix elements, $\mathcal{M} = \langle g(\boldsymbol{s}'; A', \boldsymbol{x})|\mathcal{O}|g(\boldsymbol{s}; A, \boldsymbol{x})\rangle$, of operators $\mathcal{O}$ between generating functions g of Eq. (6.19). Here we take all vectors d-dimensional. $\tilde{w}\boldsymbol{x}$ is a short-hand notation for $\sum_{i=1}^{N} w_i\boldsymbol{x}_i$. A vector product $(\boldsymbol{a} \times \boldsymbol{b})$ and a tensor product $[\boldsymbol{a} \times \boldsymbol{b}]_{2\mu}$ are defined for three-dimensional vectors $\boldsymbol{a}$ and $\boldsymbol{b}$. $B = A + A'$, $\boldsymbol{v} = \boldsymbol{s} + \boldsymbol{s}'$ and $\boldsymbol{y} = A'B^{-1}\boldsymbol{s} - AB^{-1}\boldsymbol{s}'$. P is a permutation operator and the matrix T_P is defined by $P\boldsymbol{x} = T_P\boldsymbol{x}$.

$\mathcal{O}$	$\mathcal{M}$
1	$\mathcal{M}_0 \equiv \left(\frac{(2\pi)^N}{\det B}\right)^{\frac{d}{2}} \exp(\frac{1}{2}\tilde{\boldsymbol{v}}B^{-1}\boldsymbol{v})$
$\tilde{w}\boldsymbol{x}$	$\tilde{w}B^{-1}\boldsymbol{v}\mathcal{M}_0$
$\tilde{\boldsymbol{x}}Q\boldsymbol{x}$ (Q: a symmetric matrix)	$\left\{d\mathrm{Tr}(B^{-1}Q) + \tilde{\boldsymbol{v}}B^{-1}QB^{-1}\boldsymbol{v}\right\}\mathcal{M}_0$
$\left(\tilde{w}\boldsymbol{x} \times \tilde{\zeta}\boldsymbol{x}\right)$	$\left(\tilde{w}B^{-1}\boldsymbol{v} \times \tilde{\zeta}B^{-1}\boldsymbol{v}\right)\mathcal{M}_0$
$[\tilde{w}\boldsymbol{x} \times \tilde{\zeta}\boldsymbol{x}]_{2\mu}$	$[\tilde{w}B^{-1}\boldsymbol{v} \times \tilde{\zeta}B^{-1}\boldsymbol{v}]_{2\mu}\mathcal{M}_0$
$\tilde{\zeta}\boldsymbol{\pi}$ ($\boldsymbol{\pi}_j = -i\hbar\frac{\partial}{\partial \boldsymbol{x}_j}$)	$-i\hbar\tilde{\zeta}\boldsymbol{y}\mathcal{M}_0$
$\tilde{\boldsymbol{\pi}}\Lambda\boldsymbol{\pi}$ (Λ: a symmetric matrix)	$\hbar^2\left\{d\mathrm{Tr}(AB^{-1}A'\Lambda) - \tilde{\boldsymbol{y}}\Lambda\boldsymbol{y}\right\}\mathcal{M}_0$
L_z ($\hbar\boldsymbol{L} = \sum_i(\boldsymbol{x}_i \times \boldsymbol{\pi}_i)$)	$-i\sum_{ij}(B^{-1})_{ij}(\boldsymbol{s}'_i \times \boldsymbol{s}_j)_z\mathcal{M}_0$
$\boldsymbol{L}^2$	$\left\{(d-1)\tilde{\boldsymbol{s}'}B^{-1}\boldsymbol{s} - \left(\sum_{ij}(B^{-1})_{ij}\boldsymbol{s}'_i \times \boldsymbol{s}_j\right)^2\right\}\mathcal{M}_0$
$\left(\tilde{w}\boldsymbol{x} \times \tilde{\zeta}\boldsymbol{\pi}\right)$	$-i\hbar\left(\tilde{w}B^{-1}\boldsymbol{v} \times \tilde{\zeta}\boldsymbol{y}\right)\mathcal{M}_0$
$\delta(\tilde{w}\boldsymbol{x} - \boldsymbol{r})$	$\mathcal{M}_1 \equiv (2\pi\tilde{w}B^{-1}w)^{-\frac{d}{2}} \times \exp\left\{-\frac{1}{2\tilde{w}B^{-1}w}(\boldsymbol{r} - \tilde{w}B^{-1}\boldsymbol{v})^2\right\}\mathcal{M}_0$
$\delta(\tilde{w}\boldsymbol{x} - \boldsymbol{r})\left(\tilde{w}\boldsymbol{x} \times \tilde{\zeta}\boldsymbol{\pi}\right)$	$-i\hbar\left\{\left(\boldsymbol{r} \times \tilde{\zeta}\boldsymbol{y}\right) + \frac{\tilde{\zeta}AB^{-1}w}{\tilde{w}B^{-1}w}\left(\boldsymbol{r} \times \tilde{w}B^{-1}\boldsymbol{v}\right)\right\}\mathcal{M}_1$
P	$\langle g(\boldsymbol{s}'; A', \boldsymbol{x})\|g(\widetilde{T_P}\boldsymbol{s}; \widetilde{T_P}AT_P, \boldsymbol{x})\rangle$

$$\mathcal{M} = \langle g(\boldsymbol{s}'; A', \boldsymbol{x}) | \mathcal{O} | g(\boldsymbol{s}; A, \boldsymbol{x}) \rangle, \tag{7.1}$$

where A and A' are $N \times N$ positive-definite, symmetric matrices. The integration over a number N of vectors, $\boldsymbol{x} = \{\boldsymbol{x}_1, ..., \boldsymbol{x}_N\}$, can be carried out by using Eqs. (6.112)–(6.116).

Before leaving this section, we note that the formulation with the generating function, though quite useful and powerful, is still severely restricted in its application to a many-body system by the number of particles because of the following properties of the method:

1. The number of nonlinear parameters contained in the matrix A of one basis function is $N(N+1)/2$. By increasing the number of particles the optimization of these parameters becomes prohibitively time consuming.
2. To calculate the matrix elements between the basis functions, one needs to calculate the determinant and the inverse of the $N \times N$ matrix $B = A + A'$. The number of operations needed to calculate the determinant and the inverse increases as N^3. The symmetrization postulate would require the repetition of this calculation $N!$ times for the overlap and the kinetic energy and $(N(N-1)/2)N!$ times for the potential energy matrix elements, respectively.
3. As the system gets more complex, the number of basis functions needed to represent the wave function increases considerably and more matrix elements have to be calculated.

As was shown in Chap. 4, a number of random trials are probed in the SVM to select a basis element. To calculate the energy for each trial, the determinant and the inverse of the corresponding matrix B have to be calculated many times, so the computational load mentioned in the second item would become very heavy even for systems containing a relatively small number of particles. Fortunately, when one repeats the calculation of $\det B$ and B^{-1} by changing only one element or a few elements of B, as is often the case in the SVM selection procedure, one can use the Sherman-Morrison formula [14] to reduce the load involved in the calculations. This will be explained in Complement 7.1.

7.2 Correlated Gaussians

We assume that the Hamiltonian of an N-particle system takes the form of Eq. (2.1). The set of coordinates $\tilde{\boldsymbol{x}} = (\boldsymbol{x}_1, ..., \boldsymbol{x}_{N-1})$ stands for the relative coordinates chosen to describe the basis function. As a concrete example of correlated Gaussians we assume in this section that the basis function is given by Eq. (6.27):

$$f_{KLM}(u, A, \boldsymbol{x}) = \exp\left(-\frac{1}{2}\tilde{\boldsymbol{x}}A\boldsymbol{x}\right)|\tilde{u}\boldsymbol{x}|^{2K}\mathcal{Y}_{LM}(\tilde{u}\boldsymbol{x}).$$

The function is characterized by two parameters: One is the matrix A describing the correlated motion of the particles. The other is the vector u of real numbers which defines the global vector taking care of the rotational motion. As Eq. (6.29) shows, both a change of coordinate sets and a permutation of particle labels just lead to simple transformations of u and A. We therefore only need to calculate the matrix elements between the function of Eq. (6.27). In Appendix A.2 we consider the matrix elements for another type of the correlated Gaussian which has the angular function of Eq. (6.3).

The use of Eq. (6.27) enables one to obtain the matrix element for an operator $\mathcal{O}$ through that of the generating function as follows

$$\langle f_{K'L'M'}(u', A', \boldsymbol{x})|\mathcal{O}|f_{KLM}(u, A, \boldsymbol{x})\rangle$$
$$= \frac{1}{B_{KL}B_{K'L'}}\iint d\hat{\boldsymbol{e}}d\hat{\boldsymbol{e}}'Y_{LM}(\hat{\boldsymbol{e}})Y_{L'M'}(\hat{\boldsymbol{e}}')^{*}\left(\frac{d^{2K+L+2K'+L'}}{d\lambda^{2K+L}d\lambda'^{2K'+L'}}\right.$$
$$\left.\times\langle g(\lambda'\boldsymbol{e}'u'; A', \boldsymbol{x})|\mathcal{O}|g(\lambda\boldsymbol{e}u; A, \boldsymbol{x})\rangle\right)_{\substack{\lambda=0,\ \lambda'=0\\ e=|\boldsymbol{e}|=1\\ e'=|\boldsymbol{e}'|=1}}. \tag{7.2}$$

As the matrix element $\langle g(\lambda'\boldsymbol{e}'u'; A', \boldsymbol{x})|\mathcal{O}|g(\lambda\boldsymbol{e}u; A, \boldsymbol{x})\rangle$ between the generating functions is given as a function of $\lambda, \lambda', \boldsymbol{e}$, and $\boldsymbol{e}'$ for various operators by using Table 7.1, one only needs to perform the operations prescribed in the above equation. As will be shown in the appendix, the operations can be done very systematically for many cases of practical interest for arbitrary L.

The correlated Gaussians with $L = 0$ what we call spherical Gaussians take the particularly simple form in the special case, $K = 0$. As they give a fairly good solution for a variety of few-body problems, we list the explicit form of the matrix elements for this case in the following. Because the parameter u of Eq. (6.27) is redundant in this approximation, we denote $\sqrt{4\pi}f_{000}(u, A, \boldsymbol{x})$ simply as

$$F_A(\boldsymbol{x}) = \exp\left(-\frac{1}{2}\tilde{\boldsymbol{x}}A\boldsymbol{x}\right). \tag{7.3}$$

Since this function is equal to $g(\boldsymbol{s}=0;A,\boldsymbol{x})$, a special case of the generating function (6.19), the needed matrix elements are in fact given in Table 7.1.

The overlap of the spherical Gaussians is given by

$$\langle F_{A'}|F_A\rangle = \left(\frac{(2\pi)^{N-1}}{\det(A+A')}\right)^{\frac{3}{2}}. \tag{7.4}$$

The matrix element of the kinetic energy is obtained through Eqs. (2.10) and (2.11) by

$$\langle F_{A'}|\sum_{i=1}^{N}\frac{\boldsymbol{p}_i^2}{2m_i} - T_{\rm cm}|F_A\rangle$$
$$= \frac{3}{2}\hbar^2 {\rm Tr}\Big(A(A+A')^{-1}A'\Lambda\Big)\langle F_{A'}|F_A\rangle. \tag{7.5}$$

The matrix element of the two-body potential is given by

$$\langle F_{A'}|V(\boldsymbol{r}_i-\boldsymbol{r}_j)|F_A\rangle = \langle F_{A'}|F_A\rangle\left(\frac{c_{ij}}{2\pi}\right)^{\frac{3}{2}}\int V(\boldsymbol{r})\,{\rm e}^{-\frac{1}{2}c_{ij}r^2}d\boldsymbol{r}, \tag{7.6}$$

where

$$\frac{1}{c_{ij}} = \widetilde{w^{(ij)}}(A+A')^{-1}w^{(ij)} \tag{7.7}$$

is given through the column vector $w^{(ij)}$ defined by Eq. (2.13). The matrix element of the total potential energy is easily obtained by summing Eq. (7.6) over i, j. As a by-product of Eq. (7.6), we show that the matrix element of $|\boldsymbol{r}_i-\boldsymbol{r}_j|^\alpha$ can be easily obtained because then the integral in Eq. (7.6) can be reduced to

$$\int r^\alpha\,{\rm e}^{-\frac{1}{2}cr^2}d\boldsymbol{r} = 4\pi\int_0^\infty r^{\alpha+2}\,{\rm e}^{-\frac{1}{2}cr^2}dr$$
$$= 2\pi\left(\frac{2}{c}\right)^{\frac{\alpha+3}{2}}\int_0^\infty x^{\frac{\alpha+1}{2}}\,{\rm e}^{-x}dx$$
$$= 2\pi\left(\frac{2}{c}\right)^{\frac{\alpha+3}{2}}\Gamma\Big(\frac{\alpha+1}{2}+1\Big), \tag{7.8}$$

where Γ is the Gamma function. Thus we obtain

$$\langle F_{A'} || \boldsymbol{r}_i - \boldsymbol{r}_j|^{\alpha} | F_A \rangle = \langle F_{A'} | F_A \rangle \frac{2}{\sqrt{\pi}} \left(\frac{2}{c_{ij}}\right)^{\frac{\alpha}{2}} \Gamma\left(\frac{\alpha+3}{2}\right), \tag{7.9}$$

where c_{ij} is given by Eq. (7.7).

For a spherically symmetric potential the integral

$$I(k,c) = \int_0^{\infty} V(r) r^k \, \mathrm{e}^{-\frac{1}{2}cr^2} dr \tag{7.10}$$

can be calculated analytically for a certain class of potentials. If $V(r)$ is given in the form

$$V(r) = r^n \mathrm{e}^{-ar^2 - br}, \tag{7.11}$$

then a closed expression for $I(k,c)$ becomes possible for an integer $n (n \geq -k)$ by using the following formula

$$\int_0^{\infty} r^n \mathrm{e}^{-ar^2-br} dr = \begin{cases} \sqrt{\pi}(-1)^n \left(\frac{1}{2\sqrt{a}}\right)^{n+1} \sum_{k=0}^{n} \frac{n!}{(n-k)!\,k!} f_k \, g_{n-k} & \text{for } a > 0 \\ \frac{n!}{b^{n+1}} & \text{for } {a=0 \atop b>0} \end{cases} \tag{7.12}$$

with

$$f_0 = 1, \qquad g_0 = \exp\left(\frac{b^2}{4a}\right) \operatorname{erfc}\left(\frac{b}{2\sqrt{a}}\right),$$

and for $k \geq 1$

$$f_k = \sum_{i=0}^{[\frac{k}{2}]} \frac{k!}{i!(k-2i)!} \left(\frac{b}{\sqrt{a}}\right)^{k-2i},$$

$$g_k = \frac{2}{\sqrt{\pi}} (-1)^k H_{k-1}\left(\frac{b}{2\sqrt{a}}\right). \tag{7.13}$$

Here $H_n(x)$ is the Hermite polynomial and $\operatorname{erfc}(x) = 1 - \operatorname{erf}(x)$ is the complementary error function, where $\operatorname{erf}(x)$ is the error function: $\operatorname{erf}(x) = (2/\sqrt{\pi}) \int_0^{\infty} \mathrm{e}^{-t^2} dt$. See Exercise 7.1 for the derivation of Eq. (7.12). In case where an analytic evaluation of the integral is impossible, one has to rely on a numerical integration.

Examples of application of $F_A(\boldsymbol{x})$ to various quantum-mechanical few-body problems are presented in [81]. See also Complements 11.1 and 11.3.

7.3 Correlated Gaussians in two-dimensional systems

In this section we review the trial wave function and its matrix elements in the case of two-dimensional (2D) problems. A real application of the results in this section will be made to the study on few-body problems in quantum dots in Chap. 10.

In 2D problems the motion of the particles is constrained to the xy-plane. The vector $\boldsymbol{r}$ has therefore two components $\boldsymbol{r} = (x, y)$, and all expressions for relative coordinates and other vectors introduced in the previous sections remain valid by setting $z = 0$. The scalar products and integrations over spatial variables will include only the xy components.

The main difference between the 2D and the 3D case is the orbital angular momentum. In 3D the basis function was taken as the eigenfunction of the operators $\boldsymbol{L}^2$, L_z:

$$r^l Y_{lm}(\hat{\boldsymbol{r}}). \tag{7.14}$$

In 2D $\boldsymbol{L}^2 = L_z^2$, so that one of these operators becomes redundant reflecting the fact that the system may be invariant under rotations around the z-axis only. The corresponding orbital angular functions take the form

$$r^m \mathrm{e}^{\pm im\varphi} = (x \pm iy)^m \qquad (m \geq 0), \tag{7.15}$$

that is, the states are characterized by the quantum number "m". This function, apart from normalization, can be derived from its 3D counterpart by setting $l = m$ and $z = 0$ (or $\vartheta = \pi/2$ in polar coordinate system).

The matrix elements derived for the 3D case cannot be directly used because they already include an integration over the z-component. One can, however, derive the matrix elements in a very similar way by using generating functions and exploiting the simplifications present in the 2D case. The first step toward this is to construct a generating function. Similarly to the 3D case, we assume a trial function, by using the global vector representation, in the form:

$$f_m(u, A, \boldsymbol{x}) = (v_1 + iv_2)^m \exp\left(-\frac{1}{2}\tilde{\boldsymbol{x}}A\boldsymbol{x}\right), \tag{7.16}$$

where $\boldsymbol{x}$ is the vector of relative coordinates in 2D and v_1 and v_2 are the x and y components of the vector

$$v = \sum_{i=1}^{N-1} u_i \boldsymbol{x}_i. \tag{7.17}$$

Note that the complex conjugate of $f_m(u, A, \boldsymbol{x})$ is an eigenfunction of L_z with eigenvalue $-m$.

To derive matrix elements, one can use a particularly simple generating function for $f_m(u, A, \boldsymbol{x})$:

$$f_m(u, A, \boldsymbol{x}) = \left(\frac{d^m}{dt^m} g(\boldsymbol{s}, A, \boldsymbol{x}) \right)_{t=0} \tag{7.18}$$

with

$$g(\boldsymbol{s}, A, \boldsymbol{x}) = \exp\left\{ -\frac{1}{2}\tilde{\boldsymbol{x}}A\boldsymbol{x} + t(v_1 + iv_2) \right\}, \tag{7.19}$$

where the 2D vector $\boldsymbol{s}_j$ $(j = 1, ..., N-1)$ is defined by a complex vector (tu_j, itu_j). Note that they satisfy $\boldsymbol{s}_i \cdot \boldsymbol{s}_j = 0$, $\boldsymbol{s}_i \cdot \boldsymbol{s}_j'^{\,*} = 2tt'u_iu_j'$. The matrix elements of the generating function remain valid as listed in Table 7.1 by substituting $d = 2$ and assuming 2D vectors. By using the matrix elements of the generating function all the necessary matrix elements can be easily calculated by carrying out the differentiation prescribed by Eq. (7.18). The matrix elements for some basic operators are collected in Table 7.2.

Table 7.2. Matrix elements $\mathcal{M}_{m'm} = \langle f_{m'} | \mathcal{O} | f_m \rangle$ for two-dimensional case. B is equal to $A + A'$. Λ is defined in Eq. (2.11). Q, R, ρ, $\bar{\rho}$, γ, γ', c are the same as defined in Eqs. (A.2), (A.9), and (A.18), but the factor 3 in R must be changed to 2. The integral I is defined in Eq. (7.10).

$\mathcal{O}$	$\mathcal{M}_{mm'}$
1	$\mathcal{M}_m \delta_{mm'}$ with $\mathcal{M}_m = \frac{(2\pi)^{N-1}}{\det B} m!(2\rho)^m$
$\sum_{i=1}^{N} \frac{\boldsymbol{p}_i^2}{2m_i} - T_{\rm cm} \equiv \frac{1}{2}\tilde{\boldsymbol{\pi}}\Lambda\boldsymbol{\pi}$	$\frac{1}{2}\hbar^2(R + \frac{m}{\rho}Q)\mathcal{M}_m\delta_{mm'}$
L_z	$m\mathcal{M}_m\delta_{mm'}$
$V(\lvert \boldsymbol{r}_i - \boldsymbol{r}_j \rvert)$	$\frac{(2\pi)^{N-1}c}{\det B}(m!)^2\delta_{mm'} \sum_{n=0}^{m} \frac{(2\bar{\rho})^{m-n}(\gamma\gamma')^n}{(m-n)!(n!)^2} I(2n+1, c)$

7.4 Correlated Gaussian-type geminals

We assume that the Hamiltonian of the N-particle system now takes the form of Eq. (2.2). We use the single-particle coordinates $\tilde{\boldsymbol{r}} = (\boldsymbol{r}_1, ..., \boldsymbol{r}_N)$ in this section. The transformation matrix U defined in Eq. (2.4) is now redundant or can be set equal to the unit matrix. The basis function is chosen to be the correlated Gaussian-type geminal of Eq. (6.26). The function is characterized by three parameters. The matrix A describes the correlation among the particles, the diagonal matrix B determines the spread of the single-particle wave-packets, and $\widetilde{\boldsymbol{R}} = (\boldsymbol{R}_1, ..., \boldsymbol{R}_N)$ give the centers of the packets. This type of basis function is employed in molecular physics and nuclear cluster models. The permutation P of particle indices transforms the coordinates as defined by Eqs. (2.26) and (2.27). This transformation changes the basis function to

$$Pf_{\boldsymbol{n}}(A, B, \boldsymbol{R}, \boldsymbol{r}) = f_{\widetilde{T_P}\boldsymbol{n}}(\widetilde{T_P}AT_P, \widetilde{T_P}BT_P, \widetilde{T_P}\boldsymbol{R}, \boldsymbol{r}), \tag{7.20}$$

where use is made of the relation $(T_P)^{-1} = T_{P^{-1}} = \widetilde{T_P}$ and $\widetilde{T_P}\boldsymbol{n}$ stands for $\{n_{q_{1_1}}, n_{q_{1_2}}, n_{q_{1_3}}, ..., n_{q_{N_1}}, n_{q_{N_2}}, n_{q_{N_3}})$ for

$$P^{-1} = \begin{pmatrix} p_1 & p_2 & \dots & p_N \\ 1 & 2 & \dots & N \end{pmatrix} = \begin{pmatrix} 1 & 2 & \dots & N \\ q_1 & q_2 & \dots & q_N \end{pmatrix}.$$

Thus the permutation acting on the basis function just leads to the simple transformation of the parameters, $A, B, \boldsymbol{R}$, and $\boldsymbol{n}$ and still keeps the form of the basis function invariant. It is not nice, however, that the value of $\boldsymbol{n}$ changes under the permutation.

The integral needed to evaluate the matrix element with the correlated Gaussian-type geminal is $3N$ dimensional and of multi-center type. Such integrals are extremely hard to calculate and to handle. See, e.g., [82, 83] for some developments made for those multi-center integrals which appear in molecular physics. To avoid the complication, a few simplifications are introduced in practice. For example, only one pair of particles is allowed to be correlated in each basis function. This is equivalent to assuming that the only non-vanishing elements of the matrix A are, $A_{ii} = A_{jj} = \alpha_{ij}$ and $A_{ij} = A_{ji} = -\alpha_{ij}$, if the correlation between the ith and jth particles is taken into account by $\exp\{-\alpha_{ij}(\boldsymbol{r}_i - \boldsymbol{r}_j)^2/2\}$. Another simplification is to use only primitive Gaussians, that is, the basis function with $\boldsymbol{n} = 0$. Primitive Gaussians yield solutions of good quality for few-electron systems. In this section we calculate the matrix element in the primitive Gaussian basis

$$f_0(A,B,\boldsymbol{R},\boldsymbol{r}) = \exp\Big\{-\frac{1}{2}\tilde{\boldsymbol{r}}A\boldsymbol{r} - \frac{1}{2}(\widetilde{\boldsymbol{r}-\boldsymbol{R}})B(\boldsymbol{r}-\boldsymbol{R})\Big\}$$
$$= \exp\Big(-\frac{1}{2}\tilde{\boldsymbol{R}}B\boldsymbol{R}\Big)g(B\boldsymbol{R};A+B,\boldsymbol{r}), \qquad (7.21)$$

allowing the most general type of correlations. The formulation presented here will be found to be very simple and lead to compact results for various matrix elements. The matrix elements with higher angular momenta, when needed, can be obtained from the matrix elements in the generating function by straightforward but tedious differentiation. An algorithm for a systematic evaluation of the differentiation is presented in Appendix A.3.

The overlap matrix element is readily available from Table 7.1. It reads as

$$\mathcal{M}_G \equiv \langle f_0(A',B',\boldsymbol{R}',\boldsymbol{r})|f_0(A,B,\boldsymbol{R},\boldsymbol{r})\rangle$$
$$= \left(\frac{(2\pi)^N}{\det C}\right)^{\frac{3}{2}} \exp\Big(\frac{1}{2}\tilde{\boldsymbol{v}}C^{-1}\boldsymbol{v} - \frac{1}{2}\tilde{\boldsymbol{R}}B\boldsymbol{R} - \frac{1}{2}\widetilde{\boldsymbol{R}'}B'\boldsymbol{R}'\Big), \quad (7.22)$$

where

$$C = A + B + A' + B', \qquad \boldsymbol{v} = B\boldsymbol{R} + B'\boldsymbol{R}'. \qquad (7.23)$$

Note that formulas derived in this section are valid for a general symmetric matrix B or B' provided that $A+B$ and $A'+B'$ are positive-definite.

The matrix element of the kinetic energy is also available from Table 7.1:

$$\langle f_0(A',B',\boldsymbol{R}',\boldsymbol{r})|\sum_{i=1}^{N}\frac{\boldsymbol{p}_i^2}{2m_i}|f_0(A,B,\boldsymbol{R},\boldsymbol{r})\rangle$$
$$= \frac{\hbar^2}{2}\Big\{3\mathrm{Tr}\Big((A+B)C^{-1}(A'+B')\Lambda\Big) - \tilde{\boldsymbol{y}}\Lambda\boldsymbol{y}\Big\}\mathcal{M}_G \qquad (7.24)$$

with

$$\boldsymbol{y} = (A'+B')C^{-1}B\boldsymbol{R} - (A+B)C^{-1}B'\boldsymbol{R}', \qquad (7.25)$$

where Λ is a diagonal matrix with $\Lambda_{ii} = 1/m_i$. The matrix element of the center-of-mass kinetic energy is calculated very easily as well. As Eq. (2.10) shows, it is calculated by the above formula with a simple modification: The matrix Λ is now to be taken as $\Lambda_{ij} = 1/m_{12\cdots N}$ for all i and j.

To calculate the matrix element of both one- and two-body potentials we make use of the following relation

$$\langle f_0(A', B', \boldsymbol{R}', \boldsymbol{r})|V(|\tilde{w}\boldsymbol{r} - \boldsymbol{S}|)|f_0(A, B, \boldsymbol{R}, \boldsymbol{r})\rangle$$
$$= \int V(x)\langle f_0(A', B', \boldsymbol{R}', \boldsymbol{r})|\delta(\tilde{w}\boldsymbol{r} - \boldsymbol{S} - \boldsymbol{x})|f_0(A, B, \boldsymbol{R}, \boldsymbol{r})\rangle d\boldsymbol{x}, \tag{7.26}$$

where $\tilde{w} = (w_1, ..., w_N)$ is an auxiliary parameter and $\boldsymbol{S}$ is a 3-dimensional vector. Substituting the matrix element of the δ-function in Table 7.1 into Eq. (7.26) leads to

$$\langle f_0(A', B', \boldsymbol{R}', \boldsymbol{r})|V(|\tilde{w}\boldsymbol{r} - \boldsymbol{S}|)|f_0(A, B, \boldsymbol{R}, \boldsymbol{r})\rangle$$
$$= \mathcal{M}_G\Big(\frac{2\pi}{c}\Big)^{-\frac{3}{2}} \int V(x) \exp\Big\{-\frac{c}{2}(\boldsymbol{x} + \boldsymbol{S} - \tilde{w}C^{-1}\boldsymbol{v})^2\Big\} d\boldsymbol{x}$$
$$= \mathcal{M}_G\sqrt{\frac{c}{2\pi}}\,\frac{1}{s}\,\mathrm{e}^{-\frac{1}{2}cs^2}\int_0^\infty V(x)\,x\,\mathrm{e}^{-\frac{1}{2}cx^2}(\mathrm{e}^{csx} - \mathrm{e}^{-csx})dx, \tag{7.27}$$

where

$$c^{-1} = \tilde{w}C^{-1}w \quad \text{and} \quad s = |\tilde{w}C^{-1}\boldsymbol{v} - \boldsymbol{S}| \tag{7.28}$$

were introduced.

We show that both one-body and two-body matrix elements are calculated in a unified way with an appropriate choice of w. The matrix element of the one-body potential, $V(|\boldsymbol{r}_i - \boldsymbol{S}|)$, can be obtained by choosing $w_j = \delta_{ij}$ $(j = 1, ..., N)$. If $V(x)$ takes the form of Eq. (7.11), then the above integral is obtained analytically for $n\,(n \geq -1)$. As the most important application of this formulation, we give below the matrix element for the Coulomb potential:

$$\langle f_0(A', B', \boldsymbol{R}', \boldsymbol{r})|\frac{1}{|\boldsymbol{r}_i - \boldsymbol{S}|}|f_0(A, B, \boldsymbol{R}, \boldsymbol{r})\rangle$$
$$= \mathcal{M}_G\,\frac{1}{s}\,\mathrm{erf}\Big(\sqrt{\frac{c}{2}}s\Big), \tag{7.29}$$

where c^{-1} is now defined by $(C^{-1})_{ii}$ and s is given by $|(C^{-1}\boldsymbol{v})_i - \boldsymbol{S}|$.

The two-body matrix element can also be obtained in exactly the same manner by the use of Eqs. (7.27) and (7.28). To derive the matrix element of $V(|\boldsymbol{r}_i - \boldsymbol{r}_j - \boldsymbol{S}|)$ one only needs to choose w as $w_k = \delta_{ik} - \delta_{jk}$ $(k = 1, ..., N)$. E.g., the matrix element of the Coulomb potential, $1/|\boldsymbol{r}_i - \boldsymbol{r}_j|$, can be calculated from Eq. (7.29) with the substitution of $c^{-1} = (C^{-1})_{ii} + (C^{-1})_{jj} - (C^{-1})_{ij} - (C^{-1})_{ji}$ and $s = |(C^{-1}\boldsymbol{v})_i - (C^{-1}\boldsymbol{v})_j|$. When $V(x)$ takes the Gaussian form factor $\mathrm{e}^{-\rho x^2}$, the matrix elements for both the central and the spin-orbit potentials can be expressed in particularly simple forms:

$$\langle f_0(A', B', \boldsymbol{R}', \boldsymbol{r})|\mathrm{e}^{-\rho(\tilde{w}\boldsymbol{r})^2}|f_0(A, B, \boldsymbol{R}, \boldsymbol{r})\rangle$$

$$= \mathcal{M}_G \left(\frac{c}{c+2\rho}\right)^{\frac{3}{2}} \exp\Big\{-\frac{c\rho}{c+2\rho}(\tilde{w}C^{-1}\boldsymbol{v})^2\Big\}. \tag{7.30}$$

$$\langle f_0(A', B', \boldsymbol{R}', \boldsymbol{r})|\mathrm{e}^{-\rho(\tilde{w}\boldsymbol{r})^2}(\tilde{w}\boldsymbol{r} \times \tilde{\zeta}\boldsymbol{p})|f_0(A, B, \boldsymbol{R}, \boldsymbol{r})\rangle$$

$$= -i\hbar(\tilde{w}C^{-1}\boldsymbol{v} \times \tilde{\zeta}\boldsymbol{y})\mathcal{M}_G\Big(\frac{c}{c+2\rho}\Big)^{\frac{5}{2}} \exp\Big\{-\frac{c\rho}{c+2\rho}(\tilde{w}C^{-1}\boldsymbol{v})^2\Big\}. \tag{7.31}$$

It is clear that the above formulation applies to the calculation of even N-body matrix elements as well.

7.5 Nonlocal potentials

In this section we show a simple example of the extension of the calculational technique to nonlocal potentials. The application of nonlocal potentials will be presented for positronic atoms in Sect. 8.3 and for the nucleus ^{12}C in Complement 11.4. The potential is assumed to be of separable form

$$V = \sum_{j>i=1}^{N} \sum_{a} \lambda_{ij} |\, \varphi_a(\boldsymbol{r}_i - \boldsymbol{r}_j)\rangle\langle\varphi_a(\boldsymbol{r}_i - \boldsymbol{r}_j)\,| \tag{7.32}$$

with a Gaussian form factor

$$\varphi_a(\boldsymbol{r}) = \mathrm{e}^{-\frac{1}{2}ar^2}. \tag{7.33}$$

This form is quite useful in many practical cases, but generalization to other cases is straightforward by using the same trick as for the potential functions, that is by substituting the function φ_a by a delta function $\delta(\boldsymbol{r}_i - \boldsymbol{r}_j - \boldsymbol{r})$.

The first step of the calculation of the matrix elements of the above potential is to substitute the relative coordinates $\boldsymbol{x}$ by a new set of Jacobi coordinates $\boldsymbol{y} = \{\boldsymbol{y}_1, ..., \boldsymbol{y}_{N-1}\}$. The Jacobi coordinates $\boldsymbol{y}$ are obtained from the original set of coordinates $\boldsymbol{x}$ in such a way that $\boldsymbol{y}_1 = \boldsymbol{r}_i - \boldsymbol{r}_j$ (see Sect. 2.4). The matrix of the corresponding transformation

$$\boldsymbol{y} = T^{(k)}\boldsymbol{x}, \tag{7.34}$$

whose Jacobian is $\det(\partial\boldsymbol{x}/\partial\boldsymbol{y}) = 1$, is defined in Sect. 2.4. The kth permutation in deriving $T^{(k)}$ is defined in such a way that the first Jacobi coordinate is equal to $\boldsymbol{r}_i - \boldsymbol{r}_j$.

The next step is to show the matrix elements for the generating function (6.19). The matrix element of one term of the potential reads as

$$\begin{aligned}&\langle g(\boldsymbol{s}'; A', \boldsymbol{x})|\varphi_a(\boldsymbol{r}_i - \boldsymbol{r}_j)\rangle\langle\varphi_a(\boldsymbol{r}_i - \boldsymbol{r}_j)|g(\boldsymbol{s}; A, \boldsymbol{x})\rangle \\ &\quad = \langle g(\tilde{T}\boldsymbol{s}'; \tilde{T}A'T, \boldsymbol{y})|\varphi_a(\boldsymbol{y}_1)\rangle\langle\varphi_a(\boldsymbol{y}_1)|g(\tilde{T}\boldsymbol{s}; \tilde{T}AT, \boldsymbol{y})\rangle,\end{aligned} \tag{7.35}$$

where the integration has to be carried out first over $\boldsymbol{y}_1$ in both sides and then over the remaining coordinates, $\boldsymbol{y}_i$'s. We use the same notation $\boldsymbol{y}$ to denote these remaining $(N-2)$ variables, as well. To carry out the integration implied in Eq. (7.35), we introduce the following notations:

- B: an $(N-2) \times (N-2)$ matrix obtained by suppressing the first row and the first column of the matrix $\tilde{T}AT$.
- b: an $(N-2)$-dimensional vector $\{(\tilde{T}AT)_{12}, ..., (\tilde{T}AT)_{1\,N-1}\}$.
- b_1: the first diagonal element of the matrix $(\tilde{T}AT)$, i.e., $(\tilde{T}AT)_{11}$.
- $\boldsymbol{t}$: a set of $(N-2)$ vectors $\{(\tilde{T}\boldsymbol{s})_2, ..., (\tilde{T}\boldsymbol{s})_{N-1}\}$.
- $\boldsymbol{t}_1$: the first vector $(\tilde{T}\boldsymbol{s})_1$.

The notations corresponding to the bra side are introduced in exactly the same way and distinguished with a prime symbol from the quantities of the ket side. By separating the $\boldsymbol{y}_1$-dependence explicitly from the generating function g as

$$\begin{aligned}&g(\tilde{T}\boldsymbol{s}; \tilde{T}AT, \boldsymbol{y}) \\ &\quad = g(\boldsymbol{t}; B, \boldsymbol{y}) \exp\left(-\frac{1}{2}b_1\boldsymbol{y}_1^2 - \tilde{b}\boldsymbol{y}\cdot\boldsymbol{y}_1 + \boldsymbol{t}_1\cdot\boldsymbol{y}_1\right),\end{aligned} \tag{7.36}$$

and by using the formula (6.32) one obtains

$$\langle \varphi_a(\boldsymbol{y}_1)|g(\tilde{T}\boldsymbol{s};\tilde{T}AT,\boldsymbol{y})\rangle$$
$$= \left(\frac{2\pi}{d}\right)^{\frac{3}{2}} \exp\left(\frac{1}{2d}\boldsymbol{t}_1^2\right) g\Big(\boldsymbol{t}-\frac{1}{d}b\boldsymbol{t}_1;D,\boldsymbol{y}\Big), \tag{7.37}$$

where

$$d = a + b_1, \qquad D = B - \frac{1}{d}b\tilde{b}. \tag{7.38}$$

As this function is again a generating function, the integration over the remaining variables $\boldsymbol{y}$ is just the overlap of the generating functions, and from Table 7.1 one obtains the final result as

$$\langle g(\boldsymbol{s}';A',\boldsymbol{x})|\varphi_a(\boldsymbol{r}_i-\boldsymbol{r}_j)\rangle\langle\varphi_a(\boldsymbol{r}_i-\boldsymbol{r}_j)|g(\boldsymbol{s};A,\boldsymbol{x})\rangle$$
$$= \left(\frac{(2\pi)^N}{dd'\det(D+D')}\right)^{\frac{3}{2}}$$
$$\times \exp\left(\frac{1}{2d}\boldsymbol{t}_1{}^2 + \frac{1}{2d'}\boldsymbol{t}_1'{}^2 + \frac{1}{2}\tilde{\boldsymbol{v}}_0(D+D')^{-1}\boldsymbol{v}_0\right) \tag{7.39}$$

with

$$\boldsymbol{v}_0 = \boldsymbol{t} - \frac{1}{d}b\boldsymbol{t}_1 + \boldsymbol{t}' - \frac{1}{d'}b'\boldsymbol{t}_1'. \tag{7.40}$$

The above result becomes particularly simple for the correlated Gaussian with $L = 0$ used in Sect. 7.2 and reads as

$$\langle F_{A'}|\varphi_a(\boldsymbol{r}_i-\boldsymbol{r}_j)\rangle\langle\varphi_a(\boldsymbol{r}_i-\boldsymbol{r}_j)|F_A\rangle$$
$$= \left(\frac{(2\pi)^N}{dd'\det(D+D')}\right)^{\frac{3}{2}}. \tag{7.41}$$

To calculate the matrix element for nonlocal potentials of an arbitrary form factor, we note from Eq. (7.36) that

$$\langle\delta(\boldsymbol{y}_1-\boldsymbol{r})|g(\tilde{T}\boldsymbol{s};\tilde{T}AT,\boldsymbol{y})\rangle$$
$$= g(\boldsymbol{t}-b\boldsymbol{r};B,\boldsymbol{y})\exp\left(-\frac{1}{2}b_1\boldsymbol{r}^2+\boldsymbol{t}_1\cdot\boldsymbol{r}\right), \tag{7.42}$$

which leads to the general result

$$\langle g(\boldsymbol{s}';A',\boldsymbol{x})|\delta(\boldsymbol{r}_i-\boldsymbol{r}_j-\boldsymbol{r}')\rangle\langle\delta(\boldsymbol{r}_i-\boldsymbol{r}_j-\boldsymbol{r})|g(\boldsymbol{s};A,\boldsymbol{x})\rangle$$

$$= \left(\frac{(2\pi)^{N-2}}{\det(B+B')}\right)^{\frac{3}{2}}$$
$$\times \exp\left(-\frac{1}{2}b_1\boldsymbol{r}^2 + \boldsymbol{t}_1\cdot\boldsymbol{r} - \frac{1}{2}b_1'\boldsymbol{r}'^2 + \boldsymbol{t}_1'\cdot\boldsymbol{r}' + \frac{1}{2}\tilde{\boldsymbol{v}}(B+B')^{-1}\boldsymbol{v}\right) \tag{7.43}$$

with

$$\boldsymbol{v} = \boldsymbol{t} - b\boldsymbol{r} + \boldsymbol{t}' - b'\boldsymbol{r}'. \tag{7.44}$$

By introducing the short-hand notations

$$\boldsymbol{z} = \begin{pmatrix} \boldsymbol{r} \\ \boldsymbol{r}' \end{pmatrix}, \qquad \boldsymbol{w} = \begin{pmatrix} \boldsymbol{t}_1 - \tilde{b}(B+B')^{-1}(\boldsymbol{t}+\boldsymbol{t}') \\ \boldsymbol{t}_1' - \tilde{b}'(B+B')^{-1}(\boldsymbol{t}+\boldsymbol{t}') \end{pmatrix},$$
$$C = \begin{pmatrix} b_1 - \tilde{b}(B+B')^{-1}b & -\tilde{b}(B+B')^{-1}b' \\ -\tilde{b}'(B+B')^{-1}b & b_1' - \tilde{b}'(B+B')^{-1}b' \end{pmatrix}, \tag{7.45}$$

Eq. (7.43) can be expressed in terms of g in compact form:

$$\langle g(\boldsymbol{s}'; A', \boldsymbol{x})|\delta(\boldsymbol{r}_i - \boldsymbol{r}_j - \boldsymbol{r}')\rangle\langle\delta(\boldsymbol{r}_i - \boldsymbol{r}_j - \boldsymbol{r})|g(\boldsymbol{s}; A, \boldsymbol{x})\rangle$$
$$= \left(\frac{(2\pi)^{N-2}}{\det(B+B')}\right)^{\frac{3}{2}} \exp\left(\frac{1}{2}\widetilde{(\boldsymbol{t}+\boldsymbol{t}')}(B+B')^{-1}(\boldsymbol{t}+\boldsymbol{t}')\right)$$
$$\times g(\boldsymbol{w}; C, \boldsymbol{z}). \tag{7.46}$$

7.6 Semirelativistic kinetic energy

This section shows the calculation of the matrix element of the semirelativistic kinetic energy. This will be needed in the application to subnuclear systems among others. See Chap. 9.

We assumed the nonrelativistic kinematics in Chap. 2. The single-particle kinetic energy T is then related to the momentum $\boldsymbol{p}$ by $T = \boldsymbol{p}^2/(2m)$. In the relativistic kinematics it is replaced by $T = \sqrt{c^2\boldsymbol{p}^2 + m^2c^4} - mc^2$. The speed of light c is set to unity and the term, $-mc^2$, is dropped unless otherwise stated as it just shifts the energy expectation value. The kinetic energy we consider takes the form

$$T_{\rm sr} = \sum_{i=1}^{N} \sqrt{\boldsymbol{p}_i^2 + m_i^2}. \tag{7.47}$$

The evaluation of the matrix element of the operator (7.47) is never trivial, because the separation of the center-of-mass motion requires special care. The contribution of the center-of-mass motion is removed for the nonrelativistic case in Sect. 2.1 by subtracting the center-of-mass kinetic energy from the Hamiltonian. This prescription led us to the expression (2.10) for the intrinsic kinetic energy, which, as it should, does not contain any dependence on the total momentum $\boldsymbol{\pi}_N$. However, this method no longer applies to the present case and must be replaced by a more general procedure.

We assert that the sought procedure is to evaluate the matrix element in the center-of-mass system, that is, in the system of

$$\boldsymbol{\pi}_N \equiv \sum_{i=1}^{N} \boldsymbol{p}_i = 0. \tag{7.48}$$

Note that, when applied to the nonrelativistic kinetic energy, this new procedure reduces to the previous prescription of subtracting the center-of-mass kinetic energy from the total kinetic energy as expected. This is easily seen as follows: By substituting Eq. (2.9) into $T_{\rm nr} = \sum_{i=1}^{N} \boldsymbol{p}_i^2/(2m_i)$, we obtain $T_{\rm nr} = (1/2)\sum_{i=1}^{N}\sum_{j=1}^{N} \Lambda_{ij}\boldsymbol{\pi}_i \cdot \boldsymbol{\pi}_j$ with Λ_{ij} being defined by Eq. (2.11). (The center-of-mass kinetic energy was subtracted from the beginning in Eq. (2.10), so that the suffices i and j go up to $N-1$ and Λ_{ij} was defined for $i,\, j \le N-1$. In the case where the center-of-mass kinetic energy is included, it is not difficult to see that i and j take 1 to N and Eq. (2.11) is still valid for $(i,\, j = 1, ..., N)$.) The requirement (7.48) then restricts the sum over i and j up to $N-1$ in practice instead of up to N. Thus the matrix element of the operator $T_{\rm nr}$ evaluated in the center-of-mass system is equal to that of the intrinsic kinetic energy of Eq. (2.10).

Now we will show a method of calculating the matrix element of $T_{\rm sr}$ in the center-of-mass system in two steps. The first step is to express the operator in terms of the operators defined in the relative coordinates. As Eq. (2.9) shows, the single-particle momentum operator $\boldsymbol{p}_i$ can be expressed in terms of the N momenta $\boldsymbol{\pi}_j$ $(j = 1, ..., N)$. If we use the Jacobi coordinate set $\boldsymbol{x}$ defined by the transformation matrix

(2.5), the momentum $\boldsymbol{p}_N$ becomes just $-\boldsymbol{\pi}_{N-1}$ thanks to the condition (7.48). This peculiarity was already noted in Sect. 2.4. Note that none of the other momenta takes such a simple form; they contain more than one $\boldsymbol{\pi}$'s. Let us denote this particular Jacobi coordinate set $\boldsymbol{x}$, the corresponding matrix U_J, and the momentum $\boldsymbol{\pi}_{N-1}$ defined in this coordinate set as $\boldsymbol{x}^{(N)}$, $U_J^{(N)}$, and $\boldsymbol{\pi}_{N-1}^{(N)}$, respectively. Obviously we can define $(N-1)$ other sets of Jacobi coordinates $\boldsymbol{x}^{(k)}$ $(k=1,...,N-1)$, each of which corresponds to the relative coordinate set obtained by acting with the cyclic permutation $(1,2,...,N) = \begin{pmatrix} 1 & 2 & \dots & N \\ 2 & 3 & \dots & 1 \end{pmatrix}$ k-times on the original pattern of the Jacobi set $\boldsymbol{x}^{(N)}$. See also Sect. 2.4 and Fig. 2.2. Corresponding to each set $\boldsymbol{x}^{(k)}$ a linear transformation analogous to Eq. (2.4) defines $U_J^{(k)}$. It is easy to see that $U_J^{(1)}$ can be obtained by rearranging the N column vectors of $U_J^{(N)}$ according to the permutation $(1,2,...,N)$ and at the same time by the cyclic permutation of the masses $m_1,...,m_N$. Likewise, $U_J^{(2)}$ can be obtained from $U_J^{(1)}$, $U_J^{(3)}$ from $U_J^{(2)}$, and so on. As the transformation between the single-particle and relative coordinates is always given in the form of Eq. (2.4) for any Jacobi set $\boldsymbol{x}^{(k)}$, the corresponding transformation of the single-particle and relative momenta is also given as in the form of Eq. (2.9):

$$\boldsymbol{p}_i = \sum_{j=1}^{N} \left(U_J^{(k)}\right)_{ji} \boldsymbol{\pi}_j^{(k)} \quad (i=1,...,N). \tag{7.49}$$

Using the special form of the matrix $U_J^{(k)}$ and the condition (7.48), we obtain the following useful relation

$$\boldsymbol{p}_k = -\boldsymbol{\pi}_{N-1}^{(k)} \quad (k=1,...,N). \tag{7.50}$$

That is, the kth single-particle momentum is equal to the negative of the $N-1$-th relative momentum defined in the kth Jacobi coordinate set. Therefore, with the help of Eq. (7.50) the semirelativistic kinetic energy (7.47) can be effectively replaced in the center-of-mass system by

$$T_{\rm sr} = \sum_{i=1}^{N} \sqrt{(\boldsymbol{\pi}_{N-1}^{(i)})^2 + m_i^2}. \tag{7.51}$$

Note that this operator has no cross terms such as $\boldsymbol{\pi}_i \cdot \boldsymbol{\pi}_j$. The price to be paid is that one has to transform the coordinate systems conforming to $\boldsymbol{\pi}_{N-1}^{(i)}$. There is no difficulty in evaluating the cross terms when one can work with a definite set of the relative coordinates. That is the reason why we followed that route for the nonrelativistic kinetic energy in Sect. 2.2. In the case of the semirelativistic kinetic energy the cross term has to be avoided as will be shown below.

The second step is to show how to evaluate the matrix element of the operator (7.51). One has to calculate the matrix elements term by term. For brevity we show a method only for the simplest correlated Gaussian $F_A(\boldsymbol{x})$ of Eq. (7.3). Let us assume that $\boldsymbol{x}$ denotes the last set of the Jacobi coordinates $\boldsymbol{x}^{(N)} = \{\boldsymbol{x}_1, ..., \boldsymbol{x}_{N-1}\}$. To calculate the matrix element of $\sqrt{(\boldsymbol{\pi}_{N-1}^{(i)})^2 + m_i^2}$, we have to transform the coordinate system conforming to the ith Jacobi set. This can be achieved by using the relation

$$\boldsymbol{x} = V^{(i)}\boldsymbol{x}^{(i)}, \tag{7.52}$$

where $V^{(i)}$ is an $(N-1) \times (N-1)$ matrix obtained by omitting the last row and column of $U_J^{(N)}(U_J^{(i)})^{-1}$. By using this transformation we have

$$\begin{aligned}&\langle F_{A'}(\boldsymbol{x})|\sqrt{(\boldsymbol{\pi}_{N-1}^{(i)})^2 + m_i^2}|F_A(\boldsymbol{x})\rangle\\ &\quad = (\det V^{(i)})^3 \langle F_{A'^{(i)}}(\boldsymbol{x}^{(i)})|\sqrt{(\boldsymbol{\pi}_{N-1}^{(i)})^2 + m_i^2}|F_{A^{(i)}}(\boldsymbol{x}^{(i)})\rangle,\end{aligned} \tag{7.53}$$

where $(\det V^{(i)})^3$ is a Jacobian corresponding to the change of integration variables and

$$A^{(i)} = \widetilde{V^{(i)}} A V^{(i)}, \qquad A'^{(i)} = \widetilde{V^{(i)}} A' V^{(i)}. \tag{7.54}$$

The kinetic energy operator and the relative coordinates of the correlated Gaussians are now expressed in the same coordinate set.

To cope with the square root, we go from the coordinate space to the momentum space:

$$\begin{aligned}&\langle F_{A'^{(i)}}(\boldsymbol{x}^{(i)})|\sqrt{(\boldsymbol{\pi}_{N-1}^{(i)})^2 + m_i^2}|F_{A^{(i)}}(\boldsymbol{x}^{(i)})\rangle\\ &\quad = \iint \langle F_{A'^{(i)}}|\boldsymbol{k}'\rangle\langle \boldsymbol{k}'|\sqrt{(\boldsymbol{\pi}_{N-1}^{(i)})^2 + m_i^2}|\boldsymbol{k}\rangle\langle \boldsymbol{k}|F_{A^{(i)}}\rangle \, d\boldsymbol{k}\, d\boldsymbol{k}',\end{aligned} \tag{7.55}$$

where $\boldsymbol{k} = \{\boldsymbol{k}_1, ..., \boldsymbol{k}_{N-1}\}$ and $\langle \boldsymbol{x}|\boldsymbol{k}\rangle = (2\pi)^{-\frac{3}{2}(N-1)} e^{i\tilde{\boldsymbol{k}}\boldsymbol{x}}$ is normalized as $\langle \boldsymbol{k}'|\boldsymbol{k}\rangle = \delta(\boldsymbol{k}-\boldsymbol{k}')$. The Fourier transform of the correlated Gaussian and the matrix element of the square root operator can be written in a very simple form:

$$\langle \boldsymbol{k}|F_A\rangle = \int \langle \boldsymbol{x}|\boldsymbol{k}\rangle^* \exp\Big(-\frac{1}{2}\tilde{\boldsymbol{x}}A\boldsymbol{x}\Big)d\boldsymbol{x}$$

$$= (\det A)^{-\frac{3}{2}}F_{A^{-1}}(\boldsymbol{k}), \qquad (7.56)$$

$$\langle \boldsymbol{k}'|\sqrt{(\boldsymbol{\pi}_{N-1}^{(i)})^2+m_i^2}|\boldsymbol{k}\rangle = \sqrt{\hbar^2\boldsymbol{k}_{N-1}^2+m_i^2}\,\delta(\boldsymbol{k}-\boldsymbol{k}'). \qquad (7.57)$$

By substituting these expressions into Eq. (7.55), one can reduce the integration over $\boldsymbol{k}$ as follows:

$$\langle F_{A'}(\boldsymbol{x})|\sqrt{(\boldsymbol{\pi}_{N-1}^{(i)})^2+m_i^2}|F_A(\boldsymbol{x})\rangle$$

$$= (\det V^{(i)})^3(\det A^{(i)}\det A'^{(i)})^{-\frac{3}{2}}$$

$$\times \langle F_{A'^{(i)-1}}(\boldsymbol{k})|\sqrt{(\boldsymbol{\pi}_{N-1}^{(i)})^2+m_i^2}|F_{A^{(i)-1}}(\boldsymbol{k})\rangle$$

$$= (\det V^{(i)})^3(\det A^{(i)}\det A'^{(i)})^{-\frac{3}{2}}\int\sqrt{\hbar^2\boldsymbol{q}^2+m_i^2}$$

$$\times \langle F_{A'^{(i)-1}}(\boldsymbol{k})|\delta(\boldsymbol{k}_{N-1}-\boldsymbol{q})|F_{A^{(i)-1}}(\boldsymbol{k})\rangle d\boldsymbol{q}. \qquad (7.58)$$

To calculate the matrix element of $\delta(\boldsymbol{k}_{N-1}-\boldsymbol{q})$, we rewrite it as $\delta(\tilde{w}\boldsymbol{k}-\boldsymbol{q})$ with an auxiliary $1\times(N-1)$ one-row matrix $\tilde{w}=(0,...,0,1)$ and make use of the formula in Table 7.1. Then we obtain

$$\langle F_{A'}(\boldsymbol{x})|\sqrt{(\boldsymbol{\pi}_{N-1}^{(i)})^2+m_i^2}|F_A(\boldsymbol{x})\rangle$$

$$= (\det V^{(i)})^3(\det A^{(i)}\det A'^{(i)})^{-\frac{3}{2}}\langle F_{A'^{(i)-1}}(\boldsymbol{k})|F_{A^{(i)-1}}(\boldsymbol{k})\rangle$$

$$\times \Big(\frac{c_i}{2\pi}\Big)^{\frac{3}{2}}\int \mathrm{e}^{-\frac{1}{2}c_i q^2}\sqrt{\hbar^2\boldsymbol{q}^2+m_i^2}\;d\boldsymbol{q}$$

$$= \langle F_{A'}|F_A\rangle\, f(c_i,m_i) = \left(\frac{(2\pi)^{N-1}}{\det(A+A')}\right)^{\frac{3}{2}} f(c_i,m_i), \qquad (7.59)$$

with

$$\frac{1}{c_i} = \Big(\widetilde{V^{(i)}}A(A+A')^{-1}A'V^{(i)}\Big)_{N-1\;N-1}, \qquad (7.60)$$

where the function f is defined by

$$f(x,m) = \left(\frac{x}{2\pi}\right)^{\frac{3}{2}} \int \mathrm{e}^{-\frac{1}{2}xq^2} \sqrt{\hbar^2 q^2 + m^2}\, d\boldsymbol{q}$$

$$= 4\pi \left(\frac{x}{2\pi}\right)^{\frac{3}{2}} \int_0^\infty \mathrm{e}^{-\frac{1}{2}xq^2} \sqrt{\hbar^2 q^2 + m^2} q^2\, dq, \qquad (7.61)$$

and use is made of the fact that the overlap has the following property with respect to the change of the Jacobi coordinate set:

$$\langle F_{A'}|F_A\rangle \equiv \langle F_{A'}(\boldsymbol{x})|F_A(\boldsymbol{x})\rangle$$
$$= (\det V^{(i)})^3 \langle F_{A'^{(i)}}(\boldsymbol{x}^{(i)})|F_{A^{(i)}}(\boldsymbol{x}^{(i)})\rangle$$
$$= (\det V^{(i)})^3 (\det A^{(i)} \det A'^{(i)})^{-\frac{3}{2}} \langle F_{A'^{(i)-1}}(\boldsymbol{k})|F_{A^{(i)-1}}(\boldsymbol{k})\rangle. \qquad (7.62)$$

The overlap $\langle F_{A'}|F_A\rangle$ is explicitly given by Eq. (7.4).

Complements

7.1 *Sherman-Morrison formula*

To show that the Sherman-Morrison formula [14] can be used to advantage in selecting a basis element from among many random trials, let us use the correlated Gaussians of Eq. (4.1) or Eq. (7.3) as the basis functions. The matrix elements A_{ij} or, equivalently, α_{ij} are nonlinear parameters of the basis, and they are related to each other through Eq. (2.25). As is shown in Sect. 7.2, the calculation of the matrix elements, $\langle F_{A'}|F_A\rangle$ and $\langle F_{A'}|H|F_A\rangle$, requires the evaluation of the determinant and inverse of the matrix $B = A + A'$.

Let us assume that we attempt to optimize the symmetric matrix A of nonlinear parameters by changing only one element, say $A_{kl} = A_{lk}$, or α_{ij} $(j > i)$, but by keeping all other parameters unchanged. Instead of changing all of the nonlinear parameters randomly at once, we change one particular element randomly and then proceed to other elements step by step. The latter type of optimization of A is certainly a very restricted way, but in this case the computer time required for the evaluation of the matrix elements is tremendously reduced, as will be seen below. And it is actually very useful in selecting a successful candidate from among a number of random trials.

We see from Eq. (2.25) that changing only one element α_{ij} to $\alpha_{ij} + \lambda$ produces a change in A as follows:

$$A \rightarrow A + \lambda w^{(ij)}\widetilde{w^{(ij)}}, \tag{7.63}$$

where the $(N-1)$-dimensional column vector $w^{(ij)}$ is defined in Eq. (2.13). Note that $w^{(ij)}\widetilde{w^{(ij)}}$ is an $(N-1) \times (N-1)$ matrix, whereas $\widetilde{w^{(ij)}}w^{(ij)}$ is just a number. On the other hand, changing directly only the k, l (and l, k) element $A_{kl} = A_{lk}$ of A as $A_{kl} \rightarrow A_{kl} + \lambda$ (and $A_{lk} \rightarrow A_{lk} + \lambda$) can be achieved by introducing $(N-1)$-dimensional unit vectors $e^{(m)}$ as follows:

$$A \rightarrow A + \frac{\lambda}{1+\delta_{kl}}\left(e^{(k)}\widetilde{e^{(l)}} + e^{(l)}\widetilde{e^{(k)}}\right), \tag{7.64}$$

where $e^{(m)}$ is defined by $(e^{(m)})_i = \delta_{im}$ $(i = 1, ..., N-1)$. By letting u_i and v_i stand for $(N-1)$-dimensional vectors, both of the changes given by Eqs. (7.63) and (7.64) are in general expressible as

$$A \rightarrow A + \sum_{i=1}^{p} \lambda_i u_i \widetilde{v_i}, \tag{7.65}$$

where p is either 1 or 2.

As B is equal to $A + A'$, the change of A as given by Eq. (7.65) leads to the following modification of B,

$$B \to B + \sum_{i=1}^{p} \lambda_i u_i \widetilde{v_i}. \tag{7.66}$$

Therefore the calculation of the matrix element for the above change of A results in the calculation of the inverse and determinant of the special form of matrix, $B + \sum_{i=1}^{p} \lambda_i u_i \widetilde{v_i}$. When the modification is given by just one term of the form $\lambda u\tilde{v}$, the Sherman-Morrison formula can be used to obtain

$$(B + \lambda u\tilde{v})^{-1} = B^{-1} - \frac{1}{1 + \lambda \tilde{v} B^{-1} u} \lambda B^{-1} u \tilde{v} B^{-1}, \tag{7.67}$$

and

$$\det(B + \lambda u\tilde{v}) = (1 + \lambda \tilde{v} B^{-1} u) \det B. \tag{7.68}$$

See Exercise 7.2 for the derivation of the Sherman-Morrison formula. The advantage of these formulas is apparent: By knowing B^{-1} and $\det B$ one can easily calculate the right-hand side of the equations, and the λ dependence is given in a very simple form. To change λ, therefore there is no need for the evaluation of inverses and determinants of the modified matrix B (which would require $(N-1)^3$ operations), but we get the desired results by a simple multiplication and division.

When the modification of B is given by Eq. (7.66) in full generality, then the inverse and determinant can be calculated by using the Woodbury formula [14], which is the block-matrix version of the Sherman-Morrison formula. If one wants to change a few of the α_{ij}'s or one-column (and one-row) elements of A at the same time, the summation in Eq. (7.66) has to be further extended appropriately.

Exercises

7.1. Derive Eq. (7.12).

*Solution.*The case of $a = 0$ is simple. In the case of $a > 0$, we may get the integral as follows:

$$\int_0^\infty r^n e^{-ar^2-br} dr = (-1)^n \frac{d^n}{db^n} \int_0^\infty e^{-ar^2-br} dr$$
$$= (-1)^n \frac{d^n}{db^n} \frac{1}{2}\sqrt{\frac{\pi}{a}} \exp\Big(\frac{b^2}{4a}\Big) \operatorname{erfc}\Big(\frac{b}{2\sqrt{a}}\Big). \tag{7.69}$$

By putting $x = b/(2\sqrt{a})$, the above equation becomes

$$\int_0^\infty r^n e^{-ar^2-br} dr$$
$$= \sqrt{\pi}(-1)^n \Big(\frac{1}{2\sqrt{a}}\Big)^{n+1} \sum_{k=0}^{n} \frac{n!}{(n-k)!\,k!} f_k(x)\, g_{n-k}(x), \tag{7.70}$$

where

$$f_k(x) = \exp(-x^2)\left(\exp(x^2)\right)^{(k)},$$
$$g_k(x) = \exp(x^2)\left(\operatorname{erfc}(x)\right)^{(k)}. \tag{7.71}$$

It is easy to show that $f_k(x)$ defined above is equal to the one given in Eq. (7.13). Remembering that the Hermite polynomial is given by $H_n(x) = (-1)^n e^{x^2}\left(\exp(-x^2)\right)^{(n)}$ and $\left(\operatorname{erfc}(x)\right)^{(1)} = -2e^{-x^2}/\sqrt{\pi}$, it is not difficult to check that $g_k(x)$ can be expressed as in Eq. (7.13).

7.2. Derive Eqs. (7.67) and (7.68).

Solution. Let X represent $u\tilde{v}$. Then the inverse $(B+\lambda X)^{-1}$ may be calculated as follows:

$$(B+\lambda X)^{-1} = (1+\lambda B^{-1}X)^{-1}B^{-1}$$
$$= \sum_{n=0}^{\infty}(-\lambda)^n (B^{-1}X)^n B^{-1}. \tag{7.72}$$

Because of the special form of X, $XB^{-1}X$ reduces to cX, where the constant factor c is given by $\tilde{v}B^{-1}u$. Therefore repeated use of this relation leads to the follwoing result:

$$(B^{-1}X)^n B^{-1} = B^{-1}(XB^{-1}XB^{-1}X\cdot\cdots\cdot B^{-1}X)B^{-1}$$
$$= c^{n-1}B^{-1}XB^{-1} \quad (n \geq 1). \tag{7.73}$$

By substituting this result into Eq. (7.72), we obtain

$$(B+\lambda X)^{-1} = B^{-1} - \frac{\lambda}{1+\lambda c}B^{-1}XB^{-1}, \tag{7.74}$$

which is nothing but Eq. (7.67).

To calculate the determinant of the matrix $(B+\lambda X)$ with its $(i,\, j)$ element given by $B_{ij}+\lambda u_i v_j$, let us suppose that the determinant is a function of λ: $P(\lambda) = \det(B+\lambda X)$. The function $P(\lambda)$ is a polynomial of at most degree $(N-1)$ and can be expanded as follows:

$$P(\lambda) = a_0 + a_1\lambda + \cdots + a_{N-1}\lambda^{N-1}, \tag{7.75}$$

where the coefficient a_k is calculated by $k!P^{(k)}(0)$. The rule of differentiating determinants leads us to the conclusion that $a_k = 0$ for $k \geq 2$ because of the special form of the matrix X. We have $a_0 = P(0) = \det B$. The coefficient a_1 is obtained as follows:

$$a_1 = \sum_{i=1}^{N-1} u_i \begin{vmatrix} B_{11} & B_{12} & \cdots & B_{1\,N-1} \\ \vdots & \vdots & & \vdots \\ v_1 & v_2 & \cdots & v_{N-1} \\ \vdots & \vdots & & \vdots \\ B_{N-1\,1} & B_{N-1\,2} & \cdots & B_{N-1\,N-1} \end{vmatrix}$$
$$= \sum_{i=1}^{N-1}\sum_{j=1}^{N-1} u_i v_j \Delta_{ij} = \sum_{i=1}^{N-1}\sum_{j=1}^{N-1} u_i v_j \det B (B^{-1})_{ji}$$
$$= (\tilde{v}B^{-1}u)\det B. \tag{7.76}$$

Here Δ_{ij} is the $(i,\, j)$ cofactor of the matrix B and use is made of the relation $\Delta_{ij} = \det B\,(B^{-1})_{ji}$. Thus $P(\lambda)$ is equal to $(1+\lambda\tilde{v}B^{-1}u)\det B$, which is what we want to derive.

7.3. Calculate the one-body density matrix for the generating function g:

$$\rho_i(\boldsymbol{r}', \boldsymbol{r})$$
$$= \langle g(\boldsymbol{s}'; A', \boldsymbol{x})|\delta(\boldsymbol{r}_i - \boldsymbol{x}_N - \boldsymbol{r}')\rangle\langle\delta(\boldsymbol{r}_i - \boldsymbol{x}_N - \boldsymbol{r})|g(\boldsymbol{s}; A, \boldsymbol{x})\rangle. \tag{7.77}$$

Solution. As Eq. (2.12) shows, $\boldsymbol{r}_i - \boldsymbol{x}_N$ can be expressed as $\sum_{l=1}^{N-1} w_l^{(i)} \boldsymbol{x}_l$. The argument made for the nonlocal potential in Sect. 7.5 can, therefore, be applied to this case in exactly the same way. The only necessary change is to replace $w^{(ij)}$ with $w^{(i)}$. The density matrix $\rho_i(\boldsymbol{r}', \boldsymbol{r})$ takes the same form as Eq. (7.46). When the wave function has the proper symmetry for a system of identical particles, the one-body density matrix does not depend on the suffix i.

7.4. Reproduce the nonrelativistic kinetic energy formula (7.5) by using the formulation presented in Sect. 7.6.

Solution. By using Eq. (7.50), the matrix element of the nonrelativistic kinetic energy operator in the center-of-mass system becomes

$$\begin{aligned}\langle F_{A'}(\boldsymbol{x})|\sum_{i=1}^{N} \frac{\boldsymbol{p}_i^2}{2m_i}|F_A(\boldsymbol{x})\rangle &= \langle F_{A'}(\boldsymbol{x})|\sum_{i=1}^{N} \frac{(\boldsymbol{\pi}_{N-1}^{(i)})^2}{2m_i}|F_A(\boldsymbol{x})\rangle \\ &= \langle F_{A'}|F_A\rangle \sum_{i=1}^{N} \frac{3\hbar^2}{2m_i c_i}, \end{aligned} \tag{7.78}$$

where $f(x,m) = 3\hbar^2/(2mx)$ for the nonrelativistic kinetic energy is obtained by replacing $\sqrt{\hbar^2 q^2 + m^2}$ in Eq. (7.61) with $\hbar^2 q^2/(2m)$.

To perform the summation over i in Eq. (7.78), we need to know the special matrix element $\left(V^{(i)}\right)_{k\,N-1}$:

$$\left(V^{(i)}\right)_{k\,N-1} = \sum_{l=1}^{N} \left(U_J^{(N)}\right)_{kl} \left(U_J^{(i)^{-1}}\right)_{l\,N-1}. \tag{7.79}$$

Let us recall that $U_J^{(i)}$ can be obtained by rearranging the N column vectors of $U_J^{(N)} \equiv U_J$ and at the same time by rearranging the masses $m_1, m_2, ..., m_N$ according to the operation with the cyclic permutation $(1, 2, ..., N)$ i times. From this construction it is easy to see that $U_J^{(i)^{-1}}$ can be obtained from $U_J^{(N)^{-1}}$ in exactly the same manner as $U_J^{(i)}$ is constructed from $U_J^{(N)}$ but by rearranging the row vectors instead of the column vectors. Then we obtain (see Eq. (2.6))

$$\left(U_J^{(i)^{-1}}\right)_{l\,N-1} = \begin{cases} \frac{m_i}{m_{12\cdots N}} & \text{for} \quad l \neq i \\ -(1 - \frac{m_i}{m_{12\cdots N}}) & \text{for} \quad l = i. \end{cases} \tag{7.80}$$

Using this result in Eq. (7.79) and noting the relation $\sum_{l=1}^{N}\left(U_J^{(N)}\right)_{kl}$ $= 0$ for $k \neq N$ enables us to obtain the desired matrix element as

$$\left(V^{(i)}\right)_{k\,N-1} = -\left(U_J^{(N)}\right)_{k\,i} = -\left(U_J\right)_{k\,i}. \tag{7.81}$$

The summation over i in Eq. (7.78) can easily be done to obtain

$$\sum_{i=1}^{N} \frac{3\hbar^2}{2m_i c_i} = \frac{3}{2}\hbar^2 \mathrm{Tr}\left(A(A+A')^{-1}A'\Lambda\right) \tag{7.82}$$

with the matrix Λ being defined by Eq. (2.11). This agrees with Eq. (7.5), which was obtained by explicitly subtracting the center-of-mass kinetic energy from the total kinetic energy.

This exercise serves as an indirect evidence for the validity of the formulation given in Sect. 7.6.

8. Small atoms and molecules

This chapter contains examples for the application of the method to atomic and molecular systems. The interaction between the charged particles is the Coulomb force. The long-range character of this force makes the solution of the few-body problems difficult, especially in the case of scattering. We restrict our attention to bound states, where many different methods have been elaborated in the past decades. These calculations provide an excellent possibility to test the efficiency of our method. Relativistic effects in atoms are within the reach of experimental precision of today, which calls for very accurate theoretical calculations.

8.1 Coulombic systems

The systems of charged particles can be characterized and classified by the masses $(m_1, m_2, \ldots, m_N)$ and the charges $(q_1, q_2, \ldots, q_N)$ of the constituents. One can distinguish systems formed by either equal (unit) or unequal charges, and depending on the masses of the particles one can classify the systems as adiabatic and nonadiabatic ones. In the unit charge systems of more than two particles the constituents form a molecule and the binding energy depends only on the mass ratio(s) of the particles. Atoms are good examples for systems with unequal charges. The distinction between the adiabatic and nonadiabatic cases is dictated by the possibility of a simplified treatment in the former case. In the adiabatic case the masses of a group of particles are considerably heavier than those of the rest. The classical examples are the H_2 molecule and the H_2^+ molecular ion (see Complement 8.2). In these cases the electrons move faster, while the protonic frame may rotate and vibrate by moving considerably slower. This physical picture is expressed in mathematical form as the Born-Oppenheimer approximation, where the electronic motion is first calculated by assuming the nuclei fixed in space and then the nuclei are subject to rotations and vibrations in the effective field of the fast electrons. Even though the Born-Oppenheimer method proved to be a very good and useful approximation, it has the undesirable feature that the two kinds of motion are separated. We use a unified nonadiabatic approach even in the case of infinite nuclear masses.

The variational trial function used in Sects. 8.2 and 8.3 takes the form

$$\Psi = \sum_k c_k \mathcal{A}\left\{\exp\left(-\frac{1}{2}\tilde{\boldsymbol{x}}A_k\boldsymbol{x}\right)|\boldsymbol{v}_k|^{2K+L}Y_{LM}(\widehat{\boldsymbol{v}_k})\chi_{SM_S}\right\},$$

$$\boldsymbol{v}_k = \sum_i u_{ki}\boldsymbol{x}_i. \tag{8.1}$$

The operator $\mathcal{A}$ is introduced to impose proper symmetries on identical particles (antisymmetry for fermions and symmetry for bosons) of the system. In most calculations the index K is assumed to be zero or at most zero or one. In the case of the Ps_2 molecule of Sect. 8.3, K is set equal to zero. The details of the calculations can be found in [33, 84, 85] for some of the systems presented in this chapter. The variational parameters included in each basis function are the elements of the matrix A_k and the coefficients u_{ki}, which define the global vector $\boldsymbol{v}_k$. Since the Hamiltonian used in this chapter commutes with the spin operator, the variational trial function can be chosen to have a definite spin value S. The value of S influences the symmetry of the orbital part of identical particles. Therefore, possible S values have to be tested in general to obtain the ground state. For treating the adiabatic system of small molecules in Sect. 8.4, we use combinations of the generating function g itself as the variational trial function.

8.2 Coulombic three-body systems

The Coulombic three-body systems represent a special class of three-body problems [51] where the constituents interact via the long-range Coulomb force and the correlation plays a very distinctive role. These systems have received particular attention in the last few decades, and the binding energies and other properties have been calculated with utmost accuracy by various methods (see references in Table 8.1). The accuracy of the calculations has reached such a high level, typically

13–16 figures, that the solution of the Coulombic three-body problem is one of the most useful benchmark tests to compare different methods. The accuracy pursued in Coulombic cases is not of purely academic interest, but highly motivated by the high precision of the experiments. For example, the fine-structure splitting of the $1s^2 2p^2 P_J$ state of the Li atom has been measured with an accuracy of 20 ppm (parts per million). To explain the experimental numbers one has to include relativistic and quantum electrodynamics (QED) corrections, including terms of the second- and third-order in the fine-structure constant, which require very high accuracy. In addition to the accurate reproduction of the energies, another motivation is that in the variational calculations, even though the energy is good, other physical observables might be less accurately determined. The increased accuracy of the energy, as we will demonstrate, eventually will lead to very accurate values of other observables as well.

The stability domain of a general $(m_A^+ m_B^+ m_C^-)$-type three-body system with unit charges has been thoroughly explored [86], and the requirement for stability can be phrased as an empirical rule: " like charges have to be borne by equal or nearly equal masses" [86]. In accordance with this prediction, systems such as $\mathrm{H}^-(pe^-e^-)$, $\mathrm{H}_2^+(ppe^-)$, $\mathrm{Ps}^-(e^+e^-e^-)$, $\mathrm{HD}^+(pde^-)$, $\mathrm{HT}^+(pte^-)$, or $td\mu^-$ are all bound, while $(p\bar{p}e^-)$ or (pe^+e^-) are most probably not. Here p, $\bar{p}$, d, and t are proton, antiproton, deuteron (^{2}H) and triton (^{3}H), respectively. In the latter case, the particles with opposite charges form an atom, which does not bind the third particle. Some systems, e.g., the muonic molecules such as $(tt\mu^-)$ or $(td\mu^-)$, remain bound even for $L \neq 0$ orbital angular momenta.

A second group of the Coulombic three-body problems is formed by systems where not all the particles carry unit charges. The representatives of this category are the helium atom (αe^-e^-) and the helium-like ions, where α stands for the ^{4}He nucleus. These systems often form bound states with $L \neq 0$ as well. The antiprotonic helium atom $(\alpha\bar{p}e^-)$ has been observed [87] and studied in very high orbital angular momentum states $(L = 30 - 40)$ [88, 89].

We have challenged our method to calculate the energies of some of these systems. In the calculations to be presented the number of basis functions superposed is mostly limited to be modest because our primary purpose is to demonstrate the overall performance of the correlated Gaussians and not to compete with well-established methods sharpened for these systems. We will increase the basis dimension to reach high accuracy only in cases where such calculations are considered to be important. The results of the calculations are listed in Table 8.1. Table 8.2 presents the parameters used in the calculations. The results are compared with those of other (mainly Hylleraas-type or correlated hyperspherical harmonics basis) calculations. Our results are in reasonable agreement with other calculations. In most cases a basis size of $K = 200$ was used in the SVM. The precision of the results can be improved by increasing the basis size as can be seen on the selected examples of Table 8.1. In these cases we reach almost the same precision as the other methods. The calculation extends to nonzero orbital angular momentum states as well, including an example for the $L = 31$ state of the antiprotonic helium atom or the slightly bound $^3P^e$ states of the H^- ion. These nonzero orbital angular momentum states as well as $L = 0$ states have been investigated by using the global vector representation. The recovery of the results of other calculations (which are based on several different representations of the orbital part of the wave function) convinces us of the usefulness of the global vector representation. See Complement 8.4 for a comparative calculation of the $td\mu$ molecule with the global vector representation.

To illustrate the convergence of the energy and the expectation values of average separation distances, the results at different basis sizes are tabulated in Table 8.3 for Ps^-. Actually the limit of further increasing the accuracy is the conventional precision of the computer itself. One can notice that at the basis size of $K = 100$ the energy is accurate up to six decimal digits, but only the first four figures of the separation distances can be precisely determined. By increasing the basis size, the virial coefficient falls below 10^{-10}, showing by the high accuracy of the calculation that all the digits of the reference calculation are recovered.

Quite a few very accurate methods have been developed to solve the three-body Coulomb problem. It is very difficult to go beyond their precision. This is especially true for the methods which have been elaborated for a given system only, incorporating as much physical intuition as possible into the trial function or into the solution. In contrast with these methods, we use the same trial function, which is of Gaussian nature and therefore it is not tailored to Coulomb problems at all. Still, as the examples prove, one can get a sufficiently good solution in a unified and automatic way without a priori built-in knowledge about the systems to cope with. The real power of the

Table 8.1. Energies of different Coulombic three-body systems in atomic units. K is the basis dimension. See Table 8.2 for the constants which are used in the calculations denoted superscripts a, b and c.

System	State	K	SVM	Other method	K	Ref.
Ps^-	$^1S^e$	600	−0.262005070226	−0.2620050702328	1488	[69]
$^\infty H^-$	$^1S^e$	600	−0.527710163	−0.527751016523	850	[90]
H^-	$^3P^e$	200	−0.1252865	−0.1252865	90	[91]
$tt\mu$	$^1S^e$	200	−112.97300[a]	−112.9730179[a]	500	[23]
$tt\mu$	$^1P^o$	200	−110.26210[a]	−110.2621165[a]	500	[23]
$tt\mu$	$^1D^e$	200	−105.98292[b]	−105.982930[b]	2250	[92]
$tt\mu$	$^1F^o$	200	−101.43131[c]	−101.43[c]	200	[93]
$td\mu$	$^1S^e$	200	−111.36444[a]	−111.364511474[a]	1400	[23]
$td\mu$	$^1P^o$	200	−108.17923[b]	−108.179385[b]	2662	[20]
$td\mu$	$^1D^e$	200	−103.40849[a]	−103.408481[a]	1566	[20]
$^\infty$He	$^1S^e$	600	−2.9037243769	−2.903724376984	700	[94]
He	$^1S^e$	200	−2.9037242	−2.903724372437	100	[95]
He	$^3S^e$	200	−2.1752291	−2.175293782367	700	[68]
He	$^1P^o$	200	−2.1238423	−2.123843086498	700	[68]
He	$^3P^o$	200	−2.1331635	−2.133164190779	700	[68]
He	$^1D^e$	200	−2.0556201	−2.055620732852	700	[68]
$^\infty$He	$^3D^e$	700	−2.055338993068	−2.055338993337	700	[68]
He	$^1F^o$	200	−2.03125504	−2.031255144382	700	[68]
He	$^3F^o$	200	−2.03125506	−2.031255168403	700	[68]
He	$^1G^e$	200	−2.02000069058	−2.020000710898	700	[68]
He	$^3G^e$	200	−2.02000069062	−2.020000710925	700	[68]
$\bar{p}He^+$	$L=31$	300	−3.50760	−3.50763486	1728	[89]
$^\infty Li^+$	$^1S^e$	300	−7.279913	−7.279913		[96]

Table 8.2. The constants used in the calculations. The masses are in units of the electron mass. m_α=7294.2618241, m_p=1836.1515. R_∞ is the Rydberg energy in eV.

	Set a	Set b	Set c
m_t	5496.918	5496.92158	5496.918
m_d	3670.481	3670.483014	3670.481
m_μ	206.7686	206.768262	206.769
$2R_\infty$	27.2113961	27.2113961	27.2116

Table 8.3. Energy and different separation distances for the $(e^+e^-e^-)$ Coulombic three-body system as a function of the basis dimension K. The virial ratio η is defined by $\eta = |1 + \langle V\rangle/(2\langle T\rangle)|$. See Eq. (3.50). Atomic units are used.

	SVM (K=100)	SVM (K=200)	SVM (K=600)	Hylleraas [69]
$-E$	0.26200465	0.2620050648	0.262005070226	0.2620050702328
$\langle r_{+-}\rangle$	5.489	5.48962	5.489633252	5.489633252
$\langle r_{--}\rangle$	8.548	8.54856	8.548580655	8.548580655
$\langle r_{+-}^2\rangle^{\frac{1}{2}}$	6.958	6.95832	6.95837	6.95837
$\langle r_{--}^2\rangle^{\frac{1}{2}}$	9.652	9.65284	9.65291	9.65291
η	0.46×10^{-4}	0.34×10^{-6}	0.54×10^{-10}	0.23×10^{-10}

approach will be more highlighted in the following sections, where the number of particles is more than three and the method still works almost as easily as before, while the other methods need tedious efforts.

8.3 Four or more particles

The expedition to larger Coulombic systems starts with the positronium molecule (Ps_2). This exotic molecule consists of two electrons and two positrons. The possibility that the Ps_2 molecule or in general the electron-positron system consisting of p positrons and q electrons form a bound system was originally suggested by Wheeler [97], and this question has been extensively studied since then. The existence of the positronium negative ion Ps^- ($p = 1$, $q = 2$) has experimentally been observed [98]. The binding energy of Ps_2 was first calculated by Hylleraas and Ore [99]. To date, it has not been observed yet due to the difficult experimental circumstances, and this fact has intensified the theoretical interest in solving this Coulombic four-body problem [65, 100, 69, 101–104]. Actually the positron-electron annihilation limits the lifetime of Ps_2 to few nanoseconds.

In obtaining the solution for the Ps_2 molecule, it is useful to note that the Hamiltonian for Ps_2 is invariant with respect to the charge permutation, that is, the exchange of positive and negative charges. The trial function should therefore either remain unchanged or change its sign under the charge permutation operation. The ground state of Ps_2 turns out to be even under the charge permutation.

The convergence of the energy of Ps_2 against the increase of the basis dimension is shown in Table 8.4. The fact that the best variational calculation [103] in the correlated Gaussian basis is already surpassed at the basis size of 400 illustrates the power of the random trials.

Table 8.4. The total energies (in a.u.) of the ground state and the bound excited-state of the Ps_2 molecule in atomic units. K is the basis dimension.

Method	Ps_2 $(L=0)$	Ps_2 $(L=1)$
SVM $(K=100)$	−0.516000069	−0.334376975
SVM $(K=200)$	−0.516003119	−0.334405047
SVM $(K=400)$	−0.516003666	−0.334407561
SVM $(K=800)$	−0.516003778	−0.334408177
SVM $(K=1200)$	−0.5160037869	−0.334408234
SVM $(K=1600)$	−0.516003789058	−0.3344082658
CG [103]	−0.5160024	
QMC [104]	−0.51601±0.00001	

We have examined the possible existence of bound excited-states of the Ps_2 molecule [84, 85] by taking all possible combinations of states with $L = 0, 1, 2, 3$ orbital angular momenta and $S = 0, 1, 2$ spins. By a bound excited state we mean such a state that cannot decay to any dissociation channels. The results of the calculation were negative in all but one case. In the case of $L = 1$ (with negative parity) and $S = 0$, the calculation predicts the existence of a second bound-state of the Ps_2 molecule. This unique bound-state has been found to be odd under the charge permutation operation. The convergence of the excited-state energy is shown in Table 8.4. Figure 8.1 summarizes the energy spectra of the bound states made up of two positrons and two electrons together with the relevant thresholds.

One may ask the question of why the second bound-state cannot decay to two Ps atoms in spite of the fact that it is located above the threshold of Ps(1S)+Ps(1S). Since the total spin of the state is coupled to zero, it can dissociate into two Ps (ground state) atoms provided that they have equal spins and the relative orbital angular momentum between them is $L = 1$. (Recall that the Hamiltonian preserves spin, orbital angular momentum and parity.) However, this is apparently impossible because two Ps atoms of equal spins are bosons and their relative motion can only have even L. Consequently, the Ps_2 molecule

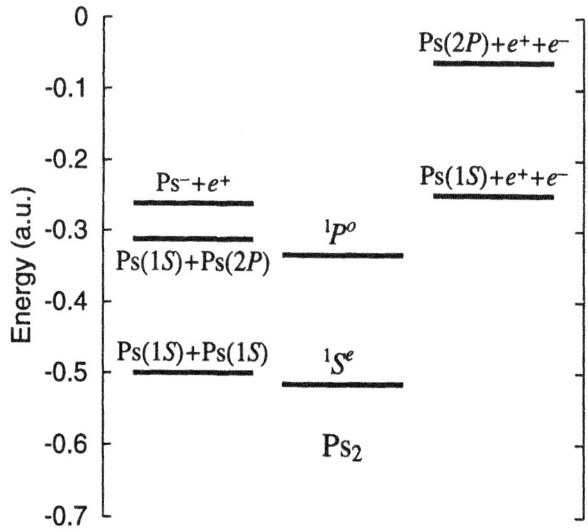

Fig. 8.1. The energy spectrum of electron and positron systems. Energies are given in atomic units.

with $L = 1$ and negative parity cannot decay into the ground states of two Ps atoms, that is, the lowest threshold of Ps(1S)+Ps(1S). Since the energy of this $L = 1$ state is calculated to be $E = -0.3344$ a.u. (see Table 8.4), which is lower than the next threshold (-0.3125 a.u.) of Ps(1S) + Ps(2P), this state is stable against autodissociation into this channel. The binding energy of this state is 0.5961 eV from this second threshold, about 40% more tightly bound than that of the ground state of Ps_2 whose binding energy is 0.4355 eV from the lowest threshold.

The expectation values of various quantities for the Ps_2 molecule are listed in Table 8.5.

In the Ps_2 molecule we deal with antiparticles, so the electron-positron pair can annihilate. The second bound-state may decay either by annihilation or by an electric dipole transition to the ground state [85]. The most dominant annihilation is accompanied by the emission of two photons with energy of about 0.5 MeV each. To have an estimate for the decay width due to the annihilation we have substituted

Table 8.5. Properties of the ground and excited states of the Ps_2 molecule. The positrons are labelled 1 and 3 and the electrons are 2 and 4. Because of charge permutation symmetry, some equalities hold, e.g. $\langle r_{12}\rangle = \langle r_{14}\rangle = \langle r_{32}\rangle = \langle r_{34}\rangle$. Atomic units are used.

	Ps_2 $(L=0)$	Ps_2 $(L=1)$
$\langle r_{12}\rangle$	4.4871530	7.56881891
$\langle r_{13}\rangle$	6.0332070	8.8575844
$\langle r_{12}^2\rangle$	29.112633	80.173836
$\langle r_{13}^2\rangle$	46.374735	96.085514
$\langle r_{12}^3\rangle$	253.04611	1041.3251
$\langle r_{13}^3\rangle$	443.85244	1226.7955
$\langle r_{12}^4\rangle$	2807.2718	15612.112
$\langle r_{13}^4\rangle$	5202.0371	17939.574
$\langle r_{12}^{-1}\rangle$	0.36839693	0.24082648
$\langle r_{13}^{-1}\rangle$	0.22079007	0.147244820
$\langle r_{12}^{-2}\rangle$	0.30310361	0.16081514
$\langle r_{13}^{-2}\rangle$	0.073444303	0.032230158
$\langle \boldsymbol{r}_{12}\cdot\boldsymbol{r}_{13}\rangle$	23.187368	48.042757
$\langle \boldsymbol{r}_{12}\cdot\boldsymbol{r}_{14}\rangle$	5.9252651	32.131079
$\langle \delta(\boldsymbol{r}_{12})\rangle$	0.0221151	0.0112091
$\langle \delta(\boldsymbol{r}_{13})\rangle$	0.0006259	0.00014591
$\langle \nabla_1^2\rangle$	−0.258001894	−0.16720401
$\langle \nabla_1\cdot\nabla_2\rangle$	0.1307732538	0.091656853
$\langle \nabla_1\cdot\nabla_3\rangle$	−0.0035446132	−0.016109693
$\lvert 1+\langle V\rangle/(2\langle T\rangle)\rvert$	0.3×10^{-9}	0.36×10^{-6}

the probability density of an electron at the position of a positron, $\langle\delta(\boldsymbol{r}_{12})\rangle$, into the formula [102]

$$\begin{aligned}\Gamma_{\text{annihi}} &= 4\pi\left(\frac{e^2}{mc^2}\right)^2 \hbar c\langle\Psi|\delta(\boldsymbol{r}_1-\boldsymbol{r}_2)|\Psi\rangle \\ &= 4\pi\left(\frac{e^2}{\hbar c}\right)^4 \hbar c a_0^{-1}\langle\delta(\boldsymbol{r}_{12})\rangle, \end{aligned} \tag{8.2}$$

where $\langle\delta(\boldsymbol{r}_{12})\rangle$, given in a.u., is equal to $a_0^3\langle\Psi|\delta(\boldsymbol{r}_1-\boldsymbol{r}_2)|\Psi\rangle$ with a_0 being the Bohr radius. Roughly speaking, the lifetime is inversely proportional to the probability. The lifetime due to the annihilation is estimated to be 0.44 ns. This is twice that of the ground state (0.22 ns).

The electric dipole transition from the excited state emits one photon with energy of 4.94 eV. The decay width Γ_{dipole} for this transition is calculated through the reduced transition probability $B(E1)$ for the electric dipole operator D_μ:

$$\Gamma_{\text{dipole}} = \frac{16\pi}{9}\left(\frac{E}{\hbar c}\right)^3 B(E1;1^- \to 0^+), \tag{8.3}$$

where

$$B(E1;1^- \to 0^+) = \sum_{\mu=-1}^{1} |\langle \Psi_{\text{gnd}}|D_\mu|\Psi_{1M}\rangle|^2, \tag{8.4}$$

with

$$D_\mu = \sum_{i=1}^{4} q_i |\boldsymbol{r}_i - \boldsymbol{x}_4| Y_{1\mu}(\widehat{\boldsymbol{r}_i - \boldsymbol{x}_4}). \tag{8.5}$$

Here $\boldsymbol{x}_4$ is the center-of-mass of the Ps_2 molecule and E is the excitation energy (4.94 eV) of the second bound-state. We calculated the $B(E1)$ value and obtained $B(E1) = 0.87e^2a_0^2$. The lifetime due to the electric dipole transition has been found to be 2.1 ns. The branching of the electric dipole transition is thus about 17 % of the total decay rate. Therefore, both branches contribute to the decay of the excited state of the Ps_2 molecule. Its lifetime is finally estimated to be about 0.37 ns. The excitation energy of 4.94 eV found for Ps_2 is different by 0.16 eV from the corresponding excitation energy (5.10 eV) of a Ps atom. This difference seems to be large enough to detect its existence, e.g. in the photon absorption spectrum of the positronium gas.

Before discussing the spatial distribution of the Ps_2 molecule, let us recall that the average distance $\langle r_{+-}\rangle$ between the positron and the electron is 3 a.u. in the ground state of the Ps atom, while it is 10 a.u. in its first excited state. The root-mean-square radius (rms), $\sqrt{\langle \sum_{i=1}^{4} (\boldsymbol{r}_i - \boldsymbol{x}_4)^2 \rangle}$, of the ground state of the Ps_2 molecule is calculated to be 3.61 a.u. The rms radius of the second bound-state is 5.66 a.u., 1.5 times larger than that of the ground state. This is not surprising if one assumes that the second bound-state is essentially a system of a Ps atom in its ground state and a Ps atom in its first (spatially extended) excited state. To check the validity of this assumption, we have restricted the model space to include only this type of configurations. This can be achieved by a special choice of the u_{ki} parameters of Eq. (8.1). The energy converged to −0.323 a.u., that is, the Ps(1S) +Ps(2P) system with zero relative orbital angular momentum forms a bound state with energy close to that of $L = 1$ state of the Ps_2 molecule, therefore this configuration is likely to be the dominant configuration in this molecule. There is a second configuration, the $Ps^- + e^+$ or $Ps^+ + e^-$ with $L = 1$ relative orbital motion, which intuitively may look important because two oppositely charged particles attract each other, but it is barely bound ($E = -0.315$ a.u).

The average distances in Table 8.5 show that in the $L = 1$ state the two atoms are well separated. In fact we can estimate the root-mean-square distance d between the two atoms by

$$d^2 = \left\langle \left(\frac{\boldsymbol{r}_1 + \boldsymbol{r}_2}{2} - \frac{\boldsymbol{r}_3 + \boldsymbol{r}_4}{2} \right)^2 \right\rangle$$

$$= \frac{1}{4} \Big(2\langle r_{12}^2 \rangle + \langle r_{13}^2 \rangle - 2\langle \boldsymbol{r}_{12} \cdot \boldsymbol{r}_{14} \rangle \Big). \tag{8.6}$$

The symmetry properties of the Ps_2 wave function are used to obtain the second equality. Using the values of Table 8.5 yields $d = 6.93$ a.u. for the $L = 1$ excited state and $d = 4.82$ a.u. for the $L = 0$ ground state. One cannot give a direct geometrical picture of the ground or excited state because the variance $\Delta r_{ij} = \sqrt{\langle r_{ij}^2 \rangle - \langle r_{ij} \rangle^2}$ is large.

The correlation function defined by

$$C(\mathbf{r}) = \langle \Psi | \delta(\mathbf{r}_i - \mathbf{r}_j - \mathbf{r}) | \Psi \rangle \tag{8.7}$$

gives more detailed information on a system than just various average distances. This quantity can be calculated for the correlated Gaussians by using Eq. (A.30) or (A.136). For the ground-state wave function Ψ with $L = 0$, $C(\mathbf{r})$ becomes a function of only r, which is called the monopole density. For the excited-state wave function with $L = 1$, $C(\mathbf{r})$ consists of the two terms of monopole and quadrupole densities.

Figure 8.2 displays the electron-electron and the electron-positron correlation functions $r^2 C(\mathbf{r})$ for the ground-state of the Ps_2 molecule. The peak position of the electron-electron correlation function is shifted to a larger distance than that of the electron-positron correlation function. The latter has much broader distribution and reaches farther in distances compared to the corresponding function of a Ps atom.

Figure 8.3 displays the electron-electron and electron-positron correlation functions for the second bound-state of the Ps_2 molecule. As mentioned above, the correlation function for the $L = 1$ state consists of the monopole and quadrupole densities and their shapes depend on the magnetic quantum number M of the wave function. Of course the M-dependence of the shapes is not independent of each other

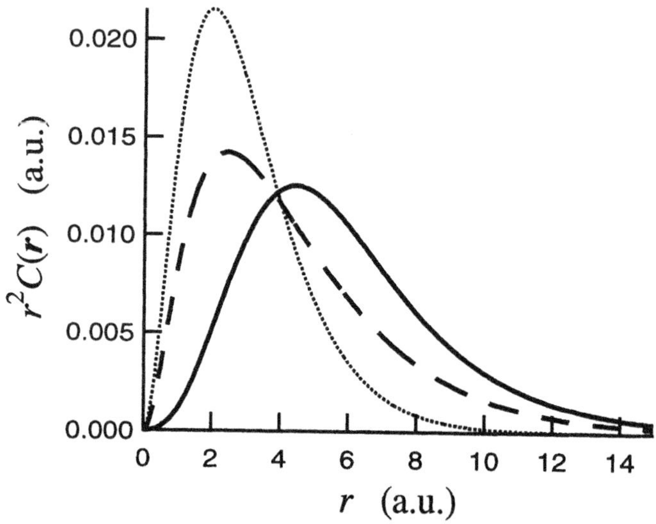

Fig. 8.2. The correlation functions $r^2C(\boldsymbol{r})$ for the ground state of the Ps_2 molecule. The solid curve denotes the electron-electron correlation and the dashed curve the electron-positron correlation. For the sake of comparison, the electron-positron correlation function for a Ps atom is drawn by the dotted curve.

but is related by the Clebsch-Gordan coefficient. See Eq. (A.136). The quadrupole density is contributed from only the P wave of the electron-positron relative motion, while the monopole density is contributed by both S and P waves. Figure 8.3 plots the correlation functions for both (a) $M = 0$ and (b) $M = 1$. As the correlation function is axially symmetric around the z axis and has a reflection symmetry with respect to the xy plane, the correlation function sliced on the xz plane is drawn as a function of $x\,(x \geq 0)$, $z\,(z \geq 0)$. The electron-electron correlation function has a peak at the point corresponding to the average distance of 7.57 a.u. The electron-positron correlation function has two peaks reflecting the fact that the basic structure of the second bound-state is a weakly coupled system of a Ps atom in the $L = 0$ state and another Ps atom in the $L = 1$ spatially extended state. The peak located at a larger distance from the origin is due to the P-wave component of the Ps_2 molecule.

The hydrogen and positronium molecules can be considered as members of the same family as both are quantum-mechanical fermionic four-body systems of two positively and two negatively charged identical particles. But they are at the opposite ends of the $(M^+M^+m^-m^-)$-

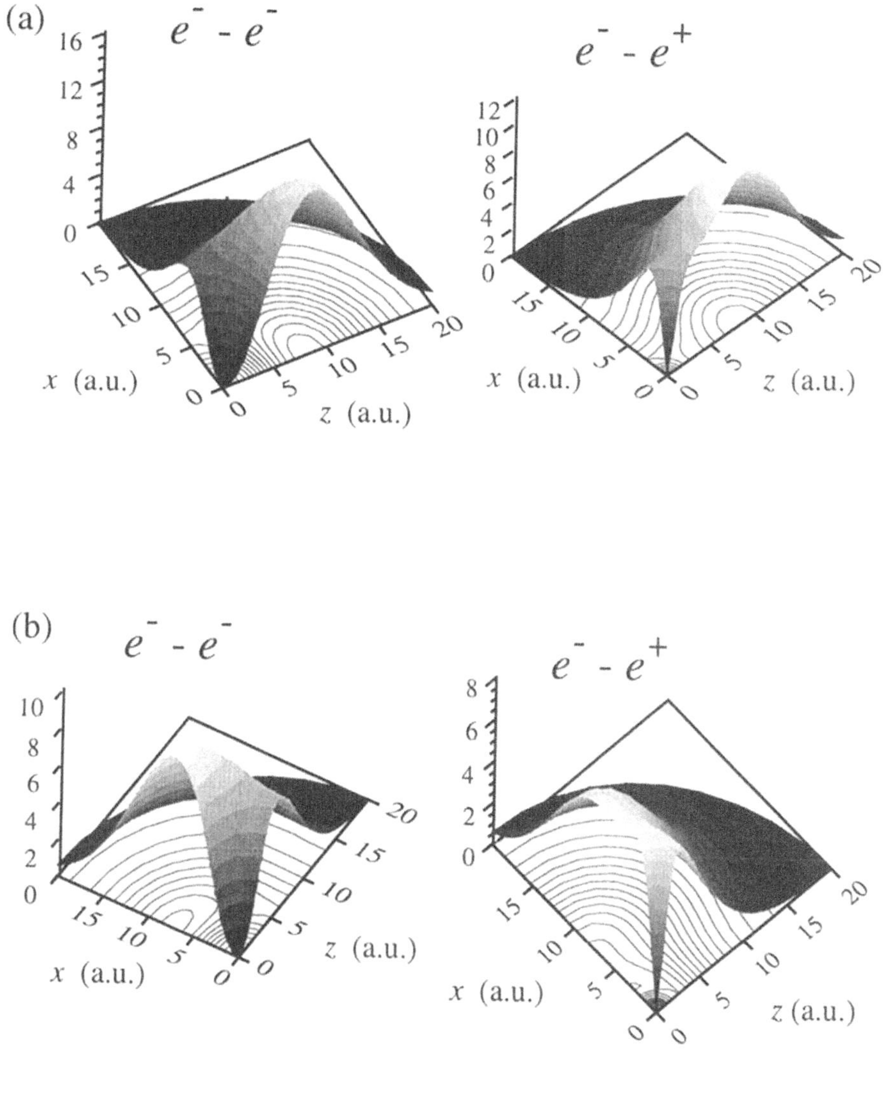

Fig. 8.3. The correlation functions $r^2C(\boldsymbol{r})$ in atomic units, multiplied by one thousand, for the second bound-state of the Ps_2 molecule. The z component of the orbital angular momentum is $M = 0$ for (a) and $M = 1$ for (b). Drawn on the xz plane are the corresponding contour maps.

type Coulombic systems called biexcitons or excitonic molecules. The biexciton (or biexciton molecule), a bound complex of two electrons and two holes, is observed in a variety of semiconductors [105, 106]. See also Sect. 10.1. The biexciton molecule is characterized by the mass ratio $\sigma = m/M$. Apart from this connection, however, their properties are radically different, e.g., H_2 is an adiabatic but Ps_2 is a highly nonadiabatic system. Moreover, while in the case of the H_2 molecule many bound excited-states have been observed experimentally and later studied theoretically, in the case of the Ps_2 molecule only the ground state and the unique excited state discussed above have so far been predicted theoretically. See also Complement 8.3 for the stability of the biexciton molecule.

Figure 8.4 displays the dependence of the binding energy of the biexciton molecule on the mass ratio $\sigma = m/M$. The changes of the binding energies in the ground state ($L = 0$) and the excited states with $L = 1$ and negative parity is similar. Both the ground and excited states become less bound by changing the mass ratio from H_2 to Ps_2, though the binding of the excited state decreases to a somewhat lesser extent. The energy of the transition from the excited state to the ground state is also shown in this figure. This transition may take place in an external field, for example.

By increasing the mass M of the positively charged particles toward infinity, one arrives at the energy of the $C\,^1\Pi_u\,2p\pi$ state of the H_2 molecule. This state is formed by an excited H-atom and a ground-state H-atom. Consequently, a statement similar to the case of the Ps_2 molecule is valid for the biexcitonic molecule: The second bound-state of the biexciton molecule is dominantly formed by an interacting pair of a ground-state exciton and an $L = 1$ excited-state exciton.

The rule that the Pauli principle forbids odd partial waves between identical bosons also applies to the biexciton with $L = 1$ and negative parity. The second bound-state of the biexciton molecule cannot decay to two ground-state excitons. A somewhat similar situation exists in the $^3P^e$ state of the H^- ion as well, where its second bound state cannot decay due to the parity conservation. By changing the mass ratio in that $(M^+m^-m^-)$ system, however, this kind of state disappears for $\sigma = 1$ and the Ps^- ion is known to have only one bound state.

Tables 8.6–8.8 show our results for various other Coulombic systems.

The investigation of the stability of positronic atoms has been motivated by the use of positrons as a tool for spectroscopy (positron

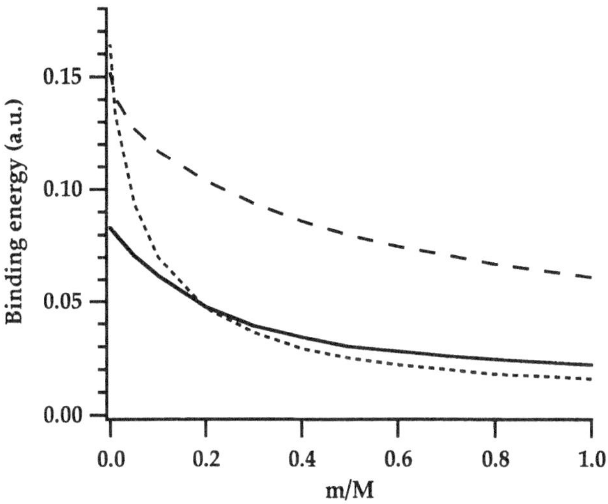

Fig. 8.4. The binding energy of the biexciton molecule as a function of the mass ratio $\sigma = m/M$. The dotted curve is the binding energy of the ground state, and the solid curve is that of the first excited state with $L = 1$ and negative parity. The dashed curve is the energy difference, multiplied by one third, between the first excited state and the ground state.

Table 8.6. Energies of different Coulombic four-body systems in atomic units. K is the basis dimension.

System	State	K	SVM	Other method	K	Ref.
Ps_2	$^1S^e$	800	−0.516003778	−0.516002	400	[102]
Ps_2	$^1P^o$	800	−0.334408112			
Li	$^1S^e$	600	−7.478058	−7.47806032	1589	[107]
Li	$^1P^o$	1000	−7.410151	−7.410156521	1715	[107]
Li	$^1D^e$	1000	−7.335520	−7.335523540	1673	[107]
$^\infty$HPs	$^1S^e$	1200	−0.7891964	−0.7891794		[103]

Table 8.7. Energies of different Coulombic five-body systems in atomic units. K is the basis dimension. The Li^- energy with $K = \infty$ is the extrapolated one [109], where $E = -7.500577$ is given by multiconfiguration Hartree-Fock calculations with $K = 2997$.

System	State	K	SVM	Other method	K	Ref.
Be	$^1S^e$	500	−14.6673	−14.667355	1200	[108]
Li^-	$^1S^e$	600	−7.50012	−7.50076	∞	[109]
$(2\pi^+, 3\pi^-)$	$^1S^e$	200	−0.5493			
$Li + e^+$	$^1S^e$	1000	−7.53218			

Table 8.8. Energies of different Coulombic six-body systems in atomic units. K is the basis dimension.

System	State	K	SVM
$(3\pi^+, 3\pi^-)$	$^1S^e$	300	−0.820
Li + Ps	$^1S^e$	600	−7.73855
$Be + e^+$	$^1S^e$	1000	−14.692

annihilation spectroscopy) in condensed matter physics. An intriguing question is whether or not a chemically stable system containing a positron or a positronium could be formed in the various targets. This question can be answered only by a sophisticated calculation or experiment because the mechanism responsible for binding the positron to the neutral atom is the polarization potential present in the atom+e^+ system. The boundness of the hydrogen positride (positronium hydride) HPs was predicted theoretically by Ore [99] in 1951 and it has recently been created and observed in collisions between positrons and methane [110]. The properties of HPs is discussed in [85]. The use of the SVM proved for the first time that the positronic lithium (Li+e^+) [111] and the positronic beryllium (Be+e^+) [112] are stable. We see from the tables that the positron separation energy of the positronic lithium is 0.054 a.u. Below the Li+e^+ channel the Li^++Ps channel is open and the binding energy of the positronic lithium is only 0.0022 a.u. against the dissociation into the Li^++Ps channel. A calculation has to be accurate at least to 10^{-3} a.u. to answer the stability of the positronic lithium. Likewise, the positron separation energy of the Be+e^+ system is only 0.025 a.u. Due to the tiny binding energies of typically about 0.01 a.u. one has to be able to reach high accuracy in these 6-particle systems. A naive picture of these systems is that the positron orbiting around the neutral atom slightly polarizes the negative electron cloud, and the positron is bound by the resulting attraction.

To extend the method (as a full N-body solution) for the investigation of the stability of much larger systems (e.g. Sodium plus positron) is out of question. One can, however, try to use a "frozen core approximation". In this approximation the positively charged core is considered to be passive (its polarization is neglected) and the problem is solved in a model space where the single-particle orbitals are orthogonal to the core orbitals. One has to solve the modified Schrödinger equation of the form

$$(H + \lambda P)\Psi = E\Psi \qquad \text{with} \quad P = \sum_{i \in \text{occupied}} |\phi_i\rangle\langle\phi_i|, \tag{8.8}$$

which produces wave functions that are orthogonal to the core orbitals provided the positive constant λ is large enough. The projection operator P is an example for the nonlocal potentials discussed in Sect. 7.5. See also Complement 11.4. One can validate this approximation by comparing it to the "exact" full N-body calculation for Li+e^+. This approximation turns out to be very accurate, reproducing the first six digits of the result of the full calculation [112]. Assuming that the accuracy holds for larger systems, one seems to find the stability of positronic sodium (Na+e^+) [112].

8.4 Small molecules

As it was mentioned at the beginning of this chapter, the molecular case demands a special treatment. The variational trial function is assumed to take combinations of the form (6.19)

$$g(\boldsymbol{s}; A, \boldsymbol{x}) = \exp\Big(-\frac{1}{2}\tilde{\boldsymbol{x}}A\boldsymbol{x} + \tilde{\boldsymbol{s}}\boldsymbol{x}\Big), \tag{8.9}$$

where A_{ij} and $\boldsymbol{s} = \{\boldsymbol{s}_1, \boldsymbol{s}_2, ..., \boldsymbol{s}_{N-1}\}$ are parameters of the basis functions. The "generator coordinates" $\boldsymbol{s}$ are chosen conforming to $\boldsymbol{x}$. Note that $\boldsymbol{x}$ is a set of relative coordinates. Our aim here is to calculate the energy of the system which can be directly compared to experiment without recourse to the adiabatic treatment like the Born-Oppenheimer approximation. Each basis function includes $N(N-1)/2+3(N-1)$ parameters to be optimized. These parameters describe various correlations. The matrix A describes the electronic correlations and motions, while the generator coordinate $\boldsymbol{s}$ makes the wave function flexible and allows us to represent several "peaks" of the density distribution when, for example in the hydrogenic limit, the holes are well separated and the electrons are on "atomic orbits" around them. By choosing $\boldsymbol{s} = 0$ the function (8.9) has its maximum at the origin and this limit is suitable around $\sigma = m/M = 1$, when the particles with nearly equal masses are moving equally fast. At the hydrogenic limit, when the motion of the heavy particles are very slow compared to the light ones, the density distribution has several peaks around the attractive centers, and to represent these configurations we need to shift the maximum of the trial functions out of the origin by choosing $\boldsymbol{s}$ appropriately.

The usefulness of the generator coordinates in the basis function (8.6) can be understood by the following example. Let us try to calculate the energy of the $^{\infty}H_2^+$ by using this basis with and without (that is by setting them to zero) the generator coordinates. The latter form corresponds to the correlated Gaussians for $L = 0$. The energy of that system is -0.6026 a.u.. Without the generator coordinates a basis of $K = 300$ Gaussians give -0.5999 a.u. for this molecule. The inclusion of the generator coordinate immediately improves the convergence and one can get -0.6024 a.u. by using $K = 10$ basis states only!

Table 8.9 shows examples of calculations for the molecules and the ions consisting of protons and electrons.

Table 8.9. Ground state energies of small molecules in atomic units. K is the basis dimension.

System	K	SVM	Other method	K	Ref.
$^{\infty}H_2^+$	50	-0.602634429	-0.602634214	160	[113]
$^{\infty}H_2$	100	-1.17445	-1.174475714	1200	[113]
$^{\infty}H_3^+$	100	-1.34351	-1.343835624	600	[114]

Complements

8.1 *The cusp condition for the Coulomb potential*

It is desirable that the trial function satisfies the proper asymptotic behavior or the special boundary condition as demanded by a given Hamiltonian. We discuss a special boundary condition, the cusp condition [115] known for the Coulomb potential, by using the hydrogen atom as an example. The local energy for the hydrogen atom is given by

$$E_{\text{loc}}(\boldsymbol{r}) = -\frac{\hbar^2}{2m}\frac{1}{\Psi}\left(\frac{\partial^2\Psi}{\partial r^2} + \frac{2}{r}\frac{\partial\Psi}{\partial r}\right) + \frac{\hbar^2}{2mr^2}\frac{1}{\Psi}\boldsymbol{l}^2\Psi - \frac{e^2}{r}, \tag{8.10}$$

where $\hbar\boldsymbol{l}$ is the angular momentum operator. As was discussed in Sect. 3.2, the local energy for the exact wave function turns out to be a constant. The Coulomb potential in the local energy gives a singular behavior at $r = 0$. For the local energy to be a constant, this singular behavior must be compensated for by the kinetic energy term. For an S wave, where the wave function Ψ has no angular dependence, $\boldsymbol{l}^2\Psi = 0$ and the constancy of E_{loc} requires that the second term in the bracket in Eq. (8.10) cancel the singular behavior of the Coulomb potential:

$$\left(\frac{1}{\Psi}\frac{\partial\Psi}{\partial r}\right)_{r=0} = -\frac{me^2}{\hbar^2} = -\frac{1}{a_0}, \tag{8.11}$$

where a_0 is the Bohr radius. It is easy to see that an exponential form of $\exp(-r/\alpha)$ for the radial part of the ground-state wave function leads to a constant local density if and only if α is chosen to be a_0. The constant of the local energy is then equal to $-\hbar^2/(2m\alpha^2) = -me^4/(2\hbar^2) = -E_I$ as expected, where E_I is the hydrogen atom ionization energy without the proton recoil effect, that is, the well-known energy of 13.6 eV.

For a general case when Ψ has angular dependence, it has to take care of both the $1/r$ and $1/r^2$ singularities. Then Ψ may be expressed as a product of radial and angular parts: $\Psi = r^s R(r) Y(\Omega)$, where s is a positive constant and $R(r)$ is assumed to be nonzero at $r = 0$. Substituting Ψ into Eq. (8.10), we obtain

$$E_{\text{loc}}(\boldsymbol{r}) = -\frac{\hbar^2}{2m}\left(\frac{1}{R}\frac{\partial^2 R}{\partial r^2} + \frac{2(s+1)}{r}\frac{1}{R}\frac{\partial R}{\partial r} + \frac{s(s+1)}{r^2}\right) + \frac{\hbar^2}{2mr^2}\frac{1}{Y}\boldsymbol{l}^2 Y - \frac{e^2}{r}. \tag{8.12}$$

As R does not vanish at $r = 0$, $1/R$ gives rise to no singularity at the origin. The condition that the local energy is constant leads to the following result:

$$\left(\frac{1}{R}\frac{\partial R}{\partial r}\right)_{r=0} = -\frac{1}{(s+1)a_0}, \qquad l^2 Y = s(s+1)Y. \tag{8.13}$$

We know that the second equation is satisfied if and only if s is a positive integer l, and then the first equation determines the correct behavior of R near the origin as $R(r) \propto \exp(-r/(l+1)a_0)$. Equations (8.11) and (8.13) are known to be the cusp conditions.

Let us attempt to solve the S-wave hydrogen atom variationally with Gaussian basis functions, $\exp[-(\alpha/2)(r/a_0)^2]$, where α is a variational parameter. We use Gaussians as an example of such basis functions that lead to a rather accurate solution but are poor in satisfying the cusp condition. When a single basis function is used, the optimal value of α is $16/(9\pi)$, giving the minimum energy of $-8/(3\pi)E_I = -0.849E_I$. A combination of a few terms approximates the ground-state energy quite well. The parameter values of α are determined by the SVM. Table 8.10 shows sample results of such calculations obtained with the code given in [81]. The calculation with five Gaussians already reproduces the energy up to three digits. The wave function obtained with ten Gaussians can reproduce both the energy and the mean values of r and $1/r$ fairly accurately. The overlap of the wave function with the exact wave function is very close to unity. We may conclude that the Gaussian basis can predict physical quantities to high accuracy. Of course, the solution does not satisfy the proper asymptotic behavior at large distances and, moreover, always gives zero for the cusp value of Eq. (8.11). The local energy displayed in Fig. 8.5 for the variational wave function indicates that with increasing K it tends to show smaller and smaller deviations from the exact wave function except for the singular points mentioned above. The local energy at large r deviates from the correct value because the Gaussian basis has the wrong asymptotic behavior.

It is possible to generalize the above arguments for the cusp value in a system of particles interacting via Coulomb potentials. Evaluating the cusp value for a pair of particles with charges q_i and q_j, we obtain

$$\frac{\langle\Psi|\delta(\boldsymbol{r}_i - \boldsymbol{r}_j)\frac{\partial}{\partial|\boldsymbol{r}_i - \boldsymbol{r}_j|}|\Psi\rangle}{\langle\Psi|\delta(\boldsymbol{r}_i - \boldsymbol{r}_j)|\Psi\rangle} = \frac{\mu_{ij}q_iq_j}{\hbar^2}, \tag{8.14}$$

where μ_{ij} is the reduced mass of the two particles. The left-hand side of Eq. (8.14) is expressed with the matrix elements involving

$\delta(\boldsymbol{r}) = \delta(r)/(4\pi r^2)$. The cusp values for a pair of particles are often used to test the quality of the variational solution at the particles' coalescence.

Table 8.10. Variational solution for the hydrogen atom with a number K of Gaussian basis functions. The last row shows the exact values. E_I is the hydrogen atom ionization energy and a_0 is the Bohr radius.

K	$\frac{E}{E_I}$	$\langle(\frac{r}{a_0})^{-2}\rangle$	$\langle(\frac{r}{a_0})^{-1}\rangle$	$\langle(\frac{r}{a_0})\rangle$	$\langle(\frac{r}{a_0})^{2}\rangle$	Overlap
1	−0.8488264	1.131774	0.8488284	1.499996	2.650706	0.9568351
3	−0.9939585	1.903352	0.9939409	1.491519	2.922759	0.9987560
5	−0.9996191	1.986219	0.9995692	1.499147	2.991284	0.9999446
10	−0.9999998	1.999700	0.9999958	1.500004	3.000014	0.9999999
	−1	2	1	1.5	3	1

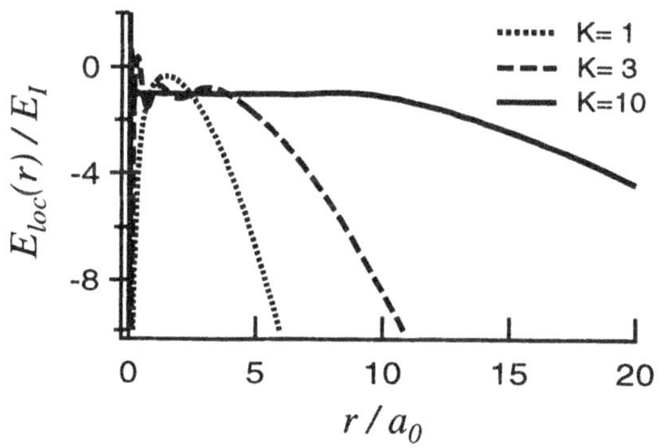

Fig. 8.5. The local energy curve plotted for the variational solution of the hydrogen atom. K denotes the number of Gaussian basis functions.

8.2 *The chemical bond: The* H_2^+ *ion*

Quantum mechanics enables us to understand the chemical bond, which is responsible for the formation of molecules from isolated atoms. The chemical bonding phenomenon involves the delocalization of electrons in an atom to gain attraction from the other nuclei when the atoms come close to each other. We take up the simplest possible molecule, the H_2^+ ion, to understand what an important role the Hellmann-Feynman and virial theorems play for clarifying the origin of the chemical bond. See [1] for detail.

The fully quantum-mechanical description of a molecule is a complex problem. This problem is usually simplified by using the Born-Oppenheimer approximation, where the electronic motion is separated from the nuclear motion, considering the fact that the electron mass is much smaller than that of the nuclei. One starts with determining the motion of the electrons for a fixed configuration $\boldsymbol{R}$ of the nuclei and obtains the ground state, of energy $U(\boldsymbol{R})$, of the electronic system. Then one assumes that, when $\boldsymbol{R}$ varies, the electronic system always remains in the ground state corresponding to $\boldsymbol{R}$, that is the electrons follow adiabatically the motion of the nuclei. The chemical bond is then determined by studying the nuclear motion in a potential energy $V(\boldsymbol{R})$ which comprises the Coulomb repulsion between the nuclei and $U(\boldsymbol{R})$.

Let $\boldsymbol{R}$ be the distance vector between the two protons of the H_2^+ ion and $\boldsymbol{r}$ be the position vector of the electron with respect to the center-of-mass of the protons. The electron motion is determined by the Hamiltonian

$$H_e(\boldsymbol{R}) = \frac{\boldsymbol{p}^2}{2\mu} - \frac{e^2}{|\boldsymbol{r} - \frac{\boldsymbol{R}}{2}|} - \frac{e^2}{|\boldsymbol{r} + \frac{\boldsymbol{R}}{2}|}. \tag{8.15}$$

Note that $\boldsymbol{R}$ is just a parameter at this stage. The reduced mass μ is equal to $2Mm/(2M+m)$, where M is the mass of the proton and m the mass of the electron. The electronic energy $U(\boldsymbol{R})$ is the lowest eigenvalue of the Hamiltonian (8.15). It is clear that $U(\boldsymbol{R})$ becomes a function of R only. The Schrödinger equation for the Hamiltonian (8.15) is completely separated in elliptical coordinates with respect to the foci, $\boldsymbol{R}/2$ and $-\boldsymbol{R}/2$. We do not need its exact solution in the following discussion, but note that it is well approximated by the variational calculation using a trial function of the form

$$\Psi \sim \psi_{1s}\Big(Z, \frac{\boldsymbol{R}}{2}\Big) + \psi_{1s}\Big(Z, -\frac{\boldsymbol{R}}{2}\Big), \tag{8.16}$$

where $\psi_{1s}(Z, \boldsymbol{s})$ is the $1s$ hydrogenic orbital of radius a_0/Z centered at $\boldsymbol{s}$. The charge Z is a variational parameter and its optimal value is determined to minimize the energy for each R. The optimal value of Z decreases from $Z = 2$ for $R = 0$ to $Z = 1$ for $R \to \infty$. At $R \to \infty$ the system will switch over to a configuration of the hydrogen atom and the proton. Between these two extremes, Z is a decreasing function of R. The energy $U(R)$ is $-4E_I(2M/(2M+m))$ for $R=0$ and approaches $-E_I(M/(M+m))$ for $R \to \infty$.

By using the virial theorem (3.49) and Eq. (3.56) with $\lambda = R_x, R_y, R_z$, we can show that the expectation values, $\langle T\rangle$ and $\langle W\rangle$, of the kinetic energy and potential energy of $H_e(\boldsymbol{R})$ satisfy the relation

$$2\langle T\rangle + \langle W\rangle + R\frac{d}{dR}U(R) = 0. \tag{8.17}$$

Here we have used the fact that $W_A + \boldsymbol{R}\cdot\frac{\partial}{\partial \boldsymbol{R}}W = -W$ and $\boldsymbol{R}\cdot\frac{\partial}{\partial \boldsymbol{R}}U(R) = R\frac{d}{dR}U(R)$. This equation, together with $\langle T\rangle + \langle W\rangle = U(R)$, enables us to express $\langle T\rangle$ and $\langle W\rangle$ in terms of the potential between the protons, $V(R) = U(R) + (e^2/R)$, as follows:

$$\langle T\rangle = -U(R) - R\frac{d}{dR}U(R) = -V(R) - R\frac{d}{dR}V(R),$$

$$\langle W\rangle = 2U(R) + R\frac{d}{dR}U(R) = 2V(R) + R\frac{d}{dR}V(R) - \frac{e^2}{R}. \tag{8.18}$$

For the chemical bond to occur in the H_2^+ system, the potential $V(R)$ must have a minimum deeper than $V(R\to\infty) \sim -E_I$ at some point R_0, that is, $V'(R_0) = 0$ and $V(R_0) < -E_I$ have to be met. Since $\langle T\rangle \sim E_I$ and $\langle W\rangle \sim -2E_I$ at $R \to \infty$ (see Eq. (3.53)), we can conclude from Eq. (8.18) that, at equilibrium ($R = R_0$), the electronic kinetic energy is increased and the electronic potential energy is decreased. The lowering of the electronic potential energy is large enough to cancel the repulsion between the protons, and that is responsible for the chemical bond. According to an exact calculation, $R_0 = 2.00a_0$ and $V(R_0) - V(R\to\infty) = -2.79$ eV.

It is challenging to perform a nonadiabatic calculation in which no separation of the electron and nuclear motion is made. The validity of the Born-Oppenheimer approximation can be tested in such a calculation. Furthermore, the development of the nonadiabatic treatment for a small molecule is of importance in its own right because the adiabaticity may be questionable when the electron is replaced with heavier particles like the muon or the pion. The excellent results obtained in Sects. 8.2-8.4 indicate that realistic nonadiabatic calculations are in fact possible in the correlated Gaussian basis.

8.3 *Stability of hydrogen-like molecules*

The existence of bound states of systems composed of particles with unit charge attracts considerable attention. We discuss this problem here by applying some of the principles discussed in Chap. 3.

The system of the hydrogen-like atom, (M^+m^-), is always bound and its binding energy is equal to $\mu/2$ in units of $e = 1$ and $\hbar = 1$, where $\mu = Mm/(M+m)$ is the reduced mass. What about the stability of a hydrogen-like molecule $(M^+M^+m^-m^-)$? This system is characterized by the mass ratio $\sigma = m/M$. Two well-known examples include the hydrogen molecule ($\sigma \ll 1$) and the positronium molecule Ps_2 ($\sigma = 1$). Another example is the biexciton molecule [105, 106]. See Sect. 8.3. The value of σ of the biexcitons can vary between the two limits, $0 < \sigma < 1$. A molecule is bound provided that the threshold of any dissociation channel is higher than the lowest energy of the system. The lowest dissociation channel is $(M^+m^-) + (M^+m^-)$ for this system, and its threshold energy is $E_{\mathrm{th}} = -\mu$. The stability of the hydrogen-like molecule has been studied numerically in Sect. 8.3 (see Fig. 8.4). However, a theoretical argument [101] makes it possible to prove that the system is bound for arbitrary values of σ, that is, the ground-state energy E of the system is lower than E_{th}. The proof relies on the scaling property of the Coulombic Hamiltonian and the stability of Ps_2. The basic point of the proof given in [101] will be shown below.

The Hamiltonian of the system $(M^+M^+m^-m^-)$ is written as

$$H = H_S + H_A, \tag{8.19}$$

$$H_S = \frac{1}{4\mu}(\boldsymbol{p}_1^2 + \boldsymbol{p}_2^2 + \boldsymbol{p}_3^2 + \boldsymbol{p}_4^2) + e^2\left(\frac{1}{r_{12}} + \frac{1}{r_{34}} - \frac{1}{r_{13}} - \frac{1}{r_{14}} - \frac{1}{r_{23}} - \frac{1}{r_{24}}\right), \tag{8.20}$$

$$H_A = \left(\frac{1}{4M} - \frac{1}{4m}\right)(\boldsymbol{p}_1^2 + \boldsymbol{p}_2^2 - \boldsymbol{p}_3^2 - \boldsymbol{p}_4^2). \tag{8.21}$$

H_S is nothing but the Hamiltonian of a system $(X^+X^+X^-X^-)$ with the mass m_{X} of a particle X being equal to 2μ. By applying a scaling transformation $\boldsymbol{r}_i \to \lambda \boldsymbol{r}_i$ ($\lambda = m_{\mathrm{e}}/(2\mu)$), we obtain

$$H_S = \frac{2\mu}{m_{\mathrm{e}}} H(\mathrm{Ps}_2), \tag{8.22}$$

where m_e is the electron mass and $H(Ps_2)$ is the Hamiltonian for the positronium molecule. This equation indicates that the ground-state energy of H_S, E_S, can be obtained from that of Ps_2, $E(Ps_2)$, by $E_S = (2\mu/m_e)E(Ps_2)$. Note that this relation can also be obtained from the Hellmann-Feynman theorem and the virial theorem. In fact the use of Eqs. (3.38) and (3.53) enables one to obtain ($m_X = 2\mu$)

$$\frac{d}{dm_X}E_S = \langle \Phi_S | \frac{\partial}{\partial m_X} H_S | \Phi_S \rangle = -\frac{1}{m_X} \langle \Phi_S | T_S | \Phi_S \rangle$$
$$= \frac{1}{m_X} E_S, \tag{8.23}$$

which leads to the solution that E_S/m_X is independent of m_X. Here Φ_S is the ground-state wave function of H_S and T_S is the kinetic energy of H_S.

According to the Ritz theorem the ground-state energy E_1 of H satisfies the following relation

$$E_1 \leq \langle \Phi_S | H_S + H_A | \Phi_S \rangle. \tag{8.24}$$

The term H_A is odd under the interchange P of the masses M and m, i.e., $PH_A = -H_A P$. Then by using that Φ_S is invariant under P we have $\langle \Phi_S | H_A | \Phi_S \rangle = \langle \Phi_S | PH_A P | \Phi_S \rangle = -\langle \Phi_S | H_A P^2 | \Phi_S \rangle = -\langle \Phi_S | H_A | \Phi_S \rangle = 0$. The right-hand side of Eq. (8.24) is thus equal to $\langle \Phi_S | H_S | \Phi_S \rangle = E_S$, and we obtain

$$E_1 - E_{\text{th}} \leq E_S - (-\mu) = \frac{2\mu}{m_e}\left(E(Ps_2) - (-\frac{m_e}{2})\right). \tag{8.25}$$

The energy of $-m_e/2$ is the threshold for Ps_2 to decay into two Ps atoms. If the stability of Ps_2 is established, that is, $E(Ps_2) + (m_e/2) < 0$, then we can conclude immediately from Eq. (8.25) that the hydrogen-like molecule is always stable. In fact Hylleraas and Ore [99] calculated the energy of Ps_2 by using a simple trial function and showed its stability.

It is of interest to examine the stability of a hydrogen-antihydrogen like molecular system, $(M^+m^+M^-m^-)$ [104]. The lowest threshold of this system is now $E_{\text{th}} = -(M/4) - (m/4)$ corresponding to the dissociation $(M^+M^-) + (m^+m^-)$ and gets deeper and deeper as M increases, which is in contrast to the case of the hydrogen-like molecule. It is unlikely that the hydrogen-antihydrogen system gets lower energy than the threshold. It is thus expected that there is a critical mass ratio M/m beyond which the system becomes unstable. According to the calculations in [104, 116], the stability limit is $M/m \lesssim 2.1$.

8.4 *Application of global vectors to muonic molecules*

The aim of this Complement is to show the utility of the global vector $\boldsymbol{v}$ of Eq. (6.4) for describing the angular part of the variational trial function. As a test example we take up a Coulombic three-body system, the $td\mu$ molecule, which has attracted much attention in relation to muon catalyzed fusion [117]. The excited P state especially plays a key role in the fusion since it lies close to the threshold for the decay to the $t\mu$ atom and the deuteron.

The basis function we use is f_{KLM} of Eq. (6.27). The Hamiltonian matrix elements for this function can be calculated with the method of Chap. 7 and the appendix. Without loss of generality the variational parameters u_1 and u_2 which define $\boldsymbol{v}$ can be normalized to satisfy $u_1^2 + u_2^2 = 1$. Each basis function for a given set of KLM values thus contains at most four nonlinear parameters, three of which come from the matrix A. The matrix A has to be positive-definite to assure the finite norm of f_{KLM}, and can in general be expressed as $A = \tilde{G}DG$, where G is a 2×2 orthogonal matrix $\begin{pmatrix} \cos\vartheta & \sin\vartheta \\ -\sin\vartheta & \cos\vartheta \end{pmatrix}$ specified by just one parameter ϑ and D is a diagonal matrix including two diagonal elements of positive d_1 and d_2 values.

The accuracy of the variational solution depends on how the parameters u_1, ϑ, d_1, and d_2 are given [33]. The most naive choice would be to take G as a unit matrix ($\vartheta = 0$), which is equivalent to using only a single set of relative coordinates $\boldsymbol{x}$, and then to try to reach convergence by including successively higher partial waves implied by an appropriate choice of K and u_1 values. Many examples show [20, 29, 30, 31], however, that this type of single-channel calculation does not work well, especially in the case where the adiabatic approximation is questionable as in the present example.

The matrix G may be chosen to let $\boldsymbol{y} = G\boldsymbol{x}$ correspond to other particular coordinate sets such as the so-called rearrangement channel [20]. (For $G\boldsymbol{x}$ to correspond to the rearrangement channel, the length scales of $\boldsymbol{x}$ and $\boldsymbol{y}$ in general have to be modified appropriately.) Three possible rearrangement channels expressed in the Jacobi coordinate set are drawn in Fig. 2.2. If the particles d, t, and μ are labelled particle 1, 2, and 3, the three patterns in Fig. 2.2 correspond to $(t\mu)d$, $(\mu d)t$, $(dt)\mu$ arrangements, respectively. If $\boldsymbol{x}$ is understood to stand for the coordinate set of the $(t\mu)d$ arrangement, the coordinate set corresponding to the other arrangement is obtained by choosing such an appropriate ϑ value that is uniquely determined by the change of the coordinates. With this choice of ϑ, we can write $\tilde{\boldsymbol{x}}A\boldsymbol{x} = \tilde{\boldsymbol{x}}\tilde{G}DG\boldsymbol{x} = \tilde{\boldsymbol{y}}D\boldsymbol{y} = d_1\boldsymbol{y}_1^2 + d_2\boldsymbol{y}_2^2$.

The following three simple types of bases were chosen:

(i) $K = 0$ and $A = \tilde{G}DG$, where G is restricted to the special matrices connecting only the three sets of rearrangement channels as explained above.
(ii) $K = 0$ or 1 and the choice of A is the same as in case (i).
(iii) $K = 0$ and $A = \tilde{G}DG$, where G is now an arbitrary orthogonal matrix.

As was mentioned in Sect. 6.2, the angular part with $K = 0$ describes only the stretched configuration, $l_1 + l_2 = L$, and therefore the basis of type (i) allows rather limited angular correlations between the particles. In fact, the possible (l_1, l_2) values are given by $(l, L - l)$, $(l = 0, ..., L)$. Basis (ii) is an extension of basis (i) to include the non-stretched coupling. With $K = 1$, possible $(l_1, l_2) = (l + 1, L - l + 1)$, $(l = 0, ..., L)$ are also allowed. Though K is set to zero in basis (iii), the angular correlation can be taken into account through the cross term of the exponential part of the basis function. In bases (i) and (ii) the factors $\exp(-\tilde{\boldsymbol{x}}A\boldsymbol{x}/2)$ are always "correlation-free" in a particular channel, that is, it contains no cross term in the exponential part. In these bases it is through the inclusion of all rearrangement channels that one takes care of the correlation. In a basis of type (iii), on the contrary, one needs to consider just any one arrangement (that is, coordinate set) explicitly; the correlations are allowed for through the general form of A.

Table 8.11 shows the results of calculations. As expected, the results with case (i) are poor compared to other bases. The correlation contained in basis (i) is too restricted to obtain a realistic description of the system. The angular correlation, which is taken into account in basis (ii), certainly improves the energy over case (i). The basis functions (iii) give even better energies. In fact they reproduce the first six figures of the most precise variational calculations [20, 23, 92] for $L = 0 - 2$ states. Even with $K = 0$, the use of the full A matrix enables one to incorporate the important correlations between the particles. A Gaussian basis of type (ii) is employed in the Coupled-Rearrangement-Channel Gaussian (CRCG) basis variational calculation of [20], where the angular part is, however, represented by the successive coupling of type (6.3). The fact that the D state energy with the type (iii) calculation of dimension 200 becomes slightly lower than that of [20] with 1566 basis functions confirms the importance of a careful optimization of the nonlinear parameters. Further increase of the dimension can improve the accuracy rather easily in the angular function of Eq. (6.4) as seen in the case of S and P states. The basis function used in [23, 92] is correlated exponential (CE) which takes the form of $\exp(-\alpha|\boldsymbol{r}_1 - \boldsymbol{r}_2| - \beta|\boldsymbol{r}_2 - \boldsymbol{r}_3| - \gamma|\boldsymbol{r}_3 - \boldsymbol{r}_1|)$ multiplied by some polynomials for the orbital motion with nonzero orbital angular momentum. The parameters α, β, γ are determined by the random tempering explained in Sect. 4.2.2. These calculations using the exponential function give very precise results.

Table 8.11. The total energies of the $td\mu$ molecule. Parameter set a of Table 8.2 is used except for the case indicated by superscript b, where set b is used. Atomic units are used.

L	Method	Dimension	E
S	SVM (i)	200	−111.29346
	SVM (ii)	200	−111.36398
	SVM (iii)	200	−111.36444
		600	−111.3645077
	CRCG[20]	1442	−111.364507
	CE[23]	1400	−111.364511474
P	SVM (i)	200	−108.13122
	SVM (ii)	200	−108.17914
	SVM (iii)	200	−108.17940
	CE[23]	1800	−108.1795424
	SVM (iii)[b]	800	−108.179361
	CRCG[20][b]	2662	−108.179385
	CE[92][b]	1900	−108.1793881
P^*	SVM (iii)[b]	800	−99.660367
	CRCG[20][b]	2662	−99.660548
	CE[92][b]	1900	−99.6605507
D	SVM (i)	200	−103.37067
	SVM (ii)	200	−103.40824
	SVM (iii)	200	−103.40849
	CRCG[20]	1566	−103.408481

9. Baryon spectroscopy

This chapter is devoted to applications of the SVM to the constituent quark model of baryons. There are several different models of baryons and this is a very hot topic nowadays. The discussion of the different models and their relative merits are far beyond the scope of this book, and our only aim here is to show the applicability of the SVM to solving three-body dynamics in constituent quark potential models.

The constituent quark model assumes that the baryons comprise three quarks with dynamical quark masses and the quarks interact via phenomelogical potentials. These potentials contain a confining term and other terms which determine the fine structure of the spectra. As there are no free quarks observed in nature, the confining interaction hinders the "ionization" of the the baryons in the constituent quark potential models. This interaction is usually taken as a power law potential ($V(r) \sim r^p$). The second part of the interaction is traditionally chosen as the one-gluon exchange potential (OGEP) which is a color analogue of the Breit interaction of quantum electrodynamics (QED). There is an alternative possibility, the meson exchange interaction, which is motivated by the fact that, despite its quite successful reproduction of the ground-state spectra of the mesons and baryons, the OGEP may not be adequate for describing the excited states as mesonic effects or quark-antiquark excitations become important. This chapter includes examples for both choices to illustrate how the SVM can be used for these systems with different interactions and with different (nonrelavistic and semirelativistic) forms of the kinetic energy.

These simple quark models have obvious limitations (e.g., large relativistic corrections can be anticipated for light quarks), but their success indicates that relativistic and field theoretical effects are somehow incorporated in the parameters. In so far as sophisticated approaches based on quantum chromodynamics (QCD) do not provide results for these systems, the application of constituent quark models is justified.

9.1 The trial function in the constituent quark model

The quark (q) carries space-, spin-, flavor-, and color degrees of freedom. The flavor is an extension of the isospin to include *up*(u), *down*(d), *strange* (s), *charm*(c), *bottom*(b), and *top*(t) quarks.

The baryons are colorless: The three quarks in a baryon form a color-singlet state, that is, they are totally antisymmetric with respect to the interchange of the color coordinates. For example, the eight baryons of spin 1/2, $N\,(p,n)$, Λ, $\Sigma\,(\Sigma^+,\Sigma^0,\Sigma^-)$, and $\Xi\,(\Xi^0,\Xi^-)$, belonging to a family of octet baryons, consist of three quarks of u, d and s flavor varieties. The product of the spin-, flavor-, and space-part of their wave functions has to be symmetric to comply with the generalized Pauli principle.

The spin part is represented by the spin functions $\chi_{(S_{12})SM_S}$ (see Complement 6.5):

$$\chi_{(1)\frac{3}{2}\frac{3}{2}} = |\uparrow\uparrow\uparrow\rangle, \qquad \chi_{(0)\frac{1}{2}\frac{1}{2}} = \frac{1}{\sqrt{2}}\Big(|\uparrow\downarrow\uparrow\rangle - |\downarrow\uparrow\uparrow\rangle\Big),$$

$$\chi_{(1)\frac{1}{2}\frac{1}{2}} = \frac{1}{\sqrt{6}}\Big(2|\uparrow\uparrow\downarrow\rangle - |\uparrow\downarrow\uparrow\rangle - |\downarrow\uparrow\uparrow\rangle\Big). \tag{9.1}$$

The flavor wave functions are shown in Table 9.1, where the baryons with the same charge are arranged in each column.

The trial wave function for baryons is a combination of the terms

$$\Psi = \mathcal{S}\Big\{\exp\Big(-\frac{1}{2}\tilde{x}Ax\Big)\chi_{(S_{12})SM}\xi_{(T_{12})TM_T}\Big\}, \tag{9.2}$$

where χ and ξ are the spin and flavor functions, and $\mathcal{S}$ is a symmetrizer.

9.2 One-gluon exchange model

The one-gluon exchange potential (OGEP) model has proved to be quite successful in describing the spectra of the ground states and various properties of the baryons [118, 119, 120]. In this model the

Table 9.1. Flavor wave functions of baryons arranged columnwise according to charge

Baryon	+2	+1	0	−1
N		uud	ddu	
Δ	uuu	uud	ddu	ddd
Λ			$\frac{1}{\sqrt{2}}(ud-du)s$	
Σ		uus	$\frac{1}{\sqrt{2}}(ud+du)s$	dds
Ξ			ssu	ssd
Ω				sss
Λ_c		$\frac{1}{\sqrt{2}}(ud-du)c$		
Σ_c	uuc	$\frac{1}{\sqrt{2}}(ud+du)c$	ddc	
Λ_b			$\frac{1}{\sqrt{2}}(ud-du)b$	
Σ_b		uub	$\frac{1}{\sqrt{2}}(ud+du)b$	ddb

quarks exchange massless but color-charged gluons and the qq potential is taken as the color analogue of the Breit interaction [3, 121]. Consequently, the simplest version of the OGEP includes a Coulomb-like $1/r$ piece and a color-magnetic ("hyperfine") piece:

$$\sum_{i<j} \frac{\pi}{6m_i m_j} (\lambda_i^C \cdot \lambda_j^C)(\boldsymbol{\sigma}_i \cdot \boldsymbol{\sigma}_j)\delta(\boldsymbol{r}_i - \boldsymbol{r}_j), \tag{9.3}$$

where the Gell-Mann matrix $\lambda_i^C(\alpha)$ ($\alpha = 1, ..., 8$) is the color SU(3) generator for the ith quark and $(\lambda_i^C \cdot \lambda_j^C)$ stands for a scalar product, $\sum_{\alpha=1}^{8} \lambda_i^C(\alpha)\lambda_j^C(\alpha)$. (In a more elaborated framework one can introduce spin-orbit, tensor and other higher order terms.) The quark masses m_i in the denominator introduce a very weak flavor dependence, but the interaction is essentially flavor-blind. In a nonperturbative treatment the contact interaction leads to a singularity. Because of this and because of the finite spatial extent of the constituent quarks the delta function should be smeared out, and it is to be substituted by a Gaussian or a Yukawa form factor.

Among the several different parametrizations of the OGEP we have chosen the AL1 potential [122] as an example:

$$V_{ij}(r) = -\frac{3}{8}(\lambda_i^C \cdot \lambda_j^C) \times \left(-\frac{\kappa}{r} + \lambda r - \Lambda + \frac{2\pi\kappa'}{3m_i m_j} \frac{\exp\left(-\frac{r^2}{\rho_{ij}^2}\right)}{\pi^{\frac{3}{2}}\rho_{ij}^3} \boldsymbol{\sigma}_i \cdot \boldsymbol{\sigma}_j \right), \tag{9.4}$$

where the length r is given in units of inverse energy (1 fm = 1/197.327 MeV^{-1}), and the smearing parameter ρ_{ij} is given in the form

$$\rho_{ij} = A\left(\frac{2m_i m_j}{m_i + m_j}\right)^{-B} . \tag{9.5}$$

The quark masses used in the calculation are $m_u = m_d = 315$ MeV, $m_s = 577$ MeV, $m_c = 1836$ MeV and $m_b = 5227$ MeV. The parameters were determined by a fit on a large sample of meson states in every flavor sector [122]: They are $\kappa = 0.5069$, $\kappa' = 1.8609$, $\lambda = 0.1653$ GeV^2, $\Lambda = 0.8321$ GeV, $B = 0.2204$ and $A = 1.6553$ GeV^{B-1}.

The matrix element of the color part can be easily obtained because the baryons are in a color singlet state. By noting the relationship between the color-exchange operator P_{ij}^C and $(\lambda_i^C \cdot \lambda_j^C)$,

$$P_{ij}^C = \frac{1}{2}(\lambda_i^C \cdot \lambda_j^C) + \frac{1}{3}, \tag{9.6}$$

the matrix element of $(\lambda_i^C \cdot \lambda_j^C)$ between the color-singlet states reduces simply to $-8/3$.

The SVM results [81, 123] are compared with the calculations of [122] in Table 9.2. The agreement between the two results is good. The experimental data [124] are included for guidance. One has to compare the energies relative to the nucleon ground state and one has to bear in mind that this potential gives good results for the mesons and no free parameter has been introduced to fit the baryons.

Table 9.2. The masses of baryons (in MeV) as predicted by the AL1 potential [122]. The numbers in parentheses are the masses relative to the nucleon mass.

Baryon	Ref. [122]	SVM	Experiment [124]
N	998	995	939
Δ		1306 (311)	1232 (293)
Λ	1154 (156)	1149 (154)	1116 (177)
Σ	1231 (233)	1228 (233)	1192 (253)
Ξ	1343 (345)	1339 (344)	1318 (379)
Ω	1674 (676)	1674 (679)	1672 (733)
Λ_c	2296 (1298)	2290 (1295)	2285 (1346)
Σ_c	2466 (1468)	2467 (1472)	2455 (1516)
Λ_b	5642 (4644)	5635 (4640)	5641±50 (4702±50)
Σ_b	5849 (4851)	5849 (4854)	
Ξ_b	5808 (4810)	5803 (4808)	

9.3 Meson-exchange model

The OGEP model is successful in describing ground-state energies and properties of baryons, but for the excited states a number of delicate problems remain unsolved. No model has been able to explain, for example, the correct level orderings in light- and strange-baryon spectra in a simple three-quark description of baryons. The source of the problem can be easily understood by the following argument: The ground state of the nucleon $N(939)$ has positive parity, followed by the positive parity $1/2^+$ Roper resonance $N(1440)$ and the negative-parity $1/2^-$ and $3/2^-$ states $N(1535)$ and $N(1520)$, respectively. The parity of the states follows as $(+,+,-,-)$. In the case of the Λ, the order of states follows a $(+,-,+,-)$ pattern. The order of parity of the spectra in a simple harmonic-oscillator quark model is $(+,-,+,-,\ldots)$. In the case of the linear confinement the level spacing changes, but the order of states remains the same. The wave function of N and Λ differs only in flavor, but the OGEP is flavor independent (though it slightly depends on flavor through the mass difference of light and strange quarks in Eq. (9.4)). Therefore it predicts the same level order for both N and Λ in contradiction to the experiments.

To resolve this problem an explicitly flavor-dependent meson-exchange interaction has been introduced [125]. The essential difference between the two approaches is that while the OGEP of type $\sum_{i<j}(\lambda_i^C \cdot \lambda_j^C)(\boldsymbol{\sigma}_i \cdot \boldsymbol{\sigma}_j)V(r_{ij})$ acts on the *color* and spin space, the meson-exchange potential $\sum_{i<j} \lambda_i^F \lambda_j^F (\boldsymbol{\sigma}_i \cdot \boldsymbol{\sigma}_j)V(r_{ij})$ acts on the *flavor* and the spin degrees of freedom.

The meson-exchange potential combined with a semirelativistic form of the kinetic energy is the second example of the application of the SVM for baryon spectroscopy. In this example no gluon-exchange mechanism is explicitly introduced to describe the qq interaction but only a meson-exchange potential is employed to generate the baryon spectra.

The three-quark Hamiltonian is a sum of the semirelativistic kinetic energy, the meson-exchange potential V and the confinement potential $V_0 + Cr$:

$$H = \sum_{i=1}^{3} \sqrt{\boldsymbol{p}_i^2 + m_i^2} + \sum_{j>i=1}^{3} \Big(V_{ij} + V_0 + C|\boldsymbol{r}_i - \boldsymbol{r}_j| \Big), \tag{9.7}$$

where the meson-exchange potential V reads as

$$V_{ij}(r) = \left\{ V_{ij}^{\pi}(r) \sum_{\alpha=1}^{3} \lambda_i^F(\alpha)\lambda_j^F(\alpha) + V_{ij}^{K}(r) \sum_{\alpha=4}^{7} \lambda_i^F(\alpha)\lambda_j^F(\alpha) \right.$$

$$\left. + V_{ij}^{\eta}(r)\lambda_i^F(8)\lambda_j^F(8) + \frac{2}{3} V_{ij}^{\eta'}(r) \right\} \boldsymbol{\sigma}_i \cdot \boldsymbol{\sigma}_j . \tag{9.8}$$

The form factor of the meson-exchange potential is parametrized as

$$V_{ij}^{\gamma}(r) = \frac{g_{\gamma}^2}{4\pi} \frac{1}{12 m_i m_j} \left\{ \mu_{\gamma}^2 \frac{e^{-\mu_{\gamma} r}}{r} - \Lambda_{\gamma}^2 \frac{e^{-\Lambda_{\gamma} r}}{r} \right\}, \tag{9.9}$$

where the coupling constant g_γ is taken to be g_8 for the pseudoscalar octet mesons (π, K, η) or g_0 for the pseudoscalar singlet meson (η'), respectively and the cut-off mass Λ_γ is assumed to be given by

$$\Lambda_\gamma = \Lambda_0 + \kappa \mu_\gamma \tag{9.10}$$

for each $\gamma = \pi$, K, η, η'. For the constituent quark masses we take the typical values, $m_u = m_d = 340$ MeV and $m_s = 500$ MeV. Parameters of the Hamiltonian are listed in Table 9.3.

Table 9.3. Parameters of the semirelativistic constituent quark model with meson-exchange potential

Fixed parameters						
Quark masses [MeV]		Meson masses [MeV]				
m_u, m_d	m_s	μ_π	μ_K	μ_η	$\mu_{\eta'}$	$\frac{g_8^2}{4\pi}$
340	500	139	494	547	958	0.67

Free parameters				
$(g_0/g_8)^2$	Λ_0 [fm^{-1}]	κ	V_0 [MeV]	C [fm^{-2}]
1.34	2.87	0.81	−416	2.33

The matrix elements of the semirelativistic kinetic energy in Eq. (9.7) can be calculated with the method presented in Sect. 7.6. The matrix elements of $\lambda_i^F(\alpha)\lambda_j^F(\alpha)$ in the flavor space can be easily calculated by using Table 9.1 and the explicit form of the Gell-Mann matrices. It is useful to know the action of the flavor-changing operator $\lambda_i^F(\alpha)\lambda_j^F(\alpha)$ on the flavor wave functions. As is summarized in Table 9.4, for example the flavor function $u_i d_j$, when acted upon by $\sum_{\alpha=1}^{3}\lambda_i^F(\alpha)\lambda_j^F(\alpha)$, transforms to $2d_i u_j - u_i d_j$.

Table 9.4. Effect of flavor-changing operators on quark pairs

Operator	uu	dd	ss	ud	us	ds
$\sum_{\alpha=1}^{3}\lambda_i^F(\alpha)\lambda_j^F(\alpha)$	uu	dd	0	$2du$-ud	0	0
$\sum_{\alpha=4}^{7}\lambda_i^F(\alpha)\lambda_j^F(\alpha)$	0	0	0	0	$2su$	$2\ sd$
$\lambda_i^F(8)\lambda_j^F(8)$	$\frac{1}{3}uu$	$\frac{1}{3}dd$	$\frac{4}{3}uu$	$\frac{1}{3}ud$	$-\frac{2}{3}us$	$-\frac{2}{3}ds$

In Fig. 9.1 the SVM predictions of this model are shown for all light- and strange-baryons with mass up to $M < 1850$ MeV; the nucleon mass is normalized to its mass of 939 MeV, which determines the value of the confinement potential parameter V_0. All masses corresponding to three- and four-star resonances in the most recent compilation of the Particle Data Group [124] are included.

The result is good, reproducing the spectra of all low-lying light and strange baryons [123, 126]. In particular, the level ordering of the lowest positive- and negative-parity states in the nucleon spectrum is reproduced correctly, with the $1/2^+$ Roper resonance $N(1440)$ falling well below the negative parity $1/2^-$ and $3/2^-$ states $N(1535)$ and $N(1520)$, respectively.

Likewise, in the Λ and Σ spectra the positive-parity $1/2^+$ excited baryons, $\Lambda(1600)$ and $\Sigma(1660)$, fall below the negative parity $1/2^-$–$3/2^-$ states, $\Lambda(1670)$–$\Lambda(1690)$, and the $1/2^-$ state $\Sigma(1750)$, respectively. In the Λ spectrum, at the same time, the negative-parity $1/2^-$–$3/2^-$ states $\Lambda(1405)$–$\Lambda(1520)$ remain the lowest excitations above the Λ ground state.

At this place, a remark is in order about the necessity of employing a semirelativistic kinetic energy operator in the three-quark Hamiltonian (9.7). Certainly, this is only an intermediate step toward a fully covariant treatment but it already allows us to include the kinematical relativistic effects. In any nonrelativistic approach these effects get compensated by the potential parameters, but there one faces a disturbing consequence of $v/c > 1$, where v is the mean velocity of the constituent quark and c is the velocity of light.

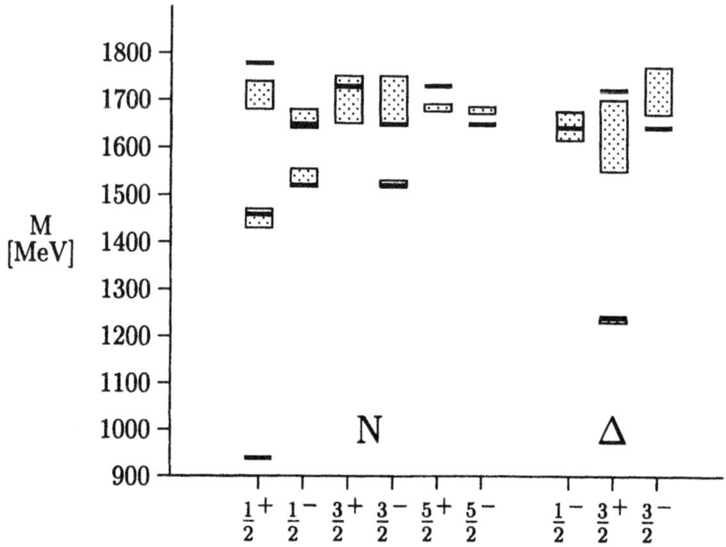

Fig. 9.1. Comparison of low-lying baryon spectra predicted by the meson-exchange potential model with experiment. The solid lines denote the calculated spectra, and the shaded boxes show the experimental masses with their error bars.

(Continued on the next page.)

One may think that the idea of the meson-exchange interaction acting between quarks is unfamiliar though it has produced the good results. To judge its merits the present calculations have to be extended into several directions (prediction of physical observables, e.g. form factors, decay widths, etc.).

The complexity of the NN interaction is basically due to the compositeness of the nucleon. As N, Λ or Σ belong to the same octet family, it is natural to attempt a unified description of the NN, ΛN or ΣN interaction from the underlying qq potentials. This goal is at present too difficult to achieve directly on the basis of QCD. In so far as QCD as yet presents no results, the effective theory formulated in a microscopic quark-cluster model [127] seems to be justified.

(Continued from Fig 9.1.)

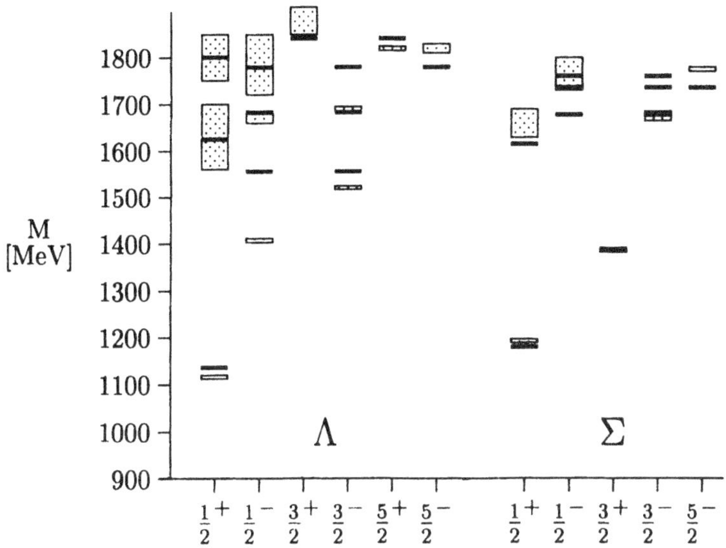

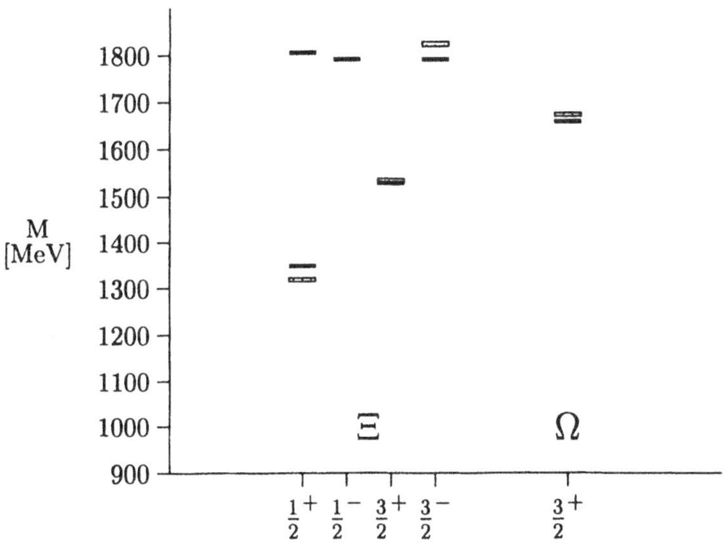

The intermediate-range attraction of the baryon-baryon interactions cannot be accounted for by the OGEP because it produces no contribution between the colorless baryons. It is not clear yet whether the meson-exchange potential introduced above leads to a realistic description of the NN interaction or not. It is probable that the OGEP combined with the meson-exchange potential is a useful, effective vehicle to obtain a reasonable description of both baryon spectra and baryon-baryon interactions at the same time. See, for example, [128] for the attempt along this line.

10. Few-body problems in solid state physics

This chapter is a collection of examples of the application of the SVM to few-body problems in the field of solid state physics. The examples include excitonic complexes, biexcitons, and quantum dots with and without magnetic field.

The biexciton (or excitonic molecules), the bound state of two holes and two electrons, provides an interesting few-body problem, since it may be thought as a positronium or a hydrogen molecule with variable electron and hole masses realized in semiconductors. The biexcitons are also often confined and can be modelled as two-dimensional (2D) systems.

In recent years there has been much excitement in the possible applications of ultrasmall systems with a length scale of 10-100 Å("nanostructures"). The technological motivation (in electronics and optoelectronics) is that the smaller components are faster. It is possible to manufacture very small nanostructures, often called "dots", which contain only a very few electrons ($N < 10$). The quantum-mechanical effects in these small systems are very important and their properties are strongly N-dependent. The energy spectra of few-particle quantum dots call for theoretical interpretation. In a nanostructure the electrons are confined in a box of sizes L_x, L_y, and L_z. If $L_x \sim L_y >> L_z$ the quantum dot is quasi 2D. The confinement is usually modelled by a harmonic-oscillator potential. This system is somewhat similar to an atom in nature, where the "confining" potential is the Coulomb attraction of the electrons to a nucleus in the atom, and therefore the quantum dots are often referred to as "artificial atoms".

Throughout this chapter we use an effective mass m_e^* for an electron. When the dielectric constant of a material is denoted by κ, the length and the energy will be measured in "atomic units" with Bohr radius $a^* = \hbar^2\kappa/(m_e^* e^2)$ and hartree (two times the Rydberg energy) $2R^* = e^2/(\kappa a^*)$, respectively.

10.1 Excitonic complexes

The excitonic complexes have considerable importance in the development of semiconductor physics and spectroscopy. The bound state of a positively charged hole with an effective mass m_h^* and an electron with an effective mass m_e^* is called an exciton in semiconductor physics. The mass ratio $\sigma = m_e^*/m_h^*$ can change between $\sigma = 0$ (hydrogenic limit) and $\sigma = 1$ (positronium limit) depending on the material and other factors. Like in the case of hydrogen, where not only the hydrogen atom but the hydrogen molecule (H_2), or the H_2^+, H^-, H_3^+ ions are bound, the system of N_e electrons and N_h holes can also be bound. The latter system is called an excitonic complex. These excitonic complexes, including the charged excitons, have been subject of intensive experimental [105, 129] and theoretical [99, 130–135] investigation. The properties and structure of these systems strongly change with the mass ratio, and, by approaching the two limiting cases, one arrives at two completely different worlds. The interest in these systems has been intensified when the advance in semiconductor technology has made possible the fabrication of artificial nanostructures with diameters comparable to atomic distances. This restricted geometry has a prominent effect on the dynamics of the excitonic species.

In the following we present the ground-state energy and some other properties of 2D excitonic complexes. The general Hamiltonian can be written as

$$H = \sum_{i=1}^{N_e} \frac{\boldsymbol{p}_i^2}{2m_e^*} + \sum_{i=N_e+1}^{N} \frac{\boldsymbol{p}_i^2}{2m_h^*} - T_{\rm cm} + \sum_{j>i=1}^{N_e} \frac{e^2}{\kappa|\boldsymbol{r}_i - \boldsymbol{r}_j|}$$
$$+ \sum_{j>i=N_e+1}^{N} \frac{e^2}{\kappa|\boldsymbol{r}_i - \boldsymbol{r}_j|} - \sum_{i=1}^{N_e} \sum_{j=N_e+1}^{N} \frac{e^2}{\kappa|\boldsymbol{r}_i - \boldsymbol{r}_j|}, \tag{10.1}$$

where κ is the dielectric constant of the material, and the position vectors are 2D vectors in the xy plane. In atomic units introduced in this chapter, the Hamiltonian becomes dimensionless and the energy depends only on σ. (Note that we do not need to specify what value of κ is used because the κ dependence is hidden in the atomic units.) This is easily understood by introducing dimensionless coordinates and momenta by $\boldsymbol{r}_i^* = \boldsymbol{r}_i/a^*$ and $\boldsymbol{p}_i^* = \boldsymbol{p}_i/(\hbar/a^*)$. Then the Hamiltonian (10.1) is reduced to

$$\frac{H}{2R^*} = \sum_{i=1}^{N_e} \frac{1}{2} \boldsymbol{p}_i^{*2} + \sum_{i=N_e+1}^{N} \frac{\sigma}{2} \boldsymbol{p}_i^{*2} - \frac{\sigma}{2(\sigma N_e + N - N_e)} \left(\sum_{i=1}^{N} \boldsymbol{p}_i^* \right)^2$$

$$+ \sum_{j>i=1}^{N_e} \frac{1}{|\boldsymbol{r}_i^* - \boldsymbol{r}_j^*|} + \sum_{j>i=N_e+1}^{N} \frac{1}{|\boldsymbol{r}_i^* - \boldsymbol{r}_j^*|} - \sum_{i=1}^{N_e} \sum_{j=N_e+1}^{N} \frac{1}{|\boldsymbol{r}_i^* - \boldsymbol{r}_j^*|}.$$

At this point, a comment is due. What we consider here is a well-defined quantum-mechanical problem to be solved. This is, however, only a model of the experimental situation. In the realistic case there are many other factors to be considered, for example the effect of other electronic bands, the possibility that the interaction is different from the pure Coulomb force due to the confinement in the z direction, etc. The discussion of the importance of these effects is beyond the scope of this book.

As the problem here is practically the same as that discussed in the chapter of small atoms and molecules (Chap. 8), the same trial function is used (see Eq. (8.1)), except that it is tailored to the 2D case, that is, the position vectors have only x and y components.

The results for $\sigma = 0$ and $\sigma = 1$ are shown in Table 10.1 for the 3D case and in Table 10.2 for the 2D case. The most striking difference between the 3D and 2D results is the large increase in the binding energy in the 2D case. E.g., the binding energy of H_2 (with $\sigma = 0$) in 2D is about 8 times of that in 3D. The increase of binding has been found experimentally as well.

Table 10.1. Energies and binding energies (in a.u.) of 3D exciton complexes of electrons and holes for two cases of the mass ratio σ. L and S are the total orbital angular momentum and spin of the exciton complex, respectively. The asterisk refers to the states which are found to be unbound. The binding energy is understood with respect to the nearest threshold.

System	(L,S)	$E\,(\sigma=0)$	$E\,(\sigma=1)$	$B\,(\sigma=0)$	$B\,(\sigma=1)$
eh	$(0,0)$	−0.500	−0.250	0.500	0.250
eeh	$(0,1/2)$	−0.528	−0.262	0.028	0.012
ehh	$(0,1/2)$	−0.602	−0.262	0.102	0.012
eehh	$(0,0)$	−1.174	−0.516	0.174	0.016
eeehh	$(0,1/2)$	*	*	*	*
eehhh	$(0,1/2)$	−1.344	*	0.169	*

Table 10.2. Energies and binding energies (in a.u.) of 2D exciton complexes of electrons and holes for two cases of the mass ratio σ. M is the z component of the total orbital angular momentum and S is the total spin. See also the caption of Table 10.1.

System	(M, S)	$E\,(\sigma=0)$	$E\,(\sigma=1)$	$B\,(\sigma=0)$	$B\,(\sigma=1)$
eh	(0, 0)	−2.00	−1.00	2.00	1.00
eeh	(0, 1/2)	−2.24	−1.12	0.24	0.12
ehh	(0, 1/2)	−2.82	−1.12	0.82	0.12
eehh	(0, 0)	−5.33	−2.19	1.33	0.19
eeehh	(0, 1/2)	*	*	*	*
eehhh	(0, 1/2)	−6.82	*	1.50	*

The average distances between the constituents are compared in Table 10.3, where a large range of σ values covers both adiabatic and nonadiabatic extremes. One needs high accuracy because these expectation values are very sensitive to the wave function. The SVM reaches this result with no bias, that is, without assuming any dominant configuration or geometrical arrangement. In Table 10.3 the first thing to be noticed is that the average distances between the particles are smaller by about 4 times in the 2D case than in the 3D case for $\sigma = 1$. As a comparison, note that the average distance between the positron and the electron in the Ps atom is 3 a.u. in 3D and 1 a.u. in 2D. The expectation values $\langle r \rangle$ and $\langle r^2 \rangle$ decrease with decreasing σ. The average distance between negative-negative and positive-positive charges, $\langle r_{--} \rangle$ and $\langle r_{++} \rangle$, are different in the hydrogenic limit ($\sigma = 0$), and gradually become the same in the limit of equal mass ($\sigma = 1$). Naturally, in order to maximize the Coulomb attraction, the distance between the positive and negative charges is smaller than that between the particles with identical charges. This is what we see in Table 10.3 except around $\sigma = 0$. Around the hydrogenic limit the average distance between positive charges becomes smaller than that between the opposite ones. This indicates that the binding mechanism has been changed. It would be very interesting to get a picture of the geometrical arrangement of these 2D systems. Our result that the relative distances between pairs of identical charges, $\langle r_{--} \rangle$ and $\langle r_{++} \rangle$, are not equal contradicts the basic assumption of [135], where the authors assume that the constituents of the biexciton form a square. This does not exclude, however, the possibility of the formation of a rhombus, for example. To have a rhombus-shaped arrangement, the relative distances have to fullfil the equality $2\langle r_{+-}\rangle = \sqrt{\langle r_{--}\rangle^2 + \langle r_{++}\rangle^2}$. Table 10.3 shows that this is not satisfied. Moreover, the fact that the uncertainties $\Delta r_{ij} = \sqrt{\langle r_{ij}^2\rangle - \langle r_{ij}\rangle^2}$ are large means that no such interpretation is possible. There is only one case where the uncertainty is very small, that is the distance between the (heavy) positive charges in the hydrogenic limit. In that case one can assume, in accordance with the adiabatic approximation, that the positive charges are fixed at the distance of 0.37 a.u. The equilibrium distance in the H_2 molecule in 3D is 1.4 a.u.

Table 10.3. Average distances in 2D biexcitons (*eehh*) as a function of the mass ratio σ. E.g., $\langle r_{--}\rangle$ is the mean distance between the two negative charges. Atomic units are used.

	$\sigma = 0$	$\sigma = 0.4$	$\sigma = 0.7$	$\sigma = 1$
$\langle r_{--}\rangle$	0.67	1.26	1.55	1.80
$\langle r_{++}\rangle$	0.37	1.14	1.49	1.80
$\langle r_{+-}\rangle$	0.47	0.93	1.16	1.38
$\langle r_{--}^2\rangle$	0.59	2.09	3.19	4.33
$\langle r_{++}^2\rangle$	0.14	1.69	2.95	4.33
$\langle r_{+-}^2\rangle$	0.31	1.29	2.05	2.88

10.2 Quantum dots

Rapid advances in semiconductor technology have led to the fabrication of nanostructures called quantum dots or artificial atoms [136, 137, 138]. In quantum dots a few electrons, typically 2 to 200 electrons, are bound at semiconductor interfaces. These few-electron systems arise when homogeneous two-dimensional (2D) electron gases of heterojunctions or metal-oxide-semiconductor (MOS) structures are laterally confined to diameters comparable to the effective Bohr radius of the host semiconductor. The interest in quantum dots arises not only from the prospect of new technological applications but also from the desire to understand the physics of a few interacting electrons in an external field. Needless to say, we concentrate on the few-electron case where the effect of electron-electron interaction and their correlation seems to be very important.

The artificial atoms we consider here are modelled by a system of N electrons confined in 2D by a harmonic-oscillator potential. Before starting with artificial atoms in 2D we present a calculation for the energies of 2D "natural" atoms where the "confining" potential is Coulombic. The results for few-electron atoms and ions are given in Table 10.4. The binding energies are again, due to the 2D geometry, much larger than in the 3D case. Otherwise the properties of these systems are rather similar to those of their 3D counterparts. For example, akin to 3D, the H, Li or Be atom can bind an extra electron, while the He atom cannot.

Table 10.4. Energies of atoms and ions confined in 2D. M is the z component of the total orbital angular momentum and S is the total spin. Atomic units are used.

System	(M, S)	Energy
H	$(0,0)$	-2.00000000
H^-	$(0,1/2)$	-2.24
He	$(0,0)$	-11.89
Li	$(0,1/2)$	-29.87
Be	$(0,0)$	-56.77

The example of quantum dots considered here is a system of 2D electrons in a quadratic confining potential. The Hamiltonian reads as

$$H = \sum_{i=1}^{N} \frac{\boldsymbol{p}_i^2}{2m^*} + \sum_{i=1}^{N} \frac{1}{2} m^* \omega_0^2 \boldsymbol{r}_i^2 + \sum_{j>i=1}^{N} \frac{e^2}{\kappa |\boldsymbol{r}_i - \boldsymbol{r}_j|}. \tag{10.2}$$

(Here and in the following section we will deal with electrons only and for simplicity the effective mass m_e^* will be abbreviated as m^*. Also, to remind the reader that we are in 2D and only the z component of the angular momentum plays a role, we use "M" for the total orbital angular momentum and "m" for the single-particle one.) The basis function with the total orbital angular momentum M , in the spirit of Sect. 7.3, is taken as

$$(v_x + iv_y)^M \exp\Big(-\frac{1}{2} \tilde{\boldsymbol{r}} A \boldsymbol{r} \Big), \tag{10.3}$$

where v_x and v_y are the x and y components of the 2D vector $\boldsymbol{v}$ defined by

$$v = \sum_{i}^{N} u_i \boldsymbol{r}_i. \tag{10.4}$$

The parameters u_i and A of the basis function are determined by the SVM.

Due to the external field (the single-particle potential), the Hamiltonian (10.2) of the system is not translationally invariant. Correspondingly, the basis function (10.3) is written in terms of single-particle coordinates instead of relative ones. This basis function depends on the center-of-mass coordinate, and the energy obtained will be the total energy of the system, unlike the previous cases where the energy of the center-of-mass motion was always subtracted. The case of the quadratic confinement is very special, because in that case the energy of the center-of-mass can be separated. In general, for an arbitrary single-particle potential, however, this cannot be done.

The case of $N = 1$ electron is trivial and its solution is a simple harmonic oscillator. For $N = 2$ an analytical solution exists for certain frequencies [139]. For example, for $\hbar\omega_0 = 1$ a.u., the ground state energy is 3 a.u., which is precisely reproduced by the numerical calculation. Not surprisingly, as is shown in Table 10.5, if the quadratic confinement is strong, the spectrum is close to that of a harmonic oscillator. In the realistic cases ($\hbar\omega_0 < 1$ a.u.), however, the eigenvalues strongly deviate from those of the harmonic oscillator. This example illustrates the importance of these systems: by changing the confinement one can "tune" the properties of these artificial atoms.

Table 10.5. Energies of N-electron quantum dots with the z component of the total orbital angular momentum M and the total spin S. The numbers in parentheses are the eigenvalues without Coulomb interaction. Atomic units are used.

N	(M,S)	$\hbar\omega_0 = 1$	$\hbar\omega_0 = 100$
2	(0,0)	3.000 (2.000)	212.198 (200.00)
3	(0,1/2)	7.220 (5.000)	525.12 (500.00)
4	(0,0)	10.600 (6.000)	654.45 (600.00)

Figure 10.1 shows the energies of the ground and excited states (in units of the energy of a harmonic-oscillator quantum) of two electrons as a function of the inverse square root of the oscillator constant, $(\hbar\omega_0)^{-1/2}$. If there would be no Coulomb interaction between the elec-

trons, the energy divided by $\hbar\omega_0$ would be given by lines parallel to the horizontal axis. The deviation from the straight lines is entirely due to the Coulomb interaction. Another interesting feature is that level crossings occur between the excited states but the ground state always remains $M = 0$.

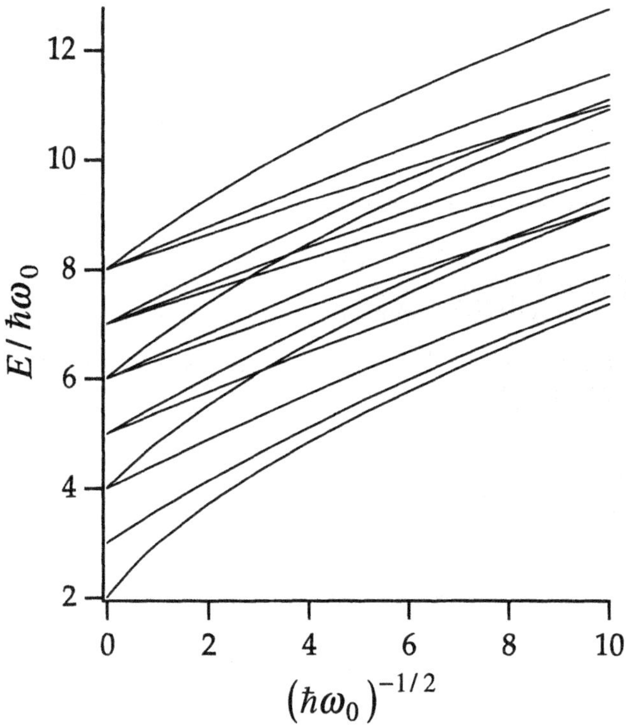

Fig. 10.1. Energy levels of two electrons confined by a harmonic-oscillator potential with an oscillator constant ω_0. Atomic units are used.

Figure 10.2 shows the radial density of the electrons for different confinement strengths. The structure of the systems is determined by the competition between the Coulomb repulsion and the strength of the confinement. For large ω_0 the electrons are close to the origin. For small ω_0 the maximum density of electrons moves out from the origin. In this case the repulsion between the electrons becomes dominant, and the distance between them becomes maximal. As is shown in Fig 10.3, the electrons practically seem to move along a ring in that case.

The change of density with an increasing number of electrons is illustrated in Fig. 10.3 in the case of a weak confining potential. The density is normalized to the number of electrons. The radius of the

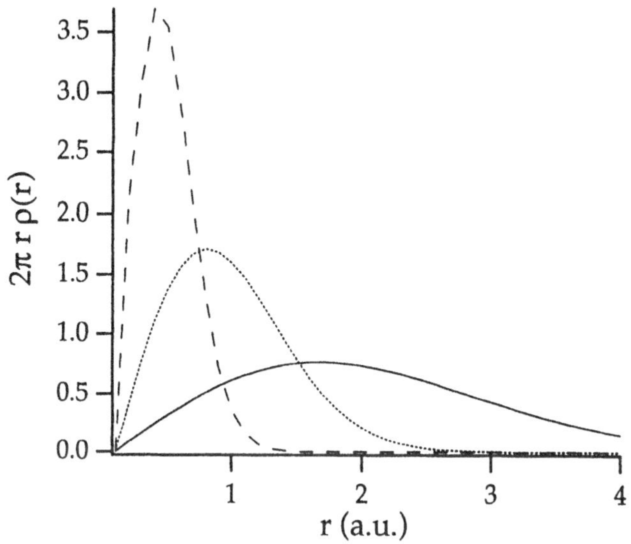

Fig. 10.2. Electron densities of two-electron quantum dots as a function of the radial distance from the origin. The solid, the dotted and the dashed curves correspond to $\hbar\omega_0 = 0.2$, $\hbar\omega_0 = 2$, and $\hbar\omega_0 = 5$, respectively. Atomic units are used.

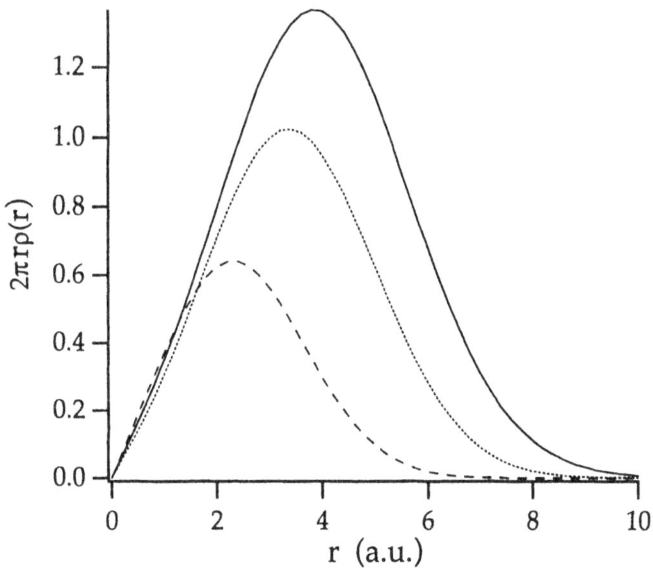

Fig. 10.3. Electron densities of two- (dashed curve), four- (dotted curve), and six-electron (solid curve) quantum dots. The oscillator frequency of the quadratic confining potential is $\hbar\omega_0 = 0.2$. Atomic units are used.

peak density inreases with the number of electrons, that is, the equilibrium configuration is realized on rings of expanding diameter. In the case of N=4 electrons, for example, the particles may move along a ring, while they are situated at the vertices of a square to minimize the Coulomb repulsion.

10.3 Quantum dots in magnetic field

A dramatic feature arises when the quantum dots are subjected to magnetic fields: The ground states are stabilized into magic-number states. When the magnitude of the magnetic field is varied, the ground state jumps from one magic number state to another. Magic numbers appear owing to the combined effect of the electron correlation and the Pauli principle [140].

The Hamiltonian of the system is given by

$$H = \sum_{i=1}^{N} \frac{1}{2m^*}(\boldsymbol{p}_i + \frac{e}{c}\boldsymbol{A}_i)^2 + \sum_{i=1}^{N} \frac{1}{2} m^* \omega_0^2 \boldsymbol{r}_i^2$$
$$+ \sum_{j>i=1}^{N} \frac{e^2}{\kappa |\boldsymbol{r}_i - \boldsymbol{r}_j|}. \tag{10.5}$$

Let us assume that a homogeneous magnetic field B is applied downward in the z direction. The above Hamiltonian can then be rewritten in the form

$$H = \sum_{i=1}^{N} \frac{\boldsymbol{p}_i^2}{2m^*} + \sum_{i=1}^{N} \frac{1}{2} m^* \omega^2 \boldsymbol{r}_i^2 - \frac{1}{2}\hbar\omega_c \sum_{i=1}^{N} l_{iz}$$
$$+ \sum_{j>i=1}^{N} \frac{e^2}{\kappa |\boldsymbol{r}_i - \boldsymbol{r}_j|}, \tag{10.6}$$

where $\omega_c = eB/(m^*c)$ is the cyclotron frequency and $\omega = \sqrt{\omega_c^2/4 + \omega_0^2}$. We ignored the magnetic interaction energy due to the electron spin. See Complement 10.1 for details. It is shown there that the part of the Hamiltonian corresponding to the single-particle motion can easily be diagonalized by the use of associated Laguerre polynomials. Depending on the strength of the magnetic field, both the single-particle motion and the correlation due to the Coulomb repulsion play decisive roles in determining the structure of the quantum dots. The trial function is the same as in the previous section.

For the two-electron case our calculation is in perfect agreement with the analytical solution given in [141]. The spectrum of two electrons in the spin-singlet state ($S = 0$) is shown in Fig. 10.4. See also Complement 10.1. The harmonic-oscillator frequency ω_0 is taken as $\hbar\omega_0 = 0.01$ a.u. The most intriguing phenomenon seen is the level crossing. Due to these level crossings, the ground state changes with the strength of the magnetic field. This is shown in a magnified scale on the right-hand side of Fig. 10.4. If there is no Coulomb interaction between the electrons, no level crossing occurs.

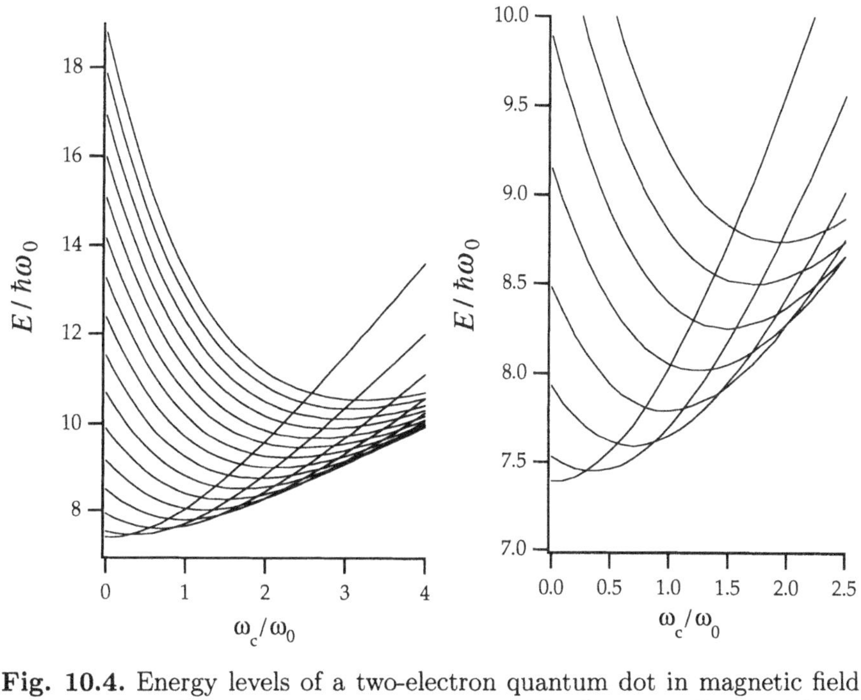

Fig. 10.4. Energy levels of a two-electron quantum dot in magnetic field as a function of the cyclotron frequency ω_c. The figure on the right-hand side magnifies the level crossing. The oscillator frequency of the quadratic confining potential is $\hbar\omega_0 = 0.01$. The total spin is $S = 0$. Atomic units are used.

Figures 10.5 and 10.6 plot the energy levels of a three-electron quantum dot (with the total spin being $S = 3/2$) against the orbital angular momentum M at magnetic fields that are typical to illustrate low and high-field behavior. In the case of low magnetic field the orbital angular momentum of the ground state is M=3. There are several sets of broadened levels. In the limit of vanishing confining potential these would be the lowest Landau levels. The general trend is that the energies increase with M because, as is shown in Complement 10.1, the single-electron energies increase with m. In the case of the low magnetic field, the ground state has $M = 3$.

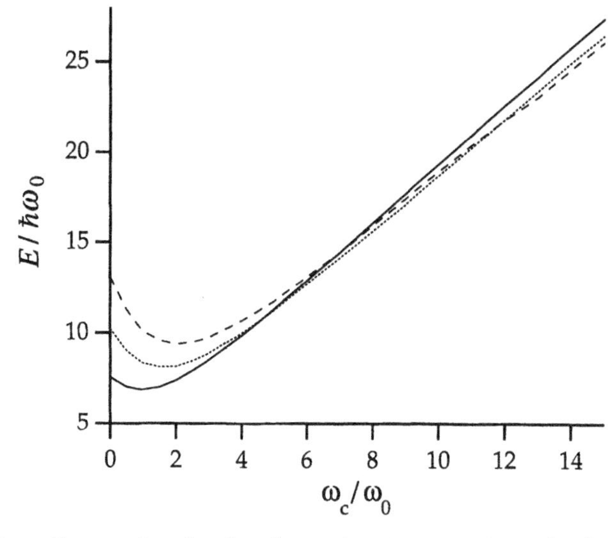

Fig. 10.5. Energy levels of a three-electron quantum dot belonging to the z component of the total orbital angular momentum $M = 3$ (solid curve), $M = 6$ (dotted curve), and $M = 9$ (dashed curve) orbital motion. The frequency of the quadratic confining potential is $\hbar\omega_0 = 1$. The total spin is $S = 3/2$. Atomic units are used.

In the case of high magnetic field the gap betwen the Landau levels is much bigger. See Fig. 10.9. The increase of energy with the orbital angular momentum is very small in this case because the energy mainly comes from the effect of the magnetic field. The ground state has M=6 as is shown in Fig. 10.7.

The single-particle energies in higher angular momentum orbits are larger because these states are subjected to stronger confinement. At the same time, the Coulomb interaction is smaller because the radii

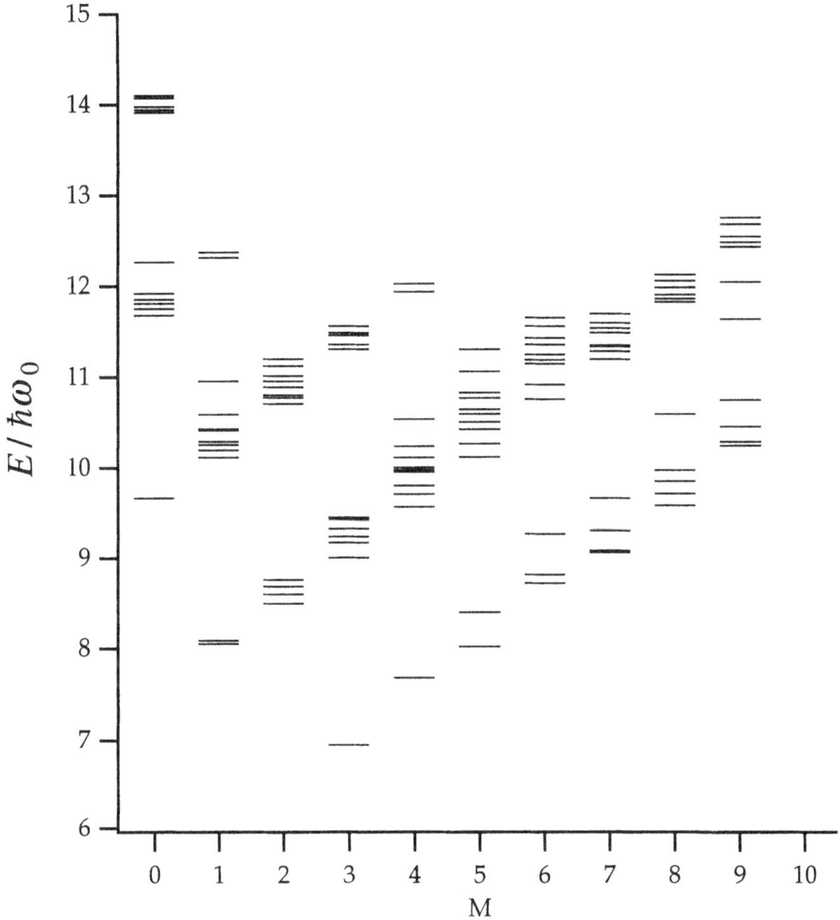

Fig. 10.6. Energy levels as a function of the z component of the total orbital angular momentum, M, of a three-electron quantum dot in a low magnetic field ($\hbar\omega_c = 0.2$, $\hbar\omega_0 = 1$). The total spin is $S = 3/2$. Atomic units are used.

of these orbits are larger (see Eq. (10.16)). The energy of the dots is determined by the interplay of the single-particle energies and the interaction energy. By increasing the strength of the magnetic field, similarly to the case of low magnetic field, the ground state does not take all possible values of M. By changing the magnetic field the states with $M = 3, 6, 9, 12, ...$ "magic numbers' become ground states.

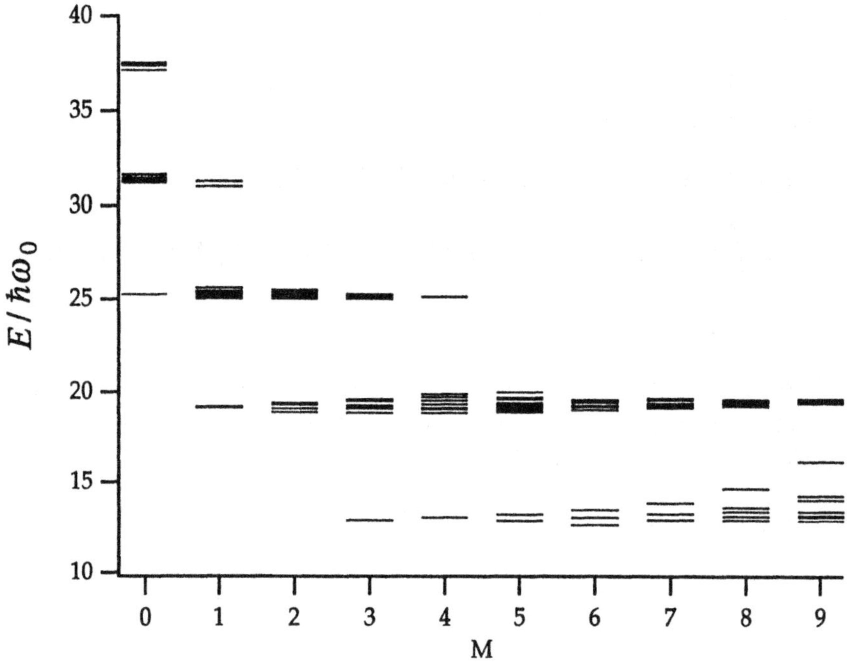

Fig. 10.7. The same as in Fig. 10.6 but for high magnetic field ($\hbar\omega_c = 6$, $\hbar\omega_0 = 1$). Atomic units are used.

The change of the ground-state angular momentum quantum numbers as a function of the magnetic-field strength is illustrated in Fig 10.5 for a three-electron quantum dot with spin $S = 3/2$. This figure shows the energies of the states with $M = 3$ (solid curve), $M = 6$ (dotted) and $M = 9$ (dashed) orbital angular momentum. The calculation shows that by changing the strength of the magnetic field the orbital angular momentum of the ground state of this system is $M = 3$ for $\omega_c/\omega \leq 5$, it is $M = 6$ for $5 \leq \omega_c/\omega \leq 11$, and then it switches over to $M = 9$, reproducing the "magic number" sequence

($M = 3, 6, 9, ..$). States with other orbital angular momentum never become the ground state and are not included in the figure.

The previous examples dealt with spin polarized electrons ($S = 3/2$). Actually, spin polarized ($S = 3/2$) and spin unpolarized ($S = 1/2$) states compete to be the ground state as the magnetic field changes. This very interesting phenomenon is illustrated in Fig. 10.8. As the magnetic field increases, the ground state changes as $(M, S) = (1, 1/2) \rightarrow (2, 1/2) \rightarrow (3, 3/2)$ and so on. At low magnetic field the lowest energy state is spin unpolarized ($S = 1/2$) and as the magnetic field increases the spin becomes polarized. The orbital angular momentum (M) changes continuously (in steps of unity).

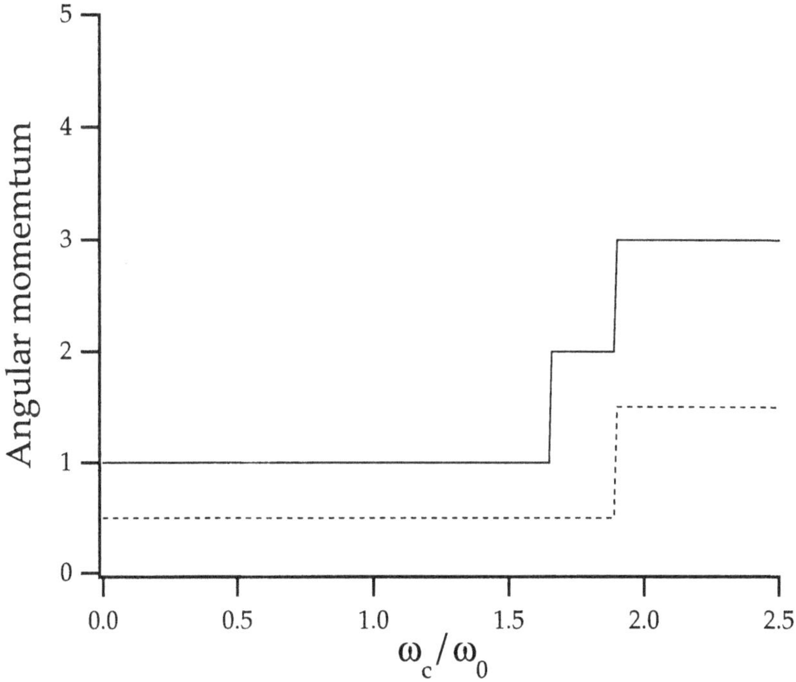

Fig. 10.8. Spin (S) and z component of the total orbital angular momentum (M) quantum numbers of the ground state of a three-electron quantum dot as a function of magnetic field strength. $\hbar\omega_0 = 1$. Solid and dotted lines denote M and S, respectively. Atomic units are used.

10.4 Quantum dots in the generator coordinate method

In Sects. 10.2 and 10.3 the correlated Gaussian basis of Eq. (10.3) is employed to obtain the solution of the quantum dots. However, this is not the only Gaussian basis function for the quantum dots. In fact, the quantum dots offer a very nice application of another basis function (8.9) that was used for the study of small molecules. That basis function is characterized by the matrix A which describes the correlated motion of the particles and also by the generator coordinates $\boldsymbol{s}$ which allow us to represent several peaks of the density distribution of the system.

In this section we consider the quantum dots in 3D and apply no magnetic field. The ground state of the quantum dot can belong to nonzero orbital angular momentum (for example, in the $N = 3$-electron case). The application of the basis function (8.9) gives a very simple way to obtain accurate energies of the dots even with nonzero orbital angular momentum, without the hassle of partial-wave expansion or angular momentum algebra. As this basis function has no definite orbital angular momentum, in principle one has to project out the component with good angular momentum quantum numbers. In practice, however, the converged wave function already belongs to the correct ground-state quantum numbers and the energy gain due to an orbital angular momentum projection is negligible. As the Hamiltonian and the $\boldsymbol{L}^2$ operator commute, the optimization of the wave function filters out the eigenstate of $\boldsymbol{L}^2$ as well. As was mentioned in Sect. 6.1, this type of calculation is called a variation before projection. In the limit that the variational trial function is chosen to be flexible enough, the solution of this type of calculation would reach the eigenstate of the conserved quantity like $\boldsymbol{L}^2$ to a good approximation.

In Table 10.6 the results of the SVM with the basis function of Eq. (8.9) is compared to that of a large scale shell-model calculation. The shell model works extremely well in this case because the confining potential is a harmonic-oscillator well. By choosing the single-particle wave functions to be the eigenfunctions of this single-particle potential, the diagonalization of the Hamiltonian with the relatively weak Coulomb interaction between the electrons is simple and gives an accurate solution. The results of the two methods are in perfect agreement. This agreement shows that the optimization of the basis of Eq. (8.9) finds the correct ground state automatically even for the case with nonzero orbital angular momentum provided that the basis size is sufficiently large. An orbital angular momentum projection may lead to a faster convergence, but as it can only be carried out by a three-dimensional numerical integration, it would prohibitively slow down the calculation.

Table 10.6. Energies of few-electron quantum dots in 3D calculated by SVM and by a large scale shell-model (SM) basis[a]. L is the total orbital angular momentum, S the total spin, and π is the parity. K is the basis dimension. $\hbar\omega_0 = 0.5$. Atomic units are used.

$N\ (L, S, \pi)$	E (SVM)	K	E (SM)	SM space (in $\hbar\omega_0$)
$3\ (1, 1/2, -)$	4.01324	100	4.01324	39 $\hbar\omega_0$
$4\ (0, 0, +)$	6.35025	300	6.3506	14 $\hbar\omega_0$
$5\ (1, 1/2, -)$	9.00331	500	9.0032	6 $\hbar\omega_0$

[a] P. Navratil, private communication.

Complements

10.1 *Two-dimensional electron motion in a magnetic field*

In the previous section we have presented a variational solution to the few-body problems of quantum dots. In this approach a correlated Gaussian basis has been used and the problem has been formulated in a relative coordinate framework. The examples from atomic physics (Chap. 8) convincingly prove that this is a very powerful way to cope with the correlation between the electrons in the field of an attractive center. This is, however, not the only possible approach. One can, alternatively, use a basis set made up from products of single-particle states to diagonalize the Hamiltonian. The properties of these single-particle states are discussed in this Complement.

The Hamiltonian of interest is given by

$$H = \sum_i \left\{ \frac{1}{2m^*}\left(\boldsymbol{p}_i + \frac{e}{c}\boldsymbol{A}_i\right)^2 + g^*\mu_{\rm B}\boldsymbol{B}\cdot\frac{1}{2}\boldsymbol{\sigma}_i \right\} + \sum_i V(\boldsymbol{r}_i) + \sum_{j>i} \frac{e^2}{\kappa|\boldsymbol{r}_i - \boldsymbol{r}_j|}, \tag{10.7}$$

where m^* is the effective electron mass, $\mu_{\rm B} = e\hbar/(2m_e c)$ the Bohr magneton of the electron, g^* the effective g-factor of the electron, and κ is the dielectric constant of the bulk material. The magnetic interaction energy due to the electron spin is included in the Hamiltonian. The magnetic field is directed downward in the z direction, i.e. $\boldsymbol{B} = (0, 0, -B)$ with $B > 0$. For high magnetic field of, e.g. $B = 10$ T, the value of $\mu_{\rm B}B$ becomes 0.5788 meV. The vectors $\boldsymbol{r}_i$ and $\boldsymbol{p}_i$ are 2D vectors, which have x and y components. The one-body potential $V(\boldsymbol{r})$ is to confine the quantum dots and is chosen to be quadratic, $m^*\omega_0^2\boldsymbol{r}^2/2$. A typical value of the confinement energy is $\hbar\omega_0 \sim 5$ meV. By taking a symmetric gauge vector potential, $\boldsymbol{A} = -\boldsymbol{r} \times \boldsymbol{B}/2 = B(y/2, -x/2, 0)$, we obtain

$$\left(\boldsymbol{p} + \frac{e}{c}\boldsymbol{A}\right)^2 = -\hbar^2\Delta + \frac{e^2B^2}{4c^2}\boldsymbol{r}^2 - \frac{e\hbar B}{c}l_z, \tag{10.8}$$

where $\Delta = \frac{\partial^2}{\partial x^2} + \frac{\partial^2}{\partial y^2}$ and l_z is the z component of the angular momentum.

To express the Hamiltonian in terms of dimensionless variables, we introduce

$$\omega_c = \frac{eB}{m^*c}, \quad \omega = \sqrt{\frac{1}{4}\omega_c{}^2 + \omega_0^2}, \quad \text{and} \quad \rho = \sqrt{\frac{\hbar}{m^*\omega}}. \tag{10.9}$$

Here ω_c is the cyclotron frequency and ρ is the magnetic length. (Note that the magnetic length is expressed by $\rho = \sqrt{2R^*/\hbar\omega}a^*$ using Bohr radius a^* and hartree $2R^*$.) Then the Hamiltonian is expressed by the dimensionless variable $\boldsymbol{r}_i$:

$$H = \hbar\omega \sum_i \left\{ -\frac{1}{2}\Delta_i + \frac{1}{2}\boldsymbol{r}_i{}^2 \right\} - \frac{1}{2}\hbar\omega_c \sum_i l_{iz} - \frac{1}{2}g^*\mu_{\rm B}B\sum_i \sigma_{iz}$$
$$+\chi \sum_{j>i} \frac{1}{|\boldsymbol{r}_i - \boldsymbol{r}_j|}, \qquad \text{with} \quad \chi = \frac{e^2}{\kappa\rho} = \frac{2R^*a^*}{\rho}. \tag{10.10}$$

Here the length is in unit of the magnetic length. The cyclotron energy is $\hbar\omega_c = (2m_e/m^*)\mu_{\rm B}B$. The ratio of m_e/m^* is of order 15 for the 2D electron system in GaAs-Ga$_x$Al$_{1-x}$As heterojunctions.

The strength of the Coulomb potential is of the order of χ, while the level spacing of the single-particle energy is determined by $\hbar\omega$. Therefore the ratio $\chi/\hbar\omega = \sqrt{2R^*/\hbar\omega}$ is the quantity which characterizes the electronic motion. If the ratio is much smaller than unity, the independent motion of the electrons becomes a good approximation. As it increases, the correlated motion of the electrons overwhelms the independent motion.

Recall that the second and third terms of the above Hamiltonian are consistent with the well-known fact that a spin 1/2 particle with mass M and charge qe in the magnetic filed $\boldsymbol{B}$ has the magnetic interaction energy $-\boldsymbol{\mu}{\cdot}\boldsymbol{B}$, where the magnetic moment $\boldsymbol{\mu}$ of the particle is in general expressed by $\boldsymbol{\mu} = g_l\boldsymbol{l} + g_s\boldsymbol{s}$ in units of $e\hbar/(2Mc)$. Here g_l and g_s are the respective gyromagnetic ratios for the orbital and spin angular momenta. The g_l value is equal to q, while the Dirac equation would give $g_s = 2q$ for a point charged particle like the electron. The precision measurement gives $|g_s| = 2.0023\,\mu_{\rm B}$ for the electron and the deviation from 2 can be very accurately computed by the higher order corrections of quantum electrodynamics. On the other hand, the experimental values for the nucleons are far from the Dirac value expected for point particles: $g_s = 5.586$ for the proton and $g_s = -3.826$ for the neutron in units of the nuclear magneton $\mu_{\rm N} = e\hbar/(2m_pc)$. This fact is ascribed to the internal structure of the nucleon.

The eigenvalue of the single-particle Hamiltonian, $H_0 = \hbar\omega\,(-\Delta/2 + r^2/2)\; -(\hbar\omega_c/2)l_z\; -(g^*\mu_B B/2)\sigma_z$, is obtained easily. Rewriting the spatial coordinate as a complex variable $z = (x - iy)/\sqrt{2}$, we define

$$a^\dagger = \frac{1}{\sqrt{2}}\left(z^* - \frac{\partial}{\partial z}\right), \qquad a = \frac{1}{\sqrt{2}}\left(z + \frac{\partial}{\partial z^*}\right),$$

$$b^\dagger = \frac{1}{\sqrt{2}}\left(z - \frac{\partial}{\partial z^*}\right), \qquad b = \frac{1}{\sqrt{2}}\left(z^* + \frac{\partial}{\partial z}\right), \tag{10.11}$$

with the commutation relations $[a, a^\dagger] = [b, b^\dagger] = 1$ and $[a, b] = [a, b^\dagger] = [a^\dagger, b] = [a^\dagger, b^\dagger] = 0$. We can then express the single-particle Hamiltonian as

$$H_0 = \hbar\omega(a^\dagger a + b^\dagger b + 1) - \frac{1}{2}\hbar\omega_c(a^\dagger a - b^\dagger b) - \frac{1}{2}g^*\mu_B B\sigma_z. \tag{10.12}$$

The operator $a^\dagger$ creates a right circular quantum of counterclockwise rotation about z axis, and $b^\dagger$ creates a left one. Let a non-negative integer N denote an eigenvalue of the total oscillator quanta $a^\dagger a + b^\dagger b$ and m denote an eigenvalue of the angular momentum $l_z = a^\dagger a - b^\dagger b$. Then the eigenvalues of $a^\dagger a$ and $b^\dagger b$ are given by $(N + m)/2$ and $(N - m)/2$, respectively. Since they should be positive integers or zero, the possible values of m for a given N are $m = N, N-2, N-4, ..., -N+2, -N$. The single-particle energy becomes

$$\begin{aligned} E_{nm} &= \hbar\omega(N+1) - \frac{1}{2}\hbar\omega_c m - g^*\mu_B B S_z \\ &= \hbar\omega(2n + |m| + 1) - \frac{1}{2}\hbar\omega_c m - g^*\mu_B B S_z \end{aligned} \tag{10.13}$$

with S_z being either $1/2$ or $-1/2$, and where the number of total quanta N is set equal to $N = 2n + |m|$ with a non-negative integer n. For a given N, possible values of n are $0, 1 ..., [N/2]$. The energies of $\hbar\omega(N + 1)$, with $N + 1$-fold degeneracy, corresponding to the 2D harmonic-oscillator are independent of m conforming to N. The single-particle energy is shown in Fig. 10.9.

In the case of no magnetic field or very weak magnetic field ($\omega_c/\omega_0 \ll 1$) the single-particle level shows a shell structure with degeneracy of $N + 1$. The single-particle level is then in the order of $(n, m) = (0, 0)$; $(0, 1)$, $(0, -1)$; $(0, 2)$, $(1, 0)$, $(0, -2)$;... The level crossing of the low-lying single-particle levels occurs at $\omega_c/\omega_0 = \sqrt{1/2} \approx 0.71$ and the level with $(n, m) = (0, 2)$ becomes lower than the one with $(0, -1)$.

If $\omega_0 = 0$, that is, there is no confining potential present or the magnetic field is very high ($\omega_c/\omega_0 \gg 1$), ω is (nearly) equal to $(1/2)\omega_c$ and the energy becomes $E_{nm} = \hbar\omega_c(n+(|m|-m)/2+1/2)-g^*\mu_B BS_z$, depending only on $n+(|m|-m)/2$. Clearly the energy becomes independent of m for positive m. This gives rise to infinitely degenerate levels called the Landau levels with the Landau level index $n+(|m|-m)/2$. The fact that the single-electron motion is quantized to the Landau levels with an infinite degeneracy number in the case of no confining potential reflects the fact that the energy is independent of the center of the cyclotron motion. In the presence of the confining potential, the single-particle energy increases with m as seen from Fig. 10.9.

By using the polar coordinates r and ϑ the eigenfunctions of the single-particle Hamiltonian are expressed as

$$\psi_{nm}(r,\vartheta) = \sqrt{\frac{n!}{\pi\rho^2(n+|m|)!}}\left(\frac{r}{\rho}\right)^{|m|} L_n^{(|m|)}\left(\frac{r^2}{\rho^2}\right)\exp\left(-\frac{r^2}{2\rho^2}\right)e^{im\vartheta}, \tag{10.14}$$

where $L_n^{(\alpha)}(x)$ is an associated Laguerre polynomial defined by

$$\begin{aligned} L_n^{(\alpha)}(x) &= \frac{e^x x^{-\alpha}}{n!}\frac{d^n}{dx^n}(e^{-x}x^{n+\alpha}) \\ &= \sum_{k=0}^{n}\frac{\Gamma(n+\alpha+1)}{k!\,(n-k)!\,\Gamma(k+\alpha+1)}(-x)^k. \end{aligned} \tag{10.15}$$

The parity operation changes ϑ to $\vartheta+\pi$, so the parity of the single-particle eigenfunction is given by $(-1)^m$.

For the lowest Landau states ($n = 0$, $m \geq 0$) the larger angular momentum state corresponds to a wave function which is more extended from the origin:

$$\langle\psi_{0m}|r^2|\psi_{0m}\rangle = (|m|+1)\rho^2. \tag{10.16}$$

The Hamiltonian (10.10) contains the Coulomb interaction between the electrons, which leads to interesting effects on quantum dots. The interplay between the magnetic field and the Coulomb interaction is an essential ingredient for studying the dots. The relative importance of various pieces of the Hamiltonian depends on m^*, κ, g^*, and B.

For two electrons the Hamiltonian can be separated in the relative and center-of-mass coordinates of the two electrons. By introducing the coordinates

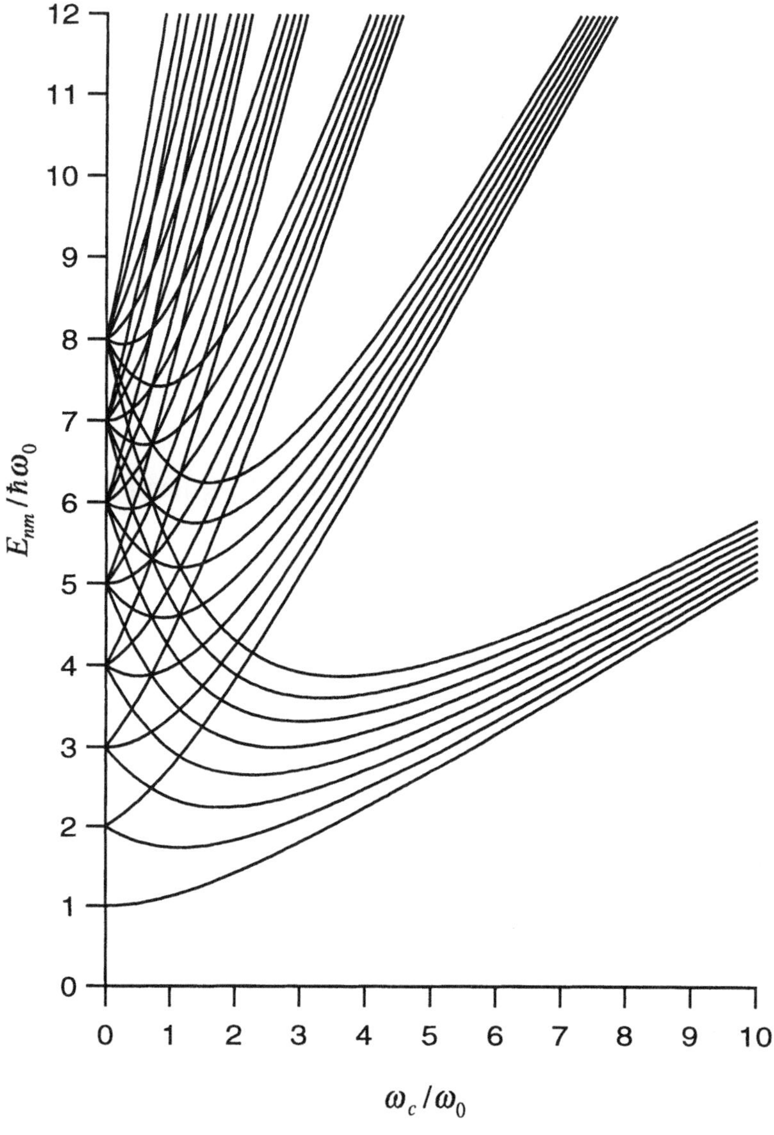

Fig. 10.9. The single-particle energy E_{nm} of a quantum dot as a function of magnetic field strength. The Zeeman spin splitting of the energy is not included.

$$\boldsymbol{R} = \frac{1}{\sqrt{2}}(\boldsymbol{r}_1 + \boldsymbol{r}_2), \qquad \boldsymbol{r} = \frac{1}{\sqrt{2}}(\boldsymbol{r}_1 - \boldsymbol{r}_2), \tag{10.17}$$

the Hamiltonian is reduced to

$$H = \hbar\omega\Big\{ -\frac{1}{2}\Delta_R + \frac{1}{2}\boldsymbol{R}^2\Big\} - \frac{1}{2}\hbar\omega_c l_{Rz} - g^*\mu_B B(S_{1z} + S_{2z})$$
$$+ \hbar\omega\Big\{ -\frac{1}{2}\Delta_r + \frac{1}{2}\boldsymbol{r}^2\Big\} - \frac{1}{2}\hbar\omega_c l_{rz} + \frac{\chi}{\sqrt{2}}\frac{1}{r}. \tag{10.18}$$

The center-of-mass motion is determined by exactly the same Hamiltonian as the single-particle Hamiltonian and its energy and eigenfunction are given by Eqs. (10.13) and (10.14). The eigenfunction $\psi(\boldsymbol{r})$ of the relative motion Hamiltonian may be obtained in polar coordinates. By separating the angular part as $\psi(\boldsymbol{r}) = u(r)\mathrm{e}^{im\vartheta}$ ($m = 0, \pm1, \pm2, \ldots$), the radial function $u(r)$ and the eigenvalue E_r are obtained from the following equation

$$\frac{d^2u}{dr^2} + \frac{1}{r}\frac{du}{dr} - \left(r^2 + \frac{\sqrt{2}\chi}{\hbar\omega}\frac{1}{r} + \frac{m^2}{r^2} - \frac{\omega_c}{\omega}m - \frac{2E_r}{\hbar\omega}\right)u = 0, \tag{10.19}$$

where $u(r)$ has to satisfy the condition that its norm, $\int_0^\infty dr r u(r)^2$, is finite.

Particular analytical solutions are available provided $\chi/\hbar\omega$ belongs to a certain enumarably infinite set of values [141]. For a general case we can obtain the eigenvalue by numerical integrations. The eigenvalue may be labelled by a quantum number $n = 0, 1, 2, \ldots$ for each m. We note the symmetry property of the eigenfunctions. The interchange of the position coordinates $\boldsymbol{r}_1$ and $\boldsymbol{r}_2$ leads to $\boldsymbol{r} \to -\boldsymbol{r}$, which is equivalent to the change of $\vartheta \to \vartheta + \pi$. Therefore the eigenfunction labelled by n and m receives a phase change $\mathrm{e}^{i\pi m} = (-1)^m$, so that it is symmetric for even m and antisymmetric for odd m. Figure 10.10 displays the energy E_r of the relative motion as a function of the cyclotron frequency. The most striking feature is the level crossing in the lowest level (ground state). With increasing magnetic field the ground state changes from $(n, m) = (0, 0)$ to $(0, 1), (0, 2), (0, 3), \ldots$, so that the angular momentum m of the ground state increases one by one. (This figure is basically the same as Fig. 10.4 which is obtained by the variational calculation, but included here to show the significant role of the Coulomb interaction.)

Finally we note that the importance of the correlated motion of electrons in a strong magnetic field has already been recognized in the quantum Hall effect [142]. The broadening of the Landau levels

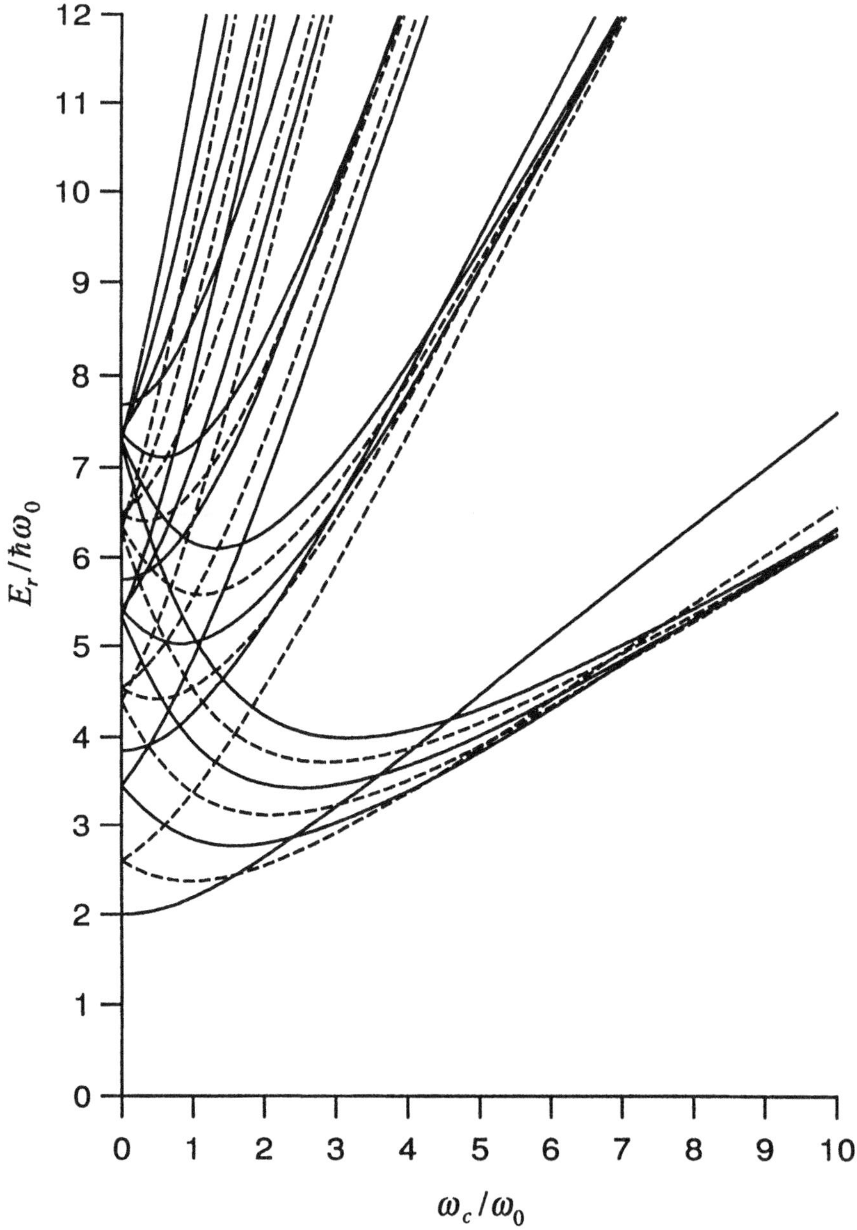

Fig. 10.10. The energy E_r of the relative motion of a two-electron quantum dot as a function of magnetic field strength. The Zeeman spin splitting of the energy is not included. The solid curves are for states with even m and the dashed curves for states with odd m. $\hbar\omega_0 = 1$. Atomic units are used.

is caused by scatterers such as impurities and lattice defects. The correlations due to the Coulomb repulsion between many electrons lead to the Anderson localization and the fractional quantum Hall effect. Laughlin [143] successfully used a correlated basis of Jastrow type to describe the electron motion.

11. Nuclear few-body systems

The stochastic variational method was originally developed to solve few-body problems in nuclear physics. In this chapter we collect a few possible applications of the SVM for nuclear systems. The nuclear force, due to the effect of the underlying quark structure and relativistic motion, is a very complicated interaction. In addition to the spin-isospin dependent central part, the two-nucleon interaction contains spin-orbit and tensor forces, and L^2- and $(LS)^2$-dependent terms. The two-body interaction designed to fit the nucleon-nucleon phase shift and the properties of the deuteron fails to reproduce the binding energies of the three- and four-nucleon nuclei, and a three-body interaction has to be introduced to get the correct binding energy. As the nucleons are not structureless point-like entities, the spatial part of the nuclear force contains a strong repulsion at short distances. These circumstances altogether make the solution of the nuclear few-body problem a formidable task.

The first application is restricted to central forces. These schematic forces are used in simple model calculations and serve for benchmark tests. Examples of more realistic problems are shown in the last section.

11.1 Introductory remark on nucleon-nucleon potentials

The nucleon (N) has spin and isospin degrees of freedom in addition to the spatial one. The operator O_{ij}^p of Eq. (2.3) is a product of these factors, one acting on each coordinate of the three types:

$$O_{ij}^p = \Big(O_{\kappa_{ij}}(\text{space}) \cdot O_{\kappa_{ij}}(\text{spin})\Big) O_{ij}(\text{isospin}), \tag{11.1}$$

where, e.g., $O_{\kappa\mu_{ij}}(\text{space})$ $(\mu = \kappa, \kappa-1, ..., -\kappa)$ is a spherical tensor operator of rank κ acting in the spatial part, and a scalar product $(T_\kappa \cdot U_\kappa)$ stands for the tensor product $(-1)^\kappa \sqrt{2\kappa+1}[T_\kappa \times U_\kappa]_{00}$, which is defined with Clebsch-Gordan coefficients by

$$[T_{\kappa_1} \times U_{\kappa_2}]_{\kappa_3\mu_3} = \sum_{\mu_1\mu_2} < \kappa_1\mu_1\kappa_2\mu_2|\kappa_3\mu_3 > T_{\kappa_1\mu_1}U_{\kappa_2\mu_2}. \tag{11.2}$$

See, e.g., [2, 70, 71, 72] for angular momentum algebra.

The NN interaction is a typical example of the strong interaction and is still not known in full detail. There are, however, several versions of the nuclear potential whose parameters have been determined fairly accurately so as to reproduce two-nucleon bound state and scattering data. The nuclear force reaches at most up to 2 fm (1 fm $= 10^{-13}$ cm). An NN potential which is consistent with the scattering data is called "realistic". The most important ingredients of the realistic potential include central, tensor, and spin-orbit components.

The central part of the potential is characterized by four components: $O_{12}^p = 1$, $\boldsymbol{\tau}_1 \boldsymbol{\tau}_2$, $\boldsymbol{\sigma}_1 \boldsymbol{\sigma}_2$, and $(\boldsymbol{\tau}_1 \boldsymbol{\tau}_2)(\boldsymbol{\sigma}_1 \boldsymbol{\sigma}_2)$. The central potential is thus expressed as

$$V_{12} = V^c(r) + V^{c\tau}(r)\boldsymbol{\tau}_1 \cdot \boldsymbol{\tau}_2 + V^{c\sigma}(r)\boldsymbol{\sigma}_1 \cdot \boldsymbol{\sigma}_2 \\ + V^{c\tau\sigma}(r)(\boldsymbol{\tau}_1 \cdot \boldsymbol{\tau}_2)(\boldsymbol{\sigma}_1 \cdot \boldsymbol{\sigma}_2). \tag{11.3}$$

These components can also be expressed by the space- (P_{12}^r), spin- $(P_{12}^\sigma = \frac{1+\boldsymbol{\sigma}_1 \boldsymbol{\sigma}_2}{2})$, and isospin-exchange $(P_{12}^\tau = \frac{1+\boldsymbol{\tau}_1 \boldsymbol{\tau}_2}{2})$ operators. The generalized Pauli principle requires the relation

$$P_{12}^r P_{12}^\sigma P_{12}^\tau = -1. \tag{11.4}$$

Therefore the central force can also be written as

$$V_{12} = V_W(r) + V_M(r)P_{12}^r + V_B(r)P_{12}^\sigma + V_H(r)P_{12}^r P_{12}^\sigma, \tag{11.5}$$

and the terms are named Wigner, Majorana, Bartlett, and Heisenberg force, respectively.

The tensor force is very strong as it is generated from the one-pion exchange which is the most important process responsible for the nuclear force. The tensor force acts only in spin-triplet states, which are symmetric with respect to the interchange of spin coordinates, and thus is given in the form

$$V_{12} = \left(V^t(r) + V^{t\tau}(r)\boldsymbol{\tau}_1 \cdot \boldsymbol{\tau}_2\right) S_{12}, \tag{11.6}$$

with the tensor operator

$$S_{12} = 3(\boldsymbol{\sigma}_1\cdot\hat{\boldsymbol{r}})(\boldsymbol{\sigma}_2\cdot\hat{\boldsymbol{r}}) - \boldsymbol{\sigma}_1\cdot\boldsymbol{\sigma}_2$$
$$= \sqrt{\frac{24\pi}{5}}\,(Y_2(\hat{\boldsymbol{r}})\cdot[\boldsymbol{\sigma}_1 \times \boldsymbol{\sigma}_2]_2)\,. \tag{11.7}$$

The deuteron consisting of a proton (p) and a neutron (n) becomes bound through the coupling of the ${}^3S_1$ and ${}^3D_1$ waves due to the tensor force. Here ${}^{2S+1}L_J$ denotes the two-nucleon channel with the spin S and the relative orbital angular momentum L being coupled to the total angular momentum J. However, no such coupling occurs in the case of two protons or two neutrons because the Pauli principle (11.4) forbids two like nucleons with $L = 0$ or 2 to be in a spin-singlet state. The interaction between like nucleons is just too weak to produce a bound state.

The spin-orbit force has non-vanishing matrix elements only in spin-triplet states, too. It is parametrized in the form

$$V_{12} = \left(V^b(r) + V^{b\tau}(r)\boldsymbol{\tau}_1\cdot\boldsymbol{\tau}_2\right)(\boldsymbol{L}\cdot\boldsymbol{S})_{12} \tag{11.8}$$

with

$$(\boldsymbol{L}\cdot\boldsymbol{S})_{12} = \left(\left(\boldsymbol{r} \times \frac{1}{2\hbar}(\boldsymbol{p}_1 - \boldsymbol{p}_2)\right)\cdot\frac{1}{2}(\boldsymbol{\sigma}_1 + \boldsymbol{\sigma}_2)\right)$$
$$= \left(\frac{1}{\hbar}(\boldsymbol{r} \times \boldsymbol{p})\cdot\frac{1}{2}(\boldsymbol{\sigma}_1 + \boldsymbol{\sigma}_2)\right) = \left(\boldsymbol{l}\cdot\frac{1}{2}(\boldsymbol{\sigma}_1 + \boldsymbol{\sigma}_2)\right), \tag{11.9}$$

where $\boldsymbol{p} = -i\hbar\frac{\partial}{\partial \boldsymbol{r}}$, $\hbar\boldsymbol{l} = (\boldsymbol{r} \times \boldsymbol{p})$, and $(\ \times\ )$ indicates a vector (outer) product. The form factors of the various components are usually expressed in terms of $Y(x) = \mathrm{e}^{-x}/x$ and its derivatives.

As is evident from the above discussion, one of the characteristic features of the realistic NN potential is its strong state-dependence, namely it differs in the four states of singlet-even, triplet-even, singlet-odd, and triplet-odd character, where even and odd refer to the parity of the orbital motion.

The tensor force makes it necessary to introduce at least D-waves in the calculations. Another important feature of the most commonly used family of NN potentials is the strong repulsion at short distances, say within 0.5 fm. This requires due care in so far as short-range correlations must be properly taken into account in theoretical models. All of these aspects of the NN interaction really make a variational calculation for a nuclear system very challenging. In these circumstances it is rather frequent to use some "effective" potentials which are not necessarily designed to reproduce the scattering data but determined to fit some bulk properties of nuclei such as the binding energy and the size. Although relativistic effects should be fairly large in nuclei ($(v/c)^2 \sim 0.08$, which is estimated from the kinetic energy $T_{\rm F} \sim 40$ MeV belonging to the Fermi level), the ambiguity of the nuclear force makes it rather difficult to extract these clearly from experimental data.

The complexity of the NN interaction is due to the quark-gluon substructure of the nucleon as was discussed in Chapt. 9 and to the compositeness of mesons exchanged between the nucleons.

11.2 Few-nucleon systems with central forces

We have tested the performance of the SVM by performing model calculations with different NN central potentials, e.g., the spin-averaged Malfliet-Tjon (MT-V) potential [144], the Volkov no. 1 (V_1) "super-soft" core potential [145], the Afnan-Tang S3 (ATS3) potential [146], the Minnesota potential [147] and the Brink-Boeker (B_1) potential [148]. Parameters of these potentials are given in Table 11.1. Some of these model problems have already been solved to high accuracy by various methods and therefore we can directly compare the solutions. The results presented in this section do not include the contribution from the Coulomb potential.

Each calculation has been repeated several times starting from different random points to check the convergence. The energy as a function of the number of basis states is shown in Fig. 11.1 for the case of ^{6}Li with the V_1 potential. The energies on different random paths approach each other after a few initial steps, and converge to the final solution. The energy difference between two random paths as well as the tangent of the curves give us some information on the accuracy of the method for a given size of the basis. The root-mean-square (rms) radius is calculated in each step and found to be rapidly convergent to its final value. By increasing the basis size the results can be arbitrarily improved when needed.

The number of basis states required to reach energy convergence increases with the number of particles, but it depends on the form of the interaction as well. This latter property is illustrated in Fig. 11.2 for the case of the alpha-particle. The soft-core Volkov (V_1) potential

Table 11.1. List of the parameters of the nucleon-nucleon central potentials. The potential consists of a few terms; each is expressed as $Vf(\mu,r)(W+MP^r+BP^\sigma+HP^rP^\sigma)$. See Eq. (11.5). For the Gaussian potentials (V_1, ATS3, Minnesota, B_1), the form factor $f(\mu,r)$ is $\exp(-\mu r^2)$, the potential strength V is in units of MeV, and the potential range μ is in units of fm^{-2}, while for the Yukawa potential (MT-V) $f(\mu,r)$ is defined by $\exp(-\mu r)/r$, V is given in units of MeV fm, and μ is in units of fm^{-1}, respectively. The Majorana mixtures m and m' are set equal to zero in the calculation. For realistic values of m and m' see the original papers. The parameter u of the Minnesota potential is set equal to unity.

Potential	V	μ	W	M	B	H
MT-V	1458.05	3.11	1	0	0	0
[144]	−578.09	1.55	1	0	0	0
V_1	144.86	0.82^{-2}	$1-m$	m	0	0
[145]	−83.34	1.60^{-2}	$1-m$	m	0	0
ATS3	1000.0	3.0	1	0	0	0
[146]	−326.7	1.05	0.5	0	0.5	0
	−166.0	0.80	0.5	0	−0.5	0
	−43.0	0.60	0.5	0	0.5	0
	−23.0	0.40	0.5	0	−0.5	0
Minnesota	200.0	1.487	$u/2$	$(2-u)/2$	0	0
[147]	−178.0	0.639	$u/4$	$(2-u)/4$	$u/4$	$(2-u)/4$
	−91.85	0.465	$u/4$	$(2-u)/4$	$-u/4$	$(-2+u)/4$
B_1	389.5	0.7^{-2}	$1-m$	m	0	0
[148]	−140.6	1.4^{-2}	$1-m'$	m'	0	0

shows rapid convergence, while the repulsive-core ATS3 interaction requires more basis states to get an accurate solution. The relatively fast convergence for the MT-V potential of a strong repulsion can be explained by the simplicity of the spin-independence of this interaction.

Table 11.2 shows the SVM results, together with results of others, for the application of the MT-V potential to N=2–7-nucleon systems. We chose $\hbar^2/m = 41.47$ MeV fm^2. In the table the ground-state energies E and point matter rms radii $\langle r^2\rangle^{1/2}$ are given. The basis dimension K of the SVM listed in the table is such that, beyond it, the energy and the radius do not change in the digits shown. For three-body systems, the solution of the Faddeev equation is known to be the method of choice (see Sect. 5.3), but the SVM can easily yield energy of the

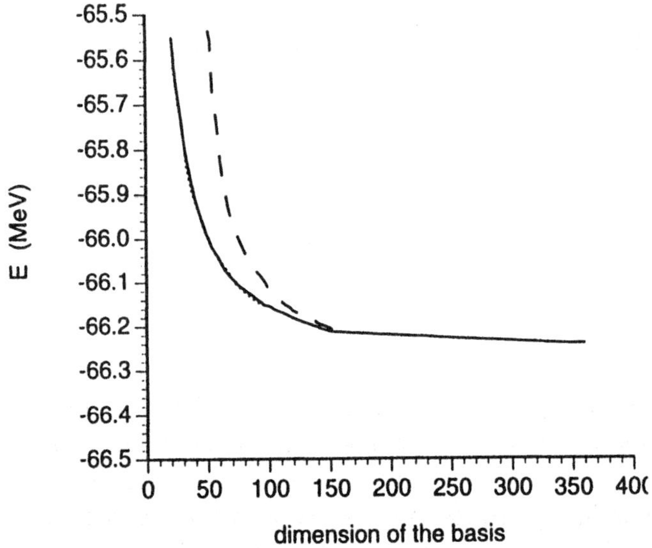

Fig. 11.1. Convergence of the ^{6}Li energy on different random paths. The Volkov (V_1) potential [145] is used.

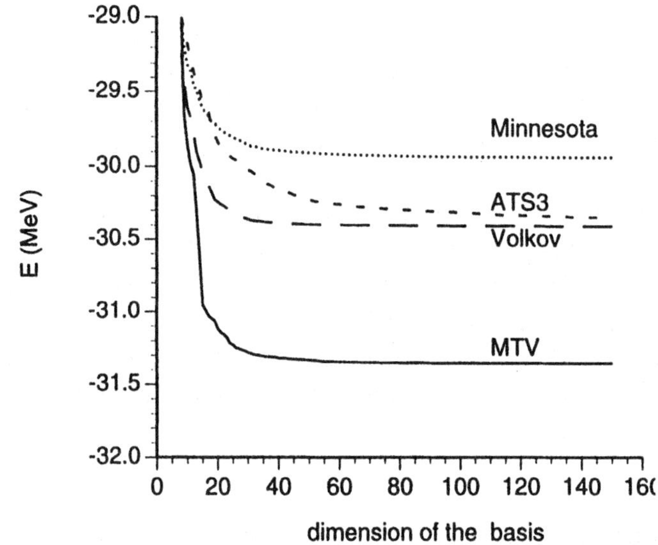

Fig. 11.2. Convergence of the alpha-particle energy for different central potentials

Table 11.2. Energies (in MeV) and root-mean-square (rms) radii (in fm) of N-nucleon systems interacting via the Malfliet-Tjon (MT-V) potential [144]. The value of K denotes the basis dimension beyond which the energies and the radii of the SVM calculation do not change in the digits shown.

N	$(L,S)J^\pi$	Method	E		$\langle r^2\rangle^{1/2}$	K
2	$(0,1)1^+$	Numerical	−0.4107		3.743	
		SVM	−0.4107		3.743	5
3	$(0,1/2)1/2^+$	Faddeev[150, 151]	−8.25273			
		ATMS[12]	−8.26	±0.01	1.682	
		CHH[152]	−8.240			
		GFMC[41]	−8.26	±0.01	1.682	
		VMC[153][a]	−8.27	±0.03	1.68	
		SVM	−8.2527		1.682	80
4	$(0,0)0_1^+$	FY[149]	−31.36			
		ATMS[12]	−31.36		1.40	
		CRCG[26]	−31.357			1000
		GFMC[41]	−31.3	±0.2	1.36	
		VMC[153][a]	−31.3	±0.05	1.39	
		SVM	−31.360		1.4087	150
	$(0,0)0_2^+$	CRCG[26]	−8.50			2000
		SVM	−8.49			150
5	$(1,1/2)3/2^-$	VMC[153][a]	−42.98	±0.16	1.51	
		SVM	−43.48		1.51	500
6	$(0,0)0^+$	VMC[153][a]	−66.34	±0.29	1.50	
		SVM	−66.68		1.52	800
7	$(1,1/2)3/2^-$	SVM	−83.4		1.68	1300

[a]Calculated with Coulomb potential, the Coulomb contribution then subtracted perturbatively. The potential strength used in the VMC [153] calculation is V_1=1458.25 and $V_2 = -578.17$ MeV, which is slightly different from that used in the calculation. See Table 11.1.

same accuracy. As the MT-V potential is a preferred benchmark test of the few-body calculations, there are numerous solutions available. Table 11.2 includes a few of the most accurate results. The nice agreement for the four-nucleon case corroborates the fact that the SVM is as accurate as the direct solution of the Faddeev-Yakubovsky (FY) equations [149], the method of the Amalgamation of Two-body correlations into Multiple Scattering (ATMS) [12] or the variational Monte Carlo (VMC) [153] and the Green's function Monte Carlo (GFMC) [41] methods. The basis used in the Coupled-Rearrangement-Channel Gaussian-basis (CRCG) variational method [26] is similar to that of the SVM, but the Gaussian parameters follow geometric progressions. The fact that the basis size needed in the SVM for both the ground and first excited states of the four-nucleon system is much smaller proves the efficiency of the selection procedure. See Chap. 4. The results of the VMC calculation for the five- and six-nucleon systems are also in good agreement with the results of the SVM. The MT-V potential has no exchange term; therefore, unlike nature, it renders the five-nucleon system bound, and the nucleus tends to collapse as the binding energy increases with the number of particles.

Neither the MT-V nor the V_1 potential contains the full set of exchange components of the central potential, so the calculation becomes rather simple. The Minnesota potential, however, contains space-, spin-, and isospin-exchange operators, and is considered to be more realistic. In fact it has been successfully used in microscopic cluster-model calculations [28, 29, 30]. An extensive calculation with this potential was carried out for N=2–6 nucleon systems in [31], where all possible spin and isospin configurations were allowed for and all angular functions that give non-negligible contributions were included. The agreement with experiment was surprisingly good. The energies and the radii of the triton and the alpha-particle converge to realistic values. The Minnesota potential, correctly, does not bind the N=5 system, but it binds ^{6}He and slightly overbinds ^{6}Li. The radius of ^{6}He is found to be much larger than that of ^{4}He, consistent with the neutron-halo structure of ^{6}He [29, 30]. The SVM calculation of [31] has made it possible to test for the first time the Minnesota force without assuming any cluster structure or restricting the model space by any other bias.

The last example is the application to Efimov states [154] to show how well the correlated Gaussian basis describes the few-body asymptotics. A three-body system with short-range forces may have several bound states. If the two-body subsystem has just one bound state whose energy is (close to) zero, the three particles can interact at long distances and an infinite number of bound states may appear in the three-body system. These "Efimov states" are extremely interesting from both experimental and theoretical points of view because of their distinctive properties. These states are very loosely bound and their wave functions extend far beyond those of normal states. By increasing the strength of the two-body interactions these states disappear.

A system of three identical bosons of nucleonic mass is considered for simplicity. The potential between the particles is taken as the Poeschl-Teller interaction because this potential is analytically solvable for the two- body case, and therefore the two-body binding energy can be easily set to zero;

$$V(r) = \frac{625.972}{\sinh(1.586\ r)^2} - \frac{1251.943}{\cosh(1.586\ r)^2} \quad \text{in MeV.} \tag{11.10}$$

Here r is in units of fm. A spatially very extended basis is needed to represent the weakly bound states. In this case we use a predefined basis. The predefined basis is created by using only the diagonal elements of the matrix A. These diagonal elements are taken as geometric progression:

$$\begin{aligned} A_{11} &= \frac{1}{(a_0 q_0^{(i-1)})^2} \qquad (i = 1, ..., n), \\ A_{22} &= \frac{1}{(a_0 q_0^{(j-1)})^2} \qquad (j = 1, ..., n). \end{aligned} \tag{11.11}$$

This construction defines an n^2-dimensional basis corresponding to $(i, j) = (1, 1), ..., (n, n)$. The parameters are chosen as $n = 20$, $a_0 = 0.1$, $q_0 = 2.4$. One can find the first three Efimov states with this basis size. The ratios of the energies of the bound states follow the $E_{i+1}/E_i = \exp\{-2\pi\}$ rule [154]. By increasing the basis size one can reveal more bound excited states. Note that the value of $a_0 q_0^{n-1}$ roughly corresponds to the spatial extension of the basis. The present basis in this example goes out up to $a_0 q_0^{n-1} = 0.1 \times 2.4^{19} = 1674990$ (fm). This extension is necessary to get the third bound state. For a comparison, to calculate the ground-state energy of -4.81 MeV it is enough to choose $n = 7$ and the basis covers only the $[0, 10]$ (fm) interval. The rms radii of the ground, first and second excited states are about 1.5, 25 and 6000 fm, respectively. As shown in Fig. 11.3, these facts illustrate the tremendous spatial extension of the Efimov

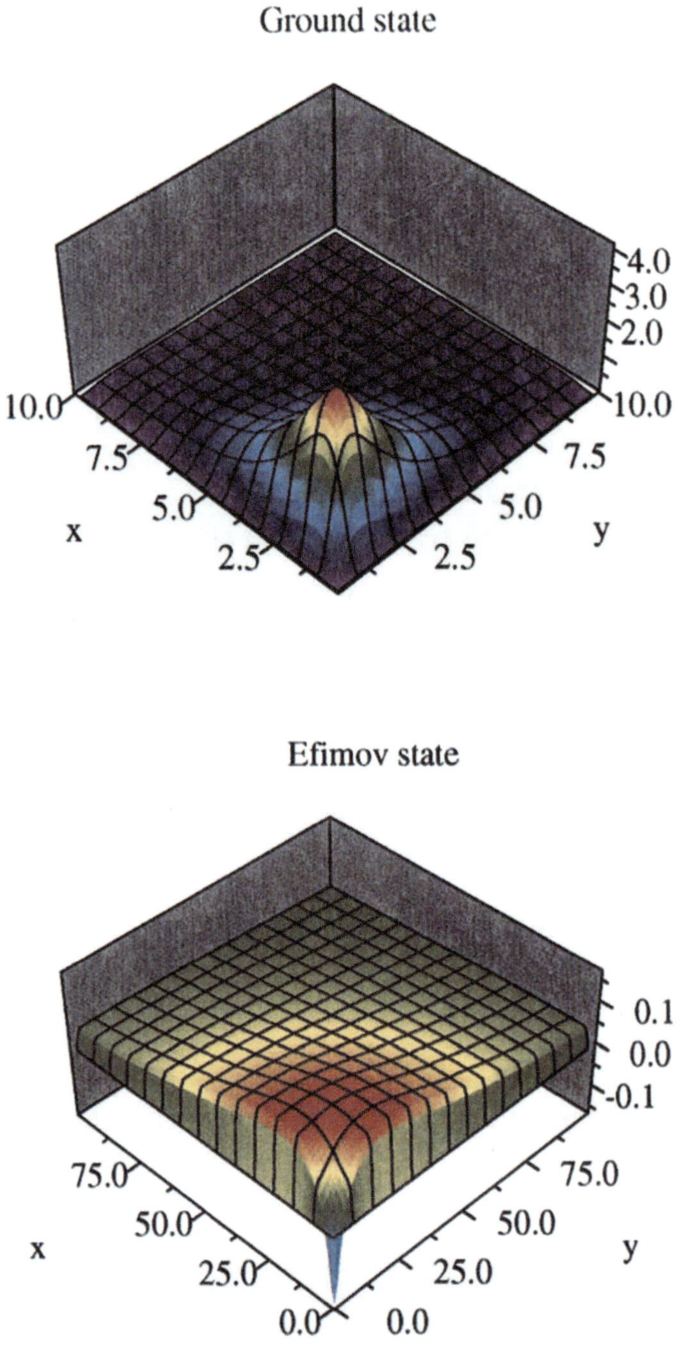

Fig. 11.3. Wave function amplitudes of the ground state and the first Efimov state. Note the different scale (in fm) of the lengths of the Jacobi coordinates x and y.

states. These results indicate that the extended correlated Gaussian basis gives a practical means to describe the asymptotics characteristics of the Efimov states. See Complement 11.4 for the possibility of Efimov states in the ^{12}C nucleus.

11.3 Realistic potentials

The trial function is given as a combination of the correlated Gaussians:

$$\psi_{(LS)JMTM_T}(\boldsymbol{x}, A) = \mathcal{A}\Big\{\exp\Big(-\frac{1}{2}\tilde{\boldsymbol{x}}A\boldsymbol{x}\Big)\,[\theta_L(\boldsymbol{x})\times\chi_S]_{JM}\,\eta_{TM_T}\Big\}, \qquad (11.12)$$

where $\boldsymbol{x} = \{\boldsymbol{x}_1, ..., \boldsymbol{x}_{N-1}\}$ is a set of relative (Jacobi) coordinates, the operator $\mathcal{A}$ is an antisymmetrizer, χ_{SM_S} is the spin function, and η_{TM_T} is the isospin function. The matrix A is an $(N-1)\times(N-1)$ symmetric, positive-definite matrix of nonlinear variational parameters. The function $\theta_{LM_L}(\boldsymbol{x})$ represents the angular part of the wave function. It is taken as a vector coupled product of partial waves $\theta_{LM_L}(\boldsymbol{x}) = [[[Y_{l_1}(\widehat{\boldsymbol{x}_1})\times Y_{l_2}(\widehat{\boldsymbol{x}_2})]_{L_{12}}\times Y_{l_3}(\widehat{\boldsymbol{x}_3})]_{L_{123}}\cdots]_{LM_L}$. This is referred to as the method of partial-wave expansion. The spin and isospin functions are also successively coupled, for example $\chi_{SM_S} = \chi_{(S_{12}S_{123}\ldots)SM_S} = [[[\chi_{1/2}\times\chi_{1/2}]_{S_{12}}\times\chi_{1/2}]_{S_{123}}\cdots]_{SM_S}$ For the triton, we used the partial-wave channels with $(l_1, l_2)L$, where $l_i \leq 2$ and $L = 0, 1, 2$. For the alpha-particle, all partial waves $((l_1, l_2)L_{12}, l_3)L$ satisfying the condition $l_1 + l_2 + l_3 \leq 4$, $l_i \leq 2$, $|l_1 - l_2| \leq L_{12} \leq l_1 + l_2$ and $L = 0, 1, 2$ have been tried. As the triton has $J = 1/2$ and $T = 1/2$, there are three spin and two isospin channels; $(S_{12})S = (0)1/2, (1)1/2, (1)3/2$ and $T_{12} = 0, 1$. For the alpha-particle ($J = 0$, $T = 0$) there are six spin and two isospin channels unless very small isospin mixing is taken into account; $(S_{12}, S_{123}) = (0, 1/2)0, (0, 1/2)1, (1, 1/2)0, (1, 1/2)1, (1, 3/2)1, (1, 3/2)2$ and $(T_{12}, T_{123}) = (0, 1/2), (1, 1/2)$.

The spin, isospin and orbital angular momentum dependence of realistic nucleon-nucleon interaction is extremely important. To extend the application of the SVM calculations to nuclear systems interacting with realistic potentials, in addition to the previously applied random selection of the nonlinear parameters of the correlated Gaussian, a random selection of the spin and orbital components of the wave function is introduced. The main motivation for the random selection is that the numbers of channels and nonlinear parameters are prohibitively large, therefore the calculation of all of the matrix elements and diagonalization including all potentially important basis states is out of the question. In the case of the alpha-particle, for example, there are 12 spin-isospin channels and 32 partial-wave channels even in our truncated-partial wave expansion. Just to show up an alternative, it is worthwhile assessing what dimensions a deterministic optimization would require. The simplest choice of the nonlinear parameters of the Gaussian basis is to use a diagonal matrix A. One of the best deterministic choices for the diagonal elements is a geometric progression [25]. To reach good accuracy, at least five values for each of A_{11}, A_{22}, and A_{33} have to be used, requiring $5^3 = 125$ functions in a given channel. The spin-isospin and space part, therefore, would result in a basis size of about one hundred thousand ($2 \times 125 \times 12 \times 32$) in such a "direct" calculation for the alpha-particle.

The literature is very rich in papers devoted to few-nucleon problems [155, 156, 157, 12, 25, 158–161]. For $N = 3$ nuclei it is possible to include enough channels and an exact Faddeev calculation can be performed, but other methods give essentially the same results. For $N = 4$ nuclei some of these methods become too complicated, and the results of the existing calculations agree only within a few hundred keV.

To test the SVM we used two, from the point of view of spatial form, rather different interactions. The first, the Argonne potential (AV6 and AV8) [162], is a well-behaved smooth function. The second, the Reid potential (RV8) [163] is more singular and includes a linear combination of $\exp(-\mu r^2)/r^k$-type terms. The latter potential can certainly cause some problems. The AV6 potential includes only central and tensor components and serves as a good test case for the inclusion of non-central forces. The AV6 and AV8 potentials are used without Coulomb potential, while in the case of the RV8 potential the Coulomb interaction is added.

To keep the number of nonlinear parameters low, we restricted the matrix A to be diagonal in one of the possible Jacobi coordinate sets. In the case of the three-nucleon system we have only one possibility of the Jacobi coordinate, $(NN) + N$. For the alpha-particle we use both possible coordinate sets, $(3N) + N$ ("K") and $(2N) + (2N)$-type ("H") Jacobi coordinates. Note that the basis function is fully antisymmetrized. This choice is also dictated by physical intuition as it is natural to consider ^{3}H+p, ^{3}He+n and $d+d$-type partitions in the alpha-particle. Test calculations show that this choice already gives satisfactory results.

As is shown in Table 11.3, the convergence is relatively fast both for the triton and the alpha-particle with the AV6 and AV8 potentials, but in the case of the RV8 potential, due to its singular nature and stronger repulsive core, the convergence is slower. About 50 basis states for the triton and about 200 basis states for the alpha-particle give fairly good binding energies. The energy of the alpha-particle with the basis size of 200 is already within 0.5 MeV of its final value. It is very satisfactory that the SVM has succeeded in reproducing a reasonably good energy with just 400 basis states.

Table 11.3. The convergence of energies (in MeV) for the triton and the alpha-particle for different potentials, AV6 [162], AV8 [162], and RV8 [163]. K is the basis dimension.

^{3}H				^{4}He			
K	AV6	AV8	RV8	K	AV6	AV8	RV8
25	−6.63	−7.36	−6.53	100	−24.37	−25.15	−23.35
50	−7.04	−7.69	−7.41	200	−25.05	−25.50	−24.15
75	−7.11	−7.74	−7.54	300	−25.28	−25.60	−24.35
100	−7.15	−7.79	−7.59	400	−25.40	−25.62	−24.49

The results of the SVM and other calculations are compared in Tables 11.4 and 11.5. We show and compare the results not only for the energy but for the rms radius and average kinetic and potential energies as well. The results remain the same by repeating the calculations several times starting from different random values. The comparison for the triton with RV8 demonstrates that the SVM results are in very good agreement with the Faddeev results. Since Faddeev calculations are considered exact for three-body systems (see Sect. 5.3), this agreement confirms that the basis selection of the SVM has been performed efficiently even for the singular potential of RV8. The correlated hyperspherical harmonics (CHH) expansion method (see Sect. 5.2) also agrees with the SVM for the triton. The results of the SVM and GFMC are very close to each other. Except for the case of the alpha-particle with AV6, the energies of the SVM are within the statistical errors of the GFMC, though the expectation values of

Table 11.4. Energies (in MeV) and radii (in fm) of the triton calculated by different methods and with different interactions. $\langle T \rangle$, $\langle V \rangle$, and $\langle V_6 \rangle$ are the expectation values of the kinetic energy, the total potential energy, and the potential energy of the central and tensor components, respectively. $\langle r^2 \rangle^{1/2}$ denotes the root-mean-square radius. The value of P_L shows the percentage of the component with the total orbital angular momentum L.

	SVM	GFMC	Faddeev	VMC	CHH
AV6					
$\langle T \rangle$	44.8				
$\langle V_6 \rangle$	−51.9	−52.0(3.00)		−43.7(1.0)	
$\langle r^2 \rangle^{1/2}$	1.76	1.75(0.10)		1.95(0.03)	
P_S	91.2				
P_P	< 0.1				
P_D	8.7				
E	−7.15	−7.22(0.12)[a]	−7.15[a]	−6.33(0.05)[a]	
AV8					
$\langle T \rangle$	46.3				
$\langle V_6 \rangle$	−52.9				
$\langle V_{LS} \rangle$	−1.2				
$\langle r^2 \rangle^{1/2}$	1.75				
P_S	91.1				
P_P	< 0.1				
P_D	8.9				
E	−7.79		−7.79[d]		−7.79[c]
RV8					
$\langle T \rangle$	52.2	54.0(0.20)	52.2		
$\langle V \rangle$	−59.8	−62.0(0.20)	−59.8		
$\langle r^2 \rangle^{1/2}$	1.75	1.68(0.07)	1.76		
P_S	90.3				
P_P	< 0.1				
P_D	9.7				
E	−7.59	−7.54(0.10)[b]	−7.59[b]	−7.44(0.03)[e]	−7.60[c]

[a]Ref. [164]. [b]Ref. [165]. [c]Ref. [157]. [d]Ref. [159]. [e]Ref. [161].

Table 11.5. Energies (in MeV) and radii (in fm) of the alpha-particle by different methods and with different interactions. The isospin mixing due to the Coulomb potential is not taken into account. See also the caption of Table 11.4.

	SVM	GFMC	FY	VMC	CHH
AV6					
$\langle T \rangle$	100.1				
$\langle V_6 \rangle$	−125.4	−122.0(3.0)		−122.0(1.0)	
$\langle r^2 \rangle^{1/2}$	1.49	1.50(0.04)		1.50(0.01)	
P_S	84.3				
P_P	0.5				
P_D	15.1				
E	−25.40	−25.50(0.20)[a]		−22.75(0.01)[a]	
AV8					
$\langle T \rangle$	98.8				
$\langle V \rangle$	−124.4	−124.20(1.0)			
$\langle V_6 \rangle$	−121.5				
$\langle V_{LS} \rangle$	−2.9				
$\langle r^2 \rangle^{1/2}$	1.50	1.51(0.01)			
P_S	85.5				
P_P	0.3				
P_D	14.2				
E	−25.62	−25.75(0.05)	−25.31[d]		−25.60[c]
RV8					
$\langle T \rangle$	111.7	109.2(0.20)			
$\langle V_6 \rangle$	−139.1	−137.5(0.20)			
$\langle V_{LS} \rangle$	2.1	2.45(0.23)			
$\langle V_{coul} \rangle$	0.75	0.71(0.02)			
$\langle r^2 \rangle^{1/2}$	1.51	1.53(0.02)			
P_S	84.1				
P_P	0.4				
P_D	15.5	15.5(0.20)			
E	−24.49	−24.55(0.13)[b]	−23.79[f]	−24.01[e]	−23.9[c]

[a]Ref. [164]. [b]Ref. [165]. [c]Ref. [157]. [d]Ref. [159]. [e]Ref. [161]. [f]Ref. [155].

the kinetic and the potential energies are somewhat different. Both of VMC and GFMC have a statistical error involved in the Monte Carlo integration. Though the error bar of VMC is influenced by the choice of the trial function as well, the error of GFMC is considered purely statistical in the limit of infinitesimal time step (see Sect. 5.1).

Variational calculations were considered to be inappropriate for realistic potentials, as they might not be able to reproduce the large cancellation between the kinetic and the potential energies [158]. Our calculation shows that, with the careful optimization of the nonlinear variational parameters, this is not the case. We can obtain accurate energies even in the case of the RV8 potential.

The experimental energies of the triton and the alpha-particle are −8.48 and −28.295 MeV, respectively. We see that the calculation using the realistic potentials underbinds about 0.9 MeV for the triton and about 3.8 MeV for the alpha-particle. This indicates that the binding energies of few-nucleon systems cannot be accounted for in two-nucleon potential models but call for other mechanism to get more attraction. One possibility for the mechanism is a three-body potential which can be produced by two-pion exchanges among three nucleons.

Once we have a basis for a given potential, one naturally expects that the same basis may give fair ground-state energy for other potentials of similar nature as well. We have checked if it is really the case. The basis optimized for the RV8 potential gives excellent results (within 100 keV) for AV6 and AV8 as well. Due to the singular nature and stronger repulsive core of RV8, however, the basis optimized for AV8 gives about 500 keV less binding for the alpha-particle with RV8 than the basis optimized for RV8 itself. This result is still not so bad and can be easily improved by refining the nonlinear parameters as explained in Sect. 4.2. Therefore, one does not have to look for an entirely new basis set for another interaction, but can use the same basis for a given system, with some "fine tuning" if necessary.

It is obviously important to calculate accurate binding energies of light nuclei with realistic forces. For example, the two-neutron separation energy of ^{6}He is about 1 MeV, so that a few hundred keV less binding is thought to change its neutron "halo" structure significantly. One of the advantages of the SVM is that it is relatively easy to extend it to $N = 5$-, 6-, 7-nucleon systems [31]. The low dimension of the bases required to solve the $N = 3$- and 4-nucleon problems confirms that the SVM is suitable for treating larger nuclei with realistic forces. The formalism and the computer code itself is general

[31], so the applicability is mostly limited by the memory and speed of the available computer. There is no difficulty in including three-body forces if necessary.

We show the utility of the global vector representation of Eq. (6.4) in describing the angular part of the system interacting via realistic potentials. We calculated the ground-state energies of the triton and the alpha-particle by using the AV6 interaction [162]. The results by the global vector representation and by the partial-wave expansion are compared to each other in Table 11.6. The model space allows for natural parity states only. The partial-wave expansion is restricted for partial waves with orbital angular momentum 0, 1 and 2. These partial waves give a good description of the ground states of these nuclei [32]. The table shows that one can reach the same results with the global vector representation as with the partial-wave expansion even in the more complicated case when the tensor force couples the different orbital angular momenta. The case of the alpha-particle is a nice illustration of the main advantage of the global vector representation over the partial-wave expansion: Even in the expansion truncated to the natural parity states, one needs 24 partial-wave combinations due to the large number of combinations of the individual partial waves, while in the global vector representation only 4 sets of (K, L) ($(K, L) = (0,0), (1,0), (0,2), (1,2)$) were used to reproduce the same result [166]. By increasing the number of particles the partial-wave expansion would contain a prohibitively large number of combinations. The number of sets (K, L), on the contrary, does not depend on the number of particles. Only the number of parameters u_i increases (linearly) by adding particles to the system.

Table 11.6. Total energies of the ground states of the triton and the alpha-particle with the AV6 interaction [162]. Only natural parity states have been taken into account for the orbital motion. Two types of angular functions are used; the global vector representation (GVR) and the partial-wave expansion (PWE). K is the basis dimension.

^{3}H			^{4}H		
Method	K	E (MeV)	Method	K	E (MeV)
GVR	100	−6.80	GVR	300	−23.29
PWE	100	−6.80	PWE	300	−23.27

Complements

11.1 *Correlations in few-nucleon systems*

To elucidate the significance of the correlation we calculate the binding energies of the triton and the alpha-particle by using the correlated Gaussians and the correlated Gaussian-type geminals. The spin-isospin parts of the triton and the alpha-particle wave functions are considered totally antisymmetric and, as the nucleon is a fermion, their spatial parts must be totally symmetric. The spatial motion in the ground state of the $N = 3$- or 4-nucleon system may thus be treated as motion of bosons. We obtain variational solutions for the triton and the alpha-partile interacting via a spin-isospin independent two-nucleon potential.

The orbital wave function is assumed to be given by combinations of the correlated Gaussians with $L = 0$, see Sect. 7.2:

$$\Psi = \sum_{k=1}^{K} c_k \mathcal{S} \exp\Big(-\frac{1}{2}\tilde{\boldsymbol{x}} A_k \boldsymbol{x} \Big). \tag{11.13}$$

Here $\boldsymbol{x}$ is the Jacobi coordinate set and $\mathcal{S}$ is a symmetrizer. Each of the positive-definite symmetric matrices A_k contains nonlinear variational parameters, three for the triton and six for the alpha-article. The parameters can be determined by the SVM. The calculation was done with the code of [81]. The number K of the basis functions has been chosen so as to reach convergence.

As another type of trial function, one may try the following correlated wave function

$$\Psi = F\Phi = F \prod_{i=1}^{N} \Big(\frac{\nu}{\pi}\Big)^{\frac{3}{4}} \exp\Big(-\frac{1}{2}\nu \boldsymbol{r}_i{}^2 \Big). \tag{11.14}$$

The function Φ, a product of the nodeless harmonic-oscillator functions (Gaussians), describes the independent single-particle motion, whereas the function F takes into account the correlations. The Jastrow-type correlation factor [47] is taken, in a single Gaussian form, as

$$\begin{aligned} F &= \prod_{j>i=1}^{N} \Big(1 + a \exp\big\{-b(\boldsymbol{r}_i - \boldsymbol{r}_j)^2\big\}\Big) \\ &= 1 + \sum_{k=1}^{K_N} \alpha_k \exp\Big(-\frac{1}{2}\tilde{\boldsymbol{r}} B_k \boldsymbol{r} \Big) \end{aligned} \tag{11.15}$$

with $K_N = 2^{\binom{N}{2}} - 1$, where α_k is given by an integer power of a, and B_k is an $N \times N$ symmetric matrix of type, $2b\sum_i u_i\widetilde{u_i}$, where u_i is an N-dimensional vector which has only two nonzero elements. Therefore the trial function Ψ becomes a combination of the correlated Gaussian-type geminals (6.26) with $\boldsymbol{n} = 0$ and $\boldsymbol{R} = 0$, and the needed formulas for matrix elements involving such special geminals were given in that section. Three parameters ν, a, and b can be varied to minimize the energy. Note that Ψ is totally symmetric and that its dependence on the center-of-mass coordinate $\boldsymbol{x}_N$ can be factored out from the intrinsic motion. This makes it possible to calculate the intrinsic energy.

We also consider a special type of the correlation factor which contains only the pair correlation

$$F_2 = 1 + \sum_{k=1}^{K} a_k \left\{ \sum_{j>i=1}^{N} \exp\Big(- b_k(\boldsymbol{r}_i - \boldsymbol{r}_j)^2\Big) \right\}. \tag{11.16}$$

Here the pair correlation can be better described with combinations of K Gaussian functions than with Eq. (11.15). It is expected that the Jastrow-type correlation becomes more useful in cases where various correlations of more than two particles play a major role. The nonlinear parameters b_k and ν and linear parameters a_k are determined to minimize the energy.

We use three potentials, the MT-V [144], V_1 [145] and B_1 [148] potentials listed in Table 11.1. The space exchange operator P^r of the potential can be set equal to unity because of the bosonic symmetry of the spatial wave function. The energy is thus independent of the Majorana mixture parameters.

Table 11.7 summarizes the results. The result for $N = 3$ and 4 [31] calculated with the correlated Gaussian basis is known to be in good agreement with other accurate calculations available in the literature. It is interesting to note that, except for the alpha-article with the MT-V potential, the pair correlation with $K = 2$ terms already gives lower energy than the Jastrow-type correlation described with a single Gaussian function. (The alpha-particle has more compact structure than the triton. Because of this property, the strong repulsion of the MT-V potential has to be fully taken care of in order to obtain the ground-state energy particularly for the alpha-particle. The correlation factor of Jastrow type appears to be more suited than just the pair correlation of Eq. (11.16) in this regard.) The small difference between $K = 2$ and $K = 3$ calculations indicates that the pair correlation can well be described with two Gaussians. A better description of the pair correlation appears to be important. The trial function with $K = 0$ is nothing but a product of the single-particle wave functions and contains no correlation. This function misses a lot of energy. This becomes more evident in more singular potentials such as the MT-V potential. Note that there is a significant energy difference between the CG and CGG results with the MT-V potential. This suggests that at least a triple-particle correlation, e.g.,

$$\sum_{k>j>i=1}^{N} \exp\Big(- b(\boldsymbol{r}_i - \boldsymbol{r}_j)^2 - b(\boldsymbol{r}_j - \boldsymbol{r}_k)^2 - b(\boldsymbol{r}_k - \boldsymbol{r}_i)^2 \Big), \quad (11.17)$$

has to be taken into account in the correlation factor for this potential to obtain accurate energy. This problem was investigated in [167], which takes into account the two-, three- and even more-particle correlations and shows the importance of triplet correlations in producing accurate energy.

Table 11.7. Energies of the triton and the alpha-particle in the correlated Gaussians (CG) of Eq. (11.13) and the correlated Gaussian-type geminals (CGG) of Eq. (11.14), which are chosen to represent correlations in one of two forms. The correlation factor F of CGG is defined by Eq. (11.15) for Jastrow and by Eq. (11.16) for the number K of pair correlations. See Table 11.1 for the nucleon-nucleon potential. The energies are in units of MeV.

	CG	CGG				
		Jastrow	$K = 0$	$K = 1$	$K = 2$	$K = 3$
MT-V [144]						
triton	−8.25	−6.00	unbound	−5.78	−7.11	−7.27
alpha	−31.36	−29.04	−6.41	−27.03	−28.47	−28.93
V_1 [145]						
triton	−8.46	−6.99	−6.66	−6.99	−8.11	−8.11
alpha	−30.42	−29.14	−27.92	−29.13	−30.06	−30.13
B_1 [148]						
triton	−11.64	−9.95	−7.00	−9.87	−11.20	−11.20
alpha	−38.34	−36.73	−28.19	−36.02	−37.60	−37.60

When the interaction has spin-isospin dependence, the correlation factor is also expected to be state-dependent. The evaluation of matrix elements with such correlation factors would then be more involved. Refer to [168] for a recent development on the application of the correlated wave functions to heavier nuclei.

11.2 *Convergence of partial-wave expansions*

Here we examine the convergence of the partial-wave expansion (PWE) in obtaining bound-state solution for few-body systems. We consider examples of three-body systems consisting of p, n, and n (triton) or of p, n, and a hyperon Λ. The Λ is a neutral baryon with mass $m_\Lambda c^2 = 1116$ MeV. It has spin 1/2, isospin 0 and strangeness −1. The $N\Lambda$ interaction is not strong enough to form a bound state between a nucleon and a Λ. When one or few Λ's are embedded in a nucleus, they form a bound system called a Λ hypernucleus. The simplest hypernucleus is the (p, n, Λ) system ${}^3_\Lambda$H, which is called a hypertriton and its ground state has $J^\pi = 1/2^+$.

The binding energy of ${}^3_\Lambda$H is 2.4 MeV from the three-particle threshold. As neither subsystem of $p\Lambda$ nor $n\Lambda$ forms a bound state, Λ couples weakly with p and n. In fact the Λ separation energy of ${}^3_\Lambda$H is only about 0.2 MeV. Most of the binding energy of ${}^3_\Lambda$H thus comes from that of the pn subsystem, namely the ground-state energy is just 0.2 MeV below the threshold of deuteron (d) + Λ. It would be natural to attempt to describe the ground-state wave function Ψ in terms of the coordinates $\boldsymbol{x}_1 = \boldsymbol{r}_p - \boldsymbol{r}_n$ and $\boldsymbol{x}_2 = (\boldsymbol{r}_p + \boldsymbol{r}_n)/2 - \boldsymbol{r}_\Lambda$.

To simplify the analysis we assume that the pn subsystem has the same spin and isospin state as d, that is spin 1 and isospin 0. This assumption, known to be a rather good approximation, leads us to the condition that Ψ must be an even function of $\boldsymbol{x}_1$. In addition we assume that the total orbital angular momentum of the system is zero, and that the three 1/2 spins are coupled to $J = 1/2$. Under these assumptions Ψ can in general be expressed as follows:

$$\Psi = f(x_1, x_2, \boldsymbol{x}_1 \cdot \boldsymbol{x}_2)\,[\chi_1(pn) \times \chi_{\frac{1}{2}}(\Lambda)]_{\frac{1}{2}M}, \tag{11.18}$$

where the space part f is a function of x_1, x_2 and $\boldsymbol{x}_1 \cdot \boldsymbol{x}_2$. For Ψ to be an even function of $\boldsymbol{x}_1$, f has to be an even function of $\boldsymbol{x}_1 \cdot \boldsymbol{x}_2$.

The Hamiltonian of the system is given by

$$\begin{aligned} H &= T + V \\ &= -\frac{\hbar^2}{2\mu_1}\frac{\partial^2}{\partial \boldsymbol{x}_1{}^2} - \frac{\hbar^2}{2\mu_2}\frac{\partial^2}{\partial \boldsymbol{x}_2{}^2} + V_{pn} + V_{p\Lambda} + V_{n\Lambda}, \end{aligned} \tag{11.19}$$

where $\mu_1 = M_N/2$, and $\mu_2 = 2M_N M_\Lambda/(2M_N + M_\Lambda)$. Only the central potential is included in the Hamiltonian. The pn interaction V_{pn} in the triplet even state is taken from the Minnesota potential [147] with $u = 1$ (see Table 11.1). The $N\Lambda$ central potential is assumed to have the form of Minnesota type

$$V_{N\Lambda} = \left(V_R + \frac{1}{2}(1+P^\sigma)V_t + \frac{1}{2}(1-P^\sigma)V_s\right) \times \left(\frac{1}{2}u + \frac{1}{2}(2-u)P^r\right) \tag{11.20}$$

with

$$V_R = V_{0R}\mathrm{e}^{-\kappa_R r^2}, \quad V_t = -V_{0t}\mathrm{e}^{-\kappa_t r^2}, \quad V_s = -V_{0s}\mathrm{e}^{-\kappa_s r^2}. \tag{11.21}$$

The u value is set equal to 1.5. We assume that the $p\Lambda$ interaction and the $n\Lambda$ interaction are the same, though there is some evidence that they are in fact slightly different. The potential parameters are determined so as to fit $A = 3, 4$ binding-energy data: The strength parameters in units of MeV are $V_{0R} = 200.0$, $V_{0t} = 109.8$, and $V_{0s} = 121.3$, and the ranges in units of fm^{-2} are $\kappa_R = 1.638$, $\kappa_t = 0.7864$, and $\kappa_s = 0.7513$, respectively. This $N\Lambda$ potential will be used in the next Complement to calculate the binding energies of s-shell hypernuclei.

The equation of motion for $f(x_1, x_2, \boldsymbol{x}_1 \cdot \boldsymbol{x}_2)$ is obtained by substituting Eq. (11.18) into the Schrödinger equation $(H - E)\Psi = 0$ and integrating over the spin coordinates. Owing to the symmetry property of the wave function and the assumption of $V_{p\Lambda} = V_{n\Lambda}$, we only need to evaluate the spin matrix element of, say $V_{n\Lambda}$. The matrix element of P^σ exchanging the spin coordinates of the $n\Lambda$ pair can be calculated by rewriting the spin function of Eq. (11.18), expressed in $(pn)\Lambda$ coupling order, to that of the $p(n\Lambda)$ order by using the Racah coefficient in unitary form (see Complement 6.3):

$$\begin{aligned}[\chi_1(pn) \times \chi_{\frac{1}{2}}(\Lambda)]_{\frac{1}{2}M} &= \sum_{S=0,1} U(\tfrac{1}{2}\tfrac{1}{2}\tfrac{1}{2}\tfrac{1}{2}; 1S)\, [\chi_{\frac{1}{2}}(p) \times \chi_S(n\Lambda)]_{\frac{1}{2}M} \\ &= \frac{\sqrt{3}}{2}\, [\chi_{\frac{1}{2}}(p) \times \chi_0(n\Lambda)]_{\frac{1}{2}M} + \frac{1}{2}\, [\chi_{\frac{1}{2}}(p) \times \chi_1(n\Lambda)]_{\frac{1}{2}M}.\end{aligned} \tag{11.22}$$

Then the desired matrix element is

$$\langle [[\chi_1(pn) \times \chi_{\frac{1}{2}}(\Lambda)]_{\frac{1}{2}}]_{\frac{1}{2}M} | P^\sigma | [[\chi_1(pn) \times \chi_{\frac{1}{2}}(\Lambda)]_{\frac{1}{2}}]_{\frac{1}{2}M} \rangle$$

$$= \sum_S (-1)^{S+1} U(\tfrac{1}{2}\tfrac{1}{2}\tfrac{1}{2}\tfrac{1}{2};1S)\, U(\tfrac{1}{2}\tfrac{1}{2}\tfrac{1}{2}\tfrac{1}{2};1S) = -\tfrac{1}{2}. \qquad (11.23)$$

The equation of motion for f reads as

$$\Bigg\{ -\frac{\hbar^2}{2\mu_1}\frac{\partial^2}{\partial \boldsymbol{x}_1{}^2} - \frac{\hbar^2}{2\mu_2}\frac{\partial^2}{\partial \boldsymbol{x}_2{}^2} + V_{pn}(x_1) + V_{p\Lambda}\Big(\Big|\boldsymbol{x}_2 + \frac{1}{2}\boldsymbol{x}_1\Big|, P^r\Big)$$

$$+ V_{n\Lambda}\Big(\Big|\boldsymbol{x}_2 - \frac{1}{2}\boldsymbol{x}_1\Big|, P^r\Big) - E \Bigg\} f(x_1, x_2, \boldsymbol{x}_1 \cdot \boldsymbol{x}_2) = 0, \qquad (11.24)$$

with the boundary condition that f has to vanish for large x_1 and x_2. Here $V_{p\Lambda}(r, P^r)$ is given by

$$V_{p\Lambda}(r, P^r) = \Big(V_R + \frac{1}{4}V_t + \frac{3}{4}V_s\Big)\Big(\frac{1}{2}u + \frac{1}{2}(2-u)P^r\Big). \qquad (11.25)$$

One has to note that P^r exchanges $\boldsymbol{r}_p$ and $\boldsymbol{r}_\Lambda$, which induces the transformation of the coordinates: $\boldsymbol{x}_1 \to \boldsymbol{x}_1/2 - \boldsymbol{x}_2$, $\boldsymbol{x}_2 \to -3\boldsymbol{x}_1/4 - \boldsymbol{x}_2/2$. The potential $V_{n\Lambda}(r, P^r)$ is defined in exactly the same manner, where P^r now transforms $\boldsymbol{x}_1$ and $\boldsymbol{x}_2$ to $\boldsymbol{x}_1/2 + \boldsymbol{x}_2$ and $3\boldsymbol{x}_1/4 - \boldsymbol{x}_2/2$, respectively.

We obtain the ground-state energy and the solution of Eq. (11.24). The variational trial function may be chosen as a combination of the correlated Gaussians (CG) of Eq. (7.3):

$$f(x_1, x_2, \boldsymbol{x}_1 \cdot \boldsymbol{x}_2) = \sum_{k=1}^{K} c_k \Big\{ F_{A_k}(\boldsymbol{x}) + F_{PA_kP}(\boldsymbol{x}) \Big\}, \qquad (11.26)$$

where the second term with $P = \begin{pmatrix} 1 & 0 \\ 0 & -1 \end{pmatrix}$ assures that the trial function is an even function of $\boldsymbol{x}_1$, i.e., for $A = \begin{pmatrix} A_{11} & A_{12} \\ A_{12} & A_{22} \end{pmatrix}$, PAP becomes $PAP = \begin{pmatrix} A_{11} & -A_{12} \\ -A_{12} & A_{22} \end{pmatrix}$.

It is interesting to analyse the partial-wave contents of the solution. For this purpose we calculate the quantity

$$c_l^2 = \langle f(x_1, x_2, \boldsymbol{x}_1 \cdot \boldsymbol{x}_2) | P_l | f(x_1, x_2, \boldsymbol{x}_1 \cdot \boldsymbol{x}_2) \rangle \qquad (11.27)$$

with

$$P_l = \big| [Y_l(\widehat{\boldsymbol{x}_1}) \times Y_l(\widehat{\boldsymbol{x}_2})]_{00} \big\rangle \big\langle [Y_l(\widehat{\boldsymbol{x}_1}) \times Y_l(\widehat{\boldsymbol{x}_2})]_{00} \big|. \qquad (11.28)$$

The result is given in Table 11.8. The admixtures of D, G and higher waves are small and the PWE is very effective in the present case. The root-mean-square (rms) radius of the hypertriton is calculated to be 4.87 fm. The rms distance between the proton and the neutron is 3.61 fm, which is slightly smaller than the corresponding distance (3.90 fm) obtained for the free deuteron, and the rms distance between the Λ and the center-of-mass of the proton and the neutron is as large as 9.98 fm. Therefore the hypertriton is barely bound by polarizing the deuteron slightly in order to increase the binding between the deuteron and the Λ.

Table 11.8. Partial-wave decomposition of the hypertriton wave function

l	0	2	4
c_l^2	0.99913	0.79×10^{-3}	0.6×10^{-4}

The trial function (11.26) gets higher partial-wave contributions from the cross term $\boldsymbol{x}_1\cdot\boldsymbol{x}_2$. The parameters A_{12} determine the appropriate weights of various partial waves. It is of course possible to solve Eq. (11.24) by expanding f in partial waves:

$$f(x_1, x_2, \boldsymbol{x}_1\cdot\boldsymbol{x}_2) = \sum_{l=\text{even}} f_l(x_1, x_2)\,[Y_l(\widehat{\boldsymbol{x}_1}) \times Y_l(\widehat{\boldsymbol{x}_2})]_{00}. \qquad (11.29)$$

Then Eq. (11.24) reduces to a coupled equation for $f_l(x_1, x_2)$'s. Refer to Appendix A.5 for a method of evaluating the matrix element with this type of basis functions in the case that $f_l(x_1, x_2)$ is expanded as a combination of Gaussians $(x_1x_2)^l e^{-ax_1^2-bx_2^2}$. Table 11.9 shows the energy convergence in the partial waves included in the calculation. In conformity with Table 11.8 the inclusion of S wave alone is a good approximation to the hypertriton and higher partial waves give modest contributions to the binding energy.

Noncentral forces necessitate the inclusion of higher partial waves and the PWE will be slowly convergent in general. Even in the case of central potentials the energy convergence in the PWE in one particular Jacobi coordinate set depends on a system and an interaction. To examine the contribution of higher partial waves in the case of central forces we consider another example of the triton (nnp), which is more tightly bound than the hypertriton. The energy has been calculated

Table 11.9. Convergence of the hypertriton energy in the partial-wave expansion (PWE). The CG indicates the result obtained with the correlated Gaussian of Eq. (11.26). The energy is from the three-particle threshold. The deuteron energy is −2.202 MeV.

		PWE		CG
	l_{max}	0	2	
E (MeV)		−2.305	−2.358	−2.378

by using the wave function of type (11.29), where $\boldsymbol{x}_1$ is the relative distance vector between the two neutrons and $\boldsymbol{x}_2$ is the relative coordinate between the proton and the center-of-mass of the neutrons. The two neutrons are assumed to be in a spin-singlet state. Two potentials are used: One is the soft-core Volkov (V_1) potential [145] and the other is the repulsive-core ATS3 potential [146]. Table 11.10 compares the PWE results with other methods, the hyperspherical harmonics (HH) expansion method (see Sect. 5.2) and the CG calculation. The PWE gives fairly fast convergence for the V_1 potential and the inclusion of S and D waves already produces the energy that is close to the accurate value by the HH method. However, for the ATS3 potential the convergence is rather slow and the partial waves of up to $l = 6$ are not sufficient to get the energy which is close to the value calculated with the CG basis (11.26). (The two neutrons are assumed to be in a spin-singlet state, but, as the ATS3 potential is spin dependent, to allow them to be in a spin-triplet state as well leads to lower energy than the result of Table 11.10. The CG calculation including this possibility gives −8.765 MeV (the rms radius is 1.67 fm) in $K = 100$ dimension, which is in perfect agreement with the energy [170] obtained by the Faddeev method (see Sect. 5.3).

Here we explain why the ATS3 potential leads to the slower convergence than the V_1 potential. The matrix element for the central potential between the two neutrons vanishes for the functions with different l values, so the potential is not responsible for the mixing-in of different partial waves. The potential between the neutron and the proton, however, induces the l-mixing. To see this, we note that the latter potential has the form $V(|a\boldsymbol{x}_1 + b\boldsymbol{x}_2|)$ (for the triton case $a = \pm(1/2)$, $b = 1$, but it is extended to a general case.), which is a function of x_1, x_2 and the angle ϑ between $\boldsymbol{x}_1$ and $\boldsymbol{x}_2$. The potential can be expanded in terms of the multipole operators as follows

Table 11.10. Convergence of the triton energy in the partial-wave expansion (PWE). The two neutrons are assumed to be in a spin-singlet state. See Table 11.1 for the nucleon-nucleon potential.

V₁ [145]						
		PWE				HH[169]
	$l_{\max}$	0	2	4	6	
E (MeV)		−8.005	−8.390	−8.447	−8.460	−8.4647
ATS3 [146]						
		PWE				CG
	$l_{\max}$	0	2	4	6	
E (MeV)		−2.948	−6.215	−7.210	−7.484	−7.616

$$V(|a\boldsymbol{x}_1 + b\boldsymbol{x}_2|) = \sum_{l=0}^{\infty} v_l(x_1, x_2) P_l(\cos\vartheta) \tag{11.30}$$

with

$$v_l(x_1, x_2) = \frac{2l+1}{2} \int_0^{\pi} V(|a\boldsymbol{x}_1 + b\boldsymbol{x}_2|) P_l(\cos\vartheta) \sin\vartheta d\vartheta. \tag{11.31}$$

For example, for the Gaussian potential of $V(r) = V_0 \mathrm{e}^{-\mu^2 r^2}$ we have (see Eq. (6.53) or Eq. (A.81))

$$v_l(x_1, x_2) = V_0(2l+1)\, \epsilon(l, ab)\, i_l(2\mu^2|ab|x_1 x_2)\, \mathrm{e}^{-\mu^2(a^2 x_1^2 + b^2 x_2^2)}, \tag{11.32}$$

where $\epsilon(l, ab) = 1$ for $ab > 0$, $\epsilon(l, ab) = (-1)^l$ for $ab \le 0$. Since the Legendre polynomial $P_l(\cos\vartheta)$ is proportional to the tensor product $[Y_l(\widehat{\boldsymbol{x}_1}) \times Y_l(\widehat{\boldsymbol{x}_2})]_{00}$ (see Eq. (6.54)) and has non-vanishing matrix elements between the functions of different partial waves, we find that the various multipole components contained in the potential bring about the mixing of the higher partial waves. The relative strength of the multipoles thus determines the extent to which each partial wave is contained in the solution. For a long-ranged potential (small μ) the magnitude of $v_l(x_1, x_2)$ is determined by the factor $\mu^{2l}/(2l-1)!!$, so that only low partial waves give main contributions to the solution. However, this is not the case for a short-ranged potential. The shorter the potential range is, the larger the contribution of the higher multipole components. E.g., in the extreme case of $V(|a\boldsymbol{x}_1 + b\boldsymbol{x}_2|) \sim$

$\delta(a\boldsymbol{x}_1 + b\boldsymbol{x}_2)$, by putting $V_0 = (\mu^2/\pi)^{3/2}$ and letting $\mu \to \infty$ in Eq. (11.32) we get $v_l(x_1, x_2) \sim ((2l+1)/(4\pi))\delta(|a|x_1 - |b|x_2)/(|ab|x_1x_2)$, which indicates that the multipole strength is equally strong for arbitrary l. The result of Table 11.10 can be understood in this way from the fact that the ATS3 potential contains the strong short-ranged repulsive-core part.

Clearly the matrix element for the central potential between the neutron and the proton vanishes for the functions of diffrent l values if they are expressed in terms of the different Jacobi coordinate set where the first coordinate is chosen to the relative coordinate between the neutron and the proton. This consideration leads us to the following ansatz for the solution

$$f = \sum_l f_l^{(1)}(x_1^{(1)}, x_2^{(1)}) \left[Y_l(\widehat{\boldsymbol{x}_1^{(1)}}) \times Y_l(\widehat{\boldsymbol{x}_2^{(1)}}) \right]_{00}$$
$$+ \sum_l f_l^{(2)}(x_1^{(2)}, x_2^{(2)}) \left[Y_l(\widehat{\boldsymbol{x}_1^{(2)}}) \times Y_l(\widehat{\boldsymbol{x}_2^{(2)}}) \right]_{00}$$
$$+ \sum_l f_l^{(3)}(x_1^{(3)}, x_2^{(3)}) \left[Y_l(\widehat{\boldsymbol{x}_1^{(3)}}) \times Y_l(\widehat{\boldsymbol{x}_2^{(3)}}) \right]_{00}, \tag{11.33}$$

where $\boldsymbol{x}_1^{(i)}$, $\boldsymbol{x}_2^{(i)}$ $(i = 1, 2, 3)$ stands for the ith Jacobi set (see Sect. 2.4). This type of functions is used in the Faddeev method and the CRCG [26] method. It is known that this expansion converges much faster than the one using one particular Jacobi set, and the inclusion of low partial waves is expected to be sufficient to get accurate solutions (see also Sect. 4.2.1 and [30]). In contrast to this approach, the CG basis of Eq. (11.26) uses only one particular Jacobi coordinate but takes account of the contribution of the high partial waves by the cross term $A_{12}\boldsymbol{x}_1 \cdot \boldsymbol{x}_2$ in the exponent of the trial function. This ensures accurate solutions in the correlated Gaussian approach. It is noted that the calculation of matrix elements with high l values is computer time consuming (see Appendix A.5), whereas in the CG basis no such PWE is employed, which makes it possible to calculate matrix elements very quickly.

11.3 *Quark Pauli effect in s-shell Λ hypernuclei*

Among a few known Λ hypernuclei, ${}^3_\Lambda$H, ${}^4_\Lambda$H, ${}^4_\Lambda$He, and ${}^5_\Lambda$He are called s-shell hypernuclei because in the simplest version of the shell model all the constituents can reside in an s orbit. By applying the $L = 0$

correlated Gaussians of Eq. (7.3), we calculate the binding energies of s-shell Λ hypernuclei in order to reveal the significant role of the quark substructure of baryons in resolving the long-standing problem of ${}^{5}_{\Lambda}$He.

Since the pioneering work of [171], the s-shell Λ hypernuclei have offered several intriguing problems, among others, the anomalously small binding of ${}^{5}_{\Lambda}$He. According to a recent survey of hypernuclear physics [172], "The anomalously small binding of ${}^{5}_{\Lambda}$He remains an enigma. Simple model calculations based upon ΛN potentials, parametrized to account for the low-energy ΛN scattering data and the binding energy of the $A = 3$, 4 Λ-hypernuclei, overbinds ${}^{5}_{\Lambda}$He by 2-3 MeV."

The trial function is given as a combination of the correlated Gaussians:

$$\Psi_{JMTM_T} = \sum_k c_k \mathcal{A}\{F_{A_k}(\boldsymbol{x})\chi_{kJM}\eta_{kTM_T}\}, \tag{11.34}$$

where the operator $\mathcal{A}$ antisymmetrizes the nucleon coordinates, χ_{kJM} is the spin function (Since $L = 0$, the spin S is equal to the total angular momentum J), and η_{kTM_T} is the isospin function. The spin and isospin functions are obtained by successive couplings, for example, $\chi_{kJM} = \chi_{(S_{12}S_{123}\ldots)JM} = [[[\chi_{1/2} \times \chi_{1/2}]_{S_{12}} \times \chi_{1/2}]_{S_{123}}\cdots]_{JM}$ with k representing a set of the intermediate spins $S_{12}, S_{123}, \ldots$. The optimal set was chosen for each matrix A_k. The matrix A_k containing nonlinear parameters was selected by the SVM.

The Minnesota potential [147] with $u = 1$ was used for the NN interaction. The $N\Lambda$ potential is the same as was used in the previous Complement. Table 11.11 lists the results of the calculation. The calculation reproduces the data well except for the case of ${}^{5}_{\Lambda}$He, for which the theory overestimates the binding energy by about 1.9 MeV, consistent with what is mentioned above. Another thing to be noticed is that ${}^{4}_{\Lambda}$He is more strongly bound than ${}^{4}_{\Lambda}$H. It would be difficult to understand this without introducing a charge-dependent component as the $N\Lambda$ interaction, because then ${}^{4}_{\Lambda}$He would be less tightly bound because of the Coulomb repulsion.

We have so far treated a baryon as a structureless particle and distinguished Λ from N. What happens if we take their quark substructure into consideration? Recently it has been shown in [173] that the anomalous binding problem of ${}^{5}_{\Lambda}$He can be, at least partly, resolved by considering the quark substructure of the baryons.

Table 11.11. Binding energies (in units of MeV) of s-shell hypernuclei. The Λ separation energy B_Λ of the hypernucleus ${}^{A}_{\Lambda}\mathrm{X}$ is defined by the difference of the binding energies, $B({}^{A}_{\Lambda}\mathrm{X})-B({}^{A-1}\mathrm{X})$.

	${}^{3}_{\Lambda}\mathrm{H}$	${}^{4}_{\Lambda}\mathrm{H}$	${}^{4}_{\Lambda}\mathrm{He}$	${}^{5}_{\Lambda}\mathrm{He}$
$B({}^{A}_{\Lambda}\mathrm{X})$	2.38	10.62	9.91	34.93
$B({}^{A-1}\mathrm{X})$	2.20	8.38	7.71	29.95
B_Λ(Cal.)	0.18	2.23	2.20	4.98
B_Λ(Exp.)	0.13±0.05	2.04±0.04	2.39±0.03	3.12±0.02

Let us assume for the sake of simplicity that a baryon is a composite particle of three quarks, $p = (u, u, d)$, $n = (u, d, d)$, $\Lambda = (u, d, s)$, and each quark moves in the $0s$ orbit in the harmonic-oscillator well with size parameter b. The spatial part of the three-quark baryon is described by the function

$$\phi_{\mathrm{B}} = (\pi b^2)^{-\frac{9}{4}} \exp\left(-\frac{1}{2b^2}(\boldsymbol{\rho}_1^2 + \boldsymbol{\rho}_2^2 + \boldsymbol{\rho}_3^2)\right)$$

$$= \left(\frac{\pi b^2}{3}\right)^{-\frac{3}{4}} \exp\left(-\frac{3}{2b^2}\boldsymbol{r}_1^2\right) \phi_{\mathrm{B}}^{(\mathrm{int})}, \qquad (11.35)$$

where $\boldsymbol{\rho}_i$ is the quark's position vector, $\boldsymbol{r}_1 = (\boldsymbol{\rho}_1 + \boldsymbol{\rho}_2 + \boldsymbol{\rho}_3)/3$ is the baryon's position vector, and $\phi_{\mathrm{B}}^{(\mathrm{int})}$ is the spatial part of the intrinsic wave function of the baryon depending on $\boldsymbol{\rho}_1 - \boldsymbol{\rho}_2$, $(\boldsymbol{\rho}_1 + \boldsymbol{\rho}_2)/2 - \boldsymbol{\rho}_3$. The value of b can be estimated to be about 0.86 fm by requiring that the function (11.35) reproduce the charge radius of the proton [174]. When the spatial part of the two-baryon wave function is described with the configuration of $\exp\{-3/(2b^2)(\boldsymbol{r}_1^2 + \boldsymbol{r}_2^2)\}$, we interpret it as indicating that all six quarks move in the $0s$ orbit, $\exp(-\boldsymbol{\rho}^2/(2b^2))$, of the common harmonic-oscillator well with size parameter b. By increasing the number of baryons, we thus expect that the many-baryon wave function may receive a special constraint arising from the quark Pauli principle that any single-particle orbit can accommodate at most six quarks (three colors and up-down spins) for each flavor.

It is easy to see that we have no apparent quark Pauli-forbidden states up to $A = 4$ hypernuclei. However, this is not the case for ${}^{5}_{\Lambda}\mathrm{He}$: Four nucleons of ${}^{5}_{\Lambda}\mathrm{He}$, when they are on top of one another, have already six u-quarks and six d-quarks in the $0s$ orbit, so neither u nor d quark of Λ can take the same $0s$ orbit. This leads us to the conclusion that the five baryons cannot take a configuration of

$\exp(-3/(2b^2)\sum_{i=1}^5 \boldsymbol{r}_i^2)$. To be more specific, the (normalized) Pauli-forbidden state for ${}^5_\Lambda$He, with its center-of-mass motion being separated, is given by

$$\Psi_{\rm PF}(\boldsymbol{x}) = \mathcal{N}^{\frac{3}{4}}\Big(\frac{\pi b^2}{3}\Big)^{-3} \exp\Big(-\frac{3}{2b^2}\sum_{i=1}^{5}(\boldsymbol{r}_i - \boldsymbol{x}_5)^2\Big), \tag{11.36}$$

where $\boldsymbol{x}_5$ is the center-of-mass coordinate of the five baryons. The normalization constant is given by $\mathcal{N} = (4m_N^2 + m_\Lambda^2)/(4m_N + m_\Lambda)^2$, where m_N is the mass of N. Other Pauli-forbidden states might possibly exist, but this is the simplest and most evident one.

The solution Ψ obtained above for ${}^5_\Lambda$He was found to contain the Pauli-forbidden component of Eq. (11.36) by only 0.44%, but its consequence is very significant: If this Pauli-forbidden component is simply subtracted from the wave function Ψ, the B_Λ value would have been changed from 4.98 MeV to 2.74 MeV, no anomalously small binding of ${}^5_\Lambda$He! The reduction is mainly due to the fact that the energy expectation value in $\Psi_{\rm F}(\boldsymbol{x})$ is very large, i.e., 513.6 MeV. However, this may be a premature conclusion from the viewpoint of the variational principle because the reduction in the binding energy is based on a calculation of the so-called variation before projection type (see Sect. 6.1), that is, the variation has been done before the elimination of the Pauli-forbidden component is made from the trial function.

To calculate the binding energy more precisely by taking into account the quark Pauli effect, we have repeated the calculation by replacing the correlated Gaussian in Eq. (11.34) with the one that has no Pauli-forbidden component:

$$\bar{F}_{A_k}(\boldsymbol{x}) = F_{A_k}(\boldsymbol{x}) - \Psi_{\rm PF}(\boldsymbol{x})\langle\Psi_{\rm PF}(\boldsymbol{x})|F_{A_k}(\boldsymbol{x})\rangle. \tag{11.37}$$

Figure 11.4 displays the Λ separation energy B_Λ of ${}^5_\Lambda$He as a function of b. The binding energy decreases rather rapidly around $b = 0.7$ fm and reaches 3.63 MeV at $b = 0.86$ fm. The discrepancy from the observed binding energy has now been reduced to just 0.5 MeV. Though this scenario may not be a final solution for resolving the long-standing problem of ${}^5_\Lambda$He, we have clearly seen that the binding mechanism of the Λ hypernuclei reveals an intriguing role of the Pauli principle which acts at the quark level.

11.4 *The* ^{12}C *nucleus as a system of three alpha-particles*

As an example of the application of nonlocal potentials, we take up a simple model for the ^{12}C nucleus, the 3α model. In this model ^{12}C

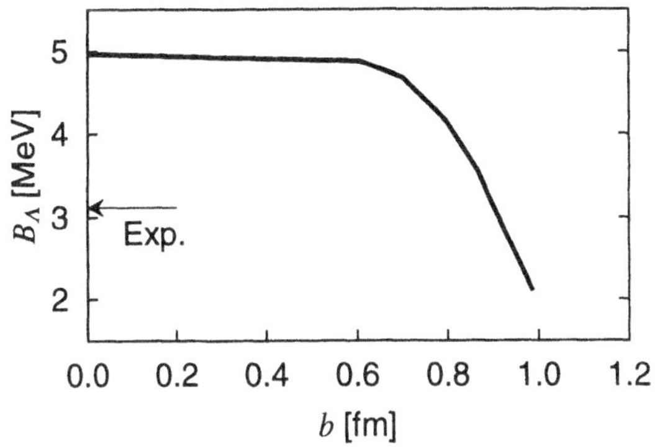

Fig. 11.4. The Λ separation energy B_Λ of ${}^5_\Lambda$H as a function of the size parameter b of three-quark baryons

containing six protons and six neutrons is approximated as a bosonic system of three alpha-particles. The physical reason behind this picture is in the fact that the alpha-particle is a strongly bound, very stable system compared to light neighboring nuclei. One needs an energy of about 20 MeV to excite the alpha-particle. Because of this unique feature, some of the light nuclei tend to form a subsystem consisting of two protons and two neutrons, which is called an alpha-cluster. The alpha-cluster would never be identical to the alpha-particle but it may happen that a model description assuming such subsystems can explain many properties of the light nuclei [60, 175, 59].

To calculate the binding energy of ^{12}C in the 3α model, one needs to know the potential between two alpha-particles. It is hard to derive a potential which acts between composite particles through the underlying interactions of the particles composing the composite particles. Thus we use a phenomenological potential which successfully reproduces the alpha-alpha scattering phase shifts. The one employed here is an l-independent local potential consisting of both the nuclear part of Gaussian form and the Coulomb part [176]:

$$V(r) = -V_0\,\mathrm{e}^{-\rho r^2} + \frac{4e^2}{r}\,\mathrm{erf}(\beta r), \tag{11.38}$$

where $V_0 = 122.6225$ MeV, $\rho = 0.22\,\text{fm}^{-2}$, and $\beta = 0.75\,\text{fm}^{-1}$. Though this potential fits the scattering data very well to about $E_{\text{cm}} = 30$ MeV and reproduces the 0^+ resonance at 92 keV quite well, we have to note that it predicts "redundant bound states", two (-72.625 and -25.617 MeV) in the $l = 0$ wave and one (-21.999 MeV) in the $l = 2$ wave. Since the two alpha-particles are known to form no bound state, these bound states are considered spurious and must be excluded from the configuration space. The existence of spurious states for a system of two nuclei is due to the Pauli principle and can be easily understood in the simplest version of the nuclear shell model. This is the basis of the orthogonality condition model [177] which succeeded to give a foundation to the deep local potential of type (11.38) from the microscopic theory of scattering between composite particles.

Let us denote the spurious states as $\varphi_{nlm}(\boldsymbol{r})$, ($n = 0$, 1 for $l = 0$, and $n = 0$ for $l = 2$). We require that the wave function Ψ for the 3α system be free from the spurious components in the alpha-alpha pairs, that is,

$$\langle \varphi_{nlm}(\boldsymbol{r}_i - \boldsymbol{r}_j)|\Psi\rangle = 0. \tag{11.39}$$

Here $\boldsymbol{r}_i$ is the position vector of ith alpha-particle. Unless this condition is imposed, the ground-state energy with the potential (11.38) would be strongly overbound because the alpha-clusters, as bosons, would occupy the lowest possible states. An alternative, convenient approach to eliminate reasonably well the spurious components is to use the nonlocal potential of projection operator type

$$\Gamma_{ij}^{nlm} = |\varphi_{nlm}(\boldsymbol{r}_i - \boldsymbol{r}_j)\rangle\langle\varphi_{nlm}(\boldsymbol{r}_i - \boldsymbol{r}_j)|, \tag{11.40}$$

and include it in the Hamiltonian as a kind of "pseudo potential" [178]

$$H = T - T_{\text{cm}} + \sum_{j>i=1}^{N} V_{ij}(|\boldsymbol{r}_i - \boldsymbol{r}_j|) + \lambda \sum_{j>i=1}^{N} \sum_{nlm} \Gamma_{ij}^{nlm}. \tag{11.41}$$

Here λ is a positive constant chosen to be very large. The idea is that if the wave function contains the spurious components, then the energy would be comparatively high. Therefore, the variational lowest solution would approach a state that has a negligible overlap with the spurious components.

The spurious bound state can be well approximated in the form

$$\varphi_{nlm}(\boldsymbol{r}) = \sum_a c_a e^{-\frac{1}{2}ar^2} \mathcal{Y}_{lm}(\boldsymbol{r}). \tag{11.42}$$

Then the nonlocal potential summed over m takes the form

$$\langle \boldsymbol{r}'| \sum_m \Gamma_{ij}^{nlm} |\boldsymbol{r}\rangle = \frac{2l+1}{4\pi}(rr')^l P_l(\cos\vartheta) \sum_{aa'} c_a{}^* c_{a'}\, \mathrm{e}^{-\frac{1}{2}ar^2-\frac{1}{2}a'r'^2}, \tag{11.43}$$

where ϑ is the angle between $\boldsymbol{r}$ and $\boldsymbol{r}'$. The trial function for the 3α system was chosen as a linear combination of the correlated Gaussians, $F_A(\boldsymbol{x})$, of Sect. 7.2. The matrix element of the nonlocal potential (11.43) in this basis can be calculated with the use of Eq. (7.46).

The high sensitivity of the energy on λ calls for numerical calculations with a high accuracy in order for the pseudo potential to play the role of the Pauli projector. It was found [179] that the projection effect is so strong that the partial waves $l = 0$ and 2 do not contain bound states, and a contribution of more than 95 % to the ground-state energy is due to the partial wave $l = 4$. Table 11.12 lists the energies of the ground state and the excited states obtained in a variational calculation [180] using the basis function of Eq. (11.33). The radial part $f_l(x_1, x_2)$ is expanded as combinations of Gaussians $x_1^l x_2^l \exp(-\alpha x_1^2 - \beta x_2^2)$. The calculation does not reproduce the experimental energies very well. In [180] the energies of the 2_1^+ and 4_1^+ states are also given. They are -3.77 and -2.25 MeV, respectively, which are compared to the experimental energies of -2.84 and 6.80 MeV. Here the discrepancy between theory and experiment is more serious: The energy of the 2_1^+ state is lower than that of the 0_1^+ state and the energy of the 4_1^+ state is much lower than the experimental energy. Note that there is no Pauli-forbidden state in the $l \geq 4$ waves, which makes the contribution of the high partial waves too much important. This calculation suggests that the 3α model with the local $\alpha\alpha$ potential has only limited success and is not very realistic to reproduce the experimental energy spectra. This doe not, however, exclude the

Table 11.12. Energies of the 0^+ states of ^{12}C calculated in the 3α model [180]. The energy is measured from the threshold of the 3α breakup.

	0_1^+	0_2^+	0_3^+
Cal. (MeV)	-3.38	-1.43	3.70
Exp. (MeV)	-7.27	0.38	3.03

usefulness of a microscopic α-cluster model [175] which explicitly employs the nucleonic degrees of freedom together with an appropriate nucleon-nuclcon potential.

The second 0^+ state of ^{12}C lies just 0.29 MeV above the ^{8}Be+α threshold. The existence of such a state was predicted in 1953 by Hoyle [181] to understand the mechanism to synthesize ^{12}C, the fourth most abundant nuclear species in the universe. Because of the instabilities of the nuclei with a mass number $A = 5$ and $A = 8$, ^{12}C has to be created using only alpha-particles by the triple-α process, $3\alpha \longrightarrow {}^{12}\mathrm{C}$:

$$\alpha + \alpha \rightleftarrows {}^{8}\mathrm{Be}, \quad \text{and} \quad {}^{8}\mathrm{Be}(\alpha,\gamma)^{12}\mathrm{C}.$$

In order for the triple-α process to produce the sufficient synthesis of ^{12}C, the second reaction of the process has to proceed through an S-wave resonance near the threshold. The experimental confirmation of Hoyle's prediction was made with success in 1957 [182]. See [183] for this subject and recent developments in nuclear astrophysics.

The second 0^+ state is one of the well-known examples which are hardly understandable in the nuclear shell model. It was discussed in [154] that the state could be a possible candidate of an "Efimov state". If a short range two-body force permits just one bound or resonant state with its energy being (very close to) zero just as the potential of the two alpha-particles, then the three-body system interacting with such a force may have a series of bound states because the constituent particles can interact attractively at long distances. In reality the Coulomb interaction between the alpha-particles makes the discussion in ^{12}C complicated, and no ideal Efimov states have so far been observed.

Appendix

Matrix elements for general Gaussians

The matrix element for the correlated Gaussian with arbitrary angular momentum is obtained in a closed form for most physically important potentials including central, tensor and spin-orbit components. A simple and straightforward method is presented to calculate the matrix element for the general correlated Gaussian-type geminals.

A.1 Correlated Gaussians

We start with Eq. (7.2) to evaluate the matrix element for a general case. The matrix element between the generating functions g is given in Table 7.1 and one has to perform the operations prescribed in Eq. (7.2).

A.1.1 Overlap of the basis functions

The overlap matrix element can be obtained through

$$\langle f_{K'LM}(u', A', \boldsymbol{x}) | f_{KLM}(u, A, \boldsymbol{x}) \rangle$$

$$= \frac{1}{B_{KL}B_{K'L}} \iint d\hat{\mathrm{e}}\, d\hat{e'} Y_{LM}(\hat{\mathrm{e}}) Y_{LM}(\hat{e'})^* \Bigg(\frac{d^{2K+L+2K'+L}}{d\lambda^{2K+L} d\lambda'^{2K'+L}}$$

$$\times \left(\frac{(2\pi)^{N-1}}{\det B} \right)^{\frac{3}{2}} \exp\left[q\lambda^2 + q'\lambda'^2 + \rho\lambda\lambda' \boldsymbol{e}\cdot\boldsymbol{e}' \right] \Bigg)_{\substack{\lambda=0,\lambda'=0 \\ e=|\boldsymbol{e}|=1 \\ e'=|\boldsymbol{e}'|=1}}, \quad \text{(A.1)}$$

where

$$B = A + A', \quad q = \frac{1}{2}\tilde{u}B^{-1}u, \quad q' = \frac{1}{2}\widetilde{u'}B^{-1}u', \quad \rho = \widetilde{u'}B^{-1}u. \tag{A.2}$$

To perform the operation prescribed in Eq. (A.1), we use the expansion

$$\exp\left[q\lambda^2 + q'\lambda'^2 + \rho\lambda\lambda' \boldsymbol{e}\cdot\boldsymbol{e}'\right]$$
$$= \sum_{n=0}^{\infty}\sum_{n'=0}^{\infty}\sum_{m=0}^{\infty} H(n,q)H(n',q')H(m,\rho)\lambda^{2n+m}\lambda'^{2n'+m}(\boldsymbol{e}\cdot\boldsymbol{e}')^m, \tag{A.3}$$

where $H(n,x)$ is introduced to simplify the notation and defined by

$$H(n,x) = \begin{cases} \frac{x^n}{n!} & \text{for non-negative integer } n \\ 0 & \text{otherwise.} \end{cases} \tag{A.4}$$

Differentiating with respect to λ and λ', followed by $\lambda = \lambda' = 0$, gives non-vanishing contribution only when $2n + m = 2K + L$ and $2n' + m = 2K' + L$, while the integration over the angles of $\boldsymbol{e}$ and $\boldsymbol{e}'$ becomes nonzero if m is equal to $L + 2k$ with non-negative integer k (see Eq. (6.18)):

$$\iint d\hat{\boldsymbol{e}}\, d\hat{\boldsymbol{e}'} Y_{LM}(\hat{\boldsymbol{e}})Y_{LM}(\hat{\boldsymbol{e}'})^*(\boldsymbol{e}\cdot\boldsymbol{e}')^{2k+L}\Big|_{e=1,e'=1} = B_{kL}. \tag{A.5}$$

Rewriting $n = K - k$ and $n' = K' - k$, we obtain the result for the overlap matrix element

$$\langle f_{K'LM}(u', A', \boldsymbol{x})|f_{KLM}(u, A, \boldsymbol{x})\rangle$$
$$= \frac{(2K+L)!(2K'+L)!}{B_{KL}B_{K'L}}\left(\frac{(2\pi)^{N-1}}{\det B}\right)^{\frac{3}{2}}$$
$$\times \sum_{k=0}^{\min(K,K')} H(K-k,q)H(K'-k,q')H(L+2k,\rho)B_{kL}. \tag{A.6}$$

The B_{nl} value is given in Eq. (6.9). Note that the values of K and K' can usually be chosen to be small in practical cases, and then the sum over k is limited to just a few terms. In particular, for the special case of $K = K' = 0$ the above result simplifies to

$$\langle f_{0LM}(u', A', \boldsymbol{x}) | f_{0LM}(u, A, \boldsymbol{x}) \rangle$$
$$= \frac{(2L+1)!!}{4\pi} \left(\frac{(2\pi)^{N-1}}{\det B} \right)^{\frac{3}{2}} \rho^L. \tag{A.7}$$

A.1.2 Kinetic energy

The kinetic energy with the center-of-mass kinetic energy subtracted is given in Eqs. (2.10) and (2.11). It is simply written as $\tilde{\boldsymbol{\pi}} \Lambda \boldsymbol{\pi}/2$. The matrix element of the kinetic energy is then easily obtained by using Table 7.1 as

$$\langle f_{K'LM}(u', A', \boldsymbol{x}) | \frac{1}{2} \tilde{\boldsymbol{\pi}} \Lambda \boldsymbol{\pi} | f_{KLM}(u, A, \boldsymbol{x}) \rangle$$
$$= \frac{\hbar^2}{2B_{KL}B_{K'L}} \left(\frac{(2\pi)^{N-1}}{\det B} \right)^{\frac{3}{2}} \iint d\hat{e}\, d\hat{e'} Y_{LM}(\hat{e}) Y_{LM}(\hat{e'})^*$$
$$\times \left(\frac{d^{2K+L+2K'+L}}{d\lambda^{2K+L} d\lambda'^{2K'+L}} \left[R + P\lambda^2 + P'\lambda'^2 + Q\lambda\lambda' \boldsymbol{e}\cdot\boldsymbol{e}' \right] \right.$$
$$\left. \times \exp\left[q\lambda^2 + q'\lambda'^2 + \rho\lambda\lambda' \boldsymbol{e}\cdot\boldsymbol{e}' \right] \right)_{\substack{\lambda=0,\lambda'=0 \\ e=1,e'=1}}, \tag{A.8}$$

where

$$R = 3\mathrm{Tr}(AB^{-1}A'\Lambda), \qquad P = -\tilde{u}B^{-1}A'\Lambda A'B^{-1}u,$$
$$P' = -\widetilde{u'}B^{-1}A\Lambda AB^{-1}u', \qquad Q = 2\,\widetilde{u'}B^{-1}A\Lambda A'B^{-1}u. \tag{A.9}$$

The matrix element of the kinetic energy can be obtained by using a manipulation similar to the overlap case:

$$\langle f_{K'LM}(u', A', \boldsymbol{x}) | \frac{1}{2} \tilde{\boldsymbol{\pi}} \Lambda \boldsymbol{\pi} | f_{KLM}(u, A, \boldsymbol{x}) \rangle$$
$$= \frac{\hbar^2}{2} \frac{(2K+L)!(2K'+L)!}{B_{KL}B_{K'L}} \left(\frac{(2\pi)^{N-1}}{\det B} \right)^{\frac{3}{2}}$$
$$\times \sum_k \Big\{ RH(K-k, q)H(K'-k, q')H(L+2k, \rho)$$
$$+PH(K-k-1, q)H(K'-k, q')H(L+2k, \rho)$$

$$+P'H(K-k,q)H(K'-k-1,q')H(L+2k,\rho)$$

$$+QH(K-k,q)H(K'-k,q')H(L+2k-1,\rho)\Big\}B_{kL}. \quad (A.10)$$

The case with $K = K' = 0$ reduces to a very simple result

$$\langle f_{0LM}(u',A',\boldsymbol{x})|\frac{1}{2}\tilde{\pi}\Lambda\pi|f_{0LM}(u,A,\boldsymbol{x})\rangle$$

$$=\frac{\hbar^2}{2}(R+LQ\rho^{-1})\frac{(2L+1)!!}{4\pi}\left(\frac{(2\pi)^{N-1}}{\det B}\right)^{\frac{3}{2}}\rho^L. \quad (A.11)$$

A.1.3 Two-body interactions

Next we derive the matrix element for the interaction of Eqs. (2.3) and (11.1). To evaluate the matrix element of the operator expressed as a tensor product, it is convenient to make use of the famous Wigner-Eckart theorem [70, 71, 72], which states that a matrix element of a spherical tensor operator $O_{\kappa\mu}$ between states with angular momenta JM and $J'M'$ can be expressed as a product of a Clebsch-Gordan coefficient and a reduced matrix element which is independent of the z-components of the angular momenta:

$$\langle J'M'|O_{\kappa\mu}|JM\rangle = \frac{\langle JM\kappa\mu|J'M'\rangle}{\sqrt{2J'+1}}\langle J' \parallel O_\kappa \parallel J\rangle. \quad (A.12)$$

Here the reduced matrix element is expressed by the so-called double-barred matrix element. The factor $1/\sqrt{2J'+1}$ is factored out because then the reduced matrix element is symmetric to the bra and ket interchange

$$\epsilon\langle J \parallel O_\kappa \parallel J'\rangle = (-1)^{J+\kappa-J'}\langle J' \parallel O_\kappa \parallel J\rangle, \quad (A.13)$$

provided that the matrix element is real and the Hermitian conjugate of $O_{\kappa\mu}$ is $(O_{\kappa\mu})^\dagger = \epsilon(-1)^{\kappa-\mu}O_{\kappa\,-\mu}$ with a phase factor ϵ ($\epsilon^2 = 1$).

By applying the Wigner-Eckart theorem we can express the matrix element for the orbital and spin angular momentum coupled wave function as follows:

$$\langle\Psi_{(L'S')JM}|V(|\boldsymbol{r}_i-\boldsymbol{r}_j|)\Big(O_{\kappa_{ij}}(\text{space})\cdot O_{\kappa_{ij}}(\text{spin})\Big)|\Psi_{(LS)JM}\rangle$$

$$=(-1)^\kappa\frac{U(L\kappa JS';L'S)}{\sqrt{(2L'+1)(2S+1)}}$$

$$\times \langle L' \parallel V(|\boldsymbol{r}_i - \boldsymbol{r}_j|) O_{\kappa_{ij}}(\text{space}) \parallel L\rangle \, \langle S' \parallel O_{\kappa_{ij}}(\text{spin}) \parallel S\rangle. \tag{A.14}$$

Here the form factor of the potential $V(\boldsymbol{r})$ is assumed to be a function of r only. The derivation of Eq. (A.14) and its extension to a general coupled tensor operator is given in Exercise A.1.

The reduced matrix element of the spatial part is obtained through

$$\begin{aligned}
&\frac{\langle LM\kappa\mu|L'M'\rangle}{\sqrt{2L'+1}} \langle L' \parallel V(|\boldsymbol{r}_i - \boldsymbol{r}_j|) O_{\kappa_{ij}}(\text{space}) \parallel L\rangle \\
&= \langle L'M'|V(|\boldsymbol{r}_i - \boldsymbol{r}_j|) O_{\kappa\mu_{ij}}(\text{space})|LM\rangle \\
&= \int V(r) \langle L'M'|\delta(\boldsymbol{r}_i - \boldsymbol{r}_j - \boldsymbol{r}) O_{\kappa\mu_{ij}}(\text{space})|LM\rangle d\boldsymbol{r}.
\end{aligned} \tag{A.15}$$

In the case where one uses the orbital and spin angular momentum uncoupled wave function, the matrix element (A.14) can be obtained by

$$\begin{aligned}
&\langle \Psi_{(L'S')JM}|V(|\boldsymbol{r}_i - \boldsymbol{r}_j|)\Big(O_{\kappa_{ij}}(\text{space})\cdot O_{\kappa_{ij}}(\text{spin})\Big)|\Psi_{(LS)JM}\rangle \\
&= (-1)^{\kappa}\sqrt{\frac{2S'+1}{2S+1}} \frac{U(L\kappa JS'; L'S)}{\langle LM_L\kappa\mu|L'M_L'\rangle\langle SM_S\kappa\nu|S'M_S'\rangle} \\
&\times \langle \Psi_{L'M_L'S'M_S'}|V(|\boldsymbol{r}_i - \boldsymbol{r}_j|) O_{\kappa\mu_{ij}}(\text{space}) O_{\kappa\nu_{ij}}(\text{spin})|\Psi_{LM_LSM_S}\rangle.
\end{aligned} \tag{A.16}$$

Since the spin matrix element is easily calculated as will be shown in Appendix A.4, we will focus on the spatial matrix element and evaluate it for the most important components of the nuclear potential, that is, central, tensor and spin-orbit components.

(i) *central and tensor interactions*

The operator $O_{\kappa\mu_{ij}}(\text{space})$ can be 1 for the central potential and $Y_{2\mu}(\widehat{\boldsymbol{r}_i - \boldsymbol{r}_j})$ for the tensor operator. See Eq. (11.7). Thus both the central and tensor components can be treated by assuming the form of $Y_{\kappa\mu}(\widehat{\boldsymbol{r}_i - \boldsymbol{r}_j})$ for $O_{\kappa\mu_{ij}}(\text{space})$. Expressing $\boldsymbol{r}_i - \boldsymbol{r}_j$ as $\widetilde{w^{(ij)}}\boldsymbol{x}$ with the $(N-1)\times 1$ column matrix $w^{(ij)}$ defined in Eq. (2.13) and using Table 7.1, we obtain

$$\langle f_{K'L'M'}(u', A', \boldsymbol{x})|\delta(\boldsymbol{r}_i - \boldsymbol{r}_j - \boldsymbol{r}) Y_{\kappa\mu}(\widehat{\boldsymbol{r}_i - \boldsymbol{r}_j})|f_{KLM}(u, A, \boldsymbol{x})\rangle$$

$$= Y_{\kappa\mu}(\hat{\boldsymbol{r}})\langle f_{K'L'M'}(u', A', \boldsymbol{x})|\delta(\widetilde{w^{(ij)}}\boldsymbol{x} - \boldsymbol{r})|f_{KLM}(u, A, \boldsymbol{x})\rangle$$

$$= \mathrm{e}^{-\frac{1}{2}cr^2} Y_{\kappa\mu}(\hat{\boldsymbol{r}}) \frac{1}{B_{KL}B_{K'L'}} \left(\frac{(2\pi)^{N-2}c}{\det B}\right)^{\frac{3}{2}}$$

$$\times \iint d\hat{\boldsymbol{e}}\, d\hat{\boldsymbol{e}}' Y_{LM}(\hat{\boldsymbol{e}}) Y_{L'M'}(\hat{\boldsymbol{e}}')^* \left(\frac{d^{2K+L+2K'+L'}}{d\lambda^{2K+L} d\lambda'^{2K'+L'}} \right.$$

$$\left. \times \exp\left[\bar{q}\lambda^2 + \bar{q}'\lambda'^2 + \bar{\rho}\lambda\lambda'\boldsymbol{e}\cdot\boldsymbol{e}' + \gamma\lambda\boldsymbol{e}\cdot\boldsymbol{r} + \gamma'\lambda'\boldsymbol{e}'\cdot\boldsymbol{r}\right] \right)_{\substack{\lambda=0,\lambda'=0 \\ e=1,e'=1}}, \tag{A.17}$$

where

$$c^{-1} = \widetilde{w^{(ij)}} B^{-1} w^{(ij)}, \qquad \gamma = c\widetilde{w^{(ij)}} B^{-1} u, \qquad \gamma' = c\widetilde{w^{(ij)}} B^{-1} u',$$

$$\bar{q} = q - \frac{1}{2c}\gamma^2, \qquad \bar{q}' = q' - \frac{1}{2c}\gamma'^2, \qquad \bar{\rho} = \rho - \frac{1}{c}\gamma\gamma'. \tag{A.18}$$

All of these quantities depend on i and j but we omit the labels i and j to simplify notations. The integration over the angles of $\boldsymbol{e}$ and $\boldsymbol{e}'$ can be performed by expanding $\exp(\bar{\rho}\lambda\lambda'\boldsymbol{e}\cdot\boldsymbol{e}' + \gamma\lambda\boldsymbol{e}\cdot\boldsymbol{r} + \gamma'\lambda'\boldsymbol{e}'\cdot\boldsymbol{r})$ in power series and using the relation

$$(\boldsymbol{e}\cdot\boldsymbol{e}')^{n_1}(\boldsymbol{e}\cdot\boldsymbol{r})^{n_2}(\boldsymbol{e}'\cdot\boldsymbol{r})^{n_3}\Big|_{e=e'=1}$$

$$= r^{n_2+n_3} \sum_{LL'\kappa} R^{n_1n_2n_3}_{LL'\kappa} \left[\left[Y_L(\hat{\boldsymbol{e}}) \times Y_{L'}(\hat{\boldsymbol{e}}')\right]_\kappa \times Y_\kappa(\hat{\boldsymbol{r}})\right]_{00} \tag{A.19}$$

with

$$R^{n_1n_2n_3}_{LL'\kappa} = (-1)^{n_1+n_2+n_3} \sum_{l_1l_2l_3} B_{\frac{n_1-l_1}{2}l_1} B_{\frac{n_2-l_2}{2}l_2} B_{\frac{n_3-l_3}{2}l_3}$$

$$\times \sqrt{\frac{(2L+1)(2\kappa+1)}{2l_2+1}} C(l_1l_2;L)C(l_1l_3;L')C(l_2l_3;\kappa)$$

$$\times U(l_1l_3L\kappa;L'l_2). \tag{A.20}$$

See Exercise A.2 for the derivation of this relation. In the defining equation for R the sum over l_1 is limited to $n_1, n_1-2, ..., 1$ or 0, and l_2 and l_3 have similar ranges, respectively. Possible values of L, L', and κ

for a given set of values of n_1, n_2, and n_3 are limited by the conditions that L takes the values $n_1+n_2, n_1+n_2-2, ..., 1$ or 0, and L' and κ take similar values given by n_1, n_3 and n_2, n_3, respectively. In addition, the sum over L, L', κ is restricted to even values of $L+L'+\kappa$. Conversely, for given values of L, L', and κ, all of the n_1+n_2-L, n_1+n_3-L', and $n_2+n_3-\kappa$ have to be non-negative and even, otherwise $R_{LL'\kappa}^{n_1n_2n_3}$ would vanish. Clearly R has the symmetry: $R_{LL'\kappa}^{n_1n_2n_3} = R_{L'L\kappa}^{n_1n_3n_2}$.

The reduced matrix element becomes

$$\begin{aligned}
&\langle f_{K'L'}(u',A',\boldsymbol{x}) \parallel V(|\boldsymbol{r}_i-\boldsymbol{r}_j|)Y_\kappa(\widehat{\boldsymbol{r}_i-\boldsymbol{r}_j}) \parallel f_{KL}(u,A,\boldsymbol{x})\rangle \\
&= (-1)^{L'}\frac{(2K+L)!(2K'+L')!}{B_{KL}B_{K'L'}}\left(\frac{(2\pi)^{N-2}c}{\det B}\right)^{\frac{3}{2}} \\
&\times \sum_{n_1n_2n_3} I(n_2+n_3+2,c)R_{LL'\kappa}^{n_1n_2n_3}H(n_1,\bar{\rho})H(n_2,\gamma)H(n_3,\gamma') \\
&\times H\Big(\frac{2K+L-n_1-n_2}{2},\bar{q}\Big)H\Big(\frac{2K'+L'-n_1-n_3}{2},\bar{q}'\Big),
\end{aligned} \tag{A.21}$$

where the integral I is defined in Eq. (7.10)

The matrix element of the potential for different pairs of particles can be calculated with the above formula by changing only $w^{(ij)}$ in Eq. (A.18).

We note some useful applications of Eq. (A.21). For example, one can calculate the matrix element of $|\boldsymbol{r}_i-\boldsymbol{r}_j|^\alpha$ simply by putting $V(r) = r^\alpha$ and $\kappa=0$. The two-particle correlation function

$$\langle f_{K'L'}(u',A',\boldsymbol{x}) \parallel \delta(|\boldsymbol{r}_i-\boldsymbol{r}_j|-a)Y_\kappa(\widehat{\boldsymbol{r}_i-\boldsymbol{r}_j}) \parallel f_{KL}(u,A,\boldsymbol{x})\rangle \tag{A.22}$$

can also be easily calculated by taking $\delta(r-a)$ for $V(r)$.

We note again that the formula (A.21) simplifies for $K=K'=0$, that is, the triple sum reduces to a single sum over, say n_1 ($n_1 = 0, 1, ..., \min(\ L, L', (L+L'-\kappa)/2)$) and n_2 and n_3 have to satisfy $n_1+n_2 = L$ and $n_1+n_3 = L'$, respectively. In addition, in this special case $R_{LL'\kappa}^{n_1n_2n_3}$ is simply calculated by a term with $l_1=n_1$, $l_2=n_2$, $l_3=n_3$ alone in Eq. (A.20). This simplicity will be used to obtain the matrix element of a density multipole operator. See Complement A.2 for the details.

In the evaluation of the matrix element explained above the integration over the angles $\hat{\boldsymbol{r}}$ has been done in the last step after the coupling of the angular momenta associated with $\boldsymbol{e}$ and $\boldsymbol{e}'$. It is of course possible to reverse the order of integration. We show in Complement A.1 that the reverse order leads to a simpler result, avoiding the angular momentum coupling.

(ii) *spin-orbit interaction*
The operator $O_{\kappa\mu_{ij}}$(space) for the spin-orbit potential is the orbital angular momentum $l_{\mu_{ij}} = ((\boldsymbol{r}_i - \boldsymbol{r}_j) \times \frac{1}{2\hbar}(\boldsymbol{p}_i - \boldsymbol{p}_j))_\mu$. See Eq. (11.9). Expressing $l_{\mu_{ij}} = \frac{1}{\hbar}\left(\widetilde{w^{(ij)}}\boldsymbol{x} \times \widetilde{\zeta^{(ij)}}\boldsymbol{\pi}\right)_\mu$ in terms of $\boldsymbol{x}$ and $\boldsymbol{\pi}$ with the use of Eqs. (2.13) and (2.14) and substituting the corresponding matrix element in Table 7.1 into Eq. (7.2), we obtain

$$\langle f_{K'L'M'}(u', A', \boldsymbol{x})|V(|\boldsymbol{r}_i - \boldsymbol{r}_j|)l_{\mu_{ij}}|f_{KLM}(u, A, \boldsymbol{x})\rangle$$

$$= \int d\boldsymbol{r} V(r) \mathrm{e}^{-\frac{1}{2}cr^2} \frac{1}{B_{KL}B_{K'L'}} \left(\frac{(2\pi)^{N-2}c}{\det B}\right)^{\frac{3}{2}}$$

$$\times \iint d\hat{\boldsymbol{e}}\, d\hat{\boldsymbol{e}}' Y_{LM}(\hat{\boldsymbol{e}}) Y_{L'M'}(\hat{\boldsymbol{e}}')^* \left(\frac{d^{2K+L+2K'+L'}}{d\lambda^{2K+L} d\lambda'^{2K'+L'}}\right.$$

$$\times \exp\left[\bar{q}\lambda^2 + \bar{q}'\lambda'^2 + \bar{\rho}\lambda\lambda'\boldsymbol{e}\cdot\boldsymbol{e}' + \gamma\lambda\boldsymbol{e}\cdot\boldsymbol{r} + \gamma'\lambda'\boldsymbol{e}'\cdot\boldsymbol{r}\right]$$

$$\left.\times i\Big\{\eta\lambda(\boldsymbol{e} \times \boldsymbol{r})_\mu - \eta'\lambda'(\boldsymbol{e}' \times \boldsymbol{r})_\mu\Big\}\right)_{\substack{\lambda=0,\lambda'=0\\ e=1,e'=1}} \tag{A.23}$$

with

$$\eta = \widetilde{\zeta^{(ij)}} A' B^{-1} u, \qquad \eta' = \widetilde{\zeta^{(ij)}} A B^{-1} u'. \tag{A.24}$$

Note that to derive Eq. (A.23) use is made of the relation

$$\int V(r)\, \mathrm{e}^{-c(\boldsymbol{r}-\boldsymbol{a})^2} (\boldsymbol{r} \times \boldsymbol{a}) d\boldsymbol{r} = 0 \tag{A.25}$$

for an arbitrary vector $\boldsymbol{a}$ provided $V(r)$ is a function of r. See Eq. (A.163).

Before performing the operation prescribed in Eq. (A.23), we note that the matrix element vanishes in the case of $L \neq L'$. Because f_{KLM} has parity $(-1)^L$ and because the spin-orbit potential does not change parity, $|L-L'|$ has to be even. On the other hand, the tensorial character of the spin-orbit potential imposes the condition $|L-L'| \le 1$. Both the conditions are met only when L is equal to L'. This special result is entirely due to the unique feature of f_{KLM} and does not always hold for general wave functions.

Combining Eq. (A.19) and the relation

$$i\Big\{\eta\lambda(\boldsymbol{e}\times\boldsymbol{r})_\mu - \eta'\lambda'(\boldsymbol{e}'\times\boldsymbol{r})_\mu\Big\}\Big|_{e=1,e'=1}$$
$$= \frac{4\sqrt{2}\pi}{3} r\Big\{\eta\lambda[Y_1(\hat{\boldsymbol{e}})\times Y_1(\hat{\boldsymbol{r}})]_{1\mu} - \eta'\lambda'[Y_1(\hat{\boldsymbol{e}'})\times Y_1(\hat{\boldsymbol{r}})]_{1\mu}\Big\} \tag{A.26}$$

yields the reduced matrix element as

$$\langle f_{K'L'}(u',A',\boldsymbol{x}) \parallel V(|\boldsymbol{r}_i-\boldsymbol{r}_j|)l_{ij} \parallel f_{KL}(u,A,\boldsymbol{x})\rangle$$
$$= \frac{4\sqrt{2}\pi}{3}\delta_{LL'}(-1)^L \frac{(2K+L)!(2K'+L)!}{B_{KL}B_{K'L}}\left(\frac{(2\pi)^{N-2}c}{\det B}\right)^{\frac{3}{2}}$$
$$\times\Bigg\{\eta\sum_{n_1n_2n_3} I(n_2+n_3+3,c)H(n_1,\bar{\rho})H(n_2,\gamma)H(n_3,\gamma')$$
$$\times H\Big(\frac{2K+L-n_1-n_2-1}{2},\bar{q}\Big)H\Big(\frac{2K'+L-n_1-n_3}{2},\bar{q}'\Big)$$
$$\times\sum_\Lambda C(\Lambda 1;L)U(L\Lambda 11;1L)R^{n_1n_2n_3}_{\Lambda L1}$$
$$+\eta'\sum_{n_1n_2n_3} I(n_2+n_3+3,c)H(n_1,\bar{\rho})H(n_2,\gamma)H(n_3,\gamma')$$
$$\times H\Big(\frac{2K+L-n_1-n_2}{2},\bar{q}\Big)H\Big(\frac{2K'+L-n_1-n_3-1}{2},\bar{q}'\Big)$$
$$\times\sum_\Lambda C(\Lambda 1;L)U(L\Lambda 11;1L)R^{n_1n_2n_3}_{L\Lambda 1}\Bigg\}. \tag{A.27}$$

The above derivation is straightforward and easy to follow. As a simple check of the above formula, the matrix element of the orbital angular momentum is calculated in Exercise A.3. It is again possible, however, to get a simpler formula by performing the $\hat{\boldsymbol{r}}$ integration first, as was done in Complement A.1. This task is reserved for Exercise A.4.

A.1.4 Density multipole operators

The matrix element of a density multipole operator plays a substantial role in investigating the properties of a system, e.g. the density, the deformation, the electromagnetic transition rates and the electron scattering form factors. The basic element of the multipole operator takes the form

$$O_{\kappa\mu_i}(\text{space}) = f(|\boldsymbol{r}_i - \boldsymbol{x}_N|)Y_{\kappa\mu}(\widehat{\boldsymbol{r}_i - \boldsymbol{x}_N}). \tag{A.28}$$

Note that the argument of the density multipole operator is not $\boldsymbol{r}_i$ but is correctly taken as $\boldsymbol{r}_i - \boldsymbol{x}_N$, which is the single-particle coordinate measured from the center-of-mass coordinate. Comparison of Eqs. (2.12) and (2.13) immediately suggests that the matrix element of the density multipole operator ought to be calculated in exactly the same manner as that of the two-body potential. In fact the reduced matrix element

$$\langle f_{K'L'}(u', A', \boldsymbol{x}) \parallel f(|\boldsymbol{r}_i - \boldsymbol{x}_N|)Y_\kappa(\widehat{\boldsymbol{r}_i - \boldsymbol{x}_N}) \parallel f_{KL}(u, A, \boldsymbol{x})\rangle \tag{A.29}$$

can be calculated by the same formula as Eq. (A.21) with the trivial replacements of $w^{(ij)} \to w^{(i)}$ in Eq. (A.18) and of $V(r) \to f(r)$ in Eq. (7.10).

Just as the matrix element for the density operator can be calculated from that of $\delta(\boldsymbol{r}_i - \boldsymbol{x}_N - \boldsymbol{r})$, the matrix element for the correlation function can also be derived from that of $\delta(\boldsymbol{r}_i - \boldsymbol{r}_j - \boldsymbol{r})$. As we have already noted in the above derivations, both of $\boldsymbol{r}_i - \boldsymbol{x}_N$ and $\boldsymbol{r}_i - \boldsymbol{r}_j$ are expressible in terms of the relative coordinates $\boldsymbol{x}$ as $\tilde{w}\boldsymbol{x}$ with $\tilde{w}$ being an appropriate $1 \times (N-1)$ row matrix. Therefore all of these matrix elements are reduced to the one of type $\langle f_{K'L'M'}(u', A', \boldsymbol{x})|\delta(\tilde{w}\boldsymbol{x} - \boldsymbol{r})|f_{KLM}(u, A, \boldsymbol{x})\rangle$. This matrix element was in fact obtained in the derivation following Eq. (A.17), so that we give the final result:

$$\begin{aligned}
&\langle f_{K'L'M'}(u', A', \boldsymbol{x})|\delta(\tilde{w}\boldsymbol{x} - \boldsymbol{r})|f_{KLM}(u, A, \boldsymbol{x})\rangle \\
&= \frac{(2K+L)!(2K'+L')!}{B_{KL}B_{K'L'}} \left(\frac{(2\pi)^{N-2}c}{\det B}\right)^{\frac{3}{2}} \frac{(-1)^{L'}}{\sqrt{2L'+1}} \mathrm{e}^{-\frac{1}{2}cr^2} \\
&\times \sum_\kappa \langle L\, M\, \kappa\, M'{-}M|L'\, M'\rangle Y_{\kappa\, M'-M}(\hat{\boldsymbol{r}})^*
\end{aligned}$$

$$\times \sum_{n_1 n_2 n_3} r^{n_2+n_3} R_{LL'\kappa}^{n_1 n_2 n_3} H(n_1, \bar{\rho}) H(n_2, \gamma) H(n_3, \gamma')$$
$$\times H\Big(\frac{2K+L-n_1-n_2}{2}, \bar{q}\Big) H\Big(\frac{2K'+L'-n_1-n_3}{2}, \bar{q}'\Big), \tag{A.30}$$

where c, γ, γ', $\bar{q}$, $\bar{q}'$, ρ are given in exactly the same manner as in Eq. (A.18) with $w^{(ij)}$ being replaced by w. Here, κ takes the values $L+L', L+L'-2, ..., |L-L'|$, and it is of course restricted by $|M'-M| \leq \kappa$. The above matrix element becomes very simple for the special case of $K = K' = 0$ as was noticed in the previous subsection. Since the matrix element has useful applications, we show in Complement A.2 that Eq. (A.30) can be simplified to just a double sum.

A.2 Correlated Gaussians with different coordinate sets

As the correlated Gaussians treated in the previous section have a very simple transformation property implied by Eq. (6.29), we assumed that they are all expressed by a particular set of the relative coordinates $\boldsymbol{x}$. However, the correlated Gaussians with the angular function $\theta_{LM}(\boldsymbol{x})$ given by Eq. (6.3) do not have such a nice property. Moreover, the use of different sets of the relative coordinates for these correlated Gaussians leads in general to a faster convergence because they allow us the possibility of describing naturally different types of correlations and asymptotics. The linear transformation of the coordinate sets, however, leads to a formidable task even for a system of only four particles, because the function $\theta_{LM}(\boldsymbol{x})$ shows no simple transformation rule. See Appendix A.5 where the matrix elements for a three-body system are explicitly evaluated by transforming the correlated Gaussians from one Jacobi coordinate set to another. The aim of this section is to outline a method of calculating the matrix elements for an N-body system using the correlated Gaussians which are expressed by different sets of the coordinates.

By making use of Eq. (6.22) to generate the correlated Gaussian, we have to evaluate the following matrix element for the operator $\mathcal{O}$:

$$\Big\langle \exp\Big(-\frac{1}{2}\tilde{\boldsymbol{x}}' A' \boldsymbol{x}'\Big) \prod_{j=1}^{N-1} \mathcal{Y}_{l_j' m_j'}(\boldsymbol{x}_j') \Big|$$

$$\times \mathcal{O}\bigg|\exp\Big(-\frac{1}{2}\tilde{\boldsymbol{x}}A\boldsymbol{x}\Big)\prod_{i=1}^{N-1}\mathcal{Y}_{l_im_i}(\boldsymbol{x}_i)\bigg\rangle$$

$$= \left(\prod_{i=1}^{N-1}\frac{1}{C_{l_im_i}}\frac{\partial^{l_i}}{\partial\alpha_i{}^{l_i}}\frac{\partial^{l_i-m_i}}{\partial\tau_i{}^{l_i-m_i}}\prod_{j=1}^{N-1}\frac{(-1)^{m'_j}}{C_{l'_j-m'_j}}\frac{\partial^{l'_j}}{\partial\alpha'_j{}^{l'_j}}\frac{\partial^{l'_j+m'_j}}{\partial\tau'_j{}^{l'_j+m'_j}}\right.$$

$$\left.\times\,\langle g(\alpha'|\boldsymbol{t}';A',\boldsymbol{x}')|\mathcal{O}|g(\alpha|\boldsymbol{t};A,\boldsymbol{x})\rangle\right)_{\substack{\alpha_i=0,\alpha'_j=0\\ \tau_i=0,\tau'_j=0}}. \tag{A.31}$$

Here use is made of the identity $(\mathcal{Y}_{lm}(\boldsymbol{x}_i))^* = (-1)^m\mathcal{Y}_{l-m}(\boldsymbol{x}_i)$ to take care of the complex conjugation in the bra side. Note that the vector $\boldsymbol{t}_i$ is defined by $(1-\tau_i{}^2, i(1+\tau_i{}^2), -2\tau_i)$ and likewise $\boldsymbol{t}'_j$ by $(1-\tau_j'^2, i(1+\tau_j'^2), -2\tau'_j)$. The vector $\boldsymbol{t}'_j$ of g in the bra side should not be complex conjugated. Assume that $\boldsymbol{x}'$ is related to $\boldsymbol{x}$ by a linear transformation T as $\boldsymbol{x}' = T\boldsymbol{x}$. With the use of Eq. (6.28) the matrix element (A.31) between two g's which are expressed in terms of different sets of relative coordinates can be reduced to

$$\langle g(\alpha'|\boldsymbol{t}';A',\boldsymbol{x}')|\mathcal{O}|g(\alpha|\boldsymbol{t};A,\boldsymbol{x})\rangle$$
$$= \langle g(\tilde{T}\alpha'|\boldsymbol{t}';\tilde{T}A'T,\boldsymbol{x})|\mathcal{O}|g(\alpha|\boldsymbol{t};A,\boldsymbol{x})\rangle, \tag{A.32}$$

and can be obtained from Table 7.1 for most operators.

For the unit operator $\mathcal{O} = 1$, the matrix element (A.32) takes, except for a trivial constant factor, the form $\mathrm{e}^{\frac{1}{2}\tilde{\boldsymbol{v}}B'\boldsymbol{v}}$, where $\boldsymbol{v} = \alpha|\boldsymbol{t} + \tilde{T}\alpha'|\boldsymbol{t}'$ and the $(N-1)\times(N-1)$ matrix B' is given by $(A+\tilde{T}A'T)^{-1}$. To simplify the notation we rename $\alpha_1\boldsymbol{t}_1 \equiv \boldsymbol{s}_1, ..., \alpha_{N-1}\boldsymbol{t}_{N-1} \equiv \boldsymbol{s}_{N-1}, \alpha'_1\boldsymbol{t}'_1 = \alpha_N\boldsymbol{t}_N \equiv \boldsymbol{s}_N, ..., \alpha'_{N-1}\boldsymbol{t}'_{N-1} = \alpha_{2N-2}\boldsymbol{t}_{2N-2} \equiv \boldsymbol{s}_{2N-2}$. Then $\tilde{\boldsymbol{v}}B'\boldsymbol{v}$ can be expressed by a $(2N-2)\times(2N-2)$ symmetric matrix B by $\tilde{\boldsymbol{s}}B\boldsymbol{s}$, where B is expressed in terms of B' and T as follows

$$B = \begin{pmatrix} B' & B'\tilde{T} \\ TB' & TB'\tilde{T} \end{pmatrix}. \tag{A.33}$$

Thus the operation prescribed in Eq. (A.31) can be reduced to the form

$$\mathcal{I} = \left(\prod_{i=1}^{2N-2}\partial_i^{l_i,m_i}\exp\left(\frac{1}{2}\tilde{\boldsymbol{s}}B\boldsymbol{s}+\tilde{\boldsymbol{w}}\boldsymbol{s}\right)\right)_{\substack{\alpha_1=0,\ldots,\alpha_{2N-2}=0,\\ \tau_1=0,\ldots,\tau_{2N-2}=0}} \tag{A.34}$$

with

$$\partial_i^{l,m} = \frac{\partial^l}{\partial {\alpha_i}^l} \frac{\partial^{l-m}}{\partial {\tau_i}^{l-m}}. \tag{A.35}$$

Here the factor $\tilde{\boldsymbol{w}}\boldsymbol{s}$ is included because it is needed in the case of the potential energy matrix elements. See Table 7.1. It is worthwhile noting that the present formulation leads to a unified prescription of Eq. (A.34), which is independent of the choice of the relative coordinate sets but requires only a very simple calculation of the matrix B. It would be extremely tedious if one were to try to rewrite the angular function of $\prod_{j=1}^{N-1} \mathcal{Y}_{l'_j m'_j}(\boldsymbol{x}'_j)$ in terms of those angular functions conforming to the coordinates $\boldsymbol{x}$.

Using Leibniz's formula we can express Eq. (A.34) as

$$\begin{aligned} \mathcal{I} = \sum_{\lambda_i \mu_i} & \left(\prod_{i=1}^{2N-2} \frac{l_i!}{\lambda_i!\,(l_i-\lambda_i)!} \frac{(l_i-m_i)!}{(\lambda_i-\mu_i)!\,(l_i-m_i-\lambda_i+\mu_i)!} \right) \\ & \times \left(\prod_{i=1}^{2N-2} \partial_i^{\lambda_i,\mu_i} \exp\left(\frac{1}{2}\tilde{\boldsymbol{s}}B\boldsymbol{s}\right) \right)_{\substack{\alpha_i=0 \\ \tau_i=0}} \\ & \times \left(\prod_{i=1}^{2N-2} \partial_i^{l_i-\lambda_i,m_i-\mu_i} \mathrm{e}^{\tilde{\boldsymbol{w}}\boldsymbol{s}} \right)_{\substack{\alpha_i=0 \\ \tau_i=0}} . \end{aligned} \tag{A.36}$$

The last factor is easily evaluated by using Eq. (6.58):

$$\begin{aligned} & \left(\prod_{i=1}^{2N-2} \partial_i^{l_i-\lambda_i,m_i-\mu_i} \, \mathrm{e}^{\tilde{\boldsymbol{w}}\boldsymbol{s}} \right)_{\substack{\alpha_i=0 \\ \tau_i=0}} \\ & \quad = \prod_{i=1}^{2N-2} C_{l_i-\lambda_i\, m_i-\mu_i} \mathcal{Y}_{l_i-\lambda_i\, m_i-\mu_i}(\boldsymbol{w}_i). \end{aligned} \tag{A.37}$$

Possible values of λ_i and μ_i are determined by the conditions that $0 \le \lambda_i \le l_i$, $0 \le \lambda_i - \mu_i \le l_i - m_i$ and $-(l_i - \lambda_i) \le m_i - \mu_i \le l_i - \lambda_i$. The last condition assures that the coefficients $C_{l_i-\lambda_i\, m_i-\mu_i}$ do not vanish. Because of a special form of $\boldsymbol{s}_i \cdot \boldsymbol{s}_j$ mentioned below, another condition that $\lambda_i - \mu_i \le 2\lambda_i$ has also to be met.

To evaluate the remaining factor, we note that $\tilde{\boldsymbol{s}}B\boldsymbol{s}$ is quadratic both in the α_i's and τ_i's because of $\boldsymbol{s}_i \cdot \boldsymbol{s}_j = -2\alpha_i\alpha_j(\tau_i - \tau_j)^2$. (Since, for each α_i, τ_i appears at most in quadratic power, the number of differentiation with respect to τ_i cannot exceed two times the number of differentiation with respect to α_i, that is, $\lambda_i-\mu_i \leq 2\lambda_i$.) Therefore, when $e^{\frac{1}{2}\tilde{s}Bs}$ is expanded in power series in both α_i's and τ_i's, and when these are set to zero after the differentiation, only the term $(\frac{1}{2}\tilde{s}Bs)^Q$ contributes provided that $\sum_{i=1}^{2N-2}\lambda_i = 2Q$ and $\sum_{i=1}^{2N-2}(\lambda_i-\mu_i) = 2Q$. Thus we have

$$\left(\prod_{i=1}^{2N-2}\partial_i^{\lambda_i,\mu_i}\exp\left(\frac{1}{2}\tilde{s}Bs\right)\right)_{\substack{\alpha_i=0\\ \tau_i=0}} = \frac{(-2)^Q}{Q!}$$

$$\times\left(\prod_{i=1}^{2N-2}\partial_i^{\lambda_i,\mu_i}\left\{\sum_{i<j}B_{ij}\alpha_i\alpha_j(\tau_i-\tau_j)^2\right\}^Q\right)_{\substack{\alpha_i=0\\ \tau_i=0}}. \tag{A.38}$$

To have non-vanishing contributions, each of Q terms must be differentiated with respect to α_i, α_j, and either twice with respect to τ_i or τ_j, or once each with respect to both τ_i and τ_j. This operation yields $\pm 2B_{ij}$, where the minus sign comes from the differentiation with respect to τ_i and τ_j. This sign is denoted ϵ_{ij}. Therefore we obtain

$$\left(\prod_{i=1}^{2N-2}\partial_i^{\lambda_i,\mu_i}\exp\left(\frac{1}{2}\tilde{s}Bs\right)\right)_{\substack{\alpha_i=0\\ \tau_i=0}}$$

$$= \delta_{\lambda_1+\cdots+\lambda_{2N-2},2Q}\,\delta_{\mu_1+\cdots+\mu_{2N-2},0}\,\frac{(-4)^Q}{Q!}$$

$$\times\sum\prod_{\substack{k=1\\ i_k<j_k}}^{Q}\epsilon_{i_kj_k}B_{i_kj_k}, \tag{A.39}$$

where the sum extends over all the possible combinations distributing the given multiple differentiation into pairs of differentiations. In fact the operation prescribed in Eq. (A.38) can be done easily by using the software *Mathematica* [184]. The case of a three-particle system is given in Complement A.3.

The matrix element for the kinetic energy can be obtained similarly. In fact the matrix element (A.32) for the kinetic energy takes the form $(\tilde{s}Cs)e^{\frac{1}{2}\tilde{s}Bs}$, but this can be dealt with easily. Using the relation

$$(\tilde{\boldsymbol{s}}C\boldsymbol{s})\exp\left(\frac{1}{2}\tilde{\boldsymbol{s}}B\boldsymbol{s}\right)=2\frac{d}{dp}\exp\left(\frac{1}{2}\tilde{\boldsymbol{s}}\left(B+pC\right)\boldsymbol{s}\right)\Bigg|_{p=0}, \tag{A.40}$$

$\left(\prod_{i=1}^{2N-2}\partial_i^{\lambda_i,\mu_i}\tilde{\boldsymbol{s}}C\boldsymbol{s}\,\mathrm{e}^{\frac{1}{2}\tilde{\boldsymbol{s}}B\boldsymbol{s}}\right)_{\alpha_i=0,t_i=0}$ can be obtained by simply replacing one of the $B_{i_kj_k}$'s in Eq. (A.39) with $C_{i_kj_k}$.

The potential energy matrix element for the spin-orbit interaction requires a slight modification. Noting that the general form of the matrix element (A.32) for this case can be expressed as $(s_k)_\mu\exp\left(\frac{1}{2}\tilde{\boldsymbol{s}}B\boldsymbol{s}+\tilde{\boldsymbol{w}}\boldsymbol{s}\right)$ (see Table 7.1), we make use of the relation

$$\begin{aligned}(s_k)_\mu\exp\left(\frac{1}{2}\tilde{\boldsymbol{s}}B\boldsymbol{s}+\tilde{\boldsymbol{w}}\boldsymbol{s}\right)&=(-1)^\mu\frac{\partial}{\partial(w_k)_{-\mu}}\exp\left(\frac{1}{2}\tilde{\boldsymbol{s}}B\boldsymbol{s}+\tilde{\boldsymbol{w}}\boldsymbol{s}\right)\\&=\left(\frac{\partial}{\partial w_k}\right)_\mu\exp\left(\frac{1}{2}\tilde{\boldsymbol{s}}B\boldsymbol{s}+\tilde{\boldsymbol{w}}\boldsymbol{s}\right),\end{aligned} \tag{A.41}$$

where the spherical components a_μ of a vector $\boldsymbol{a}$ are defined by $a_1=-\frac{1}{\sqrt{2}}(a_x+ia_y)$, $a_0=a_z$, $a_{-1}=\frac{1}{\sqrt{2}}(a_x-ia_y)$, and the spherical components $\left(\frac{\partial}{\partial a}\right)_\mu$ of a differential operator $\boldsymbol{\nabla}_a$ are given by $\left(\frac{\partial}{\partial a}\right)_1=-\frac{1}{\sqrt{2}}\left(\frac{\partial}{\partial a_x}+i\frac{\partial}{\partial a_y}\right)=-\frac{\partial}{\partial a_{-1}}$, $\left(\frac{\partial}{\partial a}\right)_0=\frac{\partial}{\partial a_z}=\frac{\partial}{\partial a_0}$, $\left(\frac{\partial}{\partial a}\right)_{-1}=\frac{1}{\sqrt{2}}\left(\frac{\partial}{\partial a_x}-i\frac{\partial}{\partial a_y}\right)=-\frac{\partial}{\partial a_1}$. By using Eqs. (A.41) and (A.34)-(A.37), the matrix element for the spin-orbit interaction, $\left(\prod_{i=1}^{2N-2}\partial_i^{\lambda_i,\mu_i}(s_k)_\mu\,\mathrm{e}^{\frac{1}{2}\tilde{\boldsymbol{s}}B\boldsymbol{s}+\tilde{\boldsymbol{w}}\boldsymbol{s}}\right)_{\alpha_i=0,\tau_i=0}$, can be obtained by replacing $\mathcal{Y}_{l_k-\lambda_k\,m_k-\mu_k}(\boldsymbol{w}_k)$ in Eq. (A.37) with $\left(\frac{\partial}{\partial w_k}\right)_\mu\mathcal{Y}_{l_k-\lambda_k\,m_k-\mu_k}(\boldsymbol{w}_k)$, which is easily calculated by the gradient formula [70, 71]

$$\begin{aligned}&\boldsymbol{r}\cdot\boldsymbol{\nabla}_a f(a)Y_{lm}(\hat{\boldsymbol{a}})\\&=\sqrt{\frac{l}{2l+1}}\left(f'(a)+\frac{l+1}{a}f(a)\right)[\boldsymbol{r}\times Y_{l-1}(\hat{\boldsymbol{a}})]_{lm}\\&-\sqrt{\frac{l+1}{2l+1}}\left(f'(a)-\frac{l}{a}f(a)\right)[\boldsymbol{r}\times Y_{l+1}(\hat{\boldsymbol{a}})]_{lm}.\end{aligned} \tag{A.42}$$

In particular we obtain

$$\begin{aligned}&\left(\frac{\partial}{\partial a}\right)_\mu\mathcal{Y}_{lm}(\boldsymbol{a})\\&=-(2l+1)\sqrt{\frac{l}{2l-1}}\langle 1\mu\, lm|l-1\, m+\mu\rangle\mathcal{Y}_{l-1\, m+\mu}(\boldsymbol{a}).\end{aligned} \tag{A.43}$$

A.3 Correlated Gaussian-type geminals

The matrix element for the general correlated Gaussian-type geminals of Eq. (6.26) can be calculated from the equation

$$\langle f_{n'}(A', B', \boldsymbol{R}', \boldsymbol{r})|\mathcal{O}|f_n(A, B, \boldsymbol{R}, \boldsymbol{r})\rangle$$
$$= \Bigg(\Bigg\{\prod_{i=1}^{N}\prod_{p=1}^{3}\prod_{j=1}^{N}\prod_{q=1}^{3}\frac{\partial^{n_{ip}}}{\partial t_{ip}{}^{n_{ip}}}\frac{\partial^{n'_{jq}}}{\partial t'_{jq}{}^{n'_{jq}}}\Bigg\}$$
$$\times \exp\Big(-\frac{1}{2}\widetilde{\boldsymbol{R}}B\boldsymbol{R} - \tilde{\boldsymbol{t}}\boldsymbol{R} - \frac{1}{2}\widetilde{\boldsymbol{R}'}B'\boldsymbol{R}' - \widetilde{\boldsymbol{t}'}\boldsymbol{R}'\Big)$$
$$\times \langle g(\boldsymbol{t}' + B'\boldsymbol{R}'; A' + B', \boldsymbol{r})|\mathcal{O}|g(\boldsymbol{t} + B\boldsymbol{R}; A + B, \boldsymbol{r})\rangle\Bigg)_{\substack{\boldsymbol{t}_i=0,\boldsymbol{t}'_i=0\\(i=1,\dots,N)}} \tag{A.44}$$

It is crucial to know the dependence of the matrix element in the generating functions on $\tilde{\boldsymbol{t}} = (\boldsymbol{t}_1, ..., \boldsymbol{t}_N)$ and $\widetilde{\boldsymbol{t}'} = (\boldsymbol{t}'_1, ..., \boldsymbol{t}'_N)$ to do multiple differentiations. As can be seen from Table 7.1, they appear as a polynomial of at most degree 2 in exponential functions for the most important operators $\mathcal{O} = 1$ and $\delta(\tilde{w}\boldsymbol{r} - \boldsymbol{S})$. The matrix element for the kinetic energy has an additional factor which is again quadratic. The basic elements of the $\boldsymbol{t}$ and $\boldsymbol{t}'$ dependence include $\boldsymbol{t}_i{\cdot}\boldsymbol{t}_j$, $\boldsymbol{t}'_i{\cdot}\boldsymbol{t}'_j$, $\boldsymbol{t}_i{\cdot}\boldsymbol{t}'_j$, and $\boldsymbol{S}{\cdot}\boldsymbol{t}_i$, $\boldsymbol{S}{\cdot}\boldsymbol{t}'_i$. As the differentiation can be made separately in each of the Cartesian coordinates t_{ip} and t'_{jq}, and as the basic elements have no cross terms in different directions of the Cartesian coordinates such as $t_{i1}t_{j2}$, it is possible to do the needed differentiation in each of the x, y, and z components.

The above considerations reduce the operation prescribed in Eq. (A.44) to the mathematical problem given in Exercise A.5. The formula developed in Exercise A.5 can be applied in each of the x, y and z components to complete the operations needed in Eq. (A.44). It is not difficult to extend the formula given in Exercise A.5 to get the matrix element for the kinetic energy which has an additional polynomial of degree 2. In fact, Eq. (A.169) can be generalized to a relation which is valid for a general function:

$$\Big(\mathcal{L}^N f(x_1, x_2, ..., x_K)\Big)_{x_1=0,\dots,x_K=0}$$

$$= \left(\frac{d^N}{d\vartheta^N} f(\lambda_1\vartheta, \lambda_2\vartheta, ..., \lambda_K\vartheta)\right)_{\vartheta=0}$$
$$= \frac{N!}{2\pi i}\oint_{|z|=1} \frac{f(\lambda_1 z, \lambda_2 z, ..., \lambda_K z)}{z^{N+1}} dz. \tag{A.45}$$

A.4 Spin matrix elements

The interaction may depend not only on position coordinates but also on intrinsic degrees of freedom such as spin and flavor. By representing the single-particle spin function by $|m = \pm 1/2\rangle$, a basic matrix element is given by

$$\langle m' \,|\, \boldsymbol{\sigma}_\mu \,|\, m\,\rangle = \sqrt{3}\,\langle \tfrac{1}{2}\, m\, 1\, \mu \,|\, \tfrac{1}{2}\, m'\rangle. \tag{A.46}$$

The matrix elements for spin operators, $\langle S'M'|\mathcal{O}_{\kappa\mu}(\text{spin})|SM\rangle$, are usually evaluated by using the Wigner-Eckart theorem (A.12). Here a reduced matrix element which is needed to evaluate the matrix element is the one for Pauli matrix and it is given by

$$\langle \tfrac{1}{2} \,\|\, \boldsymbol{\sigma} \,\|\, \tfrac{1}{2}\rangle = \sqrt{6}. \tag{A.47}$$

For the spin function discussed in Sect. 6.4 no technique involving angular momentum algebra is needed to obtain the spin matrix element. An elementary, direct method enables one to calculate the matrix element. We list the spin matrix element that appears in evaluating the nucleon-nucleon potential energy.

$$\langle m_1'\, m_2' \,|\, \boldsymbol{\sigma}_1 \cdot \boldsymbol{\sigma}_2 \,|\, m_1\, m_2\,\rangle = 3\,(-1)^{m_1'-m_1}\delta_{m_1+m_2,m_1'+m_2'}$$
$$\times \langle \tfrac{1}{2}\, m_1\, 1\, m_1'{-}m_1 \,|\, \tfrac{1}{2}\, m_1'\rangle\langle \tfrac{1}{2}\, m_2\, 1\, m_2'{-}m_2 \,|\, \tfrac{1}{2}\, m_2'\rangle. \tag{A.48}$$

$$\langle m_1'\, m_2' \,|\, [\boldsymbol{\sigma}_1 \times \boldsymbol{\sigma}_2]_{2m} \,|\, m_1\, m_2\,\rangle = 3\,\delta_{m,m_1'+m_2'-m_1-m_2}$$
$$\times \langle \tfrac{1}{2}\, m_1\, 1\, m_1'{-}m_1 \,|\, \tfrac{1}{2}\, m_1'\rangle\langle \tfrac{1}{2}\, m_2\, 1\, m_2'{-}m_2 \,|\, \tfrac{1}{2}\, m_2'\rangle$$
$$\times \langle 1\, m_1'{-}m_1\, 1\, m_2'{-}m_2 \,|\, 2\, m\rangle. \tag{A.49}$$

$$\langle m_1'\, m_2' \,|\, [\boldsymbol{r} \times (\boldsymbol{\sigma}_1 + \boldsymbol{\sigma}_2)]_{1m} \,|\, m_1\, m_2\,\rangle = \sqrt{3}\sum_{q_1=-1}^{1}\sum_{q_2=-1}^{1} r_{q_1}$$

$$\times \left\{ \delta_{m_2 m_2'} \langle \tfrac{1}{2}\, m_1\, 1\, q_2 \,|\, \tfrac{1}{2}\, m_1' \rangle + \delta_{m_1 m_1'} \langle \tfrac{1}{2}\, m_2\, 1\, q_2 \,|\, \tfrac{1}{2}\, m_2' \rangle \right\}$$
$$\times \langle 1\, q_1\, 1\, q_2 \,|\, 1\, m \rangle. \tag{A.50}$$

$$\langle m_1'\, m_2' \,|\, (\boldsymbol{r}\cdot\boldsymbol{\sigma}_2)(\boldsymbol{\sigma}_1)_m + (\boldsymbol{r}\cdot\boldsymbol{\sigma}_1)(\boldsymbol{\sigma}_2)_m \,|\, m_1\, m_2 \rangle$$
$$= 3 \Big\{ (-1)^{m_2'-m_2}\, \delta_{m,m_1'-m_1} r_{m_2-m_2'}$$
$$+ (-1)^{m_1'-m_1}\, \delta_{m,m_2'-m_2} r_{m_1-m_1'} \Big\}$$
$$\times \langle \tfrac{1}{2}\, m_1\, 1\, m_1'-m_1 \,|\, \tfrac{1}{2}\, m_1' \rangle \langle \tfrac{1}{2}\, m_2\, 1\, m_2'-m_2 \,|\, \tfrac{1}{2}\, m_2' \rangle. \tag{A.51}$$

$$\langle m_1'\, m_2' \,|\, (\boldsymbol{r}\cdot\boldsymbol{\sigma}_1)(\boldsymbol{r}\cdot\boldsymbol{\sigma}_2) \,|\, m_1\, m_2 \rangle = 3\, (-1)^{m_1'-m_1+m_2'-m_2}$$
$$\times\, r_{m_1-m_1'} r_{m_2-m_2'} \langle \tfrac{1}{2}\, m_1\, 1\, m_1'-m_1 \,|\, \tfrac{1}{2}\, m_1' \rangle \langle \tfrac{1}{2}\, m_2\, 1\, m_2'-m_2 \,|\, \tfrac{1}{2}\, m_2' \rangle. \tag{A.52}$$

$$\langle m_1'\, m_2' \,|\, \big[[\boldsymbol{r}\times\boldsymbol{\sigma}_1]_1 \times [\boldsymbol{r}\times\boldsymbol{\sigma}_2]_1 \big]_{2m} \,|\, m_1\, m_2 \rangle$$
$$= 3\, \langle \tfrac{1}{2}\, m_1\, 1\, m_1'-m_1 \,|\, \tfrac{1}{2}\, m_1' \rangle \langle \tfrac{1}{2}\, m_2\, 1\, m_2'-m_2 \,|\, \tfrac{1}{2}\, m_2' \rangle$$
$$\times \sum_{q_1=-1}^{1} \sum_{q_2=-1}^{1} r_{q_1} r_{q_2} \langle 1\, q_1{+}m_1'{-}m_1\, 1\, q_2{+}m_2'{-}m_2 \,|\, 2\, m \rangle$$
$$\times \langle 1\, q_1\, 1\, m_1'-m_1 \,|\, 1\, q_1{+}m_1'{-}m_1 \rangle \langle 1\, q_2\, 1\, m_2'-m_2 \,|\, 1\, q_2{+}m_2'{-}m_2 \rangle. \tag{A.53}$$

Here $\boldsymbol{r}$ is a 3-dimensional vector and r_m stands for its spherical component $\sqrt{4\pi/3}\, r Y_{1m}(\hat{\boldsymbol{r}})$.

Matrix elements for the isospin part can be obtained in exactly the same manner.

A.5 Three-body problem with central, tensor and spin-orbit forces

The matrix element for an N-body system can be derived from the general formulas presented in the previous sections. In order to illustrate that there exist alternative possibilities, here we use a sligthly different way of calculating the matrix elements for three-body problems with central, tensor and spin-orbit forces. In this approach the radial and angular integrations are done separately and this involves the extensive use of angular momentum algebra. Needless to say, this approach cannot be easily pushed far. One may try to attempt to derive similar formulas for the four-body case, but to go beyond that seems to be too tedious.

(i) *a choice of basis functions*
The trial function is taken in the form

$$\Psi_{JM_JTM_T}(123) = \sum_{\alpha c} C_{\alpha c} \mathcal{A}\varphi_{\alpha c}(1,23) \tag{A.54}$$

with expansion coefficients $C_{\alpha c}$ and functions $\mathcal{A}\varphi_{\alpha c}(1,23)$ that are constructed to be fully antisymmetric under the exchange of particles:

$$\begin{aligned}\mathcal{A}\varphi_{\alpha c}(1,23) &= \varphi_{\alpha c}(1,23) + \varphi_{\alpha c}(2,31) + \varphi_{\alpha c}(3,12)\\ &\quad - \varphi_{\alpha c}(1,32) - \varphi_{\alpha c}(2,13) - \varphi_{\alpha c}(3,21),\end{aligned} \tag{A.55}$$

(the construction of a totally symmetric wave function is the same, but then the last three terms should have positive sign in the above equation). The basis function $\varphi_{\alpha c}(k,pq)$ represents a specific wave function in a Jacobi channel k ($k = 1,2,3$) (see Fig. 2.2). The notation (k,pq) stands for a system where the particles p and q are first connected and then the center-of-mass of the (pq) pair is connected with the particle k. Thus the coordinate used in the kth Jacobi channel, $\boldsymbol{x}^{(k)} = \{\boldsymbol{x}_k, \boldsymbol{y}_k\}$, is given by

$$\boldsymbol{x}_k = \boldsymbol{r}_p - \boldsymbol{r}_q, \qquad \boldsymbol{y}_k = \frac{m_p\boldsymbol{r}_p + m_q\boldsymbol{r}_q}{m_p + m_q} - \boldsymbol{r}_k. \tag{A.56}$$

(Note that we use $\{\boldsymbol{x}_k, \boldsymbol{y}_k\}$ in this section, instead of $\{\boldsymbol{x}_1^{(k)}, \boldsymbol{x}_2^{(k)}\}$, to denote the first and second Jacobi coordinates.)

The basis function $\varphi_{\alpha c}(k,pq)$ is given by a product of space, spin and isospin parts:

$$\varphi_{\alpha c}(k,pq) = [\mathcal{F}_L^{\alpha}(k,pq) \times \chi_S^s(k,pq)]_{JM_J}\eta_{TM_T}^t(k,pq), \tag{A.57}$$

where the spatial part is chosen to be a correlated Gaussian of the form

$$\mathcal{F}_{LM_L}^{\alpha}(k,pq) = \exp\left(-\frac{1}{2}\widetilde{\boldsymbol{x}^{(k)}}A\boldsymbol{x}^{(k)}\right) x_k^{2\nu+\lambda} y_k^{2n+l}\, Y_{LM_L}^{(\lambda l)}(\widehat{\boldsymbol{x}_k},\widehat{\boldsymbol{y}_k}) \tag{A.58}$$

with

$$A = \begin{pmatrix} \beta & \delta \\ \delta & \gamma \end{pmatrix} \qquad (\beta\gamma - \delta^2 > 0) \tag{A.59}$$

and

$$Y_{LM_L}^{(\lambda l)}(\widehat{\boldsymbol{x}_k},\widehat{\boldsymbol{y}_k}) = [Y_\lambda(\widehat{\boldsymbol{x}_k}) \times Y_l(\widehat{\boldsymbol{y}_k})]_{LM_L}. \tag{A.60}$$

The index α stands for a set of discrete labels $\alpha = (\nu, n, \lambda, l)$ and for the matrix A which contains the nonlinear parameters β, γ, δ. Here ν, $n = 0,1,2,...$, while λ is the orbital angular momentum of the relative motion between the particles p and q, and l is the orbital angular momentum between the center-of-mass of the (pq) pair and the particle k.

The index c denotes the labels (L, s, S, t), where L is the total orbital angular momentum, S the total spin, and s and t are the spin and the isospin of the (pq) pair, respectively. (The total isospin T is here assumed to be a conserved quantity, so that no T-mixing is considered in the trial function.) The spin and isospin parts are thus given by

$$\chi_{SM_S}^s(k,pq) = [\chi_s(pq) \times \chi_{\frac{1}{2}}(k)]_{SM_S}, \tag{A.61}$$

$$\eta_{TM_T}^t(k,pq) = [\eta_t(pq) \times \eta_{\frac{1}{2}}(k)]_{TM_T}, \tag{A.62}$$

where

$$\chi_{sm}(pq) = [\chi_{\frac{1}{2}}(p) \times \chi_{\frac{1}{2}}(q)]_{sm},$$

$$\eta_{tm}(pq) = [\eta_{\frac{1}{2}}(p) \times \eta_{\frac{1}{2}}(q)]_{tm}. \tag{A.63}$$

The function $\varphi_{\alpha c}(k,qp)$ is obtained from $\varphi_{\alpha c}(k,pq)$ by changing the sign of $\boldsymbol{x}_k$ and the order of couplings in the spin and isospin functions. The sign change of $\boldsymbol{x}_k$ results in a change of δ to $-\delta$ and multiplication by the phase factor $(-1)^\lambda$, and the spin and isospin functions change as follows:

$$\chi_{sm}(qp) = (-1)^{1-s}\chi_{sm}(pq),$$
$$\eta_{tm}(qp) = (-1)^{1-t}\eta_{tm}(pq). \tag{A.64}$$

Hence the function $\varphi_{\alpha c}(k, qp)$ is obtained from $\varphi_{\alpha c}(k, pq)$ by changing the sign of δ and by multiplying by the phase factor $(-1)^{\lambda+s+t}$.

The kinetic energy operator with the center-of-mass kinetic energy subtracted can be written in terms of the relative coordinates as

$$T_{\text{rel}} \equiv \sum_{i=1}^{3}\frac{p_i^2}{2m_i} - T_{\text{cm}} = -\frac{\hbar^2}{2\mu_1^{(k)}}\Delta_{x_k} - \frac{\hbar^2}{2\mu_2^{(k)}}\Delta_{y_k} \equiv T_k, \tag{A.65}$$

where

$$\mu_1^{(k)} = \frac{m_p m_q}{m_p + m_q}, \qquad \mu_2^{(k)} = \frac{(m_p + m_q)m_k}{m_p + m_q + m_k}. \tag{A.66}$$

This expression is valid in any of the kth Jacobi coordinate set.

(ii) *transformation of the basis function*
We have to calculate the matrix elements between the basis functions expressed in terms of different coordinate sets (1,23), (2,31), ... To achieve this one has to transform the basis function from one Jacobi coordinate set to another. This can be accomplished by using the transformation

$$\begin{pmatrix} x_k \\ y_k \end{pmatrix} = T^{(kq)} \begin{pmatrix} x_q \\ y_q \end{pmatrix}, \tag{A.67}$$

where the matrices connecting the different Jacobi sets are given by

$$T^{(21)} = \begin{pmatrix} -\frac{m_2}{m_2+m_3} & 1 \\ -\frac{m_3(m_1+m_2+m_3)}{(m_2+m_3)(m_1+m_3)} & -\frac{m_1}{m_1+m_3} \end{pmatrix},$$

$$T^{(31)} = \begin{pmatrix} -\frac{m_3}{m_2+m_3} & -1 \\ \frac{m_2(m_1+m_2+m_3)}{(m_2+m_3)(m_1+m_2)} & -\frac{m_1}{m_1+m_2} \end{pmatrix},$$

$$T^{(32)} = \begin{pmatrix} -\frac{m_3}{m_1+m_3} & 1 \\ -\frac{m_1(m_1+m_2+m_3)}{(m_1+m_3)(m_1+m_2)} & -\frac{m_2}{m_1+m_2} \end{pmatrix},$$

$$T^{(12)} = \begin{pmatrix} -\frac{m_1}{m_3+m_1} & -1 \\ \frac{m_3(m_1+m_2+m_3)}{(m_3+m_1)(m_2+m_3)} & -\frac{m_2}{m_2+m_3} \end{pmatrix},$$

$$T^{(13)} = \begin{pmatrix} -\frac{m_1}{m_1+m_2} & 1 \\ -\frac{m_2(m_1+m_2+m_3)}{(m_1+m_2)(m_2+m_3)} & -\frac{m_3}{m_2+m_3} \end{pmatrix},$$

$$T^{(23)} = \begin{pmatrix} -\frac{m_2}{m_1+m_2} & -1 \\ \frac{m_1(m_1+m_2+m_3)}{(m_1+m_2)(m_3+m_1)} & -\frac{m_3}{m_3+m_1} \end{pmatrix}. \quad \text{(A.68)}$$

The space function $\mathcal{F}^{\alpha}_{LM_L}(k, pq)$, expressed in the kth Jacobi coordinate set, can be transformed to the one expressed in terms of the qth Jacobi coordinate set as follows:

$$\mathcal{F}^{\alpha}_{LM_L}(k, pq) = \sum_{\bar{\alpha}} \mathcal{B}^{(kq)}_L(\alpha\bar{\alpha})\mathcal{F}^{\bar{\alpha}}_{LM_L}(q, kp), \quad \text{(A.69)}$$

where

$$\mathcal{F}^{\bar{\alpha}}_{LM_L}(q, kp) = \exp\left(-\frac{1}{2}\widetilde{x^{(q)}}\bar{A}x^{(q)}\right) x_q^{2\bar{\nu}+\bar{\lambda}} y_q^{2\bar{n}+\bar{l}} Y^{(\bar{\lambda}\bar{l})}_{LM_L}(\widehat{x_q}, \widehat{y_q}), \quad \text{(A.70)}$$

and where $\bar{\alpha}$ stands for a set $(\bar{\nu}, \bar{n}, \bar{\lambda}, \bar{l})$, which has to satisfy the relation

$$2\bar{\nu} + \bar{\lambda} + 2\bar{n} + \bar{l} = 2\nu + \lambda + 2n + l. \quad \text{(A.71)}$$

The matrix $\bar{A}$ is uniquely determined from A by the transformation (A.67) as

$$\bar{A} = \widetilde{T^{(kq)}} A T^{(kq)} \equiv \begin{pmatrix} \bar{\beta} & \bar{\delta} \\ \bar{\delta} & \bar{\gamma} \end{pmatrix}. \quad \text{(A.72)}$$

The $\mathcal{B}^{(kq)}_L(\alpha\bar{\alpha})$ in Eq. (A.69) is the transformation coefficient of the polynomial part:

$$x_k^{2\nu+\lambda} y_k^{2n+l} Y^{(\lambda l)}_{LM_L}(\widehat{x_k}, \widehat{y_k}) = \sum_{\bar{\alpha}} \mathcal{B}^{(kq)}_L(\alpha\bar{\alpha}) x_q^{2\bar{\nu}+\bar{\lambda}} y_q^{2\bar{n}+\bar{l}} Y^{(\bar{\lambda}\bar{l})}_{LM_L}(\widehat{x_q}, \widehat{y_q}). \quad \text{(A.73)}$$

To get this expression one has to make use of Eq. (6.7) and to recouple the product of the two angular functions (see Complement 6.3) by

$$[Y_{l_{12}}^{(l_1 l_2)}(\hat{\boldsymbol{x}}, \hat{\boldsymbol{y}}) \times Y_{l_{34}}^{(l_3 l_4)}(\hat{\boldsymbol{x}}, \hat{\boldsymbol{y}})]_{LM}$$
$$= \sum_{l_{13} l_{24}} E_{l_{13} l_{24}}^{l_1 l_2 l_{12} l_3 l_4 l_{34} L} Y_{LM}^{(l_{13} l_{24})}(\hat{\boldsymbol{x}}, \hat{\boldsymbol{y}}), \tag{A.74}$$

where the coefficient E is given by a unitary $9j$ symbol and the coefficient C of Eq. (6.11) as

$$E_{l_{13} l_{24}}^{l_1 l_2 l_{12} l_3 l_4 l_{34} L} = \begin{bmatrix} l_1 & l_2 & l_{12} \\ l_3 & l_4 & l_{34} \\ l_{13} & l_{24} & L \end{bmatrix} C(l_1\, l_3; l_{13})\, C(l_2\, l_4; l_{24}). \tag{A.75}$$

Note that the coefficient E may be defined by the reduced matrix element

$$\sqrt{2L+1}\, E_{l_{13} l_{24}}^{l_1 l_2 l_{12} l_3 l_4 l_{34} L}$$
$$= \sqrt{2L+1}\, \langle Y_{LM}^{(l_{13} l_{24})}(\hat{\boldsymbol{x}}, \hat{\boldsymbol{y}}) | [Y_{l_{12}}^{(l_1 l_2)}(\hat{\boldsymbol{x}}, \hat{\boldsymbol{y}}) \times Y_{l_{34}}^{(l_3 l_4)}(\hat{\boldsymbol{x}}, \hat{\boldsymbol{y}})]_{LM} \rangle$$
$$= (-1)^{l_{12}+l_{34}-L} \langle Y_L^{(l_{13} l_{24})}(\hat{\boldsymbol{x}}, \hat{\boldsymbol{y}}) \parallel Y_{l_{12}}^{(l_1 l_2)}(\hat{\boldsymbol{x}}, \hat{\boldsymbol{y}}) \parallel Y_{l_{34}}^{(l_3 l_4)}(\hat{\boldsymbol{x}}, \hat{\boldsymbol{y}}) \rangle$$
$$= (-1)^{l_3+l_4} \langle Y_{l_{12}}^{(l_1 l_2)}(\hat{\boldsymbol{x}}, \hat{\boldsymbol{y}}) \parallel Y_L^{(l_{13} l_{24})}(\hat{\boldsymbol{x}}, \hat{\boldsymbol{y}}) \parallel Y_{l_{34}}^{(l_3 l_4)}(\hat{\boldsymbol{x}}, \hat{\boldsymbol{y}}) \rangle, \tag{A.76}$$

which is easily verified by using Eq. (A.147) and the relation

$$\langle Y_l(\hat{\boldsymbol{x}}) \parallel Y_k(\hat{\boldsymbol{x}}) \parallel Y_{l'}(\hat{\boldsymbol{x}}) \rangle = (-1)^{l'} \sqrt{2k+1}\, C(l l'; k)$$
$$= \sqrt{2l+1}\, C(k l'; l). \tag{A.77}$$

The transformation coefficient $\mathcal{B}_L^{(kq)}(\alpha\bar{\alpha})$ is then expressed as

$$\mathcal{B}_L^{(kq)}(\alpha\bar{\alpha}) = \sum_{\substack{\nu_1 \lambda_1 \nu_2 \lambda_2 \\ n_1 l_1 n_2 l_2}} D_{\nu_1 \lambda_1 \nu_2 \lambda_2}^{\nu\lambda} D_{n_1 l_1 n_2 l_2}^{nl} E_{\bar{\lambda} \bar{l}}^{\lambda_1 \lambda_2 \lambda l_1 l_2 l L}$$
$$\times \left(T_{11}^{(kq)}\right)^{2\nu_1+\lambda_1} \left(T_{12}^{(kq)}\right)^{2\nu_2+\lambda_2} \left(T_{21}^{(kq)}\right)^{2n_1+l_1}$$
$$\times \left(T_{22}^{(kq)}\right)^{2n_2+l_2}, \tag{A.78}$$

where the coefficient D is given in Eq. (6.8) and where the summation is constrained by the conditions

$$2\nu_1 + \lambda_1 + 2\nu_2 + \lambda_2 = 2\nu + \lambda, \qquad 2n_1 + l_1 + 2n_2 + l_2 = 2n + l,$$

$$2\nu_1 + \lambda_1 + 2n_1 + l_1 = 2\bar{\nu} + \bar{\lambda}, \qquad 2\nu_2 + \lambda_2 + 2n_2 + l_2 = 2\bar{n} + \bar{l}, \tag{A.79}$$

and by the triangular inequality for the angular momentum among $(\lambda_1, \lambda_2, \lambda)$, (l_1, l_2, l), $(\lambda_1, l_1, \bar{\lambda})$, and $(\lambda_2, l_2, \bar{l})$, respectively.

The transformation to the pth Jacobi coordinate set can be performed as well in a similar manner.

The spin part of the wave function transforms as

$$\begin{aligned}\chi^{s}_{SM_S}(k, pq) &= \sum_{\bar{s}} (-1)^{\frac{1}{2}+s-S} U(\tfrac{1}{2}\tfrac{1}{2}S\tfrac{1}{2}; \bar{s}s)\chi^{\bar{s}}_{SM_S}(q, kp) \\ &= \sum_{\bar{s}} (-1)^{\frac{1}{2}+\bar{s}-S} U(\tfrac{1}{2}\tfrac{1}{2}S\tfrac{1}{2}; s\bar{s})\chi^{\bar{s}}_{SM_S}(p, qk).\end{aligned} \tag{A.80}$$

The transformation of the isospin part is given in exactly the same way.

The radial and angular integrations are separated in this approach. However, no such separation is made in the factor $\mathrm{e}^{-\delta \boldsymbol{x}_k \cdot \boldsymbol{y}_k}$ which appears in the exponent of the function $\mathcal{F}^{\alpha}_{LM_L}(k, pq)$. Therefore in the integration one has to expand this term into partial waves (see Eq. (6.53)):

$$\mathrm{e}^{-w\boldsymbol{x}\cdot\boldsymbol{y}} = 4\pi \sum_{l} \sqrt{2l+1}\, \epsilon(l, w)\, i_l(|w|xy) Y^{(ll)}_{00}(\hat{\boldsymbol{x}}, \hat{\boldsymbol{y}}), \tag{A.81}$$

where $i_l(x)$ is the modified spherical Bessel function of the first kind (6.48) and $\epsilon(l, w) = 1$ for $w > 0$, $\epsilon(l, w) = (-1)^l$ for $w \le 0$. In the radial part one encounters an integral of the form

$$\begin{aligned}I(n, l, v, |w|) &\equiv \int_0^\infty y^{2n+l+2} \mathrm{e}^{-\frac{1}{2}vy^2} i_l(|w|y)dy \\ &= \sqrt{\frac{\pi}{2}} \frac{(2n)!!|w|^l}{v^{n+l+\frac{3}{2}}} \exp\left(\frac{w^2}{2v}\right) L_n^{(l+\frac{1}{2})}\left(-\frac{w^2}{2v}\right),\end{aligned} \tag{A.82}$$

where n is a non-negative integer and $L_n^{(\alpha)}(x)$ is the associated Laguerre polynomial (10.15). We define the following integral

$$\begin{aligned}&I(\nu, n, l, u, v, |w|; V) \\ &\equiv \int_0^\infty \int_0^\infty x^{2\nu+l+2} y^{2n+l+2} \mathrm{e}^{-\frac{1}{2}ux^2 - \frac{1}{2}vy^2} V(x) i_l(|w|xy)dxdy\end{aligned}$$

$$= \int_0^\infty x^{2\nu+l+2} e^{-\frac{1}{2}ux^2} V(x) I(n,l,v,|w|x)dx. \tag{A.83}$$

The one-dimensional integration in the above equation, depending on the functional form of $V(x)$, can be analytically or numerically evaluated. In the special case when $V(x) = 1$ we obtain

$$I(\nu,n,l,u,v,|w|;V{=}1) = \sqrt{\frac{\pi}{8}}(2n)!!\Gamma\left(n+l+\tfrac{3}{2}\right)\frac{|w|^l}{v^{n+l+\frac{3}{2}}}$$

$$\times \sum_{k=0}^{n} \frac{\Gamma(k+\nu+l+\frac{3}{2})}{k!(n-k)!\Gamma(k+l+\frac{3}{2})} \frac{\left(\frac{w^2}{2v}\right)^k}{\left[\frac{1}{2}\left(u-\frac{w^2}{v}\right)\right]^{k+\nu+l+\frac{3}{2}}}, \tag{A.84}$$

where Γ is the Gamma function given in Eq. (7.8).

(iii) *overlap matrix elements*

Now we are ready to calculate the matrix elements. The overlap between the functions in different Jacobi sets k and q can be calculated separately in the space, spin and isospin parts:

$$\begin{aligned}\langle k|q\rangle &\equiv \langle \varphi_{\alpha c}(k,pq)|\varphi_{\alpha' c'}(q,kp)\rangle \\ &= \delta_{LL'}\delta_{SS'}\langle \mathcal{F}^{\alpha}_{LM_L}(k,pq)|\mathcal{F}^{\alpha'}_{LM_L}(q,kp)\rangle \\ &\quad\times \langle \chi^{s}_{SM_S}(k,pq)|\chi^{s'}_{SM_S}(q,kp)\rangle\langle \eta^{t}_{TM_T}(k,pq)|\eta^{t'}_{TM_T}(q,kp)\rangle.\end{aligned} \tag{A.85}$$

The spin part is easily obtained by using Eq. (A.80) as

$$\langle \chi^{s}_{SM_S}(k,pq)|\chi^{s'}_{SM_S}(q,kp)\rangle = (-1)^{\frac{1}{2}+s-S}U(\tfrac{1}{2}\tfrac{1}{2}S\tfrac{1}{2};s's). \tag{A.86}$$

Likewise, the isospin part becomes

$$\langle \eta^{t}_{TM_T}(k,pq)|\eta^{t'}_{TM_T}(q,kp)\rangle = (-1)^{\frac{1}{2}+t-T}U(\tfrac{1}{2}\tfrac{1}{2}T\tfrac{1}{2};t't). \tag{A.87}$$

The integration of the space part can be performed by transforming the function in the bra to the one depending on the qth Jacobi coordinate. By substituting Eqs. (A.69) and (A.70) we obtain (the suffix q to label the coordinate set is suppressed in the integral)

$$\langle \mathcal{F}^{\alpha}_{LM_L}(k,pq)|\mathcal{F}^{\alpha'}_{LM_L}(q,kp)\rangle = \sum_{\bar{\alpha}} \mathcal{B}^{(kq)}_L(\alpha\bar{\alpha})$$

$$\times \iint x^{2\bar{\nu}+\bar{\lambda}+2\nu'+\lambda'} y^{2\bar{n}+\bar{l}+2n'+l'} e^{-\frac{1}{2}ux^2-\frac{1}{2}vy^2-w\boldsymbol{x}\cdot\boldsymbol{y}}$$

$$\times \left(Y_{LM_L}^{(\bar{\lambda}\bar{l})}(\hat{\boldsymbol{x}},\hat{\boldsymbol{y}})\right)^* Y_{LM_L}^{(\lambda' l')}(\hat{\boldsymbol{x}},\hat{\boldsymbol{y}})d\boldsymbol{x}d\boldsymbol{y} \tag{A.88}$$

with

$$u=\bar{\beta}+\beta', \qquad v=\bar{\gamma}+\gamma', \qquad w=\bar{\delta}+\delta'. \tag{A.89}$$

The angular integration can be done easily by the use of Eqs. (A.81) and (A.74) as

$$\iint e^{-w\boldsymbol{x}\cdot\boldsymbol{y}}\left(Y_{LM_L}^{(\bar{\lambda}\bar{l})}(\hat{\boldsymbol{x}},\hat{\boldsymbol{y}})\right)^* Y_{LM_L}^{(\lambda' l')}(\hat{\boldsymbol{x}},\hat{\boldsymbol{y}})d\hat{\boldsymbol{x}}d\hat{\boldsymbol{y}}$$

$$= 4\pi\sum_{\kappa}\sqrt{2\kappa+1}\,\epsilon(\kappa,w)\,E_{\bar{\lambda}\bar{l}}^{\kappa\kappa0\lambda' l' LL}\, i_{\kappa}(|w|xy). \tag{A.90}$$

One can easily verify that $\bar{l}+l'-\kappa$ has to be even and non-negative, and then one can use the integral formula (A.84) to obtain

$$\langle\mathcal{F}_{LM_L}^{\alpha}(k,pq)|\mathcal{F}_{LM_L}^{\alpha'}(q,kp)\rangle$$

$$= 4\pi\sum_{\bar{\alpha}}\mathcal{B}_L^{(kq)}(\alpha\bar{\alpha})\sum_{\kappa}\sqrt{2\kappa+1}\,\epsilon(\kappa,w)\,E_{\bar{\lambda}\bar{l}}^{\kappa\kappa0\lambda' l' LL}$$

$$\times I\left(\bar{\nu}+\nu'+\frac{\bar{\lambda}+\lambda'-\kappa}{2},\bar{n}+n'+\frac{\bar{l}+l'-\kappa}{2},\kappa,u,v,|w|;V=1\right). \tag{A.91}$$

(iv) *matrix elements of kinetic energy operator*
The next step is the calculation of the matrix elements of the kinetic energy operator

$$\langle k\,|T_{\rm rel}|\,q\rangle = \langle\varphi_{\alpha c}(k,pq)\,|T_q|\,\varphi_{\alpha' c'}(q,kp)\rangle. \tag{A.92}$$

In the above expression we used the fact that the kinetic energy operator can be expressed in different relative coordinate sets (see Eq. (A.65)), and thus we defined $T_{\rm rel}$ in conformity with the coordinate set in the ket. To calculate the matrix elements of the kinetic energy operator one first notes that

$$\Delta_x\left(x^{2\nu+\lambda}y^{2n+l}e^{-\frac{1}{2}\beta x^2-\frac{1}{2}\gamma y^2-\delta\boldsymbol{x}\cdot\boldsymbol{y}}Y_{LM}^{(\lambda l)}(\hat{\boldsymbol{x}},\hat{\boldsymbol{y}})\right)$$

$$= \Big(-3\beta x^2 + \beta^2 x^4 + \delta^2 x^2 y^2 - 2\beta(2\nu+\lambda)x^2 + 2\nu(2\nu+2\lambda+1) \Big)$$

$$\times\, x^{2\nu+\lambda-2} y^{2n+l} \mathrm{e}^{-\frac{1}{2}\beta x^2 - \frac{1}{2}\gamma y^2 - \delta \boldsymbol{x}\cdot\boldsymbol{y}} Y_{LM}^{(\lambda l)}(\hat{\boldsymbol{x}},\hat{\boldsymbol{y}})$$

$$-2\delta x^{2\nu+\lambda-1} y^{2n+l+1} \mathrm{e}^{-\frac{1}{2}\beta x^2 - \frac{1}{2}\gamma y^2 - \delta \boldsymbol{x}\cdot\boldsymbol{y}} \sqrt{\frac{4\pi}{3}}$$

$$\times \Bigg\{ \sqrt{\frac{\lambda}{2\lambda+1}} (2\nu + 2\lambda + 1 - \beta x^2)$$

$$\times \sum_{\kappa} U(\lambda\!-\!1\,1\,L\,l;\lambda\,\kappa) C(1l;\kappa) Y_{LM}^{(\lambda-1\,\kappa)}(\hat{\boldsymbol{x}},\hat{\boldsymbol{y}})$$

$$-\sqrt{\frac{\lambda+1}{2\lambda+1}} (2\nu - \beta x^2)$$

$$\times \sum_{\kappa} U(\lambda\!+\!1\,1\,L\,l;\lambda\,\kappa) C(1l;\kappa) Y_{LM}^{(\lambda+1\,\kappa)}(\hat{\boldsymbol{x}},\hat{\boldsymbol{y}}) \Bigg\}. \qquad \text{(A.93)}$$

A proof of this relation is given in Exercise A.6. A similar expression holds for the second relative coordinate $\boldsymbol{y}$. By the help of the above equation the effect of the kinetic energy operator applied on the basis function is given by a linear combination of terms akin to the basis function. The matrix element of the kinetic energy can therefore be given by repeatedly using the expression obtained for the overlap of the basis function.

(v) *matrix elements of central potentials*
The central two-body interaction can be written as

$$V = V(|\boldsymbol{r}_2 - \boldsymbol{r}_3|) + V(|\boldsymbol{r}_1 - \boldsymbol{r}_3|) + V(|\boldsymbol{r}_1 - \boldsymbol{r}_2|)$$
$$= V(x_1) + V(x_2) + V(x_3), \qquad \text{(A.94)}$$

that is, the potential is a sum of terms depending on the first relative coordinate in different Jacobi sets. In calculating the matrix element for the case

$$\langle k\,|V(x_p)|\,q\rangle \equiv \langle \varphi_{\alpha c}(k,pq)|V(x_p)|\varphi_{\alpha' c'}(q,kp)\rangle$$
$$= \delta_{LL'}\delta_{SS'} \langle \mathcal{F}^{\alpha}_{LM_L}(k,pq)|V(x_p)|\mathcal{F}^{\alpha'}_{LM_L}(q,kp)\rangle$$
$$\times \langle \chi^{s}_{SM_S}(k,pq)|\chi^{s'}_{SM_S}(q,kp)\rangle \langle \eta^{t}_{TM_T}(k,pq)|\eta^{t'}_{TM_T}(q,kp)\rangle,$$

(A.95)

the spin and isospin parts can be calculated as before but the spatial functions in the bra and the ket have to be transformed into the pth Jacobi coordinate set in which the potential is defined.

The space part of the matrix element is calculated by using the transformations $k \to p$ and $q \to p$:

$$\langle \mathcal{F}^{\alpha}_{LM_L}(k,pq)|V(x_p)|\mathcal{F}^{\alpha'}_{LM_L}(q,kp)\rangle = \sum_{\bar{\alpha}} \mathcal{B}^{(kp)}_L(\alpha\bar{\alpha}) \sum_{\bar{\alpha}'} \mathcal{B}^{(qp)}_L(\alpha'\bar{\alpha}') \times \langle \mathcal{F}^{\bar{\alpha}}_{LM_L}(p,qk)|V(x_p)|\mathcal{F}^{\bar{\alpha}'}_{LM_L}(p,qk)\rangle . \tag{A.96}$$

Here the matrices corresponding to the transformations are defined by

$$\bar{A} = \widetilde{T^{(kp)}} A T^{(kp)} \equiv \begin{pmatrix} \bar{\beta} & \bar{\delta} \\ \bar{\delta} & \bar{\gamma} \end{pmatrix},$$

$$\bar{A}' = \widetilde{T^{(qp)}} A' T^{(qp)} \equiv \begin{pmatrix} \bar{\beta}' & \bar{\delta}' \\ \bar{\delta}' & \bar{\gamma}' \end{pmatrix}, \tag{A.97}$$

respectively. By introducing

$$u = \bar{\beta} + \bar{\beta}', \qquad v = \bar{\gamma} + \bar{\gamma}', \qquad w = \bar{\delta} + \bar{\delta}', \tag{A.98}$$

one obtains

$$\langle \mathcal{F}^{\bar{\alpha}}_{LM_L}(p,qk)|V(x_p)|\mathcal{F}^{\bar{\alpha}'}_{LM_L}(p,qk)\rangle = 4\pi \sum_{\kappa} \sqrt{2\kappa+1}\, \epsilon(\kappa,w)\, E^{\kappa\kappa 0 \bar{\lambda}' \bar{l}' LL}_{\bar{\lambda}\bar{l}} \times I\left(\bar{\nu}+\bar{\nu}'+\frac{\bar{\lambda}+\bar{\lambda}'-\kappa}{2}, \bar{n}+\bar{n}'+\frac{\bar{l}+\bar{l}'-\kappa}{2}, \kappa, u, v, |w|; V\right). \tag{A.99}$$

(vi) *matrix elements of tensor potentials*

The tensor interaction is defined through the tensor operator (11.7)

$$S_p = \sqrt{\frac{24\pi}{5}}\, (Y_2(\widehat{\boldsymbol{x}_p}) \cdot [\boldsymbol{\sigma}_q \times \boldsymbol{\sigma}_k]_2) . \tag{A.100}$$

To calculate the matrix element of the tensor interaction one has to follow the same steps as for the central interaction, except that now the spin part of the wave function has to be taken into account, and the spin part should also be transformed into the same coordinate system as the potential. The use of the Wigner-Eckart theorem (A.14) enables one to obtain

$$\begin{aligned}
\langle k|V(x_p)S_p|q\rangle &\equiv \langle \varphi_{\alpha c}(k,pq)\,|V(x_p)S_p|\,\varphi_{\alpha' c'}(q,kp)\rangle \\
&= \sqrt{\frac{24\pi}{5}}\frac{U(L'2JS;LS')}{\sqrt{(2L+1)(2S'+1)}} \\
&\quad\times \langle \mathcal{F}_L^{\alpha}(k,pq)\parallel V(x_p)Y_2(\widehat{\boldsymbol{x}_p})\parallel \mathcal{F}_{L'}^{\alpha'}(q,kp)\rangle \\
&\quad\times \langle \chi_S^{s}(k,pq)\parallel [\boldsymbol{\sigma}_q\times\boldsymbol{\sigma}_k]_2\parallel \chi_{S'}^{s'}(q,kp)\rangle \\
&\quad\times \langle \eta_{TM_T}^{t}(k,pq)|\eta_{TM_T}^{t'}(q,kp)\rangle .
\end{aligned} \tag{A.101}$$

Here the spatial part of the matrix element is calculated as before by transforming the basis functions to the pth Jacobi coordinate set

$$\begin{aligned}
&\langle \mathcal{F}_L^{\alpha}(k,pq)\parallel V(x_p)Y_2(\widehat{\boldsymbol{x}_p})\parallel \mathcal{F}_{L'}^{\alpha'}(q,kp)\rangle \\
&\quad= \sum_{\bar\alpha}\mathcal{B}_L^{(kp)}(\alpha\bar\alpha)\sum_{\bar\alpha'}\mathcal{B}_{L'}^{(qp)}(\alpha'\bar\alpha') \\
&\qquad\times \langle \mathcal{F}_L^{\bar\alpha}(p,qk)\parallel V(x_p)Y_2(\widehat{\boldsymbol{x}_p})\parallel \mathcal{F}_{L'}^{\bar\alpha'}(p,qk)\rangle ,
\end{aligned} \tag{A.102}$$

where the last term is obtained by using Eqs. (A.81) and (A.82) as

$$\begin{aligned}
&\langle \mathcal{F}_L^{\bar\alpha}(p,qk)\parallel V(x_p)Y_2(\widehat{\boldsymbol{x}_p})\parallel \mathcal{F}_{L'}^{\bar\alpha'}(p,qk)\rangle \\
&\quad= 4\pi\sum_{\kappa}\sqrt{2\kappa+1}\,\epsilon(\kappa,w) \\
&\qquad\times \langle Y_L^{(\bar\lambda\bar l)}(\hat{\boldsymbol{x}},\hat{\boldsymbol{y}})\parallel Y_0^{(\kappa\kappa)}(\hat{\boldsymbol{x}},\hat{\boldsymbol{y}})Y_2(\hat{\boldsymbol{x}})\parallel Y_{L'}^{(\bar\lambda'\bar l')}(\hat{\boldsymbol{x}},\hat{\boldsymbol{y}})\rangle \\
&\qquad\times I\left(\bar\nu+\bar\nu'+\frac{\bar\lambda+\bar\lambda'-\kappa}{2},\bar n+\bar n'+\frac{\bar l+\bar l'-\kappa}{2},\kappa,u,v,|w|;V\right).
\end{aligned} \tag{A.103}$$

Here u,v,w are defined in Eq. (A.98). Note that $\bar l+\bar l'-\kappa$ has to be even and non-negative.

The reduced matrix element of the angular part is calculated by using the angular momentum recoupling technique (see Complement 6.3) and Eq. (A.76). One obtains

$$\begin{aligned}
&\langle Y_L^{(\bar{\lambda}\bar{l})}(\hat{\boldsymbol{x}},\hat{\boldsymbol{y}}) \parallel Y_0^{(\kappa\kappa)}(\hat{\boldsymbol{x}},\hat{\boldsymbol{y}})Y_2(\hat{\boldsymbol{x}}) \parallel Y_{L'}^{(\bar{\lambda}'\bar{l}')}(\hat{\boldsymbol{x}},\hat{\boldsymbol{y}})\rangle \\
&\quad = \sum_{\kappa'} U(2\kappa 2\kappa;\kappa' 0)\, C(2\kappa;\kappa') \\
&\qquad \times \langle Y_L^{(\bar{\lambda}\bar{l})}(\hat{\boldsymbol{x}},\hat{\boldsymbol{y}}) \parallel Y_2^{(\kappa'\kappa)}(\hat{\boldsymbol{x}},\hat{\boldsymbol{y}}) \parallel Y_{L'}^{(\bar{\lambda}'\bar{l}')}(\hat{\boldsymbol{x}},\hat{\boldsymbol{y}})\rangle \\
&\quad = (-1)^{\bar{\lambda}'+\bar{l}'} \sum_{\kappa'} \sqrt{\frac{2\kappa'+1}{2\kappa+1}}\, C(2\kappa;\kappa') E_{\kappa'\kappa}^{\bar{\lambda}\bar{l}L\bar{\lambda}'\bar{l}'L'2}. \qquad \text{(A.104)}
\end{aligned}$$

The spin matrix element is also easily calculated by transforming the spin functions to the pth Jacobi system and by using Eqs. (A.147) and (A.149) as

$$\begin{aligned}
&\langle \chi_S^s(k,pq) \parallel [\boldsymbol{\sigma}_q \times \boldsymbol{\sigma}_k]_2 \parallel \chi_{S'}^{s'}(q,kp)\rangle \\
&\quad = \sum_{\bar{s}} (-1)^{\frac{1}{2}+\bar{s}-S} U(\tfrac{1}{2}\tfrac{1}{2}S\tfrac{1}{2};s\bar{s}) \sum_{\bar{s}'} (-1)^{\frac{1}{2}+s'-S'} U(\tfrac{1}{2}\tfrac{1}{2}S'\tfrac{1}{2};\bar{s}'s') \\
&\qquad \times \langle \chi_S^{\bar{s}}(p,qk) \parallel [\boldsymbol{\sigma}_q \times \boldsymbol{\sigma}_k]_2 \parallel \chi_{S'}^{\bar{s}'}(p,qk)\rangle, \qquad \text{(A.105)}
\end{aligned}$$

where

$$\begin{aligned}
&\langle \chi_S^{\bar{s}}(p,qk) \parallel [\boldsymbol{\sigma}_q \times \boldsymbol{\sigma}_k]_2 \parallel \chi_{S'}^{\bar{s}'}(p,qk)\rangle \\
&\quad = (-1)^{\bar{s}'-S'+S-\bar{s}} \sqrt{\frac{2S+1}{2\bar{s}+1}}\, U(\tfrac{1}{2}\bar{s}'S2;S'\bar{s}) \\
&\qquad \times \langle (\tfrac{1}{2}\tfrac{1}{2})\bar{s} \parallel [\boldsymbol{\sigma}_q \times \boldsymbol{\sigma}_k]_2 \parallel (\tfrac{1}{2}\tfrac{1}{2})\bar{s}'\rangle \qquad \text{(A.106)}
\end{aligned}$$

with

$$\begin{aligned}
&\langle (\tfrac{1}{2}\tfrac{1}{2})\bar{s} \parallel [\boldsymbol{\sigma}_q \times \boldsymbol{\sigma}_k]_2 \parallel (\tfrac{1}{2}\tfrac{1}{2})\bar{s}'\rangle \\
&\quad = \sqrt{\frac{2\bar{s}+1}{4}} \begin{bmatrix} \frac{1}{2} & \frac{1}{2} & \bar{s}' \\ 1 & 1 & 2 \\ \frac{1}{2} & \frac{1}{2} & \bar{s} \end{bmatrix} (\sqrt{6})^2 = 2\sqrt{5}\,\delta_{\bar{s}1}\delta_{\bar{s}'1}. \qquad \text{(A.107)}
\end{aligned}$$

(vii) *matrix elements of spin-orbit potentials*
The spin-orbit force is given through the operator (11.9)

$$\boldsymbol{L}_p\cdot\boldsymbol{S}_p \quad \text{with} \quad \boldsymbol{L}_p = -i\boldsymbol{x}_p\times\boldsymbol{\nabla}_p, \quad \boldsymbol{S}_p = \frac{1}{2}(\boldsymbol{\sigma}_q+\boldsymbol{\sigma}_k). \tag{A.108}$$

As in the case of the tensor interaction, with the use of the Wigner-Eckart theorem (A.14) one obtains

$$\begin{aligned}\langle k|V(x_p)\boldsymbol{L}_p\cdot\boldsymbol{S}_p|q\rangle &\equiv \langle\varphi_{\alpha c}(k,pq)\,|V(x_p)\boldsymbol{L}_p\cdot\boldsymbol{S}_p|\,\varphi_{\alpha'c'}(q,kp)\rangle\\ &= -\frac{U(L'1JS;LS')}{\sqrt{(2L+1)(2S'+1)}}\langle\mathcal{F}_L^{\alpha}(k,pq)\parallel V(x_p)\boldsymbol{L}_p\parallel\mathcal{F}_{L'}^{\alpha'}(q,kp)\rangle\\ &\quad\times\langle\chi_S^s(k,pq)\parallel\boldsymbol{S}_p\parallel\chi_{S'}^{s'}(q,kp)\rangle\langle\eta_{TM_T}^t(k,pq)|\eta_{TM_T}^{t'}(q,kp)\rangle.\end{aligned} \tag{A.109}$$

The spatial part of the matrix element is calculated as before by transforming the basis functions to the pth Jacobi coordinate set

$$\begin{aligned}&\langle\mathcal{F}_L^{\alpha}(k,pq)\parallel V(x_p)\boldsymbol{L}_p\parallel\mathcal{F}_{L'}^{\alpha'}(q,kp)\rangle\\ &\quad=\sum_{\bar\alpha}\mathcal{B}_L^{(kp)}(\alpha\bar\alpha)\sum_{\bar\alpha'}\mathcal{B}_{L'}^{(qp)}(\alpha'\bar\alpha')\\ &\qquad\times\langle\mathcal{F}_L^{\bar\alpha}(p,qk)\parallel V(x_p)\boldsymbol{L}_p\parallel\mathcal{F}_{L'}^{\bar\alpha'}(p,qk)\rangle.\end{aligned} \tag{A.110}$$

When the orbital angular momentum operator acts on the basis function (provided both the basis function and the operator are expressed in the same coordinate system) one obtains

$$\begin{aligned}&\boldsymbol{L}\left(x^{2\nu+\lambda}y^{2n+l}\mathrm{e}^{-\frac{1}{2}\beta x^2-\frac{1}{2}\gamma y^2-\delta\boldsymbol{x}\cdot\boldsymbol{y}}\,Y_{LM}^{(\lambda l)}(\hat{\boldsymbol{x}},\hat{\boldsymbol{y}})\right)\\ &\quad=i\delta(\boldsymbol{x}\times\boldsymbol{y})x^{2\nu+\lambda}y^{2n+l}\mathrm{e}^{-\frac{1}{2}\beta x^2-\frac{1}{2}\gamma y^2-\delta\boldsymbol{x}\cdot\boldsymbol{y}}\,Y_{LM}^{(\lambda l)}(\hat{\boldsymbol{x}},\hat{\boldsymbol{y}})\\ &\quad+x^{2\nu+\lambda}y^{2n+l}\mathrm{e}^{-\frac{1}{2}\beta x^2-\frac{1}{2}\gamma y^2-\delta\boldsymbol{x}\cdot\boldsymbol{y}}\,\boldsymbol{L}Y_{LM}^{(\lambda l)}(\hat{\boldsymbol{x}},\hat{\boldsymbol{y}}).\end{aligned} \tag{A.111}$$

Note that the outer product $i(\boldsymbol{x}\times\boldsymbol{y})_\mu$ is equal to the tensor product $\sqrt{2}[\boldsymbol{x}\times\boldsymbol{y}]_{1\mu} = (4\sqrt{2}\pi/3)xyY_{1\mu}^{(11)}(\hat{\boldsymbol{x}},\hat{\boldsymbol{y}})$. Therefore the spatial part of the matrix element, $\langle\mathcal{F}_L^{\bar\alpha}(p,qk)\parallel V(x_p)\boldsymbol{L}_p\parallel\mathcal{F}_{L'}^{\bar\alpha'}(p,qk)\rangle$, can be obtained in exactly the same way as in the central potential. It is given by a sum of two terms: One is

$$\begin{aligned}&4\pi\frac{4\sqrt{2}\pi}{3}\,\delta\sum_{\kappa}\sqrt{2\kappa+1}\,\epsilon(\kappa,w)\\ &\quad\times\langle Y_L^{(\bar\lambda\bar l)}(\hat{\boldsymbol{x}},\hat{\boldsymbol{y}})\parallel Y_0^{(\kappa\kappa)}(\hat{\boldsymbol{x}},\hat{\boldsymbol{y}})Y_1^{(11)}(\hat{\boldsymbol{x}},\hat{\boldsymbol{y}})\parallel Y_{L'}^{(\bar\lambda'\bar l')}(\hat{\boldsymbol{x}},\hat{\boldsymbol{y}})\rangle\end{aligned}$$

$$\times I\left(\bar{\nu}+\bar{\nu}'+\frac{\bar{\lambda}+\bar{\lambda}'+1-\kappa}{2},\bar{n}+\bar{n}'+\frac{\bar{l}+\bar{l}'+1-\kappa}{2},\kappa,u,v,|w|;V\right),$$

(A.112)

and the other is

$$4\pi\sum_{\kappa}\sqrt{2\kappa+1}\,\epsilon(\kappa,w)\langle Y_L^{(\bar{\lambda}\bar{l})}(\hat{\boldsymbol{x}},\hat{\boldsymbol{y}})\parallel Y_0^{(\kappa\kappa)}(\hat{\boldsymbol{x}},\hat{\boldsymbol{y}})\boldsymbol{L}\parallel Y_{L'}^{(\bar{\lambda}'\bar{l}')}(\hat{\boldsymbol{x}},\hat{\boldsymbol{y}})\rangle$$

$$\times I\left(\bar{\nu}+\bar{\nu}'+\frac{\bar{\lambda}+\bar{\lambda}'-\kappa}{2},\bar{n}+\bar{n}'+\frac{\bar{l}+\bar{l}'-\kappa}{2},\kappa,u,v,|w|;V\right).$$

(A.113)

Here u, v, w are defined in Eq. (A.98). Note that in the first term $\bar{l}+\bar{l}'+1-\kappa$ has to be even and non-negative, and likewise in the second term $\bar{l}+\bar{l}'-\kappa$ has to be even and non-negative.

The angular part in the first term can be obtained by using Eqs. (A.74) and (A.76) as

$$\langle Y_L^{(\bar{\lambda}\bar{l})}(\hat{\boldsymbol{x}},\hat{\boldsymbol{y}})\parallel Y_0^{(\kappa\kappa)}(\hat{\boldsymbol{x}},\hat{\boldsymbol{y}})Y_1^{(11)}(\hat{\boldsymbol{x}},\hat{\boldsymbol{y}})\parallel Y_{L'}^{(\bar{\lambda}'\bar{l}')}(\hat{\boldsymbol{x}},\hat{\boldsymbol{y}})\rangle$$
$$=(-1)^{\bar{\lambda}'+\bar{l}'}\sqrt{3}\sum_{pq}E_{pq}^{\kappa\kappa01111}E_{pq}^{\bar{\lambda}\bar{l}L\bar{\lambda}'\bar{l}'L'1}. \qquad (A.114)$$

The angular part in the second term becomes

$$\langle Y_L^{(\bar{\lambda}\bar{l})}(\hat{\boldsymbol{x}},\hat{\boldsymbol{y}})\parallel Y_0^{(\kappa\kappa)}(\hat{\boldsymbol{x}},\hat{\boldsymbol{y}})\boldsymbol{L}\parallel Y_{L'}^{(\bar{\lambda}'\bar{l}')}(\hat{\boldsymbol{x}},\hat{\boldsymbol{y}})\rangle$$
$$=E_{\bar{\lambda}\bar{l}}^{\kappa\kappa0\bar{\lambda}'\bar{l}'LL}\langle Y_L^{(\bar{\lambda}'\bar{l}')}(\hat{\boldsymbol{x}},\hat{\boldsymbol{y}})\parallel\boldsymbol{L}\parallel Y_{L'}^{(\bar{\lambda}'\bar{l}')}(\hat{\boldsymbol{x}},\hat{\boldsymbol{y}})\rangle, \qquad (A.115)$$

where

$$\langle Y_L^{(\bar{\lambda}'\bar{l}')}(\hat{\boldsymbol{x}},\hat{\boldsymbol{y}})\parallel\boldsymbol{L}\parallel Y_{L'}^{(\bar{\lambda}'\bar{l}')}(\hat{\boldsymbol{x}},\hat{\boldsymbol{y}})\rangle$$
$$=(-1)^{L-L'}\sqrt{\frac{2L+1}{2\bar{\lambda}'+1}}\,U(\bar{l}'\bar{\lambda}'L1;L'\bar{\lambda}')\langle Y_{\bar{\lambda}'}(\hat{\boldsymbol{x}})\parallel\boldsymbol{L}\parallel Y_{\bar{\lambda}'}(\hat{\boldsymbol{x}})\rangle$$

(A.116)

with

$$\langle Y_l(\hat{\boldsymbol{x}})\parallel\boldsymbol{L}\parallel Y_l(\hat{\boldsymbol{x}})\rangle=\sqrt{l(l+1)(2l+1)}. \qquad (A.117)$$

The spin matrix element can be calculated as in the tensor interaction. The result is

$$\begin{aligned}
&\langle \chi_S^s(k,pq) \parallel \boldsymbol{S}_p \parallel \chi_{S'}^{s'}(q,kp)\rangle \\
&\quad = \sum_{\bar{s}} (-1)^{\frac{1}{2}+\bar{s}-S} U(\tfrac{1}{2}\tfrac{1}{2}S\tfrac{1}{2};s\bar{s}) \sum_{\bar{s'}} (-1)^{\frac{1}{2}+s'-S'} U(\tfrac{1}{2}\tfrac{1}{2}S'\tfrac{1}{2};\bar{s'}s') \\
&\qquad \times \langle \chi_S^{\bar{s}}(p,qk) \parallel \boldsymbol{S}_p \parallel \chi_{S'}^{\bar{s'}}(p,qk)\rangle,
\end{aligned} \tag{A.118}$$

where

$$\begin{aligned}
&\langle \chi_S^{\bar{s}}(p,qk) \parallel \boldsymbol{S}_p \parallel \chi_{S'}^{\bar{s'}}(p,qk)\rangle \\
&\quad = \frac{1}{2}(-1)^{\bar{s'}-S'+S-\bar{s}} \sqrt{\frac{2S+1}{2\bar{s}+1}}\, U(\tfrac{1}{2}\bar{s'}S1;S'\bar{s}) \\
&\qquad \times \langle (\tfrac{1}{2}\tfrac{1}{2})\bar{s} \parallel \boldsymbol{\sigma}_q + \boldsymbol{\sigma}_k \parallel (\tfrac{1}{2}\tfrac{1}{2})\bar{s'}\rangle
\end{aligned} \tag{A.119}$$

with

$$\begin{aligned}
&\langle (\tfrac{1}{2}\tfrac{1}{2})\bar{s} \parallel \boldsymbol{\sigma}_q + \boldsymbol{\sigma}_k \parallel (\tfrac{1}{2}\tfrac{1}{2})\bar{s'}\rangle \\
&\quad = \sqrt{\frac{2\bar{s}+1}{2}}\, U(\tfrac{1}{2}\tfrac{1}{2}\bar{s}1;\bar{s'}\tfrac{1}{2})\sqrt{6}\left((-1)^{\bar{s}-\bar{s'}}+1\right) \\
&\quad = 2\sqrt{6}\,\delta_{\bar{s}1}\delta_{\bar{s'}1}.
\end{aligned} \tag{A.120}$$

Complements

A.1 *Matrix elements of central potentials*

The matrix element of the central and tensor potentials can be obtained through a step which is slightly different from the procedure presented in Appendix A.1.3. Then the formula becomes simpler than Eq. (A.21). Here we outline the step which leads to this simpler result.

Instead of using Eq. (A.19), we first perform the integration over the angle $\hat{\boldsymbol{r}}$ in Eq. (A.17), which leads to

$$\langle f_{K'L'M'}(u',A',\boldsymbol{x})|\frac{\delta(|\boldsymbol{r}_i-\boldsymbol{r}_j|-r)}{r^2}Y_{\kappa\mu}(\widehat{\boldsymbol{r}_i-\boldsymbol{r}_j})|f_{KLM}(u,A,\boldsymbol{x})\rangle$$

$$= \mathrm{e}^{-\frac{1}{2}cr^2}\frac{4\pi}{B_{KL}B_{K'L'}}\left(\frac{(2\pi)^{N-2}c}{\det B}\right)^{\frac{3}{2}}\iint d\hat{\boldsymbol{e}}\,d\hat{\boldsymbol{e}'}$$

$$\times Y_{LM}(\hat{\boldsymbol{e}})Y_{L'M'}(\hat{\boldsymbol{e}'})^*\left(\frac{d^{2K+L+2K'+L'}}{d\lambda^{2K+L}d\lambda'^{2K'+L'}}\right.$$

$$\left.\times\exp\left[\bar{q}\lambda^2+\bar{q}'\lambda'^2+\bar{\rho}\lambda\lambda'\boldsymbol{e}\cdot\boldsymbol{e}'\right]i_\kappa(rv)Y_{\kappa\mu}(\hat{\boldsymbol{v}})\right)_{\substack{\lambda=0,\lambda'=0\\ e=1,e'=1}} \qquad \text{(A.121)}$$

with

$$\boldsymbol{v}=\gamma\lambda\boldsymbol{e}+\gamma'\lambda'\boldsymbol{e}'. \qquad \text{(A.122)}$$

If either γ or γ' vanishes, v is independent of the scalar product $\boldsymbol{e}\cdot\boldsymbol{e}'$, and the integration and the differentiation prescribed in Eq. (A.121) becomes very easy. When neither γ nor γ' is zero, the term in the exponent can be expressed in terms of $\boldsymbol{v}^2$ as

$$\bar{q}\lambda^2+\bar{q}'\lambda'^2+\bar{\rho}\lambda\lambda'\boldsymbol{e}\cdot\boldsymbol{e}'$$

$$=\frac{\bar{\rho}}{2\gamma\gamma'}\boldsymbol{v}^2+\left(\bar{q}-\frac{\gamma}{2\gamma'}\bar{\rho}\right)\lambda^2+\left(\bar{q}'-\frac{\gamma'}{2\gamma}\bar{\rho}\right)\lambda'^2. \qquad \text{(A.123)}$$

One first has to expand $\exp(\frac{\bar{\rho}}{2\gamma\gamma'}v^2)i_\kappa(rv)$ in terms of powers of v (see Eq. (6.48) for the expansion). Secondly, its general term $v^{2n+\kappa}$ multiplied by the spherical harmonic $Y_{\kappa\mu}(\hat{\boldsymbol{v}})$ must be expanded in the coupled tensor form of $[Y_L(\hat{\boldsymbol{e}})\times Y_{L'}(\hat{\boldsymbol{e}'})]^{(\kappa)}_\mu$. The second step can be done with the use of Eq. (6.7). In this derivation one can avoid recoupling the angular momenta, and hence would get the simplest possible formula for the matrix element.

In the following we work out the details for the central potential with $\kappa = 0$. However, we shall not follow the above route exactly but make a little detour in order to make transparent the relation to the overlap matrix element. To this end we rewrite the left-hand side of Eq. (A.123) as follows:

$$\bar{q}\lambda^2 + \bar{q}'\lambda'^2 + \bar{\rho}\lambda\lambda'\boldsymbol{e}\cdot\boldsymbol{e}' = q\lambda^2 + q'\lambda'^2 + \rho\lambda\lambda'\boldsymbol{e}\cdot\boldsymbol{e}' - \frac{1}{2c}\boldsymbol{v}^2. \quad \text{(A.124)}$$

The term $\mathrm{e}^{-\frac{v^2}{2c}} i_0(rv)$ is expanded in powers of v with the use of the Hermite polynomials:

$$\exp\Big(-\frac{v^2}{2c}\Big) i_0(rv) \equiv \exp\Big(-\frac{v^2}{2c}\Big)\frac{1}{2rv}(\mathrm{e}^{rv} - \mathrm{e}^{-rv})$$
$$= \frac{1}{\sqrt{2cr}}\sum_{n=0}^{\infty}\frac{1}{(2n+1)!}H_{2n+1}\Big(\sqrt{\frac{c}{2}}r\Big)\Big(\frac{v^2}{2c}\Big)^n. \quad \text{(A.125)}$$

Next we expand $\exp\big(q\lambda^2 + q'\lambda'^2 + \rho\lambda\lambda'\boldsymbol{e}\cdot\boldsymbol{e}'\big)\left(\frac{v^2}{2c}\right)^n$ as follows:

$$\exp\Big(q\lambda^2 + q'\lambda'^2 + \rho\lambda\lambda'\boldsymbol{e}\cdot\boldsymbol{e}'\Big)\left(\frac{v^2}{2c}\right)^n = \frac{n!}{(2c)^n}\sum_{pp'lmm'}$$
$$\times H(p-m, q\lambda^2)H(p'-m', q'\lambda'^2)H(l-n+m+m', \rho\lambda\lambda'\boldsymbol{e}\cdot\boldsymbol{e}')$$
$$\times H(m, \gamma^2\lambda^2)H(m', \gamma'^2\lambda'^2)H(n-m-m', 2\gamma\gamma'\lambda\lambda'\boldsymbol{e}\cdot\boldsymbol{e}')$$
$$= \sum_{pp'l} F^n_{pp'l}(c, q, q', \rho, \gamma, \gamma')\lambda^{2p+l}\lambda'^{2p'+l}(\boldsymbol{e}\cdot\boldsymbol{e}')^l \quad \text{(A.126)}$$

with

$$F^n_{pp'l}(c, q, q', \rho, \gamma, \gamma') = \frac{n!}{c^n}\sum_{mm'} H(p-m, q)H(p'-m', q')$$
$$\times H(l-n+m+m', \rho)\frac{\gamma^{n+m-m'}\gamma'^{n-m+m'}}{2^{m+m'}m!\,m'!\,(n-m-m')!}. \quad \text{(A.127)}$$

Here p, p', l, m, and m' are all non-negative integers and the ranges of their values are determined by the conditions $p + p' + l \geq n$ and $m \leq p$, $m' \leq p'$, $n - l \leq m + m' \leq n$. The operation in Eq. (A.121) is now easily done, leading to the result

$$\langle f_{K'L'M'}(u', A', \boldsymbol{x})|V(|\boldsymbol{r}_i - \boldsymbol{r}_j|)|f_{KLM}(u, A, \boldsymbol{x})\rangle$$

$$= \delta_{LL'}\delta_{MM'} \frac{(2K+L)!(2K'+L)!}{B_{KL}B_{K'L}} \left(\frac{(2\pi)^{N-1}}{\det B}\right)^{\frac{3}{2}} \sum_{n=0}^{K+K'+L} J(n,c)$$

$$\times \sum_{k=0}^{\min(K,K')} F^{n}_{K-k\,K'-k\,L+2k}(c,q,q',\rho,\gamma,\gamma')\, B_{kL}, \tag{A.128}$$

where $J(n,c)$ is the integral defined by

$$J(n,c) = \frac{1}{\sqrt{\pi}(2n+1)!}$$

$$\times \int_0^\infty V\left(\sqrt{\frac{2}{c}}x\right) \mathrm{e}^{-x^2} H_1(x) H_{2n+1}(x) dx. \tag{A.129}$$

In the case of $V(r) = 1$ the formula (A.128) reduces to that of the overlap matrix element (A.6), because then the integral $J(n,c)$ becomes nonzero if and only if $n = 0$ and $F^{n=0}_{K-k\,K'-k\,L+2k}(c,q,q',\rho,\gamma,\gamma')$ reduces to $H(K-k,q)H(K'-k,q')H(L+2k,\rho)$. Though this relationship is not apparent in Eq. (A.21), both equations should give the same result for the central potential.

For the special case with $K = K' = 0$, Eq. (A.128) reduces to a much simpler form which essentially includes a double sum:

$$\langle f_{0LM}(u',A',\boldsymbol{x})|V(|\boldsymbol{r}_i - \boldsymbol{r}_j|)|f_{0LM}(u,A,\boldsymbol{x})\rangle$$

$$= \langle f_{0LM}(u',A',\boldsymbol{x})|f_{0LM}(u,A,\boldsymbol{x})\rangle$$

$$\times \sum_{n=0}^{L} \frac{L!}{(L-n)!} \left(\frac{\gamma\gamma'}{c\rho}\right)^n J(n,c), \tag{A.130}$$

where the overlap $\langle f_{0LM}(u',A',\boldsymbol{x})|f_{0LM}(u,A,\boldsymbol{x})\rangle$ is given by Eq. (A.7). The above formula has a particularly nice feature for power law potentials $V(r) = r^k$, because then the c-dependence of $J(n,c)$, now denoted $J_k(n,c)$, is factored out as follows:

$$J_k(n,c) = \left(\frac{2}{c}\right)^{\frac{k}{2}} \frac{-2}{\sqrt{\pi}(2n+1)!} \int_0^\infty x^{k+1} \left(\frac{d^{2n+1}}{dx^{2n+1}} \mathrm{e}^{-x^2}\right) dx$$

$$= \left(\frac{2}{c}\right)^{\frac{k}{2}} \frac{1}{\sqrt{\pi}} \sum_{m=0}^{n} \frac{(-1)^m 2^{2n-2m+1}}{m!(2n-2m+1)!} \Gamma\left(n-m+\frac{k+3}{2}\right). \tag{A.131}$$

Since the particle indices i and j are contained only in c and $\gamma\gamma'/c\rho$, this factorization makes the summing up of Eq. (A.130) over i and j much faster. Especially for the Coulomb force ($k = -1$), we get

$$J_{-1}(n,c) = \sqrt{\frac{2c}{\pi}}\frac{(-1)^n}{(2n+1)n!}. \tag{A.132}$$

For other types of potentials the route given below Eq. (A.123) leads us to the following simpler result including only a single sum:

$$\begin{aligned}&\langle f_{0LM}(u',A',\boldsymbol{x})|V(|\boldsymbol{r}_i-\boldsymbol{r}_j|)|f_{0LM}(u,A,\boldsymbol{x})\rangle\\ &=\frac{(2L+1)!!}{4\pi}\left(\frac{(2\pi)^{N-1}}{\det B}\right)^{\frac{3}{2}}\sum_{n=0}^{L}\frac{L!}{n!(L-n)!}\left(\frac{\gamma\gamma'}{c}\right)^n\bar{\rho}^{L-n}\\ &\times\frac{2^{n+2}}{\sqrt{\pi}(2n+1)!!}\int_0^\infty V\left(\sqrt{\frac{2}{c}}x\right)x^{2n+2}\mathrm{e}^{-x^2}dx.\end{aligned} \tag{A.133}$$

A particular simplicity occurs for a Gaussian potential as one can then perform the n sum easily. For $V(r)=\mathrm{e}^{-\kappa r^2}$, we obtain

$$\begin{aligned}&\langle f_{0LM}(u',A',\boldsymbol{x})|\exp\{-\kappa(\boldsymbol{r}_i-\boldsymbol{r}_j)^2\}|f_{0LM}(u,A,\boldsymbol{x})\rangle\\ &=\frac{(2L+1)!!}{4\pi}\left(\frac{(2\pi)^{N-1}}{\det B}\right)^{\frac{3}{2}}\left(1+\frac{2\kappa}{c}\right)^{-\frac{3}{2}}\left(\bar{\rho}+\frac{\gamma\gamma'}{c}\frac{1}{1+\frac{2\kappa}{c}}\right)^{L}.\end{aligned} \tag{A.134}$$

A.2 *Matrix elements of density multipoles*

Here we show that Eq. (A.30) takes a simple form for the special case of $K = K' = 0$. As we have seen in the chapters of applications, the correlated Gaussian with $K = 0$ is, with the use of its full general A matrix and u vector, a very useful basis function. The simplified form of the matrix elements was in fact used in our calculations. In the case of $K = 0$, $L-n_1-n_2$ has to be non-negative and even in order to have a non-vanishing contribution from the term $H((2K+L-n_1-n_2)/2,\bar{q})$. Likewise $L'-n_1-n_3$ has to be non-negative and even. On the other hand, the recoupling coefficient $R^{n_1n_2n_3}_{LL'\kappa}$ of Eq. (A.20) becomes non-vanishing provided that n_1+n_2-L, n_1+n_3-L' and $n_2+n_3-\kappa$ are all non-negative and even. Therefore we are led to the result that the summation in Eq. (A.30) is restricted to the case of $n_1+n_2=L$ and $n_1+n_3=L'$. By renaming n_1 as n, we obtain $n_2=L-n$ and

$n_3 = L' - n$, where n runs from 0 to $\min(L, L')$. Possible values of κ are limited by the condition that $n_2 + n_3 - \kappa = L + L' - 2n - \kappa$ be non-negative and even and also $\kappa \geq \max(|L - L'|, |M - M'|)$ due to the triangular relation of the Clebsch-Gordan coefficient $\langle L\, M\, \kappa\, M'-M|L'\, M'\rangle$.

Furthermore, we note that under these conditions $R^{n_1 n_2 n_3}_{LL'\kappa} = R^{n\, L-n\, L'-n}_{LL'\kappa}$ is contributed by a single term with $l_1 = n_1 = n$, $l_2 = n_2 = L - n$, $l_3 = n_3 = L' - n$ in Eq. (A.20). This is because $l_1 \leq n_1$, $l_2 \leq n_2$, $l_3 \leq n_3$ and $l_1 + l_2 \geq L$, $l_1 + l_3 \geq L'$. Using the well-known formulas for the Clebsch-Gordan coefficients $\langle l\, 0\, l'\, 0|L\, 0\rangle$ and the Racah coefficient [75]

$$U(n\; L'-n\; L\; \kappa; L'\; L-n) = \sqrt{\frac{(2L-2n+1)!\,(2L'-2n)!\,(L+L'-\kappa)!\,(L+L'+\kappa+1)!}{(2L+1)!\,(2L')!\,(L+L'-2n-\kappa)!\,(L+L'-2n+\kappa+1)!}}, \tag{A.135}$$

we obtain the following result:

$$\begin{aligned} &\langle f_{0L'M'}(u', A', \boldsymbol{x})|\delta(\tilde{w}\boldsymbol{x} - \boldsymbol{r})|f_{0LM}(u, A, \boldsymbol{x})\rangle \\ &= \left(\frac{(2\pi)^{N-2}}{\det B}\right)^{\frac{3}{2}} c^{\frac{3}{2}}\, \mathrm{e}^{-\frac{1}{2}cr^2} \sum_{n=0}^{\min(L,L')} \bar{\rho}^n \gamma^{L-n} \gamma'^{L'-n} r^{L+L'-2n} \\ &\times \sum_{\kappa=\max(|L-L'|,|M-M'|)}^{L+L'-2n} Z_{LL'n\kappa} \langle L\, M\, \kappa\, M'-M|L'\, M'\rangle \\ &\times\; Y_{\kappa\, M'-M}(\hat{\boldsymbol{r}})^*, \end{aligned} \tag{A.136}$$

where $\kappa + L + L'$ has to be even and the coefficient $Z_{LL'n\kappa}$ is given by

$$\begin{aligned} Z_{LL'n\kappa} &= (-1)^{\frac{1}{2}(L-L'+\kappa)} \sqrt{\frac{(2L+1)(2\kappa+1)}{4\pi}} \\ &\times \frac{\sqrt{(L+L'-\kappa)!\,(L+L'+\kappa+1)!}}{(2n)!!\,(L+L'-2n-\kappa)!!\,(L+L'-2n+\kappa+1)!!} \\ &\times \sqrt{\frac{(L-L'+\kappa-1)!!\,(L'-L+\kappa-1)!!}{(L-L'+\kappa)!!\,(L'-L+\kappa)!!}}. \end{aligned} \tag{A.137}$$

With an appropriate choice of the $(N-1)$-dimensional vector w one can calculate various matrix elements. To calculate the matrix element of $\delta(\boldsymbol{r}_i - \boldsymbol{x}_N - \boldsymbol{r})$, one only needs to set w equal to $w^{(i)}$ which is defined in Eq. (2.12). The values of c, $\bar{\rho}$, γ, γ' are obtained from Eq. (A.18) with $w^{(ij)}$ being replaced by $w^{(i)}$. In exactly the same manner, one can also calculate the matrix element of $\delta(\boldsymbol{r}_i - \boldsymbol{r}_j - \boldsymbol{r})$ by employing the vector $w^{(ij)}$, defined in Eq. (2.13), for w. By using the well-known formula

$$\delta(\tilde{w}\boldsymbol{x} - \boldsymbol{r}) = \frac{\delta(|\tilde{w}\boldsymbol{x}| - r)}{r^2} \sum_{\kappa\,\mu} Y_{\kappa\mu}(\widehat{\tilde{w}\boldsymbol{x}}) Y_{\kappa\mu}(\hat{\boldsymbol{r}})^*, \tag{A.138}$$

Eq. (A.136) immediately enables one to obtain the matrix element of the density multipole operator $\delta(|\tilde{w}\boldsymbol{x}| - r) Y_{\kappa\mu}(\widehat{\tilde{w}\boldsymbol{x}})$.

The matrix element for the central potential, $V(|\boldsymbol{r}_i - \boldsymbol{r}_j|)$, is obtainable from Eq. (A.136) as well. By integrating the equation multiplied by $V(r)$ over $\boldsymbol{r}$, we see that only the term with $\kappa = 0$ contributes to the matrix element of the central potential. It is easy to see that the integration leads us to the same result as Eq. (A.133).

A.3 *Overlap matrix elements of the correlated Gaussians for a three-particle system*

We tabulate in Table A.1 results of the calculation for the quantity

$$\left(\prod_{i=1}^{2N-2} \frac{\partial^{l_i}}{\partial {\alpha_i}^{l_i}} \frac{\partial^{l_i - m_i}}{\partial {\tau_i}^{l_i - m_i}} \exp\left(\frac{1}{2}\tilde{\boldsymbol{s}} B \boldsymbol{s}\right)\right)_{\substack{\alpha_1 = 0, \ldots, \alpha_{2N-2} = 0 \\ \tau_1 = 0, \ldots, \tau_{2N-2} = 0}} \tag{A.139}$$

for the three-particle system $(N = 3)$ assuming $l_i \leq 2$. The vector $\boldsymbol{s}_i$ is defined by $\alpha_i(1 - {\tau_i}^2, i(1 + {\tau_i}^2), -2\tau_i)$. The calculation was done with *Mathematica* [184]. The quantity apparently has a symmetry property with respect to the interchange of $l_i, m_i \leftrightarrow l_j, m_j$. E.g., if one calculates this quantity for a given set of angular momenta, then the quantity corresponding to a set $(l_1, m_1, l_3, m_3, l_2, m_2, l_4, m_4)$ can easily be obtained from it by simply interchanging the suffices 2 and 3 of the B matrix. Note that the table can be used for the four-particle system as well if one of the angular momenta is restricted to zero.

Table A.1. Tabulation of Eq. (A.139) for the three-particle system. The vector s_i is defined by $\alpha_i(1-\tau_i{}^2, i(1+\tau_i{}^2), -2\tau_i)$. The matrix B is a 4×4 symmetric matrix and its elements are denoted $a = B_{12}$, $b = B_{13}$, $c = B_{14}$, $d = B_{23}$, $e = B_{24}$, $f = B_{34}$. The tenth to twelfth columns give the types and the coefficients of terms needed to construct the solution, where each coefficient must be multiplied by a factor given in the ninth column of the corresponding entry. For example, in the case of $l_1 = m_1 = l_2 = m_2 = 2$, $l_3 = 2$, $m_3 = -2$, $l_4 = 2$, $m_4 = -2$, the quantity of Eq. (A.139) is given by $12288 \times (3b^2e^2 + 3c^2d^2 + 12bcde)$.

l_1	m_1	l_2	m_2	l_3	m_3	l_4	m_4				
								4	f		
0	0	0	0	1	1	1	−1		−1		
0	0	0	0	1	0	1	0		1		
								192	f^2		
0	0	0	0	2	2	2	−2		1		
0	0	0	0	2	1	2	−1		−1		
0	0	0	0	2	0	2	0		1		
								32	ef		
0	0	1	1	1	1	2	−2		6		
0	0	1	1	1	0	2	−1		−3		
0	0	1	1	1	−1	2	0		1		
0	0	1	0	1	0	2	0		2		
0	0	1	0	1	−1	2	1		−1		
0	0	1	−1	1	−1	2	2		1		
								16	af	be	cd
1	1	1	1	1	−1	1	−1			1	1
1	1	1	0	1	0	1	−1				−1
1	0	1	0	1	0	1	0		1	1	1

(Continued on the next page.)

(Continued from Table A.1.)

								1536	def		
0	0	2	2	2	0	2	−2		−2		
0	0	2	2	2	−1	2	−1		3		
0	0	2	1	2	1	2	−2		2		
0	0	2	1	2	0	2	−1		−1		
0	0	2	0	2	0	2	0		2		
								256	af^2	bef	cdf
1	1	1	1	2	0	2	−2			−6	−6
1	1	1	1	2	−1	2	−1			9	9
1	1	1	0	2	1	2	−2				6
1	1	1	0	2	0	2	−1			3	−6
1	1	1	−1	2	2	2	−2		−3		−6
1	1	1	−1	2	1	2	−1		3		3
1	1	1	−1	2	0	2	0		−3	−1	−1
1	0	1	0	2	2	2	−2		3		
1	0	1	0	2	1	2	−1		−3	−3	−3
1	0	1	0	2	0	2	0		3	4	4
1	0	1	−1	2	2	2	−1				3
1	0	1	−1	2	1	2	0			1	−2
1	−1	1	−1	2	2	2	0			−1	−1
1	−1	1	−1	2	1	2	1			1	1
								12288	a^2f^2 $abef$	b^2e^2 $acdf$	c^2d^2 $bcde$
2	2	2	2	2	−2	2	−2			3	3 12
2	2	2	1	2	−1	2	−2				−3 −6
2	2	2	0	2	0	2	−2		 2	 2	3 2
2	2	2	0	2	−1	2	−1		 −3	 −3	 6
2	1	2	1	2	0	2	−2		 −2	 −2	 4
2	1	2	1	2	−1	2	−1		 3	3 3	3 6
2	1	2	0	2	0	2	−1		 4	 −5	−3 −5
2	0	2	0	2	0	2	0		3 6	3 6	3 6

Exercises

A.1. Derive Eq. (A.14).

Solution. Noting the definition of the scalar product in Eqs. (11.1) and (11.2), expanding the angular momentum coupled states of bra and ket functions, e.g.,

$$|(LS)JM\rangle = \sum_{m_1 m_2} \langle Lm_1 Sm_2|JM\rangle |Lm_1\rangle |Sm_2\rangle, \tag{A.140}$$

and using the Wigner-Eckart theorem (A.12), we have

$$\begin{aligned}
&\langle (L'S')JM|\Big(O_\kappa(\text{space})\cdot O_\kappa(\text{spin})\Big)|(LS)JM\rangle \\
&= \sum_{\mu m_1 m_2 m_1' m_2'} (-1)^\mu \langle Lm_1 Sm_2|JM\rangle \langle L'm_1' S'm_2'|JM\rangle \\
&\quad \times \frac{\langle Lm_1 \kappa\mu|L'm_1'\rangle}{\sqrt{2L'+1}} \langle L' \parallel O_\kappa(\text{space}) \parallel L\rangle \\
&\quad \times \frac{\langle Sm_2 \kappa\, {-\mu}|S'm_2'\rangle}{\sqrt{2S'+1}} \langle S' \parallel O_\kappa(\text{spin}) \parallel S\rangle .
\end{aligned} \tag{A.141}$$

The use of a symmetry property of Clebsch-Gordan coefficients

$$\langle Sm_2 \kappa\, {-\mu}|S'm_2'\rangle = (-1)^{\kappa-\mu} \sqrt{\frac{2S'+1}{2S+1}} \langle \kappa\mu S'm_2'|Sm_2\rangle \tag{A.142}$$

enables one to rewrite the above matrix element as

$$\begin{aligned}
&\langle (L'S')JM|\Big(O_\kappa(\text{space})\cdot O_\kappa(\text{spin})\Big)|(LS)JM\rangle \\
&= (-1)^\kappa \frac{\langle L' \parallel O_\kappa(\text{space}) \parallel L\rangle \langle S' \parallel O_\kappa(\text{spin}) \parallel S\rangle}{\sqrt{(2L'+1)(2S+1)}} \\
&\quad \times \sum_{\mu m_1 m_2 m_1' m_2'} \langle Lm_1 \kappa\mu|L'm_1'\rangle \langle L'm_1' S'm_2'|JM\rangle \\
&\quad \times \langle Lm_1 Sm_2|JM\rangle \langle \kappa\mu S'm_2'|Sm_2\rangle .
\end{aligned} \tag{A.143}$$

The sum over $\mu, m_1, m_2, m_1', m_2'$ of the product of the four Clebsch-Gordan coefficients is just the recoupling coefficient $U(L\kappa JS'; L'S)$ (see Eq. (6.67)). This completes the derivation.

A generalization to the matrix element of a coupled tensor operator can be made similarly. Let us calculate the matrix element

$$\begin{aligned}&\langle (L'S')J'M'|\,[O_l(\text{space})\times O_\lambda(\text{spin})]_{\kappa\nu}\,|(LS)JM\rangle\\&=\sum_{m_1'm_2'm_1m_2m\mu}\langle L'm_1'S'm_2'|J'M'\rangle\langle Lm_1Sm_2|JM\rangle\langle lm\lambda\mu|\kappa\nu\rangle\\&\times\frac{\langle Lm_1lm|L'm_1'\rangle}{\sqrt{2L'+1}}\langle L'\parallel O_l(\text{space})\parallel L\rangle\\&\times\frac{\langle Sm_2\lambda\mu|S'm_2'\rangle}{\sqrt{2S'+1}}\langle S'\parallel O_\lambda(\text{spin})\parallel S\rangle.\end{aligned}\tag{A.144}$$

The sum over $m_1', m_2', m_1, m_2, m, \mu$ of the product of the five Clebsch-Gordan coefficients

$$\begin{aligned}&\sum_{m_1m_2m\mu m_1'm_2'}\langle Lm_1Sm_2|JM\rangle\langle lm\lambda\mu|\kappa\nu\rangle\langle Lm_1lm|L'm_1'\rangle\\&\qquad\times\langle Sm_2\lambda\mu|S'm_2'\rangle\langle L'm_1'S'm_2'|J'M'\rangle\end{aligned}\tag{A.145}$$

can be taken by means of the unitary $9j$ symbol as

$$\langle JM\kappa\nu|J'M'\rangle\begin{bmatrix}L & S & J\\ l & \lambda & \kappa\\ L' & S' & J'\end{bmatrix}.\tag{A.146}$$

To derive this result multiply Eq. (A.145) by $\langle JM\kappa\nu|J'M'\rangle$ and sum over M and ν with M' being fixed, then what one has is a sum of the product of six Clebsch-Gordan coefficients involving nine angular momenta, and it is nothing but the $9j$ symbol in unitary form given in Eq. (A.146) (see Eq. (6.78)). Then with the help of the orthogonality relation of the Clebsch-Gordan coefficients Eq. (A.145) itself has to be equal to (A.146). Substituting this result into Eq. (A.144) and using the definition of the reduced matrix element one obtains

$$\begin{aligned}&\langle (L'S')J'\parallel[O_l(\text{space})\times O_\lambda(\text{spin})]_\kappa\parallel(LS)J\rangle\\&=\sqrt{\frac{2J'+1}{(2L'+1)(2S'+1)}}\begin{bmatrix}L & S & J\\ l & \lambda & \kappa\\ L' & S' & J'\end{bmatrix}\\&\quad\times\langle L'\parallel O_l(\text{space})\parallel L\rangle\langle S'\parallel O_\lambda(\text{spin})\parallel S\rangle.\end{aligned}\tag{A.147}$$

In special cases one can set $\lambda = 0$ or $l = 0$ in the above formula and by using

$$\langle S' \parallel 1 \parallel S\rangle = \delta_{SS'}\sqrt{2S+1} \tag{A.148}$$

derive useful formulas as given below:

$$\begin{aligned}&\langle (L'S')J' \parallel O_l(\text{space}) \parallel (LS)J\rangle \\ &\quad = \delta_{SS'}(-1)^{L-J+J'-L'}\sqrt{\frac{2J'+1}{2L'+1}}\,U(SLJ'l;JL') \\ &\qquad \times \langle L' \parallel O_l(\text{space}) \parallel L\rangle. \end{aligned} \tag{A.149}$$

$$\begin{aligned}&\langle (L'S')J' \parallel O_\lambda(\text{spin}) \parallel (LS)J\rangle \\ &\quad = \delta_{LL'}\sqrt{\frac{2J'+1}{2S'+1}}\,U(LSJ'\lambda;JS')\langle S' \parallel O_\lambda(\text{spin}) \parallel S\rangle. \end{aligned} \tag{A.150}$$

A.2. Derive Eq. (A.19).

Solution. Using Eq. (6.18), we have

$$\begin{aligned}&(\boldsymbol{e}\cdot\boldsymbol{e}')^{n_1}(\boldsymbol{e}\cdot\boldsymbol{r})^{n_2}(\boldsymbol{e}'\cdot\boldsymbol{r})^{n_3}\Big|_{e=e'=1} \\ &\quad = r^{n_2+n_3}\sum_{2k_1+l_1=n_1}\ \sum_{2k_2+l_2=n_2}\ \sum_{2k_3+l_3=n_3} B_{k_1l_1}B_{k_2l_2}B_{k_3l_3} \\ &\qquad \times (-1)^{l_1+l_2+l_3}\sqrt{(2l_1+1)(2l_2+1)(2l_3+1)}\,[Y_{l_1}(\hat{\boldsymbol{e}}) \times Y_{l_1}(\hat{\boldsymbol{e}'})]_{00} \\ &\qquad \times [Y_{l_2}(\hat{\boldsymbol{e}}) \times Y_{l_2}(\hat{\boldsymbol{r}})]_{00}\,[Y_{l_3}(\hat{\boldsymbol{e}'}) \times Y_{l_3}(\hat{\boldsymbol{r}})]_{00}. \end{aligned} \tag{A.151}$$

As was done in Exercise 6.1 (see Eq. (6.106)), we recouple the product

$$\begin{aligned}&[Y_{l_2}(\hat{\boldsymbol{e}}) \times Y_{l_2}(\hat{\boldsymbol{r}})]_{00}\,[Y_{l_3}(\hat{\boldsymbol{e}'}) \times Y_{l_3}(\hat{\boldsymbol{r}})]_{00} \\ &\quad = \sum_\kappa \begin{bmatrix} l_2 & l_2 & 0 \\ l_3 & l_3 & 0 \\ \kappa & \kappa & 0 \end{bmatrix} C(l_2\,l_3;\kappa)\,[[Y_{l_2}(\hat{\boldsymbol{e}}) \times Y_{l_3}(\hat{\boldsymbol{e}'})]_\kappa \times Y_\kappa(\hat{\boldsymbol{r}})]_{00} \\ &\quad = \sum_\kappa \sqrt{\frac{2\kappa+1}{(2l_2+1)(2l_3+1)}}\,C(l_2\,l_3;\kappa)\end{aligned}$$

$$\times [[Y_{l_2}(\hat{e}) \times Y_{l_3}(\hat{e'})]_\kappa \times Y_\kappa(\hat{\boldsymbol{r}})]_{00}. \tag{A.152}$$

The last step is to combine the two scalar products as follows:

$$\begin{aligned}
&[Y_{l_1}(\hat{e}) \times Y_{l_1}(\hat{e'})]_{00}\,[[Y_{l_2}(\hat{e}) \times Y_{l_3}(\hat{e'})]_\kappa \times Y_\kappa(\hat{\boldsymbol{r}})]_{00} \\
&= [[[Y_{l_1}(\hat{e}) \times Y_{l_1}(\hat{e'})]_0 \times [Y_{l_2}(\hat{e}) \times Y_{l_3}(\hat{e'})]_\kappa]_\kappa \times Y_\kappa(\hat{\boldsymbol{r}})]_{00} \\
&= \sum_{LL'} E_{LL'}^{l_1 l_1 0 l_2 l_3 \kappa\kappa}\,[[Y_L(\hat{e}) \times Y_{L'}(\hat{e'})]_\kappa \times Y_\kappa(\hat{\boldsymbol{r}})]_{00},
\end{aligned} \tag{A.153}$$

where the coefficient E is given by Eq. (A.75). When one of the angular momenta is zero, the $9j$ symbol is reduced to the Racah coefficient (see Eq.(6.86)). Using this simplification in the coefficient E and substituting the above results completes the derivation.

A.3. Calculate the matrix element of the orbital angular momentum $l_{\mu_{ij}}$ for the correlated Gaussians.

Solution. The matrix element of $l_{\mu_{ij}}$ can be obtained from that between the generating functions of Table 7.1 or by performing the integration over $\boldsymbol{r}$ in Eq. (A.23) with $V(r)$ being set to unity. The result is

$$\begin{aligned}
&\langle f_{K'L'M'}(u', A', \boldsymbol{x}) | l_{\mu_{ij}} | f_{KLM}(u, A, \boldsymbol{x}) \rangle \\
&= \frac{1}{B_{KL} B_{K'L'}} \left(\frac{(2\pi)^{N-1}}{\det B} \right)^{\frac{3}{2}} \frac{1}{c} (\eta\gamma' + \eta'\gamma) \\
&\times \iint d\hat{e} d\hat{e'} Y_{LM}(\hat{e}) Y_{L'M'}(\hat{e'})^* \Bigg(\frac{d^{2K+L+2K'+L'}}{d\lambda^{2K+L} d\lambda'^{2K'+L'}} \\
&\times \lambda\lambda' \exp\left[q\lambda^2 + q'\lambda'^2 + \rho\lambda\lambda' \boldsymbol{e}\cdot\boldsymbol{e}' \right] i(\boldsymbol{e} \times \boldsymbol{e}')_\mu \Bigg)_{\substack{\lambda=0,\lambda'=0 \\ e=1, e'=1}}
\end{aligned} \tag{A.154}$$

The integration over $\hat{e}$ and $\hat{e'}$ can be done by using the formula

$$\begin{aligned}
&\iint d\hat{e} d\hat{e'} Y_{LM}(\hat{e}) Y_{L'M'}(\hat{e'})^* \Big(i(\boldsymbol{e} \times \boldsymbol{e}')_\mu (\boldsymbol{e}\cdot\boldsymbol{e}')^n \Big)_{e=1, e'=1} \\
&= 4\sqrt{\frac{2}{3}}\pi \frac{1}{\sqrt{2L'+1}} \langle LM\, 1\mu | L'M' \rangle \sum_{2k+l=n} B_{kl} \sqrt{2l+1}\, E_{LL'}^{ll01111}.
\end{aligned}$$

(A.155)

Clearly l has to be $L \pm 1$ and also $L' \pm 1$. In order that a possible value of l exists, either $L = L'$ or $|L - L'| = 2$ has to be satisfied. The latter case is obviously impossible because the condition $|L - L'| \leq 1$ has to be met. Thus we reach the same conclusion that L and L' must be equal, otherwise the integral would vanish. Using

$$C(L-1\,1;L) = \sqrt{\frac{3L}{4\pi(2L+1)}}, \quad C(L+1\,1;L) = -\sqrt{\frac{3(L+1)}{4\pi(2L+1)}},$$

$$U(L-1\,1L1;L1) = \sqrt{\frac{L+1}{2L}}, \quad U(L+1\,1L1;L1) = -\sqrt{\frac{L}{2L+2}},$$

(A.156)

together with the relation

$$B_{k\,L-1} = \frac{2k+2L+1}{2k+L} B_{k\,L}, \qquad B_{k-1\,L+1} = \frac{2k}{2k+L} B_{k\,L}, \quad \text{(A.157)}$$

we can show that

$$\iint d\hat{e}d\hat{e'} Y_{LM}(\hat{e}) Y_{L'M'}(\hat{e'})^* \Big(i(\boldsymbol{e} \times \boldsymbol{e'})_\mu (\boldsymbol{e}\cdot\boldsymbol{e'})^n \Big)_{e=1,e'=1}$$

$$= \delta_{LL'} \sqrt{L(L+1)} \langle LM1\mu|LM'\rangle \frac{1}{n+1} B_{\frac{n+1-L}{2}\,L}. \quad \text{(A.158)}$$

In order for the integral not to vanish, $n+1-L$ must be non-negative and even.

Employing this result in Eq. (A.154) and comparing its result with Eq. (A.6) leads to a solution

$$\langle f_{K'L'M'}(u', A', \boldsymbol{x}) | l_{\mu_{ij}} | f_{KLM}(u, A, \boldsymbol{x}) \rangle$$

$$= \delta_{LL'} \sqrt{L(L+1)} \langle LM1\mu|LM'\rangle$$

$$\times \frac{1}{\rho c} (\gamma\eta' + \gamma'\eta) \langle f_{K'LM}(u', A', \boldsymbol{x}) | f_{KLM}(u, A, \boldsymbol{x}) \rangle. \quad \text{(A.159)}$$

As a check of the above formula, one may use this to calculate the matrix element of the total orbital angular momentum $L_\mu = \frac{2}{N} \sum_{i<j} l_{\mu_{ij}}$ (see Eq. (2.22)). The summation, over $i < j$, of $\frac{1}{c}(\gamma\eta' + \gamma'\eta)$ can be easily done with the use of Eq. (2.15), yielding just $\frac{N}{2}\rho$. Therefore, we expect that the desired matrix element becomes

$$\langle f_{K'L'M'}(u',A',\boldsymbol{x})|L_\mu|f_{KLM}(u,A,\boldsymbol{x})\rangle = \delta_{LL'}\sqrt{L(L+1)} \\ \times \langle LM1\mu|LM'\rangle\langle f_{K'LM}(u',A',\boldsymbol{x})|f_{KLM}(u,A,\boldsymbol{x})\rangle. \qquad (A.160)$$

On the other hand, we can reach the same result through the Wigner-Eckart theorem:

$$\langle\alpha'L'M'|L_\mu|\alpha LM\rangle = \delta_{LL'}\frac{\langle LM\,1\mu|LM'\rangle}{\sqrt{2L+1}}\langle\alpha'L\parallel \boldsymbol{L}\parallel\alpha L\rangle$$

$$= \delta_{LL'}\frac{\langle LM\,1\mu|LM'\rangle}{\sqrt{2L+1}}\sqrt{L(L+1)}\langle\alpha'L\parallel 1\parallel\alpha L\rangle$$

$$= \delta_{LL'}\langle LM\,1\mu|LM'\rangle\sqrt{L(L+1)}\langle\alpha'LM|\alpha LM\rangle, \qquad (A.161)$$

where the labels α and α' stand for the other labels of the wave functions. Here we used the fact that L_μ does not change the magnitude of the angular momentum of the wave function because it is a generator of the rotation group.

A.4. Calculate the matrix element of the spin-orbit potential following the method of Complement A.1.

Solution. The integration over $\hat{\boldsymbol{r}}$ is first done in Eq. (A.23), leading to

$$\langle f_{K'L'M'}(u',A',\boldsymbol{x})|V(|\boldsymbol{r}_i-\boldsymbol{r}_j|)l_{\mu_{ij}}|f_{KLM}(u,A,\boldsymbol{x})\rangle$$

$$= (\gamma\eta'+\gamma'\eta)\int_0^\infty dr r^2 V(r)\,\mathrm{e}^{-\frac{1}{2}cr^2}\frac{4\pi}{B_{KL}B_{K'L'}}\left(\frac{(2\pi)^{N-2}c}{\det B}\right)^{\frac{3}{2}}$$

$$\times\iint d\hat{e}\,d\hat{e'}Y_{LM}(\hat{e})Y_{L'M'}(\hat{e'})^*\left(\frac{d^{2K+L+2K'+L'}}{d\lambda^{2K+L}d\lambda'^{2K'+L'}}\right.$$

$$\left.\times\lambda\lambda'\exp\left[\bar{q}\lambda^2+\bar{q}'\lambda'^2+\bar{\rho}\lambda\lambda'\boldsymbol{e}\cdot\boldsymbol{e}'\right]i(\boldsymbol{e}\times\boldsymbol{e}')_\mu\frac{r}{v}i_1(rv)\right)\Bigg|_{\substack{\lambda=0,\lambda'=0\\ e=1,e'=1}}. \qquad (A.162)$$

Here v is the length of the vector $\boldsymbol{v}$ defined by Eq. (A.122) and use is made of the formula

$$\int \mathrm{e}^{\boldsymbol{a}\cdot\boldsymbol{r}}(\boldsymbol{b}\times\boldsymbol{r})_\mu d\hat{\boldsymbol{r}} = 4\pi\frac{r}{a}i_1(ar)(\boldsymbol{b}\times\boldsymbol{a})_\mu, \qquad (A.163)$$

which can be proved by using Eqs. (6.53) and (6.54) and noting $(\boldsymbol{b}\times\boldsymbol{r})_\mu = -\sqrt{2}i[\boldsymbol{b}\times\boldsymbol{r}]_{1\mu} = -\sqrt{\frac{8\pi}{3}}ir[\boldsymbol{b}\times Y_1(\hat{\boldsymbol{r}})]_{1\mu}$.

The term $e^{-\frac{v^2}{2c}}\frac{r}{v}i_1(rv)$ is expanded as was done in Complement A.1:

$$\exp\Big(-\frac{v^2}{2c}\Big)\frac{r}{v}i_1(rv) = \frac{1}{(2c)^{\frac{3}{2}}r}\sum_{n=0}^{\infty}\frac{2n+2}{(2n+3)!}\Big(\frac{v^2}{2c}\Big)^n$$
$$\times\Big\{H_{2n+3}\Big(\sqrt{\frac{c}{2}}r\Big)+2(2n+3)H_{2n+1}\Big(\sqrt{\frac{c}{2}}r\Big)\Big\}. \tag{A.164}$$

By using the expansion (A.126) and integrating over $\hat{\boldsymbol{e}}$ and $\hat{\boldsymbol{e}'}$ with the use of Eq. (A.158), we obtain

$$\langle f_{K'L'}(u',A',\boldsymbol{x}) \parallel V(|\boldsymbol{r}_i-\boldsymbol{r}_j|)l_{ij} \parallel f_{KL}(u,A,\boldsymbol{x})\rangle$$
$$= \delta_{LL'}\sqrt{L(L+1)(2L+1)}\frac{(2K+L)!(2K'+L)!}{B_{KL}B_{K'L}}\Big(\frac{(2\pi)^{N-1}}{\det B}\Big)^{\frac{3}{2}}$$
$$\times\frac{1}{c}(\gamma\eta'+\gamma'\eta)\sum_{n=0}^{K+K'+L-1}\frac{n+1}{\sqrt{\pi}(2n+3)!}\int_0^\infty V\Big(\sqrt{\frac{2}{c}}x\Big)e^{-x^2}$$
$$\times H_1(x)\Big\{H_{2n+3}(x)+2(2n+3)H_{2n+1}(x)\Big\}dx$$
$$\times\sum_{k=0}^{\min(K,K')}F^n_{K-k\,K'-k\,L+2k-1}(c,q,q',\rho,\gamma,\gamma')\frac{1}{L+2k}B_{kL}. \tag{A.165}$$

In the case when $V(r)=1$ only the term with $n=0$ contributes to the matrix element. The function $F^{n=0}_{K-k\,K'-k\,L+2k-1}(c,q,q',\rho,\gamma,\gamma')$ then becomes $\frac{L+2k}{\rho}H(K-k,q)H(K'-k,q')H(L+2k,\rho)$. Therefore we can confirm that Eq. (A.165) reduces to Eq. (A.159).

A.5. For a given function $f(x_1,x_2,...,x_K)$ of the form

$$f(x_1,x_2,...,x_K)=\frac{1}{2}\tilde{x}Ax+\tilde{b}x,$$

where $\tilde{x}=(x_1,...,x_K)$, $\tilde{b}=(b_1,...,b_K)$, and A is a $K\times K$ symmetric matrix, calculate

$$\left(\prod_{i=1}^{K}\frac{\partial^{n_i}}{\partial {x_i}^{n_i}}\mathrm{e}^{f(x_1,x_2,\ldots,x_K)}\right)_{x_1=0,\ldots,x_K=0}. \tag{A.166}$$

Solution. A solution can be obtained with the use of Leibniz's formula as was done in A.2. Here we show another way to solve this problem. Let an operator $\mathcal{L}$ denote

$$\mathcal{L}=\lambda_1\frac{\partial}{\partial x_1}+\lambda_2\frac{\partial}{\partial x_2}+\cdots+\lambda_K\frac{\partial}{\partial x_K}, \tag{A.167}$$

where $\tilde{\lambda}=(\lambda_1,\ldots,\lambda_K)$ is an auxiliary parameter. Then we have

$$\begin{aligned}\mathrm{e}^{\vartheta\mathcal{L}}\mathrm{e}^{f(x_1,\ldots,x_K)} &= \mathrm{e}^{f(x_1+\vartheta\lambda_1,\ldots,x_K+\vartheta\lambda_K)}\\ &= \mathrm{e}^{f(x_1,\ldots,x_K)}\exp\left(\frac{1}{2}\tilde{\lambda}A\lambda\vartheta^2+\tilde{\lambda}b\vartheta+\tilde{\lambda}Ax\vartheta\right). \end{aligned} \tag{A.168}$$

Let N be equal to $\sum_{i=1}^{K} n_i$. Expanding the above equation in power series of ϑ, collecting the terms of the Nth power in ϑ, and putting $x_i=0$ $(i=1,\ldots,K)$ leads to

$$\begin{aligned}&\left(\mathcal{L}^N\mathrm{e}^{f(x_1,\ldots,x_K)}\right)_{x_1=0,\ldots,x_K=0}\\ &=\sum_{k=0}^{[\frac{N}{2}]}\frac{N!}{k!\,(N-2k)!}\left(\frac{1}{2}\tilde{\lambda}A\lambda\right)^k(\tilde{\lambda}b)^{N-2k}.\end{aligned} \tag{A.169}$$

The problem of finding a solution to Eq. (A.166) is thus reduced to a simpler problem of expanding the right-hand side of Eq. (A.169) in the λ_i's. To this end $(\tilde{\lambda}b)^{N-2k}$ and $(\frac{1}{2}\tilde{\lambda}A\lambda)^k$ are expanded as follows:

$$(\tilde{\lambda}b)^{N-2k}=\sum_{l_i}\frac{(N-2k)!}{l_1!\,l_2!\cdots l_K!}(\lambda_1 b_1)^{l_1}(\lambda_2 b_2)^{l_2}\cdots(\lambda_K b_K)^{l_K},$$

$$\begin{aligned}\left(\frac{1}{2}\tilde{\lambda}A\lambda\right)^k&=\left(\frac{1}{2}\right)^k\sum_{m_{ij}}\frac{k!}{m_{11}!\,m_{12}!\cdots m_{KK}!}(A_{11}{\lambda_1}^2)^{m_{11}}\\ &\quad\times(2A_{12}\lambda_1\lambda_2)^{m_{12}}\cdots\times(A_{KK}{\lambda_K}^2)^{m_{KK}}.\end{aligned} \tag{A.170}$$

Collecting those terms which have powers of ${\lambda_1}^{n_1}{\lambda_2}^{n_2}\cdots{\lambda_K}^{n_K}$ we obtain a solution

$$\left(\prod_{i=1}^{K}\frac{\partial^{n_i}}{\partial {x_i}^{n_i}}\mathrm{e}^{f(x_1,x_2,\ldots,x_K)}\right)_{x_1=0,\ldots,x_K=0}$$

$$=\sum_{k=0}^{[\frac{N}{2}]}\left(\frac{1}{2}\right)^k\sum_{m_{ij}}\left\{\prod_{i=1}^{K}n_i!H(m_{ii},A_{ii})H(l_i,b_i)\right\}$$

$$\times\prod_{j>i=1}^{K}H(m_{ij},2A_{ij}), \tag{A.171}$$

where m_{ij}'s and l_i's are all non-negative integer and must satisfy the following relations

$$\sum_{j\geq i=1}^{K}m_{ij}=k,\quad m_{ii}+\sum_{j=1}^{K}m_{ji}+l_i=n_i\quad (i=1,...,K). \tag{A.172}$$

A.6. Derive Eq. (A.93).

Solution. First we note that

$$\Delta_x\left(x^{2\nu+\lambda}\mathrm{e}^{-\frac{1}{2}\beta x^2-\delta \boldsymbol{x}\cdot\boldsymbol{y}}\,Y_{LM}^{(\lambda l)}(\hat{\boldsymbol{x}},\hat{\boldsymbol{y}})\right)$$

$$=\left(\Delta_x\mathrm{e}^{-\frac{1}{2}\beta x^2-\delta \boldsymbol{x}\cdot\boldsymbol{y}}\right)x^{2\nu+\lambda}Y_{LM}^{(\lambda l)}(\hat{\boldsymbol{x}},\hat{\boldsymbol{y}})$$

$$+\,\mathrm{e}^{-\frac{1}{2}\beta x^2-\delta \boldsymbol{x}\cdot\boldsymbol{y}}\left(\Delta_x x^{2\nu+\lambda}Y_{LM}^{(\lambda l)}(\hat{\boldsymbol{x}},\hat{\boldsymbol{y}})\right)$$

$$+\,2\left(\boldsymbol{\nabla}_x\mathrm{e}^{-\frac{1}{2}\beta x^2-\delta \boldsymbol{x}\cdot\boldsymbol{y}}\right)\cdot\left(\boldsymbol{\nabla}_x x^{2\nu+\lambda}Y_{LM}^{(\lambda l)}(\hat{\boldsymbol{x}},\hat{\boldsymbol{y}})\right). \tag{A.173}$$

The differentiation in the second term of the right-hand side of the above equation can be performed by using Eq. (6.59) as

$$\left(\Delta_x x^{2\nu+\lambda}Y_{LM}^{(\lambda l)}(\hat{\boldsymbol{x}},\hat{\boldsymbol{y}})\right)$$

$$=2\nu(2\nu+2\lambda+1)x^{2\nu+\lambda-2}Y_{LM}^{(\lambda l)}(\hat{\boldsymbol{x}},\hat{\boldsymbol{y}}). \tag{A.174}$$

The third term can be reduced, by means of the gradient formula (A.42), to

$$\left(\boldsymbol{\nabla}_x\mathrm{e}^{-\frac{1}{2}\beta x^2-\delta \boldsymbol{x}\cdot\boldsymbol{y}}\right)\cdot\left(\boldsymbol{\nabla}_x x^{2\nu+\lambda}Y_{LM}^{(\lambda l)}(\hat{\boldsymbol{x}},\hat{\boldsymbol{y}})\right)$$

$$= x^{2\nu+\lambda-1} e^{-\frac{1}{2}\beta x^2 - \delta \boldsymbol{x}\cdot\boldsymbol{y}} \Bigg(\sqrt{\frac{\lambda}{2\lambda+1}} (2\nu+2\lambda+1) \left[[(-\beta\boldsymbol{x}-\delta\boldsymbol{y}) \times Y_{\lambda-1}(\hat{\boldsymbol{x}})]_\lambda \times Y_l(\hat{\boldsymbol{y}})\right]_{LM}$$
$$-\sqrt{\frac{\lambda+1}{2\lambda+1}} 2\nu \left[[(-\beta\boldsymbol{x}-\delta\boldsymbol{y}) \times Y_{\lambda+1}(\hat{\boldsymbol{x}})]_\lambda \times Y_l(\hat{\boldsymbol{y}})\right]_{LM} \Bigg). \tag{A.175}$$

By using the angular momentum algebra the first term in the round bracket of the above equation is reduced to

$$\left[[(-\beta\boldsymbol{x}-\delta\boldsymbol{y}) \times Y_{\lambda-1}(\hat{\boldsymbol{x}})]_\lambda \times Y_l(\hat{\boldsymbol{y}})\right]_{LM}$$
$$= -\sqrt{\frac{4\pi}{3}} \Bigg(\beta x\, C(1\,\lambda{-}1;\lambda) Y_{LM}^{(\lambda l)}(\hat{\boldsymbol{x}},\hat{\boldsymbol{y}})$$
$$+ \delta y \sum_\kappa U(\lambda{-}1\,1\,L\,l;\lambda\,\kappa)\, C(1l;\kappa) Y_{LM}^{(\lambda-1\,\kappa)}(\hat{\boldsymbol{x}},\hat{\boldsymbol{y}}) \Bigg)$$
$$= -\sqrt{\frac{\lambda}{2\lambda+1}} \beta x Y_{LM}^{(\lambda l)}(\hat{\boldsymbol{x}},\hat{\boldsymbol{y}})$$
$$-\sqrt{\frac{4\pi}{3}} \delta y \sum_\kappa U(\lambda{-}1\,1\,L\,l;\lambda\,\kappa)\, C(1l;\kappa) Y_{LM}^{(\lambda-1\,\kappa)}(\hat{\boldsymbol{x}},\hat{\boldsymbol{y}}), \tag{A.176}$$

where use is made of Eq. (A.156) for $C(1\,\lambda-1;\lambda) = C(\lambda-1\,1;\lambda)$. Similarly the second term becomes

$$\left[[(-\beta\boldsymbol{x}-\delta\boldsymbol{y}) \times Y_{\lambda+1}(\hat{\boldsymbol{x}})]_\lambda \times Y_l(\hat{\boldsymbol{y}})\right]_{LM}$$
$$= \sqrt{\frac{\lambda+1}{2\lambda+1}} \beta x Y_{LM}^{(\lambda l)}(\hat{\boldsymbol{x}},\hat{\boldsymbol{y}})$$
$$-\sqrt{\frac{4\pi}{3}} \delta y \sum_\kappa U(\lambda{+}1\,1\,L\,l;\lambda\,\kappa) C(1l;\kappa) Y_{LM}^{(\lambda+1\,\kappa)}(\hat{\boldsymbol{x}},\hat{\boldsymbol{y}}). \tag{A.177}$$

The contribution of the first term is easily obtained by using

$$\left(\Delta_x \mathrm{e}^{-\frac{1}{2}\beta x^2-\delta \boldsymbol{x}\cdot\boldsymbol{y}}\right)$$

$$= \big(-3\beta+\beta^2x^2+\delta^2y^2+2\beta\delta\boldsymbol{x}\cdot\boldsymbol{y}\big)\,\mathrm{e}^{-\frac{1}{2}\beta x^2-\delta\boldsymbol{x}\cdot\boldsymbol{y}}, \qquad \text{(A.178)}$$

where the last term involving the scalar product $\boldsymbol{x}\cdot\boldsymbol{y}$ can be reduced to

$$\boldsymbol{x}\cdot\boldsymbol{y}Y_{LM}^{(\lambda l)}(\hat{\boldsymbol{x}},\hat{\boldsymbol{y}}) = -\sqrt{3}[\boldsymbol{x}\times\boldsymbol{y}]_{00}Y_{LM}^{(\lambda l)}(\hat{\boldsymbol{x}},\hat{\boldsymbol{y}})$$

$$= -\sqrt{3}\sum_{\kappa=\lambda\pm1}U(11\lambda\lambda;0\kappa)\big[[\boldsymbol{y}\times[\boldsymbol{x}\times Y_\lambda(\hat{\boldsymbol{x}})]_\kappa]_\lambda\times Y_l(\hat{\boldsymbol{y}})\big]_{LM}$$

$$= \sqrt{\frac{\lambda}{2\lambda+1}}x\,\big[[\boldsymbol{y}\times Y_{\lambda-1}(\hat{\boldsymbol{x}})]_\lambda\times Y_l(\hat{\boldsymbol{y}})\big]_{LM}$$

$$-\sqrt{\frac{\lambda+1}{2\lambda+1}}x\,\big[[\boldsymbol{y}\times Y_{\lambda+1}(\hat{\boldsymbol{x}})]_\lambda\times Y_l(\hat{\boldsymbol{y}})\big]_{LM}. \qquad \text{(A.179)}$$

The two terms of the right-hand side of the above equation appear in Eqs. (A.176) and (A.177), respectively.

Combining these results we obtain the desired equality (A.93)

$$\Delta_x\left(x^{2\nu+\lambda}\mathrm{e}^{-\frac{1}{2}\beta x^2-\delta\boldsymbol{x}\cdot\boldsymbol{y}}\,Y_{LM}^{(\lambda l)}(\hat{\boldsymbol{x}},\hat{\boldsymbol{y}})\right)$$

$$= \Big(-3\beta x^2+\beta^2x^4+\delta^2x^2y^2-2\beta(2\nu+\lambda)x^2$$

$$+2\nu(2\nu+2\lambda+1)\Big)\,x^{2\nu+\lambda-2}\,\mathrm{e}^{-\frac{1}{2}\beta x^2-\delta\boldsymbol{x}\cdot\boldsymbol{y}}\,Y_{LM}^{(\lambda l)}(\hat{\boldsymbol{x}},\hat{\boldsymbol{y}})$$

$$-2\delta x^{2\nu+\lambda-1}y\,\mathrm{e}^{-\frac{1}{2}\beta x^2-\frac{1}{2}\gamma y^2-\delta\boldsymbol{x}\cdot\boldsymbol{y}}\sqrt{\frac{4\pi}{3}}$$

$$\times\Bigg\{\sqrt{\frac{\lambda}{2\lambda+1}}\,(2\nu+2\lambda+1-\beta x^2)$$

$$\times\sum_\kappa U(\lambda{-}1\,1\,L\,l;\lambda\,\kappa)C(1l;\kappa)Y_{LM}^{(\lambda-1\,\kappa)}(\hat{\boldsymbol{x}},\hat{\boldsymbol{y}})$$

$$-\sqrt{\frac{\lambda+1}{2\lambda+1}}\,(2\nu-\beta x^2)$$

$$\times\sum_\kappa U(\lambda{+}1\,1\,L\,l;\lambda\,\kappa)C(1l;\kappa)Y_{LM}^{(\lambda+1\,\kappa)}(\hat{\boldsymbol{x}},\hat{\boldsymbol{y}})\Bigg\}. \qquad \text{(A.180)}$$

References

1. C. Cohen-Tannoudji, B. Diu and F. Laloë, *Quantum Mechanics* (A Wiley-Interscience publication, Hermann and John Wiley & Sons, Paris, 1977).
2. A. Messiah, *Quantum Mechanics* (North Holland, Amsterdam, 1961).
3. H. A. Bethe and E. E. Salpeter, *Quantum Mechanics of One and Two Electron Atoms* (Springer-Verlag, Berlin, 1957).
4. A. Theophilou, J. Phys. C**12**, 5419 (1979).
5. R. G. Parr and W. Yang, *Density-Functional Theory of Atoms and Molecules* (Oxford University Press, 1989).
6. D. H. Weinstein, Proc. Nat. Acad. Sci. **20**, 529 (1934).
7. S. Goedecker and K. Maschke, Phys. Rev. B**44**, 10 365 (1991).
8. G. Temple, Proc. Roy. Soc. (London) A**211**, 204 (1952).
9. M. A. Abdel-Raouf, Phys. Rep. **84**, 163 (1982).
10. C. J. Umrigar, K. G. Wilson and J. K. Wilkins, Phys. Rev. Lett. **60**, 1719 (1988).
11. J. A. Carlson and R. B. Wiringa, in *Computational Nuclear Physics 1*, ed. K. Langanke, J. A. Maruhn and S. E. Koonin, (Springer-Verlag, Berlin, 1991), p. 171.
12. Y. Akaishi, International Review of Nuclear Physics, Vol. 4, *Cluster Models and Other Topics* (World Scientific, Singapore, 1986), p. 259.
13. N. Metropolis, A. W. Rosenbluth, M. N. Rosenbluth, A. H. Teller and E. Teller, J. Chem. Phys. **21**, 1087 (1953).
14. W. H. Press, S. A. Teukolsky, W. T. Vetterling and B. P. Flanney, *Numerical Recipes in FORTRAN*, 2nd edition, (Cambridge University Press, New York, 1992).
15. D. Cvijovic and J. Klinowski, Science **267**, 664 (1995).
16. P. Serra, F. Stanton and S. Kais, Phys. Rev. E**55**, 1162 (1997).
17. S. Kirkpatrick, C. D. Gelett. Jr. and M. P. Vecchi, Science **220**, 671 (1983).
18. D. E. Goldberg, *Genetic Algorithms in Search, Optimization, and Machine Learning* (Addison Wesley, Reading, Massachusetts, 1989).
19. M. W. Schmidt and K. Ruedenberg, J. Chem. Phys. **71**, 3951 (1979).
20. M. Kamimura, Phys. Rev. A**38**, 621 (1988).
21. A. J. Thakker and V. H. Smith, Jr., Phys. Rev. A**15**, 1 (1977); A**15**, 16 (1977).
22. S. A. Alexander, H. J. Monkhorst and K. Szalewicz, J. Chem. Phys. **85**, 5821 (1986); **87**, 3976 (1987); **89**, 355 (1988).

23. S. A. Alexander and H. J. Monkhorst, Phys. Rev. A**38**, 26 (1988).
24. V. I. Kukulin, V. M. Krasnopol'sky, V. T. Voronchev and P. B. Sazonov, Nucl. Phys. A**417**, 128 (1984).
25. H. Kameyama, M. Kamimura and Y. Fukushima, Phys. Rev. C**40**, 974 (1989).
26. M. Kamimura and H. Kameyama, Nucl. Phys. A**508** 17c, (1990).
27. V. I. Kukulin and V. M. Krasnopol'sky, J. Phys. G: Nucl. Part. Phys. **3**, 795 (1977).
28. K. Varga, Y. Suzuki and R. G. Lovas, Nucl. Phys. A**371**, 447 (1994).
29. K. Arai, Y. Suzuki and K. Varga, Phys. Rev. C**51**, 2488 (1995).
30. K. Varga, Y. Suzuki and I. Tanihata, Phys. Rev. C**52**, 3013 (1995).
31. K. Varga and Y. Suzuki, Phys. Rev. C**52**, 2885 (1995); Phys. Rev. A**53**, 1907 (1996).
32. K. Varga, Y. Ohbayasi and Y. Suzuki, Phys. Lett. B**396**, 1 (1997).
33. Y. Suzuki, J. Usukura and K. Varga, J. Phys. B: At. Mol. Opt. Phys. **31**, 31 (1998).
34. W. Cencek and J. Rychlewski, J. Chem. Phys. **98**, 1252 (1993).
35. R. Bukowski, B. Jeziorski, S. Rybak and K. Szalewicz, J. Chem. Phys. **102**, 888 (1995); B. Jeziorski, R. Bukowski and K. Szalewicz, Int. J. Quant. Chem. **61**, 769 (1997).
36. R. Krivec, M. I. Haftel and V. B. Mandelzweig, Phys. Rev. A**44**, 7158 (1991).
37. J. Baker, R. N. Hill and J. D. Morgan III, in *Relativistic, Quantum Electrodynamic and Weak Interaction Effects in Atoms*, Proceedings of the Program held on Relativistic, Quantum Electrodynamic and Weak Interaction Effects in Atoms at the Institute of Theoretical Physics, Santa Barbara, CA, 1988, AIP Conf. Proc. No. **189**, ed. W. Johnson, P. Mohr and J. Sucher (AIP, New York, 1988), p. 123.
38. H. De Raedt and M. Frick, Phys. Rep. **231**, 109 (1993).
39. A. Mushinski and M. P. Nightingale, J. Chem. Phys. **101**, 8831 (1994).
40. M. H. Kalos, D. Levesque and L. Verlet, Phys. Rev. A**9**, 2178 (1974).
41. J. G. Zabolitzky and M. H. Kalos, Nucl. Phys. A**356**, 114 (1981); J. G. Zabolitzky, K. E. Schmidt and M. H. Kalos, Phys. Rev. C**25**, 1111 (1982).
42. J. Carlson, Phys. Rev. C**36**, 2026 (1987).
43. W. von der Linden, Phys. Rep. **220**, 53 (1992).
44. B. L. Hammond, W. A. Lester, Jr. and P. J. Reynolds, *Monte Carlo Methods in Ab Initio Quantum Chemistry* (World Scientific, Singapore, 1994).
45. Y. Alhassid, D. J. Dean, S. E. Koonin, G. Lang and W. E. Ormand, Phys. Rev. Lett. **72**, 613 (1994).
46. R. Krivec and V. B. Mandelzweig, Phys. Rev. A**42**, 3779 (1990).
47. R. Jastrow, Phys. Rev. **79**, 389 (1950).
48. R. Krivec, M. I. Haftel and V. B. Mandelzweig, Phys. Rev. A**47**, 911 (1993).
49. M. Fabre de la Ripelle, Ann. of Phys. **147**, 281 (1983).
50. J. Avery, *Hyperspherical Harmonics, Application in Quantum Theory* (Kluwer, Dordrecht, 1989).

51. C. D. Lin, Phys. Rep. **257**, 1 (1995).
52. L. D. Faddeev, Sov. Phys. JETP **12**, 1014 (1961); *Mathematical Aspects of the Three-Body Problem in the Quantum Scattering Theory* (Steklov Mathematical Institute, Leningrad, 1963 [English translation: Israel Program for Scientific Translation, Jerusalem, 1965] and John Wiley & Sons, Paris, 1977).
53. E. W. Schmid and H. Ziegelmann, *The Quantum Mechanical Three-Body Problem* (Pergamon, Oxford, 1974).
54. W. Glöckle, *The Quantum Mechanical Few-Body Problem* (Springer-Verlag, Berlin, Heidelberg, 1983).
W. Glöckle, H. Witała, D. Hüber, H. Kamada and J. Golak, Phys. Rep. **274**, 107 (1996).
55. Y. Wu, S. Ishikawa and T. Sasakawa, Few-Body Systems **15**, 145 (1993).
56. Z. Papp and W. Plessas, Phys. Rev. C**54**, 50 (1996).
57. D. L. Hill and J. A. Wheeler, Phys. Rev. **89**, 1102 (1953).
58. P. Ring and P. Schuck, *The Nuclear Many-Body Problem* (Springer-Verlag, New York, 1980).
59. Y. C. Tang, *Microscopic Description of the Nuclear Cluster Theory,* Lecture Notes in Physics, Vol. 145 (Springer-Verlag, Berlin, 1981), p. 571.
60. D. Brink, *Proceedings of the International School of Physics "Enrico Fermi"*, Course **36** (1965), ed. C. Bloch, (Academic Press, New York and London, 1966), p.247.
61. S. F. Boys, Proc. R. Soc. London Ser. A**258**, 402 (1960).
62. K. Singer, Proc. R. Soc. London Ser. A**258**, 412 (1960).
63. S. Fantoni, L. Panattoni and S. Rosati, Nuovo Cimento A**69**, 81 (1970).
64. P. M. Kozlowski and L. Adamowicz, J. Chem. Phys. **95**, 6681 (1991).
65. D. B. Kinghorn and R. D. Poshusta, Phys. Rev. A**47**, 3671 (1993).
66. B. J. Persson and P. R. Taylor, J. Chem. Phys. **105**, 5915 (1996).
67. H. F. King, J. Chem. Phys. **46**, 705 (1967).
68. G. W. F. Drake and Zong-Chao Yan, Phys. Rev. A**46**, 2378 (1992).
69. Y. K. Ho, Phys. Rev. A**48**, 4780 (1993).
70. A. R. Edmonds, *Angular Momentum in Quantum Mechanics*, (Princeton University Press, Princeton, 1957).
71. M. E. Rose, *Elementary Theory of Angular Momentum*, (John Wiley & Sons, New York, 1957).
72. D. M. Brink and G. R. Satchler, *Angular Momentum* (Clarendon Press, Oxford, 1968).
73. R. A. Sack, J. Math. Phys. **5**, 252 (1964).
74. K. G. Kay, H. D. Todd and H. J. Silverstone, J. Chem. Phys. **51**, 2359 (1969).
75. D. A. Varshalovich, A. N. Moskalev and V. K. Khersonskii, *Quantum Theory of Angular Momentum* (World Scientific, Singapore, 1988), p. 168.
76. V. Bargmann, Commun. Pure and Appl. Math. **14**, 187 (1961).

77. V. Bargmann, in *Analytic Methods in Mathematical Physics*, ed. R. P. Gilbert and R. G. Newton (Gordon and Breach, New York, 1968), p. 27.
78. K. T. Hecht and Y. Suzuki, J. Math. Phys. **24**, 785 (1983).
79. B. Klahn and W. A. Bingel, Theoret. Chim. Acta **44**, 9 (1977).
80. R. N. Hill, Int. J. Quant. Chem. **68**, 357 (1998).
81. K. Varga and Y. Suzuki, Comp. Phys. Commun. **106**, 157 (1997).
82. P. Schwerdtfeger and H. Silberbach, Phys. Rev. A**37**, 2834 (1988).
83. B. Dunlap, Phys. Rev. A**42**, 1127 (1990).
84. K. Varga, J. Usukura and Y. Suzuki, Phys. Rev. Lett. **80**, 1876 (1998).
85. J. Usukura, K. Varga and Y. Suzuki, Phys. Rev. A, in press (1998).
86. A. Martin, J.-M. Richard and T. T. Wu., Phys Rev. A**52** (1992) 3695.
87. M. Iwasaki, S. N. Nakamura, K. Shigaki, Y. Shimizu, H. Tamura, T. Ishikawa, R. S. Hayano, E. Takada, E. Widman, H. Outa, M. Aoki, P. Kitching and T. Yamazaki, Phys. Rev. Lett. **67**, 1246 (1991).
T. Yamazaki *et al.*, Nature **361**, 238 (1993).
88. I. Shimamura, Phys. Rev. A**46**, 3776 (1992).
89. V. I. Korobov, Phys. Rev. A **54** R1749 (1996).
90. A. M. Frolov and V. H. Smith, Phys. Rev. A**49**, 3580 (1994).
91. A. K. Bhatia, Phys. Rev. A**2**, 1667 (1970).
92. A. M. Frolov, V. H. Smith, Jr. and D. M. Bishop, Phys. Rev. A**49**, 1686 (1994).
D. M. Bishop, A. M. Frolov and V. H. Smith, Jr., Phys. Rev. A**51**, 3636 (1995).
93. S. I. Vinitskiĭ, V. S. Melezhik, L. I. Ponomarev, I. V. Puzynin, T. P. Puzynina, L. N. Somov and N. F. Truskova, Sov. Phys. JETP **52**, 353 (1980).
94. A. Yeremin, A. M. Frolov and E.B. Kutukova, Few-Body Systems **4**, 111 (1988).
95. N. M. Cann and A. J. Thakkar, Phys. Rev. A**46**, 5397 (1992).
96. C. L. Pekeris, Phys. Rev. **112**, 1649 (1958).
97. J. A. Wheeler, Ann. N. Y. Acad. Sci. **48**, 219 (1946).
98. A. P. Mills, Phys. Rev. Lett. **46**, 717 (1981).
99. E. A. Hylleraas and A. Ore, Phys. Rev. **71**, 493 (1947); A. Ore, *ibid.* **71**, 913 (1947); **83**, 665 (1951).
100. P. M. Kozlowski and. L. Adamowicz, Phys. Rev. A**48**, 1903 (1993).
101. J. M. Richard, Phys. Rev. A**49**, 3573 (1994).
102. A. M. Frolov, S. I. Kryuchkov and V. H. Smith, Jr., Phys. Rev. A **51**, 4514 (1995).
103. A. M. Frolov and V. H. Smith, Jr., J. Phys. B: At. Mol. Opt. Phys. **29**, L433 (1996); Phys. Rev. A**55**, 2662 (1997).
104. D. Bressanini, M. Mella and G. Morosi, Phys. Rev. A**55**, 200 (1997).
105. R. C. Miller, D. A. Kleinman, A.C. Gossard and O. Muntenau, Phys. Rev. B**25**, 6545 (1982).
106. E. I. Rashba and M. D. Sturge, *Excitons* (North-Holland, New York, 1982).
107. Zhong-Chao Yan and G. W. F. Drake, Phys. Rev. A**52**, 3711 (1995).
108. J. Komasa, W. Cencek and J. Rychlewski, Phys Rev. A**52**, 4500 (1995).

109. C. Froese Fisher, J. Phys. B: At. Mol. Opt. Phys. **26**, 855 (1993).
110. D. M. Schrader, F. M. Jacobsen, N-P. Frandsen and U. Mikkelsen, Phys. Rev. Lett. **69**, 57 (1992).
111. G. G. Ryzhikh and J. Mitroy, Phys. Rev. Lett. **79**, 4124 (1997).
112. G. G. Ryzhikh, J. Mitroy and K. Varga, J. Phys. B: At. Mol. Opt. Phys. **31**, L265 (1998).
113. W. Cencek and W. Kutzelnigg, J. Chem. Phys. **105**, 5878 (1996).
114. W. Cencek, J. Komasa and J. Rychlewski, Chem. Phys. Letters **246**, 417 (1995).
115. T. Kato, Commun. Pure and Appl. Math. **10**, 151 (1957).
R. T. Pack and W. B. Brown, J. Chem. Phys. **45**, 556 (1966).
D. P. Chong and D. M. Schrader, Mol. Phys. **16**, 137 (1969).
116. Y. Suzuki, K. Varga and J. Usukura, Nucl. Phys. A**631**, 91c (1998).
117. L. I. Ponomarev, Atomic Phys. **10**, 197 (1987).
118. A. de Rujula, H. Georgi and S. L. Glashow, Phys. Rev. D**12**, 147 (1975).
119. N. Isgur and G. Karl, Phys. Rev. D**18**, 4187 (1978); D**19**, 2653 (1979); D**20**, 1191 (1979).
120. F. E. Close, *An Introduction to Quarks and Partons* (Academic Press, New York, 1979).
121. V. B. Beretetskii, E. M. Lifshitz and L. P. Pitaevskii, *Relativistic Quantum Mechanics* (Pergamon Press, Oxford, 1971).
122. B. Silvestre-Brac, Few-Body Sytems **20**, 1 (1996); B. Silvestre-Brac and C. Gignoux, Phys. Rev. D**32**, 743 (1985).
123. L. Ya. Glozman, Z. Papp, W. Plessas, K. Varga and R. F. Wagenbrunn, Nucl. Phys. A**623**, 90c (1997); Phys. Rev. C**57**, 3406 (1998).
124. Particle Data Group, Phys. Rev. D**54**, 1 (1996).
125. L. Ya. Glozman and D. O. Riska, Phys. Rep. **268**, 263 (1996).
126. L. Ya. Glozman, W. Plessas, K. Varga and R. F. Wagenbrunn, Phys. Rev. D**58**, 094030 (1998).
127. M. Oka and K. Yazaki, in *Quarks and Nuclei*, ed. W. Weise (World Scientific, Singapore, 1984), p. 489.
128. Y. Fujiwara, C. Nakamoto and Y. Suzuki, Phys. Rev. C**54**, 2180 (1996).
129. J. R. Haynes, Phys. Rev. Lett. **17**, 860 (1966).
130. W. F. Brinkmann, T. M. Rice and B. Bells, Phys. Rev. B**8**, 1570 (1973).
131. M. A. Lee, P. Vashishta and R. K. Kalia, Phys. Rev. Lett. **51**, 2422 (1983).
132. D. A. Kleinmann, Phys. Rev. B**28**, 871 (1983).
133. B. Stebe and A. Ainane, Superlatt. Microstuct. **5**, 545 (1989).
134. A. Thilagam, Phys. Rev. B**55** 7804, (1997).
135. J. Singh, D. Birkedal, V. G. Lyssenko and J. M. Hvam, Phys. Rev. B**53**, 15909 (1996).
136. M. A. Kastner, Phys. Today **46** No. 1, 24 (1993).
137. R. C. Ashoori, H. L. Stormer, J. S. Weiner, L. N. Pfeiffer, K. W. Baldwin and K. W. West, Phys. Rev. Lett. **71**, 613 (1993).

138. L. Jacak, P. Hawrylak and A. Wójs, *Quantum Dots* (Springer-Verlag, Berlin, Heidelberg, 1998).
139. M. Taut, Phys. Rev. A**48**, 3561 (1993).
140. P. A. Maksym, Physica B**184**, 385 (1993).
141. M. Taut, J. Phys. A: Math. Gen. **27**, 1045 (1994).
142. *The Quantum Hall Effect*, ed. R. E. Prange and M. Girvin (Springer-Verlag, New York, 1987).
143. R. B. Laughlin, Phys. Rev. Lett. **50**, 1395 (1983); see also [142], p. 233.
144. R. A. Malfliet and J. A. Tjon, Nucl. Phys. A**127**, 161 (1969).
145. A. B. Volkov, Nucl. Phys. **74**, 33 (1965).
146. I. R. Afnan and Y. C. Tang, Phys. Rev. **175**, 1337 (1968).
147. D. R. Thompson, M. LeMere and Y. C. Tang, Nucl. Phys. A**286**, 53 (1977); I. Reichstein and Y. C. Tang, Nucl. Phys. A**158**, 529 (1970).
148. D. M. Brink and E. Boeker, Nucl. Phys. A**91**, 1 (1967).
149. H. Kamada and W. Glöckle, Nucl. Phys. A**548**, 205 (1992).
150. J. L. Friar, B. F. Gibson and G. L. Payne, Phys. Rev. C**24**, 2279 (1981); C**25**, 2279 (1982).
151. N. W. Schellingerhout and L. P. Kok, Nucl. Phys. A**508**, 299c (1990).
152. A. Kievsky, M. Viviani and S. Rosati, Nucl. Phys. A**501**, 503 (1989); A**551**, 241 (1993); A**577**, 511 (1994).
153. R. B. Wiringa, private communication.
154. V. Efimov, Phys. Lett. B**33**, 563 (1970).
155. W. Glöckle and H. Kamada, Phys. Rev. Lett. **71**, 971 (1993) and private communication. In the case of RV8 potential, Coulomb energy of 0.75 MeV is added. For the alpha particle with RV8, the calculation includes maximum two-body angular momentum j_{max} up to 3.
156. B. S. Pudliner, V. R. Pandharipande, J. Carlson and R. B. Wiringa, Phys. Rev. Lett. **74**, 4396 (1995).
157. M. Viviani, A. Kievsky and S. Rosati, Few-Body Systems **18**, 25 (1995) and private communication.
158. J. Carlson, Nucl. Phys. A**508**, 141c (1990).
159. W. Glöckle, H. Kamada, H. Witała, D. Hüber, J. Golak, K. Miyagawa and S. Ishikawa, Few-Body Systems, Suppl. **8**, 9 (1995).
160. C. R. Chen, G. L. Payne, J. L. Friar and B. F. Gibson, Phys. Rev. C**31**, 2266 (1985).
161. A. Arriaga, V. R. Pandharipande and R. B. Wiringa, Phys. Rev. C**52**, 2362 (1995).
162. R. B. Wiringa, R. A. Smith and T. L. Ainsworth, Phys. Rev. C**29**, 1207 (1984).
163. R .B. Wiringa and V. R. Pandharipande, Nucl. Phys. A**317**, 1 (1979).
164. J. Carlson, Phys. Rev. C**36**, 2026 (1987).
165. J. Carlson, Phys. Rev. C**38**, 1879 (1988).
166. K. Varga, Y. Suzuki and J. Usukura, Few-Body Systems **24**, 81 (1998).
167. R. F. Bishop, E. Buendía, M. F. Flynn and R. Guadiola, J. Phys. G: Nucl. Part. Phys. **19**, 1163 (1993).
168. R. Guardiola, P. I. Moliner, J. Navarro, R. F. Bishop, A. Puente and N. R. Walet, Nucl. Phys. A**609**, 218 (1996).

169. G. Erens, J. L. Visschers and R. van Wageningen, Ann. Phys. (N.Y.) **67**, 461 (1971).
170. J. L. Friar, quoted in [42].
171. R. H. Dalitz, R. C. Herndon and Y. C. Tang, Nucl. Phys. B**47**, 109 (1972).
172. B. F. Gibson and E. V. Hungerford III, Phys. Rep. **257**, 349 (1995).
173. H. Nemura, Y. Suzuki, Y. Fujiwara and C. Nakamoto, to be submitted.
174. G. G. Simon, Ch. Schmitt, F. Borkowski and V. H. Walther, Nucl. Phys. A**333**, 381 (1980).
M. McCord *et al.*, Nuclear Instruments and Methods in Physcs Research B**56/57**, 496 (1991).
175. Y. Fujiwara, H. Horiuchi, K. Ikeda, M. Kamimura, K. Katō, Y. Suzuki and E. Uegaki, Prog. Theor. Phys. Suppl. **68**, 29 (1980).
176. B. Buck, H. Friedrich and C. Wheatley, Nucl. Phys. A**275**, 246 (1977).
177. S. Saito, Prog. Theor. Phys. **41**, 705 (1969).
178. V. I. Kukulin and V. N. Pomenertsev, Ann. Phys. (N. Y.) **111**, 330 (1978).
179. D. V. Rutgauzer, Sov. J. Nucl. Phys. **51**, 787 (1990).
180. E. M. Tursunov, European Conference on Advances in Nuclear and Related Areas, Thesaloniki, Greece, 8-12 July, 1997, Abstracts, p. 194.
181. F. Hoyle, D. N. F. Dunbar, W. A. Wenzel and W. Whaling, Phys. Rev. **92**, 1095 (1953).
182. C. W. Cook, W. A. Fowler, C. C. Lauritsen and T. Lauritsen, Phys. Rev. **107**, 508 (1957).
183. C. E. Rolfs and W. S. Rodney, *Cauldrons in the Cosmos*, (The University of Chicago Press, Chicago and London, 1988).
184. S. Wolfram, *Mathematica: A System for Doing Mathematics by Computer*, 2nd Edition, (Addison-Wesley, Reading, 1991).

Index

Lecture Notes in Physics

For information about Vols. 1–478
please contact your bookseller or Springer-Verlag

Vol. 479: H. Latal, W. Schweiger (Eds.), Perturbative and Nonperturbative Aspects of Quantum Field Theory. Proceedings, 1996. X, 430 pages. 1997.

Vol. 480: H. Flyvbjerg, J. Hertz, M. H. Jensen, O. G. Mouritsen, K. Sneppen (Eds.), Physics of Biological Systems. From Molecules to Species. X, 364 pages. 1997.

Vol. 481: F. Lenz, H. Grießhammer, D. Stoll (Eds.), Lectures on QCD. VII, 276 pages. 1997.

Vol. 482: X.-W. Pan, D. H. Feng, M. Vallières (Eds.), Contemporary Nuclear Shell Models. Proceedings, 1996. XII, 309 pages. 1997.

Vol. 483: G. Trottet (Ed.), Coronal Physics from Radio and Space Observations. Proceedings, 1996. XVII, 226 pages. 1997.

Vol. 484: L. Schimansky-Geier, T. Pöschel (Eds.), Stochastic Dynamics. XVIII, 386 pages. 1997.

Vol. 485: H. Friedrich, B. Eckhardt (Eds.), Classical, Semi-classical and Quantum Dynamics in Atoms. VIII, 341 pages. 1997.

Vol. 486: G. Chavent, P. C. Sabatier (Eds.), Inverse Problems of Wave Propagation and Diffraction. Proceedings, 1996. XV, 379 pages. 1997.

Vol. 487: E. Meyer-Hofmeister, H. Spruik (Eds.), Accretion Disks – New Aspects. Proceedings, 1996. XIII, 356 pages. 1997.

Vol. 488: B. Apagyi, G. Endrédi, P. Lévay (Eds.), Inverse and Algebraic Quantum Scattering Theory. Proceedings, 1996. XV, 385 pages. 1997.

Vol. 489: G. M. Simnett, C. E. Alissandrakis, L. Vlahos (Eds.), Solar and Heliospheric Plasma Physics. Proceedings, 1996. VIII, 278 pages. 1997.

Vol. 490: P. Kutler, J. Flores, J.-J. Chattot (Eds.), Fifteenth International Conference on Numerical Methods in Fluid Dynamics. Proceedings, 1996. XIV, 655 pages. 1997.

Vol. 491: O. Boratav, A. Eden, A. Erzan (Eds.), Turbulence Modeling and Vortex Dynamics. Proceedings, 1996. XII, 245 pages. 1997.

Vol. 492: M. Rubí, C. Pérez-Vicente (Eds.), Complex Behaviour of Glassy Systems. Proceedings, 1996. IX, 467 pages. 1997.

Vol. 493: P. L. Garrido, J. Marro (Eds.), Fourth Granada Lectures in Computational Physics. XIV, 316 pages. 1997.

Vol. 494: J. W. Clark, M. L. Ristig (Eds.), Theory of Spin Lattices and Lattice Gauge Models. Proceedings, 1996. XI, 194 pages. 1997.

Vol. 495: Y. Kosmann-Schwarzbach, B. Grammaticos, K.M. Tamizhmani (Eds.), Integrability of Nonlinear Systems. VII, 380 pages. 1997.

Vol. 496: F. Lenz, H. Grießhammer, D. Stoll (Eds.), Lectures on QCD. VII, 483 pages. 1997.

Vol. 497: J. P. Greve, R. Blomme, H. Hensberge (Eds.), Stellar Atmospheres: Theory and Observations. VIII, 352 pages. 1997

Vol. 498: Z. Horváth, L. Palla (Eds.), Conformal Field Theories and Integrable Models. Proceedings, 1996. X, 251 pages. 1997.

Vol. 499: K. Jungmann, J. Kowalski, I. Reinhard, F. Träger (Eds.), Atomic Physics Methods in Modern Research, IX, 448 pages. 1997.

Vol. 500: D. Joubert (Ed.), Density Functionals: Theory and Applications, XVI, 194 pages. 1998.

Vol. 501: J. Kertész, I. Kondor (Eds.), Advances in Computer Simulation. VIII, 166 pages. 1998.

Vol. 502: H. Aratyn, T. D. Imbo, W.-Y. Keung, U. Sukhatme (Eds.), Supersymmetry and Integrable Models. Proceedings, 1997. XI, 379 pages. 1998.

Vol. 503: J. Parisi, S. C. Müller, W. Zimmermann (Eds.), A Perspective Look at Nonlinear Media. From Physics to Biology and Social Sciences. VIII, 372 pages. 1998.

Vol. 504: A. Bohm, H.-D. Doebner, P. Kielanowski (Eds.), Irreversibility and Causality. Semigroups and Rigged Hilbert Spaces. XIX, 385 pages. 1998.

Vol. 505: D. Benest, C. Froeschlé (Eds.), Impacts on Earth. XVII, 223 pages. 1998.

Vol. 506: D. Breitschwerdt, M. J. Freyberg, J. Trümper (Eds.), The Local Bubble and Beyond. Proceedings, 1997. XXVIII, 603 pages. 1998.

Vol. 507: J. C. Vial, K. Bocchialini, P. Boumier (Eds.), Space Solar Physics. Proceedings, 1997. XIII, 296 pages. 1998.

Vol. 508: H. Meyer-Ortmanns, A. Klümper (Eds.), Field Theoretical Tools for Polymer and Particle Physics. XVI, 258 pages. 1998.

Vol. 509: J. Wess, V. P. Akulov (Eds.), Supersymmetry and Quantum Field Theory. Proceedings, 1997. XV, 405 pages. 1998.

Vol. 510: J. Navarro, A. Polls (Eds.), Microscopic Quantum Many-Body Theories and Their Applications. Proceedings, 1997. XIII, 379 pages. 1998.

Vol. 511: S. Benkadda, G. M. Zaslavsky (Eds.), Chaos, Kinetics and Nonlinear Dynamics in Fluids and Plasmas. Proceedings, 1997. VIII, 438 pages. 1998.

Vol. 512: H. Gausterer, C. Lang (Eds.), Computing Particle Properties. Proceedings, 1997. VII, 335 pages. 1998.

Vol. 513: A. Bernstein, D. Drechsel, T. Walcher (Eds.), Chiral Dynamics: Theory and Experiment. Proceedings, 1997. IX, 394 pages. 1998.

Vol. 514: F. W. Hehl, C. Kiefer, R. Metzler, Black Holes: Theory and Observation. Proceedings, 1997. XV, 519 pages. 1998.

Vol. 515: C.-H. Bruneau (Ed.), Sixteenth International Conference on Numerical Methods in Fluid Dynamics. Proceedings. XV, 568 pages. 1998.

Vol. 516: J. Cleymans, H. B. Geyer, F. G. Scholtz (Eds.), Hadrons in Dense Matter and Hadrosynthesis. Proceedings, 1998. XII, 253 pages. 1999.

Monographs

For information about Vols. 1–10
please contact your bookseller or Springer-Verlag

Vol. m 11: A. D. Yaghjian, Relativistic Dynamics of a Charged Sphere. XII, 115 pages. 1992.

Vol. m 12: G. Esposito, Quantum Gravity, Quantum Cosmology and Lorentzian Geometries. Second Corrected and Enlarged Edition. XVIII, 349 pages. 1994.

Vol. m 13: M. Klein, A. Knauf, Classical Planar Scattering by Coulombic Potentials. V, 142 pages. 1992.

Vol. m 14: A. Lerda, Anyons. XI, 138 pages. 1992.

Vol. m 15: N. Peters, B. Rogg (Eds.), Reduced Kinetic Mechanisms for Applications in Combustion Systems. X, 360 pages. 1993.

Vol. m 16: P. Christe, M. Henkel, Introduction to Conformal Invariance and Its Applications to Critical Phenomena. XV, 260 pages. 1993.

Vol. m 17: M. Schoen, Computer Simulation of Condensed Phases in Complex Geometries. X, 136 pages. 1993.

Vol. m 18: H. Carmichael, An Open Systems Approach to Quantum Optics. X, 179 pages. 1993.

Vol. m 19: S. D. Bogan, M. K. Hinders, Interface Effects in Elastic Wave Scattering. XII, 182 pages. 1994.

Vol. m 20: E. Abdalla, M. C. B. Abdalla, D. Dalmazi, A. Zadra, 2D-Gravity in Non-Critical Strings. IX, 319 pages. 1994.

Vol. m 21: G. P. Berman, E. N. Bulgakov, D. D. Holm, Crossover-Time in Quantum Boson and Spin Systems. XI, 268 pages. 1994.

Vol. m 22: M.-O. Hongler, Chaotic and Stochastic Behaviour in Automatic Production Lines. V, 85 pages. 1994.

Vol. m 23: V. S. Viswanath, G. Müller, The Recursion Method. X, 259 pages. 1994.

Vol. m 24: A. Ern, V. Giovangigli, Multicomponent Transport Algorithms. XIV, 427 pages. 1994.

Vol. m 25: A. V. Bogdanov, G. V. Dubrovskiy, M. P. Krutikov, D. V. Kulginov, V. M. Strelchenya, Interaction of Gases with Surfaces. XIV, 132 pages. 1995.

Vol. m 26: M. Dineykhan, G. V. Efimov, G. Ganbold, S. N. Nedelko, Oscillator Representation in Quantum Physics. IX, 279 pages. 1995.

Vol. m 27: J. T. Ottesen, Infinite Dimensional Groups and Algebras in Quantum Physics. IX, 218 pages. 1995.

Vol. m 28: O. Piguet, S. P. Sorella, Algebraic Renormalization. IX, 134 pages. 1995.

Vol. m 29: C. Bendjaballah, Introduction to Photon Communication. VII, 193 pages. 1995.

Vol. m 30: A. J. Greer, W. J. Kossler, Low Magnetic Fields in Anisotropic Superconductors. VII, 161 pages. 1995.

Vol. m 31 (Corr. Second Printing): P. Busch, M. Grabowski, P.J. Lahti, Operational Quantum Physics. XII, 230 pages. 1997.

Vol. m 32: L. de Broglie, Diverses questions de mécanique et de thermodynamique classiques et relativistes. XII, 198 pages. 1995.

Vol. m 33: R. Alkofer, H. Reinhardt, Chiral Quark Dynamics. VIII, 115 pages. 1995.

Vol. m 34: R. Jost, Das Märchen vom Elfenbeinernen Turm. VIII, 286 pages. 1995.

Vol. m 35: E. Elizalde, Ten Physical Applications of Spectral Zeta Functions. XIV, 224 pages. 1995.

Vol. m 36: G. Dunne, Self-Dual Chern-Simons Theories. X, 217 pages. 1995.

Vol. m 37: S. Childress, A.D. Gilbert, Stretch, Twist, Fold: The Fast Dynamo. XI, 406 pages. 1995.

Vol. m 38: J. González, M. A. Martín-Delgado, G. Sierra, A. H. Vozmediano, Quantum Electron Liquids and High-Tc Superconductivity. X, 299 pages. 1995.

Vol. m 39: L. Pittner, Algebraic Foundations of Non-Com-mutative Differential Geometry and Quantum Groups. XII, 469 pages. 1996.

Vol. m 40: H.-J. Borchers, Translation Group and Particle Representations in Quantum Field Theory. VII, 131 pages. 1996.

Vol. m 41: B. K. Chakrabarti, A. Dutta, P. Sen, Quantum Ising Phases and Transitions in Transverse Ising Models. X, 204 pages. 1996.

Vol. m 42: P. Bouwknegt, J. McCarthy, K. Pilch, The W3 Algebra. Modules, Semi-infinite Cohomology and BV Algebras. XI, 204 pages. 1996.

Vol. m 43: M. Schottenloher, A Mathematical Introduction to Conformal Field Theory. VIII, 142 pages. 1997.

Vol. m 44: A. Bach, Indistinguishable Classical Particles. VIII, 157 pages. 1997.

Vol. m 45: M. Ferrari, V. T. Granik, A. Imam, J. C. Nadeau (Eds.), Advances in Doublet Mechanics. XVI, 214 pages. 1997.

Vol. m 46: M. Camenzind, Les noyaux actifs de galaxies. XVIII, 218 pages. 1997.

Vol. m 47: L. M. Zubov, Nonlinear Theory of Dislocations and Disclinations in Elastic Body. VI, 205 pages. 1997.

Vol. m 48: P. Kopietz, Bosonization of Interacting Fermions in Arbitrary Dimensions. XII, 259 pages. 1997.

Vol. m 49: M. Zak, J. B. Zbilut, R. E. Meyers, From Instability to Intelligence. Complexity and Predictability in Nonlinear Dynamics. XIV, 552 pages. 1997.

Vol. m 50: J. Ambjørn, M. Carfora, A. Marzuoli, The Geometry of Dynamical Triangulations. VI, 197 pages. 1997.

Vol. m 51: G. Landi, An Introduction to Noncommutative Spaces and Their Geometries. XI, 200 pages. 1997.

Vol. m 52: M. Hénon, Generating Families in the Restricted Three-Body Problem. XI, 278 pages. 1997.

Vol. m 53: M. Gad-el-Hak, A. Pollard, J.-P. Bonnet (Eds.), Flow Control. Fundamentals and Practices. XII, 527 pages. 1998.

Vol. m 54: Y. Suzuki, K. Varga, Stochastic Variational Approach to Quantum-Mechanical Few-Body Problems. XIV, 324 pages. 1998.

Vol. m 55: F. Busse, S. C. Müller, Evolution of Spontaneous Structures in Dissipative Continuous Systems. X, 559 pages. 1998.

GPSR Compliance

The European Union's (EU) General Product Safety Regulation (GPSR) is a set of rules that requires consumer products to be safe and our obligations to ensure this.

If you have any concerns about our products, you can contact us on ProductSafety@springernature.com

In case Publisher is established outside the EU, the EU authorized representative is:

Springer Nature Customer Service Center GmbH
Europaplatz 3
69115 Heidelberg, Germany

Batch number: 10584688

Printed by Printforce, the Netherlands